Space, Time, and Aliens

For Terry
Greg & Jenna
Anthony & Elizabeth

One Last Time

With Love

Preface: Cosmos and Culture

While this may seem at first glance an eclectic collection, all of the chapters in this volume are connected by a single theme: the impact of astronomy and space exploration on culture. Although such was not the intent when most of the articles were originally written, in retrospect this theme not only unites them but also constitutes an important story in its own right. In short, this book is more than the sum of its parts, and I hope highlights the idea that culture and cosmos are inextricably intertwined, not in an astrological way that all too many people have believed throughout history, but in innumerable other ways not often recognized. The 40 years of my writings offered here, while not exhaustive on this theme, nevertheless represent a fair sample of the sometimes surprising connections between the heavens and the Earth.

The claim of a connection between cosmos and culture as embodied in this volume requires some explanation. Parts I and II need no justification in this regard; they encompass one of the most popular ideas in Western civilization—that life might exist beyond the Earth. Where this worldview of "the biological universe" fits into the history of science and how it impacts culture may hold the key to human destiny. In film, literature, science, and art, it is a subject that engages the general public as few other topics in astronomy. In short it is an integral and pervasive part of Western culture. The implications of discovering life beyond Earth, whether in the form of simple, complex, or intelligent life, has only relatively recently been taken up by scholars, though often in obscure places. The general public deserves to know more about this work, especially since the World Economic Forum has identified the discovery of life beyond Earth as one of five "X factors" affecting humanity's future. Parts I and II provide an entrée into that literature and its sometimes "strange seas of thought," to borrow a phrase from William Wordsworth's characterization of Isaac Newton.

The exploration of space as represented in Part III has already changed the view of ourselves and our place in the universe, while also yielding practical applications such as the Global Positioning System, communications and weather satellites, and the reconnaissance of the Earth for both military and civilian purposes. From a practical standpoint, daily life on our planet would be very different without the sometimes unforeseen benefits of the Space Age. The Apollo images of Earthrise and the "Blue Marble," as well as the Voyager image of our "Pale Blue Dot" set against the stark background of space, have provided a new perspective on our fragile planet, even if it has not always translated into environmental action and even though the images of borderless continents have certainly not brought world peace. Part III, featuring articles written during my time as NASA Chief Historian, begins by highlighting some of humanity's greatest achievements in space exploration, attempting to place that exploration in the context of world history. The chapters on the Hubble Space Telescope remind us how robotic space exploration has also shaped our view of the universe and our place in it. At the same time, these chapters demonstrate the fragile hold we have had on the telescope through five servicing missions, the last of which was cancelled and then revived with results we still see today. Additionally, Part III looks at the role of imagination in spaceflight and introduces the reader to one aspect of "astroculture," a new field of historical inquiry that analyzes how humans have come to terms with the universe of which they are a part. During the 1960s the so-called Space Race highlighted the international stakes in what scholars have detailed as the geopolitical connections between the heavens and the Earth. Human spaceflight, and particularly the Apollo program, captured the human imagination and held out the unrealized hope that seen from a distance, Earthlings might finally see themselves as one human family. In any case, we argue that an agenda of bold exploration, continually looking forward to the new, is important to any vibrant society and is a choice that every society continually must make.

The subject of Part IV may at first seem esoteric by comparison and far removed from our daily lives. Although positional astronomy, or "astrometry" as it is more formally known, is in many ways a highly technical science, it has historically had many benefits for society, ranging from time determination, calendar reform, navigation, and surveying, to spacecraft navigation, geodesy, and astrophysics. Historically national observatories often carried out that work, and the rise of national observatories in the form of Greenwich, Paris, Pulkovo, and the U. S. Naval Observatory, among others, is largely a story of practical applications undertaken through public funding. I wrote the papers in this section while working almost 25 years at the U. S. Naval Observatory, having experienced firsthand the promise, problems and difficulties of this science, and having met and collaborated with many of its practitioners. Finally, the two colorful international astronomical expeditions discussed in Part IV demonstrate the necessity of international cooperation in science, even as they were affected by their own societies in terms of politics, funding, and popular support.

Part V emphasizes the importance of discovery in astronomy and again demonstrates how, since time immemorial, the heavens have affected culture in the form of meteor storms and other astronomical spectacles such as eclipses and comets, often seen as omens good or bad. In quite another arena, occasionally the public becomes enmeshed in astronomical issues such as the status of Pluto as a planet, or discoveries such as the moons of Mars, or the fate of the Hubble Space Telescope. At still another level, the twentieth-century leap from the claustrophobic but widely accepted universe of A. R. Wallace around 1900 to the expansive universe of Edwin

Hubble only two decades later, emphasizes how drastically our view of the universe has changed over the last century. Astronomical worldviews, whether geocentric, heliocentric, galactocentric, or biocentric, have always affected culture in multiple ways, providing the very framework for human existence.

In Part VI we contemplate another theme that unites many of the chapters in this volume: the philosophy of astronomy, cosmology, and astrobiology. As a subject functioning at the limits of science, astrobiology is in a particularly strong position to shed light on the scientific enterprise in general. And although astrobiology is the discipline where I first came to realize that science is not always as straightforward as sometimes depicted, astronomy and cosmology in general share many of its metaphysical conundrums, and cosmology brings these concerns to the ultimate problems in determining the structure and extent of the universe and our place in it. Part VI is a call for philosophers, historians, scientists, and science studies scholars to join in the new endeavor of philosophy of astronomy and cosmology, just as other sciences have been enriched by the systematic study of its philosophical aspects. Such a field adds a new dimension to astronomical endeavors.

Finally, in Part VII we contemplate not only the role of the cosmos for humanity on Earth, but also the future of humanity in space. Using the philosopher Olaf Stapledon's landmark essay "Interplanetary Man?" as a springboard, I argue that the prospect of "interstellar humanity" during the next millennium is likely to have an effect on all branches of terrestrial endeavor, whether religion, philosophy, science, or the arts. The stage of human drama will be vastly expanded. Despite many religious worldviews that claim to have a monopoly on the fate of humanity, astronomy holds the ultimate key to human destiny, for that destiny is entwined with the fate of the universe in which we live. The final chapter argues that "the consolation of astronomy" and the cosmic perspective are well worth the journey embodied in this book, just as the sixth-century philosopher Boethius argued for the consolation of philosophy in his book by that title.

Many of these chapters consist of articles delivered at a variety of meetings and venues, and that therefore first appeared in exotic places ranging from the Proceedings of the 5th International Conference on Bioastronomy (Chap. 4) and Proceedings of the Society of Amateur Radio Astronomers (Chap. 40) to deliberations at the International Astronomical Union (Chap. 28) and the International Academy of Astronautics (Chap. 6). Articles in such proceedings tend to be obscure for a reason: they are usually aimed at a narrow audience of specialists. However, in my case as a historian of science my role has often been to cover broad themes, and to place the scientific discoveries announced at these meetings into context. This is an endeavor that should be of interest to the general public as well as scholars. Some of the chapters have not been published before (Chaps. 15, 22, 27, 37, and 42) or were written for this volume, as in the final Chap. 42. The original sources and permissions are listed in Appendix 2.

All articles are reprinted substantially as they originally appeared, except for formatting changes, the addition of abstracts where they were lacking, and minor corrections to the bibliographies where some titles listed as "forthcoming" have now been published. Among the features of the book are the "Commentary 2020" sections added at the end of many chapters. The latter are necessary not only for updating the science, but also for updating the history, since interpretations of historical events inevitably change with more insight. I have also taken the opportunity in those sections to provide additional bibliographic references. The venues where the papers were originally presented are also given in those sections, since context is important, whether in the developing field of astrobiology or in the broader field of astronomy. I have not hesitated to add a human element to these sections, mainly in the form of my personal experiences at these meetings. Too often we forget that science and history are very human endeavors, a fact often not reflected in scholarly papers.

Upon finishing this book, it serendipitously turned out to consist of 42 chapters. Forty-two, as all fans of Douglas Adams's *The Hitchhiker's Guide to the Galaxy* know, is "the ultimate answer to the question of life, the universe, and everything," as calculated by the supercomputer Deep Thought. I do not fancy that this volume contains ultimate answers, but will be content if it stimulates thought on the many issues associated with cosmos and culture.

Finally, I wish to thank my editors at Springer, Maury Solomon and Hannah Kaufman, for their advocacy, advice, and diligence in seeing this rather hefty collection through the press.

Ashburn, VA, USA Steven J. Dick
January, 2020

This was the man who once was free
To climb the sky with zeal devout
To contemplate the crimson sun,
The frozen fairness of the moon –
Astronomer once used in joy
To comprehend and to commune
With planets on their wandering ways.
—Boethius, *The Consolation of Philosophy*, 6th century AD

Contents

Part IV Measuring the Universe: Goals, Institutions, Techniques

Part I
The Biological Universe

Part I Frontispiece Cosmic evolution, depicting the formation of galaxies, stars, planets, and life. The epic of cosmic evolution provides a kind of Genesis for the twenty-first century, an origin story that all cultures can hold in common based on modern science. It provides the cosmic context for the possibility of life on other worlds, the biological universe. (From the Roadmap for NASA's Office of Space Science Origins Theme, 1997)

The clear galaxy
Shorn of its hoary luster, wonderful
Distinct and vivid with sharp points of light
Blaze within blaze, an unimagin'd depth
And harmony of planet-girded suns
And moon-encircled planets, wheel in wheel,
Arch'd the wan sapphire. Nay – the hum of men
Or other things talking in unknown tongues
And notes of busy life in distant worlds
Beat like a far wave on my anxious ear.

Tennyson, Timbuctoo, 1829

That the idea of life other worlds has been a persistent theme in Western civilization is now well documented, and evident in our daily lives. From sending spacecraft to Mars and the water worlds of the gas giant planets, to the discovery of exoplanets, searches for extraterrestrial intelligence, alien science fiction literature, and the great UFO debate, we are immersed in a worldview that I have called the *biological universe*, the idea that life is abundant throughout the cosmos. Only in the last four decades have both historians and popular writers documented the pervasiveness of the theme, its relation to science and natural philosophy, and how unusual this obsession is by comparison with non-Western cultures (Dick 1982, 1996; Achenbach 1999; Crowe 1986; Guthke 1990). Why this should be so is still a subject of active research.

In Part I of this volume we examine the outlines of this theme, explore how it fits into the history of science, and illuminate one of its central icons, the Drake Equation. Chapter 1 presents in broad brushstrokes the checkered history of the idea from the ancient Greeks through most of the twentieth century, just prior to the discovery that exoplanets are common throughout the universe, now seen as a landmark in the debate. One of its central themes is that what has often been seen as an eccentric idea has actually been tied to major traditions of natural philosophy, including the ancient atomist, Aristotelian, Copernican, Cartesian and Newtonian traditions. After being dominated for much of the nineteenth century by philosophical and theological explorations, it received its scientific underpinnings with the Darwinian theory of evolution by natural selection and the rise of astronomical spectroscopy. The former eventually provided the basis for a discussion of the evolution of life under differing conditions beyond the Earth, while the latter provided a tool for studying the nature of the planets and stars in ever-increasing detail, demonstrating that the building blocks for matter and life were alike throughout the universe.

Chapter 2 zeroes in on the twentieth century debate and distinguishes four of its major themes: the role of planetary science, the search for planets beyond the solar system, research on the origins of life, and the Search for Extraterrestrial Intelligence (SETI). We describe the birth of exobiology/astrobiology as a new discipline, emphasize the discovery of cosmic evolution as the proper context for the debate, and suggest that it is best seen as a worldview comparable to the great worldviews of the past. The next three chapters place the theme in the context of the history of science, arguing that the move from the physical world to the biological universe constitutes a fundamental shift in our cosmological worldview, and that cosmic evolution may have resulted in a third possible shift, from a biological to a postbiological universe. We argue that the confirmation of one of these worldviews will have profound implications for human destiny.

Chapter 6 surprisingly demonstrates just how long the actual search for extraterrestrial intelligence has been contemplated, extending back to the earliest days of radio science. Prior to what has been known since the 1960s as SETI, there first came a period dubbed the era of "interstellar communication," dominated by radio pioneers such as Nikola Tesla and Guglielmo Marconi, and including astronomers such as Harvard Observatory Director Donald Menzel. This era is filled with parallels, contrasts, and lessons for those interested in the survival of SETI in its current incarnation.

This sets us up for the final chapter of Part I, which addresses one of the icons of the SETI movement—the Drake Equation, originated in connection with the first modern search for extraterrestrial intelligence by Frank Drake in Project Ozma in 1960. A method for estimating the number of radio communicative civilizations in our galaxy, it was a product of its time, embodying contemporary ideas about star formation, the frequency of planetary systems, the origins of life and intelligence, and the lifetimes of technological civilizations. We examine the origins and development of the equation, place it in the context of the science of its times, examine some of its hidden assumptions, and analyze criticisms and variations of it. Sixty years after it was originated, it still embodies our cultural hopes and fears.

Part I also incidentally demonstrates key changes in terminology, typical of a new discipline. What began as the *plurality of worlds* tradition in the Middle Ages morphed into *exobiology* at the beginning of the Space Age, became formally known as *bioastronomy* among astronomers at their meetings of the International Astronomical Union, and in the mid-1990s transformed into *astrobiology*, as biologists became the dominant players. Often, astronomers saw the subject as a branch of astronomy, while biologists saw it as a new part of biology. Similarly, what was most often known as SETI in the United States was often referred to as CETI (Communication with Extraterrestrial Intelligence) in the Soviet Union. This implied actual communication rather than just searching, a bridge too far for some in the United States who were seeking government funding. More recently, as we shall see in Chap. 15, "active SETI," sometimes referred to as METI (Messaging Extraterrestrial Intelligence), has been the focus of considerable attention. This endeavor to actually initiate messages has been even more controversial because it raises issues of who speaks for Earth and who should be consulted regarding the message content and targets, along with a Pandora's Box of other issues. Each change in terminology reflects the topic's cultural and scientific context at the time.

References

Achenbach, Joel. 1999. *Captured by Aliens: The Search for Life and Truth in a Very Large Universe*, Simon and Schuster, New York.

Crowe, Michael. J. 1986. *The Extraterrestrial Life Debate, 1750-1900: The Idea of a Plurality of Worlds from Kant to Lowell*. Cambridge, Cambridge University Press.

Dick, Steven. J. 1982. *Plurality of Worlds: The Origins of the Extraterrestrial Life Debate from Democritus to Kant*. Cambridge, Cambridge University Press.

Dick, Steven. J. 1996. *The Biological Universe: The Twentieth Century Extraterrestrial Life Debate and the Limits of Science*. Cambridge, Cambridge University Press

Guthke, Karl. S. 1990. *The Last Frontier: Imagining other Worlds from the Copernican Revolution to Modern Science Fiction*. Ithaca: Cornell University Press

Chapter 1
Plurality of Worlds: A Persistent Theme in Western Civilization

Abstract This chapter argues that the plurality of worlds tradition originated with and was sustained by cosmological worldviews through the middle of the eighteenth century, was dominated for the following century by philosophical explorations, and in the late nineteenth century received its scientific foundations in modern terms. Even so, the extraterrestrial life debate in the twentieth century remains an example of science functioning at its limits. This very fact garners a large amount of interest among historians and philosophers of science.

1.1 Introduction

Plurality of worlds (*plures mundi; Mehrheit der Welten; pluralité des mondes*) is the term historically used by many cultures for the concept of other worlds beyond the Earth. In ancient Greek times this meant a plurality of ordered world systems, referred to as *kosmoi*. Beginning in the seventeenth century the term came to mean Earth-like planets, complete with intelligent inhabitants. In the twentieth century the tradition has become known as the extraterrestrial life debate, a pursuit that biologists have labeled *exobiology* and astronomers have termed *bioastronomy*. Because it is intrinsically difficult to verify the existence of other worlds and extraterrestrial life, the debate has always incorporated a good measure of philosophy. Critics have called it "a science without a subject," and some even have questioned whether it is a scientific issue at all. Depending on the period under consideration, historians have concluded both that the debate has been conducted primarily in philosophical terms (Lovejoy 1936) and that it has been conducted primarily in scientific terms (Dick 1982). To an extent, these conclusions depend on how one defines *science*; any account must take into consideration the changing nature of science over the millennia during which the subject of multiple worlds has been discussed.

First published as "Plurality of Worlds," in Norriss S. Hetherington, ed., *Encyclopedia of Cosmology: Historical, Philosophical and Scientific Foundations of Modern Cosmology*, Garland Publishing (New York and London, 1993), pp. 502-512.

S. J. Dick, *Space, Time, and Aliens*, https://doi.org/10.1007/978-3-030-41614-0_1

1.2 The Cosmological Connection

The idea of other worlds already was present in the earliest of humankind's cosmological worldviews—ancient atomism. Constructed in the fourth and fifth centuries BC by Leucippus, Democritus, and Epicurus, this system held that.

> There are infinite worlds both like and unlike this world of ours. For the atoms being infinite in number, as was already proved, are borne on far out into space. For those atoms which are of such a nature that a world could have been created by them ... have not been used up either on one world or a limited number of worlds ... So that there nowhere exists an obstacle to the infinite number of worlds.

The important point here is that the atomists directly tied their cosmology to the physical principles of the atomist system; it was no half-hearted afterthought, but an integral part of the theory. Moreover, it is remarkable that this infinite number of worlds (*aperoi kosmoi*) existed completely beyond the human senses, for the entire visible world by the Greek definition composed a single *kosmos*. It was this non-empirical aspect that caused some seventeenth-century critics to complain that atomists should not posit infinite unseen worlds when so little was understood about the world we did see.

The Roman poet Lucretius spread the atomist doctrine of an infinite number of worlds, together with the rest of atomism, throughout Europe in his *De rerum natura* (*On the Nature of Things*). The atomist system, however, was not destined to win the day; it was almost two thousand years before it would be revived in the sixteenth and seventeenth centuries with the birth of modern science. In the meantime, a far more elaborate cosmology was constructed by Aristotle, whose life overlapped that of Epicurus by two decades and who gave new meaning to the word *kosmos*. Aristotle's cosmology placed the earth at the center of a nested hierarchy of celestial spheres, from the spheres of the Moon and planets to the sphere of the fixed stars (Fig. 1.1). The Earth in this system was more than a physical center—it was also the center of motion. According to one of the basic tenets of Aristotle's cosmology, the doctrine of natural motion and place, everything in the cosmos moved with respect to that single center: the element earth moved naturally toward the Earth, and the element fire moved naturally away, while the elements air and water assumed intermediate natural places. Aristotle's belief in the impossibility of more than a single *kosmos* was directly tied to this basic tenet. In his cosmological treatise *De caelo* (*On the Heavens*) he reasoned that if there were more than one world:

> It must be natural therefore for the particles of earth in another world to move towards the center of this one also, and for the fire in that world to move toward the circumference of this. This is impossible, for if it were to happen the earth would have to move upwards in its own world and the fire to the center; and similarly earth from our own world would have to move naturally away from the center, as it made its way to the center of the other, owing to the assumed situation of the worlds relative to each other.

The issue of a plurality of worlds thus was reduced to a confrontation with the most basic assumptions of Aristotle's system. Either he must reject his doctrine of natural motion and place (on which he had built his entire physics) and reject as well his

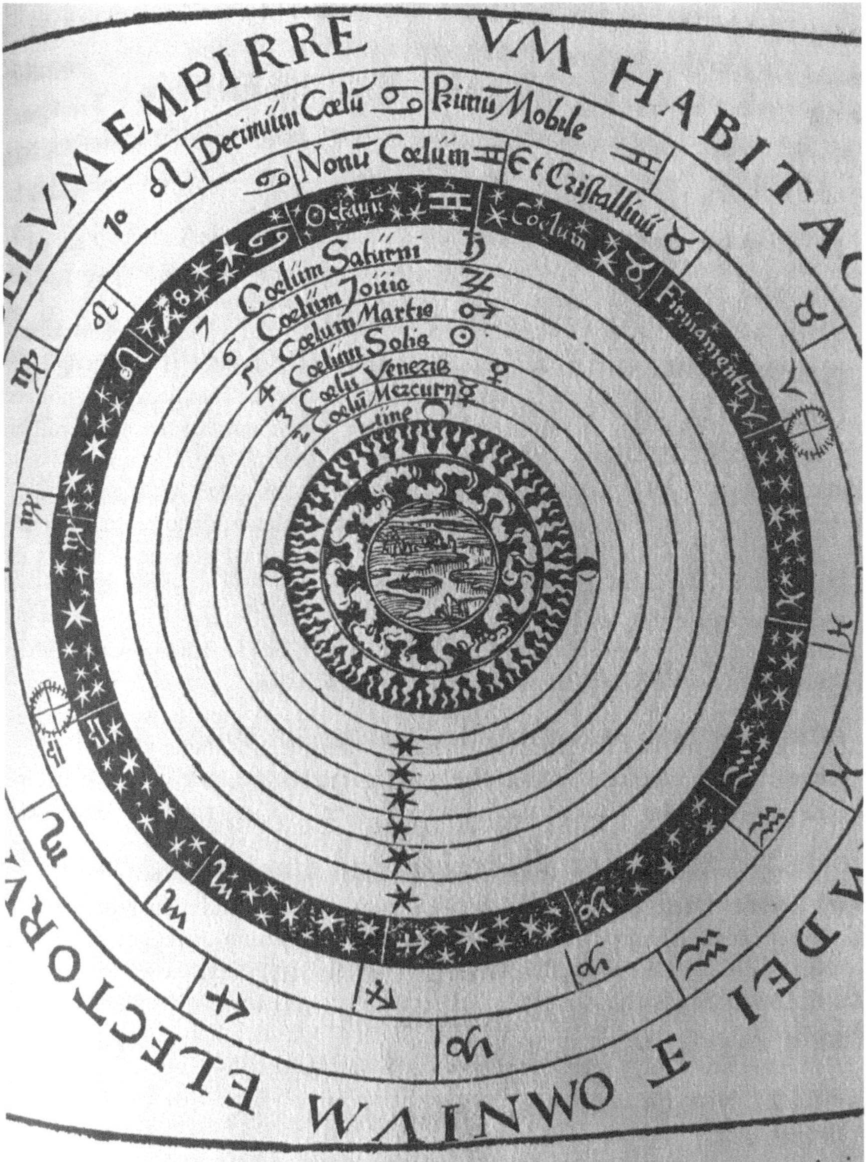

Fig. 1.1 A representation of the medieval *kosmos*, based on Aristotle, from Peter Apianus, *Cosmographicus liber Petri Apiani mathematici studiose collectus* (1524). Although the number of spheres sometimes varied, the framework of the medieval view was always the same, with the central Earth the point from which all motion was defined, according to Aristotle's physics. Based primarily on his definition of motion, Aristotle held that there was only one such *kosmos*, or world

belief in four elements (on which his theory of matter rested), or he must conclude that the world was unique. The choice was not a difficult one; indeed, he must have taken comfort in reaching a conclusion so diametrically opposed to the atomists, whose system differed from his in so many other ways. Aristotle also proposed auxiliary metaphysical arguments for a single world, but this physical argument was the central argument.

It was Aristotle's system that was transmitted to the Latin West, where it was commented upon again and again, but now in the context of the Christian system. The problem that Christianity had with the doctrine of a plurality of worlds was as follows: suppose God wished to create another world. How could He do so given the principles of Aristotle? Either Aristotle was wrong, and by his own admission wrong in some very basic principles, or God's power was severely limited. This dilemma was handled in various ways by commentators in the Middle Ages. Thomas Aquinas found God's perfection and omnipotence in the unity of the world rather than in its plurality. By late in the thirteenth century several commentators at Paris and Oxford universities argued that the plurality of worlds was not theologically impossible because God can act beyond the Aristotelian laws of nature. In the fourteenth century Jean Buridan and William of Ockham argued that the elements in each world would return to the natural place within their own world, either supernaturally or naturally. By 1377 the Paris master Nicole Oresme had completely reformulated the doctrine of natural place to state in no uncertain terms that other worlds were possible without any supernatural intervention.

The transition to the more modern plurality of worlds tradition, however, was not made through successive rebuttals to Aristotle's doctrine of a single world. Rather, it stemmed from the complete overthrow of Aristotle's geocentric universe and its replacement with the Copernican system of the world. By placing the sun in the center of the system of planets, and making the earth one of those planets, Copernicus gave birth to a new tradition where the term *world* (*mundus*) was now redefined to be an Earth-like planet, and each of these Earthlike planets took on the kinematic or motion-related functions of the single Earth in the old geocentric system. Just as the kinematic implications of the decentralization of the Earth led to the birth of a new physics, so the implications of that move for the physical nature of the planets led to the concept of inhabited worlds. All discussions of life on other worlds since the Copernican Revolution recall the argument set in motion by Copernicus: If the Earth is a planet, then the planets may be Earths; if the Earth is not central, then neither is man.

Copernicus himself did not pursue the implications of his system for planetary physics, but the Italian philosopher Giordano Bruno, an avowed Copernican, showed just how far such implications might go. Although in his *De l' infinito universo e mondi* (*On the Infinite Universe and Worlds*) of 1584 Bruno pointed primarily to metaphysical ideas such as the unity and plenitude of nature as the source of his belief in an infinite number of worlds, this was also the view toward which the Copernican system inexorably led. Even before the invention of the telescope, the young astronomer Johannes Kepler, already a convinced Copernican under the influence of his teacher Michael Maestlin, would ascribe inhabitants to the Moon.

The telescope accelerated this trend: Galileo in his *Siderius nuncius* (*The Sidereal* [or *Starry] Messenger*) of 1610 noted that the surface of the Moon was "not unlike the face of the earth." Because of theological difficulties, Galileo himself sought to downplay the similarities between Earth and Moon, and he admitted in 1632 in his *Dialogue on the Two Chief World Systems* only that if there was lunar life, it would be "extremely diverse and far beyond all our imaginings." The Copernican tide, however, could not be stemmed. Six years later, in the less repressive atmosphere of Anglican England, Bishop John Wilkins penned his *Discovery of a World in the Moone*, in which Galilean caution was thrown to the wind (Fig. 1.2). Copernicanism was not synonymous with inhabited planets, but it did give theoretical underpinning to habitable planets. The proof or disproof of this implication remained a goal of astronomers until the Viking landers touched down on Mars late in the twentieth century.

All-important cosmological worldviews of the seventeenth century onward incorporated the Copernican system as a basic truth. Such was the case with the first complete physical system proposed since Aristotle—that of the French philosopher Rene Descartes. His 1644 *Principia Philosophiae* (*Principles of Philosophy*), greatly influenced by a revived atomism, offered a mechanical philosophy in which atoms in motion once again formed the basis for a rational cosmology. For the plurality of worlds tradition it did even more, as it was through the Cartesian cosmology that the quest for a biological universe was first carried to other solar systems, and in a fashion so graphic that it has remained an ingrained concept to the present day. Unlike the void space of his atomist predecessors (and his Newtonian successors), Descartes proposed that the universe was a plenum, filled with atoms in every nook and cranny. A consequence of this was that, once set in motion by God, the particles of the plenum formed into vortices (systems analogous to our Solar System) centered around every star. Though Descartes himself, again for religious reasons, was careful not to specify that these vortices consisted of inhabited planets, his application of Cartesian laws to the entire universe and the graphic vortex cosmology was plain for all to see. Descartes's followers were not slow to realize the implications, none more boldly than his countryman Bernard le Bovier de Fontenelle. His *Entretriens sur la pluralité des mondes* of 1686 exploits both the Copernican and Cartesian theories to shed light on the question of life on other worlds (Fig. 1.3). Fontenelle asked: "If the fix'd Stars are so many Suns, and our Sun the centre of a Vortex that turns round him, why may not every fix'd Star be the Centre of a Vortex that turns round the fix'd Star? Our Sun enlightens the Planets; why may not every fix'd Star have Planets to which they give light?" In the same year, the Dutch astronomer Christiaan Huygens began to formulate very similar ideas, published posthumously in his *Cosmotheoros* in 1698. Although Cartesian vortices would be swept away by the Newtonian system, the general idea of planetary systems would not.

It is ironic that of all cosmological worldviews, the scientific principles of the Newtonian worldview entailed extraterrestrial life least of all. Although a mechanical philosophy like that of Descartes, Newton's atoms and void, with each body subject to universal gravitation according to fixed laws, did not necessarily imply other solar systems. No mechanical necessity dictated the formation of solar

Fig. 1.2 This elaboration of the title page from John Wilkins' *Discovery of a World in the Moone* (1638) appeared as the title page to the combined edition of the *Discovery and the Discourse concerning a New Planet* (1640). At left, Copernicus offers his heliocentric worldview, while at right, Galileo offers his telescope and behind him Kepler wishes for wings that he might visit the new world. The Sun says "I give light, heat, and motion to all"

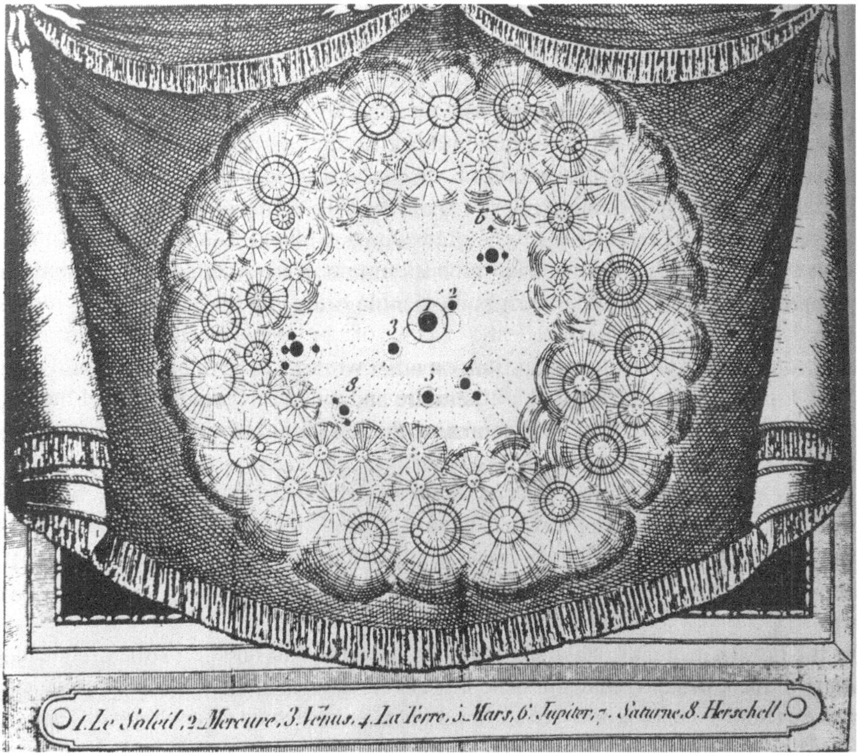

Fig. 1.3 A similar version of this frontispiece adorned many editions of Fontenelle's *Entretriens sur la pluralité des mondes*, including the first edition of 1686. It clearly indicates Fontenelle's belief in the plurality of planetary systems, and clearly depicts planets circling other fixed stars. The depiction of the planet Uranus (discovered by William Herschel in 1781) in this 1821 French edition is an indication of how Fontenelle's work was often updated by new astronomical discoveries long after Descartes' vortex cosmology was abandoned

systems as in Descartes's worldview; indeed, under Newtonian principles the whole question of other solar systems has proved to be one of greatest complexity even to the present day. Newton himself declined to expound any rational cosmogony that might shed light on the question. He insisted only that the formation of ordered systems was contingent upon God's will, contenting himself with the observation in the second edition of the *Principia* in 1713 that "if the fixed stars are the centres of other like systems, these, being formed by the like wise counsel, must be all subject to the dominion of the One." The major effect of the Newtonian worldview on the plurality of worlds tradition was to incorporate it into the tradition of natural theology, where it assumed the role of an important counterbalance in a system that had lessened the need for a Deity to keep the universe running. In one Newtonian treatise after another, the theological view of an inhabited universe was joined to the physical principles of Newton's system. Again and again, a universe full of inhabited solar systems was applauded as one "far more magnificent, worthy of, and becoming the infinite Creator, than any of the other narrower schemes." Once this argument had been made, overwhelming all Scriptural objections, other arguments such as teleology could be adduced in its favor.

This satisfying vision of the universe, operated by Newtonian laws and reflecting the power of the Deity by spreading intelligence through the universe, was passed on to the modern world. The proof of other solar systems by observation, and the proof of their likely formation by Newtonian principles, remained a desired goal in the centuries to follow. But the basic predisposition toward a universe of inhabited solar systems was set, almost within the lifetime of Newton himself. Philosopher-cosmologists such as Thomas Wright, Immanuel Kant, and Johann Lambert—committed Newtonians all—spread the vision of an inhabited universe in their eighteenth-century treatises. In the new Newtonian cosmological worldview (Fig. 1.4), the plurality of worlds tradition was joined to Christianity and natural theology. That conjunction would take a central place in the plurality of worlds debate in the nineteenth century.

1.3 Philosophical Explorations

Following the triumph of the Newtonian system in the middle of the eighteenth century, the extraterrestrial-life debate was waged not so much on a cosmological scale as on a scale of worldviews a level or more below the cosmological. Though sometimes discussed by the elaboration of Newtonian science (such as the Laplacian nebular hypothesis), more often the extraterrestrial life debate fell into the domain of philosophical explorations, both secular and religious. If cosmological worldviews gave birth to the idea of extraterrestrial life, then philosophy and literature, in their traditional role of examining the human condition, explored the ramifications of the idea borne of that cosmological context.

In particular, much of the plurality of worlds debate late in the eighteenth century and into the nineteenth century, at least in the West, may be understood as a struggle with that widespread philosophical world view known as Christianity. If in the Newtonian system the plurality of worlds concept was reconciled with theism through natural theology, this was not equivalent to a reconciliation with Christianity; as Professor Michael Crowe succinctly states in his study of the nineteenth-century plurality of worlds tradition: "structures of insects or solar systems may evidence God's existence, but they are mute as to a Messiah" (Crowe 1986).

Three choices were logically open to Christians who pondered the question of other worlds: they could reject other worlds, reject Christianity, or attempt to reconcile the two. Historically, all three of these possibilities came to pass in the eighteenth and nineteenth centuries. Although the Scriptural and doctrinal problems of the issue had been widely discussed throughout the seventeenth century, only to be overwhelmed by natural theology, no one more forcefully expressed the continuing difficulties of the plurality of worlds doctrine for Christianity than did Thomas Paine. In 1793 in his influential *Age of Reason*, Paine bluntly stated that "To believe

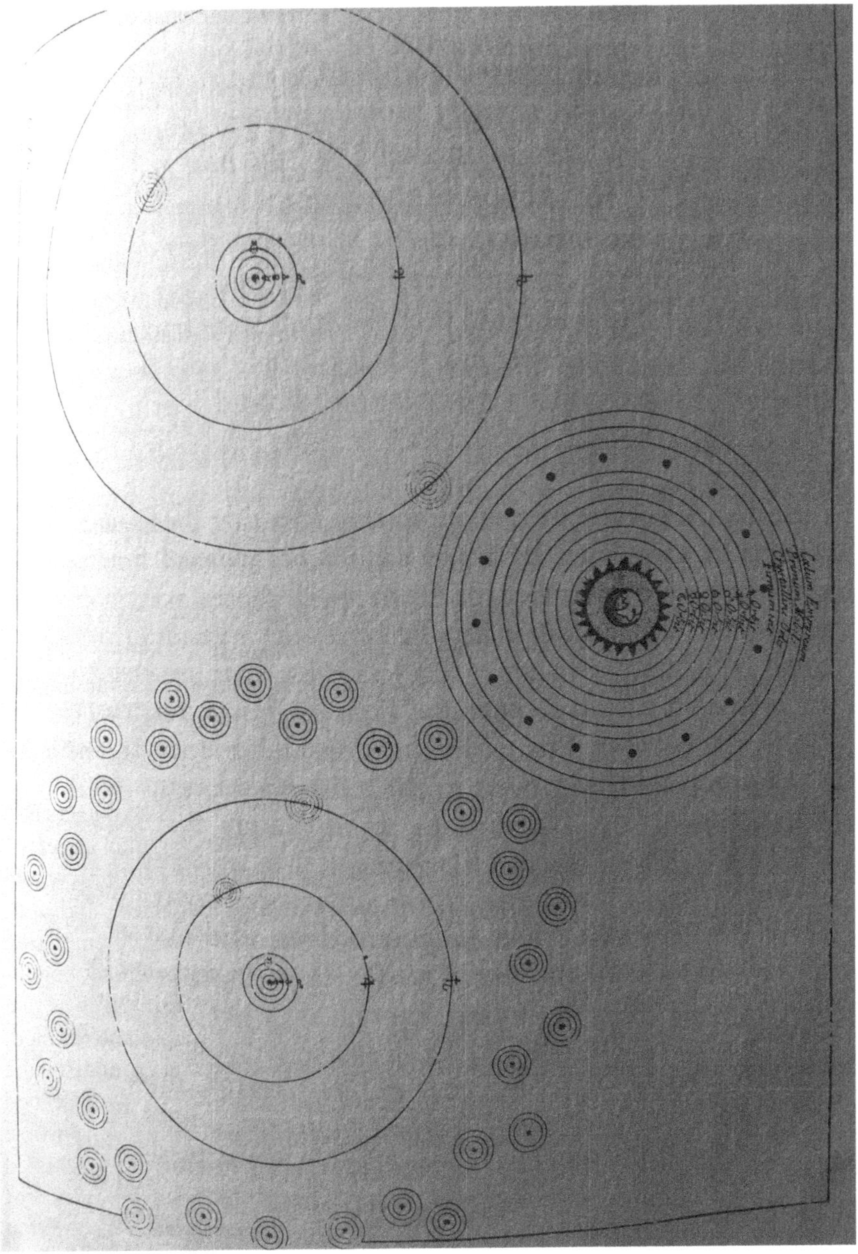

Fig. 1.4 The Ptolemaic, Copernican, and the "new system" according to William Derham, *Astro-Theology* (1715). The new system as depicted at the bottom indicates Derham's belief in a plurality of solar systems, a concept that plays a significant role in his work

that God created a plurality of worlds at least as numerous as what we call stars, renders the Christian system of faith at once little and ridiculous and scatters it in the mind like feathers in the air. The two beliefs cannot be held together in the same mind; and he who thinks that he believes in both has thought but little of either." Pointing to the Christian doctrines of Redemption and Incarnation, and to the absurdity of a planet-hopping Savior, Paine rejected Christianity.

Though few would reject Christianity because of Paine's argument, few would reject plurality of worlds either, a testimony to its entrenchment by the end of the eighteenth century. This left but one alternative: the two systems and all they implied would have to coexist. That other worlds could be incorporated into Christianity, despite Paine, was demonstrated by the Scottish theologian Thomas Chalmers. His 1817 *Astronomical Discourses* incorporated plurality of worlds into evangelical religion, and his countryman Thomas Dick made it a staple of Christianity in a number of works during the first half of the nineteenth century. Even astronomers such as John Herschel were strongly influenced by philosophical arguments on this issue.

Paine's objections, however, would not disappear. By mid-century the consonance of plurality of worlds with Christianity was once again called into serious question in one of the most interesting intellectual disputes of the nineteenth century. The instigator was William Whewell, philosopher, scientist, and Master of Trinity College, Cambridge. Influenced by Chalmers, Whewell was a pluralist from at least 1827. By 1850, as Professor Crowe has shown in his analysis of an unpublished Whewell manuscript, Whewell opposed pluralism. Then in 1853, Whewell's treatise *Of the Plurality of Worlds: An Essay*—the most learned, radical, and influential anti-pluralist treatise of the century—appeared anonymously.

In his *Essay* Whewell confirmed that the existence of other Earthlike planets and solar systems was commonly accepted. It is clear that his own Christian concerns were the source of Whewell's treatise, in particular the overwhelming reinforcement other worlds gave to the cry of the Psalmist "What is man, that thou art mindful of him?" and to the doctrines of Redemption and Incarnation. Before contemplating radical changes to Christianity, Whewell insisted, one should first examine the plurality of worlds doctrine. This was the purpose of the *Essay*, and the result was that Whewell would argue that it was pluralism, not Christianity, that should be rejected. Quickly disposing of one of the chief philosophical arguments of the pluralists, the teleological argument that the vast space must have some purpose, Whewell argued that confining intelligence to the "atom of space" that was the Earth was no worse than confining humanity to the "atom of time"that geology revealed it had existed on the earth. On the more empirical side, Whewell argued that no proof existed of other solar systems, that the stars might not be exactly similar to our Sun, and that in any case many of them were binary stars whose putative planets would therefore not have conditions conducive for life. In our own Solar System, only Mars approached the conditions of Earth, and it was just as likely as not that Mars was still in a condition of "preintelligence." Finally, Whewell cautioned against the unbridled use of the analogy argument in science. Altogether, Whewell's was a serious challenge to a doctrine that had come to be cherished by both science and natural theology.

Whewell's treatise generated a tremendous amount of debate, but in the end it did little to weaken support for a plurality of worlds among scientists or the religious. Professor Crowe documents twenty books and some 54 articles and reviews in response to Whewell; of these about two-thirds still favored pluralism despite Whewell's arguments (Crowe 1986). Treatises such as Sir David Brewster's *The Creed of the Philosopher and the Hope of the Christian* continued to be driven by an attachment to teleology and natural theology. Reconciliation with the doctrines of Incarnation and Redemption was never achieved. The claim that Christ's incarnation on Earth was of great enough force to save extraterrestrials prevented a planet-hopping Christ, but strained credulity. The concept of a plurality of worlds even became a central doctrine for at least two nineteenth-century religions: the Mormons and the Seventh Day Adventists. Yet another religion, the Swedenborgians, had held it as one of their beliefs since the middle of the eighteenth century.

Thus Christianity holds the distinction of being the philosophical worldview that most influenced the plurality of worlds doctrine in the nineteenth century, at least in the Western world. Secular philosophies also interacted with the concept of other worlds, though none so strongly or persistently as Christianity. Already in the seventeenth century the British empiricist John Locke had pointed out in his1689 *Essay Concerning Human Understanding* that human ideas are limited by the human senses, and that extraterrestrials might have no such limitations, or at least different ones. The German philosopher Gottfried Wilhelm Leibniz, well known for his belief that ours is the best of all possible worlds, may have been influenced in that belief by the possibility of actual worlds. This view was satirized by the most famous of the philosophers, Voltaire, who nevertheless also made use of extraterrestrials in his writings.

By contrast, many German philosophers of the nineteenth century were opposed to a plurality of worlds, not because of science or religion, but because of their anthropocentrism. G. W.F. Hegel held that the Earth is the most excellent of all planets, and several of his students argued strongly against the pluralist position. Friedrich Schelling and his disciple Heinrich Steffens, though not themselves Hegelians, also opposed with anthropocentric arguments the idea of other worlds. The ever-pessimistic Arthur Schopenhauer, though he accepted the existence of extraterrestrials, also believed that man was at the pinnacle of creation.

Literature, through its emotional exploration of human purpose, played an important role in responding to the challenge of other worlds (Guthke 1990). While John Milton in seventeenth-century England had cautioned, "Dream not of other worlds; what creatures there live in what state, condition or degree," Alexander Pope's *Essay on Man* suggested that he who contemplated the "worlds on worlds" of the universe might "tell why Heaven has made us as we are." A century later Tennyson expressed similar sentiments, and the Romantic poets Byron, Shelley, and Coleridge used other worlds in a religious context. In prose no less than in poetry, the extraterrestrial perspective became entrenched. From the cosmic-voyage genre of the seventeenth century to the science fiction of H.G. Wells at the end of the nineteenth century, in which implications of extraterrestrials and their worlds attained the status of a classic theme, the implications began to be explored at an increasing pace in ever more detail.

By the last quarter of the nineteenth century, plurality of worlds clearly had won the day, but based more on philosophical rather than scientific arguments. The Copernican system remained in the background as a driving force, and science kept chipping away at the problem with its limited empirical tools, until two unexpected developments laid the scientific foundations for all subsequent discussion of life beyond the earth.

1.4 Scientific Foundations

As the Whewell debate reached its height, two fundamental developments in science profoundly affected the plurality of worlds debate: in 1859 Charles Darwin and A.R. Wallace published their theory of them origin of species and of evolution by natural selection, and in the early 1860s the new technique of spectroscopy began to be applied to astronomy. Although these and less sweeping developments in other fields did not effect an immediate and radical change in the character of the plurality of worlds debate, they did signal the beginnings of a long-term change that would bring the subject of other worlds increasingly under the purview of modern science. Natural selection not only provided the basis for a discussion of the evolution of life under differing conditions beyond the Earth, but also gave impetus to the idea of the physical evolution of the universe. And spectroscopy provided a tool for studying the nature of the planets and stars in ever increasing detail, as well as a means of proving the truth of the evolutionary universe, prerequisites for determining the possibility of life.

Of the two developments, spectroscopy would have the more immediate and profound impact on the debate over other worlds. Although the arguments of analogy and uniformity of nature had for a long time given credence to the belief that the building blocks for matter and life were alike throughout the universe, now for the first time this great truth could be observationally proven. Many of the spectroscopic pioneers, including Sir William Huggins, did not fail to see the connection of their research to life in the universe. Huggins and his collaborator William Miller wrote in 1864 that their work contributed toward an "experimental basis" that the stars were "energising centres of systems of worlds adapted to be the abode of living beings." Huggins's early attempts to probe planetary atmospheres spectroscopically was the first step toward yet another research program that would become increasingly central to the extraterrestrial life debate in the future. Other pioneers in the new science (including Angelo Secchi, Heinrich Schellen, Jules Janssen, John and Henry Draper, and S.P. Langley) did not fail to make similar connections.

The Darwinian theory of evolution had a more gradual effect, but one eventually no less significant. Its earliest effect was in its general application to the idea of the physical evolution of the universe. This is evident already in one of the most prominent treatises on the plurality of worlds: Richard A. Proctor's 1870 *Other Worlds than Ours*, subtitled *The Plurality of Worlds Studied Under the Light of Recent Scientific Researches*. Like Proctor, his French counterpart Camille Flammarion professed to take a scientific approach to the problem of other worlds, though it is clear already from his 1862 *La pluralité des mondes habites* and subsequent works that far more than empirical science drove Flammarion to his belief in other worlds. (He had read pluralist authors, and also Jean Reynaud, who advocated the transmigration of souls from planet to planet, progressively improving at each stage.) By the end of the century it was not Flammarion's radical pluralism but Proctor's more limited version that became prevalent in the writings of British and American writers alike.

Other less encompassing theories born during this era also were destined to play an important, if delayed role in the plurality of worlds debate. G. Johnstone Stoney undertook the application of the kinetic theory of gases to planetary atmospheres as early as 1869. Laplace's nebular hypothesis was elaborated in more subtle form late in the nineteenth century, before being eclipsed for a while by the theory of close stellar encounters as a means of producing solar systems. Based on theories of solar system formation, the rarity or abundance of planetary systems would play an extremely important role in the debate over life.

Not all of these approaches achieved fruition in the nineteenth century, but as the old century transformed into the new, there was an increasing consensus that if there was to be an answer at all, it could only emerge from a scientific approach. At the turn of the century, in 1903, A.R. Wallace himself wrote a treatise, *Man's Place in the Universe: A Study of the Results of Scientific Research in Relation to the Unity or Plurality of Worlds*. The problem was that the intrinsic observational difficulties allowed scientific research to give no definitive answer; yet so great was the urge to resolve this age-old question that many scientists plunged ahead nevertheless. The result is a case study of how science and scientists function at their limits (Chap. 35).

The twentieth-century extraterrestrial life debate has quite naturally been dominated by the relatively nearby search for life in our Solar System, but increasingly in the second half of the century it has been joined by three other scientific components: the search for other planetary systems, experiments on the origin of life as applied to life beyond the Earth, and finally the Search for Extraterrestrial Intelligence (SETI) by means of radio telescopes (see Chap. 2). Each component of the debate has risen in turn to prominence, and in the 1960s these components began to converge toward a new scientific discipline, known first as exobiology and then as bioastronomy. Characterized by a coherent research program, federal funding, a tight-knit community of scientists, and formal institutionalization into the structure of scientific organizations (such as the International Astronomical Union), what began early in the century as a loose set of ideas ended as a protoscience with a broad-based constituency.

The search for life in the Solar System has been focused on Mars ever since Percival Lowell founded the Lowell Observatory in 1894 especially to search for life on that tantalizing planet (Hoyt 1976). The very next year—much too quickly, critics said—Lowell published his book *Mars*, in which he claimed to have mapped

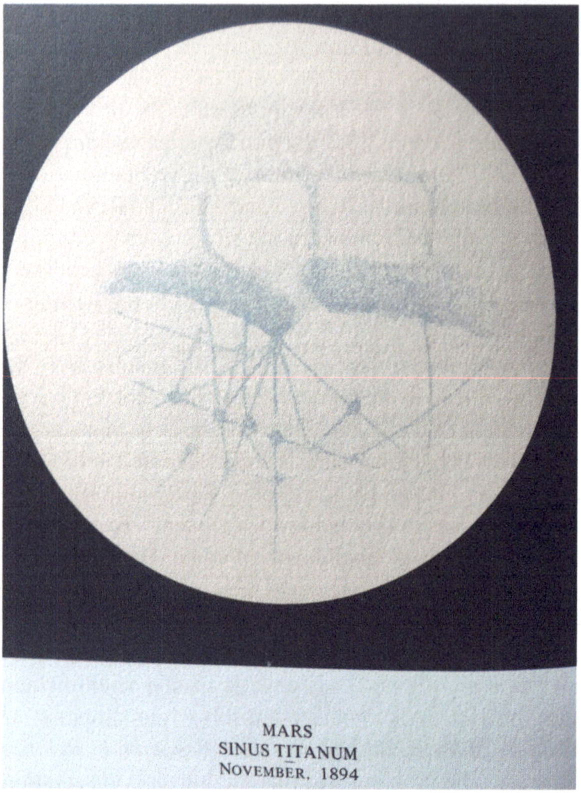

Fig. 1.5 Frontispiece to Percival Lowell's *Mars* (1895). This visual drawing depicts markings he interpreted as "canals" built by an intelligent Martian civilization channeling water on a dying planet. Other astronomers also claimed to see such markings, but many could not, initiating a controversy that lasted decades and illustrated the limits of observation and interpretation in science

"canals," long straight markings crisscrossing the planet in a system that he argued had been engineered by intelligent Martians (Fig. 1.5). Lowell was not the first to claim the existence of such markings on Mars. The Italian astronomer Giovanni Schiaparelli had mapped *canali* beginning in 1877, leaving their origin unexplained. But the connection that Lowell made between canals and intelligence began a controversy that peaked about 1910, and continued sporadically even after Lowell's death in 1916. The chief issues were observational, revolving around the ability of telescopes to detect fine detail on planetary surfaces and the ability of the human eye and brain to interpret this detail. Further observational issues centered around the very difficult determination of atmospheric and surface conditions on Mars.

The resolution of these issues was complicated by the varying distance of Mars, which approached relatively close to the Earth every 15 years only to recede again. Nevertheless, during the close approach of 1909 the French astronomer E.M. Antoniadi, using the 33-inch telescope at the Meudon Observatory in France, resolved some canals into dark splotches, a feat that began the downfall of Lowell's theory. However, this achievement left unexplained why Lick Observatory astronomers, who were critical of Lowell's observations, did not avert the whole episode by resolving canals during observations made in 1888 with their even larger 36-inch refractor. Even worse, modern spacecraft observations have shown little correlation between Martian surface markings and the canals that Lowell mapped, indicating that most of the objects of the debate, including Antoniadi's resolved canals, were illusory. The whole episode raises fundamental questions about observation and evidence in science. It had a profound effect on planetary astronomy during the rest of the century, causing some to leave the field of planetary astronomy because of the fierce debate and inspiring others to enter the field in order to discover the real nature of Mars.

After Lowell's death, with the close approach of Mars in 1924, attention focused on the possibility of Martian vegetation rather than intelligence. In one particularly important case this was still tied to the old visual method and canals; using the 36-inch refractor, Lick astronomer Robert J. Trumpler concluded that the canals were the result of natural topography, but that vegetation caused the dark Martian areas and made the canals visible. But the mid-1920s marked a new era in Martian studies: physical methods of spectroscopy and infrared astronomy now came into widespread use in the attempt to determine temperature and atmospheric conditions. Respected scientists such as W.W. Coblentz of the National Bureau of Standards, C.O. Lampland at the Lowell Observatory, and Edison Petit and Seth Nicholson at the Mount Wilson Observatory, pioneering in the field of infrared astronomy, determined that the temperature conditions on Mars were adequate for some form of Martian vegetation. Using spectroscopic techniques, others found evidence of oxygen and water vapor in the Martian atmosphere, but in increasingly minute amounts, now known to be spurious. Despite the desert conditions revealed by the new physical methods, by 1957 and the dawn of the Space Age the existence of hardy, perhaps lichen-like Martian vegetation was widely accepted, especially in the wake of William Sinton's claims in that year to have discovered infrared bands in the Martian spectrum that were unique to vegetation.

These hopes were partially dashed in the early 1960s when the Sinton bands were found to be due to deuterated water in the Earth's own atmosphere, and the water content of the Martian atmosphere was lowered almost to the vanishing point (see Chap. 39). But hopes were completely dashed two decades into the Space Age, when the Viking landers demonstrated in 1976 not only the lack of vegetation on Mars but also the complete absence of any organic molecules (Fig. 1.6). Although one of the three prime biology experimenters on the Viking project still maintains that his results were compatible with life, the consensus today is that life is absent on the planet Mars. Thus the twentieth century has seen the question of life on Mars progress from intelligence to vegetation to organic molecules—all having been disproven. With the discovery at mid-century that Venus is a victim of the greenhouse effect, with temperatures consequently at the 800 degrees Fahrenheit level, the Viking results left a Solar System largely bereft of life beyond Earth, though organics are still considered possible on some of the moons of the outer gaseous giant

Fig. 1.6 Viking 2 lander image showing the spacecraft and part of Utopia Planitia on Mars, looking south, November 2, 1976. The horizon is about 2 miles distant. The area is a region of fractured plains. The Viking landers found no organic molecules at their landing sites Mars down to parts per billion, though the results remain controversial. NASA/JPL

planets. But because Mars in particular had been widely viewed as a test case for life in the universe, the absence of life there was a correspondingly great blow to the concept of a universe filled with life.

Nevertheless, it is not surprising that long before the Viking results were in hand, attention had turned beyond the Solar System to the possibility of the existence of other planetary systems—a prerequisite for life in the realm of the stars. Belief in such systems has been greatly affected throughout the century by theories of their origin (Jaki 1978). The nebular hypothesis of Laplace, whereby planetary systems originated from rotating gas clouds that formed the stars themselves, indicated that planets were a natural by-product of star formation, and therefore very abundant. At the turn of the century, however, this theory was under heavy attack. In its place, the geologist T.C. Chamberlin and the astronomer F.R. Moulton proposed that solar systems originated by the close encounters of stars, which resulted in the tidal ejection of matter which cooled to form small planetesimals, which in turn accreted to form planets. This *planetesimal hypothesis*, elaborated and modified by the British astronomer James Jeans from 1916 almost until his death in 1946, implied that solar systems were extremely rare, since stellar collisions in the vastness of space were extremely rare. For this reason, during the 1920s and 1930s belief in extraterrestrial life was at a low point; it was difficult to conceive of life without planets.

But the fifteen years between 1943 and 1958 saw once again a complete turnabout in opinion. In 1943 two astronomers independently claimed they had observed the gravitational effects of planets orbiting the stars 61 Cygni and 70 Ophiuchi. Although these observations were proven spurious decades later, they filled a need at the time. Doubts expressed about Jeans's stellar encounter hypothesis by the American astronomer Henry Norris Russell in 1935 had grown to a crisis point by the early 1940s. Carl Friedrich von Weizsäcker began the revival of a modified nebular hypothesis in 1944, and the theoretical basis once again was laid for abundant planetary systems.

Although the new nebular hypothesis has been elaborated in ever more subtle form since that time, attempts to pin down the abundance of planetary systems have proven very difficult. Observationally, the search has been dominated by the astrometric method, whereby the proper motions of stars are studied for the gravitational effects of planetary systems. Since the 1960s several claims have been made by Peter van de Kamp and others for planetary systems around several stars. In the 1980s another method for determining planetary effects on stars—this time, on their radial velocities—came into use. As with the astrometric method, at the distances of even the nearest stars these effects are so small as to be at the limits of observation, and therefore are still controversial. Theory has therefore still predominated in the debate. Aside from the general fact that, according to the nebular hypothesis, solar systems are a normal occurrence during stellar evolution, subsidiary arguments also have been important. Especially since the 1950s, the knowledge that stars of the F spectral type exhibit a greatly slowed rotation rate has been used as an argument that they may have lost their angular momentum to planetary systems, as is the case in our own Solar System. Circumstellar material also has been observed that may represent solar systems in formation. Although such arguments lead astronomers to believe planetary systems are abundant, and despite numerous attempts to detect them observationally, no other solar systems have yet been unambiguously confirmed beyond our own.

As the idea of abundant planetary systems was being revived in the 1950s, work also was progressing on the biological question of the origins of life, a crucial factor in the question of extraterrestrial life. In particular, the work of Urey and Miller showed how organics could be produced under simulated primitive atmospheric conditions. However, since that time a better appreciation of the difficulties of the many steps in the origin of life has somewhat tempered optimism among biologists. Whereas astronomers focus on the enormous size of the universe and the likelihood of planets emerging from an abundance of stars, biologists concentrate on the extremely complex steps in the origin and evolution of life. Thus some dichotomy of opinion has developed between astronomers and biologists, further widened by the biologists' recognition that the evolution of life beyond the Earth might lead to forms of life and intelligence very different from the humanoid form and alien to the human concept of intelligence.

Given the inherent difficulties in the search for planetary systems and the uncertainties in our knowledge of the origin and evolution of life, interest since 1959 has focused on the detection of intelligent signals of extraterrestrial origin, the so-called Search for Extraterrestrial Intelligence (SETI), a detection that would leapfrog many of these uncertain arguments. The physicists Giuseppe Cocconi and Philip Morrison proposed in that year a search in the radio region of the spectrum using the 21-cm hydrogen line, and the American radio astronomer Frank Drake independently undertook the first search for such signals at the National Radio Astronomy Observatory in 1960. It was in the context of a meeting in 1961 in the wake of this search that the so-called Drake equation was formulated (Chap. 7). A general equation embodying the various factors of star and planet formation, the likelihood of the origin and evolution of life and intelligence, and the lifetimes of technical civilizations, it came to serve in the last third of the century as a paradigm for discussion of the issues. Although almost everyone acknowledges that the parameters of the equation are not well known—values range from one planet in our galaxy with intelligence (our own) to a hundred million or more planets with intelligence—this uncertainty has not prevented use of the equation as a basis for discussion of the prevalence of technological civilizations in the galaxy. Many radio searches have been undertaken worldwide since 1960, with NASA sponsoring the largest program now underway.

Although no radio searches have been successful, the existence of extraterrestrial intelligence is widely accepted in the scientific community, as well as among the public. Leaving aside the motivations of the public, much of which is swayed by nonscientific considerations such as unidentified flying objects, the scientific acceptance is an interesting commentary on the methodology of scientists, many of whom have preferred not to reject a theory that seems plausible on general grounds, even as it has awaited empirical confirmation for centuries.

To return full circle to the cosmological connection with which we began, the twentieth-century view of a universe full of life may perhaps best be seen as a cosmology in its own right, a *biophysical cosmology* that asserts the importance of both the physical and biological components of the universe (Dick 1989, and see Chap. 4). Like all cosmologies, it makes a claim about the large-scale nature of the universe; its claim is that life is not only a possible implication, but also a basic property of the universe. Like all cosmologies, the biophysical cosmology redefines our place in the universe. And most important, like other cosmologies, the biophysical cosmology has become increasingly testable in the twentieth century; this is the role and the importance of modern SETI programs. Viewed in this light, the transition from the physical world to the biological universe (Chap. 3) is one of the great revolutions in Western thought, no less profound than the move from the closed world to the infinite universe described by the French historian of science Alexandre Koyré more than three decades ago (Koyré 1957).

1.5 Commentary 2020

While the cosmological connections presented in Sect. 1.2 are now widely accepted, this was not always the case. The idea of a cosmological connection was first set out in my doctoral dissertation at Indiana University on the plurality of worlds debate from the ancient Greeks to the mid-eighteenth century. In a sign of the times, my history of astronomy professor was appalled when I proposed this as a dissertation topic: in a history and philosophy of science department, he said, it had two strikes against it—it wasn't science and it had no intellectual history worth documenting. Luckily, the Department included the distinguished medieval historian Edward Grant. Professor Grant was much more a history of ideas scholar, was well aware of the importance of the medieval plurality of worlds tradition, and was delighted to be my dissertation advisor. The dissertation was eventually published by Cambridge University Press (Dick 1982), the first dissertation published out of this Department.

My thesis was this: that far from being a subject undertaken by crackpots, minor thinkers, or those out of the mainstream, life on other worlds was explored by some of the leading thinkers of their time. More than that, the idea was explored in the context of the science or *natural philosophy* of the time. And even more than that, the idea was not only associated with, but entirely dependent upon, the major cosmological worldviews from Democritus to Kant. Specifically, as summarized in this chapter, those worldviews were (1) the ancient atomist tradition of Democritus, Epicurus, Lucretius and their followers; (2) the Aristotelian tradition that argued for a single *kosmos*, an idea repeatedly taken up by the medieval Scholastic commentators like Thomas Aquinas, John Buridan, William of Ockham and Nicole Oresme, before being transformed by the likes of Nicholas of Cusa and Giordano Bruno; (3) the Copernican tradition, which changed the meaning of world from *kosmos* to *mundus*, a tradition that made the Earth a planet and the planets potential Earths; (4) the Cartesian tradition, with its vortices interpreted by many Cartesians as solar systems; and (5) the Newtonian tradition, in which solar systems were possible due to a mix of scientific and natural theology arguments.

This argument was quite the opposite of the historian of ideas Arthur O. Lovejoy, who in *The Great Chain of Being* (Lovejoy 1936) had penned an entire chapter arguing that the metaphysical principle of plenitude was the chief argument that drove the idea of other inhabited worlds. In its most general form, the principle states that the fecundity of Nature or of God (depending on one's philosophical bent), demands that an idea or process that is possible be realized eventually in actuality. As our world gives ample evidence of the potentiality of world formation, the principle of plenitude demands that as many worlds are being made as can be made, and that they be inhabited.

My conclusions were quite different from Lovejoy's, namely, that "the emergence of the concept of extraterrestrial intelligence into the mainstream of European consciousness occurred only after complex, painstaking, and reasoned analysis, led by natural philosophers who drew upon a tradition of thought stretching back to antiquity. That tradition, as all natural philosophy, drew upon both metaphysical and

Fig. 1.7 This image released in 2014 shows a protoplanetary disk around the young star HL Tau. The disk is composed of multiple rings and gaps as the emerging planets sweep their orbits clear of dust and gas—a striking confirmation of the nebular hypothesis. The system, located about 450 light years from Earth, is less than 1 million years old. (Credit: ALMA (ESO/NAOJ/NRAO))

physical principles. From the infinite number of *kosmoi* of the ancient atomists to the infinity of stellar systems expounded by Newtonians such as Immanuel Kant, metaphysical concepts such as Divine omnipotence, teleology, and plenitude played a significant role. But they played this role only in the larger context of contemporary physical theory, and in consistency with whatever observational evidence was available" (Dick 1982).

Forty years on I still believe that conclusion is valid, with some interesting exceptions such as Giordano Bruno. If there was a mix of metaphysics and theory in Bruno, the proportion was highly tilted in favor of metaphysics (Chap. 42). But the point remains that one cannot judge the past based on the science of the present. Most historians would agree that it is unsurprising, and uninteresting, that the past does not measure up to the definition of science in the present. Rather, the past must be judged on its own terms—one of the reasons I found George Basalla's recent history of the extraterrestrial life debate flawed (Basalla 2006). Having said that, we must admit that metaphysics has not been banished from science even today, as the extraterrestrial life debate itself clearly shows (Fry 2012, 2015).

Even as I was working on the history of this subject through 1750, Professor Michael Crowe at the University of Notre Dame was researching the debate from 1750 to 1900 (Crowe 1986). I consider our collaboration and correspondence during this period a model of scholarly cooperation, and many of the insights of Sect. 1.3 are from his work, also published by Cambridge. I completed what we consider the "Cambridge trilogy" on the history of the extraterrestrial life debate with my book on the twentieth century debate (Dick 1996). Also important is the work of Harvard cultural historian Karl Guthke, who looked at the subject from a cultural and literary point of view Guthke (1990). And as the following chapters demonstrate, in the last two decades research from multiple disciplines have increased our understanding of the role of this idea in history, philosophy, and theology, and its potential role in the future.

In addition to the historiographic issues, illustrating that our understanding of history and historical interpretations continually change with new information and perspectives, a great many scientific advances have occurred since this article was published in 1993, affecting the last section. Some of these updates will be apparent in Chap. 2 and following chapters, especially the discovery of exoplanets and planetary systems in formation. In light of the long history described in this chapter, it is nothing short of amazing that we can now image these systems (Fig. 1.7). But the most basic points of this chapter remain valid: both Darwinian theory and astronomical spectroscopy are the foundations for all current work in what is today known as astrobiology.

References

Basalla, George. 2006. *Civilized Life in the Universe: Scientists on Intelligent Extraterrestrials*. Oxford, Oxford University Press

Crowe, Michael J. 1986. *The Extraterrestrial Life Debate, 1750–1900: The idea of a Plurality of Worlds from Kant to Lowell*. Cambridge: Cambridge University Press.

Dick, Steven J. 1982. *Plurality of Worlds: The Origins of the Extraterrestrial Life Debate from Democritus to Kant*. Cambridge: Cambridge University Press.

Dick, Steven J. 1989. The concept of extraterrestrial intelligence – An emerging cosmology, *Planetary Report*, 9: 13–17.

Dick, Steven J. 1996. *The Biological Universe: The Twentieth Century Extraterrestrial Life Debate and the Limits of Science*. Cambridge: Cambridge University Press.

Dick, Steven J. 2015. *The Impact of Discovering Life Beyond Earth*. Cambridge: Cambridge University Press

Fry, Iris. 2012. "Is Science Metaphysically Neutral?" *Studies in the History and Philosophy of Biological and Biomedical Sciences*, 43, 665–673.

Fry, Iris. 2015. "The Copernican and Darwinian Presuppositions," in Dick (2015), 23–37.

Guthke, Karl S. 1990. *The Last Frontier: Imagining other Worlds from the Copernican Revolution to Modern Science Fiction*. Ithaca: Cornell University Press.
Hoyt, William G. *Lowell and Mars*. 1976. Tucson: Univ. of Arizona Press
Jaki, S. L. 1978. *Planets and Planetarians: A History of Theories of the Origin of Planetary Systems*. Edinburgh, Scottish Academic Press.
Koyré, Alexandre, 1957. *From the Closed World to the Infinite Universe,* Johns Hopkins University Press, Baltimore.
Lovejoy, A. O. 1936. *The Great Chain of Being*. Cambridge, Mass., Harvard University Press

Chapter 2
The Twentieth Century History of the Extraterrestrial Life Debate: Major Themes

Abstract In this chapter we provide an overview of the extraterrestrial life debate since 1900, drawing largely on the major histories of the subject during this period, *The Biological Universe*, *Life on Other Worlds*, and *The Living Universe*, as well as other published works. We outline the major components of the debate, including (1) the role of planetary science, (2) the search for planets beyond the Solar System, (3) research on the origins of life, and (4) the Search for Extraterrestrial Intelligence (SETI). We describe the birth of exobiology/astrobiology as a new discipline, emphasize the discovery of cosmic evolution as the proper context for the debate, and suggest that it is best seen as a worldview comparable to the great worldviews of the past.

2.1 Introduction

When the twentieth century began, the idea of a universe filled with life was widely accepted, completely unproven, and heavily burdened with a long and checkered history that finally held the promise of more successful scientific scrutiny. The challenge was to bring new data to bear on an age-old controversy. The infamous episode of Percival Lowell and the canals of Mars, resolved to the satisfaction of most astronomers by 1912 (see Chap. 1), demonstrated just how difficult that challenge could be. Difficulties notwithstanding, the search for life would continue not only in our Solar System with tools ranging from ground-based telescopes to *in situ* observations on Mars, but also in the realm of the stars with the search for extrasolar planets, in laboratories and environments on Earth performing research bearing on the origins of life, and with the radio search for signals from extraterrestrial intelligence. We now examine the major themes of each of these areas in turn.

First published as part of "The Twentieth Century Extraterrestrial Life Debate: Major Themes and Lessons Learned," in *Astrobiology, History and Society: Life Beyond Earth and the Impact of Discovery*, Douglas A. Vakoch, ed. (Springer: Heidelberg, 2013), pp. 133-175.

S. J. Dick, *Space, Time, and Aliens*, https://doi.org/10.1007/978-3-030-41614-0_2

2.2 Major Themes of the Debate

2.2.1 Planetary Science

In the wake of the demise of the idea of canals on Mars, the red planet remained a focus for the search for life in the Solar System. After Lowell's death in 1916, with the close approach of Mars in 1924 attention focused on the possibility of Martian vegetation rather than intelligence. In one particularly important case this was still tied to the old visual method and canals; using the 36-inch Lick Observatory refractor astronomer Robert J. Trumpler concluded that the canals were the result of natural topography but that vegetation caused the dark Martian areas and made the canals visible (Trumpler 1927). But the mid-1920s mark a new era in Martian studies: physical methods of spectroscopy and infrared astronomy now came into widespread use in the attempt to determine temperature and atmospheric conditions. Respected scientists like W. W. Coblentz of the National Bureau of Standards, C. O. Lampland of Lowell Observatory, and Edison Petit and Seth Nicholson at Mt. Wilson Observatory, pioneering in the field of infrared astronomy, determined that the temperature conditions on Mars were adequate for some form of Martian vegetation (Coblentz and Lampland 1924; Petit and Nicholson 1924).

The belief in a harsher, but Earth-like Mars with vegetation was still very much alive at mid-century. At that time astronomers believed Mars had an atmospheric pressure of about 85 millibars at its surface, ten times thinner than Earth's. In 1949 the Dutch-American astronomer Gerard Kuiper had used early near-infrared techniques to discover carbon dioxide, one of the principle gases in the process of photosynthesis (Kuiper 1949). Seasonal vegetation across parts of Mars was commonly accepted, based on visual and photographic observations showing unmistakable seasonal changes on the surface as the polar caps melted, spreading a wave of darkening (Slipher 1927; Barabashev 1952). The second edition of the standard astronomy textbook of the time was pessimistic about the existence of even primitive animal life, but asserted that the existence of vegetation was "more likely than not" (Russell et al. 1945). Meanwhile, in the Soviet Union the astronomer Gavriil A. Tikhov assumed the mantle of the Russian Lowell, with a passion for Martian vegetation rather than Martian canals. In a career spanning many decades, Tikhov used reflection spectra to study the optical properties of terrestrial vegetation in harsh climates and applied the results to Martian observations, claiming a new science of *astrobotany* (Tikhov 1955, 1960). Tikhov's work, like Lowell's, provoked great criticism in his own country as well as abroad.

Using spectroscopic techniques, others found evidence of oxygen and water vapor in the Martian atmosphere, but in increasingly minute amounts, now known to be largely spurious (Spinrad et al. 1963). Despite the desert conditions revealed by the new physical methods, by 1957 and the dawn of the Space Age the existence of hardy, perhaps lichen-like Martian vegetation was widely accepted, especially in the wake of William Sinton's claims in that year to have discovered infrared bands in the Martian spectrum that were unique to vegetation (Sinton 1957, 1959).

These hopes were partially dashed in the early 1960s when the Sinton bands were found to be caused by deuterated water in the Earth's own atmosphere, and the water content of the Martian atmosphere was lowered almost to the vanishing point. But hopes were completely dashed two decades into the Space Age when the Viking orbiters and landers in 1976 seemed to demonstrate not only the lack of vegetation on Mars, but also the complete absence of any organic molecules at the two landing sites (Dick 1996, 153). And they showed an average atmospheric surface pressure of only 6 millibars. As we shall see in the next section, the Viking results on organic molecules—the *sine qua non* for life—have been questioned, and in the decades since that time other spacecraft have shown evidence of abundant water flow on Mars in the past. The Mars Global Surveyor and Mars Odyssey missions have both indicated that water ice still exists in plentiful amounts just below the surface, and the Mars Exploration Rovers have found strong evidence for plentiful liquid water below and on the surface in the past.

Nonetheless, evidence for life itself has not been found on Mars. The tantalizing seasonal changes were shown not to be due to vegetation, but to seasonal wind-blown sand. With the discovery at mid-century that Venus was a victim of the greenhouse effect, with temperatures consequently at the 800 °F level, it appeared that the Solar System was bereft of life beyond Earth. Hope of microbial life in the Solar System has not totally disappeared, due especially to the possibility that organics exist on some of the moons of the outer gas giants, notably Europa, Ganymede, Callisto and Titan. But because Mars had been viewed as a test case for life in the universe, the apparent absence of life there was a correspondingly great blow to the concept of a universe filled with life.

2.2.2 Planetary Systems

Long before the Viking results were in hand, attention had turned beyond the Solar System to the possibility of the existence of other planetary systems—a prerequisite for life in the realm of the stars. Since they could not be directly observed, belief in such systems was greatly affected for most of the century by theories of their origin (Dick 1996). The nebular hypothesis of Laplace, whereby planetary systems were theorized to originate from the same rotating gas clouds that formed the stars themselves, indicated that planets were a natural byproduct of star formation and, therefore, very abundant (Brush 1996). At the turn of the century, however, this theory was under heavy attack. In its place the geologist T. C. Chamberlin and the astronomer F. R. Moulton, both at the University of Chicago, proposed that solar systems originated by the close encounters of stars, which resulted in the tidal ejection of matter, which then cooled to form small planetesimals, which in turn accreted to form planets (Chamberlin and Moulton 1900). This *planetesimal hypothesis*, elaborated and modified by the British astronomer James Jeans from 1916 almost until his death three decades later (Jeans 1917), implied that solar systems were extremely rare, since stellar collisions in the vastness of space were extremely rare. For this reason, during the 1920s and 1930s belief in extraterrestrial life was at a low point; it was difficult to conceive of life without planets.

But the 15 years from 1943 to 1958 saw once again a complete turnabout in opinion (see Table 3.1). In 1943 two astronomers independently claimed they had observed the gravitational effects of planets orbiting the stars 61 Cygni and 70 Ophiuchi (Reuyl and Holmberg 1943). Although these observations were proven spurious decades later, they filled a need at the time. Doubts expressed in 1935 about Jeans's stellar encounter hypothesis by the dean of American astronomers, Henry Norris Russell, had grown to a crisis point by the early 1940s. Carl Friedrich von Weizsäcker began the revival of a modified nebular hypothesis in 1944, and the theoretical basis was once again laid for abundant planetary systems. The turnabout involved not only possible planetary companions and the revived nebular hypothesis, but also arguments from binary star statistics and stellar rotation rates. Helping matters along was Russell, whose *Scientific American* article "Anthropocentrism's demise" enthusiastically embraced numerous planetary systems (Russell 1943). Definitive evidence, however, would be much more elusive, for it turned out that Russell's declaration was 50 years premature.

Even as the nebular hypothesis has been elaborated in ever more subtle form, attempts to pin down the abundance of planetary systems proved very difficult. Through the 1960s and 1970s the search was dominated by the astrometric method, whereby the proper motions of stars are studied for the gravitational effects of planetary systems. In the 1960s Peter van de Kamp and others made claims for planetary systems around other stars (Van de Kamp 1963). In the 1980s another method for determining planetary effects on stars—this time utilizing their line-of-sight *radial velocities*—came into use. At the same time the Infrared Astronomical Satellite spacecraft discovered circumstellar disks, initially interpreted as protoplanetary disks (now believed to be debris disks left over after planet formation). But it was only in 1995 that the radial velocity method proved unambiguously successful, when the Swiss astronomers Michel Mayor and Didier Queloz discovered a planet around the star 51 Pegasi (Mayor and Queloz 1995). The American astronomers Geoff Marcy and Paul Butler confirmed the discovery almost immediately, and after that the floodgates were opened for more discoveries. They came not only from the radial velocity method, but also from the *photometric method*, whereby millimagnitude dips in stellar brightness were measured as a planet passed in front of its parent star. It was this method that the Kepler spacecraft used beginning in 2009, discovering more than 2000 planetary candidates by 2012. Of these almost 900 are Earth- or Super-Earth-sized, 1200 are Neptune sized, and about 250 are Jupiter sized or larger. Forty-eight planet candidates were found in the habitable zones of their stars, and it is estimated that at least 5% of all Sun-like stars host Earth-sized planet candidates.

Fig. 2.1 Stanley Miller with one of his laboratory flasks enclosing a simulated Earth atmosphere, February, 1970. (Courtesy Stanley Miller)

2.2.3 Origins of Life

Even as the idea of abundant planetary systems was being revived in the 1950s, work was also progressing on the biological question of the origins of life, a crucial factor in the question of extraterrestrial life (Fry 2000). In the 1920s the Russian biochemist Aleksandr Ivanovich Oparin (Oparin 1924, 1936) and the British biologist J. B. S. Haldane had independently suggested that life originated on Earth by chemical evolution in a hot dilute soup under conditions of a primitive Earth atmosphere. The experiments of Harold Urey and Stanley Miller in 1953 (Fig. 2.1) showed how amino acids could be produced under just such conditions, believed at the time to be highly "reducing" atmosphere, rich in hydrogen compounds such as methane and ammonia. Their success set off numerous experiments around the world in chemical evolution as related to the origins of life. The major thrust of

NASA's exobiology program, begun in the early 1960s, was to undertake such experiments on the origin of life, as well as to research life detection methods for spacecraft headed to Mars (Dick and Strick 2004).

Since the original Miller-Urey experiments, a better appreciation of the difficulties of the many steps in the origin of life—as well as uncertainty about the nature of the primitive Earth atmosphere—has somewhat tempered optimism among biologists. Whereas astronomers focus on the enormous size of the universe and the likelihood of planets emerging from an abundance of stars, biologists point to the extremely complex steps in the origin and evolution of life. Thus a dichotomy of opinion has developed between astronomers and biologists, further widened by the biologists' recognition that the evolution of life beyond Earth might lead to forms of life and intelligence very different from the humanoid form and alien to the human concept of intelligence.

Over the past quarter century theories of the origin of life have proliferated, with various implications for exobiology. Furthermore, the discovery of life in extreme environments—around deep sea hydrothermal vents, in deep underground rock, and in conditions of great salinity and acidity, has fostered a new appreciation for the tenacity of life and broadened our idea of the conditions under which life might originate on another planet or on Earth. As the possibilities of panspermia have become more widely accepted, spurred on by the Mars rock controversy (discussed in the next section) and by the realization that material does transfer between planets, some researchers believe that so-called *exogenous delivery* of organic compounds may be the key to the origin of life on Earth.

The question of the origin of life on Earth and in space shared many philosophical issues. Old problems such as chance, necessity, and the nature of life—already recognized in the terrestrial realm—were magnified in the extraterrestrial realm. The crucial question for exobiology was whether life would arise wherever it could, or whether the Earth was a fluke. The contingency or necessity of life would be one of the greatest scientific and philosophical questions of the extraterrestrial life debate. The two points of view are classically represented by the French biologist and Nobelist Jacques Monod on the one hand, and the Belgian-American biochemist and Nobelist Christian deDuve on the other. In his classic work *Chance and Necessity*, Monod (1971) argued "the universe was not pregnant with life, nor the biosphere with man. Our number came up in the Monte Carlo game." Nor was Monod the only one to favor chance; the astronomer Fred Hoyle agreed that the chance of a random shuffling of amino acids producing a workable set of enzymes was miniscule, and went one step further in asserting that life must have been assembled by a "cosmic intelligence," though not necessarily the supernatural intelligence of Christianity (Hoyle 1983). DeDuve, on the other hand, argued just the opposite, declaring Monod wrong and viewing life as a "cosmic imperative," while evolutionary biologist Richard Dawkins argued that "climbing Mt. Improbable" was not impossible (De Duve 1995; Dawkins 1997).

2.2.4 Search for Extraterrestrial Intelligence

All these questions in the origin of life arena are multiplied when it comes to the nature of consciousness, mind, and intelligence. In many ways defining "intelligence" remains more problematic than defining "life," with many different possible approaches undertaken in a very large literature (Sternberg 2000, 2002). To frame it another way, there is no "general theory of intelligence" or even of human brain function, much less a general theory of intelligence in a cosmic context. Carl Sagan argued in his *Dragons of Eden* that "once life has started in a relatively benign environment and billions of years of evolutionary time are available, the expectation of many of us is that intelligent beings would develop. The evolutionary path would, of course, be different from that taken on Earth … But there should be many functionally equivalent pathways to a similar end result. The entire evolutionary record on our planet, particularly the record contained in fossil endocasts, illustrates a progressive tendency toward intelligence" (Sagan 1977, p. 230).

That conclusion embodies many assumptions that others have questioned. Evolutionists such as George Gaylord Simpson and Theodosius Dobzhansky, for example, had already argued just the opposite (Simpson 1964; Dobzhansky 1972), and Harvard evolutionist Ernst Mayr also differed strongly with Sagan, arguing that intelligence (by his definition) had emerged only once on Earth (Mayr 1985, 1988). Outspoken Harvard evolutionist Stephen Jay Gould agreed with the non-prevalence of *humanoid* intelligence, arguing in an entire book on the Burgess Shale fossils of the Cambrian explosion that if we "Wind back the tape of life to the early days of the Burgess Shale; let it play again from an identical starting point, and the chance becomes vanishingly small that anything like human intelligence would grace the replay." By contrast, evolutionary paleobiologist Simon Conway Morris (Conway Morris 1998, 2003) has argued from the same evidence, and others, that evolutionary convergence applies not only to morphology, but also to intelligence, if only the conditions are present. He is, however, skeptical that the proper conditions often obtain, summarizing his position in the subtitle of his 2003 book *Life's Solution: Inevitable Humans in a Lonely Universe*. In this he reached the same conclusion as had Peter Ward and Donald Brownlee (Ward and Brownlee 2000), who famously argued that complex life and thus intelligence in the universe will be rare, not from a lack of convergence but because so many factors must come together in order for it to exist.

These problems are leapfrogged to some extent by the radio search for extraterrestrial intelligence, or, to put it more accurately, the search for extraterrestrial technology. In 1959 the physicists Giuseppe Cocconi and Philip Morrison, both at Cornell, proposed a search in the radio region of the spectrum using the 21-cm hydrogen line (Cocconi and Morrison 1959). The radio astronomer Frank Drake independently undertook the first search of such signals at the National Radio Astronomy Observatory in 1960. It was in the context of a meeting in 1961 in the wake of this search that the so-called Drake equation was formulated (see Chap. 7). A general equation embodying the various factors of star and planet formation, the

Fig. 2.2 The Allen Telescope Array, a dedicated facility in California for SETI observations. (Photo credit: Seth Shostak)

likelihood of the origin and evolution of life and intelligence, and the lifetimes of technical civilizations, it came to serve in the last third of the century as a paradigm for discussion of the issues (Dick, 1996, pp. 431–454). Although almost everyone acknowledges that the parameters of the equation are not well known, resulting in values ranging from one planet in our galaxy with intelligence (our own) to 100 million or more, this uncertainty has not prevented its use as a basis for discussion of the abundance of technological civilizations in the galaxy. Many radio searches have been undertaken worldwide since 1960, all unsuccessful (Fig. 2.2).

2.3 Birth of a New Discipline

In the 1950s and 1960s these four scientific fields—planetary science, the search for planetary systems, origin of life studies, and SETI—converged to give birth to the field of exobiology (Dick, 1996). At first quite separate in terms of researchers, techniques, and goals, these fields over four decades gradually became integrated, in large measure because of the scientific and public desire to search for life beyond Earth. NASA served as the most important patron for the new field. By 1963 NASA's life sciences expenditures (including exobiology) had reached $17 million. The $100 million spent on the Viking biology experiments was closely related to origin of life issues, since an informed search for life required a definition of life and knowledge of its origins. Even though exobiology saw a slump in the 1980s in terms of space missions in the aftermath of the Viking results, NASA kept the program more than alive with a grant program of about $5 to $10 million per year, funding research on such broad topics as deep ocean hydrothermal vents and their associated archaea, the primitive Earth atmosphere, the Gaia hypothesis, mass extinctions, exogenous delivery of organic compounds, and the RNA world (Dick and Strick 2004). At the same time NASA also operated the largest exobiology laboratory in the world at its Ames Research Center in California.

In 1995 a deep organizational restructuring at NASA precipitated a rebirth of the field under a new name, *astrobiology*. NASA's strategic plan for 1996 used the term astrobiology for the first time anywhere in a NASA document (though it had been sporadically used elsewhere as much as 50 years earlier). Astrobiology under NASA was "the study of the living Universe" to be sure, but in particular it was seen as providing the scientific foundation for studying the origin and distribution of life in the universe, the role of gravity in living systems, and the study of the Earth's atmosphere and ecosystems. In 1998 an astrobiology "roadmap" laid out three specific questions: How does life begin and evolve? Does life exist elsewhere in the universe? And what is life's future on Earth and beyond? Specific goals were set to answer these questions (Des Marais et al. 2008).

The contrast between the exobiology and astrobiology programs was quite striking. They both shared the core concerns of origin of life research and the search for life beyond Earth. But astrobiology placed life in the context of its planetary history, encompassing the search for planetary systems, the study of biosignatures, and the past, present and future of life (Fig. 2.3). Astrobiology added new techniques and concepts to exobiology's repertoire, raised multidisciplinary work to a new level, and included the study of the history of Earth's life and present organisms. Today astrobiology is a robust field, a worldwide effort supported especially by NASA, but also by other international research-funding agencies.

All of this did not occur without skepticism, extending even to the period 50 years ago when exobiology was born. In 1964 George Gaylord Simpson, pointing to the long history of the debate, wrote that "There is even increasing recognition of a new science of extraterrestrial life, sometimes called exobiology—a curious development in view of the fact that this 'science' has yet to demonstrate that its subject matter exists!" Simpson noted that this supposed new science was very expensive, and called exobiology "a gamble at the most adverse odds in history," resembling "more a wild spree more than a sober scientific program" (Simpson 1964). Simpson concluded with a plea "that we invest just a bit more of our money and manpower, say one-tenth of that now being gambled on the expanding space program," on studying the systematic and evolution of earthly organisms—that is to say, his own field! An interesting case of the rhetoric of science, clearly Simpson had an ulterior motive in declaring that exobiology was not a science. But with Isaac Asimov's article in the *New York Times Magazine* the following year entitled "A Science in Search of a Subject" (Asimov 1965), the phrase was too good to ignore as a kind of mindless meme deployed innumerable times in the course of the following decades, despite the article's positive assessment of exobiology (Strick 2004).

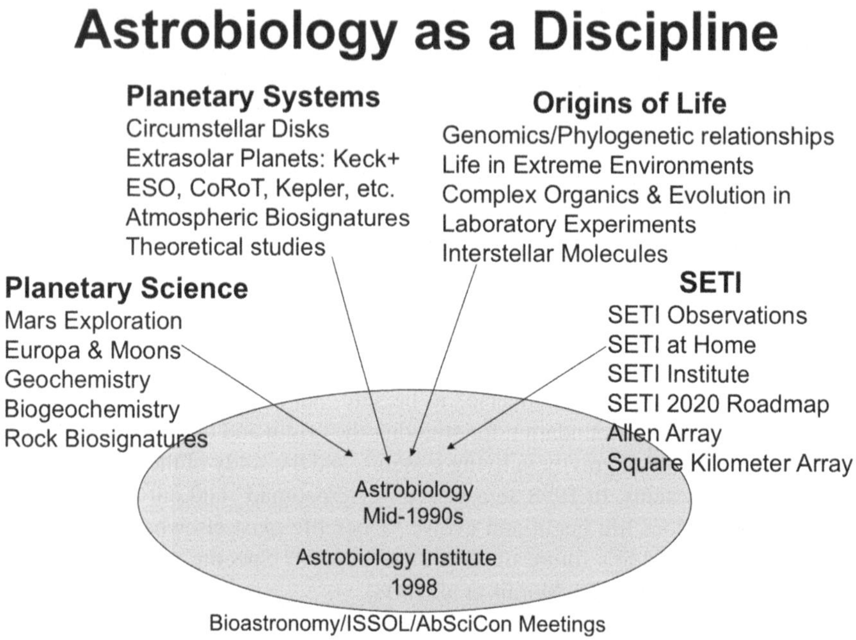

Fig. 2.3 Astrobiology as a discipline. Astrobiology as it developed in the mid-1990s at NASA was much broader than exobiology as practiced over the previous 50 years. In addition to origins of life studies, astrobiology embraced planetary science and planetary systems science. While SETI was not a programmatic element of astrobiology at NASA at this time, it remains a central intellectual element of the field. Whether astrobiology is, or should be, a separate discipline, is open to discussion. ISSOL and AbSciCon at the bottom refer to the International Society for the Study of the Origins of Life, and the biennial Astrobiology Science Conferences that began in 2000. Slowly the social sciences, humanities, and philosophy were added to the discipline, and more attention was paid to the importance of the media, education and public outreach. Compare to Fig. 16.1

Even a minimal consideration of this idea suffices to show that it is a misrepresentation of science, even if admittedly a catchy phrase. One could say the search for gravitational waves, or the Higgs boson, or planetary systems, are, or were, "sciences without a subject." But this hardly seems a productive way of approaching the problem. Every science is looking for a subject until it finds it (planetary systems), thinks it may have found it (the Higgs boson), or does not find it (gravitational waves, at least so far). From an epistemological point of view, the methods of astrobiology are as empirical as in any historical science such as astronomy or geology (Cleland 2001, 2002), though it is true that astrobiological observations and experiments are often especially difficult, and the inferences more tenuous. With the broad array of research now being undertaken in astrobiology, the "science without a subject" meme has outlived its usefulness.

Although Simpson criticized the pioneer in the field, Joshua Lederberg, by claiming that exobiology was not strictly biology because its techniques differed (Wolfe 2002), certainly astrobiologists today would be surprised to learn they are not doing science; from their point of view their endeavors constitute not only science, but cutting-edge science. While more than one practitioner early on heralded astrobiology or its equivalent as a new scientific discipline (Shklovskii 1965; Billingham 1981), these claims may have been premature (Dick 1996, pp. 475–478). Moreover, being labeled a discipline may be good or bad in terms of "Balkanization" and isolation from broader parent fields, such as was contemplated, but did not happen, in the case of radio astronomy in relation to astronomy as a whole (Sullivan III 2009, pp. 435–438). An historical comparison of discipline formation in other fields such as biochemistry (Kohler 1982), molecular biology (Abir-Am 1992), and geophysics (Good 2000) would help illuminate the problem for astrobiology.

2.4 Cosmic Evolution as the Context for Astrobiology

The concerns of astrobiology—the origins and evolution of life, intelligence and culture—are embedded in the larger process of cosmic evolution, the 13.7 billion year Master Narrative of the Universe (Fig. 2.4). The concept has its roots in the eighteenth and nineteenth centuries, but only became widely accepted and a major driver for research programs in the last half of the twentieth century (Dick 2009; Zakariya 2017). I have argued elsewhere (see Chap. 12) that the outcome of cosmic evolution may result in a physical, biological or postbiological universe, in other words, a physical universe composed of planets, stars and galaxies in which life is a fluke; a biological universe full of carbon-based life; or a postbiological universe in which cultural evolution has resulted in a universe full of artificial intelligence (Dick 2003). These outcomes determine the long-term destiny of humanity, and because the scope of astrobiology as set down in the Astrobiology Roadmap applies not only to the past and present, but also the future, the destiny of humanity falls within the purview of the philosophy of astrobiology.

2.5 The Biological Universe as Worldview

The twentieth-century view of a universe full of life may perhaps best be seen as a cosmology in its own right, a *biophysical cosmology* that asserts the importance of both the physical and biological components of the universe (see Chap. 4). Like all cosmologies, it makes a claim about the large-scale nature of the universe, and its claim is that life is not only a possible implication, but also a basic property of the universe. Over the last four decades some scientists have come to question why the laws of nature and the physical constants appear to be *biofriendly*, giving rise to what has been termed the *anthropic principle*. The principle has many variants, all having to do with the apparent fine-tuning of the physical constants for life (Carter 1974; Barrow and Tipler 1986; Carr 2007). The phrase is a spectacular misnomer, and the term *biocentric principle* is much preferred, since in the context of

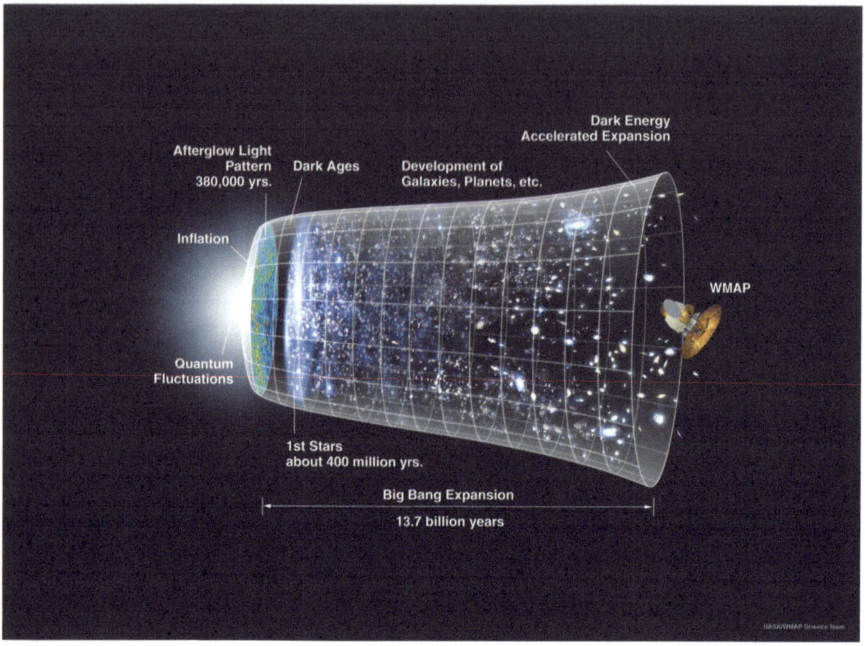

Fig. 2.4 The master narrative of the universe, 13.7 billion years of cosmic evolution, as depicted by the Wilkinson microwave anisotropy probe (WMAP) program, which narrowed the age of the universe to within 100 million years. The current model has the universe beginning with the Big Bang, stars forming within the first few hundred million years, followed by the development of galaxies, planets and life. The concerns of astrobiology must be seen within this framework, which encompasses physical, biological and cultural evolution (see Chap. 8). (Courtesy NASA/WMAP Science team)

astrobiology the universe appears to be friendly to life, and the very question to be answered is whether humans are the only intelligent life (Davies 2007).

The prospect of a fine-tuned universe has given rise to the idea of an ensemble of universes, termed a *multiverse*, as an explanation for why we happen to be in a universe particularly suited for life (Carr 2007). Whether or not we invoke the multiverse, the physicist Freeman Dyson has suggested that the prospects are bright for a future-oriented science, joining together in a disciplined fashion the resources of biology and cosmology (Dyson 1988). In such a *cosmic ecology*, life and intelligence would play a central role in the evolution of the universe, no less than its physical laws.

Like other cosmologies the biophysical cosmology redefines our place in the universe. And most importantly, like other cosmologies in the twentieth and twenty-first centuries the biophysical cosmology has become increasingly testable; this is the role and the importance of modern astrobiology and SETI programs. Viewed in this light, the transition from the physical world to the biological universe is one of the great revolutions in Western thought, no less profound that the move from the closed world to the infinite universe described by the French historian of science

Alexandre Koyré almost a half century ago (Koyré 1957). That transition has already occurred to some extent in the minds of most people. Whether the biological universe exists in reality, and what its effect will be on culture when and if it extraterrestrial life is actually discovered, remains to be seen.

2.6 Commentary 2020

Although this chapter was written only a few years ago, discoveries in astrobiology continue apace, even while still falling far short of the actual discovery of life. The "water worlds" or "ocean worlds" of our Solar System continue to spur research. NASA's Europa Clipper mission to Jupiter's icy moon was advanced to the final design phase in August, 2019, with a projected launch date of 2025. The search for life on Mars and other bodies of the Solar System also continues, with the biggest news being the NASA Curiosity Rover's discovery of organic molecules on Mars. Seasonal variations of methane on Mars are also tantalizing, but it is not yet determined whether the methane is biogenic.

Meanwhile, research on the origins of life continues in many areas, but it is not known whether life on Earth originated single or multiple times, in single or multiple locations, in Darwin's warm little pond or in deep sea hydrothermal vents, or via panspermia from outer space. Almost 70 years after the Miller-Urey experiments, the precise pathways and mechanisms of life's origin remain unknown. What *is* known is that hundreds of types of interstellar organic molecules, the building blocks of life, have been found in the molecular clouds out of which stars and planets form. In short, origins of life research has not yet advanced to the stage where we can determine if life is a *cosmic imperative* or a lucky accident restricted to Earth.

After 9 years of operation and an extended mission dubbed "K2," the Kepler spacecraft ceased operation on October 30, 2018. As of late 2019 Kepler had discovered 2345 confirmed planets with another 2420 candidates yet to be confirmed. Altogether, more than 4000 planets have been confirmed through both spacecraft and ground-based methods. Among these more than 600 are multiplanet systems, which are very diverse compared to our own. Many of them seem to contain only planets with five to ten Earth masses, rather than gas giants, and so are scaled-up versions of the systems around the gas giants in our Solar System rather than versions of our Solar System itself. Some discoveries, such as GJ 3512b consisting of a giant planet around an M dwarf star, challenge our ideas of how planets form. The search continues with spacecraft such as the Transiting Exoplanet Survey Satellite (TESS) and more to come. And on the ground 23 new high-precision spectrographs are completed or near completion for use in the radial velocity method for planet detection. The Swiss astronomers Michel Mayor and Didier Queloz received the 2019 Nobel Prize for pioneering this method with their first discovery in 1995. It is safe to say that the search for planetary systems, including Earth-like planets, will be a continuing research program that has burgeoned since its first discoveries 25 years ago. The latest Kepler news is at https://www.nasa.gov/mission_pages/

kepler/main/index.html and the latest exoplanet counts, exoplanet data, and much more are at the NASA Exoplanet Archive https://exoplanetarchive.ipac.caltech.edu/docs/counts_detail.html. TESS research can be followed at https://www.nasa.gov/tess-transiting-exoplanet-survey-satellite

The Search for Extraterrestrial Intelligence (SETI) has also been revitalized in recent years (see Chap. 7).

References

Abir-Am, P. G. 1992. "The Politics of Macromolecules: Molecular Biologists, Biochemists and Rhetoric." *Osiris* 7: 164–191

Asimov, Isaac. 1965. "A Science in Search of a Subject." *New York Times Magazine*, 23 May, pp. 52–58.

Barabashev, Nicholas P. 1952. *A Study of the Physical Conditions of the Moon and Planets*. Kharkov, Russia: Kharkov University Press.

Barrow, John D. and Frank J. Tipler. 1986. *The Anthropic Cosmological Principle*. Oxford: Oxford University Press.

Billingham, John. 1981. *Life in the Universe*. Cambridge, Mass.: MIT Press.

Brush, Stephen. 1996. *Nebulous Earth: The Origin of the Solar System and the Core of the Earth from Laplace to Jeffreys*. Cambridge: Cambridge University Press.

Carr, Bernard, editor. 2007. *Universe or Multiverse*? Cambridge: Cambridge University Press.

Carter, Brandon. 1974. Large number coincidences and the anthropic principle in cosmology, in M. S. Longair, ed. *Confrontation of Cosmological Theories with Observational Data*. Dordrecht: Reidel, 291–298.

Chamberlin, Thomas C. and Forest R. Moulton, 1900. "Certain Recent Attempts to Test the Nebular Hypothesis," *Science*, 12, 201–208.

Cleland, Carol E. 2001. "Historical Science, Experimental Science, and the Scientific Method," *Geology* 29, 978–990.

Cleland, Carol E. 2002. "Methodological and Epistemic Differences between Historical Science and Experimental Science," *Philosophy of Science*, 69: 474–496.

Coblentz, William W. and C. O. Lampland. 1924. "New Measurements of Planetary Radiation," *Science*, 60, 295.

Cocconi, Giuseppe and Philip Morrison. 1959. "Searching for Interstellar Communications," *Nature*, 184, 844.

Conway Morris Simon, . 1998. *The Crucible of Creation*, Oxford: Oxford University Press.

Conway Morris Simon. 2003. *Life's Solution: Inevitable Humans in a Lonely Universe*, Cambridge: Cambridge University Press.

Davies, Paul C. W. 2007. *Cosmic Jackpot: Why Our Universe Is Just Right for Life,* Boston: Houghton Mifflin.

Dawkins, Richard. 1997. *Climbing Mount Improbable*, W. W. Norton, New York.

Des Marais, David, J. A. Nuth, III, Louis Allamandola. 2008. "The NASA Astrobiology Roadmap," *Astrobiology*, 8: 715–730.

Dick, Steven J. 1996. *The Twentieth Century Extraterrestrial Life Debate: A Study of Science at its Limits*, Cambridge: Cambridge University Press.

Dick, Steven J. 2003. "Cultural Evolution, the Postbiological Universe and SETI," *International Journal of Astrobiology,* 2: 65–74; reprinted as Bringing Culture to Cosmos: The Postbiological Universe, in Dick and Lupisella (2009), pp. 463-488.

Dick, Steven J. 2009. "Cosmic Evolution: History, Culture and Human Destiny," in Dick and Lupisella (2009), pp. 25–59.

Dick, Steven J. and James E. Strick. 2004. *The Living Universe: NASA and the Development of Astrobiology*, New Brunswick, NJ: Rutgers University Press.

Dobzhansky, Theodosius. 1972. "Darwinian Evolution and the Problem of Extraterrestrial Life," *Perspectives in Biology and Medicine*, 15: 157–175

Christian De Duve. 1995. *Vital Dust: Life as a Cosmic Imperative*. New York: Basic Books.

Dyson, Freeman. 1988. *Infinite in All Directions*, New York: Harper and Row.

Fry, Iris. 2000. *The Emergence of Life on Earth: A Historical and Scientific Overview*. New Brunswick, New Jersey: Rutgers University Press.

Good, Gregory. 2000. "The Assembly of Geophysics: Scientific Disciplines as Frameworks of Consensus," *Studies in the History and Philosophy of Modern Physics*, 31, 259–292.

Hoyle, Fred. 1983. *The Intelligent Universe: A New View of Creation and Evolution*. New York: Holt, Rinehart and Winston.

Jeans, James. 1917. *The Motion of Tidally Distorted Masses*, London: Memoirs of the Royal Astronomical Society.

Koyré, Alexandre. 1957. *From the Closed World to the Infinite Universe*. Baltimore: Johns Hopkins University Press.

Kohler, Robert. E. 1982. *From Medical Chemistry to Biochemistry*. Cambridge: Cambridge University Press.

Kuiper, Gerard P., ed. 1949. *The Atmospheres of the Earth and Planets*. Chicago: University of Chicago Press.

Mayor, Michel. and Didier Queloz. 1995. "A Jupiter-Mass Companion to a Solar-Type Star." *Nature*, **378**, 355.

Mayr, Ernst. 1985. "The Probability of Extraterrestrial Intelligent Life," in Mayr (1988), pp. 67–74.

Mayr, Ernst. 1988. *Toward a New Philosophy of Biology: Observations of an Evolutionist*, Cambridge, MA: Harvard University Press.

Monod, Jacques. 1971. *Chance and Necessity*. New York: Vintage Books.

Oparin, Alexander I. 1924. *Proiskhozhdenie zhinzy*, translated by Ann Synge as "The Origin of Life," in J. D. Bernal, *The Origin of Life* (Cleveland and New York, 1957), and reprinted in D. W. Deamer and G. R. Fleischaker, *Origins of Life: The Central Concepts* (Boston, 1994), 31–71.

Oparin, Alexander I. 1936. *Vozhiknovenie zhizny na aemle*, translated as *The Origin of Life* (London, 1938), republished (New York, 1952) by Dover.

Peter D. Ward and Donald Brownlee. 2000. *Rare Earth: Why Complex Life is Uncommon in the Universe*. New York: Springer.

Petit, Edward and Seth B. Nicholson. 1924. "Radiation Measures on the Planet Mars," *Publications of the Astronomical Society of the Pacific*, 36, 269–272.

Reuyl, Dirk. and Erik Holmberg. 1943. "On the Existence of a Third Component in the System 70 Ophiuchi." *Astrophysical Journal*, **97**, 41–45.

Russell, Henry N. 1943. "Anthropocentrism's Demise." *Scientific American (July, 1943), 18–19.*

Russell, Henry N., R. S. Dugan, and J. Q. Stewart. 1945. *Astronomy: A Revision of Young's Manual of Astronomy*. Boston and New York: Ginn and Company.

Sagan, Carl. 1977. *The Dragons of Eden: Speculations on the Evolution of Human Intelligence*, New York: Random House.

Shklovskii, Ioseph S. 1965. "Multiplicity of Inhabited Worlds and the Problem of Interstellar Communications," In *Extraterrestrial Civilizations*, ed. G. V. Tovmasyan. Erevan: Akademii Nauk.

Simpson, George G. 1964. "The Non-Prevalence of Humanoids," *Science*, 143, February 21, 769–775, reprinted in *This View of Life: The World of an Evolutionist*, Harcourt, Brace & World, New York, pp. 253–271.

Sinton, William. 1957. "Spectroscopic Evidence for Vegetation on Mars." *The Astrophysical Journal* 126:231.

Sinton, William. 1959. "Further Evidence of Vegetation on Mars: The presence of large organic molecules is indicated by recent infrared-spectroscopic tests". *Science* **130** (3384):1234–1237.

Slipher, Edward C. 1927 "Atmospheric and Surface Phenomena on Mars," *Publications of the Astronomical Society of the Pacific*, **39** (August, 1927), 209–216.

Spinrad, Hyron, Guido Münch, and L. D. Kaplan. 1963. "The Detection of Water Vapor on Mars," *Astrophysical Journal*, **137** (1963), 1319–1321.

Sternberg, Robert J. editor. (2000), *Handbook of Intelligence*, Cambridge: Cambridge University Press.

Sternberg, Robert J. 2002. The search for criteria: why study the evolution of intelligence. In *The Evolution of Intelligence*, edited by R. J. Sternberg and J. C. Kaufman, Lawrence Erlbaum Associates, Mahwah, N. J, pp. 1–8

Strick, James E. (2004) 'Creating a Cosmic Discipline: The Crystallization and Consolidation of Exobiology, 1957–1973,' *Journal of the History of Biology*, 37: 131–80.

Sullivan, Woodruff T., III. 2009. *Cosmic Noise: A History of Early Radio Astronomy*, Cambridge: Cambridge University Press.

Tikhov, Gavriil A. 1955. "Is Life Possible on Other Planets?," Journal of the British Astronomical Association, 65 (1955), 193–204.

Tikhov, Gavriil A., 1960. *Principal Works: Astrobotany and Astrophysics, 1912–1957*. Washington, DC

Trumpler, Robert J. 1927. "Observations of Mars at the Opposition of 1924," *Lick Observatory Bulletin*, 387, 19–45.

Van de Kamp, Peter. 1963. "Astrometric Study of Barnard's Star from Plates Taken with the 24-inch Sproul Refractor," Astronomical Journal, 68 (September, 1963), 515–521.

Wolfe, Audra. 2002. "Germs in Space: Joshua Lederberg, Exobiology, and the Public Imagination, 1958-1964," *Isis*, 93, 183–205.

Zakariya, Nasser. 2017. *A Final Story*: *Science, Myth and Beginnings*. Chicago: University of Chicago Press.

Chapter 3
From the Physical World to the Biological Universe: Historical Developments Underlying the Search for Extraterrestrial Intelligence (SETI)

Abstract The SETI endeavor represents a test for a fundamental shift in cosmological worldview, from the physical world to the biological universe. I define the "biological universe" as the scientific worldview that holds that life is widespread throughout the universe. This chapter is meant to be a contribution to the ongoing endeavor to understand where the extraterrestrial life debate fits in the history of science.

3.1 Introduction

More than 30 years ago, the French historian of science Alexandre Koyré (1957) wrote his classic volume *From the Closed World to the Infinite Universe,* in which he argued that a fundamental shift in worldview had taken place in seventeenth-century cosmology. Between Nicholas of Cusa in the fifteenth century and Newton and Leibniz in the seventeenth, he found that the very terms in which humans thought about their universe had changed. These changes he characterized broadly as the destruction of the closed finite cosmos and the geometrization of space. The occasion of the International Bioastronomy Symposium in France is an especially appropriate time to argue that the SETI endeavor represents a test for a similar fundamental shift in cosmological world view, from the physical world to the biological universe. I define the *biological universe*, equivalent to what I have called before the *biophysical cosmology* (Dick 1989), as the scientific worldview that holds that life is widespread throughout the universe. In this case the biological universe does not necessarily supersede the physical universe, but a universe filled with life would certainly fundamentally alter our attitude toward the universe and our place in it. Although Koyré mentioned life beyond the Earth as an adjunct to the revolution from the closed world to the infinite universe, only in the 1980s has the history of science begun to give full treatment to the subject.

First published in *Bioastronomy: The Search for Extraterrestrial Life—The Exploration Broadens,* Proceedings of the Third International Symposium on Bioastronomy Held at Val Cenis, France, 18–23 June, 1990, Jean Heidmann and Michael Klein, eds. (Springer-Verlag, Berlin, 1991).

S. J. Dick, *Space, Time, and Aliens*, https://doi.org/10.1007/978-3-030-41614-0_3

The modern era in the extraterrestrial life debate is normally dated from Cocconi and Morrison's paper in 1959, and though one can always find precursors (see Chap. 6), this in my view is a valid perception. Cocconi and Morrison gave definite form to SETI, Frank Drake independently first carried out the experiment, a network of interested scientists began to form and met in Green Bank in November 1961, and the most distinctive part of the modern era of the extraterrestrial life debate—the Search for Extraterrestrial Intelligence by means of radio telescopes—was off and running. In this paper, after briefly reviewing some of the long-term steps toward the biological universe, I would like to examine the immediate precursors to this modern era in the 1940s and 1950s.

3.2 Long-Term Developments

By the 1950s there was, of course, a long tradition behind the debate over life on other worlds. Dick (1980, 1982) and Crowe (1986) have now well documented this history through the nineteenth century, and work on twentieth century history is now in progress (Dick 1996). It is now widely accepted that the extraterrestrial life debate began and was sustained by a "cosmological connection" stretching back to the ancient atomists and Aristotle in the fourth century B.C. and revitalized by the Copernican, Cartesian and Newtonian traditions in the seventeenth century. Copernicus made the planets Earth-like in theory, Descartes and Newton proposed other solar systems as part of their cosmologies, and the Newtonians added the crucial factor for them that a universe filled with life was in agreement with an omnipotent God, the God of natural theology. Philosophical principles such as plenitude and purpose also played a role, but, I would argue, a subordinate role given meaning only in the context of these cosmologies. So in a real sense, the mental adjustment from the physical world to the biological universe was, for many natural philosophers, made by the beginning of the eighteenth century.

But of course proof was another matter. Though much was learned about our own planetary system in the next two centuries, most of the nineteenth century was spent in explorations of the philosophical or religious implications of this idea. Thomas Paine (1793) said that he who believed in Christianity and plurality of worlds had thought but little of either, and himself came down on the side of other worlds. At the opposite extreme William Whewell (1853), Master of Trinity College Cambridge, rejected plurality of worlds in favor of Christianity. And as a middle ground Thomas Chalmers (1817) and others attempted to reconcile religion and extraterrestrial life.

Two great scientific developments in the latter half of the nineteenth century gave credence to extraterrestrial life: the rise of astronomical spectroscopy and Darwin's theory of evolution. Again and again in twentieth century discussions of life beyond the Earth, we see reference to these two nineteenth-century achievements; astrophysics shows the elements of matter to be the same throughout the universe, and Darwin's theory of biological evolution by natural selection not only would hold for organisms throughout the universe but also may be viewed as the end product of physical evolution in the universe. It is notable that Percival Lowell was much influenced by the idea of planetary evolution, not only in his books on Mars but also in *The Evolution of Worlds* (1909). In *The Study of Stellar Evolution* (1907), George Ellery Hale pointed out that while evolution was not a new idea to astronomers in the nineteenth century, "it has occupied a more important position since Darwin published his great work." It did not escape Hale's notice that in 1859, the very year of the publication of the *Origin of Species,* Kirchoff began his experiments aimed at determining the chemical composition of the Sun, launching the field of stellar evolution. Despite the extreme skepticism of Lowell's claims of canals on Mars, prominent astronomers in the early twentieth century such as W. W. Campbell (1920) could point to the results of spectroscopy to support the broader claim of life in the universe: "If there is a unity of materials, unity of laws governing those materials throughout the universe, why may we not speculate somewhat confidently upon life universal?" he asked. He even spoke of "other stellar systems … with degrees of intelligence and civilization from which we could learn much, and with which we could sympathize." Such a general argument was enough to carry the day for many astronomers up to 1920.

3.3 Short-Term Developments

With the birth of SETI in 1959—the centennial year of Darwin's *Origin* and Kirchoff's identification of elements in the Sun—all of these general steps to the biological universe lay in the background. From the viewpoint of 1959, the more immediate steps in the emergence of the biological universe stretched back less than a generation. In fact the 1950s was emerging from a 25-year period of extreme skepticism regarding life in the universe. It is significant that the general principles of the uniformity of nature and stellar evolution had not been enough for most scientists to accept life on other worlds in the first half of the twentieth century in the face of contrary theories. Just about the time Campbell wrote his article in 1920, James Jeans (1919, 1923) argued that the Solar System may be unique, or at the very least "astronomy … begins to whisper that life must necessarily be somewhat rare" in the universe. This whisper grew to a crescendo by the 1930s. Harvard Observatory Director Harlow Shapley (1923), just fresh from his triumphant use of globular clusters to show the eccentric position of the Solar System in the galaxy, held that planetary systems were unlikely and habitable planets very uncommon. And Henry Norris Russell et al. (1926) agreed that planetary systems were infrequent and habitable planets pure speculation. From 1920 to about 1945 we see the idea of extraterrestrial life at a low point—the biological universe was in danger of extinction. The reason is to be found almost totally in a shift in theories of planetary formation, from the nebular hypothesis to the *close encounter* or *tidal hypothesis.* Developed by Chamberlin and Moulton at the University of Chicago around 1900 (Brush 1978), this theory in the hands of Jeans gave a pessimistic view of the

Table 3.1 Estimates of the frequency of planetary systems, 1920–1961

Author	Argument	No. of planetary systems in galaxy	No. of habitable planets in galaxy
Jeans (1919, 1923)	Tidal theory	Unique	1
Shapley (1923)	Tidal theory	"Unlikely"	"Uncommon"
Russell et al. (1926)	Tidal theory	"Infrequent"	"Speculation"
Jeans(1942a)	No. of stars	10^2	–
Jeans (1942a, b)	Improved tidal	One in six stars	Abundant
Russell (1943)	Companions	Very large	$>10^3$
Page (1948)	Weizsäcker	$>10^9$	$>10^6$
Hoyle (1950)	Supernovae	10^7	10^6
Kuiper (1951)	Binary star statistics	10^9	–
Hoyle (1955)	Stellar rotation	10^{11}	–
Shapley (1958)	Nebular hypothesis	10^6–10^9	–
Huang (1959)	Stellar rotation	10^9	10^9
Hoyle (1960)	Stellar rotation	10^{11}	10^9
Struve (1961)	Stellar rotation	$>10^9$	–

Adapted from Dick (1996:199)

possibility of planetary systems, since stellar encounters would be very rare. Jeans' ideas were widely accepted in the scientific community, and his numerous popularizations of this idea spread it far and wide. As long as this idea held sway, planetary systems were freak occurrences divorced from normal stellar evolution.

At least four factors may be discerned in the modern reemergence of the biological universe. First, and arguably most importantly, a radical shift occurred once again in the estimation of the likelihood of planetary systems, from both observational and theoretical points of view. Dynamical objections to his theory led Jeans (1942a, b) in the final years of his life to postulate a much larger primordial Sun, and therefore to conclude that stellar collisions might not be so rare after all, perhaps forming planets around one in six stars. But new developments quickly moved beyond his tidal theory; I would date the turning point for planetary systems at 1943 (Table 3.1). In that year Russell (1943) spoke of "a radical change—indeed practically a reversal—of the view which was generally held a decade or two ago," regarding the scarcity of planetary systems. He specifically referred to the apparent discovery of planetary companions by Strand (1943) around 61 Cygni, and by Reuyl and Holmberg (1943) around 70 Ophiuchi. Both used the technique of photographic astrometry to detect perturbations in the orbits of these double stars.

Although their discoveries would eventually prove spurious, at the time Russell undoubtedly took them as vindication of his earlier analysis (Russell 1935) that there were grave angular momentum problems with the close encounter hypothesis, and that some other theory must replace it. By 1944 Carl F. von Weizsäcker had come up with the beginnings of such a theory, a modified nebular hypothesis, which he elaborated (von Weizsäcker 1951) and which opened the floodgates to similar theories(Brush 1981, 1990; Jaki 1978). The Swedish physicist Alfvén, Hoyle in Britain, Kuiper in the U.S. and others elaborated their own forms of the nebular hypothesis. Planetary systems were returned to the realm of stellar evolution. Hoyle estimated by 1950 ten million planetary systems in the galaxy and a million habitable planets. Kuiper (1951), on the basis of binary star separation statistics, estimated a billion planetary systems in the galaxy. The new cosmology, with its vastly expanded universe full of galaxies, also supported many planetary systems. Shapley (1958), formerly so skeptical, detailed the arguments here and also concluded for billions of planetary systems in the galaxy. Struve's estimate in 1961 of billions of planetary systems in the galaxy was therefore quite common by that time.

Observational research by Otto Struve also lent support to the view of many planetary systems. Struve's work (1930, 1950) on stellar rotation showed that there was a discontinuity at the F spectral type where stellar rotation slowed. Although several braking mechanisms were possible, by the 1950s Struve (1952), his student Su-Shu Huang (1957, 1959) and others were plausibly surmising that the angular momentum might have gone into planetary systems. In Huang's words "... planetary systems emerge as axial rotation declines. According to this view, planets are formed around the main sequence stars of spectral types later than F5. Thus, planets are formed just where life has the highest chance to flourish. Based on this view we can predict that nearly all single stars of the main sequence below F5 and perhaps above K5 have a fair chance of supporting life on their planets. Since they compose a few percent of all stars, life should indeed be a common phenomenon in the universe." By 1952 Struve (1952) even published a "Proposal for a Project of High-Precision Stellar Radial Velocity Work," designed to detect planets at the level of a few 100 m/s. It is interesting that although both Kuiper's binary star separation statistics and Struve's F5 stellar rotation discontinuity were known in the 1930s, it was not until the 1950s—after the downfall of the rare encounter hypothesis—that they were used as arguments for many planetary systems.

Whether or not one accepted these specific theoretical and observational arguments for planetary systems, with increasing knowledge of stars and stellar evolution, one could still argue more generally, as Struve himself did in 1955, that the physical properties of the Sun resembled in every respect other stars of similar type, right down to axial rotation. He argued that we must infer that this similarity also extends to star formation, and accompanying planets. "Since we cannot adduce a proof one way or the other, we must rely upon what seems to be the most logical hypothesis. And this is without doubt the assumption that all, or at least most, dwarf stars of the solar type have planetary systems. The total number of planets in the Milky Way may thus be counted in the billions" (Struve 1955, 146). Taking the Solar System as an example, where one of nine planets clearly has life, one (Mars) may have had life, and one (Venus) may have life in the future, Struve concluded that the number of planets in the Milky Way with some form of life might also number in the billions. Planetary systems, as supported by renewed forms of the nebular hypothesis and theories of stellar evolution, by the apparent indirect observations of planets, by the facts of stellar rotation, and by the new cosmology, were thus the first and primary factor in the reemergence of the biological universe.

Secondly, the evidence for life in the Solar System grew increasingly positive, feeding hope that this was an indication of the case for the broader universe. It is true that Rupert Wildt's postulation in 1940 of a greenhouse effect on Venus was gradually accepted and finally eliminated that planet from consideration as a biological habitat. But conditions on Mars were still believed to be acceptable, if harsh, for life, and in 1947 at the famous University of Chicago conference on planetary atmospheres Kuiper (1949) postulated plants similar to lichens on Mars, and in 1957 and 1959 Sinton gave his widely accepted spectroscopic proof of vegetation on Mars. Again and again, astronomers reasoned that if life had developed on two sites in our Solar System, then it was most likely common throughout the universe.

Thirdly, regarding the crucial question of the origin of life, the idea of chemical evolution was gaining widespread acceptance in the 1950s. Oparin's work, begun in the 1920s in Russia, was first published in English in 1953. In the same year Miller (1953), stimulated by Harold Urey's conclusion that the primitive Earth must have had a reducing atmosphere, published his results of their first experiments on the formation of organic compounds under conditions of a reducing atmosphere. In 1957 the first International Symposium on the Origin of Life was held in Moscow, and that field was off and running just prior to the launching of SETI. Melvin Calvin, a representative of this chemical evolution tradition, participated in discussions of life beyond the Earth at the Lunar and Planetary Exploration Colloquia at least as early as 1959 (Calvin 1959), and was present at the 1961 Green Bank meeting, during which he was notified he had received the Nobel Prize.

Finally, we should not forget that it was a technological development—the emergence of radio astronomy—that made SETI possible for the first time over large scales. Whatever other influences they may have felt, it is certain that Drake and his SETI successors were most directly influenced by the emergence and development of this field (Drake 1960, 1961; Morrison et al. 1977, and many others). They may or may not have been influenced by some of the other developments we have mentioned, though the influence of Struve on Drake, at least to the extent of allowing Drake to perform the experiment at the NRAO, was obvious and direct. Bracewell also emphasized the importance of Struve in his own development and that of other SETI pioneers, and Calvin acknowledged the influence of Shapley (Swift 1990). The importance of these ideas in leading to the use of radio telescopes for SETI is obvious, but without that radio technology SETI would not be the well-developed program it is today.

3.4 Summary

As Cocconi and Morrison wrote their landmark paper in 1959 and Drake made his first radio search for extraterrestrial life in 1960, they had behind them—whether they knew it or not—a widespread acceptance of the idea of such life dating from the seventeenth century, of the principles of evolution and uniformity of nature dating from the nineteenth century, but only 15 years of the reemerging tradition that planetary systems and life were likely. Belief in vegetation on Mars was at a high point due largely to the work of Kuiper and Sinton. Belief in abundant planetary systems was widespread due to the work of Russell, von Weizsäcker, Struve, Hoyle and others. And belief in the ability of life to develop via chemical evolution was on the rise due to the work of Urey and Miller on synthesis of amino acids under conditions of a primitive atmosphere. Moreover, at least two prominent astronomers explicitly acknowledged at an early stage the potentially revolutionary character of these ideas. By 1958 Shapley characterized the existence of extraterrestrial life as a possible "Fourth Adjustment"that humanity would have to make in its overall view of the universe. In 1961 Struve wrote that astronomy has had three great revolutions in the past 400 years: Copernicus'removal of the Earth from the center of the Solar System, Shapley's removal of the Solar System from the center of the galaxy, and the revolution occurring now, embodied in the question "Are we alone in the universe?"

While the biological universe has been widely debated for more than two millennia, and widely accepted for more than two centuries, it has fallen to the last half of the twentieth century to provide observational proof for the hypothesis that there is more to the universe than matter in motion. Although difficult, and although labeled by some a pseudoscience (Tipler 1987, 1988), that task in my view falls squarely in the tradition of the history of science, which frames hypotheses and attempts to test them. In a significant local result, the Viking test for life on Mars failed. It is up now to SETI programs to test the hypothesis on the cosmological scale, and to determine whether we, or future generations, will really need to make the shift to a new scientific worldview, from the physical world to the biological universe. If we do make that shift, I predict the effect on astronomy and culture will be even more profound than the move from the closed world to the infinite universe three centuries ago.

3.5 Commentary 2020

This paper was given at the Third International Symposium on Bioastronomy, held in the region of Val Cenis, Savoie, in the French Alps in June, 1990 (Heidmann and Klein 1991). Although the definitive discovery of planets beyond our Solar System was still 5 years in the future, the program was rich in papers about methods for detecting such planets by radial velocity methods, astrometry, and occultation. As we saw in Chap. 2, the first method was the one used for the detection a planet around 51 Pegasi announced in 1995. The latter, eventually termed the photometric method, was the method used by the Kepler spacecraft, responsible for the vast majority of exoplanets detected to date. The astrometric method has thus far detected only one exoplanet. Struve's 1952 "Proposal for a Project of High-Precision Stellar Radial Velocity Work" mentioned in this chapter, designed to detect planets at the level of a few 100 m/s, was indeed prophetic. Radial velocities are now routinely measured to less than 1 m/s.

In addition to its scientific components, the meeting was notable for its wide-ranging character, reflecting the interests of its organizers, the French astronomer Jean Heidmann, the American astronomer Michael J. Klein, and the scientific organizing committee. Among the astronomers and biologists were experts on consciousness and cognition (William Calvin and Irene Pepperberg), the anthropologist Ben Finney, sociologist David Swift, science fiction writer David Brin, and one historian of science (myself). In addition, a special session of the meeting discussed the post-detection protocols in the event of a SETI detection. All of this was a foreshadowing of deeper interest in the societal implications, evident in Part II of this volume.

References

Brush, Stephen. 1978. *Journal for the History of Astronomy*, 9, 1–41, 77–104.

Brush, Stephen. 1981. *Space Science Comes of Age,* ed. Paul Hanle and Von del Chamberlain, Smithsonian Press: Washington, D.C.

Brush, Stephen. 1990. *Reviews of Modern Physics*, 62, 43–112.

Calvin, Melvin. 1959. *Proceedings of the Lunar and Planetary Exploration Colloquium*, April 25, 1959, 1, no. 6, 8–18.

Campbell, W. W. 1920. *Science*, 52 (December 10, 1920), 550.

Chalmers, Thomas.1817. *A Series of Discourses on the Christian Revelation, Viewed in Connexion with Modern Astronomy,* Edinburgh.

Cocconi, Giuseppe and Philip Morrison.1959. *Nature*, 184, 844.

Crowe, Michael. 1986. *The Extraterrestrial Life Debate, 1750–1900: The Idea of a Plurality of Worlds from Kant to Lowell,* Cambridge University Press, Cambridge.

Dick, Steven J. 1980. *Journal of the History of Ideas*, 1–27.

Dick, Steven J. 1982. *Plurality of Worlds: The Origins of the Extraterrestrial Life Debate from Democritus to Kant,* Cambridge University Press, Cambridge.

Dick, Steven J. 1989. *The Planetary Report*, March–April, 13–17.

Dick, Steven J. 1996. *The Twentieth Century Extraterrestrial Life Debate: A Study of Science at its Limits,* Cambridge University Press, Cambridge.

Drake, Frank D. 1960. *Sky and Telescope*, 19, 140–43.

Drake, Frank D.1961. *Physics Today*, 14, 40.

Hale, George Ellery. 1907. *The Study of Stellar Evolution,* University of Chicago Press, Chicago, 2.

Heidmann, Jean and Michael J. Klein. 1991. *Bioastronomy: The Search for Extraterrestrial Life—The Exploration Broadens*. Berlin, Springer.

Hoyle, Fred. 1950. *The Nature of the Universe,* 26, 101.

Hoyle, Fred. 1955. *Frontiers of Astronomy,* 83, 104–05.

Hoyle, Fred. 1960. *The Nature of the Universe,* 2ndd ed., 32, 81, 90.

Huang, Su-Shu. 1957, *Publications of the Astronomical Society of the Pacific,* 69, 427.

Huang, Su-Shu. 1959. *Publications of the Astronomical Society of the Pacific*, 71, 421.

Jaki, Stanley L. 1978. *Planets and Planetarians: A History of Theories of the Origin of Planetary Systems,* Scottish Academic Press, Edinburgh.

Jeans, James. 1919. *Problems of Cosmogony and Stellar Dynamics*, Cambridge University Press, Cambridge, 290.

Jeans, James. 1923. *The Nebular Hypothesis and Modern Cosmogony*, Cambridge University Press, Cambridge, 30.
Jeans, James. 1942a. *Science*, 95, 589.
Jeans, James: 1942b. *Nature*, 149, 695.
Koyré, Alexandre. 1957. *From the Closed World to the Infinite Universe*, Johns Hopkins University Press, Baltimore.
Kuiper, Gerard P. 1949. in *Atmospheres of the Earth and Planets*, University of Chicago Press, Chicago.
Kuiper, Gerard P.: 1951, ch. 8 in *Astrophysics*, (McGraw Hill, New York) ed. J. A. Hynek, 416–417.
Lowell, Percival. 1909. *The Evolution of Worlds* (New York).
Miller, Stanley L. 1953. *Science*, 117, 528–29.
Morrison, Philip, John Billingham and John Wolfe. 1977. *The Search for Extraterrestrial Intelligence*NASA: Washington.
Page, Thornton. 1948. *Physics Today* (October, 1948), 12–24.
Paine, Thomas. 1793. *Age of Reason.*
Reuyl, Dirk and Erik Holmberg. 1943. *Astrophysical Journal*, 97, 41–45.
Russell, Henry Norris. 1935. *The Solar System and Its Origin.*
Russell, Henry Norris. 1943. *Scientific American* (July, 1943), 18–19.
Russell, Henry Norris, R. S. Dugan and J. Q. Stewart. 1926. *Astronomy*, vol. 1, 468.
Shapley, Harlow. 1923. *Harper's Monthly Magazine*, 146, 716–22.
Shapley, Harlow. 1958. *Of Stars and Men*, Beacon Press, Boston, 104–114.
Sinton, William. 1957. *Astrophysical Journal*, 126, 231–39.
Sinton, William. 1959. *Science* (6 November, 1959), 1234.
Strand, Kaj. 1943. *Publications of the Astronomical Society of the Pacific*, 55, 29–32.
Struve, Otto. 1930. *Astrophysical Journal*, 72, l.
Struve, Otto. 1950. *Stellar Evolution*, 150–51, 231–39.
Struve, Otto. 1952. *Observatory*, 72, 199–200.
Struve, Otto. 1955. *Sky and Telescope*, 14, 137–140, 146.
Struve, Otto. 1961. *The Universe*, MIT Press, Cambridge, Mass., 157–59.
Swift, David. 1990. *SETI Pioneers*, University of Arizona Press, Tucson, 54–85; 130, 141-44.
Tipler, Frank. 1987. *Physics Today*, December, 1987, 92.
Tipler, Frank. 1988. *Physics Today*, September, 1988, 14–15 and 142–44.
von Weizsäcker, C. F.. 1944. Zeitschrift für Astrophysik, 22, 319–55.
von Weizsäcker, C. F.. 1951. *Astrophysical Journal*, 114, 165–86.
Whewell, William. 1853. *Of the Plurality of Worlds: An Essay*, London.

Chapter 4
The Biophysical Cosmology: The Place of Bioastronomy in the History of Science

Abstract The recent discoveries of planets around Sun-like stars, possible primitive Martian fossil life, and conditions on Europa conducive to microbial life, render more urgent the question of the place of bioastronomy in the history of science. This paper argues that the tenets of bioastronomy constitute a *biophysical cosmology*, a scientific worldview that holds that life is common throughout the universe. Many of the activities of the field of bioastronomy are tests of this cosmology. Like other cosmologies, the biophysical cosmology bears strongly on humanity's place in the universe. Cosmological status may also be useful in discussing the implications of contact, when one considers the response to other cosmologies as partial, if imperfect, analogues.

4.1 Introduction

The field of bioastronomy, which encompasses both the search for microbial life often subsumed under the name *exobiology* and the Search for Extraterrestrial Intelligence (SETI), has a long and checkered history that now approaches four decades. During this time its status as science has often been questioned. Evolutionary biologist George Gaylord Simpson (1964) called exobiology "a science without a subject." Twenty years later physicist Frank Tipler (1987) labeled bioastronomy a pseudoscience. And shortly before U.S. government funding was cut for SETI, Harvard evolutionary biologist Ernst Mayr (1993) called the radio search a "highly dubious endeavor" and a waste of taxpayers' money, based on his calculation of the low odds of the evolution of extraterrestrial intelligence. These claims, and the spectacular discoveries now being made in the field, require that we place bioastronomy in a proper context in the history of science.

First published in *Astronomical and Biochemical Origins and the Search for Life in the Universe*, C.B. Cosmovici, S. Bowyer and D. Werthimer, eds. (Editrice Compositori, Bologna, 1997), 785–88.

S. J. Dick, *Space, Time, and Aliens*, https://doi.org/10.1007/978-3-030-41614-0_4

4.2 Bioastronomy as Cosmology

To state my conclusion first: *bioastronomy* (a term used to describe both a discipline and its tenets, used here in the latter sense) must be conferred the status of a cosmological worldview. This worldview has a long history of acceptance (Dick 1982; Crowe 1986; Guthke 1990), but has only recently been subject to significant scientific testing. Although the idea of life in the universe has sporadically been seen as a worldview by some of its pioneers, bioastronomy as cosmology has not received the attention it deserves. As early as 1958, astronomer Harlow Shapley saw extraterrestrial life as the "Fourth Adjustment" in humanity's view of itself, after the geocentric, heliocentric and galactocentric revolutions. Otto Struve (1962) held a similar view, calling the search for life in the universe the third great revolution in astronomy, after Copernicanism in the sixteenth century and Shapley's "galactocentric" revolution in the early twentieth century. And Barney Oliver and Billingham (1972), in his influential Cyclops Report spoke of a "biocosmology." All these pioneers implicitly recognized one fact: that the concept of abundant life in the universe is more than an idea, more than another theory or hypothesis, more, indeed, than most scientific disciplines can hope to be; it is sufficiently comprehensive to qualify as a worldview, comparable to other overarching worldviews in the past. Moreover, because it is testable it is a scientific worldview, one that makes an assumption about the basic nature of the universe, and thus a cosmology. Because it encompasses both physical and biological elements, we term it here the *biophysical cosmology* (Dick 1989). Stated another way, if astronomical science since the ancient Greeks has been largely an exploration of the physical world, the quickening pace and more solid results of the extraterrestrial life debate in the twentieth century herald a transition in cosmological thought from the physical world to the biological universe (Dick 1991, 1996).

Seen in this light, bioastronomy as cosmology yields a unified picture of the activities and goals of the field. The search for life on Mars culminating with the Viking project in 1976 was not only a driving force behind the American space program (Ezell and Ezell 1984), but it was also clearly a local test for the biophysical cosmology, the view that life is widespread throughout the universe. And if that test failed to satisfy aspirations for a universal biology, this failure may be partially compensated if the claims of past Martian life prove true. In either case the outcome has ramifications for bioastronomers well beyond Mars. Meanwhile, the search for interstellar organics, for extrasolar planets, and for artificial radio signals of extraterrestrial origin are direct or indirect tests for this cosmology on a larger scale. And much research in origins of life is now supported by the NASA Exobiology Program (Klein 1986; Space Studies Board 1990), not only because of the possible extraterrestrial origin of life on Earth by impact delivery of organics, but also because experiments on chemical evolution under primitive Earth conditions may shed light on estimates of the abundance of extraterrestrial life.

4.3 Role of Cosmic Evolution

The concept of cosmic evolution, invoked sporadically throughout history and given impetus as Darwinian ideas spread to the universe at large, is the central assumption of the biophysical cosmology. Percival Lowell understood at the beginning of the century that the Solar System was evolving; his picture of a dying Mars whose inhabitants desperately tried to channel their water resources, was the epitome not just of an overactive imagination, but also of his concept of a Solar System constantly subject to change. That Solar Systems might exist beyond our own was an implication of the nebular hypothesis, and even if the tidal close-encounter theory of James Jeans and others temporarily eclipsed this hypothesis, both theories encompassed a dynamically evolving universe. If extrasolar planets did exist, it was postulated again and again in the second half of our century that chemical evolution and the origins and evolution of life had taken their course, according to Darwinian principles, on each planet. The Drake Equation incorporating all these parameters, for all its uncertainties and consequent eight order-of-magnitude variations in the estimates for the number of technological civilizations in the galaxy, served for the last four decades of the twentieth century as the symbol and the icon of cosmic evolution (Chap. 7). Even if it did not render definitive answers in our century, it may serve as bioastronomy's research agenda for the twenty-first century.

4.4 Science at its Limits

The work of bioastronomy, under the guiding concept of cosmic evolution, has not been carried out without grave difficulties. In fact, one must agree that tests for the biological universe have been carried out at the very limits of science. William Whewell noted this characteristic of the debate almost 150 years ago in his treatise *Of a Plurality of Worlds*, where he stated that "the discussions in which we are engaged belong to the very boundary regions of science, to the frontier where knowledge, at least astronomical knowledge end, and ignorance begins." Although our century has pushed back the ignorance considerably, most practitioners in the field still acknowledge that we are treading at the limits of science. But this striving for knowledge at the boundaries, aside from being one method by which new discoveries must be made, is precisely one of the compelling reasons the extraterrestrial life debate is of such interest to the history of science.

Again, one need not look far in the debate for examples of science attempting to function at its limits. The most notorious case was the extended debate over Lowell's claim of artificial canals on Mars, carried out beginning in 1895, reaching a peak at the 1909 opposition, and taking many years after that to taper off. The Martian canal controversy was by no means unique in the difficulty of its resolution; later claims for Martian vegetation, and even the results of the Viking experiments conducted *in situ* were controversial. Similarly, the search for planetary systems is a long chronicle of spurious claims beginning well before the well-known announcement of Van de Kamp in 1963 that one or two planets exist around Barnard's star, a claim now believed also to be spurious.

In the case of the origin and evolution of life, the difficulties were of a different nature. No one could observe the primitive Earth atmosphere or the origin of life. The constituents of that atmosphere were assumed, and even if one accepted them, the results were only suggestive. Moreover, the pioneering observations of complex organic molecules in space, and the discovery of amino acids in meteorites, were still only suggestive as far as the abundance of life is concerned. And yet, it should not be forgotten that progress *was* made in the debate, albeit at an excruciatingly slow pace. First intelligence, then vegetation, then organic molecules were banished from the Martian surface, only to have a possible detection of Martian fossil life (McKay et al. 1996). Despite all the spurious claims, once the technology was refined, planetary systems were rapidly discovered. And even the seemingly endless task of the SETI programs was given some hope of resolution by the construction of multichannel spectrum analyzers.

4.5 Implications

As with any cosmology, the concerns of the biophysical cosmology extend well beyond the scientific. Much science fiction literature and film, as well as unsubstantiated claims of UFOs, are attempts to deal with the new cosmology in popular culture (Dick 1996). The huge popular response to movies such as *ET, Close Encounters of the Third Kind,* and *Independence Day,* is a partial measure of the pervasiveness of interest in the subject. And as with previous cosmologies, the philosophical and religious implications of the world view must be addressed—a process that began in the seventeenth century, peaked in the 19th, and may once again be given impetus by the discoveries of the late twentieth century (Dick 1996).

Finally, to the extent that the discovery of extrasolar planets and possible Martian fossil life increase the likelihood of extraterrestrial intelligence, the concept of the biophysical cosmology may also help provide guidelines as we contemplate the implications of contact in the event of success in SETI. Although historical analogues must be used with caution, and although comparisons with physical culture contact on Earth are unlikely to be useful in assessing impact of long-range extraterrestrial signals, the reactions to massive change in world view such as Copernicanism and Darwinism may hold clues to the reaction of the discovery of ETI (Dick 1995).

4.6 Summary

Like all cosmologies, the biophysical cosmology makes a claim about the large-scale nature of the universe: that life is not only a possible implication, but a basic property, of the universe, an outcome of cosmic evolution distinct from other cosmologies in that it extends well beyond the physical evolution of planets, stars and galaxies. Like all cosmologies, it redefines our place in the universe. And most importantly, like other cosmologies, in the twentieth century the biophysical cosmology has become increasingly testable—even if it still embodies philosophical assumptions along with scientific theory and observation, and even if it functions at the limits of science.

Finally, bioastronomy as cosmology accounts for the passionate nature of the extraterrestrial life debate, as both proponents and opponents realize that much more than a scientific theory is at stake. Worldviews do not change rapidly, nor without compelling reason. Nevertheless, vociferous critics notwithstanding, both science and the public have perhaps never before been so willing to accept a new worldview. All that remains is compelling evidence, which will be followed by intense debate over the uncertain implications for humanity.

4.7 Commentary 2020

This paper was given at the fifth Triennial International Conference on Bioastronomy, held on the island of Capri, Italy from July 1–5, 1996 (Cosmovici et al. 1997). Attendees included about 200 astronomers, biologists, chemists, physicists, and myself representing the humanities and social sciences and attempting to place the extraterrestrial life debate in historical context. Previous bioastronomy conferences were held in Boston (1984) Lake Balaton, Hungary (1987), Val Cenis, France (1990), and Santa Cruz (1993). The meetings represented another advance in the new discipline: they were organized by the International Astronomical Union Commission 51 on Bioastronomy, which had been newly created in 1982—another recognition that the field was becoming more coherent as it advanced. These meetings proved important in legitimizing the subject of exobiology and SETI. It is notable that the astronomers considered their subject a branch of astronomy, thus "bioastronomy" rather than exobiology or astrobiology, as the field would later become known as biologists dominated the program.

The meeting followed an established pattern where the origins of life, the Solar System, the search for extrasolar planets, and SETI were discussed in turn. The date of the meeting proved crucial because the previous October the Swiss astronomers Michel Mayor and Didier Queloz had announced the first unambiguous detection of planets beyond the Solar System, the now-famous 51 Pegasi b. This was a landmark event—the beginning of the discovery of thousands of planets that had previously only been theorized but not observed.

One final connection went largely unremarked at the time. The Isle of Capri is only about 25 miles off the Adriatic coast from Naples, a 2-hour boat ride on the hydrofoil across the Gulf of Naples. And Naples itself is only 20 miles west of a little town called Nola. My attempts to visit Nola were fruitless; everyone said "nothing to see in Nola." But not only is Nola the place where Augustus Caesar died in 14 AD (and his father before him), it is also the birthplace of Giordano Bruno, sometimes called "the Nolan," in 1548. Bruno (see Fig. 42.3 and Chap. 42) was a philosopher (and some would say troublemaker) burned at the stake in 1600 in Rome by the Roman Inquisition, at least in part for his belief in other inhabited worlds (Martinez 2018). So this bioastronomy conference was held less than 50 miles from the birthplace of a philosopher who gave his life partly for his belief in a subject that had now become a robust scientific discipline.

The volume of proceedings of the meeting (Cosmovici et al. 1997) was dedicated to Carl Sagan, who was not present at the meeting; he was 6 months away from his death the following December.

References

Cosmovici, Cristiano, Stuart Bowyer and Dan Werthimer. 1997. *Astronomical and Biochemical Origins and the Search for Life in the Universe.* Editrice Compositori: Bologna, Italy.

Crowe, M.J. 1986. *The Extraterrestrial Life Debate 1750-1900: The Idea of a Plurality of Worlds from Kant to Lowell,* New York and Cambridge: Cambridge University Press.

Dick, S.J. 1982. *Plurality of Worlds: The Origins of the Extraterrestrial Life Debate from Democritus to Kant,* New York and Cambridge: Cambridge University Press.

Dick, S.J. 1989. The Concept of extraterrestrial intelligence: an emerging cosmology?, *The Planetary Report,* 9, 13–17.

Dick, S.J. 1991. From the physical world to the biological universe: historical developments underlying SETI, in *Bioastronomy—the Search for Extraterrestrial Intelligence: The Exploration Broadens* J. Heidmann and M. J. Klein, Eds., Berlin and New York: Springer, 356–363.

Dick, S.J. 1995. Consequences of success in SETI: lessons from the history of science, in *Progress in the Search for Extraterrestrial Life,* San Francisco, Astronomical Society of the Pacific (G. Seth Shostak, Ed.), 521–532.

Dick, S.J. 1996. *The Biological Universe: The Twentieth Century Extraterrestrial Life Debate and the Limits of Science,* New York and Cambridge: Cambridge University Press.

Ezell, E. C. and Ezell, L. N. 1984. *On Mars: Exploration of the Red Planet; 1958-1978,* Washington, D.C.: NASA.

Guthke, K.S. 1990. *The Last Frontier: Imagining Other Worlds from the Copernican Revolution to Modern Science Fiction,* Ithaca: Cornell University Press.

Klein, H.P. 1986. Exobiology Revisited, *Advances in Space Research,* 6 (12), 187–92.

Martinez, Alberto A. 2018. *Burned Alive: Giordano Bruno, Galileo and the Inquisition.* London, Reaktion Books.

Mayr, E. 1993. The search for intelligence, *Science, **259,*** 1522.

McKay, D.S., Gibson, E. *et al.* 1996. Search for past life on Mars: possible relic biogenic activity in Martian meteorite ALH84001, *Science,* 273, 924–930.

Oliver, B. and Billingham, J., eds. 1972. *Project Cyclops,* NASA Ames Research Center.

Shapley, H. 1958. *Of Stars and Men: The Human Response to the Expanding Universe,* Boston: Beacon Press, ch. 8, "The Fourth Adjustment"

Simpson, G. G. 1964. The non-prevalence of Humanoids, *Science,* **143,** 769.

Space Studies Board. 1990. Committee on Planetary Biology and Chemical Evolution, *The Search for Life's Origins: Progress and Future Directions in Planetary Biology and Chemical Evolution,* National Academy Press: Washington, D. C.

Struve, O. 1962. *The Universe,* Cambridge, Mass.: MIT Press.

Tipler, F. 1987. Review of M. D. Papagiannis, *The Search for Extraterrestrial Life: Recent Developments, Physics Today* (December, 1987), 92

Chapter 5
The Biological Universe Revisited

Abstract Cosmic evolution has been seen as leading to two possible worldviews: a physical universe in which life is rare or unique to Earth, and a biological universe, in which the processes of cosmic evolution commonly end in life. These two worldviews now hang in the balance, in the same way that the heliocentric and geocentric worldviews were in the balance 400 years ago when Galileo wrote his *Dialogue on the Two Chief World Systems* (1632). Astrobiology is the science that will decide which of the two modern astronomical worldviews is true. A third worldview, the postbiological universe, is also possible and deserves more discussion. The confirmation of one of these worldviews will have profound implications for human destiny.

5.1 Introduction

Almost ten years ago I documented the twentieth-century history of what I described as a major cosmological worldview, the biological universe—the idea that the universe is full of life (Dick 1996). In this paper I want to revisit that claim, and suggest there is another possibility beyond the biological universe. To put it another way, I want to claim that cosmic evolution harbors at least three vastly different possibilities for the universe. The ultimate product of cosmic evolution may be only planets, stars and galaxies—a *physical universe* in which we are unique or extremely rare. By contrast, cosmic evolution through biological evolution may commonly result in life, mind and intelligence, an outcome that I term the *biological universe.* Finally, there is another possibility not often discussed, but that I wish to argue needs to be taken seriously. Taking a long-term view, cultural evolution on other planets may have already produced artificial intelligence, constituting a *postbiological universe.*

Seen within this framework, these possible outcomes are not just speculation—they are the result of taking cosmic evolution seriously in all three components of the Drake Equation: astronomical, biological and cultural (see Chap. 7 and Table 12.2). Just as the outcome of astronomical evolution was once speculative, and just

First published in *The New Astronomy: Opening the Electromagnetic Window and Expanding our View of Planet Earth*, Wayne Orchiston, ed. (Springer: 2005), 15-26.

S. J. Dick, *Space, Time, and Aliens*, https://doi.org/10.1007/978-3-030-41614-0_5

as the outcome of biological evolution in the universe still is speculative, so the outcome of cultural evolution is also speculative—even on Earth. But given the existence of extraterrestrial intelligence, the fact of cultural evolution beyond Earth is not speculative, and must be taken into account.

5.2 Cosmic Evolution: Three Possible Outcomes

Because cosmic evolution is the basis of my claims, I want to elaborate briefly how we came to the idea of cosmic evolution. A century ago, it is safe to say, cosmic evolution was not the accepted worldview. The worldview of that time was graphically captured in A. R. Wallace's volume on the plurality of worlds (Wallace 1903). Wallace, cofounder with Darwin of the theory of natural selection, argued that the universe was only 3600 light years in extent, a common view at the time. He noted how all the conditions for life on Earth had to be just right—the same kind of arguments recently used to claim that Earth-like planets are rare (Ward and Brownlee 2000). Wallace concluded that humans were alone in the universe, that the Solar System was nearly at its center, and that humans were its ultimate purpose. Despite all the contributions he made to the study of biological evolution on Earth, Wallace believed in the physical universe, not a biological universe. And it was devoid of cosmic evolution in any significant sense (see Chap. 35).

A century later, our view of the universe has immensely enlarged. We now know that the universe is about 13.7 billion light years in extent and full of galaxies, as graphically captured by the Hubble Space Telescope. But size has not been the most important change in our ideas about the universe over the last century. We now detect not only the expanding universe, but also the accelerating universe; we also know about space-time, inflationary cosmology, and dark energy. But arguably no concept has been so radical as cosmic evolution, which encompasses all these new concepts, and in fact embraces all we know about the universe. Despite his belief in biological evolution, Wallace's universe was static. Ours today is evolving, and cosmic evolution is the guiding principle for all of astronomy. Thus, in contrast to a century ago, we can speculate on the destiny of life—on Earth and in the universe—based on what we now know about cosmic evolution.

The intellectual basis for the guiding principle of cosmic evolution had its roots in the nineteenth century when a combination of Laplace's nebular hypothesis and Darwinian evolution gave rise to the first tentative expressions of parts of this worldview (Dick and Strick 2004). But cosmic biological evolution first had the potential to become a research program in the 1950s and 1960s, when its cognitive elements had developed enough to become experimental and observational sciences, and when the researchers in these disciplines first realized they held the key to a larger problem that could not be resolved by any one part, but only by all of them working together. Harvard College Observatory Director Harlow Shapley was an early modern proponent of this concept, which he spoke of in the 1950s in now familiar terms. The Earth and its life, he asserted, are "on the outer fringe of one galaxy in a universe of millions of galaxies. Man becomes peripheral among the billions of stars in his own Milky Way; and according to the revelations of paleontology and geochemistry he is also exposed as a recent, and perhaps an ephemeral manifestation in the unrolling of cosmic time" (Shapley 1958). Shapley went on to elaborate his belief in billions of planetary systems, where "life will emerge, persist and evolve." Shapley's belief in life was unproven then and remains to be proven today. The transition from belief to proof is tantamount to discovering whether cosmic evolution commonly ends with planets, stars and galaxies, or with life, mind and intelligence. Put another way, does cosmic evolution produce not only a physical universe, but also a biological universe? In recent years, JoAnn Palmeri has even found that Shapley in his correspondence used the term "biological universe," unbeknownst to me when I titled my book *The Biological Universe* (Palmeri 2001).

Already as the Space Age began, then, the concept of cosmic evolution—the connected evolution of planets, stars, galaxies *and* life—provided the grand context within which the enterprise of exobiology was undertaken. The idea of cosmic evolution spread rapidly over the next 40 years, both as a guiding principle within the scientific community and as an image familiar to the general public (Chaisson 1981; Reeves 1981; Sagan 1980). NASA enthusiastically embraced, elaborated and spread the concept of cosmic evolution from the Big Bang to intelligence as part of its SETI and exobiology programs in the 1970s and 1980s. And when in 1997 NASA published its Origins program Roadmap, it described the goal of the program as "following the 15 billion year long chain of events from the birth of the universe at the Big Bang, through the formation of chemical elements, galaxies, stars, and planets, through the mixing of chemicals and energy that cradles life on Earth, to the earliest self-replicating organisms—and the profusion of life" (NASA 1997). With this proclamation of a new Origins program, cosmic evolution became the organizing principle for most of NASA's space science effort, and the concept continues to be elaborated today in ever more subtle form (Chaisson 2001; Delsemme 1998).

Today, the Big Question remains—how far does cosmic evolution commonly go? Does it end with the evolution of matter, the evolution of life, or the evolution of intelligence? In this sense two astronomical world views hang in the balance in modern astronomy, just as they did four centuries ago when Galileo wrote his *Dialogue on the Two Chief World Systems* (Dick 2000). The two chief world systems in 1600, of course, were the geocentric and the heliocentric (Fig. 5.1). The two chief world systems today are the physical universe and the biological universe. But even NASA's early SETI discussions hinted at a third world view opened up by cosmic evolution—the postbiological universe based on cultural evolution. That is a worldview that deserves a great deal more attention than it has heretofore received.

Fig. 5.1 Worldviews in the balance in this frontispiece to Giovanni Riccioli's *Almagestum Novum* (1651). The geocentric system lies discarded at bottom right, while the heliocentric system and a hybrid system developed by Tycho Brahe and modified by Riccioli hang in the balance. In Riccioli's view at mid-seventeenth century, the modified Tychonic system clearly outweighs the heliocentric system. Telescopic observations, which necessitated discarding the geocentric system, are illustrated at top right, including the moons of Jupiter, the rings of Saturn, and surface features on the Moon

5.3 The Physical Universe

I will say only a few words about the first possible outcome of cosmic evolution—the physical universe—because almost all of the history of astronomy, from Stonehenge through much of the twentieth century, deals with the people, the concepts, and the techniques that gave rise to our knowledge of the physical universe. Babylonian and Greek models of planetary motion, medieval commentaries on Aristotle and Plato, the astonishing advances of Galileo, Kepler, Newton and their comrades in the Scientific Revolution, thermodynamics, the physics of stellar energy and stellar evolution, the elegant results of modern astronomy—all these and more address the physical universe. The physical universe has been the subject of astronomy for millennia, and it now boasts a whole bestiary of objects unknown a century ago—blazers and quasars, pulsars and black holes, and more familiar objects like planetary nebulae, which have now been beautifully rendered in detail undreamt of before, thanks to space telescopes such as Hubble and Chandra. The quest for a biological universe should in no way obscure the fact that the physical universe—the domain of the entire field of astronomy and astrophysics—is in itself truly amazing. Thousands of astronomers worldwide are working on understanding its dynamics, structure, and composition. The choice of worldviews I have given certainly does not deny that there is a physical universe—the distinction comes when one considers the endpoints of cosmic evolution.

5.4 The Biological Universe

The second possible outcome of cosmic evolution is the biological universe—the universe in which cosmic evolution commonly ends in life. Ideas about a possible biological universe date back to ancient Greece, in a history that is now well known (Crowe 1986; Dick 1982, 1996, 1998; Guthke 1990). The Copernican Revolution, which made the Earth a planet and the planets potential Earths, provided the theoretical underpinnings for the concept of extraterrestrial life.

Unlike the physical universe, we have addressed in a substantive empirical way this new worldview of the biological universe only over the last four decades. Despite false starts like Lowell's canals of Mars, only in the 1950s and 1960s did four intellectual elements—planetary science, the search for planetary systems, origin of life studies, and the Search for Extraterrestrial Intelligence (SETI)—converge to give birth to the field of exobiology. At first quite separate in terms of researchers, techniques and common goals, these fields over four decades gradually became integrated. Early signs of the potential marriage of astronomy and biology occurred in the 1950s, for example, with what was billed "the first American symposium on astrobiology" in 1957 (Wilson 1958).

More significant was what was thought to be the first empirical evidence of extraterrestrial life, in particular William Sinton's claim of spectroscopic evidence of life on Mars (Dick 1996, 122). Although it was soon disproved, this finding played an important role as the Space Age began. The beginning of the Space Age offered the means to actually go to Mars, and thus NASA became an important patron of the new science of "exobiology." Also in the early 1960s, Peter van de Kamp's claim of a planet around Barnard's star raised great excitement—we now know it was about three decades premature. Meanwhile, with the ideas of A. I. Oparin and J. B. S. Haldane in the background, origin of life studies took a giant leap forward with the Urey-Miller experiment in 1953. Already in 1959 Urey and Miller saw the relevance of space to their work, arguing that the discovery of life beyond Earth was a testbed for theories of the origin of life. The following year Drake undertook project Ozma, and began SETI, another thread in the new discipline of exobiology. So by the mid-1960 practitioners began to declare the beginnings of a new discipline (Dick 1996).

The Viking landers were the highlight of NASA's early foray into exobiology. The negative result for life on Mars caused a period of decline in the field in terms of the *in situ* search for life beyond Earth, even as NASA's exobiology program supported path-breaking work in life in extreme environments, the Earth's primitive atmosphere, Lovelock's Gaia hypothesis, Carl Woese's three domains of life, among other areas (Dick and Strick 2004). By the 1990s many events conspired to revitalize exobiology's search for life in the Solar System: the Mars rock ALH84001, the Mars Global Surveyor observations of the gullies of Mars and the Mars Odyssey detection of water near the surface, and the Galileo observations of Europa indicating a possible ocean. The discoveries of circumstellar matter, extrasolar planets, life in extreme environments such as deep sea hydrothermal vents, and increasingly complex interstellar organics fueled the possibilities of life beyond the Solar System. All these elements fed into NASA's new astrobiology program, which emerged from a deep organizational restructuring at NASA in 1995 (Dick and Strick 2004). Astrobiology involved much more than renaming a discipline; it was much more broadly defined than exobiology, and was to include research in cosmochemistry, chemical evolution, the origin and evolution of life, planetary biology and chemistry, formation of stars and planets, and expansion of terrestrial life into space. Astrobiology today is a much more robust science than exobiology was 40 years ago. Despite all the activity, the circumstantial evidence that the universe may be *biofriendly*, and the recent Mars Exploration Rovers discovery of likely past standing water on Mars, the biological universe remains to be proven (Darling 2001; Goldsmith and Owen 2001; Jakosky 1998; Koerner and LeVay 2000).

5.5 The Postbiological Universe

Although the biological universe remains unproven, the two chief world views today are the physical universe and the biological universe, with many believing it is only a matter of time until proof comes for the latter. I have only skimmed the surface of a subject that has been documented in detail. We now come to the third option, distinct from the physical and the biological universe, an option that thus far has not been taken seriously. But if we take seriously physical and biological cosmic evolution, we also need to take seriously cultural evolution as an integral part of cosmic evolution and the Drake Equation. Those familiar with the vast sweep of time in Olaf Stapledon's *Last and First Men* (1930) and *Star Maker* (1937) will know what I mean when I say that we need to think in Stapledonian terms (see Chap. 12). While astronomers are accustomed to thinking on cosmic time scales for physical processes, even they do not commonly think on cosmic time scales for biology and culture. But cultural evolution now completely dominates biological evolution on Earth. Given the age of universe, and if intelligence is common, it may have evolved far beyond us. I have recently argued in the *International Journal of Astrobiology* and elsewhere (Dick 2003a; Dick 2003b; Chap. 12 in this volume) that cultural evolution over thousands or millions of years will likely result in a "postbiological universe" populated by artificial intelligence, with sweeping implications for SETI strategies and for our worldview.

Let me just give you the outlines of this idea. MacGowan and Ordway III (1966), Davies (1995) and Shostak (1998), among others, have broached the subject, but it has not been given the attention it is due, nor has it been carried to its logical conclusion. The two methodological principles are those I have already mentioned: that long-term Stapledonian thinking is a necessity if we are to understand the nature of intelligence in the universe today, and that cultural evolution must be seen as an integral part of cosmic evolution and the Drake equation. The three scientific premises are (1) that the maximum age of extraterrestrial intelligence (ETI) is several billion years; (2) the lifetime of a technological civilization is greater than 100 years and probably much larger; and (3) in the long term cultural evolution will supersede biological evolution, and produced something far beyond biological intelligence. Let us look at each of these premises in turn.

It is widely agreed that the maximum age of extraterrestrial intelligence, if it exists, is billions of years. Recent results from the Wilkinson Microwave Anisotropy Probe (WMAP) place the age of the universe at 13.7 billion years, with a 1% uncertainty, and confirm that the first stars formed at about 200 million years after the Big Bang. The oldest Sun-like stars probably formed within about a billion years, or 12.5 billion years ago. By that time enough heavy element generation and interstellar seeding had taken place for the first rocky planets to form. Then, if Earth's history is any guide, it may have taken another 5 billion years for intelligence to evolve. In a universe 13.7 billion years old, this means that the first intelligence could have evolved 7.5 billion years ago. Norris (2000), Tough (2000) Livio (1999), and Kardashev (1997) have all argued that extraterrestrial civilizations could be billions of years old, and this assumption is commonly accepted among SETI practitioners.

But what about the second premise, that the lifetime of a technological civilization (denoted as L in the Drake Equation, and defined as starting when a civilization becomes radio communicative), could be billions of years? It is true that the only data point we have is ourselves. Sagan, Drake, and others generally assigned L values in the neighborhood of a million years, and even some pessimists admit

10,000 years is not unlikely. Of course there are a variety of natural and societal catastrophes that could prevent civilizations from reaching ages of millions or billions of years. But the key point is the age of extraterrestrial intelligence does not have to be large for cultural evolution to do its work. Even at our low current value of L on Earth, biological evolution by natural selection is already being overtaken by cultural evolution, which is proceeding at a vastly faster pace than biological evolution (Dennett 1996). Technological civilizations do not remain static; even the most conservative technological civilizations on Earth have not done so, and could not, given the dynamics of technology and society. Unlike biological evolution, L need only be thousands of years for cultural evolution to have drastic effects on civilization.

But how can we possibly predict the course of cultural evolution? We certainly cannot *predict* anything, least of all cultural evolution on Earth, much less in the universe. Darwinian models of cultural evolution have been the subject of much recent study (Lalande and Brown 2002), but they are fraught with problems and controversy—we need only think of the controversies generated by sociobiology, behavioral ecology, evolutionary psychology, gene-culture coevolution and memetics.

While theoretical and empirical studies of cultural evolution hold hope for a science of cultural evolution, lacking a robust theory of cultural evolution to at least guide our way, we are reduced at present to the extrapolation of current trends supplemented by only the most general evolutionary concepts. Several fields are most relevant, including genetic engineering, biotechnology, nanotechnology, and space travel. But one field—artificial intelligence—may dominate all other developments in the sense that other fields can be seen as subservient to intelligence. Biotechnology is a step on the road to AI, nanotechnology will help construct efficient AI and fulfill its goals, and space travel will spread AI. Genetic engineering may eventually provide another pathway toward increased intelligence, but it is limited by the structure of the human brain. In sorting priorities, I adopt what I term the central principle of cultural evolution, which I refer to as the Intelligence Principle: *the maintenance, improvement and perpetuation of knowledge and intelligence is the central driving force of cultural evolution, and that to the extent intelligence can be improved, it will be improved.* The Intelligence Principle implies that, given the opportunity to increase intelligence (and thereby knowledge), whether through biotechnology, genetic engineering or AI, any society would do so, or fail to do so at its own peril. I have elsewhere attempted to justify this principle (Dick 2003a, b), but what it comes down to is this: culture may have many driving forces, but none can be so fundamental, or so strong, as intelligence itself.

The field of AI is a striking example of the Intelligence Principle of cultural evolution. Although there is much controversy over whether artificial intelligence can be constructed that is equivalent or superior to human intelligence—the so-called Strong AI argument—several AI experts have come to the conclusion that AI will eventually supersede human intelligence on Earth. Moravec (1988) spoke of "a world in which the human race has been swept away by the tide of cultural change, usurped by its own artificial progeny." Kurzweil (1999) also sees the takeover of biological intelligence by AI, not by hostility, but by willing humans who have their brains scanned uploaded to a computer, and live their lives as software running on machines. Tipler (1994), well known for his work on the anthropic principle and the Fermi paradox, concluded that machines may not take over, but will at least enhance our wellbeing. But the self-reproducing von Neumann machines that Tipler foresaw in his explanation of the Fermi paradox may well exist if his view of the Fermi paradox is wrong.

It may be that Moravec, Kurzweil and their proponents underestimate the moral and ethical brakes on technological inertia. But such objections fail to take into account cultural evolution, and may lose their impact over the longer term, as the Intelligence Principle asserts itself. When one considers the accelerating pace of cultural evolution as we enter the third millennium of our era, radical change of the sort foreseen by Moravec and Kurzweil does not seem so farfetched.

Thus, it is possible that L need not be billions or millions of years for a postbiological universe scenario. It is possible that such a universe would exist if L exceeds a few hundred or a few thousand years, where L is defined as the lifetime of a technological civilization that has entered the electronic computer age (which on Earth approximately coincides with the usual definition of L as a radio communicative civilization).

The postbiological universe cannot mean a universe totally devoid of biological intelligence, since we are an obvious counterexample. Nor does it mean a universe devoid of lower life forms, as advocated by Ward and Brownlee (2000). Rather, the postbiological universe is one in which the majority of intelligent life has evolved beyond flesh and blood. The argument makes no more, and no fewer, assumptions about the probability of the evolution of intelligence or its abundance than standard SETI scenarios; it argues only that if such intelligence does arise, cultural evolution must be taken into account, and that this may result in a postbiological universe.

Although some may consider this a bold argument, its biggest flaw is probably that it is not bold enough. It is a product of our current ideas of AI, which in themselves may be parochial. It is possible after a few million years, cultural evolution may result in something even beyond AI.

5.6 Summary

The new universe, driven by the astronomical, biological and cultural components of cosmic evolution, may result in any of the three outcomes described here: the physical universe, the biological universe, or the postbiological universe. Which of the three the universe has produced in reality we do not yet know. But we can say that these three possible outcomes of cosmic evolution have very different consequences for human destiny. If life is limited to Earth in this physical universe, the destiny of life is for humans, or their robotic ancestors, to populate the universe. In such a universe, where we are unique or very rare, stewardship of our rare pale blue dot takes on special significance. The destiny of human life in a biological universe is quite different from that in a physical universe. Rather than populating a universe empty of life, the destiny of humanity is perhaps to interact with extraterrestrials, to join what has been called a "galactic club" whose goal is to enhance knowledge. The destiny of human life in a postbiological universe is as a fleeting stage in cosmic evolution prior to being lifted to a higher nonbiological intelligence. If other civilizations in the universe are already the postbiological outcome of cultural evolution, it is possible that we will find in them the future destiny of life on Earth. Whether or not postbiological humans remain human I leave as an exercise for the reader.

5.7 Commentary 2020

This paper was given in June, 2004 at the 60th birthday symposium for Woodruff ("Woody") T. Sullivan, an astronomer at the University of Washington in Seattle (Orchiston 2005). Woody had an unusual breadth of interests, ranging from sundials and history of astronomy to astrobiology. The meeting opened with Chris Chyba and me speaking on broad aspects of astrobiology before moving on to the history of astronomy and sundials. Woody had done pioneering research in the history of radio astronomy, and so many of the papers addressed that area, including the fascinating history of the discovery of Sagittarius A∗, the black hole at the center of our Milky Way Galaxy.

The "biological universe revisited" refers to the idea not just of a biological universe, but a possible postbiological universe, first laid out in detail the previous year, reprinted in this volume as Chap. 12, where the subsequent fate of the idea is discussed in the commentary section.

References

Chaisson, E. 1981. *Cosmic Dawn: The Origins of Matter and Life*. Little, Brown and Co., Boston.

Chaisson, E., 2001. *Cosmic Evolution: The Rise of Complexity in Nature*. Harvard University Press, Cambridge, Mass.

Crowe, M. J., 1986. *The Extraterrestrial Life Debate, 1750-1900: The Idea of a Plurality of Worlds from Kant to Lowell*. Cambridge University Press, Cambridge; Dover reprint, 1999.

Darling, D., 2001. *Life Everywhere: The Maverick Science of Astrobiology*. Basic Books, New York.

Davies, P., 1995. *Are We Alone? Philosophical Implications of the Discovery of Extraterrestrial Life*. Basic Books, New York.

Delsemme, A. 1998. *Our Cosmic Origins: From the Big Bang to the Emergence of Life and Intelligence*. Cambridge University Press, New York.

Dennett, D., 1996. *Darwin's Dangerous Idea*. Simon and Schuster, New York.

Dick, S. J., 1982. *Plurality of Worlds: The Extraterrestrial Life Debate from Democritus to Kant*. Cambridge University Press, Cambridge.

Dick, S. J., 1996. *The Biological Universe: The Twentieth Century Extraterrestrial Life Debate and the Limits of Science*. Cambridge University Press, Cambridge.

Dick, S. J., 1998. *Life on Other Worlds*. Cambridge University Press, Cambridge.
Dick, S. J., 2000. *Extraterrestrial Life and Our Worldview at the Turn of the Millennium*. Smithsonian Institution, Washington, D.C.
Dick, S. J., 2003a. Cultural Evolution, the Postbiological Universe and SETI. *International Journal of Astrobiology*, **2**: 65–74.
Dick, S. J., 2003b. They Aren't Who You Think. *Mercury*, 32, 18–26.
Dick, S. J. and Strick, J. 2004. *The Living Universe: NASA and the Development of Astrobiology*. Rutgers University Press, New Brunswick.
Goldsmith, D. and Owen, T. 2001. The Search for Life in the Universe, 3rdd edition. University Science Books, Sausalito, CA.
Guthke, K. S., 1990. *The Last Frontier: Imagining other Worlds from the Copernican Revolution to Modern Science Fiction*. Cornell University Press, Ithaca, N.Y.
Jakosky, B. 1998. *The Search for Life on Other Planets*. Cambridge University Press, Cambridge.
Kardashev, N. S., 1997. Cosmology and civilizations, *Astrophysics and Space Science*, 252: 25–40.
Koerner, D. and LeVay, S., 2000. *Here Be Dragons: The Scientific Quest for Extraterrestrial Life*. Oxford University Press, New York.
Kurzweil, R., 1999. *The Age of Spiritual Machines: When Computers Exceed Human Intelligence*. Penguin Books, New York.
Lalande, K. N. and Brown, G. R., 2002. *Sense & Nonsense: Evolutionary Perspectives on Human Behaviour*. Oxford University Press, Oxford.
Livio, M., 1999. How rare are extraterrestrial civilizations and when did they emerge?, *Astrophysical Journal*, **511**: 429–431.
MacGowan, R. and Ordway, F.I., III (1966). *Intelligence in the Universe*. Prentice-Hall, Englewood Cliffs, NJ.
Moravec, H., 1988. *Mind Children: The Future of Robot and Human Intelligence*. Harvard U. Press: Cambridge, Mass.
NASA, 1997. *Origins: Roadmap for the Office of Space Science Origins Theme*. Pasadena: NASA/JPL; revised ed., 2000.
Norris, R. P. , 2000. How old is ET?, in Tough (2000), pp. 103–105.
Orchiston, W. ed. 2005. *The New Astronomy: Opening the Electromagnetic Window and Expanding our View of Planet Earth*. Dordrecht, Springer.
Palmeri, J., 2001. Popular and Pedagogical Uses of Cosmic Evolution, session on Evolution and Twentieth Century Astronomy, History of Science Society meeting, Denver, Colo., 8 November, 2001.
Reeves, H. 1981. Patience dans l'azur: L'evolution cosmique, Editions du Seuil, Paris; translation *Atoms of Silence: An Exploration of Cosmic Evolution*, MIT Press, Cambridge.
Sagan, C. 1980. *The Cosmic Connection*. Random House, New York.
Shapley, H., 1958. *Of Stars and Men*. Beacon Press: Boston.
Shostak, S., 1998. *Sharing the Universe: Perspectives on Extraterrestrial Life*. Berkeley Hills, Berkeley, Ca., 103–109.
Tipler, F. , 1994. *The Physics of Immortality*. Doubleday, New York.
Tough, A. ed., 2000. *When SETI Succeeds: The Impact of High-Information Contact*, Foundation for the Future, Bellevue, Wash.
Wallace, A. R., 1903. *Man's Place in the Universe: The Results of Scientific Research in Relation to the Unity or Plurality of Worlds*. Macmillan, New York.
Ward, P. and Brownlee, D., 2000. *Rare Earth: Why Complex Life is Uncommon in the Universe*. Copernicus, New York.
Wilson, A. G. 1958. Problems Common to the Fields of Astronomy and Biology, *Publications of the Astronomical Society of the Pacific*, 70, 41–78.

Chapter 6
Back to the Future: SETI before the Space Age

Abstract The modern era of the Search for Extraterrestrial Intelligence (SETI) was inaugurated some 35 years ago, with the seminal paper by Giuseppe Cocconi and Philip Morrison in 1959 (Cocconi and Morrison. Nature 184:844, 1959) and the Project Ozma search by Frank Drake in 1960. But even many SETI enthusiasts do not realize that this era of *interstellar communication*, as it was originally called, was preceded by a colorful era of *interplanetary radio communication*, involving radio pioneers including Nikola Tesla and Guglielmo Marconi. This era was filled with parallels, contrasts, and lessons for those interested in the survival of SETI in its current incarnation.

6.1 The Radio Pioneers: Tesla and Marconi

Although the idea that visual signals might be sent to the Moon or Mars was common in the nineteenth century (Crowe 1986), it was the idea of radio communication between Earth and these bodies in space that caught the public fancy and the early interest of several radio pioneers. Heinrich Hertz, the German physicist who first demonstrated the existence of radio waves, died in 1894, too early to see the application of his work to even terrestrial communication. But two of his contemporaries, Nikola Tesla and Guglielmo Marconi, not only foresaw the use of radio technology for communication beyond Earth, but believed they had actually detected signals of intelligent origin.

Tesla knew of the electrical disturbances produced by the Sun, the aurora borealis and Earth itself; the new signals were more regular than any of these. "It was some time afterward," he reported in 1901, "when the thought flashed upon my mind that the disturbances I had observed might be due to an intelligent control. Although I could not decipher their meaning, it was impossible for me to think of them as having been entirely accidental. The feeling is constantly growing on me that I had been the first to hear the greeting of one planet to another" (Dick 1996, 401; Tesla 1901, 359).

First published in *The Planetary Report* (January–February, 1995).

S. J. Dick, *Space, Time, and Aliens*, https://doi.org/10.1007/978-3-030-41614-0_6

Tesla, the Serbian-born American physicist and engineer, was the first to publish this idea in 1901—and under remarkable circumstances. During experiments at his laboratory in Colorado Springs, Colorado in 1899–1900, Tesla observed unusual electrical disturbances that "positively terrified me, as there was present in them something mysterious, not to say supernatural."

Tesla went on to a busy and then a reclusive life and never followed up on this idea. Astronomers were skeptical, and the revival of interplanetary radio communication in a more sustained form came not from them, but from the great Marconi himself, by now world famous for his work in radio communication. On January 20, 1919, *The New York Times* ran a front-page article with the headline "Radio to Stars, Marconi's Hope." Asked during an interview whether he believed waves of ether were eternal, Marconi replied, "Yes, I do. Messages that I sent off 10 years ago have not yet reached some of the nearest stars. When they arrive there why should they stop?" (New York Times 1919a, 1).

Comparing the weakening radio signals to a repeating decimal that never comes to an end, Marconi said this property of radiation "is what makes me hope for a very big thing in the future … Communication with intelligences on other stars may some day be possible, and as many of the planets are much older than ours the beings who live there ought to have information for us of enormous value." Then, hesitating slightly, Marconi admitted he had often "received strong signals out of the ether which seemed to come from some place outside the earth and which might conceivably have proceeded from the stars" (New York Times 1919a, 1).

These ideas were repeated in more detail in the January 27, 1920 *New York Times*, where Marconi reported that Morse code letters occurred often in these signals but no message was decipherable. Because the signals occurred simultaneously in the London and New York receiving stations and because they were of equal intensity, Marconi inferred they originated at very great distances. "We have not yet the slightest proof of their origin," he noted, saying that it could be the Sun (New York Times 1920a). It was not Marconi but the press that raised the contentious question about other planets being the source, to which Marconi replied, "I would not rule out the possibility of this, but there is no proof. We must investigate the matter much more thoroughly before we venture upon a definite explanation" (New York Times 1920a, 7).

For the next few weeks, *The New York Times* followed up on the story almost daily, sometimes on the front page (New York Times 1920b). Radio engineers were quoted as being highly skeptical that the signals emanated from another planet, especially from intelligence, and felt that they were atmospheric disturbances induced by the Sun. The United States Navy Department, with its advanced radio communications system, was reported to be keeping an open mind, with interest manifestly outweighing skepticism. Charles Steinmetz, the famous inventor and engineer, denied that the signals came from Mars but held that "if the United States … should go into the effort to send messages to Mars with the same degree of intensity and thoroughness with which we went into the war it is not at all improbable that the plan would succeed." C. G. Abbot, director of the Smithsonian Astrophysical Observatory, suggested that Venus was a much more likely source than Mars, since Mars was too cold and lacked water. And Elmer Sperry, head of the Sperry Gyroscope Company, boasted that his company could send a message to Mars using 150 or 200 Sperry searchlights adding up to a billion candlepower (New York Times 1920c).

The debate reached continental Europe when the French Academy of Sciences agreed to judge a competition for a 100,000-franc prize "for the best means of making a sign to a heavenly body and the receipt of a reply." And no less a scientific icon than Albert Einstein was quoted as believing that Mars and other planets might be inhabited, but that Marconi's signals were due either to atmospheric disturbances or to experiments with other wireless systems. If intelligent beings on other planets attempted to communicate with Earth, Einstein added, he would expect them to use rays of light, which are more easily controlled. Einstein was thus an early proponent of optical SETI (New York Times 1920d)!

Two weeks later, *Scientific American* argued that while Marconi's conjectures should not be dismissed, there was absolutely no proof that Martians existed; it was unlikely that they would develop a Morse code as on Earth; they could not transmit over the 80-million-kilometer (50-million-mile) distance separating Earth and Mars; and radio stations at the Eiffel Tower and elsewhere, including those of the U. S. Navy, had not heard the Marconi signals "although they have searched for them" (Scientific American 1920, 156). Suspecting even the Japanese or "the Russian Bolsheviki, who have turned to radio as a convenient means of propagating their cause at home and abroad," the editors concluded nonetheless that "this matter deserves careful study when a scientist of Mr. Marconi's standing takes it so seriously." The following month *Scientific American* featured an article entitled "What Shall We Say to Mars?", which attempted to determine how knowledge might be communicated by dots and dashes in the absence of a common language (Fig. 6.1) (Nieman and Nieman 1920).

Marconi's interest in interplanetary communication apparently peaked during a trip from Southampton, England to New York City aboard his floating laboratory, the yacht *Elettra,* from May 23 to June 16, 1922. *The New York Times* reported that he "spent the time crossing the Atlantic performing many electrical experiments, principally by listening for signals from Mars." Marconi admitted that "they might have come from any region in the universe where electrons are in vibration" (New York Times 1922, 19). It is notable, however, that when Marconi addressed the respected Institute of Radio Engineers and the American Institute of Electrical Engineers on the subject of radiotelegraphy a few days after his arrival in New York, he discussed long-distance radio communication but had nothing to say on the subject of interplanetary radio communication.

Of all the reactions to Marconi's statements, none was more poignant than *The New York Times'* editorial "Let the Stars Alone," which argued that "even if it could be done one doubts it would be desirable … Quite possibly there are even yet more things in heaven and earth than are dreamed of in our philosophy, and it would be better to find them out in our own slow, blundering way rather than have knowledge for which we are unprepared precipitated on us by superior intelligences" (New York Times 1919b, p. 8). This viewpoint also finds its parallel in the modern era.

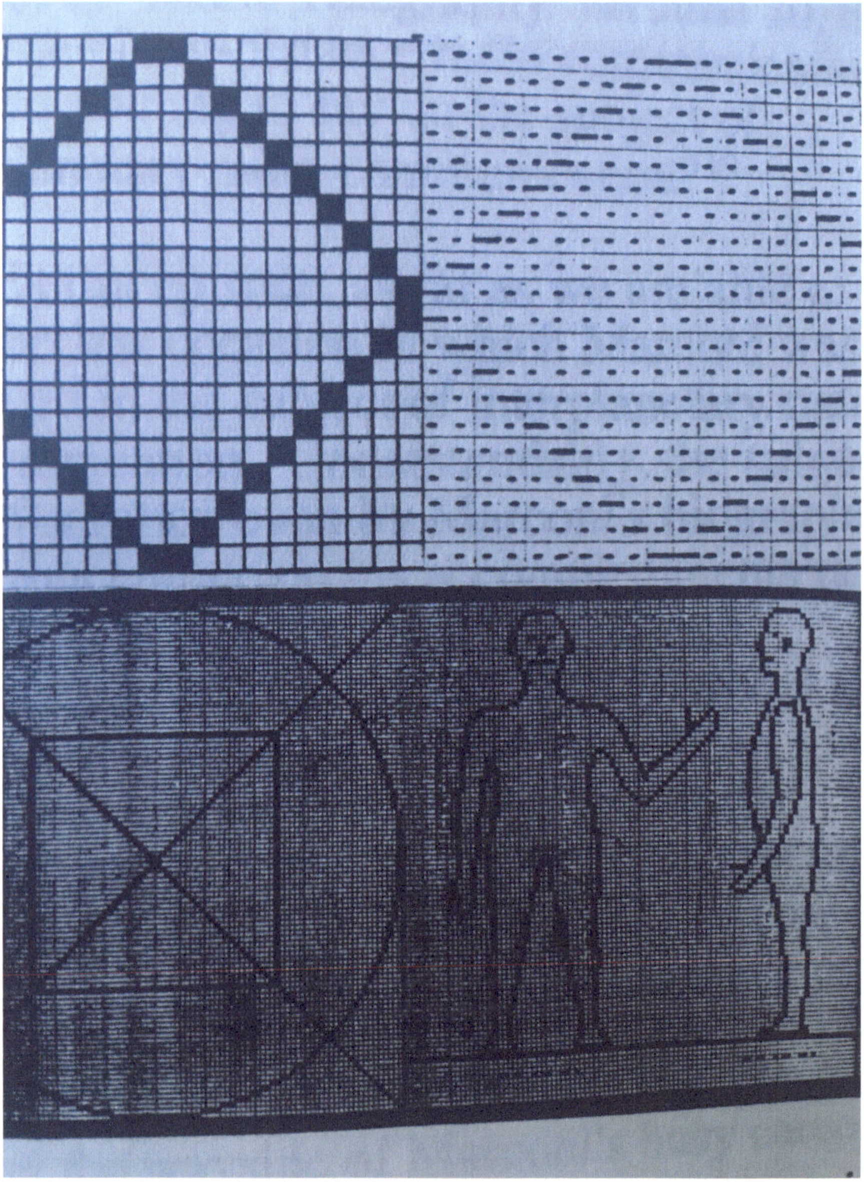

Fig. 6.1 1920 proposal from *Scientific American* for communication by dots and dashes arranged in blocks. Top: using strips of telegraph tape and plotted on graph paper. Bottom: A larger number of blocks allows more complicated messages. Note the similarity to idea that later arose with regard to interstellar communication of determining blocks of pulses and spaces using prime numbers to determine the dimensions of the array

Some were even more skeptical. When asked during a visit to Philadelphia about the possibility of radio communication with Mars, radio pioneer Sir Oliver Lodge's replied that it was "nonsense." For reasons that are unclear, he preferred the visual signal tradition, with a twist: "It would be possible to describe an immense geometrical figure, say, on the Sahara Desert ... and then, if the inhabitants of Mars are of a high order of intelligence, it is possible that with their powerful means of magnification they might be able to distinguish the figure and determine that it had been the work of man. Geometry is a universal science and it is not unlikely that if they are of a higher order of intelligence than we they would interpret the gigantic figure as an effort at communication" (New York Times 1920e).

6.2 David P. Todd, Balloon SETI and Other Schemes

In the midst of the Marconi flap, another thread of the story was developing: the idea of wireless interplanetary communication from a balloon. The prime mover in this daring enterprise, which combined an imaginative idea with a bold technology, was the well-known astronomer David P. Todd, director of the observatory at Amherst College from 1881 to 1920. As early as 1909, Todd had suggested that Martians might communication with Earth using Hertzian waves, and that the most sensitive wireless receivers should be taken up in a balloon to diminish atmospheric effects. A skeptical *Scientific American,* anticipating in rudimental form the problem of radio frequency interference (RFI) that frustrates modem SETI searchers, pointed out that about 2000 wireless stations were scattered over the surface of Earth, that the Sun and Earth's atmosphere might also be sources of electrical signals, and that it would thus be difficult to pinpoint the source of any supposed Martian signals (Scientific American 1909a, b).

Nevertheless, in 1920 *The New York Times* reported that Todd, "after more than 5 years of preparation, during which time he has studied the proposition from every conceivable angle," had set the date of April 23 for a balloon ascent to try to communicate with Mars (New York Times 1920f). Alas, we can only wonder what became of Todd's ambitious expedition. In subsequent days the *Times* reported only on a separate ground-based attempt at radio communication with Mars and offered a cryptic statement that while construction of Todd 's balloon was progressing, the "experiment will be held in abeyance, however, until sanctioned by the U.S. Government" (New York Times 1920g). Once again, shades of recent SETI events! Though unsuccessful with his balloon experiments, by 1924 Todd pressed yet another bold project related to interplanetary communication. The *New York Times* reported that Todd had obtained informal assurances from the U. S. Army and Navy that they would observe, as much as possible, a period of radio silence on August 22 and 23, when Mars was at closest approach. In an effort to obtain worldwide cooperation, Todd also discussed radio silence with the State Department and several embassies. In addition, Army and Navy radio operators would "listen in" for

Fig. 6.2 The U. S. Army listens for Martian radio signals, according to the plan of D. P. Todd, as pictured in *Radio Age* for October, 1924

signals from Mars, and be "ready to translate any peculiar messages that might come by radio from Mars (Fig. 6.2).

Department of Commerce officials said they too were ready to cooperate if asked. "Although officials were strongly skeptical as to success, they seemed to take the attitude that there could be no objection to giving communication with Mars a fair trial under the best possible conditions," the *Times* reported (New York Times 1924). Some experts were even more skeptical—too much so, we now know; the chief of the radio laboratory of the National Bureau of Standards declared that Earth's atmosphere would prevent a radio signals from reaching the ground!

On August 21 the Point Grey wireless station in Vancouver, British Columbia reported having received unusual signals during the preceding week. Although the frequency was not given, "Four distinct groups of four dashes each came through the ether today," one operator had stated. The signals "were in no known code, starting on a low note and ending with a 'zipp,'" and neither a spark nor a continuous wave could have been responsible for the sounds. The following day the signals were heard again in Vancouver and reported to have been heard at the same time of day for more than 4 weeks. Moreover, the Associated Press reported that British wireless experts, using "a twenty-four tube set erected on a hill at Dulwich" had heard at a wavelength of 30 km (18.6 miles) sounds "likened to harsh dots," but they could not be interpreted as any known code. Frank Drake and others have speculated that Tesla, Marconi and others might have been hearing "whistlers," low-frequency waves generated by lightning flashes that propagate along Earth's magnetic lines of force. These early attempts at interplanetary interpretations for unknown signals thus lend credence to the "Occam's razor" rule that mundane explanations should always be given priority over exotic ones.

6.3 Donald Menzel, Radio Amateurs and Radio Astronomy

In 1932, 10 years after Marconi's words on radiotelegraphy were published in the *Proceedings of the Institute of Radio Engineers,* Karl G. Jansky of Bell Telephone Laboratories reported in the same journal that he had detected a strange radio static that he could not attribute to any known source. This he interpreted in the following year as coming from beyond the Solar System, a claim that was greeted with skepticism by most astronomers. Occam's razor doesn't always work!

Interest in radio communication with the planets, however, had not quite run its course. "The question of radio communication with distant planets still holds supreme charm for all red-blooded radio experimenters," the editor of *Short Wave and Television* magazine (none other than science fiction pioneer Hugo Gernsback) wrote in the December 1937 issue. One of those radio amateurs happened to be Harvard astronomer Donald Menzel, who in the same issue wrote the article, "Can We Signal Mars by Shortwave?" (Fig. 6.3). Menzel noted that "the general consensus of opinion is that no very high degree of intelligent life exists in our solar system," and the thrust of his article was therefore not to propose the transmission or receipt of actual signals, but to explore the question of whether we might in principle be able to communication information via radio signals sent from beyond Earth (Menzel 1937).

If we received a radio message from Mars, Menzel argued, radio technology implied knowledge of mathematics and physical science, and mathematics was a natural starting point for communication. Using dots and dashes, one could begin by transmitting arithmetical problems and answers, to which the Martians would reply with their own. One could then advance to abstract numbers like pi, and the relative distances of the planets from the Sun. The alphabet could be transmitted by means

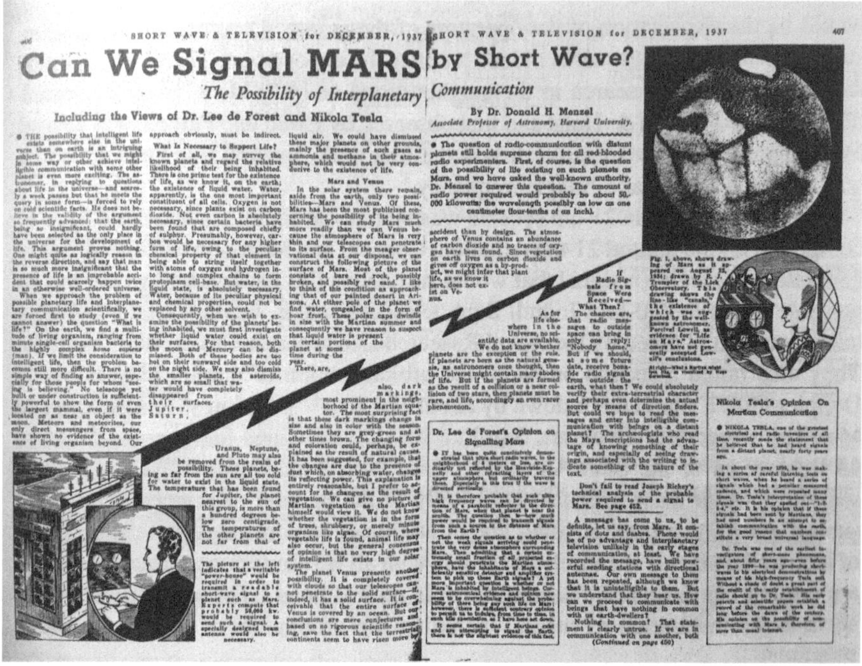
SHORT WAVE & TELEVISION for DECEMBER, 1937

SHORT WAVE & TELEVISION for DECEMBER, 1937

Can We Signal MARS by Short Wave?

The Possibility of Interplanetary Communication

Including the Views of Dr. Lee de Forest and Nikola Tesla

By Dr. Donald H. Menzel

Associate Professor of Astronomy, Harvard University.

The question of radio-communication with distant planets still holds supreme charm for all red-blooded radio experimenters. First, of course, is the question of the possibility of life existing on such planets as Mars, and we have asked the well-known authority, Dr. Menzel to answer this question. The amount of radio power required would probably be about 50,000 kilowatts; the wavelength possibly as low as one centimeter (four-tenths of an inch).

Fig. 6.3 Interplanetary signaling by radio was a topic often discussed in the first half of the twentieth century, paralleling the later interest in interstellar communication. This article, by Harvard astronomer Donald Menzel, is notable for including the views of radio pioneers Lee de Forest and Nikola Tesla, as well as for Menzel's ideas about what might be communicated. It appeared in the magazine *Short Wave and Television* for December, 1937

of a series of paired numbers, coordinates on a graph. Proceeding in this way to more and more complex information, Menzel saw "no obvious limit to the information that could be exchanged."

The same issue of *Short Wave and Television* included the cautious opinion of radio pioneer Lee De Forest, the more imaginative interpretations of Tesla, and a remarkable analysis by American Telephone and Telegraph (AT&T) staffer Joseph L. Richey of the optimal wavelengths and the power required to send a signal to Mars (Richey 1937). Although Richey proposed that radiation between the infrared and 10-m radio (as well as optical) wavelengths would be optimal for penetrating Earth's atmosphere, he concluded that communication with Mars was "economically and technically not feasible with present-day equipment." Such conclusions did not dampen the enthusiasm of radio amateurs, including pioneer Hiram Percy Maxim, whose book *Life's Place in the Cosmos* (Maxim 1933, 148–160) gave prominent coverage to *interstellar* communication. A generation ahead of his time in this proposal, Maxim simply had faith that some day the technology would be developed.

6.4 Two Eras, Two Outcomes?

Like the modern era of interstellar communication, the era of interplanetary radio communication was led by innovative scientists with broad but practical interests. Tesla, Marconi, Todd, and Menzel were all originals in their own way, not afraid to vent controversial ideas. All were well grounded in technical interests, and the interplanetary communication era was woven from those interests, though based on only the slightest of evidence. With time, the technical concerns increased in sophistication, beginning simply with unexplained signals and ending with concerns about optimal wavelengths, the effects of Earth's atmosphere, and power requirements. Prospective interplanetary communicators anticipated not only technical problems, but also philosophical, linguistic, and cultural factors, only a few of which we have mentioned here.

But there are also major differences. The degree of technical sophistication is certainly one, as well as the manner in which the current debate is carried out. While a few papers appeared in *Scientific American,* debate was largely carried out not in professional journals but in newspapers. Though interplanetary listening projects were discussed and governmental cooperation was even secured in 1924, there was no real attempt to expend public funds on any project, a political process that dominated the NASA SETI program and led to its demise. Finally, we must not lose sight of the fact that despite their interesting discussions, Tesla, Marconi, and others were mistaken in their interpretation of radio signals as artificial, a pattern that would-be interstellar communicators hope not to repeat. Needless to say, modern researchers would like to avoid the rather precipitous announcements of Tesla and Marconi to the media; history shows the inevitable result all too clearly. Our journey back to the future also demonstrates that while technical considerations are important, they should not act as an absolute constraint on thinking. Unbridled imagination is a dangerous thing, but it may also lead to the truth.

If we consider the era of interplanetary communication to be bracketed by Tesla in 1901 and Menzel in 1937, its lifespan is 36 years, almost the same duration as the current era of interstellar communication. It is true that now the entire universe awaits, rather than our own parochial Solar System. But whether lack of detections, political will, or funding results in a limited lifespan for modern SETI, and whether a century from now it will be seen as only a curious episode in the history of science like its predecessor, only the future will tell. Either way, both eras are a part of the venerable tradition of the search for humanity's place in nature.

References

Crowe, M. 1986. *The Extraterrestrial Life Debate 1750-1900: The Idea of a Plurality of Worlds from Kant to Lowell.* Cambridge: Cambridge University Press.

Dick, S. J. 1996. *The Biological Universe: The Twentieth Century Extraterrestrial Life Debate and the Limits of Science.* Cambridge: Cambridge University Press.

Maxim, H. P. 1933. *Life's Place in the Cosmos*. New York and London: D. Appleton and Company.
Menzel, D. 1937. Can We Signal Mars? *Short Wave & Television* (December), 406–407, 450–451.
New York Times. 1919a. Radio to Stars, Marconi's Hope, January 20, 1919, p. 1.
New York Times. 1919b. Let the Stars Alone, January 21, 1919, p. 8.
New York Times. 1920a. Marconi Still at Sea on Mysterious Sounds, January 27, 1920, p. 7.
New York Times. 1920b. Astronomer Thinks Mars Could Signal, January 28, 1920, p. 5; Marconi Testing His Mars Signals, Jan. 29, p. 1.
New York Times. 1920c. First Mars Message Would Cost Billion, Jan. 30, 1920, p. 18; Opposing Views on Mars Signals, Jan 31, p. 24.
New York Times. 1920d. Might Talk to Mars on Waves of Light, February 1, 1920, p. 1; Offers a $20,000 Prize for Sign to a Planet—French Academy of Sciences to Decide Winner—Einstein Would Use Rays of Light, Feb. 2, p. 24.
New York Times. 1920e. Lodge's Signal to Mars, Feb. 4, 1920, p. 13.
New York Times. 1920f. Effort to Signal Mars, 15 April, 1920, p.10; To Try This Week to Talk to Mars, 18 April, 1920, sect. 2, p. 1.
New York Times. 1920g. Radio Expert Hopes to Get Mars Signal, 21 April, 1920, p. 17; Listens for Mars Signal, 22 April, p. 2; No Sounds from Mars Greets Experimenters, 23 April, p.17.
New York Times. 1922. No Mars Message Yet, Marconi Radios; Ends Yacht Trip 'Listening In' on Planet Today, 16 June, 1922, p. 19, reprinted in D. Goldsmith, The Quest for Extraterrestrial Life (Mill Valley, 1980), p. 80.
New York Times. 1924. Asks Air Silence When Mars is Near: Prof. Todd Obtains Official Aid in Washington Despite Doubts of Its Efficacy, 21 August, 1924, p. 11; Listening for Mars: Heard Anything?, 22 August, 1924, p. 12; Radio Hears Things as Mars Nears Us, 23 August, 1924, p. 1.
Nieman, H. W. and Nieman, C. W.. 1920. What Shall We Say to Mars, *Scientific American* (20 March, 1920), 122, 298.
Richey, J. L. 1937. Communicating with Mars—A Few Technical Considerations, Short Wave and Television (December, 1937), 452–454.
Scientific American. 1909a. More about Signalling to Mars (May 15, 1909), 371.
Scientific American. 1909b. Prof. David Todd's Plan of Receiving Martian Messages, 100 (June 5, 1909), 423.
Scientific American. 1920. Those Martian Radio Signals, 122 (Feb. 14, 1920), 156.
Tesla, N. 1901. Talking with the Planets, *Collier's Weekly*, 26, 19, 4, reprinted in *Current Literature* (March 1901), 359–360.

Chapter 7
The Drake Equation in Context

Abstract The Drake Equation, a method for estimating the number of communicative civilizations in our galaxy, was a product of its time in several important ways. After a period of several decades during which the idea of life on other planets had reached a low point due to rise of the *rare collision hypothesis* for planet formation, by the 1950s the nebular hypothesis was once again in favor, whereby planets would form as a common byproduct of stellar evolution. The Miller-Urey experiments in the early 1950s produced complex organic molecules under simulated primitive Earth conditions, indicating life might easily originate given the proper conditions. And while little was known about the gap between primitive life and intelligent life, and a sophisticated understanding of intelligence was lacking, the Lowellian Mars still lingered in the cultural background and, along with contemporary astronomical advances, stimulated consideration of aliens in the scientific imagination. The original emphasis on "radio communicative" reflected the new era of radio astronomy, exemplified by the radio telescopes under construction at the newly founded National Radio Astronomy Observatory (NRAO) in Green Bank, West Virginia, where Frank Drake was working. Drake's ability to undertake such a controversial subject, including the first radio search for extraterrestrial intelligence in 1960, was aided by senior scientists Lloyd Berkner and Otto Struve. Assessments of the probabilities of extraterrestrial life and intelligence had been sporadically undertaken in the course of the twentieth century, but most particularly by former Harvard Observatory Director Harlow Shapley in his book *Of Stars and Men* (1958); Drake had recently graduated from the Harvard astronomy program and had cited the book. Here we look at the origins and development of the equation over time, including its significant variations; examine positive and negative views of its epistemological status and utility ranging from scientists to popular authors such as Michael Crichton; and attempt to tease out the scientific and metaphysical assumptions behind the equation. We conclude by discussing the future of the equation, and the cultural hopes and fears it embodies.

First published as the Introduction to *The Drake Equation: Estimating the Prevalence of Extraterrestrial Life Through the Ages*, Douglas Vakoch and Matthew Dowd, eds. (Cambridge University Press: Cambridge, 2015), pp. 1–20.

S. J. Dick, *Space, Time, and Aliens*, https://doi.org/10.1007/978-3-030-41614-0_7

7.1 Origins of the Equation

The Drake Equation was born during an informal conference on "Extraterrestrial Intelligent Life" held on November 1–2, 1961 at the nascent National Radio Astronomy Observatory (NRAO) in Green Bank, West Virginia. The meeting, sponsored by the Space Science Board of the National Academy of Sciences, was held in the wake of the excitement generated by Project Ozma, the first search for interstellar communications, conducted at the NRAO by Frank Drake, a young astronomer on its staff (Fig. 7.1). The 200-hour search, with the Observatory's 85-foot Tatel radio telescope in April, 1960 (Fig. 7.2), targeted only two nearby Sun-like stars, Tau Ceti and Epsilon Eridani, around the 21-cm line of neutral hydrogen (Drake and Sobel 1992). Although it failed to detect any extraterrestrial civilizations, the project captured the imaginations of scientists and public alike. The Ozma search (though independently conceived by Drake) followed the landmark publication by Giuseppe Cocconi and Philip Morrison arguing on theoretical grounds that such a search should be undertaken (Cocconi and Morrison, 1959). The Green Bank meeting was

Fig. 7.1 Frank Drake at the National Radio Astronomy Observatory, 1962, where he had conducted Project Ozma 2 years before. Drake, recently graduated from Cornell and Harvard, had been interested in extraterrestrial life from an early age, and had been influenced during his Cornell years by a lecture on planetary systems given by Otto Struve (Credit: NRAO/AUI/NSF)

Fig. 7.2 Drake at the 85-ft Tatel telescope during a visit to give the October, 1999 Jansky Lecture. The Tatel telescope was completed in early 1959 at the nascent NRAO, and is named after its designer, Howard E. Tatel (Credit: NRAO/AUI/NSF)

therefore the last in a troika of events from 1959 to 1961 that launched the modern era of the Search for Extraterrestrial Intelligence (Dick 1996, 1998).

The standard story, even from Drake himself, is that press coverage of Project Ozma triggered the interest of the National Academy (Drake 1992, pp. 14–15; Drake and Sobel 1992, p. 46). But National Academy records demonstrate that the

Fig. 7.3 The author with the Drake Equation, as depicted on a plaque in the room at the National Radio Astronomy Observatory (now the Green Bank Observatory) where the equation was first written during a meeting in 1961 convened by the National Academy of Sciences

immediate cause was actually a lecture Drake gave on the subject at the Philosophical Society of Washington on March 10, 1961 (Drake 1961a, b; Pearman 1961). In the audience was biologist J. P. T. Pearman of the National Academy's Space Science Board staff, who that very night after the lecture discussed with Drake the possibility of such a meeting (Pearman 1961). By March 13 Drake replied to Pearman with a letter stating that NRAO Director Otto Struve not only approved such a meeting but also offered to hold it at NRAO. NRAO had living accommodations for about 30 people, Drake noted, and "the isolation of Green Bank would also help solve the problem of keeping the symposium quiet and scientific" (Drake 1961a, b).

The National Academy records indicate Pearman immediately set to work, handling much of the logistics for the meeting. But the organization of the scientific content fell largely to Drake. Thinking in the days before the meeting about how to proceed, Drake decided to arrange the discussions of extraterrestrial intelligence around an equation that concisely represented the relevant factors. Thus appeared for the first time the formulation that would be used repeatedly in the following decades in attempts to determine the likelihood of radio communicative civilizations in our galaxy—and thus the likelihood of success in any such search.

The original form in which Drake wrote the equation was $N = R* \; f_p \; n_e \; f_l \; f_i \; f_c \; L$, where each symbol on the right side of the equation represents a factor bearing on the number of radio communicating civilizations in the galaxy (N) (Fig. 7.3). The first three factors were astronomical, estimating respectively the rate of star formation in the galaxy, the fraction of stars with planets, and the number of planets per star with environments suitable for life. The fourth and fifth factors were biological: the fraction of suitable planets on which life developed, and the fraction of those life bearing planets on which intelligence evolved. The last two factors were social: the fraction of civilizations that were radio communicative over interstellar distances, and the lifetime (L) of radio communicative civilizations. The uncertainties, already shaky enough for the astronomical factors, nevertheless increased as one progressed from the astronomical to the biological to the social. Taken together, they represented cosmic evolution writ large.

Although Drake was the first to put these factors in simple equation form, he was not the first to ask the question in terms of probabilities. Assessments of the probabilities of extraterrestrial life and intelligence had been sporadically undertaken in the course of twentieth-century discussions of the subject. On the eve of the events of 1959–61, former Harvard Observatory Director Harlow Shapley had calculated the number of intelligent civilizations in the universe based on probabilities, but had not discussed interstellar communication (Shapley 1958, 73–74). Drake had recently graduated from the Harvard astronomy program, and had cited Shapley's calculations prior to the Green Bank meeting (Drake 1959). Probabilities had also been used by radio astronomer Ronald Bracewell in another early discussion of the number of advanced communities in the Galaxy (Bracewell 1960, 670). Bracewell, however, had couched his discussion in graphical rather than equation form. And astronomer Sebastian von Hoerner had concluded using probabilities that one in three million stars might have a technical civilization, but that the longevity of a technical civilization (a concept he credited to Bracewell) might be very limited (Von Hoerner 1961).

When Drake began the Green Bank meeting by writing his equation on the board, he could not have known that he was establishing a paradigm for SETI discussions that would last into the twenty-first century. But by considering in turn astrophysical, biological, and social factors he did just that, and Green Bank was only the first of many occasions where experts would discuss the factors that Drake proposed. In the wake of the Green Bank meeting discussions centered on the likelihood of communicative extraterrestrial civilizations utilizing radio technology. The calculations of N varied wildly, over a range not seen before in the history of science. One could take this as an indication of a very unsettled protoscience, though one that held promise for the future.

In the task of calculating the number of radio communicative civilizations, the compelling nature of an equation—even one whose parameters were not well known—was not to be denied, since an equation is a symbol of science and lends authority to any scientific discussion. The meteoritic career of the Drake Equation, rather than one of the other probabilistic assessments, is evidence of such authority. Only a month after the Green Bank meeting in November, 1961, Philip Morrison used a similar equation in a NASA lecture (Morrison 1962). The equation first saw print not in an article by Drake, but in Pearman's account of the Green Bank conference published in a 1963 volume of collected articles on the subject entitled *Interstellar Communication* (Cameron 1963a; Pearman 1963). In the same volume its editor, the astrophysicist A. G. W. Cameron, used a similar equation (Cameron 1963b, c). Sagan was also among the first to publish the equation (Sagan 1963), and Drake himself used it in a paper presented at a JPL symposium on exobiology in February, 1963 (Drake 1965). The equation was thus not known at first as the Drake Equation, and only after a period of uncertainty when it was called the Sagan Equation or the Green Bank Equation did the originator receive due credit (Drake 1992). Perhaps the decisive events in the spread of the Drake Equation were Walter Sullivan's popularized account of it in *We Are Not Alone* (Sullivan 1964) and Sagan's incorporation of it into his translation and expansion of Russian astrophysicist Joseph Shklovskii's book *Intelligent Life in the Universe* (Shklovskii and Sagan 1966), which became the Bible of the SETI movement. These books assured the rapid diffusion of the Drake equation to the public and interested scientists alike.

Although not immediately used in the Soviet Union, the Drake Equation, with its emphasis on radio communication, focused attention on the electromagnetic radio search paradigm. Already by 1966 this concept, and all of the assumptions that went with it, was sufficiently entrenched that physicist Freeman Dyson labeled it the "orthodox view" of interstellar communication, characterized not by interstellar travel, but by "a slow and benign exchange of messages, a contact carrying only information and wisdom around the galaxy, not conflict and turmoil" (Dyson, 1966). As anyone who read science fiction knew, this was not the only possible view of the universe. But it was a practical method, a logical extension of the new field of radio astronomy, and one that at least some of its practitioners were keen to carry out. For these reasons the discussion of rationale and strategy within the radio search paradigm continued its upward climb.

By 1971, 10 years after its origin, the Drake Equation was the centerpiece for the first international SETI meeting, held at the Byurakan Astrophysical Observatory in Yerevan, the Soviet Union (Sagan 1973). This time the organizers of the meeting, sponsored by the Academies of Sciences of both the United States and the Soviet Union, included not only Drake, but also Carl Sagan and Philip Morrison of the United States, as well as Victor Ambartsumian, Nikolai Kardashev, Shklovskii and Troitskii of the Soviet Union. Instead of the 11 participants at the Green Bank meeting in 1961, 28 Soviets, 15 Americans, and 4 scientists from other nations participated. They concluded that perhaps a million technical civilizations existed in the galaxy. SETI, though still a small endeavor by science standards, was growing, and the Drake Equation was its central icon.

7.2 The Equation in Context

The Drake Equation was a product of its time, triggered by the ability of radio telescopes to search for artificial signals from nearby stars (Drake and Sobel 1992, Chap. 2). Drake has given us an inside look at the 11 participants as they gathered

Fig. 7.4 Otto Struve, director of the NRAO, July 1, 1959 through December, 1961. Neither Project Ozma nor the Green Bank meeting on interstellar communication could have been undertaken without his enthusiastic acceptance (Credit: NRAO/AUI/NSF)

at the Green Bank meeting (Drake and Sobel 1992, Chap. 3): Drake himself was the expert young radio astronomer. His boss, Otto Struve (Fig. 7.4), and Struve's former student Su-Shu Huang, were the experts on planetary systems. Other participants were specifically recruited for their expertise in a particular factor in the Drake Equation. Collectively, they represented most of the factors in the equation, but not all. Notably, no social science or humanities experts were present to discuss the number of civilizations or their lifetimes, in part a reflection of the gulf between the two cultures of science and the humanities in the early 1960s.

At the time of the Green Bank meeting, the idea of extraterrestrial life was gaining momentum. After a period of several decades during which the idea of life on other planets had reached a low point due to rise of the *rare collision hypothesis* for planet formation, by 1960 the *nebular hypothesis* was once again in favor, whereby planets would be a common byproduct of stellar evolution. At the Green Bank meeting Struve was enthusiastic about the number of planetary systems based primarily on his work on stellar rotation, and was supported by Huang, who had concluded from his own research on habitable zones around stars that the number of planets in the galaxy suitable for life was indeed very large (Dick 1996).

In the wake of NASA's founding in 1958, planetary science was also on the upswing, with the real possibility of sending spacecraft to study planetary surfaces and atmospheres. Indeed, that is precisely what happened, with the search for life on Mars often in the forefront as a driver of space science. Although no one represented NASA at the meeting, the young planetary scientist Carl Sagan, already involved in many planetary projects, was in attendance and well aware of NASA's planetary efforts. Joshua Lederberg, who had just coined the word *exobiology*, had some input to the meeting, but could not attend (Lederberg 1961).

The origins of life was also a hot topic at the time. The 1953 Miller-Urey experiments simulating life under primitive Earth conditions indicated life might easily originate given proper conditions. Melvin Calvin, an expert on chemical evolution, argued at the meeting that the origin of life was a common and even inevitable step in planetary evolution, and his already formidable credentials were given another boost when he received notification during the meeting that he would be awarded the Nobel Prize for his work on the chemical pathways of photosynthesis. And while little was known about the gap between primitive life and intelligent life, or even the definition of intelligence, the Lowellian Mars of artificially constructed canals still lingered in the background stimulating imaginative scientists about aliens.

It was a very large step from the origins of life to intelligence, and the concept of "intelligence" was not well defined or understood—still the case even today. Perhaps not surprisingly in this environment, the organizers looked for a participant doing practical research in the field. John Lilly, who had just come out with his controversial book *Man and Dolphin*, met that criterion and argued at Green Bank that dolphins were an intelligent species with a complex language, and that we might even be able to communicate with them. Dolphins thus became a kind of symbol for interspecies communication.

The equation's emphasis on "radio communicative" reflects the new era of radio astronomy, exemplified by the radio telescopes being built at the newly founded NRAO. All this early history and subsequent events are elaborated in detail in Dick (1996, 1998), and there is no need to repeat it here. Summarizing the results of their discussions, the members of the conference concluded that depending on the average lifetime for a civilization, the number of communicative civilizations in the galaxy might range from less than 1000 to 1 billion. Opting for optimism (likely an unfounded bias based on their interest in the subject), most of the members felt the higher number was likely to be closer to the truth.

7.3 Hidden Assumptions

Even as it grew in popularity, the Drake Equation embodied many hidden assumptions, perhaps responsible for both confusion and its enduring legacy. Nowhere is this more true than it its first and last factors, $R*$ and L, which are the only parameters with dimensions (stars forming per year and number of years). It is often forgotten that Drake's formulation was an eminently practical exercise, driven by Project Ozma and the desire to estimate the chances of its success by estimating the number of communicative civilizations existing *now*. This explains why the first parameter in the equation was not simply the number of stars existing today in the galaxy, whose formation began some 11 or 12 billion years ago. Nor was it even the number of stars existing 4.5 billion years ago, since they were all in different stages of development; if those stars had birthed civilizations, they would all be in different stages of development. Rather, Drake was interested in civilizations that were communicating now and at about the same stage of development as our civilization. He therefore used as the first parameter of the equation a rate of star formation rather than a number of stars. And he used L because it was the bottleneck that restricted technological civilizations to those communicating *now*.

The rate of star formation in our galaxy was the best-known quantity in the equation, and by the evidence of the time, it was calculated in a straightforward "quick and dirty," way, not taking into account current theories of star formation. In Pearman's account of the meeting the calculation went as follows: "If stars of solar type only are considered, a rough estimate of R∗ is given by the total number of such stars in the galaxy divided by their average lifetime. Thus $R* = 10^{10}/10^{10} = 1$ per year. This is perhaps a conservative estimate and less restrictive considerations permitting the inclusion of some Population II stars would give values as high as 10 stars per year." (Pearman 1963, 289). In other words, estimating 10 billion solar-type stars in the galaxy, each with a lifetime of about 10 billion years, yields one star forming per year. Including Population II stars (still Sun-like stars but older than our Sun) one could raise this estimate to 10 per year. Thus the often-used estimate in the Drake Equation of 1–10 stars forming in our galaxy per year.

Needless to say, this assumes a uniform rate of star formation over the lifetime of the galaxy, which we know today not to be the case. The same can be said for the calculation sometimes used, employing the number of solar type stars in the galaxy divided by the age of the galaxy. Strictly speaking R∗ today is defined not as the rate of star formation over the lifetime of the galaxy, but as the rate of star formation 4.5 billion years ago when our Sun and its planets were formed. At least that is the way Drake defines it. Responding to an inquiry about this shift in definition of R∗, Drake wrote:

> I prefer it because it more accurately quantifies the process by which current intelligent technology-using life came about. There are two versions of the equation which occur in various textbooks, etc. One uses number of stars/age of galaxy. The other uses R∗. The first conceals a somewhat important aspect of the whole picture, since it implies that the relevant star formation rate is the mean rate during the existence of the galaxy. But that is not the one which applies to the calculation of how many technology civilizations are out there to be found now. That number is governed by not the mean rate of star formation, but the rate of star formation which existed at the time stars of about the same age as the sun were formed, namely about 4.5 billion years ago. (Drake 2014a)

This formulation assumes that extraterrestrial technological civilizations develop at about the same rate as on Earth—a very large assumption indeed. Drake fully recognizes the assumption, but finds it necessary considering our ignorance:

> What we really need to know is the statistics of star formation over a substantial period 4.5 billion years ago, since the process of producing an intelligent species will take some range of time intervals. We won't know that until SETI succeeds. However, the rate of star forma-

> tion 4.5 billion years ago is the best estimate we can use in our current state of knowledge … the rate of star formation started high early on, and has been gradually decreasing over the history of the galaxy. So, to be as rigorous as possible, I always use the form of the equation with R∗. (Drake 2014a)

How different was the rate of star formation over the history of the galaxy? In particular how different was it 4.5 billion years ago compared to now? The first two chapters in this volume address the history of astronomer's ideas about star formation. But what about planets that could have formed much earlier in the history of the galaxy, say, as much as 8 billion years ago according to current theories (Dick 2003; Larson and Bromm 2003; Lemonick 2014)? Drake's answer is that they would not be at the same stage of civilization as ours. Drake emphasizes that his use of R∗ corresponding to about 4.5 billion years ago "is based on a quiet, unwritten rule, we use in SETI. 'Only assume phenomena you know exist.' Of course this is limiting, because history tells us that there must be powerful inventions yet to be made. But without such a rule, wild speculation can run rampant, and where do you draw the line?" (Drake 2014b). One such example is that civilizations that old might even be postbiological (Dick 2003). Moreover, Drake points out that older civilizations might no longer be detectable by radio:

> Our civilization, in an effort save resources and money (probably fair to believe all civilizations practice economies), is moving to communication techniques which release minimal energy into space, therefore releasing minimal energy to serve as a sign of our existence, and wasting minimal energy. Prime examples are cable TV and direct-to-home TV from satellites. Therefore, it appears that our detectability may last only a few hundred years, unless, of course, much more sensitive search systems such as using a solar gravitational lens are developed. (Drake 2014b)

This brings us full circle to L, which we need to remember is not the lifetime of a technological civilization, but the lifetime of a *communicating* technological civilization. Despite the limitation mentioned above, there remains the possibility of civilizations continuing to send out beacons, possibly altruistically for the benefit of others. Drake also recognizes this possibility:

> Of course, we should always have in the back of our head the thought that maybe there are possibly a small fraction of civilizations which are altruistic and maintain a bright, easily detectable, signal for very long times to enrich the knowledge of other civilizations. This would change the value of N a lot. If just one percent of civilizations maintained a "contact" beacon for a billion years (not crazy!), then L would be ten million years! A byproduct of this scenario is that the right strategy is to search in the directions where you will test the maximum number of stars for signals, which is in directions close to the galactic plane. (Drake 2014b)

The extent of extraterrestrial altruism might seem unknowable, but an entire volume of essays has been written on just this subject (Vakoch 2014).

The same kinds of hidden assumptions are present in the dimensionless factors sandwiched between R∗ and L. Referring to the third factor in the Equation, the number of planets that can potentially support life, Drake points out:

> …as another example of a complicated parameter, the ecosphere is a loose concept because in the simplest form usually given, the ecosphere is bounded by the boiling and freezing

> points of water. But we know of ways by which it extends to much larger distances. A deep atmosphere, and/or lots of CO_2, or having ice layer covering an ocean, all extend it out to much greater distances. A surface like that of Mars or Earth extends it way out because life can exist at suitable depths beneath the surface, since the temperature of a solid surface always rises with depth. In fact, just about everything planets can have – solid body, atmosphere, ocean, extends the ecosphere outwards. Think Enceladus. (Drake 2014a)

One more example of a hidden assumption involves L again. "Is L the total time a technical civilization exists?", Drake asks, then goes on to say:

> Using that definition gives you how many such civilizations there are. This L might be a billion years. But if you want to know how many *detectable* civilizations there are, then L is the length of time civilizations manifest themselves in some way a plausible detection system (another thing which has a wide range of possibilities) could detect. This could be only a couple of hundred years, if we are an average example, and our radio transmissions are the detectable signs of our existence. If we go to nothing but cable TV, and direct-to-home TV from satellites, our L will be measured in hundreds of years or less. On the other hand, if civilizations really can build useful telescopes using their star as a gravitational lens, then L possibly becomes a billion years! That is a big uncertainty and exciting possibility! (Drake 2014a)

The Drake Equation is again quite conservative in this sense: "when it comes to L, we are stuck with using ourselves as a model, which is all we have to go on until we discover another civilization." (Drake 2014b). To many, the Drake Equation parameters between R∗ and L immediately make sense as fractional factors winnowing the possibilities of communicating civilizations. But the meaning of the product of R∗ and L is not so intuitive. What does the rate of star formation have to do with the lifetime of a radio communicative civilization? Given all we have said above about R∗ and L, the bottom line is that if one radio communicative civilization is forming per year, and if L is 100, then any observer can only see such civilizations in the radio domain for 100 years. So, on the conservative view that we are only talking about *radio communicative* civilizations, and on the admittedly shaky assumption that those civilizations are developing at the same rate as we are, one only needs to look at 100 years around the star-forming time domain 4.5 billion years ago because all the other civilizations would have blinked out. Many more technological civilizations *could* exist—we just cannot see them because they are no longer communicative, at least in the radio spectrum. To put it another way, L acts as a kind of gateway that rejects all years during which a civilization does not communicate or have detectable radiation in the radio spectrum, no matter how many civilizations are forming. Drake likes the analogy of a Christmas tree, where each light blinks only once at various times (Drake 2014c). This is the source of the often repeated phrase in connection with the Drake Equation (and a source of license plates among SETI pioneers), that "N EQLS L," or N approximates L.

In the context of the lifetimes of civilizations, one must remember that the early 1960s were the height of Cold War. Estimates for L were remarkably optimistic for a time when civilization on Earth might have been wiped out by nuclear war at any moment. L is at one and the same time the potential bottleneck for the success of all SETI searches and conservative in its own way. As radio astronomer Seth Shostak points out, L for a radio communicative civilization might be shorter than L in the optical region of the spectrum or for communication modes based on other technology, or for a civilization not communicating at all. If we cannot detect a civilization, L is interesting for sociological reasons, but not for SETI reasons (Shostak 2009). Moreover, L should not be based on the lifetime of any particular civilization on Earth, such as Chinese, Greek, or Roman (Denning 2011). If Earth is any guide, the civilization in question will not be global; all that matters is that it be detectable.

The parameters between R∗ and L, while straightforward as members of the Drake Equation, are increasingly unknown as one moves to the right. One begins to see how the equation does an excellent job stimulating thought and discussion, and will likely continue to do so into the foreseeable future.

7.4 Criticisms and Variations

Not everyone has praised the Drake Equation. In fact, it has been highly criticized by people ranging from fiction writers to real scientists. Representative of the former is Michael Crichton, the best-selling author of books and movies such as *The Andromeda Strain, Jurassic Park*, *Sphere,* and *Prey*, which feature the failures of technology in society. In 2003, in the context of denying global warming arguments, Crichton used the Drake Equation as another example of bad science:

> This serious-looking equation gave SETI a serious footing as a legitimate intellectual inquiry. The problem, of course, is that none of the terms can be known, and most cannot even be estimated. The only way to work the equation is to fill in with guesses. And guesses – just so we're clear – are merely expressions of prejudice. Nor can there be "informed guesses." If you need to state how many planets with life choose to communicate, there is simply no way to make an informed guess. It's simply prejudice. The Drake equation can have any value from "billions and billions" to zero. An expression that can mean anything means nothing. Speaking precisely, the Drake equation is literally meaningless, and has nothing to do with science. I take the hard view that science involves the creation of testable hypotheses. The Drake equation cannot be tested and therefore SETI is not science. SETI is unquestionably a religion. (Crichton 2003)

SETI proponents would argue that not only could the results of the Drake Equation be tested, they had been tested with Project Ozma and could be tested with improved telescopes searching various targets at a range of frequencies. But the point about shaky values for individual parameters is indisputable. In this respect Crichton was not the first, nor the last, to take the equation too seriously. Most of the creators and users of the equation realized, and often explicitly stated, its limitations; Drake himself was amazed at its popularity, and SETI pioneer Bernard Oliver referred to the equation in his Project Cyclops report as "a way of compressing a large amount of ignorance into a small space" (Oliver 1971, p. 26). No one claimed the Drake Equation had the status of a scientific law such as $F = ma$ or $E = mc^2$. Although criticism is always welcome in science, Crichton's declaration seems suspiciously harsh, indicating he may have had some ideological agenda, not unusual in the context of climate change. His outburst brought responses ranging from bloggers to scientists. One journalist, William M. Briggs, argued that since we exist, it is not improbable that extraterrestrials exist, unless we were created in single unique event, which is a religious explanation, not science (Briggs 2008). Scientists assuredly cannot assume what they are trying to prove, but they are allowed to make probability estimates. The equation must be seen primarily as a useful heuristic, a way of contemplating the problem, not as a law of nature. As such it cannot be seen as strong justification for SETI searches, but rather as suggestive that such searches *could* be successful.

Viewed in this way, as an organized method for stimulating discussion, the equation has been a smashing success. This is evident not only in its appearance in numerous textbooks, lectures, and TV presentations, but also in the number of variations it has generated (which, it must be pointed out, may also take the equation too seriously). Already in 1971 J. G. Kreifeldt noted the equation was defective from a temporal point of view, in the sense that it did not allow for the time dependence of its terms. He went on to present a dynamic formulation that took into account different star generation rates, civilization lifetimes, and so on, resulting in an expression for the number of communicative civilizations in the galaxy as a function of time, including a variance for this estimate (Kreifeldt 1971; Wallenhorst (1981) elaborated on this temporal deficiency, and concluded that as a result N might only be 100 rather than one million. In a similar vein Ćirković (2004) argued that the galaxy was not habitable during its entire lifetime, since time is required for heavy elements to form the terrestrial planets. One result of this consideration is that civilizations are more concentrated in a given period of the galaxy's history.

A second type of variation on the Drake Equation involves taking interstellar colonization into account, in the wake of claims made by Michael Hart (1975) that colonization of the galaxy would take place over relatively short time scales, thus leading to the *Fermi Paradox*. The latter, which brought a crisis to SETI community thinking in the 1970s, states that if there are so many civilizations they would have been here, so "where are they?" In 1980 Walters, Hoover and Kotra suggested adding a new parameter "C" to the Drake Equation, taking into account the fraction of civilizations that wish to colonize, the fraction of stars with planets suitable for colonization, and the ability to reach those stars. If no civilizations wish to colonize, C is 1, reverting to the original Drake Equation. Based on a variety of considerations, the authors concluded that C would be less than 10, not the devastating impact on SETI that Hart had suggested. In 1983 astrophysicist and science fiction writer David Brin took a similar tack with the concept of a "contact cross section" to explain what he called "The Great Silence" (Brin 1983). While these variations may or may not give a more accurate picture of what is really going on, the more complex forms of the equation are not likely to appear in popular lectures. Its simplicity remains one of its enduring features, and is in part responsible for its longevity and continued utility (Drake 2013).

A third line of reasoning relevant to the Drake Equation was begun by Gonzalez et al. (2001), who argue that the Galactic Habitable Zone (GHZ) further narrows the possibilities for life. The GHZ, analogous to the circumstellar habitable zone often used in estimates of the fraction of planets where the conditions arise necessary for life, is the region in the galaxy where planets can retain liquid water on their surfaces and provide a long-term habitat for complex life. The central concept here is *metallicity*, the existence of heavy elements in a stellar nursery in amounts high enough to build a rocky terrestrial planet, a condition that does not occur everywhere in the galaxy. These arguments fed into Ward and Brownlee's bestselling book *Rare Earth*, the title of which indicates their conclusion (Ward and Brownlee 2000). Lineweaver and colleagues (Lineweaver et al. 2004) have further elaborated this concept, taking into account not only metallicity but also an environment free of life-extinguishing supernova and other factors.

The Drake Equation has also spawned discussion in more exotic directions, including modes of communication, difficulties of communication, and epistemological considerations (Hetesi and Regaly 2006), the messaging (METI) factor (Zaitsev 2005), and a "statistical Drake Equation" that assumes some standard deviation for each of the seven parameters (Maccone 2010). Again, any of the ideas behind these elaborations could have been undertaken without the equation, and do not stand or fall with the equation, but the fact is they were stimulated by the equation.

Does all this mean the original Drake Equation is obsolete? That was the argument of the Canadian futurist, science writer, and ethicist George Dvorsky, whose article "The Drake Equation is Obsolete" pulled no punches (Dvorsky 2007). He argues it is arbitrary, does not account for cosmological changes over time, and that the radio window on Earth is closing and so only accounts for a narrow class of civilizations radiating in the radio spectrum.

7.5 Future of the Equation

The Drake Equation continues to inspire discussion, as evident in a special issue devoted to it in the *International Journal of Astrobiology* (International Journal of Astrobiology 2013). Yet, others have concluded that the Equation was being left behind by events (Burchell 2006). In an article cleverly titled, "W(h)ither the Drake Equation?," Burchell concludes that with the rise of astrobiology, in which SETI is only one small intellectual part (and no federally funded programmatic part), the Drake Equation has become less important. The vast bulk of research in astrobiology today applies to microbes, and many consider not only that microbes may be the first extraterrestrial life to be discovered but also that microbes, not intelligence, may rule the universe. This may well be true. But it also true that the existence of extraterrestrial intelligence remains a major scientific question, one whose funding has suffered mainly due to congressional politics in the United States. SETI may rise again and even become an integral part of astrobiology (Dick 2013), and when it does, the Drake Equation will remain as relevant as ever.

Some of the criticisms of the Drake Equation are constructive and well taken. But existential threats to its existence are exaggerated. The continued utility of this type of equation may be seen in the fact that it has also inspired other equations of the same type. Most recently MIT astrophysicist Sara Seager adapted it to the current effort to search for biosignatures in exoplanet atmospheres. The *Seager Equation*, written as $N = N_{*}F_{Q}F_{HZ}F_{O}F_{L}F_{S}$, estimates N, the number of planets with detectable biosignature gases, where N_{*} is the number of stars within the sample, F_Q the fraction of quiet stars, F_{HZ} the fraction with rocky planets in the habitable zone, F_O the fraction of observable systems, F_L the fraction with life, and F_S the fraction with detectable spectroscopic signatures. The new equation (Seager 2013) is sometimes referred to as a "revised Drake Equation," although it is not primarily addressed to the search for radio communicative civilizations. But like the Drake Equation, it is driven by a practical consideration: if a system is built to detect biosignatures, what are the chances of success? The Transiting Exoplanet Survey Satellite (TESS) and the James Webb Space Telescope (JWST) are systems currently under development to address just this problem, among others, and the Seager Equation was developed for this purpose. Such biosignatures might range from simple microbial biosignatures to the most complex, technological biosignatures. The Seager Equation is subject to many of the same uncertainties as the Drake Equation, but it is seen to be useful in any case.

To summarize, we may state the following: (1) It is important to specify what assumptions are made when putting numbers into the Drake Equation. Part of the charm of the equation is that the parameters may be defined slightly differently, if one wants to know the number of civilizations, or the number of radio communicative civilizations, or the number of civilizations communicative in some other region of the spectrum, and so on. All such calculations are possible as long as one defines how each term is being used. Even with such definitions, hidden assumptions abound, which some see as a weakness, but may also be a strength in the sense that they generate even more discussion. (2) As is often stated, most of the parameters themselves are wildly uncertain, even at the beginning of the twenty-first century. Nevertheless, they hold the promise of improved estimates over time, as demonstrated by the current daily improvement in our knowledge of the fraction of stars with planets, based on both ground-based observations and the Kepler spacecraft results. Still, the equation should not be taken too seriously; its epistemological status is as a heuristic device rather than a law. (3) The equation itself is quite conservative in many ways, calculating only the number of radio communicative civilizations, a number that may be much smaller than the total number of civilizations. (4) The equation was a product of its time, but its utility is evident in its longevity, and in the birth of new equations of the same kind such as the Seager Equation, also a product of its time.

When all is said and done, of course, the equation remains only a guideline. This, after all, is all that Drake intended it to be when he first wrote the equation on the board at Green Bank meeting in 1961, in an attempt to estimate the chances of success for any SETI search using radio telescopes. As in all areas of science, in the end there is no substitution for observations, no matter what the theory or expectations are. Even given its inherent limitations, the future of the original Drake Equation remains bright, precisely because it is simple but begs for more rigorous elaboration taking into account hidden assumptions, increasingly accurate parameters based on observation, and our subjective hopes and fears. Drake's image of the Christmas tree is a vivid and haunting one, each light representing a civilization blinking on, for a period short or long depending on its lifetime, then blinking off again, perhaps forever, whether due to a change in technology, self-destruction, or some other unfathomed reason. That, too, is a reflection of the age during which the equation was constructed, with the threat of nuclear war blinking out one light on the tree of civilizations.

7.6 Commentary 2020

The Search for Extraterrestrial Intelligence (SETI) has been revitalized in recent years, with the Breakthrough Listen project, funded at the level of 100 million dollars. Moreover, NASA—its previous SETI program having been terminated by congressional action in 1993—is showing renewed interest. It sponsored a workshop on technosignatures in October 2018, foreshadowing more to come. In 2019 the SETI pioneer Jill Tarter launched a web-based archive of all SETI searches from 1960 to the present. Dubbed "Technosearch" and hosted at the SETI Institute in Mountain View, California, the archive can be found at https://technosearch.seti.org. The Drake Equation continues to be a guiding tool for all discussions of the likelihood of intelligent life in the universe.

Frank Drake (Fig. 7.5) continues to be actively involved in matters pertaining to SETI. And Green Bank Observatory, where his original observations for Project Ozma were undertaken, also continues to be active in SETI as one of the sites for the Breakthrough Listen observations. In addition to many other research projects, it continues to highlight its role as a pioneering institution in SETI, holding a conference on the 50th anniversary of Ozma in 2010 "From Project Ozma to the Starship Enterprise: A Conversation About the Next 50 Years of SETI," and another one in 2019 on the occasion of the 50th anniversary of the Apollo 11 Moon landing, titled "Moonshots and Earthshots in the Search for Life Beyond Earth." Both meetings were notable not only for the science talks, but also for a strong component on the humanities and social sciences. Also present and participating at both talks was David Tatel, a Judge on the U. S. Court of Appeals for the District of Columbia, whose father designed the radio telescope that Drake used for Project Ozma. More information and videos of the talks is available at http://www.gb.nrao.edu/OZMA@50/SETI_Attendees.shtml and https://greenbankobservatory.org/science/meetings-and-workshops/moonshots-and-earthshots/

Acknowledgments I am thankful to the ever-helpful archivists at the National Academy of Sciences, the National Radio Astronomy Observatory, the Smithsonian Institution and the Library of Congress (Carl Sagan papers) for locating relevant documents at their institutions. I am grateful to David H. DeVorkin for comments on the paper, and to Ken Kellermann for consultation on the origin of the Drake Equation.

Fig. 7.5 Frank Drake (with the author) signing a copy of his book *Intelligent Life in Space* (1962) at a meeting of Scientific Advisory Board of the SETI Institute, December 17, 2018. In the background are Nathalie Cabrol, Director of the Carl Sagan Center for the Study of Life in the Universe at the SETI Institute, and John Rummel, Chair of the Advisory Board

References

Bracewell, Ronald. 1960. "Communications from Superior Galactic Communities," *Nature*, 186, 670, reprinted in Cameron (1963a), 243–248

Briggs, William M. 2008. "Michael Crichton and SETI," online at http://wmbriggs.com/blog/?p=249

Brin, David. 1983. "The 'Great Silence': The Controversy Concerning Extraterrestrial Intelligent Life," QJRAS, 24, 283–309.

Burchell, Mark J. 2006. "W(h)ither the Drake Equation," IJA, 5, 243–250.

Cameron, A. G. W. 1963a. *Interstellar Communication*. New York, W. A. Benjamin.

Cameron, A. G. W. 1963b. "Future Research on Interstellar Communication," in Cameron (1963a), 309–315.

Cameron, A. G. W. 1963c. "Communicating with Intelligent Life on Other Worlds, *Sky and Telescope* (Nov, 1963), 258–261.
Ćirković, Milan. 2004. "On the Temporal Aspect of the Drake Equation and SETI," *Astrobiology*, 4, 225–231.
Cocconi, Giuseppe and Philip Morrison. 1959. "Searching for Interstellar Communications." *Nature* 184 (4690):844–846.
Crichton, Michael. 2003. "Aliens Cause Global Warming," https://stephenschneider.stanford.edu/Publications/PDF_Papers/Crichton2003.pdf
Denning, Kathryn. 2011. "L on Earth," in Vakoch (2011), 74–83
Dick, Steven J. 1996. *The Biological Universe: The Twentieth Century Extraterrestrial Life Debate and the Limits of Science*. Cambridge: Cambridge University Press.
Dick, Steven J. 1998. *Life on Other Worlds*. Cambridge: Cambridge University Press.
Dick, Steven J. 2003. "Cultural evolution, the postbiological universe and SETI," *International Journal of Astrobiology*, 2: 65–74; reprinted as "Bringing culture to cosmos: The postbiological universe," in Dick and Lupisella (2009), pp. 463–488.
Dick, Steven J. 2013. Congressional testimony before U. S. House Committee on Science, online at https://science.house.gov/hearing/full-committee-hearing-astrobiology-search-biosignatures-our-solar-system-and-beyond
Dick, Steven J. and Mark Lupisella. 2009. *Cosmos and Culture: Cultural Evolution in a Cosmic Context*. Washington, NASA SP-2009-4802.
Drake, Frank, 1959. "How Can We Detect Radio Transmissions from Distant Planetary Systems?," *Sky and Telescope*, 19, 140, reprinted in Cameron (1963), pp. 165–175.
Drake, Frank, 1961a. PSW lecture in Smithsonian Institution archives
Drake, Frank, 1961b. Drake to Pearman, March 13, 1961, NAS-NRC Archives: SSB Records Group: ADM: C&B: SSB: Conferences: Extraterrestrial Intelligent Life: Green Bank (West Virginia): 1961–1962.
Drake, Frank. 1965. "The Radio Search for Intelligent Extraterrestrial Life," in Current Aspects of Exobiology, ed., G. Mamikunian and M. H. Briggs. Oxford, 323–345.
Drake, Frank. 1992. Oral History Interview by Steven Dick, May 29 and May 30, 1992. p. 17 (NASA HQ History Office).
Drake, Frank. 2013. "Reflections on the Equation." International Journal of Astrobiology, Special Issue on the Drake Equation, 12, 3, 173–176.
Drake, Frank. 2014a. Drake to Steven J. Dick, private communication, March 6, 2014.
Drake, Frank. 2014b. Drake to Dick, private communication, March 10, 2014
Drake, Frank. 2014c. Drake to Dick, private communication, March 17, 2014.
Drake, Frank and Dava Sobel. 1992. *Is Anyone Out There?* New York. Delacorte Press.
Dvorsky, George. 2007. "The Drake Equation is Obsolete," online at http://www.sentientdevelopments.com/2007/05/drake-equation-is-obsolete.html
Dyson, F. 1966. "The Search for Extraterrestrial Technology." In *Perspectives in Modern Physics*, ed. R. E. Marshak, 641–55. New York: Interscience Publishers.
Goldsmith, Donald. 1980. *The Quest for Extraterrestrial Life*. Mill Valley, CA.
Gonzalez, Guillermo, Donald Brownlee, and Peter Ward. 2001. "The Galactic Habitable Zone: Galactic Chemical Evolution," *Icarus*, 152, 185–200.
Hart, Michael. 1975. "An Explanation for the Absence of Extraterrestrial Life on Earth," QJRAS. 16. 128–135, reprinted in Goldsmith (1980).
Hetesi, Zsolt and Zsolt Regaly. 2006. "A New Interpretation of the Drake Equation," JBIS, 59, 11–14
International Journal of Astrobiology. 2013. Special Issue on the Drake Equation, 12, 3.
Kreifeldt, J. G.. 1971. "A Formulation for the Number of Communicative Civilizations in the Galaxy," *Icarus*, 14, 419–430.
Larson, R. and V. Bromm. 2003. "The First Stars in the Universe," *Scientific American*, 285, 64–71.
Lederberg, Joshua. 1961. Lederberg to Pearman, April 11, 1961. NAS-NRC Archives: SSB Records Group: ADM: C&B: SSB: Conferences: Extraterrestrial Intelligent Life: Green Bank (West Virginia): 1961–1962.
Lemonick, Michael D. 2014. "The First Starlight," *Scientific American*, 310 (April), 38–45.

Lineweaver, C. H., Y. Fenner, and B. K. Gibson. 2004. 'The Galactic Habitable Zone and the Age Distribution of Complex Life in the Milky Way," *Science*, 303, 59–62.

Maccone, Claudio. 2010. "The Statistical Drake Equation," *Acta Astronautica*, 67, 1366–1383.

Morrison, Philip. 1962. "Interstellar Communication," *Bulletin of the Philosophical Society of Washington*, 16, 58, reprinted in Cameron, 1963a, 249–271.

Oliver, B. M. 1971. *Project Cyclops: A Design Study of a System for Detecting Extraterrestrial Intelligent Life*. NASA Ames, Moffett Field, CA.

Pearman, J. P. T. 1961. Pearman to O. G. Villard, Jr., Dept. of Engineering, Stanford University, NAS-NRC Archives: SSB Records Group: ADM: C&B: SSB: Conferences: Extraterrestrial Intelligent Life: Green Bank (West Virginia): 1961–1962.

Pearman, J.P. T. 196. "Extraterrestrial Intelligent Life and Interstellar Communication: An Informal Discussion," in Cameron, 1963.

Sagan, Carl. 1963. "Direct contact Among Galactic Civilizations by Relativistic Interstellar Spaceflight," *Planetary and Space Science*, 11, 485, reprinted in Goldsmith (1980), pp. 205–213

Sagan, Carl, ed. 1973. *Communication with Extraterrestrial Intelligence* (CETI) Cambridge, Mass.: MIT Press.

Seager, Sara. 2013. "An Equation to Estimate the Probability of Identifying An Inhabited World Within the Next Decade," online at http://www.cfa.harvard.edu/events/2013/postkepler/Exoplanets_in_the_Post_Kepler_Era/Program_files/Seager.pdf

Shapley, Harlow. 1958. "An Inquiry Concerning Other Worlds," *Of Stars and Men*, Boston: Beacon Press. pp. 53–75

Shklovskii, I. S. and Sagan, C. 1966. *Intelligent Life in the Universe*, Holden-Day, San Francisco.

Shostak, Seth. 2009. "The Value of 'L' and the Cosmic Bottleneck," in Dick and Lupisella, *Cosmos and Culture*, pp. 399–414.

Sullivan, Walter. 1964. *We Are Not Alone*. New York.

Vakoch, Douglas. 2014. *Extraterrestrial Altruism: Evolution and Ethics in the Cosmos*. Berlin: Springer.

Wallenhorst S. G. 1981. "The Drake Equation Reexamined," QJRAS, 22, 380.

Walters, C., R. A. Hoover and R. K. Kotra. 1980. "Interstellar Colonization: A New Parameter for the Drake Equation?" *Icarus* 41, 193–197.

Ward, Peter and Donald Brownlee. 2000. *Rare Earth: Why Complex Life is Uncommon in the Universe*.

Zaitsev, A. 2005. "Messaging to Extra-Terrestrial Intelligence", http://arxiv.org/abs/physics/0610031

Part II
Cosmic Evolution and Implications of Alien Life

Part II Frontispiece Isaac Newton, forever voyaging through strange seas of thought, alone. The implications of alien life leads us through many strange seas of thought. Newton himself spoke of "worlds of several sorts in several parts of the universe." (Photo of Newton at Trinity College, Cambridge, by Steven Dick)

Near me hung Trinity's loquacious clock,
Who never let the quarters, night or day,
Slip by him unproclaimed, and told the hours
Twice over with a male and female voice
Her pealing organ was my neighbor too;
And from my pillow, looking forth by light
Of moon or favoring stars, I could behold
The antechapel where the statue stood
Of Newton with his prism and silent face,
The marble index of a mind forever
Voyaging through strange seas of Thought, alone.

William Wordsworth, The Prelude (1850), Book III (Residence at Cambridge), 60–64

I have always warmed to these lines from William Wordsworth, with their image of the great poet looking across the green of Trinity College, Cambridge toward the statue of Isaac Newton (Fig. 1), whom Wordsworth fancied had spent a lifetime "Voyaging through strange seas of Thought, alone." When it comes to the implications of finding life beyond Earth, we indeed encounter many strange seas of thought ranging from astrotheology and astroethics to artificial intelligence and the postbiological universe. Increasingly, however, the encounter is not a lonely one, as a bevy of scholars from philosophy, theology, the humanities, and social sciences begin to analyze the implications of finding life whether microbes or intelligence.

In Part II of this volume, we tackle those implications of alien life, beginning with a chapter on cosmic evolution, which I consider the context for astrobiology and its cultural implications. Chapters 9 and 10 then provide early general overviews of cultural implications of astrobiology as visualized around the turn of the millennium when astrobiology was becoming a more coherent and expansive discipline in the wake of NASA roadmaps on the subject. Chapter 9, my first foray into the subject of implications, was a paper delivered at the Bioastronomy Symposium in Santa Cruz, California in 1993. It examines historical analogs for first contact, and argues that the history of science offers deeper insights than political history or anthropology, since the contact will likely not be physical. The transmission of science to the Latin West in the twelfth and thirteenth centuries, and the reception of scientific worldviews such as the Copernican and Darwinian, are offered as analogs to receipt of an extraterrestrial intelligent signal. Chapter 10 was written in the aftermath of my role in what was at the time (1998) considered a renegade group that argued NASA's astrobiology roadmap needed to include cultural aspects. In the end, although the study of societal impact was not incorporated as a specific goal of the NASA Astrobiology Roadmap (NASA 1998; Des Marais et al. 1998), a broad societal interest in the implications was incorporated as Principle 3 of it four operating principles. I argued that astrobiology, already an interdisciplinary field in terms of the physical and biological sciences, should embrace the humanities and the social and behavioral sciences in order to explore its cultural implications. The latest *Astrobiology Strategy* document (NASA 2015), makes the same case.

The next six chapters look at individual issues in astrobiology and society. Chapter 11 explores how anthropology and SETI can be mutually beneficial, an example of how the social sciences should become not a peripheral, but an integral, part of astrobiology. Following up on the implications of SETI and cultural evolution over the next two decades led me to consider two rather bold ideas explored in Chaps. 12 and 13: the postbiological universe and cosmotheology. Chapter 12 argues that that we may in fact live in a postbiological universe, one that has evolved beyond flesh and blood intelligence to artificial intelligence (AI), a product of cultural rather than biological evolution. Chapter 13 lays out a cosmotheology, defined as a theology that takes into account what we know about the universe based on science. Eschewing supernaturalism, we present six principles of cosmotheology, including the idea that we are not physically, biologically, cognitively, or morally central in the universe; that any concept of God must be grounded in naturalistic cosmic evolution; that it must have an expansive moral dimension—an *astroethics* extending to all life in the universe; and that while a human destiny linked to cosmic evolution rather than supernaturalism is a radical departure from the past, it is in the end beneficial and liberating. We argue that such a worldview, which might also be held by extraterrestrials since it is based on naturalistic principles, resolves many ancient theological problems. Following these ideas also leads to astroethics as discussed briefly in Chap. 14. I detail the origin of these ideas in the commentary section of each chapter.

In Chap. 15 we examine the concept of Messaging Extraterrestrial Intelligence (METI). Although directly messaging ETI raises many ethical questions, we argue that it should be undertaken for a variety of reasons. One of the most controversial issues is the extent that prior consultation should be undertaken, and if so consultation with whom. After examining the Asilomar process that took place in the 1970s in connection with biotechnology, we argue that such consultations should take place at the level of the practitioners, supplemented by an array of scholars, in conjunction with an organization like the International Academy of Astronautics in order to give it more force. But we need to have clarity about the purpose of the consultations and much else, since this is a problem that may affect all of humanity.

Chapter 16, the final chapter of Part II, is an indication of how much the field of astrobiology and society has advanced over the last three decades since NASA inaugurated its SETI observations and contemplated cultural implications in its Cultural Aspects of SETI (CASETI) workshops in 1991 and 1992. Chapter 16 comes back full circle from Chaps. 9 and 10 in order to describe the ongoing work in what is rapidly becoming a new field of astrobiology and society. It is only a glimpse at the work now undertaken by scholars from many fields. In my own case this research culminated in two volumes, an edited volume representing 22 scholars in *The Impact of Discovering Life Beyond Earth* (2015), and my own take on the subject in *Astrobiology, Discovery and Societal Impact* (2018). Both volumes were the result of my time as the Baruch S. Blumberg NASA/Library of Congress Chair in Astrobiology, itself an indication of how seriously this subject is now taken, to the extent that it funded by NASA and also of interest to the U. S. Congress (see Appendix 1).

References

Des Marais, David, J. A. Nuth, III, Louis Allamandola et al. 2008. "The NASA Astrobiology Roadmap," *Astrobiology*, 8: 715–730.

Dick, Steven J., ed. 2015. *The Impact of Discovering Life Beyond Earth*. Cambridge: Cambridge University Press.

Dick, Steven J. 2018. *Astrobiology, Discovery, and Societal Impact*. Cambridge: Cambridge University Press.

NASA, 1998. "Astrobiology Roadmap," Ames Research Center, Moffet Field, CA. online at https://nai.nasa.gov/media/roadmap/1998/

NASA. 2015. *Astrobiology Strategy* online at https://nai.nasa.gov/media/medialibrary/2016/04/NASA_Astrobiology_Strategy_2015_FINAL_041216.pdf

Chapter 8
Cosmic Evolution: History, Culture, and Human Destiny

Abstract Astrobiology must be seen in the context of cosmic evolution, the 13.7 billion-year master narrative of the universe. The idea of an evolving universe dates back only to the nineteenth century, and became a guiding principle for astronomical research only in the second half of the twentieth century. The modern synthesis in evolutionary biology hastened the acceptance of the idea in its cosmic setting, as did the confirmation of the Big Bang theory for the origin of the universe. NASA programs such as Origins incorporated it as a guiding principle. Cosmic evolution encompasses physical, biological and cultural evolution, and may result in a physical, biological or postbiological universe, each with its own implications for long-term human destiny, and each imbuing the meaning of life with different values. It has the status of an increasingly accepted worldview that is beginning to have a profound effect not only in science but also in religion and philosophy.

8.1 Introduction

During the course of the twentieth century a powerful new idea gradually entered human consciousness and culture: that we are part of a cosmos billions of years old and billions of light years in extent, that all parts of this cosmos are interconnected and evolving, and that the stories of our galaxy, our Solar System, our planet, and ourselves are part and parcel of the ultimate master narrative of the universe—a story we now term *cosmic evolution*. Even as in some quarters of popular culture heated debate continues over Darwinian evolution 150 years after the idea was published, over the last 50 years the much more encompassing idea that Carl Sagan embodied in the phrase the *cosmic connection* (Sagan 1973, 2000) has become more and more a part of our daily lives, and will be even more in the future as our cosmic consciousness increases.

Cosmic evolution provides the proper universal context for biological evolution, revealing that the latter is only a small part of the bigger picture, in which everything is evolving, including life and culture. The more we know about science, the more we know culture and cosmos are connected, to such an extent that we can now see the cosmos is inextricably intertwined with human destiny, both in the short term

First published in Steven J. and Mark Lupisella, eds. *Cosmos & Culture: Cultural Evolution in a Cosmic Context*. Washington, DC: NASA, 2009

S. J. Dick, *Space, Time, and Aliens*, https://doi.org/10.1007/978-3-030-41614-0_8

and the long-term, impinging on (and arguably essential to) questions normally reserved for religion and philosophy. It is the purpose of this chapter to uncover the historical evolution of this new understanding of the cosmos, describe the effects on culture so far, and outline the potentially far-reaching impact on the future of humanity.

8.2 Cosmic Evolution and History

The idea of cosmic evolution implies a continuous evolution of the constituent parts of the cosmos from its origins to the present. Planetary evolution, stellar evolution, and the evolution of galaxies could in theory be seen as distinct subjects, in which one component evolves but not the other, and in which the parts have no mutual relationships. Indeed, in the first half of the twentieth century scientists treated the evolution of planets, stars and galaxies for the most part as distinct subjects, and historians of science still tend to do so (Hale 1908; Lowell 1909). But the amazing and stunning idea that overarches these separate histories is that the entire universe is evolving, that all of its parts are connected and interact, and that this evolution applies not only to inert matter, but also to life, intelligence, and culture. Physical, biological, and cultural evolution are the essence of the universe. This overarching idea is what is called cosmic evolution, and the idea has itself evolved to the extent that some modern scientists even talk of a cosmic ecology, the "life of the cosmos," and the "natural selection" of universes (Dyson 1988; Smolin 1997).

Although the question of extraterrestrial life is very old, the concept of a full-blown cosmic evolution—the connected evolution of planets, stars, galaxies, and life on Earth and beyond—is much younger. As historian Michael Crowe has shown in his study of the plurality of worlds debate, in the nineteenth century a combination of ideas—the French mathematician Pierre Simon Laplace's *nebular hypothesis* for the origin of the Solar System, the British naturalist Robert Chamber's application of evolution to other worlds, and Darwinian evolution on this world—gave rise to the first tentative expressions of parts of this world view (Crowe 1986; Schaffer 1989; Zakariya 2010). The philosophy of Herbert Spencer extended it to the evolution of society, although not to extraterrestrial life or society. But some Spencerians, notably Harvard philosopher John Fiske in his *Outlines of a Cosmic Philosophy Based on the Doctrine of Evolution* (1875), did extend evolutionary principles to life on other planets (Strick 2000).

Neither astronomers nor biologists tended to embrace such a broad philosophical, and empirically unsupported, concept as full-blown cosmic evolution. Influenced by Darwin, nineteenth century astronomers and popularizers did occasionally propound the rudiments of the idea. In England, Richard A. Proctor proposed an evolutionary view in which all planets would attain life in due time (Proctor 1870). In France, Camille Flammarion argued that life began by spontaneous generation, evolved via natural selection by adaptation to its environment, and was ruled by survival of the fittest, wherever it was found in the universe (Flammarion 1872). In this scheme of cosmic evolution, anthropocentrism was banished; the Earth was not unique, and humans were in no sense the highest form of life. Thus were the general outlines of the idea of cosmic evolution spread to the populace, not only by these forerunners of Carl Sagan, but also by a variety of Victorian popularizers of science (Lightman 2007).

But such a set of general ideas is a long way from a research program. In the first half-century of the post-Darwinian world, cosmic evolution did *not* find fertile ground among astronomers, who were hard-pressed to find evidence for it. Spectroscopy, which displayed the distinct "fingerprints" of each of the chemical elements, did reveal to astronomers that those same elements were found in the terrestrial and celestial realms. This confirmed the widely assumed idea of "uniformity of nature"—that both nature's laws and its materials were everywhere the same. Astronomers recognized and advocated parts of cosmic evolution, as in William Herschel's ruminations on the classification of nebulae, the British astrophysicist Norman Lockyer's work on the evolution of the elements, or the American astronomer George Ellery Hale's *Study of Stellar Evolution* (1908). In their published writings, however, Hale and his colleagues stuck very much to the techniques for studying the evolution of the physical universe. Even Percival Lowell's *Evolution of Worlds* (1909) spoke of the evolution of the physical universe, not a *biological universe* full of life, his arguments for Martian canals built by an alien intelligence notwithstanding. Although Lowell was a Spencerian, had been influenced by Fiske at Harvard, and had addressed his graduating class on "The Nebular Hypothesis" 2 years after Fiske's *Cosmic Philosophy*, he did not apply the idea of advanced civilizations to the universe at large (Strauss 2001).

Even in the first half of the twentieth century, astronomers had to be content with the uniformity of nature argument confirmed by spectroscopy. In an article in *Science* in 1920, the American astronomer W. W. Campbell (a great opponent of Lowell's canalled Mars) enunciated exactly this general idea of widespread life via the uniformity of nature argument: "If there is a unity of materials, unity of laws governing those materials throughout the universe, why may we not speculate somewhat confidently upon life universal?" he asked. He even spoke of "other stellar systems … with degrees of intelligence and civilization from which we could learn much, and with which we could sympathize" (Campbell 1920).

That was about all the astronomers of the time could say. As Helge Kragh concluded in his history of the Big Bang cosmology, "during the nineteenth century the static clockwork universe of Newtonian mechanics was replaced with an evolutionary worldview. It now became accepted that the world has not always been the same, but is the result of a natural evolution from some previous state probably very different from the present one. Because of the evolution of the world, the future is different from the past—the universe acquired a history." But the nineteenth century went only so far: "The Victorian conception of the universe was, in a sense, evolutionary, but the evolution was restricted to the constituents of the universe and did not, as in the world models of the twentieth century, cover the universe in its entirety" (Toulmin and Goodfield 1982; Kragh 1996).

For the most part, biologists were also reluctant cosmic evolutionists even at the beginning of the twentieth century. The British naturalist Alfred Russel Wallace, cofounder with Darwin of the theory of natural selection, wrote in 1903 that "Our position in the material universe is special and probably unique, and … it is such as to lend support to the view, held by many great thinkers and writers today, that the supreme end and purpose of this vast universe was the production and development of the living soul in the perishable body of man" (Wallace 1903a, b). While he believed in a modicum of physical evolution in his small Solar System-centric universe, he concluded that intelligence beyond Earth was highly improbable, calculating the physical, cosmic, and evolutionary improbabilities against the evolution of an equivalent moral or intellectual being to man, on any other planet, as a hundred million million to one. Clearly, for this pioneer in evolution by natural selection there was no cosmic evolution in its fullest sense, no biological universe (Dick 2008a).

Similarly, Lawrence J. Henderson, a professor of biological chemistry at Harvard, wrote 10 years after Wallace:

> There is … one scientific conclusion which I wish to put forward as a positive statement and, I trust, fruitful outcome of the present investigation. The properties of matter and the course of cosmic evolution are now seen to be intimately related to the structure of the living being and to its activities; they become, therefore, far more important in biology than has been previously suspected. For the whole evolutionary process, both cosmic and organic, is one, and the biologist may now rightly regard the universe in its very essence as biocentric. (Henderson 1913)

Clearly, Henderson grasped essential elements of cosmic evolution, used its terminology, and believed his research into the fitness of the environment pointed in that direction (Fry 1996). Yet, although he had a productive career at Harvard until his death in 1942, Henderson never enunciated a full-blown concept of cosmic evolution, nor did any of his astronomical colleagues.

Henderson's idea of a biologically robust cosmic evolution in 1913 was largely stillborn, perhaps in part because just a few years later the British astronomer James Jeans' theory of the formation of planetary systems by close stellar encounters convinced the public, and most scientists, that planetary systems were extremely rare (Dick 1996). The idea remained entrenched until the mid-1940s. Without planetary systems, cosmic evolution was stymied at the level of the innumerable stars, well short of the biological universe. In the absence of evidence, cosmic evolution was left to science fiction writers like Olaf Stapledon, whose *Last and First Men* and *Star Maker* novels in the 1930s embraced it in colorful terms. But Henderson had caught the essence of a great idea—that life and the material universe were closely linked, a fundamental tenet of cosmic evolution that would lay dormant for almost a half century.

The humble and sporadic origins of the idea of cosmic evolution demonstrate that it did not have to become what is now the leading overarching principle of twentieth century astronomy (Zakariya 2010). But it did, helped along by the Big Bang cosmology featuring a universe with a beginning slowly unfolding over time. The history of the Big Bang cosmology therefore parallels to some extent the history of cosmic evolution in its grandest sense, and Edwin Hubble's empirical observations of galaxies consistent with the concept of an expanding universe added a further dimension to the new world view (Kragh 1996). Almost all astronomers today view cosmic evolution as a continuous story from the Big Bang to the evolution of intelligence, accepting as proven the evolution of the physical universe while leaving open the still-unproven question of the biological universe, whose sole known exemplar remains planet Earth. Today, the central question remains how far cosmic evolution commonly proceeds. Does it end with the evolution of matter, the evolution of life, the evolution of intelligence, or the evolution of culture? But today, by contrast with 1950, cosmic evolution is the guiding conceptual scheme for a substantial research program.

When and how did astronomers and biologists come to believe in cosmic evolution as a guiding principle for their work, and how did it become a serious research program? In her pioneering book *Unifying Biology: The Evolutionary Synthesis and Evolutionary Biology*, historian Betty Smocovitis has emphasized that with the rise of the Modern Synthesis in biology, by mid-century evolution had become a unifying theme for biology, with Julian Huxley and others also extolling its place in cosmic evolution. By the 1940s, Smocovitis wrote, "cosmic, galactic, stellar, planetary, chemical, organic evolution and cultural evolution emerged as a continuum in a 'unified' evolutionary cosmology" (Smocovitis 1996). But it was only in the 1950s and 1960s that the cognitive elements—planetary science, planetary systems science, origin of life studies, and the Search for Extraterrestrial Intelligence (SETI)—combined to form a robust theory of cosmic evolution, as well as to provide an increasing amount of evidence for it. Only then, and increasingly thereafter, were serious claims made for disciplinary status for a field known as exobiology, astrobiology, and bioastronomy—the biological universe component of cosmic evolution. And only then did government funding become available, as the search for life became one of the prime goals of space science, and cosmic evolution became public policy.

We have already hinted at why this coalescence had not happened earlier, Spencerian philosophy, and the ideas of Flammarion, Proctor, and Henderson notwithstanding. Although the idea of the physical evolution of planets and biological evolution of life on those planets in our Solar System had been around for a while—and even some evidence in the form of seasonal changes and spectroscopic evidence of vegetation on Mars—not until the space program did the technology become available, resulting in large amounts of government funding poured into planetary science so that these tentative conclusions could be further explored. Moreover, if evolution was truly to be conceived as a cosmic phenomenon, planetary systems outside our Solar System were essential. Only in the 1940s, when the nebular hypothesis came back into vogue, could an abundance of planetary systems once again be postulated. During a 15-year period from 1943 to 1958, the commonly accepted frequency of planetary systems in the galaxy went from 100 to 1 billion, a difference of seven orders of magnitude (Dick 1996). The turnaround involved many arguments, from the observations of a few possible planetary companions in 1943, to binary star statistics, the nebular hypothesis, and stellar rotation rates. Helping matters along was the dean of American astronomers,

Henry Norris Russell, whose 1943 *Scientific American* article "Anthropocentrism's Demise" enthusiastically embraced numerous planetary systems based on just a few observations by Kaj Strand and others (Russell 1943). By 1963 the American astronomer Peter van de Kamp announced his discovery of a planet around Barnard's star, and the planet chase was on, to be truly successful only at the end of the century (van de Kamp 1963).

Thus was one more step in cosmic evolution made plausible by mid-century, even though it was a premature and optimistic idea, since only in 1995 were the first planets found around Sun-like stars, and those were gas giants like Jupiter. But how about life? That further step awaited developments in biochemistry, in particular the Oparin-Haldane theory of chemical evolution for the origin of life. The first paper on the origins of life by the Russian biochemist Aleksandr Ivanovich Oparin was written in 1924, elaborated in the 1936 book *Origin of Life*, and reached the English world in a 1938 translation (Fry 2000). By that time the British geneticist and biochemist J. B. S. Haldane had provided a brief independent account of the origin of life similar to Oparin's chemical theory. By 1940, when the British Astronomer Royal Sir Harold Spencer Jones wrote *Life on Other Worlds*, he remarked, "It seems reasonable to suppose that whenever in the Universe the proper conditions arise, life must inevitably come in to existence" (Spencer Jones 1940).

The contingency or necessity of life would be one of the great scientific and philosophical questions of cosmic evolution, but in any case the Oparin-Haldane chemical theory of origin of life provided a basis for experimentation, beginning with the famous experiment of Stanley Miller and Harold Urey in 1953, in which amino acids—the building blocks of proteins and life—were synthesized under possible primitive Earth conditions. By the mid-1950s, another step of cosmic evolution was coming into focus—the possibility of primitive life. Again, the optimism was premature, but the point is that it set off numerous experiments around the world to verify another step in cosmic evolution. Already in 1954 Harvard biochemist George Wald proclaimed the Oparin-Haldane process a natural and inevitable event, not just on our planet, but on any planet similar to ours in size and temperature (Wald 1954). By 1956 Oparin had teamed with Russian astronomer V. Fesenkov to write *Life in the Universe*, which expressed the same view of the inevitability of life as had Wald (Oparin and Fesenkov 1961).

What remained was the possible evolution of intelligence in the universe. Although hampered by a lack of understanding of how this had happened on Earth, discussion of the evolution of intelligence in the universe was spurred on by the famous paper by the American physicists Giuseppe Cocconi and Philip Morrison in *Nature* in 1959. "Searching for Interstellar Communications" showed how the detection of radio transmissions was feasible with radio telescope technology already in hand. In the following year astronomer Frank Drake, a recent Harvard graduate, undertook just such a project (Ozma) at the National Radio Astronomy Observatory (NRAO), ushering in a series of attempts around the world to detect such transmissions. And in 1961 Drake, supported by NRAO director Otto Struve, convened the first conference on interstellar communication at Green Bank, West Virginia. Although a small conference attended by only 11 people including Struve, representatives were present from astronomy, biology and physics, already hinting at the interdisciplinary nature of the task (Dick 1996). Thus by 1961, the elements of the full-blown cosmic evolution debate were in place.

It was at the Green Bank meeting that the now-famous Drake Equation was first formulated. The eq. $N = R_* \times f_p \times n_e \times f_l \times f_i \times f_c \times L$—purporting to estimate the number (N) of technological civilizations in the galaxy—eventually became the icon of cosmic evolution—showing in one compact equation not only the astronomical and biological aspects of cosmic evolution, but also its cultural aspects. The first three terms represented the number of stars in the Galaxy that had formed planets with environments suitable for life; the second two terms narrow the number to those on which life and intelligence actually develop; and the final two represent radio communicative civilizations. "L", representing the lifetime of a technological civilization, embodied the success or failure of cultural evolution. Unfortunately, depending on who assigned values to the parameters of the equation, it yielded numbers ranging from one (Earth) to many millions of technological civilizations in the Galaxy. Drake and most others in the field recognized then, and recognize even now almost 50 years later, that this equation is a way of organizing our ignorance. At the same time, progress has been made on at least one of its parameters; the fraction of stars with planets (f_p) is now known to be between 5% and 10% for gas giant planets around solar type stars.

The adoption of cosmic evolution was by no means solely a Western phenomenon. On the occasion of the fifth anniversary of Sputnik, Soviet radio astronomer Joseph Shklovskii wrote *Universe, Life, Mind* (1962). When elaborated and published in 1966 as *Intelligent Life in the Universe* by Carl Sagan, it became the Bible for cosmic evolutionists interested in the search for life (Shklovskii and Sagan 1966). Nor was Shklovskii's book an isolated instance of Russian interest. As early as 1964 the Russians convened their own meetings on extraterrestrial civilizations, funded their own observing programs, and published extensively on the subject (Tovmasyan 1965).

Thus, cosmic biological evolution first had the potential to become a research program in the early 1960s when its cognitive elements had developed enough to become experimental and observational sciences, and when the researchers in these disciplines first realized they held the key to a larger problem that could not be resolved by any one part, but only by all of them working together. At first this was a very small number of researchers, but it has expanded greatly over the last 40 years, especially under NASA patronage. The idea was effectively spread beyond the scientific community by a variety of astronomers. As early as 1958, cosmic evolution was being popularized by Harvard astronomer Harlow Shapley in *Of Stars and Men*, and Shapley used it thereafter in many of his astronomical writings emphasizing its impact on culture (Palmeri 2000, 2009). The idea was spread much more by Sagan's *Cosmos* (1980) and astronomer Eric Chaisson's works (Chaisson 1981; Chaisson 1987; Chaisson 2001; Chaisson 2006), and in France by Hubert Reeves *Patience dans l'azur: L'evolution cosmique* (1981), among others. By the end of the century cosmic evolution was viewed as playing out on an incomparably larger stage than conceived by A. R. Wallace a century ago.

The catalyst for the unified research program of cosmic evolution—and for the birth of a new scientific discipline—was the Space Age. No one would claim that a field of extraterrestrial life studies, or cosmic evolution, existed in the first half of the twentieth century. Even by 1955, when Otto Struve pondered the use of the word "astrobiology" to describe the broad study of life beyond the Earth, he explicitly decided against a new discipline: "The time is probably not yet ripe to recognize such a completely new discipline within the framework of astronomy. The basic facts of the origin of life on Earth are still vague and uncertain; and our knowledge of the physical conditions on Venus and Mars is insufficient to give us a reliable background for answering the question" of life on other worlds (Struve 1955). But the imminent birth of "exobiology" was palpable in 1960 when Joshua Lederberg coined the term and set forth an ambitious but practical agenda based on space exploration in his article in *Science*, "Exobiology: Experimental Approaches to Life Beyond the Earth" (Lederberg 1960) Over the next 20 years numerous such proclamations of a new discipline were made. By 1979, NASA's SETI chief, John Billingham, wrote that "over the past 20 years, there has emerged a new direction in science, that of the study of life outside the Earth, or exobiology. Stimulated by the advent of space programs, this fledgling science has now evolved to a stage of reasonable maturity and respectability" (Billingham 1981).

The extent to which NASA had served as the chief patron of cosmic biological evolution is evident in its sponsorship of many of the major conferences on extraterrestrial life, although the Academies of Science of the United States and the USSR were also prominent supporters. It was NASA that adopted exobiology as one of the prime goals of space science, and it was from NASA that funding would come, despite an early but abortive interest at the National Science Foundation (Appel 2000). Pushed by prominent biologists such as Joshua Lederberg, beginning already in the late 1950s soon after its origin, NASA poured a small but steady stream of money into exobiology and the life sciences in general. By 1976 $100 million had been spent on the Viking biology experiments designed to search for life on Mars from two spacecraft landers. Even as exobiology saw a slump in the 1980s in the aftermath of the Viking failure to detect life on Mars unambiguously, NASA kept exobiology alive with a grant program at the level of $10 million per year, the largest exobiology laboratory in the world at its Ames Research Center, and evocative images of cosmic evolution (Fig. 8.1). Cosmic evolution's potential by the early 1960s to become a research program was converted to reality by NASA funding.

This is true not only of NASA's exobiology laboratory and grants program, but also of its SETI program. Born at Ames in the late 1960s quite separate from the exobiology program, NASA SETI expended some $55 million prior to its termination by Congress in 1993 (Dick 1996; Garber 1999). It was the NASA SETI program that was the flag bearer of cosmic evolution. As it attempted to determine how many planets might have evolved intelligent life, all of the parameters of cosmic evolution, as encapsulated in the Drake Equation, came into play.

With the demise of a publicly funded NASA SETI program in 1993, the research program of cosmic evolution did not end. The remnants of the NASA SETI program were continued with private funding, and similar, if smaller SETI endeavors are still

Fig. 8.1 Cosmic evolution is depicted in this image from the exobiology program at NASA Ames Research Center, 1986. *Upper left:* the formation of stars, the production of heavy elements, and the formation of planetary systems, including our own. At *left* prebiotic molecules, RNA, and DNA are formed within the first billion years on the primitive Earth. At *center* the origin and evolution of life leads to increasing complexity, culminating with intelligence, technology, and astronomers, *upper right,* contemplating the universe. The image was created by David DesMarais, Thomas Scattergood, and Linda Jahnke at NASA Ames in 1986 and reissued in 1997

carried out around the world. Within NASA, a program of cosmic evolution research continued, with its images subtly changed. In 1995 NASA announced its Origins program, which 2 years later it described in its Origins Roadmap as "following the 15 billion year-long chain of events from the birth of the universe at the Big Bang, through the formation of chemical elements, galaxies, stars, and planets, through the mixing of chemicals and energy that cradles life on Earth, to the earliest self-replicating organisms—and the profusion of life." Any depiction of "intelligence" is conspicuously absent from the new imagery (see the frontispiece to Part I), for due to Congressional action, programmatically it could no longer be supported with public funding. With this proclamation of a new Origins program, cosmic evolution became the organizing principle for most of NASA's space science effort. In a broad sense, most of NASA's space science program can be seen as filling in the gaps in the story of cosmic evolution.

In 1996 the "Astrobiology" program was added to NASA's lexicon. The NASA Astrobiology Institute, centered at NASA's Ames Research Center, funds numerous centers nationwide for research in astrobiology at the level of several tens of millions of dollars (Dick and Strick 2004). Its paradigm is also cosmic evolution, even if it also tends to avoid mention of extraterrestrial intelligence due to Congressional disapproval stemming from cancellation of the NASA SETI program in 1993. No such restriction is evident at the SETI Institute in Mountain View California, headed by Frank Drake. The Institute has under its purview tens of millions of dollars in grants, all geared to answering various parameters of the Drake Equation, the embodiment of cosmic evolution, including the search for intelligence.

As we enter the twenty-first century there is no doubt about the existence of a robust cosmic evolution research program. NASA is its primary patron, and even many scientists without government funding now see their work in the context of this research program. Other agencies, including the European Space Agency, are also funding research essentially in line with the Origins and Astrobiology programs, not to mention their spacecraft that help to fill in the gaps in the grand narrative of cosmic evolution. Within the last 40 years, all the elements of a new discipline gradually came into place: the cognitive elements, the funding resources, and the community and communications structures common to new disciplines. As we enter the twenty-first century, cosmic evolution is a thriving enterprise, providing the framework for an expansive research program, drawing in young talent sure to perpetuate a new field of science that a half-century ago was nonexistent.

8.3 Cosmic Evolution and Culture

Since Darwin propounded his theory of evolution by natural selection, evolution has been much more than a science. It has been a worldview that has affected culture in numerous ways, and different cultures in diverse ways (Greene 1981; Bowler 1983). As we have noted, in her history of the modern evolutionary synthesis in biology, historian Betty Smocovitis found that by the late 1950s and early 1960s, the wider culture was "permeated with evolutionary science" and "resonated with evolutionary themes" (Barlow 1995; Smocovitis 1996). The leaders of that evolutionary synthesis, including Julian Huxley, Theodosius Dobzhansky, Ernst Mayr, and George Gaylord Simpson espoused an "evolutionary humanism," a secular progressive vision of the world that for Huxley at least, was "the central feature of his worldview and of his scientific endeavors" (Simpson 1949; Huxley 1964; Dobzhansky 1969). In books and articles, each of these scientists addressed the future of mankind in evolutionary terms. Huxley (grandson of Darwin's chief defender T. H. Huxley) "offered an inquiry … into an ethical system, an ethos, grounded in evolution, now a legitimate science, with its fundamental principle of natural selection, verifiable and testable through observation and experiment." Cosmic evolution was part of this worldview, even if Mayr and Simpson would later express serious doubts about the chances for success of exobiology and SETI programs (Dick 1996).

In the 1950s and 1960s Harlow Shapley was a prime example of a cosmic evolution evangelist from the astronomical side, being among the first to popularize the cosmic evolutionary perspective with "missionary zeal." In Shapley's view this perspective inspired a religious attitude, needed to be incorporated into current religious traditions, and went beyond those traditions in questioning the need for the supernatural. He even spoke of a "stellar theology," a view that had broader implication for ethics. Cosmic evolution has also been used to bolster the idea of biological evolution, though apparently with little impact to this day among skeptical Americans (Palmeri 2009). Shapley's books *Of Stars and Men: The Human Response to an Expanding Universe* (1958), *The View from a Distant Star* (1963), and *Beyond the Observatory* (1967) spread these ideas worldwide.

During the second half of the twentieth century, then, the evolutionary view of the universe was not only fully in place both from the point of view of at least some astronomers and biologists, but was also spreading to the broader culture. Instead of the small and relatively static universe accepted at the turn of the twentieth century, humanity was now asked to absorb the idea of an expanding (now known to be accelerating) universe 13.7 billion light years in extent, full of billions of evolving galaxies floating in an Einsteinian space-time with no center. The Big Bang theory, though still in competition in the 1950s with Fred Hoyle's Steady State theory that denied an overarching linear cosmic evolution, would receive increasing confirmation through the detection of the cosmic microwave background in 1965, and its study at ever-finer resolution through the COBE and WMAP satellites (Kragh 1996). The Hubble Space Telescope and other spacecraft brought the impact of this worldview directly to the people, through spectacular imagery of objects in the evolutionary narrative, and through more global images such as the Hubble Deep Field (for example see Fig. 18.5). The biological universe full of life was conjectured, but not proven, though SETI and astrobiology programs received much popular attention, particularly in the case of the supposed fossil life found in the Mars rock, evidence hotly contested, in part because of the high stakes for broader worldviews (Sawyer 2006).

In seeking the impact of the new universe on culture in the modern era, we need to remember that "culture" is not monolithic and that "impact" is a notoriously vague term. Thus it is no surprise that the new universe and its master narrative of cosmic evolution evoked different meanings for different groups. Cosmic consciousness in the form of a biological universe was expressed in many forms in popular culture, some of them unpalatable to most scientists: belief in UFOs and extraterrestrial abduction, space-oriented religious cults, and ever more elaborate alien scenarios in science fiction. Indeed, all three of these developments may be seen as ways that popular culture attempts to work out the worldview implied by the new universe. The UFO debate and alien science fiction both had their predecessors in the late nineteenth century, but only in the second half of the twentieth century did they come into their own as major elements of popular culture. During this time evolutionary themes became common in science fiction, notably in Arthur C. Clarke's work such as *Childhood's End.* Some of the most popular films of all times featured aliens, among them *Star Wars, Close Encounters of the Third Kind, ET: The Extraterrestrial*, *War of the Worlds*, and *Men in Black.* Obviously, and understandably, popular culture became preoccupied with whether the biological universe is hostile or friendly (Dick 1996).

Although human reactions to the new universe and cosmic evolution have not been monolithic, certain underlying themes are pervasive. In the eyes of many astronomers the increased awareness of the new universe and the possibility of a biological universe largely dashed any remaining hopes for an anthropocentric universe, with all that implies for religion and philosophy (van de Kamp 1965; Berenzden 1975). Even though the idea that the universe was made for humans survives in the form of the elegantly misnamed *anthropic principle*, in fact that principle is (to use L. J. Henderson's term from 1913 mentioned earlier), a *biocentric principle* of the fine-tuning of universal laws that points to the possible abundance of life in the universe in many forms, rather than in human form only (Barrow and Tipler 1986; Carr 2007; Dick 2008b). And if life is common throughout the universe, then our religions, philosophies, and other human endeavors are too parochial and will need to be significantly altered, expanded, or discarded. As physicist Paul Davies has said, "if it turns out to be the case that the universe is biofriendly … then … the scientific, theological and philosophical implications will be extremely significant" (Davies 2000).

The religious and philosophical implications of astronomical discoveries have been discussed especially since the time of the Copernican revolution, which made the Earth a planet and the planets potential Earths (Blumenthal 1987; Dick 1996). These implications were reflected by a few farsighted thinkers in the early twentieth century. Much to the chagrin of the Catholic Church, the French Jesuit priest, philosopher, and paleontologist Pierre Teilhard de Chardin famously made the evolution of the cosmos the central theme of his posthumous book *The Phenomenon of Man* (King 1996; Teilhard de Chardin 2002; Aczel 2007). Here he embraced cosmic evolution and argued for a teleological evolution in which man would end in a collective consciousness called the "noosphere," which would ultimately lead to the Omega Point, the maximum level of consciousness, which he also identified with God. Though the idea was not accepted within the Catholic church, a few have followed in Teilhard's footsteps, including the Catholic priest Thomas Berry and physicist Brian Swimme, whose book *The Universe Story* emphasizes the religious significance of cosmic evolution (Berry and Swimme 1994).

The new universe of the late twentieth century has spawned renewed analysis of the relation of humans to the cosmos, both inside and outside established religions (Dick 2000a; Bertka 2010). Biologist Ursula Goodenough argues in *The Sacred Depths of Nature* that cosmic evolution is a shared worldview capable of evoking an abiding religious response. "Any global tradition," she writes, "needs to begin with a shared worldview—a culture-independent, globally accepted consensus as to how things are" (Goodenough 1998). She finds this consensus in "our scientific account of Nature, an account that can be called The Epic of Evolution. The Big Bang, the formation of stars and planets, the origin and evolution of life on this planet, the advent of human consciousness and the resultant evolution of cultures—this is the story, the one story, that has the potential to unite us, because it happens to be true." She calls her elaboration of the religious implications "religious naturalism."

Similarly, but with the Christian tradition, the British biochemist and Anglican priest Sir Arthur Peacocke has called cosmic evolution "Genesis for the third millennium." He believes that "any theology – any attempt to relate God to all-that-is – will be moribund and doomed if it does not incorporate this perspective [of cosmic evolution] into its very bloodstream" (Peacocke 2000). Michael Dowd and Connie Barlow, who consider themselves, "evangelists of cosmic evolution," have proposed "evolutionary Christianity"—very different from Huxley's evolutionary humanism, but both featuring evolution as a central concept. Evolutionary Christianity embraces cosmic evolution, variously termed "the Great Story" and the "epic of evolution," much more than did Huxley's original evolutionary humanism, undoubtedly because cosmic evolution has been so much more developed over the last 50 years, complete with evocative images from the Hubble Space Telescope (Barlow 1995; Dowd 2008).

While Freeman Dyson among others have argued that the age-old mystery of God will be little changed by human attempts to read his mind, others argue that the new universe not only could, but should, lead to a new *cosmotheology*, or a new *cosmophilosophy*. Among the elements such a cosmotheology must take into account are (1) that humanity is in no way physically central to the universe, but located on a small planet circling a star on the outskirts of the Milky Way galaxy; (2) that humanity is probably not central biologically, even if our morphology may be unique; (3) that humanity is likely somewhere near the bottom, or at best midway, in the great chain of being, a likelihood that follows from the age of the universe and the youth of our species; (4) that we must be open to radically new conceptions of God, grounded in cosmic evolution, including the idea of a "natural" rather than a "supernatural" God; and (5) that it must have a moral dimension, a reverence and respect for life that includes all species in the universe (Dick 2000b).

Each of these elements of cosmotheology provides vast scope for elaboration. Perhaps the most radical consequences stem from the fourth principle, which states that we must be open to new conceptions of God, stemming from our advancing knowledge of cosmic evolution and the universe in general. As the God of the ancient Near East stemmed from ideas of supernaturalism, our concept of a modern God could stem from modern ideas divorced from supernaturalism. The billions of people attached to current theologies may consider this no theology at all, for a transcendent God above and beyond nature is the very definition of their theology. The supernatural God "meme," which we should remember is a historical idea the same as any other, has been very efficient in spreading over the last few thousand years, picking up new memes such as those accepted by Christianity and other religions. Nonetheless, the idea of a "natural" God in the sense of a superior intelligence is appealing to some (Hoyle 1983; Harrison 1995; Gardner 2003, 2007). A natural God need not intervene in human history, nor be the cause for religious wars such as witnessed through human history. It remains an open question whether a natural God fulfills the apparent need that many have for "the Other;" such a "God" is different enough from tradition concepts that some may wish to call it a "cosmophilosophy" rather than a cosmotheology. In any case some will see it as an important part of religious naturalism.

Over the next centuries or millennia religions will likely adjust to these cosmotheological principles. The adjustment will be most wrenching for those monotheistic religions that see man in the image of God (Judaism, Christianity, and Islam), a one-to-one relationship with a single Godhead. It will be less wrenching for Eastern religions that teach salvation through individual enlightenment (Buddhism and Hinduism) rather than through a Savior, or that are this-worldly (Confucianism) rather than otherworldly. The adjustment will be not be to the physical world, as in Copernicanism, nor to the biological world, as in Darwinism, where man descended from the apes but still remained at the top of the terrestrial world. Rather the adjustment will be to the biological, or even postbiological, universe, in which intelligences are likely to be superior to us.

Even the possibility of life beyond Earth raises such theological questions, but particularly intriguing are impact scenarios in the event of the actual discovery of such life. The impact would undoubtedly very much depend on how the discovery was made and the nature of the discovery. Finding microbial life and even complex, but non-sentient life, might be of more interest to science than to philosophy or theology, as scientists probed the nature of the newfound life and determined whether it was based on the same DNA structure and biochemistry as life on Earth. The discovery of intelligent life, on the other hand, would be of immediate interest not only to science, but to such age-old philosophical problems as the nature of objective knowledge (would we perceive the universe in the same way as extraterrestrials?) and theology, typically meaning the relationship between man and God, but now recast as the relationship between all intelligent beings in the universe and God. In general the urgency of the societal implications of extraterrestrial intelligence would depend on whether physical contact was made (considered unlikely to the extent that evidence for UFOs is weak), or if contact was made via a remote radio signal through a SETI program. If the latter, a great deal would depend on the message received, if indeed it were decipherable.

While all of these scenarios are interesting to contemplate, most compelling and most discussed is the problem of how the discovery of clear evidence of a signal from extraterrestrial intelligence would affect theology on Earth, even if no message were deciphered. This is still a complex question, because there are many terrestrial theologies and they would undoubtedly be affected in different ways. And there would be much discussion, and perhaps no consensus, even within a particular theology. We know this will be the case because the discussion has already been underway for over 500 years (Dick 1982, 1996; Crowe 1986, 1997; Randolph et al. 1997). As Michael J. Crowe, one of the premier historians of the extraterrestrial life debate, has emphasized, extraterrestrials have already influenced life on Earth and the history of ideas in many areas, in the sense that the possibility of their existence and the implications of their discovery have been the subject of discussion for centuries.

Real SETI programs in the twentieth century, however, made the problem more real, even if the same concerns were raised again and again (O'Meara 1999; Peters 2003). Ernan McMullin (a priest and philosopher at the University of Notre Dame) and George Coyne (the Jesuit director of the Vatican Observatory) are among those who have recently provided reflections from within the Catholic tradition. McMullin related the problem to that faced by sixteenth-century Europeans discovering the peoples of Mesoamerica. Fully aware of Thomas Paine's objections to Christianity in the late eighteenth century, McMullin noted that "the proven reality of ETI might even more effectively encourage a broadening among the theologians and religious believers generally of the realization that the Creator of a galactic universe may well choose to relate to creatures made in the Creator's own image in ways and on grounds as diverse as those creatures themselves." The problems of such a broadening of Christian doctrine related for McMullin to three issues: original sin, soul and body, and incarnation. He speculated that an omnipotent Creator might want "to try more than once the fateful experiment of allowing freedom to a creature," such as the Eve/apple event in the Garden of Eden. He pointed to the possibility that aliens might or might not have souls; if they did, "God also might elect to become incarnate in their nature or to interact in some other way with them," depending on their response to an Eden-like challenge. Regarding Incarnation, which he calls "the defining doctrine of the Christian tradition," McMullin suggests that conflicting theological interpretations of that doctrine would influence anyone faced with the ETI situation. Thus the discovery of ETI would result in a range of answers from Christian theologians with regard to whether Christ would become incarnate on another world, ranging from "certainly yes" to "certainly no." McMullin's own answer is "maybe" (McMullin 2000).

George Coyne, at that time Director of the Vatican Observatory, posed similar reservations about a definitive answer. He concluded that with the discovery of ETI, "theologians must accept a serious responsibility to rethink some fundamental realities within the context of religious belief" (Coyne 2000). Among those realities are the nature of a human being, and whether Jesus Christ could exist on more than one planet at one time. While theologians are limited in their ability to answer such questions, varying interpretations of Christian doctrines suggests that were a discovery of ETI actually made, a way would be found for Christian doctrine to absorb it, though perhaps not easily. The alternative would be extinction, and Christianity has shown its ability to adapt to scientific discovery, if very slowly at times.

The extraterrestrial life debate has also stimulated Jewish thought about the implications of ETI. Rabbi Norman Lamm, for example, noted that "this challenge must be met forthrightly and honestly," and called those who shrink from pursuing it "parochial and provincial." Citing astronomers who emphasize our peripheral place in the new universe, Rabbi Lamm noted that "Never before have so many been so enthusiastic about being so trivial." Cautioning that extraterrestrial life is far from proven, Lamm explored "a Jewish exotheology" and concluded that "A God who can exercise providence over one billion earthmen can do so for then billion times that number of creatures throughout the universe" (Lamm 1978).

The case where an extraterrestrial message is decoded is even more startling. Astronomer Jill Tarter, a pioneer in the field of SETI, believes an extraterrestrial message, unambiguously decoded, might be "a missionary campaign without precedent in terrestrial history," leading to the replacement of our diverse collection of terrestrial religions by a "universal religion" (Tarter 2000). Alternatively, a message that indicates long-lived extraterrestrials with no need for God or religion might undermine our religious worldview completely.

If there was any consensus, it was that terrestrial religions would adjust to extraterrestrials, an opinion echoed in late twentieth-century studies of religious attitudes toward the problem (Ashkenazi 1992; Dick 1996). As McMullin and others have pointed out, various extraterrestrial theological scenarios have also been worked out in detail in science fiction, including C. S. Lewis's *Perelandra* and Walter Miller's *Canticle for Leibowitz*. More recently, Mary Doria Russell has taken up these questions in her novels *The Sparrow* and *Children of God*. These fictional scenarios nevertheless represent deep thought about a problem that has now been with us for 500 years in hypothetical form, and that will be given greater urgency as soon as a discovery is made.

The impact of the new cosmos and its master narrative of cosmic evolution need not be couched solely in terms of theology. Mark Lupisella and John Logsdon have proposed a *cosmocentric ethic*, which they characterize as one which "(1) places the universe at the center, or establishes the universe as the priority in a value system, (2) appeals to something characteristic of the universe (physical and/or metaphysical) which might then (3) provide a justification of value, presumably intrinsic value, and (4) allow for reasonably objective measurement of value" (Lupisella and Logsdon 1997). A cosmocentric ethic would have some of the same concerns as cosmotheology, devoid of the theological implications. For example, a cosmocentric ethic would dictate that we treat extraterrestrial life forms, whether primitive or intelligent, taking into account not only our own homocentric interests, but also the interests of the other life forms. The prospect of terraforming entire planets also raises the question of whether questions of terrestrial environmental ethics should be extended to the cosmic stage. In the context of spaceflight, human interaction in general, whether among ourselves or with other intelligence, would seem to demand a reorientation toward a cosmic rather than a geocentric perspective. Lupisella has recently expanded on the theme of life and the creation of cosmic value (Lupisella 2009).

Quite aside from theological and philosophical implications, cosmic evolution provides humanity a cosmic context in time, allowing us to place humanity in the 13.7-billion-year history of the universe. Although it is difficult to grasp that span of time, attempts have been made for several decades using the *cosmic calendar*, which conflates the history of the universe into a single year, showing humans arising in the last 1.5 h of the last day of cosmic history, with the European Age of Discovery taking place 1 s ago (Sagan 1977). More substantively, a small but increasing discipline known as *Big History* seeks to incorporate human history into cosmic history in a more systematic way (Spier 1996; Christian 2004, 2009). Big History links our understanding of human history with our understanding of other historical sciences, such as cosmology, geology, and biology. It allows us to appreciate the emergent properties of culture in the same way as the emergent properties along the earlier path of cosmic evolution. And it highlights our unique collective learning ability and capacity for symbolic thought that results in our need to find meaning. In short, it reintegrates humans with the long history of the cosmos whence they sprang.

Finally, cosmic evolution integrates humans into the cosmos quite literally by teaching us that we are all "star stuff." Once again Harlow Shapley was an early proponent of this perspective. "Mankind is made of star stuff," he wrote already in 1963, "ruled by universal laws. The thread of cosmic evolution runs through this history, as through all phases of the universe—the microcosmos of atomic structures, molecular forms, and microscopic organisms, and the macrocosmos of higher organisms, planets, stars, and galaxies. Evolution is still proceeding in galaxies and man—to what end, we can only vaguely surmise" (Shapley 1963; Palmeri 2009). The colorful terminology of star stuff and "starfolk" was picked up by Carl Sagan among others; its integration of humans into the cosmos encourages us to be "at home in the universe" in the felicitous phrase used by several distinguished scientists in the late twentieth century (Kauffman 1995; Wheeler 1996). We now know that the atoms in our bodies were forged in nuclear reactions in stellar furnaces, spewed into the universe in supernovae explosions, and incorporated into our bodies through the long process of the evolution of life over the last 3.8 billion years on Earth. We recognize that after death our bodily atoms will be dispersed once again through the universe, recycled to once again become star stuff in a cycle of events that will end only with the death of the universe itself. We are part and parcel of the universe, and at the hour of our death when we return to the universe, the old phrase from the *Book of Common Prayer* based on Genesis and often used in burial ceremonies—"earth to earth, ashes to ashes, dust to dust"—need only be slightly altered to "earth to earth, ashes to ashes, stardust to stardust," to be literally true. Cosmic evolution provides us with a master narrative in which our own birth, life, and death are integral parts of the universe, without recourse to the supernatural. In the end, that may be the ultimate message of the new universe and cosmic evolution.

While only a small portion of humanity yet realizes the implications of the new universe and cosmic evolution, the incorporation of these ideas into educational curricula and the general reawakening to our place in the universe ensure these ideas an increasingly important role in culture. Such educational curricula have emerged from the astrobiology and SETI programs, and are reaching an increasing number of students. The SETI Institute's "Life in the Universe" curriculum "Voyages Through Time" provides standards-based materials for a 1-year high school integrated science course using cosmic evolution as its unifying theme. Its six modules include Cosmic Evolution, Planetary Evolution, Origin of Life, Evolution of Life, Hominid Evolution, and Evolution of Technology. The Wright Center for Science Education at Tufts University is also a valuable educational resource directly centered on "Cosmic Evolution: From Big Bang to Humankind," not surprising since the Center's director until recently was Eric Chaisson.

Following in the tradition of Shapley's *Of Stars and Men* (1958), a variety of popular books are also bringing cosmic evolution to a broader audience, including Neil DeGrasse Tyson's *Origins: Fourteen Billion Years of Cosmic Evolution* (also a Nova special on PBS); *The Universe Story: From the Primordial Flaring Forth to the Ecozoic Era--A Celebration of the Unfolding of the Cosmos* by physicist Brian Swimme and theologian Thomas Berry; *Children of the Stars: Our Origin, Evolution and Destiny* by astronomer Daniel Altschuler; and *Atoms of Science: An Exploration of Cosmic Evolution,* by astrophysicist Hubert Reeves. In short, an increasing number of people around the world are seeing for the first time their place within this naturalistic worldview. This recognition represents for humanity a return to the cosmos, a more sophisticated integration of culture and cosmos that humans possessed when cultures began, ranging from Stonehenge and the ancient civilizations such as Sumer and Egypt to native American Indians and the Australian aborigines (Krupp 1983).

8.4 Cosmic Evolution and Human Destiny: Three Scenarios

In addition to the impact of the new universe on culture, cosmic evolution also provides a window on long-term human destiny. Although historians are understandably loathe to use the word "destiny," associating it with the misguided "Manifest Destiny" doctrine in which American colonists viewed it as their inherent right to expand westward and seize territory from the native Americans, the word can and must be dissociated from that historical event. In fact, the concept of "destiny" has often been used in the context of theological discussion. A little over a month after the outbreak of World War II in 1939, theologian Reinhold Niebuhr began his Gifford Lectures on "Human Destiny," published in 1941 under the title *The Nature and Destiny of Man*, in which he concluded that human destiny must lie outside of history, outside of nature, in the supernatural realm espoused by Christianity. In 1947, just after the War's end, the French biophysicist and philosopher Pierre Lecomte du Noüy published his volume *Human Destiny*, which espoused confidence in the broad scope of evolution in the universe, but ultimately found human destiny in God. And as we have seen, human destiny was explicit in Teilhard de Chardin's works, written in the first half of the twentieth century.

In the realm of the natural world, in the broadest sense we have only a limited number of destinies, whether we like it or not. Cosmic evolution provides at least three vastly different scenarios of what the long-term human future may be. The ultimate product of cosmic evolution may be only planets, stars and galaxies—a *physical universe* in which life is extremely rare. This has, in fact, been our chief worldview for the last several millennia, the plurality of world tradition notwithstanding. Almost all of the history of astronomy, from Stonehenge through much of the twentieth century, encompasses the people, the concepts and the techniques that gave rise to our knowledge of the physical universe. Babylonian and Greek models of planetary motion, medieval commentaries on Aristotle and Plato, the astonishing advances of Galileo, Kepler, Newton and their comrades in the Scientific Revolution, the details of planetary, stellar and galactic evolution—all these and more address the physical universe. The physical universe is truly amazing in its own right, boasting a whole bestiary of remarkable objects.

For millennia, our perceptions of the destiny of human life on Earth were tied to the physical universe as represented by the geocentric system associated with Aristotle, with the Earth at the center and the heavens above. This cosmological world view provided the very reference frame for daily life, religious and intellectual. Writers from Claudius Ptolemy to Dante Alighieri touted it as the true system of the world in which humans sought meaning. The heliocentric system of Copernicus changed all that, making the Earth and planet and the planets potential Earths. Societal uproar followed this daring new cosmological worldview. Since then the history of modern astronomy has been one of the increasing decentralization of humanity. In 1920 Harlow Shapley showed our Solar System is at the periphery of our Milky Way Galaxy rather than its center, and since then billions of galaxies have been discovered beyond our own.

In the physical universe scenario, all is not lost with respect to the status of humanity. In a universe in which life on Earth is unique or rarely duplicated, humans may still have an important role. Indeed, in such a universe stewardship of our pale blue dot takes on special significance, for life in the universe depends on our actions over long periods of time bounded only by physical reality. In 2 billion years the Sun will have increased in brightness enough to induce a runaway greenhouse effect on our home planet. Long before that we will likely have escaped to another star, offering our species us a much longer longevity. The process will repeat, until star formation in galaxies halts in 100 trillion years (Adams and Laughlin 2000). Assuming we don't remain Earthbound, the destiny of life in the physical universe is for humans, sooner or later, to populate the universe. Many options exist for humans in a universe devoid of life, and many scenarios in science fiction address this possibility. Isaac Asimov has played out one scenario in his *Foundation* series, and the philosopher John Leslie has addressed some of the philosophical implications (Leslie 1996).

The second possible outcome of cosmic evolution reveals quite a different destiny. The biological universe—the universe in which cosmic evolution commonly ends in life, mind and intelligence—means that we will almost certainly interact with extraterrestrials. Ideas about a possible biological universe date back to ancient Greece, in a history that is now well known (Dick 1982, 1996, 1998; Crowe 1986; Guthke 1990). It is the universe that astrobiology and SETI program are attempting to prove (Dick and Strick 2004). There is again no lack of ideas about human-extraterrestrial interaction in such a universe. Science fiction is filled with possibilities, from the horrors of a war of the worlds to warm and fuzzy ETs. Arthur C. Clarke, author of *Childhood's End*, *Rendezvous with Rama*, *2001: A Space Odyssey* and its sequels, among much other "alien literature," is the prophet of this worldview replete with extraterrestrials. In such a universe, humanity may join what has been called a "galactic club" whose goal is to enhance knowledge.

Taking a long-term view not often discussed, cosmic evolution may have already resulted in a third scenario. Cultural evolution in a biological universe may have already produced, or replaced, biologicals with artificial intelligence, constituting what I have called a *postbiological universe* (Dick 2003). This idea requires us to take cultural evolution just as seriously as astronomical and biological evolution. It requires us to contemplate cultural evolution on cosmic "Stapledonian" time scales, as did Olaf Stapledon in his novels *Last and First Men* (1930) and *Star Maker* (1937). While astronomers are accustomed to thinking in these terms for physical processes, they are not accustomed to thinking on cosmic time scales for biology and culture. But cultural evolution now completely dominates biological evolution on Earth. Given the age of the universe, and if intelligence is common, it may have evolved far beyond us. If intelligence is highly valued for its evolutionary advantage, extraterrestrials will long ago have sought the best way to improve their intelligence, and it is likely to involved artificial intelligence, yielding the postbiological universe. Nor does L need to be millions of years for such a scenario. It is possible that such a universe would exist if L exceeds a few hundred or a few thousand years, where L is defined as the lifetime of a technological civilization that has entered the electronic computer age (which on Earth approximately coincides with the usual definition of L as a radio communicative civilization). Indeed, some predict the Earth will be postbiological in a few generations (Moravec 1988, 1999; Kurzweil 1999, 2006).

Such a postbiological universe would have sweeping implications for SETI strategies, for our worldview, and for the destiny of life on Earth if it has already happened throughout the universe. We may see our own future in the evolution of extraterrestrial civilizations, perhaps another motivation for searching. How such postbiologicals—whether terrestrial or extraterrestrial—would use their knowledge and intelligence is a value question at present unanswerable. Whether one relishes or opposes the idea of a universe dominated by machines, the transition to such a universe presents many moral dilemmas and raises with renewed urgency the ancient philosophical question of destiny and free will.

In short, both in our relationship with extraterrestrials and with God, however conceived, human destiny would be quite different in a universe full of biologicals or postbiologicals than if we are alone. If extraterrestrial intelligence is abundant, it will be our destiny to interact with that intelligence, whether for good or ill, for life identifies with life. It is here that the fifth cosmotheological principle, or the cosmocentric ethic, comes into play. The moral dimension—a reverence and respect for extraterrestrial intelligence that may be morphologically very different from terrestrial life forms—will surely challenge a species that has come to blows over superficial racial and national differences. If we are wise, humanity will realize that our species is one, a necessary realization before we have any hope of dealing with extraterrestrial beings in a morally responsible way.

Although the physical, biological, and postbiological universe may be facts that the universe imposes on us, humans will still have great scope for choice and free will within these broad scenarios. The founders of the modern evolutionary synthesis emphasized this point already at the middle of the twentieth century. George Gaylord Simpson for one, echoing Huxley's evolutionary humanism, wrote that "it is another unique quality of man that he, for the first time in the history of life, has increasing power to choose his course and to influence his own future evolution. It would be rash, indeed, to attempt to predict his choice. The possibility of choice can be shown to exist. This makes rational the hope that choice may sometime lead to what is good and right for man. Responsibility for defining and for seeking that end belongs to all of us" (Simpson 1949).

Whether intelligence is rare or abundant, whether extraterrestrial life is of a lower order or a higher order than *homo sapiens,* human destiny is intimately connected with cosmic evolution. Driven by the astronomical, biological, and cultural components of cosmic evolution, the universe may have generated any of the three outcomes described here: the physical universe, the biological universe, or the postbiological universe. Which of the three the universe has produced in reality we do not yet know—this is one of the many challenges of astrobiology with its goal of analyzing the future of life as well as its past and present. Ours may be a cosmos in which humanity is not central, yet where it can be at home in the universe in which it plays its role. Whatever its long-term destiny, it is surely the destiny of humanity in the near future to follow the trail of scientific evidence wherever it may lead, even if it means abandoning old scientific, philosophical, and theological ideas. Humans have always known intuitively that culture and cosmos are intertwined. We are just now beginning to realize what this coevolution may mean.

Acknowledgments I wish to thank Jorge Horvath, Douglas Galante, and all the organizers of the Sao Paolo Advanced School for Astrobiology (SPASA). This is a modified version of an article that appeared in Dick and Lupisella (2009).

References

Aczel, A. 2007. *The Jesuit & the Skull: Teilhard de Chardin, Evolution, and the Search for Peking Man*. New York, Penguin.

Adams, F. & Laughlin, G. 2000. *The Five Ages of the Universe: Inside the Physics of Eternity*. The Free Press, New York.

Appel, T. 2000. *Shaping Biology: The National Science Foundation and American Biological Research, 1952–1975*. Baltimore, Johns Hopkins University Press, p. 132.

Ashkenazi, M. 1992. 'Not the Sons of Adam: Religious Response to ETI.' *Space Policy*, 8, 341–350.

Barlow, C. 1995. *Evolution Extended: Biological Debates on the Meaning of Life*. MIT Press, Cambridge, Mass.

Barrow, J. & Tipler, F. 1986. *The Anthropic Cosmological Principle*, New York, Oxford University Press.

Berenzden, R. 1975. 'Geocentric to Heliocentric to Galactocentric to Acentric: The Continuing Assault on the Egocentric,' in Arthur Beer and K. Aa. Strand, eds., *Copernicus: Yesterday and Today*. Oxford, Pergamon Press, pp. 65–83.

Berry, T. & Swimme, B. 1994. *The Universe Story: From the Primordial Flaring Forth to the Ecozoic Era-A Celebration of the Unfolding of the Cosmos*. New York, Harper.

Bertka, C., ed. 2010. *Exploring the Origin, Extent, and Future of Life: Philosophical, Ethical and Theological Perspectives*. Cambridge University Press, Cambridge

Billingham, J., ed. 1981. *Life in the Universe*. Cambridge, MA, MIT Press.

Blumenthal, H. 1987. *The Genesis of the Copernican World*. Cambridge, MA., MIT Press.

Bowler, P. J. 1983. *Evolution: The History of an Idea*. Berkeley and Los Angeles, University of California Press.

Campbell, W. W. 1920. 'The Daily Influences of Astronomy,' *Science*, 52 (December 10, 1920), 543–552: 540.

Carr, B., ed. 2007. *Universe or Multiverse?* Cambridge, Cambridge University Press.

Chaisson, E. 1981. *Cosmic Dawn: The Origins of Matter and Life*. New York, Little Brown & Company.

Chaisson, E. 1987. *The Life Era: Cosmic Selection and Conscious Evolution*. New York, W.W. Norton.

Chaisson, E. 2001. *Cosmic Evolution: The Rise of Complexity in Nature*. Cambridge, Mass., Harvard University Press

Chaisson, E. 2006. *Epic of Evolution: Seven Ages of the Cosmos*. New York: Columbia University Press.

Christian, D. 2004. *'Maps of Time': An Introduction to 'Big History'*. Berkeley, CA, University of California Press.

Christian, D. 2009. 'History and Science after the Chronometric Revolution,' in Dick and Lupisella 2009, pp. 441–462.

Coyne, G. V. 2000. 'The Evolution of Intelligent Life on Earth and Possibly Elsewhere: Reflections from a Religious Tradition,' in Dick, (2000a), pp. 177–190.

Crowe, M. J. 1986. *The Extraterrestrial Life Debate, 1750–1900: The Idea of a Plurality of Worlds from Kant to Lowell*. Cambridge, Cambridge University Press.

Crowe, M. J. 1997. 'A History of the Extraterrestrial Life Debate.' *Zygon*, 32, 147–162.

Davies, P. 2000. 'Biological Determinism, Information Theory, and the Origin of Life,' in Dick 2000a, pp. 15–28.

Dick, S. 2008a. 'The Universe and Alfred Russel Wallace,' in Charles H. Smith and George Beccaloni, eds., *Natural Selection and Beyond: The Intellectual Legacy of Alfred Russel Wallace*, Oxford: Oxford University Press, pp. 320–340.

Dick, S. J. 1982. *Plurality of Worlds: The Origins of the Extraterrestrial Life Debate from Democritus to Kant*. Cambridge, Cambridge University Press.

Dick, S. J. 1996. *The Biological Universe: The Twentieth Century Extraterrestrial Life Debate and the Limits of Science*. Cambridge, Cambridge University Press

Dick, S. J. 1998. *Life on Other Worlds*. Cambridge, Cambridge University Press.

Dick, S. J., ed. 2000a. *Many Worlds: The New Universe, Extraterrestrial Life and the Theological Implications*. Philadelphia: Templeton Foundation Press.

Dick, S. J. 2000b. 'Cosmotheology: Theological Implications of the New Universe,' in Dick (2000a), pp. 191–21

Dick, S. J. 2003. 'Cultural Evolution, the Postbiological Universe, and SETI.' *Int. J. Astrobiol.*, 2, 65–74. Reprinted in Dick and Lupisella (2009).

Dick, S. J. 2008b. 'Cosmology and Biology.' *Proceedings of the 2008 Conference of the Society of Amateur Radio Astronomers*, National Radio Astronomy Observatory & American Radio Relay League, Green Bank, West Virginia, pp. 1–16.

Dick, S. J. & Lupisella, M. 2009 *Cosmos and Culture: Cosmic Evolution in a Cosmic Context*. Washington, DC, NASA

Dick, S. J. & Strick, J. E. 2004. *The Living Universe: NASA and the Development of Astrobiology*. New Brunswick, Rutgers University Press.

Dobzhansky, T. 1969. *Mankind Evolving: The Evolution of the Human Species, New Haven and London*. Yale University Press; first edition, 1962.

Dowd, M. 2008. *Thank God for Evolution: How the Marriage of Science and Religion Will Transform Your Life and Our World*. New York, Viking.

Dyson, F. 1988. *Infinite in All Directions*. New York, Harper and Row, p. 51.

Flammarion, C. 1872. *La pluralité des mondes* habités. Paris, Didier.

Fry, I. 1996. 'On the Biological Significance of the Properties of Matter: L. J. Henderson's Theory of the Fitness of the Environment.' *J. History of Biology*, 29, pp. 155–196.

Fry, I. 2000. *The Emergence of Life on Earth: A Historical and Scientific Overview*. New Brunswick, N. J., Rutgers University Press.

Garber, S. J. 1999. 'Searching for Good Science: The Cancellation of NASA's SETI Program.' *J. Brit. Interplanetary Society*, 52, 3–12.

Gardner, J. 2003. *Biocosm: The New Scientific Theory of Evolution: Intelligent Life is the Architect of the Universe*. Makawao, Maui, Hawaii: Inner Ocean Publishing.

Gardner, J. 2007. *The Intelligent Universe: AI, ET and the Emerging Mind of the Cosmos*. Franklin Lakes, N.J., New Page Books, p. 159

Goodenough, U. 1998. *The Sacred Depths of Nature*. Oxford, Oxford University Press.

Greene, J. C. 1981. *Science, Ideology and World View: Essays in the History of Evolutionary Ideas*. Berkeley, University of California Press.

Guthke, K. S. 1990. *The Last Frontier: Imagining Other Worlds from the Copernican Revolution to Modern Science Fiction*. Ithaca, NY, Cornell University Press.

Hale, G. E. 1908. *The Study of Stellar Evolution*. Chicago, University of Chicago Press.

Harrison, E. W. 1995. 'The Natural Selection of Universes Containing Intelligent Life.' *QJRAS*, 36, no. 3, 193.

Henderson, L. J. 1913. *The Fitness of the Environment*. New York, MacMillan Company, reprinted with an Introduction by George Wald, Gloucester, Mass.: Peter Smith, 1970, p. 312.
Hoyle, F. 1983. *The Intelligent Universe: A New View of Creation and Evolution*. New York, Holt, Rinehart and Winston.
Huxley, J. 1964. *Evolutionary Humanism*, Buffalo, NY: Prometheus Books, 1992.
Kauffman, S. 1995. *At Home in the Universe: The Search for the Laws of Self- Organization and Complexity*. Oxford: Oxford University Press.
King, U. 1996. *The Life and Vision of Teilhard de Chardin*. New York, Orbis Books.
Kragh, H. 1996. *Cosmology and Controversy: The Historical Development of Two Theories of the Universe*. Princeton, Princeton University Press, p. 4.
Krupp, E. C. 1983. *Echoes of the Ancient Skies: The Astronomy of Lost Civilizations*. Harper & Row, Cambridge.
Kurzweil, R. 1999. *The Age of Spiritual Machines: When Computers Exceed Human Intelligence*. New York, Penguin Books.
Kurzweil, R. 2006. *The Singularity is Near: When Humans Transcend Biology*. New York, Penguin Books.
Lamm, N. 1978. 'The Religious Implications of Extraterrestrial Life,' in *Challenge: Torah Views on Science and its Problems*, Aryeh Carmell and Cyril Domb, eds. Feldheim, Jerusalem and New York, 354–398.
Lederberg, J. 1960. 'Exobiology: Approaches to Life Beyond the Earth,' *Science*, 132, 393–400.
Leslie, J. 1996. *The End of the World: the Science and Ethics of Human Extinction*. London and New York, Routledge.
Lightman, B. 2007. *Victorian Popularizers of Science: Designing Nature for New Audiences*. Chicago, University of Chicago Press.
Lowell, P. 1909. *The Evolution of Worlds*. New York, Macmillan Company.
Lupisella, M. 2009. 'Cosmocultural Evolution: The Coevolution of Culture and Cosmos and the Creation of Cosmic Value,' in Dick and Lupisella (2009), pp. 321–359.
Lupisella, M. & Logsdon, J. 1997. 'Do We need a Cosmocentric Ethic?' International Astronautical Congress paper IAA-97-IAA.9.2.09, October 6–10, 1997, p. 1.
McMullin, E. 2000. 'Life and Intelligence Far from Earth: Formulating Theological Issues,' in Dick (2000a), pp. 151–176: 162, 169, 171
Moravec, H. 1988. *Mind Children: The Future of Robot and Human Intelligence*. Cambridge, Harvard University Press.
Moravec H. 1999. *Robot: Mere Machine to Transcendent Mind*. Oxford. Oxford University Press.
O'Meara, T. F. 1999. 'Christian Theology and Extraterrestrial Intelligent Life.' *Theological Studies*, 60, 3–30.
Oparin, A. I. & Fesenkov, V. G. 1961. *Life in the Universe*. New York, 1961.
Palmeri, J. 2000. 'An Astronomer Beyond the Observatory: Harlow Shapley as Prophet of Science,' University of Oklahoma dissertation
Palmeri, J. 2009. 'Bringing Cosmos to Culture: Harlow Shapley and the Uses of Cosmic Evolution,' in Dick and Lupisella (2009), pp. 489–521.
Peacocke, A. 2000. 'The Challenge and Stimulus of the Epic of Evolution to Theology,' in Dick (2000a).
Peters, T. 2003. 'Exotheology: Speculations on Extraterrestrial Life,' in *Science, Theology and Ethics*. Burlington: Ashgate.
Proctor, R. A. 1870. *Other Worlds than Ours*. London.
Randolph, R. O., Race, M. S., & McKay, C. P. 1997. 'Reconsidering the Theological and Ethical Implications of Extraterrestrial Life,' *CTNS* [Center for Theology and the Natural Sciences] *Bulletin*, 17, no. 3, 1–8
Russell, H. N. 1943. 'Anthropocentrism's Demise.' *Scientific American* (July, 1943), 18–19.
Sagan, C. 1973. *The Cosmic Connection: An Extraterrestrial Perspective*. Doubleday, Garden City, N.Y.
Sagan, C. 1977. *The Dragons of Eden*, Random House, New York.

Sagan, C. 1980. *Cosmos*. New York, Random House.
Sagan, C. 2000. *Carl Sagan's Cosmic Connection: An Extraterrestrial Perspective*. Cambridge and New York, Cambridge University Press.
Sawyer, K. 2006. *The Rock from Mars: A Detective Story on Two Planets*. New York, Random House.
Schaffer, S. 1989. 'The Nebular Hypothesis and the Science of Progress,' in *History, Humanity and Evolution: Essays for John C. Greene*, J. R. Moore, ed. Cambridge, Cambridge U. Press, pp. 131–164.
Shapley, H. 1958. *Of Stars and Men: Human Response to an Expanding Universe*. Boston, Beacon Press.
Shapley, H. 1963. *The View from a Distant Star: Man's Future in the Universe*. New York, Basic Books, p. 5.
Shapley, H. 1967. *Beyond the Observatory*. New York, Scribner.
Shklovskii, J. & Sagan, C. 1966. *Intelligent Life in the Universe*, San Francisco: Holden-Day.
Simpson, G. G. 1949. *The Meaning of Evolution: A Study of the History of Life and of its Significance for Man*. New Haven, Yale University Press, p. 349.
Smocovitis, V. B. 1996. *Unifying Biology: The Evolutionary Synthesis and Evolutionary Biology*. Princeton, Princeton University Press, p. 165.
Smolin, L 1997. *The Life of the Cosmos*. New York and Oxford, Oxford University Press.
Spencer Jones, H. 1940. *Life on Other Worlds*, New York, Macmillan, p. 57.
Spier, F. 1996. *The Structure of Big History: From the Big Bang Until Today*. Amsterdam, Amsterdam University Press.
Strauss, D. 2001. *Percival Lowell: The Culture and Science of a Boston Brahmin*. Cambridge, Mass., Harvard University Press, pp. 97–165.
Strick, J. 2000. *Sparks of Life: Darwinism and the Victorian Debates over Spontaneous Generation*. Cambridge, Mass.: Harvard University Press, pp. 94–95.
Struve, O. 1955. 'Life on Other Worlds,' *Sky and Telescope*, 14, 137–146.
Tarter, J. 2000. 'SETI and the Religions of the Universe,' in Dick, (2000a), pp. 143–150.
Teilhard de Chardin, P. 2002. *The Phenomenon of Man*. New York, Harper Collins, 2002, first published in French in 1955, in English in 1959.
Toulmin, S. & Goodfield, J. 1982. *The Discovery of Time*. Chicago, University of Chicago Press.
Tovmasyan, G. M., ed. 1965. *Vnezemnye tsivilizatsii*. Erevan, Akademii Nauk Armyanskoi SSR; English translation *Extraterrestrial Civilizations*, Jerusalem, Israel Program for Scientific Translations, 1967.
Van de Kamp, P. 1963. 'Astrometric Study of Barnard's Star from Plates Taken with the 24-inch Sproul Refractor, *Astron. Journal*, 68, 515–521.
Van de Kamp, P. 1965. 'The Galactocentric Revolution: A Reminiscent Narrative.' *Publ. Astronomical Society of the Pacific*, 77 (October, 1965), 325–335.
Wald, G. 1954. 'The Origin of Life.' *Scientific American* (August, 1954), 44.
Wallace, A. R. 1903a. 'Man's Place in the Universe.' *The Independent* (New York), vol. 55, 473–383: 474.
Wallace, A. R. 1903b. *Man's Place in the Universe*. London and New York, McClure, Phillips & Co.
Wheeler, J. 1996. *At Home in the Universe*. New York, Springer.
Zakariya, N. 2010. *Towards a Final Story: Time, Myth and the Origins of the Universe*. PhD dissertation, Harvard University.

Chapter 9
Consequences of Success in SETI: Lessons from the History of Science

Abstract The consequences of receipt of a dial tone or information flow from an extraterrestrial civilization are considered in light of historical analogues. It is argued that the history of science offers deeper insights than political history or anthropology, since the contact would be intellectual and not physical. Specific cases of the transmission of knowledge across terrestrial cultures, and of the reception of scientific worldviews, are offered as analogues to receipt of an extraterrestrial intelligent signal. Particularly apt analogues are the transmission of Greek science to the Latin West via the Arabs in the twelfth and thirteenth centuries, and reception of the Copernican and Darwinian worldviews. A rich literature awaits those wishing to study the impact of success in SETI based on such analogues.

9.1 Introduction: The Relevance of History of Science

With the inauguration of NASA's High Resolution Microwave Survey (HRMS) in 1992, the continuing improvements in ongoing programs such as the Planetary Society's META and Berkeley's Project SERENDIP, and the worldwide contemplation of new projects in the Search for Extraterrestrial Intelligence (SETI), it is only prudent that attention should increasingly turn to the societal implications in the event of success in SETI. Many approaches may be taken in discussing such implications. A rich repository of ideas in the literature of science fiction envisions possible consequences of extraterrestrial contact, though much of this is centered on face-to-face rather than radio contact. Hoyle and Elliot (1962), Brown and Zerwick (1968), Gunn (1972), and Sagan (1985) are notable fictional treatments authored by scientists and dealing with the consequences of radio contact.

On the more sober side, a series of three workshops on the Cultural Aspects of SETI (CASETI), sponsored by NASA in 1991–92, set forth broad approaches from the point of view of history, behavioral science, policy, and education (Billingham et al. 1999). Here we consider how analogues in history might help to envision and assess the impact following successful detection of a signal from extraterrestrial intelligence.

First published in *Progress in the Search for Extraterrestrial Life*, Seth Shostak, ed. (Proceedings of Santa Cruz meeting on SETI, August, 1993; ASP Conference series: San Francisco, 1995), 521–532.

S. J. Dick, *Space, Time, and Aliens*, https://doi.org/10.1007/978-3-030-41614-0_9

We recognize at the outset that the societal impact will depend strongly on the circumstances of radio contact. A "dial tone" signal, only giving evidence of intelligence, will be quite different in impact from the decipherment of significant amounts of information. If the latter is achieved, the impact will in turn depend on the nature of the information. Moreover, there are likely to be both short-term and long-term impacts. In this paper we further narrow our focus by considering the long-term impact of the receipt of an intelligent "dial tone" signal (itself a powerful bit of information), or of the flow of information from a deciphered signal, without regard to the information content (which in any case can only multiply speculation). Such a focus, it is hoped, may at least yield a first-order approximation to the general reaction to discovery of intelligent life in the universe. A study of short-term impact, probably dominated by media reaction and political maneuvering in addition to scientific study, requires quite a different analysis not attempted here.

The consequences of detection of a signal from extraterrestrial intelligence have sometimes been compared to physical contact between cultures on Earth. Readers have been invited to consider Cortez and the Aztecs, Pizarro and the Incas, or Europeans and the American Indians. The European clashes with the Ottoman Empire in the sixteenth and seventeenth centuries, the British Raj in India, the attempts of Peter the Great at the Westernization of Russia in the eighteenth century, and Matthew Perry and the opening of Japan in the mid-nineteenth century are other examples taken only from the Western tradition of expansionism. We propose here, however, that the signal obtained from SETI programs is not analogous to physical contact, but rather to intellectual contact and the diffusion of ideas among cultures. This suggests that analogues should be drawn not from political history and anthropology, but from the history of ideas, and in particular the history of science, since the discovery will be a scientific event whose consequences may best be compared to past scientific breakthroughs. As the CASETI participants well realized, even these analogues must not be taken as predictors of action, but only as useful guides to thinking. Analogues "are invoked because so much about SETI is conjectural," the CASETI historians wrote. We follow their principle that "Where ignorance forces conjecture, analogy is a useful (and perhaps the only) guide."

Neither, however, should we underestimate the force of analogy. Scientists as well as philosophers of science (Giere 1977; Harre 1972; Hempel 1965; Hesse 1963) recognize analogy as an essential tool of science, both as an aid to thought and more substantially as transferring cognitive content from one scientific problem to another. Historical examples from natural science include the propagation of waves on the surface of a pond and the propagation of light waves, fluid flow and the flow of electricity, and domestic selection and natural selection. One might well argue that analogy may also be a useful tool in social and behavioral science, certainly as an aid to thought, and perhaps in transferring principles of human experience between similar events. The underlying assumption for such a suggestion is that while cultural circumstances vary enormously over time and across cultures, human nature does not. At the same time, no analogy is ever perfect: correspondences need not be one-to-one, and any two analogous systems will have differences that may be called "negative analogies" (Hesse 1967).

In this paper, three models from the history of science are offered as analogues to the impact of a SETI detection: the transmission of Greek science to the Latin West via the Arabs in the twelfth and thirteenth centuries, the reception of great cosmological ideas such as the Copernican theory of the sixteenth century and the "galactocentric revolution" of the early twentieth century, and the reception of Darwinian evolution. The first is a model for the consequences of transmission of a broad array of knowledge from one culture to another. The last two provide models for reaction to specific scientific worldviews similar in nature and scope to the new worldview that a successful SETI detection entails. I have called this elsewhere the *biophysical cosmology* (Dick 1989, 1991).

9.2 The Transmission of Science to the West in the Twelfth and Thirteenth Centuries

Assuming that a SETI signal is deciphered and significant information is transmitted, the flow of information between terrestrial civilizations across time is a tantalizing analogue from previous human experience (Fig. 9.1). The transmission of Greek science via the Arabs to the Latin West in the twelfth and thirteenth centuries is an example of what historian Arnold Toynbee called "encounters between civilizations in time"(Toynbee 1954). Such encounters—which in the case at hand resulted in a renaissance of learning in Europe in the twelfth century (Haskins 1927)—are particularly apt comparisons because they deal with the transmission of knowledge from non-contemporary civilizations across time. Because of the finite speed of microwave communications, non-contemporaneity will hold for extraterrestrial contact in direct proportion to the distance between the communicative civilizations.

One of the best-known episodes of human history is the political fallout from the end of the Roman Empire in the Latin West by 500 AD. Less widely appreciated is that this political dissolution brought with it the loss of Greek learning in science as well as other areas of knowledge. While the flame of knowledge flickered in a few places, only in the twelfth century was it recovered, via the Arabs. Scholars, especially in Spain, where the Islamic civilization had been transported in the eighth century, began to translate Greek treatises into Latin, either from the Greek originals or from the Arabic, which by this time had added its own gloss on the ancient knowledge. "First a trickle and eventually a flood," one historian of science recently wrote, the new material "radically altered the intellectual life of the West." Western Europe, which had been struggling to keep the intellectual flame from being extinguished, now had to assimilate a torrent of new ideas (Grant 1971; Lindberg 1978, 1992).

While we do not fancy our civilization analogous to the Middle Ages, the torrent of new ideas would be analogous to a significant flow of information from an extraterrestrial civilization to one probably less knowledgeable but eager to learn. The army of translators involved in the recovery of lost learning may find its analogy in the legions of scientists, cryptographers, linguists and others sure to participate in

Fig. 9.1 Although SETI programs are designed only to detect a "dial tone" initially, eventually they may seek to exchange information. This process and its consequences may parallel in some ways the translation and dissemination of information from one culture to another on Earth. A prime example is the transmission of Greek knowledge via the Islamic civilization to the Latin West in the twelfth and thirteenth centuries, leading to the Renaissance in western science, art, and culture. This possibility is symbolized here by this 1332 fresco by Tommaso da Modena of Albert the Great, thirteenth-century teacher of Thomas Aquinas and leader in the spread of the new knowledge. From Lindberg (1992) (Credit: Alinari/Art Resource, NY)

any attempt to decipher an extraterrestrial signal. However, as Lindberg (1978) has pointed out, the medieval translators acted as individuals; there was no central bureau of translations. By contrast, one would hope for some centralization among the many nations likely to be working on the decipherment of a SETI signal.

The result of the newly recovered knowledge is a matter of record. The thirteenth century—the century of Thomas Aquinas, Albert the Great, and Roger Bacon, among other luminaries—was characterized by the attempt of its best scholars to reconcile the new Greek and Arabic knowledge with Christianity. Surely the century following an extraterrestrial signal bearing significant knowledge will also be marked by the attempts of scholars to reconcile terrestrial and extraterrestrial knowledge in many areas, ranging from science to religion and the arts. One can only hope that it will not be marked by religious attempts to restrict access to the new knowledge. Even if it is, we note that Aristotelian science was not suppressed forever. It was, however, eventually superseded, leading one to wonder by analogy if terrestrial and extraterrestrial knowledge will be mutually exclusive, coexist with minimal interaction, or blend to become part of a long-sought "objective knowledge" (Popper 1979).

Many other interesting threads might be followed in this analogy. It is difficult to find in terrestrial history a more appropriate analogy in which the impact of new knowledge was not also accompanied by physical contact or occupation—a complicating element we do not expect from the current radio searches for extraterrestrial intelligence. In summary, in our present model the Greeks are the extraterrestrials, the Arabs the knowledge-bearing medium through which the information is passed, and the medieval translators and commentators those who will bring the new knowledge to the masses. In order to anticipate what might happen then—or to assess what might happen if simply a "dial tone" confirms the extraterrestrial worldview—we now turn to the reception of scientific worldviews in terrestrial history.

9.3 Cosmology as an Analogue

More than three decades ago, astronomers were already drawing analogies between great changes in cosmological worldview and the impact of discovering extraterrestrial intelligence. Shapley (1958) suggested such a discovery would be *the Fourth Adjustment* in mankind's view of itself, after the shift to the geocentric, heliocentric, and *galactocentric* worldviews. The latter was precipitated by Shapley's own work showing that the Solar System was at the periphery of our galaxy. Struve (1961) agreed that astronomy had undergone three great revolutions in the past four centuries: the removal of the Earth from the center of the Solar System by Copernicus (Fig. 9.2), Shapley's removal of the Solar System from the center of the galaxy, and the revolution embodied in the question "Are we alone in the universe?" Struve seemed to think the latter revolution was already underway, and he may have been right, but surely it will be accelerated by the actual discovery.

What, then, might we learn from the reception of these cosmological worldviews? Quite different lessons, in fact. The gradual acceptance of the Copernican theory, followed by its triggering of the Scientific Revolution and indeed its impact in all areas of human thought, has now been studied extensively (Beer and Strand 1975; Kuhn 1957; Blumenberg 1987; Stimson 1972; Westman 1975). The Copernican theory eventually gave birth to a new physics, caused wrenching controversy in theology, and made the Earth a planet and the planets potential earths. Few other revolutions in history have had such broad, if delayed, consequences.

The galactocentric revolution, on the other hand, is an example of a silent revolution. Astronomers celebrated the discovery, the press routinely reported it, and the general population went about its business as usual, despite humanity's slide from the center to the edge of the halaxy (Bok 1974; Berenzden et al. 1976; Smith 1982). Berenzden (1975) has noted this dichotomy in the reactions to certain worldviews. The proof of the galactocentric universe by 1924, he found, "caused almost no discussion whatsoever" in the press, as opposed to Hubble's announcement of the expanding universe 5 years later. In attempting to explain the difference between silent and noisy revolutions, Berenzden notes that by the mid-1920s, revolutions including Darwin, Einstein, and Freud (not to mention Copernicanism) had inured

Fig. 9.2 Copernicus (1473–1543), made the Earth a planet and the planets potential Earths—one of the major assumptions that underlie astrobiology. This image is believed to be a modern reworking based on Pierre Gassendi's Copernicus biography of 1654

the public to marginalization. Moreover, he notes that the basis for the galactocentric conclusion was technical and conceptually non-trivial, lessening its public impact. Finally, and perhaps most importantly, Berenzden emphasized that the galactocentric model did not pose a threat to societal institutions. "When scientific revolutions impinge upon metaphysics or social theory," he wrote, "they are likely to become unusually polemical and possibly unacceptable." These hints, and a voluminous literature on the history and structure of scientific revolutions (Kuhn 1962; Cohen 1985), lead us to conclude that the discovery of extraterrestrial intelligence is more likely to follow the Copernican rather than the galactocentric pattern.

At the same time, a more general lesson may be learned from the reception of worldviews. Each of these revolutions (indeed all revolutions) follow stages that may be roughly identified as periods of motivation, presentation, elaboration and refinement, exploration of implications, opposition, acceptance, and definitive confirmation or rejection. An examination of Table 9.1 shows the specifics of these stages in the development of previous cosmological worldviews. We note that the geocentric worldview held sway for more than 2000 years before its rejection, that a century-and-a-half passed before the heliocentric worldview was widely accepted (and almost an equal time interval before stellar parallax proved Earth motion around the Sun), and that the galactocentric worldview, by contrast, was accepted very quickly. In many ways, the extraterrestrial, or biophysical, worldview has been widely accepted since the mid-eighteenth century, despite the lack of direct evidence. In a sense the stages of elaboration, opposition, and exploration of implications have historically already been passed (Crowe 1986; Dick 1982). But alas, there has been no empirical evidence of the actual existence of extraterrestrials, nor of course any final confirmation. When such evidence is at hand, the biophysical worldview is likely to repeat the series of stages, at which time the arguments of its historical predecessor—arguments based only on the hypothetical existence of life beyond the Earth—will take on renewed and more immediate meaning. In any event, a closer study of the stages of past worldviews is sure to illuminate the general path of the putative extraterrestrial worldview.

Nor need such analogues be confined to physical worldviews. To the contrary, because the evolution of life in the universe has an obvious connection with Darwinism—and may even be viewed as an extension of it—the reception of Darwinism may be the best analogue of all in terms of assimilation of worldviews.

Table 9.1 Stages in World View Development

Stage	Geocentric	Heliocentric	Galactocentric	Extraterrestrial/ Biophysical
Motivation	Motion of planets Anthropocentrism	Motion of planets Neoplatonism Geocentric problems	Globular cluster Distribution	Copernican theory Cosmic evolution
Presentation Based on Observation	Eudoxus/Aristotle Fourth century BC	Copernicus 1543	Shapley 1917	Kepler (disproven) Lowell (disproven) Pulsars (disproven) Radio signal?
Elaboration	Ptolemy et al.	Galileo Kepler Newton et al.	Trumpler Oort et al.	Range of scientists
Opposition	Anti-rationalists	Geocentrists Religious	Curtis et al.	Religious Scientific Philosophical
Exploration of implications Outside field	Anthropocentric Religions & Philosophies	Philosophical Literary Scientific	Further proof of non-anthropocentrism	All aspects of human knowledge
General acceptance	Fourth century BC	1700	1930s	Widely accepted 1750s
Final confirmation	Disproven	1838 stellar parallax = earth motion	1950s radio maps of galaxy	Deciphered signal? UFO identified? Discovery of ET?

9.4 Darwinian Evolution as an Analogue

The Darwinian revolution provides a compelling analogue of the trajectory of a biological worldview that bears directly on humanity's place in nature. There are, of course, differences in the cognitive status of a theory of evolution by natural selection and a discovery of an artificial signal from outer space. Yet, like the Darwinian theory, the interpretation of an extraterrestrial signal is likely to be ambiguous and debatable, and the diverse reaction to such a signal may therefore be comparable (Fig. 9.3).

The details of the Darwinian revolution are well-known. Indeed, no event in the history of science has been the subject of so much analysis as Darwin's theory and its impact, especially since the centenary of the *Origin of Species* in 1959 spawned what has aptly been called the "Darwin industry." From the early general historical treatments of Darwinism (Eiseley 1958; Greene 1959; Himmelfarb 1959) to recent historical, philosophical, and scientific analyses, the Darwin industry itself provides a model of scholarship likely to be precipitated by a discovery of extraterrestrial intelligence.

Fig. 9.3 Charles Darwin (1809–1882) gave rise to a revolution that may bear similarities to the reaction if extraterrestrial life is discovered. This image showing Darwin late in life is attributed to the British photographer Julia Margaret Cameron

These studies show that although there was a long prehistory of the idea of biological evolution (as of extraterrestrial life), the reaction to the *Origin of Species*—and to the independent work of A. R. Wallace on natural selection—was immediate, widespread, and felt at many levels of society. The first printing of the *Origin* was sold out in a day, and was followed by many more printings in many languages. In England, T. H. Huxley and his allies championed Darwin's cause against all opponents, including Bishop Wilberforce at the infamous BAAS meeting in 1860 at which Huxley announced his preference for the ancestry of an ape rather than the ancestry of a person who, though endowed with intelligence, had Wilberforce's ignorance of science. The debates over Darwinism raged over Europe and the Western world, and eventually over the entire world. Studies have shown how Darwin's theory had distinctive impacts over the short term (Vorzimmer 1970) and the long term (Bowler 1989), and among scientists (Hull 1973), theologians and other segments of the population.

Nor were the battle lines drawn only between broad disciplines such as science and theology. A diversity of opinion existed even within disciplines. Many scientists, while accepting evolution, rejected natural selection, the centerpiece of the *Origin*. Although this has led some (Bowler 1992) to the view of a "non-Darwinian revolution," Mayr (1991) has emphasized that Darwinism meant many things to many people but the revolution was Darwinian nonetheless. While Darwinism spurred scientific research and gave birth to biology as a unified science (Smocovitis 1992), it was eclipsed for many years until the second Darwinian revolution in the 1930s and 1940s incorporated genetics in the "evolutionary synthesis"(Mayr 1982, 1988; Mayr and Provine 1980). And on the theological side, the debate has never ended. It spawned the Scopes trial in 1925, and is still with us today in the form of "creation science." In a broader sense, Darwin unconsciously spawned aberrant theories like social Darwinism.

Many of these characteristics are likely to be mimicked by the discovery of extraterrestrial intelligence: an immediate strong reaction despite a long prehistory of the idea, the short-term heated controversies, the spur to scientific research punctuated by periods of relative neglect, the diversity of opinion among and within groups, the widespread effect on areas of society that we cannot now predict, and above all, the transformation of the way in which we view our place in nature. Huxley's *Man's Place in Nature* (1863), in which he discussed man's place among the apes, is likely to be extended to a discussion of humanity's place in the universal chain of being, giving specifics to a long historical tradition (Lovejoy 1971).

The aptness of the Darwinism analogy is heightened by the fact that Wallace himself already saw a connection between evolution and extraterrestrials in his volume *Man's Place in the Universe: The Results of Scientific Research in Relation to the Unity or Plurality of Worlds* (1903). Wallace, however, used an anthropocentric cosmology to conclude that the Earth was the only inhabited planet in the universe, and was also the first evolutionist to argue that the complexity of life and the principles of natural selection would never lead to man or intelligence on another planet. This is a view that some evolutionists such as Simpson (1964) and Mayr (1985, 1993) continue to espouse. For our purposes here it matters not whether one accepts

Darwinism as an argument for or against extraterrestrial life. What matters is that the validity of Darwinism as an analogy for reaction to the extraterrestrial worldview is greatly strengthened by the fact that Darwin's theory applies to the terrestrial evolution of life, while the extraterrestrial worldview deals with the cosmic evolution of life. The discovery of extraterrestrial intelligence will surely shed light on the universality of Darwinian principles; conversely, terrestrial Darwinism may become a subset of the extraterrestrial worldview. Instead of humankind absorbing the impact of a position at the pinnacle of the apes, it may be forced to deal with its position among civilizations millions of years older.

9.5 Conclusions

In this paper we have presented only a bare outline of specific events in the history of science most relevant to understanding the impact of the successful detection of extraterrestrial intelligence. A rich literature in that discipline awaits those interested in further studying the impact of the transmission of knowledge across cultures, as well as in assessing the imp act of new scientific worldviews on culture in relation to the possible impact of a successful SETI detection. One must always take precautions in using such analogues. Nonetheless, they may well serve not only as useful guides to thinking but also as real, if imperfect, indicators of likely human reaction to future events corresponding to past human experience.

We do not wish to imply that other disciplines may not also be useful in discussing human reaction to the discovery of extraterrestrial intelligence. Anthropology and history may contribute in other ways. For example Finney (1990) uses culture contact to argue cogently that communication between terrestrial and extraterrestrial cultures may be much more difficult than we anticipate—a point that may well be valid. But for the long-term reaction to extraterrestrial intelligence considered as a worldview with serious implication s for humanity's place in the universe, the history of science offers unparalleled analogues .

As a model for the transmission of broad body of knowledge, the renaissance of the twelfth century, and the assimilation of the new knowledge in the thirteenth and fourteenth centuries, serves as an optimal terrestrial analogue. As a model for the reception of a new worldview, reaction to the Copernican and galactocentric cosmologies offer contrasting models, while the reception of Darwinism—a biological worldview with clear implications for humanity's place in nature—may be the best model of all. It is widely assumed, in fact, that the evolution of extraterrestrial intelligence will occur by Darwinian natural selection, so that the discovery of life in the universe may be viewed as an extension of the Darwinian revolution. The reaction to that worldview may also extend to extraterrestrials. Close scrutiny of detailed studies of these terrestrial analogues might repay substantial dividends for those interested in cultural aspects of SETI.

Whatever model is taken as the best terrestrial analogue, much may be gained from an analysis of the general course of scientific worldviews. In its broad outlines, the discovery of extraterrestrial intelligence is likely to follow the same general course as have other scientific worldviews. As Copernicus eventually had his Galileo and Newton, and as Darwin had his Huxley, so will the biophysical worldview have its defenders. As Copernicus had his religious and scientific critics and Darwin had his Wilberforce, so will extraterrestrials. The intellectual turmoil following the twelfth-century renaissance, and the Copernican and Darwinian worldviews, is sure to be duplicated. But eventually—if the evidence bears scrutiny—there will be final confirmation that over the long term will overwhelm the skeptics. Surely, history teaches us that the impact of a successful detection of extraterrestrial intelligence will vary with different levels of society, and may be absorbed over a lengthy time period.

This much is predictable from human nature, which has remained largely unchanged over the millennium of analogues discussed here, despite scientific and technological advancements. By contrast, we know nothing of extraterrestrial nature or the extent of their knowledge. Therefore, whether the course of the extraterrestrial revolution will bring us, in Arthur C. Clarke's concept, to "Childhood's End," we cannot now say.

9.6 Commentary 2020

This paper was presented at the Fourth Bioastronomy Symposium in Santa Cruz, California, August 16–20, 1993, and published in the Proceedings edited by Shostak (1995). As the text indicates, this meeting came at a crucial time, namely during the 1 year that NASA's SETI program was operational before termination by Congress for political reasons (Garber 2014). As mentioned in Sect. 9.1 the program at the time was termed "High Resolution Microwave Survey (HRMS)," which some claimed was a way to hide what NASA was doing. If so, the attempt did not work, and portions of the government program were taken over by the SETI Institute.

As mentioned in the Introduction to Part II, this article was my first foray into the subject of implications of extraterrestrial life following NASA's CASETI meetings in the early 1990s, described in the commentary section of the next chapter. It was followed by many more, some not represented in this volume. Less than a year after the Santa Cruz meeting, the first in a series of conferences on the Inspiration of Astronomical Phenomena (INSAP) was held under the auspices of the Vatican Observatory on June 27–July 2, 1994. Conferees met at the retreat house "Mondo Migliore" on the Rocca di Papa above the Lago Albano, and also across the lake at the Vatican Observatory at Castel Gandolfo, the Pope's summer residence. The former is in Italy, the latter is an outlying portion of the Vatican State. The subjects varied widely, so much so that the results were published in two quite different journals, *Vistas in Astronomy* (39, part IV 1995), and *Leonardo: Journal of the International Society for the Arts, Sciences, and Technology* (29, no. 2, 1996). My own contribution, on the cultural significance of the extraterrestrial life debate, appeared in *Leonardo* (Dick 1996). The INSAP meetings have proven so rich that they have continued every 3 years or so, the latest, INSAP X, being held in 2017 in Santiago de Compostela, Spain. And the reconnaissance of the implications of finding life beyond Earth has proven so rich that I continued giving papers at a number of venues, ranging from the Dibner lecture at the Smithsonian (Dick 2000) to the program of Dialogue on Science, Ethics, and Religion (DoSER) of the American Association for the Advancement of Science (Dick 2009). As evidence in the papers in Part II, scholars from a wide array of disciplines have now joined in this study, with a quick overview in Chap. 16 and a more detailed analysis in Dick (2019).

The idea of analogs as an aid to discussing the societal impacts of discovering life beyond Earth has since been elaborated in several venues, most recently Dick (2015) and Dick (2018). The latter includes a much more sophisticated analysis of the promise and peril of analogy than given in this chapter, as well as its role in scientific reasoning, and the latest research being done on this topic.

Acknowledgments As a member of the CASETI workshops, the author acknowledges the stimulating discussions with the CASETI history group, including John Heilbron, Jill Conway, Kent Cullers, Ben Finney, Karl S. Guthke, and Ken Keniston, as well as cross-fertilization with the other groups of the workshop.

References

Beer, A. and Strand, K. Aa. eds. 1975. "Copernicus Yesterday and Today," *Vistas in Astronomy,* 17, Pergamon Press, Oxford

Berenzden, R. 1975, in Beer and Strand, pp. 65–83

Berenzden, R., Hart, R. and Seeley, D. 1976. *Man Discovers the Galaxies,* Science History Publications (New York)

Bertka, Constance, ed. 2009. *Exploring the Origin, Extent, and Future of Life: Philosophical, Ethical and Theological Perspectives*. Cambridge: Cambridge University Press.

Billingham, J., Heyns, R., Milne, D. et al. 1999. *Social Implications of Detecting an* Extraterrestrial Civilization: A Report of the Workshop on the Cultural Aspects of SETI

Blumenberg, H. 1987. *The Genesis of the Copernican World,* trans. R. M. Wallace, MIT Press, Cambridge, MA

Bok, B. 1974. in *The Heritage of Copernicus: Theories "Pleasing to the Mind,"* J. Neyman, ed., MIT Press, Cambridge, pp. 26–62

Bowler, P. J. 1989. *Evolution: The History of an Idea,* University of California Press, Berkeley, ch. 7

Bowler, P. J. 1992. *The Non-Darwinian Revolution: Reinterpreting a Historical Myth,* Johns Hopkins University Press, Baltimore

Brown, H. and Zerwick, C. 1968. *The Cassiopeia Affair,* Modern Literary Editions, New York

Cohen, I. B. 1985. *Revolution in Science,* Harvard University Press, Cambridge

Crowe, M. J. 1986. *The Extraterrestrial Life Debate, 1750–1900: The Idea of a Plurality of Worlds from Kant to Lowell,* Cambridge University Press, Cambridge

Dick, S. J. 1982. *Plurality of Worlds: The Origins of the Extraterrestrial Life Debate from Democritus to Kant,* Cambridge University Press, Cambridge

Dick, S. J. 1989. *The Planetary Report,* March–April, 13–17

Dick, S. J. 1991. in *Bioastronomy: The Search for Extraterrestrial Life—The Exploration Broadens,* eds. J. Heidmann and M. J. Klein, Springer, Berlin, 356–363

Dick, S. J. 1996. "Other Worlds: The Cultural Significance of the Extraterrestrial Life Debate," Vatican Conference on the Inspiration of Astronomical Phenomena, 27 June–2 July, 1994, *Leonardo*, 29, no. 2 (1996), pp. 133–137.
Dick, S. J. 2000. "Extraterrestrial Life and our World View at the Turn of the Millennium," Dibner Library Lecture, Smithsonian Institution, Washington, D. C, online at https://library.si.edu/digital-library/book/extraterrestria00dick
Dick, Steven J. 2009. "A Historical Perspective on the Extent and Search for Life," in (Bertka, 2009), pp. 167–185.
Dick, Steven J., ed. 2015. *The Impact of Discovering Life Beyond Earth*. Cambridge: Cambridge University Press.
Dick, Steven J. 2018. *Astrobiology, Discovery, and Society*. Cambridge: Cambridge University Press.
Dick, Steven J. 2019. "Humanistic Implications of Discovering Life Beyond Earth," in *Handbook of Astrobiology*, Vera Kolb, ed., CRC Press, 741–756
Eiseley, L. 1958. *Darwin's Century,* Doubleday, New York
Finney, B. 1990. in "SETI Post-Detection Protocol," J. C. Tarter and M. Michaud, eds., *Acta Astronautica,* 21, 2, 47–51
Garber, Stephen J. 2014. "A Political History of NASA's SETI Program," in Vakoch (2014).
Giere, R. 1977. *Understanding Scientific Reasoning,* Holt, Rinehart and Winston, New York, 79–80
Grant, E. 1971. *Physical Science in the Middle Ages,* Wiley, New York
Greene, J. C. 1959. *The Death of Adam: Evolution and Its Impact on Western Thought,* Iowa State University Press, Ames
Gunn, J. E. 1972. *The Listeners,* Charles Scribner's Sons, New York
Harre, R. 1972. *The Philosophies of Science,* Oxford University Press, Oxford, 168–183
Haskins, C. 1927. *The Renaissance of the Twelfth Century,* Harvard University Press Cambridge
Hempel, C. G. 1965. *Aspects of Scientific Explanation,* The Free Press, New York, 433–447
Hesse, M.B. 1963. *Models and Analogies in Science,* Sheed and Ward, London
Hesse, M. B. 1967. in *Encyclopedia of Philosophy,* P. Edwards, ed., New York, 5, 354–359
Himmelfarb, G. 1959. *Darwin and the Darwinian Revolution*, W. W. Norton, New York
Hoyle, F. and Elliot, J. 1962. *A For Andromeda,* Fawcett, Greenwich
Hull, D. 1973. *Darwin and His Critics: The Reception of Darwin's Theory of Evolution by the Scientific Community,* Harvard University Press, Cambridge
Kuhn, T. S. 1957. *The Copernican Revolution,* Harvard University Press, Cambridge
Kuhn, T. S. 1962. *The Structure of Scientific Revolutions,* University of Chicago Press Chicago, 2nd ed., 1970
Lindberg, D. C. 1978. *Science in the Middle Ages,* D. Lindberg, ed., University of Chicago Press, Chicago, 52–90
Lindberg, D.C. 1992. *The Beginnings of Western Science,* University of Chicago Press, Chicago
Lovejoy, A. O., 1971. *The Great Chain of Being,* Harvard University Press, Cambridge, 1st ed., 1936
Mayr, E. 1982. *The Growth of Biological Thought,* Harvard University Press, Cambridge
Mayr, E. 1985. in *Extraterrestrials: Science and Alien Intelligence,* E. Regis, Jr., ed., Cambridge University Press, Cambridge, 23–30
Mayr, E. 1988. *Toward a New Philosophy of Biology,* Harvard University Pres, Cambridge
Mayr, E. 1991. *One Long Argument: Charles Darwin and the Genesis of Modern Evolutionary Thought,* Harvard University Press, Cambridge
Mayr, E. 1993. *Science,* ***259,*** 12 March, 1522–23
Mayr, E. and Provine, W. B., eds. 1980. *The Evolutionary Synthesis: Perspectives on the Unification of Biology,* Harvard University Press, Cambridge
Popper, K. R. 1979. *Objective Knowledge: An Evolutionary Approach,* The Clarendon Press, Oxford
Sagan, C. 1985. *Contact,* Simon and Schuster, New York
Shapley, H. 1958. *Of Stars and Men: Human Response to an Expanding Universe,* Beacon Press Boston, 104–114

Shostak, S., ed. 1995. *Progress in the Search for Extraterrestrial Life*. Astronomical Society of the Pacific: San Francisco.

Simpson, G. G. 1964. *This View of Life,* Harcourt, Brace and World, New York, 253–271

Smith, R. 1982. *The Expanding Universe: Astronomy's "Great Debate," 1900–1931,* Cambridge University Press, Cambridge

Smocovitis, V. B. 1992. *Journal of the History of Biology,* **25,** no. 1, 1–65

Stimson, D. 1972. *The Gradual Acceptance of the Copernican Theory of the Universe,* Peter Smith Gloucester

Struve, O. 1961. *The Universe,* MIT Press, Cambridge, 157–159

Toynbee, A. 1954. *A Study of History,* vol. 9, Oxford University Press, London

Vakoch, Douglas., ed., 2014. *Archaeology, Anthropology and Interstellar Communication.* Washington, D.C: NASA.

Vorzimmer, P. J. 1970. *Charles Darwin, the Years of Controversy: The Origins of Species and It Critics, 1859–82,* Temple University Press, Philadelphia

Westman, R. S., ed. 1975. *The Copernican Achievement,* University of California Press, Berkeley

Chapter 10
Cultural Aspects of Astrobiology: A Preliminary Reconnaissance at the Turn of the Millennium

Abstract Among the four operating principles of the NASA Astrobiology Roadmap, Principle 3 recognizes broad societal interest for the implications of astrobiology. Although several meetings have been convened in the past decade to discuss the implications of extraterrestrial intelligence, none have addressed the broader implications of astrobiology as now defined at NASA. Here we survey these societal questions and argue that they deserve further serious study, in accordance with the National Aeronautics and Space Act of 1958. Astrobiology, already an interdisciplinary field in terms of the physical and biological sciences, should now embrace the humanities and the social and behavioral sciences in order to explore its cultural implications. Such study is part of the general need for better dialogue between science and society.

10.1 Justification for Study of Cultural Questions

Astrobiology, as defined within the NASA Astrobiology Roadmap (NASA 1999), seeks to answer three fundamental questions: (1) How does life begin and evolve? (2) Does life exist elsewhere in the universe? and (3) What is life's future on Earth and beyond? Because the answers to these questions bear on fundamental human concerns, I argue here that NASA's Astrobiology Program, as well as exobiologists and bioastronomers in general, should address the cultural impact of their work. In doing so, they should encourage input from specialists in the humanities and the social and behavioral sciences.

It is important at the outset to define what we mean by "culture." For anthropologists, culture is "the total way of life of a discrete society—its religion, myths, art, technology, sports, and all the other systematic knowledge transmitted across generations." Put another way, "culture is a product; is historical; includes ideas, patterns, and values; is selective; is learned; is based upon symbols; and is an abstraction from behavior and the products of behavior" (Wilson 1998, p. 130).

First published in *Bioastronomy 99: A New Era in Bioastronomy*, Guillermo Lemarchand and Karen Meech, eds. (San Francisco, 2000).

S. J. Dick, *Space, Time, and Aliens*, https://doi.org/10.1007/978-3-030-41614-0_10

According to Harvard biologist E. O. Wilson, each society creates culture, and is created by it. Our inquiry, then, is to determine the potential impact of astrobiology on this symbolic communal and evolving world view that each society creates—a tall order indeed, but one that multidisciplinary study may systematically tackle in increments.

The study of the cultural impact of astrobiology is justified from many points of view. Primarily, it is an interesting and important problem that adds another dimension to astrobiology. It is well to remember that cosmic evolution does not end with astronomy or biology, but with culture; the evolution of human culture, and possibly cultures beyond the Earth, is not only part of cosmic evolution, but arguably the most interesting part. Such study is also entirely in keeping with the National Aeronautics and Space Act of 1958, in which one of the eight objectives of the U. S. space program is "the establishment of long-range studies of the potential benefits to be gained from, the opportunities for, and the problems involved in the utilization of aeronautical and space activities for peaceful and scientific purposes" (Logsdon et al. 1995). Though the Space Act has been amended many times, this objective has remained unchanged. It has also remained largely unfulfilled, aside from a NASA-sponsored Brookings Institution study (U. S. Congress 1961), a NASA-sponsored study at Boston University (Berenzden 1973), and a series of NASA workshops in 1991–1992 (Billingham et al. 1999). There is interest, however, both among the public and at the highest levels of government, as evidenced by the Vice President's Space Science Symposium convened in December, 1996 in the wake of the Mars rock (Fig. 10.1), especially to discuss the cultural implications of that discovery (Lawler 1996a, b; Macilwain 1996). We are thus faced with a golden opportunity. With the inauguration of NASA's Astrobiology Program, the time has come to focus on this objective once again.

One of the exciting aspects of astrobiology, and one of the features that distinguishes it from the earlier exobiology program in NASA, is that the Astrobiology Roadmap recognizes the cultural dimensions of its work from the beginning. One of the groups at the Roadmap meeting formulated a "Question 7" in addition to the scientific questions: "How will astrobiology affect and interact with human societies and cultures?" The third of the four operating principles of the Roadmap "integral to the entire Astrobiology Program" states that "Astrobiology recognizes a broad societal interest in our subject, especially in areas such as the search for extraterrestrial life and the potential to engineer new life forms adapted to live on other worlds." This principle, as distinct from Principle 4 on education and public outreach, was presumably formulated with a view toward action, no less than the Roadmap's scientific aspects. In this paper we make a first reconnaissance of the scope of the cultural aspects of astrobiology as defined above, and issue a call for action.

Fig. 10.1 In the wake of the claim that the Mars rock ALH84001 contained nanofossils, a high-level discussion of the implications of finding life beyond Earth took place on December 11, 1996, in the Indian Treaty room of the Old Executive Office Building adjacent to the White House. From right to left: Vice President Al Gore, NASA Administrator Dan Goldin, astronomers Anneila Sargent and John Bahcall, historian of science Steven Dick, theoretical biologist Stuart Kauffman, biologist Lynn Margulis, astrobiologist David McKay, theologian Joan Brown Campbell, and NASA Associate Administrator Wes Huntress. Among those not visible are journalist Bill Moyers, Harvard biologist Stephen Jay Gould, and Presidential science advisor Jack Gibbons, who was seated to Gore's left (U. S. Government photo)

10.2 Astrobiology's Three Fundamental Questions and their Implications

Although a good deal of thought has been given to the cultural impact of a successful SETI (Search for Extraterrestrial Intelligence) program, the impact of astrobiology, as encapsulated in its three fundamental questions, is much broader. In addition to intelligent life, we are interested in the quite different implications of microbial life and life that may be complex, but not intelligent. Moreover, astrobiology has a large historical dimension in that we are also interested in life's past, and it has a forward-looking dimension because we want to explore life's future on Earth and beyond. These questions give astrobiology a breadth that exobiology never had, with correspondingly broader implications.

If we take each of the three questions in turn, and ask what the implications will be for society, we end up with an enormous two-dimensional matrix. Here we concentrate on only a small part of the matrix, the philosophical, ethical, and theological implications (Table 10.1), which are also identified in the Table with each of the Astrobiology Roadmap Goals. These parts of the matrix, in my opinion, are particularly important because they form our world view, and thus affect all other parts of society. Several aspects of this matrix merit emphasis: (1) The listing of representative studies indicates that some thought has been given to these issues; the point is that the entire problem has not been approached systematically. The study of the cultural implications of astrobiology is at a stage where exobiology was 40 years ago, with sporadic individual interest but little dialogue and thus little progress in the sense of systematic study. (2) We must recognize a third dimension to the matrix: different societies will be affected differently because they each have different cultures. Thus, the theological effects of contact with extraterrestrial

Table 10.1 Astrobiology roadmap questions and their cultural implications

Questions	Implications			
Roadmap scientific/ representative cultural	Philosophical	Ethical	Theological	General Societal
Q1 Origin and evolution of life (Goals 1–4)				
Our place in history of life	Schopf (1999)			
Nature of life	Davies (1998)			
A cosmic imperative?	De Duve (1995)			
Chance & necessity	Monod (1971)			
Q2. Life in the universe (Goals 5–8)				
A. Primitive				
B. Intelligent				
Contact	Davies (1995)	Ruse (1985)		Billingham et al. (1999) Harrison (1997)
Epistemology	Rescher (1985) Minksy (1985)			
Relation to God			McMullin (2000) Coyne (2000)	
Q3. Future of life on earth and beyond (Goals 9–10)				
Planetary protection	Randolph et al. (1997)			
Environmental change & ecosystems				
Artificial life/bioengineering				
Terraforming		McKay (1990)		
Space exploration				McCurdy (1997)
Space colonization				

intelligence would be very different for Chinese religions as contrasted with the Christianity embraced by much of the Western World. (3) An important feature of the matrix is the policy dimension: the study of cultural implications is not purely academic, but is undertaken with the idea of informing policy. For a national policy strategy, the matrix would be considerably smaller, but for global policy we can see the complexity involved. Thus we envision a very large three-dimensional matrix as the structure for our study, of which we address only a very small part in this paper.

10.2.1 Origin and Evolution of Life

Astrobiology's emphasis on the origin and evolution of life (Roadmap Goals 1–4) recalls the statement of T. H. Huxley in the context of Darwinism that "The question of questions for mankind—the problem which underlies all others, and is more deeply interesting than any other, is the ascertainment of the place which Man occupies in nature and of his relations to the universe of things (Huxley 1863)." Surely one of the overarching results of origin and evolution of life studies will be a better understanding of our temporal place in the history of life on Earth. Surely, the discovery of the ancient origins of life some 3.8 billion years ago has already had an effect on human culture, as has the demonstration that bacteria ruled the Earth for the vast majority of that period. The relatively recent rise of the genus *homo*, much less *homo sapiens*, surely has lessons for our world view. Exactly what they are should be the subject of further research.

Aside from illuminating our place in nature, origin of life studies force us to ask further questions such as "What is life?", "Is there a cosmic imperative for life imbedded in the laws of Nature?", and "What is the role of chance and necessity in the origin and evolution of life?" Research on molecular biology has already produced considerable discussion on the latter (Monod 1971), but the answer to this question and others will depend on which of the three or four theories of origin of life, or what combination of them, prove to be true (Davies 1998). Life arising from panspermia will have quite different implications than if it arose on Earth, whether in Darwin's warm pond, in hydrothermal vents, or in the hot deep biosphere (Gold 1999). Scientists have been asking these questions for years; it is time to engage the broader scholarly world as well.

10.2.2 Life in the Universe

The question of life in the universe (Roadmap Goals 5–8) brings another set of concerns. In any discussion of the cultural implications of life in the universe we immediately need to distinguish primitive from intelligent life. Given the history of life on Earth—ruled by bacteria for more than 2 billion years—we perhaps need to consider that the universe is full of bacteria. Anyone who thinks this has no implications for society should recall the reaction 3 years ago to the claim of Martian fossils. The media was full of speculations about their meaning; the Vice President specifically convened a seminar of experts to discuss the societal implications, and funding was provided in no small part leading to the Origins and Astrobiology programs we have today. Undoubtedly part of the excitement had to do with the implications for the abundance of extraterrestrial intelligence, but the existence of extraterrestrial bacteria, possibly with their own biochemistries, would have its own set of implications. The cultural impact of primitive life, however, has received no serious study.

The impact of intelligent life, by contrast, has been the subject of much speculation, and some serious study. Different approaches to the long-term problems of contact have been explored by Almar (1995), Billingham et al. (1999), Dick (1995), Harrison (1997), and Tough (1991), among others. The short-term reaction in the event of contact has been discussed in considerable detail (Tarter and Michaud 1990), and policy issues regarding a response to an extraterrestrial communication are under consideration (Michaud 1998). The problem of objective knowledge, or "extraterrestrial epistemology," has been broached by Rescher (1985) and Minksy (1985), while Ruse (1985) and Randolph et al. (1997) have outlined ethical considerations. Theological issues are coming more to the fore, and are discussed in Dick (1996, 1998, 2000a, 2000b), Crowe (1986), Coyne (2000), McMullin (2000), and Peters (1994), among others. From this small sample, one can glimpse the scope of the problem of the cultural implications of extraterrestrial intelligence. Social scientists have only begun to think about how these problems might be addressed (Harrison et al. 1998).

One of the conclusions of the studies thus far is that the discovery of extraterrestrial intelligence will be very much scenario-dependent. Any serious study of the impact of extraterrestrial intelligence must categorize the types of contact; a very general categorization of scenarios as terrestrial or extraterrestrial, and direct or remote is given in Table 10.2, together with examples from science fiction. Although terrestrial modes of contact are not currently in favor among most scientists, they are logical possibilities and the subject of both science (Bracewell 1975; Tough 1998) and science fiction. (There is also a considerable popular following in the case of UFOs and alien abductions.) Direct extraterrestrial contact is also currently

Table 10.2 Modes of contact with extraterrestrial intelligence and some representative science fiction scenarios

	Terrestrial	Extraterrestrial
Direct	Wells, *War of the Worlds*	Clarke, *Rendezvous with Rama*[a]
	Clarke, *Childhood's End*	Bradbury, *Martian Chronicles*
	ET: The Extraterrestrial	*Alien* and its sequels
Indirect	Clarke, *2001: A Space Odyssey*[a]	Gunn, *The Listeners*
	McCollum, *Lifeprobe*	
	Hoyle, *The Black Cloud*	Sagan, *Contact*[a]

[a]More than one mode of contact takes place

considered unlikely, but again the subject of much science fiction. Indirect contact by radio, optical, or other electromagnetic means is currently the favored scenario, and the one to which most attention has been given in terms of implications. But clearly each of the four types of contact would have its own set of implications for each of the elements in the cultural matrix. Even a brief consideration of the cultural implications of SETI demonstrates that the subject is complex, involving matrices embedded within matrices, but that these complexities may be approached systematically in discrete parts.

10.2.3 Future of Life on Earth and beyond

The future of life on Earth and beyond, the subject of Roadmap Goals 9 and 10, has implications best known today in terms of planetary protection, and the problems of contamination and back contamination. These indeed are important and have been given prominent attention because the problems are immediate, and the potential implications catastrophic. The ethical questions, however, have only begun to be explored (Randolph et al. 1997). Moreover, astrobiology's third question raises many other cultural issues. Moving beyond the planet may mean producing artificial life for bioengineering ecosystems, in its grandest vision known as terraforming. Probably in the lifetimes of our children, certainly in the twenty-first century, the issue of terraforming Mars will become real; it behooves us to begin to consider the philosophical, ethical, and broader cultural implications now. Similarly, Goal 9's emphasis on the interplay of environmental change and ecosystems raises broad questions that society has already had to tackle. As McKay (1990) points out, we may soon be faced with extending the principles of environmental ethics to Mars.

Movement off of planet Earth (Finney and Jones 1984) also raises the entire spectrum of issues associated with space exploration, in terms of manned or unmanned, the problems and opportunities of space colonization, and societal spending priorities. Perhaps more than the other two questions, question 3 raises the issue of where our species wants to go in its cultural evolution, and emphasizes that to a large extent human cultural evolution is in our own hands. I stress again that cultural evolution must be viewed as part of cosmic evolution; indeed it is indisputable that the pace of cultural evolution now rapidly outpaces biological evolution (though genetic engineering may change that in the future). An understanding of human cultural evolution is essential to understanding ourselves and our future, and it will be essential for mutual understanding in the event of extraterrestrial contact. Viewed as a part of cosmic evolution, cultural evolution fits squarely in the context of astrobiology and the famous "L" (lifetime of communicative civilizations) parameter of the Drake Equation. Indeed, many have pointed out that the number of communicative civilizations in our Galaxy (N) approximates L; since L depends in large part on the success or failure of cultural evolution, an understanding of human cultural evolution is one of the few ways we have at present to study L, and better determine N. The humanities and social sciences are in a position to make significant contributions to this study.

10.3 Approaches and Goals

Social scientists must ask how we can *systematically* approach these difficult questions, these questions of the "benefits, opportunities, and problems" of Astrobiology, in the spirit of the National Aeronautics and Space Act of 1958 and in the interest of encouraging dialogue between science and society. Of course we cannot *predict* the short- or long-term implications of astrobiology. But is there any systematic way for at least discussing them? Given the fact that different scenarios imply different implications, let me suggest three approaches that might guide us in our thinking about implications.

First, we must make use of the humanities, for the humanities study the elements that drive cultural evolution. History may be seen as a vast set of social experiments, conducted under many conditions. Surely, the record of these experiments must be used in any assessment of the effect of astrobiology on human cultural evolution. As a start, we should ask what effect space exploration has had on human cultural evolution in the last 40 years (McCurdy 1997). Next, we might ask how humans have reacted to particular ideas or events. The historical record of public reaction to past false alarms of extraterrestrial life, whether the canals of Mars, reaction to the Orson Welles broadcast of "War of the Worlds" in 1938, reaction to UFOs, the discovery of pulsars, and the now well-known the history of the extraterrestrial life debate (Crowe 1986; Dick 1982, 1996, 1998; Guthke 1990), are among the events that should prove relevant. The idea of life beyond Earth, whether termed exobiology, astrobiology, or bioastronomy, has exercised a peculiarly strong lure in American culture, a phenomenon that should itself be studied.

More generally, the humanities provide us with analogues of possible futures. An analogue is no more than a model, a concept very successfully used in the natural sciences, less so in the humanities and social sciences. Astrobiologists do not hesitate to use, with caution, Antarctica and Lake Vostok as analogues to conditions on Mars and Europa, respectively. In the case of SETI, to take a well-known example of analogical reasoning, one hears a good deal about physical culture contacts on Earth. But most scientists in the SETI field think direct physical culture contacts are unlikely, though contact with an alien probe in the vicinity of Earth must be considered a logical possibility (Bracewell 1975; Tough 1998). In the typical radio SETI scenario, a simple "dial tone" would provide evidence of a technological civilization, while decoding a message would initiate intellectual contact. For the latter, a much better analogue in Earth history is the transmission of knowledge from the ancient Greeks to the Latin West via the Arabs in the twelfth and thirteenth centuries (Dick 1995), an event that led to the European Renaissance.

More generally still, I have argued elsewhere that the idea of a universe with abundant life constitutes a worldview, analogous to the Copernican and Darwinian world views. If one accepts the claim that the biological universe is very different from the physical universe, we can study what effect changing worldviews have had on society (Dick 1995). Worldviews traverse various stages, from motivation to evidence to opposition and confirmation or rejection, and there are very rich studies in the history of science elaborating the short- and long-term implications of world-views like Copernicanism or Darwinism. So the humanities offer a number of approaches to the cultural implications of astrobiology.

Secondly, aside from history and the humanities, one should use the tools of the social and behavioral sciences, which admittedly are not as robust as the natural sciences, but which should play a role in the multidisciplinary science that is astrobiology. If, as E. O. Wilson and others have argued, there is such a thing as gene-culture co-evolution, it offers a starting point for studying cultural evolution based on the natural sciences. If, as Dawkins (1976) has argued, the "meme" is the unit of cultural evolution equivalent to the gene in biology, human cultural evolution including movement off of planet Earth may be studied using this increasingly developed concept (Blackmore 1999). Alternatively, Albert Harrison's recent book, *After Contact: The Human Response to Extraterrestrial Life* has led the way in showing how fields such as psychology, sociology, and anthropology can be used as an aid to thinking about implications of contact, an approach that may be generalized to astrobiology. In particular he advocates a kind of systems approach, called *Living Systems Theory*, in which what we know about organisms, societies and supranational systems on Earth can be used to discuss the outer space analogues of aliens, alien civilizations and the galactic club. Yet another approach envisions an "alien anthropologist" who could apply the tools of anthropology to the Earth from an alien perspective.

Third, in addition to the humanities and social sciences, human imagination, so colorfully rendered in science fiction literature, is a rich resource for studying the implications of astrobiology. Authors such as Arthur C. Clarke have given considerable thought to the consequences of contact in fictional form. *Childhood's End*, *Rendezvous with Rama*, and *2001: A Space Odyssey* all provide engaging explorations of different contact scenarios. Carl Sagan's *Contact* asks probing questions about the relation of science and religion. At the other extreme of extraterrestrial morality, we have the genre of H. G. Wells *War of the Worlds*, *Aliens*, *Independence Day*, and *Starship Troopers*. Perhaps more realistically, Stanislaw Lem's *Solaris* is a haunting story of contact with intelligence beyond our understanding. On the issue of extraterrestrial contamination, Michael Crichton's *The Andromeda Strain* is a thought-provoking exercise. Other science fiction authors have explored extraterrestrial environmental ethics, terraforming, and the problems of space colonization. Although one can argue about which scenarios are more likely, there is an enormous database of thought in the best science fiction that should not be ignored.

Undoubtedly a group of interdisciplinary specialists would produce a more robust list of approaches to the cultural implications of astrobiology's three questions. Some may consider such study premature, since we do not yet know whether or not life exists beyond Earth, or when we will terraform planets or colonize outer space. But I think it fair to say the scientific consensus is that extraterrestrial life is likely, and that problems like terraforming and colonization will face us as real problems in the twenty-first century. As the anthropologist Ashley Montagu said 25 years ago at the NASA-sponsored Symposium at Boston University, it is important that we think about the cultural impact of exobiology well in advance of the discovery of extraterrestrial life. With the increasing attention now given to astrobiology, that sentiment may be reiterated, and extended to all of astrobiology's broader implications.

10.4 Conclusions

In closing, I would argue that it is prudent and essential for the Astrobiology Program to support research on the implications of its work. The public supports NASA's Astrobiology Program with its tax dollars; it is interested in the implications of this research, which deserves nothing less than systematic study by experts from many fields. The need for study of the implications of science has been explicitly recognized, for example, in the Human Genome Project, which devotes 3–5% of its budget to ethical, legal, and social issues. While the genomic issues are admittedly more pressing, it may surely be argued that a small percentage of Astrobiology funding should be allocated to studying cultural implications, in accordance with the National Aeronautics and Space Act of 1958.

The study of the cultural aspects of astrobiology, however, need not confine its hopes to the success of similar studies such as the Human Genome Project. As Finney (2000) argues, astrobiology is strategically placed at the boundaries between disciplines—whether of the natural sciences, the social sciences, or the humanities—and so is in a unique position to cultivate the unity of knowledge in the deep sense that E. O. Wilson has elaborated in his recent book *Consilience* (Wilson 1998). Even if life is not discovered beyond the Earth, a fundamental role in bringing about the unity of knowledge would be a stunning success for astrobiology in the twenty-first century and the third millennium. Exobiology has already brought together the physical and biological sciences in unprecedented cooperation. I urge NASA and the astrobiology/bioastronomy communities to broaden their interdisciplinary scope yet again, this time to the humanities and social sciences, and to take up the broader challenges sure to come as astrobiology moves forward with its scientific goals.

10.5 Commentary 2020

This paper was given at the sixth Bioastronomy Symposium, held at the Hapuna Prince Hotel, Kohala Coast, Hawaii from August 2–6. 1999. The Proceedings (Lemarchand and Meech 2000) heralded "A New Era in Bioastronomy," perhaps justified in NASA's new astrobiology program, roadmap, and funding, as well as in the science itself in the form of increasing numbers of exoplanets. This Hawaii meeting was also important because it sparked a follow-up meeting on cultural aspects of astrobiology sponsored by the Foundation for the Future (Tough 2000). The latter, founded by aerospace businessman Walter Kistler, had been holding meetings at its Seattle Headquarters looking forward to a one-thousand-year horizon. It had decided that the discovery of life beyond Earth was one of the wild-cards that could transform humanity, thus the meeting to focus on this idea while experts were present in Hawaii. It was organized by University of Toronto futurist Allen Tough, with Proceedings and other relevant articles published in Tough (2000).

In addition to the landmark Astrobiology Roadmap meeting described in this chapter, several events from the 1990s only briefly mentioned here also deserve elaboration. The first is when John Billingham, the head of NASA's SETI program at its Ames Research Center, convened a series of workshops on "The Cultural Aspects of SETI" (CASETI) on the eve of the inauguration of NASA SETI observations in 1992. When in October, 1991 the attendees gathered for the first of three workshops in Santa Cruz, California, we could not have known that exactly 2 years later the U. S. Congress would cancel the entire NASA SETI program (Garber 2014). We gathered at the Chaminade Conference Center confident, not that the search would be immediately successful, but that it would be an ongoing research program of long duration with some chance of eventual success—enough chance that we needed to gauge the cultural impact. Alas, petty politics intervened, and the search was halted in the same session of Congress that cancelled the Superconducting Supercollider in Texas. But the intimate gathering of two dozen scholars was a model of interdisciplinary brainstorming, with astronomers including Frank Drake and Jill Tarter, anthropologists represented by Ben Finney and Michael Ashkenazi, religious scholars and historians including Georgetown's Langdon Gilkey and Harvard's Karl Guthke, several representatives from media studies, and even two diplomats, represented by Michael Michaud from the State Department and Nandasiri Jasentuliyana, the Director of the Office of Outer Space Affairs at the United Nations (Fig. 10.2). The gathering was a *de facto* recognition that this was a broad-based problem not to be solved by scientists alone. While the publication of the results (Billingham et al. 1999) was delayed almost a decade by the cancellation of the scientific program, its recommendations are still valuable for contemplating the aftermath of any successful SETI program.

The second event was Vice President Gore's extraordinary symposium on the cultural implications in the wake of the claim of possible fossil life in the Mars rock, also mentioned in Sect. 10.1 and depicted in Fig. 10.1. In August, 1996 scientists at NASA's Johnson Space Center and their colleagues made the spectacular claim that they had discovered nanofossils in a Mars rock. Surprisingly to most people, Mars rocks do indeed land on Earth, and are usually found in the Antarctic, where they are then taken back to laboratories for study. Dick and Strick (2004) describe the announcement and its fallout in detail. Because my book, *The Biological Universe* had just come out, I received an invitation to this high-level meeting held on December 11, 1996. The meeting, which ran almost 3 hours, took place in the Indian Treaty Room of the Old Executive Office Building adjacent to the White House, the building where the Vice President has his office. The meeting included many luminaries, including Stephen Jay Gould, Lynn Margulis, John Bahcall, Bill Moyers, some of the scientists involved (David McKay and Everett Gibson), and theologians, a total of perhaps 20 seated around the table with the Vice President, the NASA Administrator (Dan Goldin) and the President's science advisor, Jack Gibbons.

Fig. 10.2 A rare photograph of the participants in the 1991–1992 interdisciplinary workshops on the Cultural Aspects of SETI (CASETI) in Santa Cruz, California. The meeting was held just before the inauguration of the NASA SETI program, and was the first of the meetings listed in Table 16.1. First row, left to right: David Milne, John Billingham (NASA Ames), education specialist Julia Koppich, Roger Heyns, Kent Cullers (SETI Institute), anthropologist Ben Finney. Second row: historian of science John Heilbron, JPL astronomer Michael Klein, SETI Institute CEO Tom Pierson. Third row: Vera Buescher (SETI Institute), historian of science Steven Dick, astronomer Julie Lutz, astronomer Jill Tarter, Alex Inkeles, Vivian Sobchak, lawyer Steve Doyle. Fourth row: Gary Coulter (NASA HQ), Amahl Drake, astronomer Andrew Fraknoi, anthropologist Michael Ashkenazi, John Rummel (NASA HQ), Harvard cultural historian Karl Guthke, student Alison Tucher, theologian Langdon Gilkey. Top row: astronomer Frank Drake, Bob Arnold, Michael Michaud (U.S. State Department) (Credit: Seth Shostak)

The purpose of the meeting was to discuss the societal implications if the nanofossils claim proved to be true. The meeting has been briefly described in a number of books, including Kathy Sawyer's *The Rock From Mars,* and my recent book *Astrobiology, Discovery, and Societal Impact*. Gore was very impressive with his interest and knowledge of science, and very nearly became President 4 years later.

Although a FOIA request turned up no transcript of the discussions at the meeting, it did reveal supporting material and photographs, including Fig. 10.1, which can be accessed at the National Archives in Washington, DC, Presidential Materials Division.

Carl Sagan was invited to attend the meeting, but was very ill and less than 2 weeks from death. This led to a third little-known event, which did not have wide impact but is interesting as a case where writing about extraterrestrials led to a court case. To wit, only a few weeks after his death the director Francis Ford Coppola sued the Sagan estate for intellectual property rights infringement. The case revolved around the book *Contact*, mentioned in Sect. 10.3, which Sagan published in 1985, made into a movie by the same name released in 1997 starring Jodie Foster. On December 26, 1996, while *Contact* was still filming and Sagan had just died, Coppola filed suit against the Sagan estate claiming Sagan had used numerous ideas in *Contact* that Sagan had originally written under a March, 1975 contract to Coppola for a television documentary titled *First Contact*. Coppola went on to produce *Apocalypse Now* (1979) and many other projects, and the *First Contact* documentary was never produced. For this case I was hired by the law firm Gibson Dunn as an expert witness on the side of the defendant, on the recommendation of John Billingham, by then the former head of the SETI program at NASA Ames. In 1997 I flew to Century City in Los Angeles to be deposed. Following depositions as expert witnesses by myself, Jill Tarter, and Andrew Fraknoi, in February 1998 a Los Angeles Superior Court judge issued a summary judgment in favor of the Sagan estate and Warner Brothers. Coppola appealed, and in April 2000 a California Court of Appeal also dismissed the suit, saying it was brought too late. In any case I had shown in my deposition that all the ideas Coppola claimed were his intellectual property rights had been put forth by Sagan or others prior to 1975. The episode is not well known but is mentioned in William Poundstone's biography of Sagan (Poundstone 1999). According to Poundstone, Coppola had written Sagan in November, 1995 stating he expected to receive a share of Sagan's profits from *Contact*, for which Sagan had received a highly publicized two million dollar advance. Sagan wrote a letter stating Coppola's claim was without merit, which went unanswered during Sagan's remaining year of life.[1]

Since this chapter was written the NASA Astrobiology Roadmap has been published in several iterations (Des Marais et al. 2008), the latest being the NASA *Astrobiology Strategy* (NASA 2015). The Harrison et al. (1998) citation below has since been published in Harrison et al. (2000), and the NASA Ames workshop is Harrison and Connell (2001).

In summary, the 1990s not only saw the first NASA astrobiology roadmaps, but also began to demonstrate a sporadic but serious interest in the cultural aspects of astrobiology. How this interest had expanded 20 years later is described in Chap. 16.

Acknowledgements The author thanks members of the "Question 7" Group at the Astrobiology Roadmap meeting at NASA Ames in 1998, and particularly Lynn Harper, Kathleen Connell, and Ken Rose for active discussions on the subject since that time.

Note

1. My papers on this case, including detailed analysis of the differences between *Contact* and the Coppola documents, are in the Adler Planetarium Archives in Chicago. See also http://variety.com/1998/film/news/coppola-s-contact-claim-is-dismissed-1117467799/ and http://variety.com/2000/film/news/coppola-loses-contact-1117780544/

References

Almar, I. 1995. "The Consequences of a Discovery: Different Scenarios", in G. Seth Shostak, ed., *Progress in the Search for Extraterrestrial Life*, San Francisco: Astronomical Society of the Pacific, 499–505.

Berenzden, R. 1973. *Life Beyond Earth and the Mind of Man*. NASA, Washington, D.C..

Billingham, J. *et al.* 1999. *Social Implications of the Detection of an Extraterrestrial Civilization*. SETI Press; Mountain View, CA.

Blackmore, S. 1999. *The Meme Machine*. Oxford University Press, Oxford.

Bracewell, R. 1975. *The Galactic Club*. San Francisco, 69–83.

U. S. Congress, 1961. *Proposed Studies on the Implications of Peaceful Space Activities for Human Affairs*, a Report of the Committee on Science and Astronautics, U. S. House of Representatives, 87th Congress, First Session, prepared for NASA by the Brookings Institution. Washington: Government Printing Office).

Coyne, G. 2000, "The Evolution of Intelligent Life on Earth and Possibly Elsewhere: Reflections from a Religious Tradition," in Steven J. Dick, ed., *Many Worlds: The New Universe and its Theological Implications*. Templeton Press: Philadelphia

Crowe, M. 1986. *The Extraterrestrial Life Debate, 1750–1900: The Idea of a Plurality of Worlds from Kant to Lowell*. Cambridge University Press, Cambridge.

Davies, P. 1995. *Are We Alone? Philosophical Implications of the Discovery of Extraterrestrial Life* Basic Books, New York.

Davies, P. 1998. *The Fifth Miracle: The Search for the Origin of Life*. Penguin Press, London

Dawkins, R. 1976. *The Selfish Gene*. Oxford University Press, Oxford.

De Duve, C. 1995. *Vital Dust: Life as a Cosmic Imperative*. Basic Books, New York

Des Marais, David., J. A. Nuth, III, Louis Allamandola et al. 2008. "The NASA Astrobiology Roadmap,"*Astrobiology*, 8: 715–730.

Dick, S. J. 1995. "Consequences of Success in SETI: Lessons from the History of Science," in G. Seth Shostak, ed., *Progress in the Search for Extraterrestrial Life*. Astronomical Society of the Pacific, San Francisco, 521–532.

Dick, S. J. 1996. *The Biological Universe: The Twentieth Century Extraterrestrial Life Debate and the Limits of Science*, Cambridge University Press, Cambridge.

Dick, S. J. 1998. *Life on Other Worlds: The Twentieth Century Extraterrestrial Life Debate*. Cambridge University Press, Cambridge.

Dick, S. J. ed. 2000a. *Many Worlds: The New Universe and its Theological Implications*. Templeton Press, Philadelphia

Dick, S. J. 2000b. "Cosmotheology", in Steven J. Dick, ed., *Many Worlds: The New Universe and its Theological Implications*. Templeton Press, Philadelphia

Dick, S. J. and James Strick. 2004. *NASA and the Development of Astrobiology*. Rutgers University Press, New Brunswick.

Dick, Steven J. 1982. *Plurality of Worlds: The Origins of the Extraterrestrial Life Debate from Democritus to Kant*. Cambridge University Press, Cambridge.

Finney, B. 2000. "SETI, Consilience and the Unity of Knowledge," in Lemarchand and Meech (2000), 641–648.

Finney, B. and E. M. Jones, 1984. *Interstellar Migration and the Human Experience*. University of California Press, Berkeley.

Garber, Stephen J. 2014. "A Political History of NASA's SETI Program," in Vakoch (2014).

Gold, T. 1999. *The Hot Deep Biosphere*. Springer Verlag, New York.

Guthke, K. 1990. *The Last Frontier: Imagining Other Worlds from the Copernican Revolution to Modern Science Fiction*. Cornell University Press, Ithaca, N.Y.

Harrison, A. 1997. *After Contact: The Human Response to Extraterrestrial Life*. Plenum, New York.

Harrison, A., J., Billingham, S. J. Dick *et al.* 1998. "Increasing the Role of Social Science in SETI," unpublished paper prepared for the SETI Committee of the International Academy of Astronautics, Melbourne.

Harrison, Albert A., John Billingham, Steven Dick, et al. 2000. "The Role of the Social Sciences in SETI," in Tough (2000), pp. 71–85.

Harrison, Albert A. and Kathleen Connell. 2001. *Workshop on the Societal Implications of Astrobiology*. Moffett Field: NASA Ames Research Center. Online at http://www.astrosociology.org/Library/PDF/NASA-Workshop-Report-Societal-Implications-of-Astrobiology.pdf

Huxley, T. H. 1863. *Man's Place in Nature*. London, 71.

Lawler, A, 1996a. "Building a Bridge Between the Big Bang and Biology," *Science*, 274 (8 November, 1996), 912.

Lawler, A. 1996b. "Origins Researchers Win Gore's Ear, Not Pocketbook," *Science*, 274 (20 December, 1996), 2003.

Lemarchand, Guillermo and Karen Meech. 2000. *Bioastronomy '99: A New Era in Bioastronomy*. Astronomical Society of the Pacific, San Francisco.

Logsdon, J. *et al.* 1995. *Exploring the Unknown: Selected Documents in the History of the U. S. Civil Space Program*, vol I, "Organizing for Exploration," NASA: Washington, DC.

Macilwain, C. 1996. "Goldin Wants More NASA Biologists as Gore is Briefed on Space Plans," *Nature*, 384 (19/26 December, 1996), 601.

McCurdy, H. 1997. *Space and the American Imagination*. Smithsonian Institution Press, Washington.

McKay, C. 1990. "Does Mars Have Rights? An Approach to the Environmental Ethics of Planetary Engineering," in *Moral Expertise*, ed. D. MacNiven. Routledge: New York, 184–197.

McMullin, E. 2000. "Life and Intelligence far From Earth: Formulating Theological Issues," in Steven J. Dick, ed., *Many Worlds: The New Universe and its Theological Implications*. Templeton Press, Philadelphia.

Michaud, M. A. G. 1998. "Policy Issues in Communicating with ETI," *Space Policy*, 14, 173–178.

Minksy, M. 1985. "Why Intelligent Aliens will be Intelligible", in Edward Regis, Jr. ed., *Extraterrestrials: Science and Alien Intelligence*, Cambridge University Press, Cambridge, 117–128.

Monod, J. 1971. *Chance and Necessity*. New York.

NASA, 1999. "Astrobiology Roadmap." Ames Research Center, Moffet Field, CA.

NASA. 2015. *Astrobiology Strategy* online at https://nai.nasa.gov/media/medialibrary/2016/04/NASA_Astrobiology_Strategy_2015_FINAL_041216.pdf

Peters, T. 1994. "Exo-Theology: Speculations on Extra-Terrestrial Life," *Center for Theology and the Natural Sciences Bulletin*, vol. 14, no. 3, 1–9.

Poundstone, William. 1999. *Carl Sagan: A Life in the Cosmos*. New York: Henry Holt, p. 375.

Randolph, R., M. Race, and C. McKay. 1997. "Reconsidering the Theological and Ethical Implications of Extraterrestrial Life," *Center for Theology and Natural Sciences Bulletin*, vol. 17, no. 3, 1–8, Berkeley.

Rescher, N. 1985. "Extraterrestrial Science," in Edward Regis, Jr. ed., *Extraterrestrials: Science and Alien Intelligence*, Cambridge University Press, Cambridge, 83–116.

Ruse, M., 1985. "Is Rape Wrong on Andromeda?," in Edward Regis, Jr. ed., *Extraterrestrials: Science and Alien Intelligence*, Cambridge University Press, Cambridge, 43–78.

Schopf, J. W. 1999. *The Cradle of Life*. Princeton University Press, Princeton.

Tarter, J. C. and M. A. G. Michaud, eds, 1990. "SETI Post-Detection Protocol," *Acta Astronautica*, 21 (February, 1990), 69–154.

Tough, A. 1991. *Crucial Questions About the Future*. Lanham, MD.

Tough, A. 1998. "Small Smart Interstellar Probes," *Journal of the British Interplanetary Society*, 51 (May, 1998), 167–174.

Tough, Allen, ed. 2000. *When SETI Succeeds: The Impact of High-Information Contact*. Foundation for the Future, Bellevue, Washington

Vakoch, Douglas, ed., 2014. *Archaeology, Anthropology and Interstellar Communication*. Washington, D.C: NASA.

Wilson, E. O. 1998. *Consilience: The Unity of Knowledge*. Knopf, New York

Chapter 11
The Role of Anthropology in SETI: A Historical View

Abstract This paper examines what role anthropology has historically played in SETI, and how the two intellectual cultures of natural scientists and social scientists made contact in this field. I argue that these historical interactions bode well for beneficial mutual interactions between anthropology and SETI in the future. What has been lacking is a *systematic* approach applying anthropology to the search for extraterrestrial intelligence. There is considerable evidence that such study would repay both disciplines.

11.1 Introduction

Three events mark the beginning of the modern era of the Search for Extraterrestrial Intelligence (SETI): (1) the publication of the landmark paper by Giuseppe Cocconi and Philip Morrison, "Searching for Interstellar Communications," in *Nature* in 1959, suggesting that a search be carried out at the 21-cm radio wavelength; (2) Frank Drake's Project Ozma in 1960, which carried out the first such search at Green Bank, West Virginia; and (3) a small but now legendary conference at Green Bank in 1961, where the feasibility of a search was discussed, and the Drake Equation was proposed as a way of estimating the number of communicative civilizations in our Milky Way Galaxy. Modern SETI was born during those 3 years, 1959–1961, setting the agenda in the field for much of the next 50 years (Dick 1996).

By the 1960s when modern SETI began, anthropology as a discipline was about a century old. As the Greek roots of the word indicate, the discipline is meant to encompass the study of humans. One might well ask, then, why it should apply to the extraterrestrial life debate, which obviously deals with non-humans. The answer is that in its broadest sense, anthropology has developed a set of approaches and methods to analyze cultures and cultural evolution. If there is intelligence beyond the Earth, it has likely developed culture. If, as many SETI proponents expect, those cultures are millions of years old, cultural evolution will have taken place, with all that implies for development, communication, cultural diffusion, and so on. All of these are areas of study that anthropologists, along with their social and behavioral science colleagues, have refined over the last century for terrestrial cultures (Smith 1997).

First published in *Anthropology Today*, 22 (April, 2006), 3–7.

S. J. Dick, *Space, Time, and Aliens*, https://doi.org/10.1007/978-3-030-41614-0_11

11.2 Beginnings

It would seem that the social sciences, and anthropology in particular, have the potential to illuminate a subject whose central concerns are, after all, societies and cultural evolution, even if the setting happens to be extraterrestrial. Yet, the historical record shows that the social sciences played no important role in SETI's first decade. This undoubtedly reflects a variety of factors, including C. P. Snow's "two cultures" phenomenon, increasing specialization already in full swing in the early 1960s, and plenty of problems on Earth for social scientists to tackle. Thus, while the Green Bank conference included astronomers, physicists, a biochemist, an engineer, and even a specialist on dolphin communication (John Lilly), no one represented the social sciences or humanities.

There appear to be only two cases in the 1960s where anthropology was discussed in relation to SETI. One was an article published in 1962 entitled "Interstellar Communication and Human Evolution," authored by Robert and Marcia Ascher, respectively an anthropologist and mathematician at Cornell, the home institution of Cocconi and Morrison. The authors suggested an "analogy between prehistoric contact and exchange, and hypothesized extraterrestrial contact and exchange." In early prehistory, when biologically distinct hominid populations existed, they pointed out that contact and exchange "occurred between technologically similar but biologically diverse populations. In later prehistory contact was usually initiated by those populations with advanced techniques and equal exchange was rare" (Ascher and Ascher 1962). This history, they suggested, might shed light on the nature of contact with extraterrestrial civilizations. Secondly, a NASA-commissioned study, published in 1961, had broached another possible role for social sciences in SETI—assessing the impact of the discovery of extraterrestrial intelligence. In a statement often cited since, the authors warned that substantial contact could trigger a foreboding effect: "Anthropological files contain many examples of societies, sure of their place in the universe, which have disintegrated when they had to associate with previously unfamiliar societies espousing different ideas and different life ways; others that survived such an experience usually did so by paying the price of changes in values and attitudes and behavior" (U.S. House of Representatives 1961).[1] This statement begs for elaboration and documentation. Over the last four decades, anthropology has certainly tackled the problem of culture contact for terrestrial societies. But it has not systematically studied the implications for extraterrestrial contact.

Already by the early 1960s, then, two roles had been identified for anthropology in the context of SETI: the study of human evolution models as analogies to extraterrestrial contact, and the study of the impact of such contact. Both roles embedded the problems and the promise of analogical thinking.[2]

11.3 Early SETI Overtures to Social Science

These ideas lay mostly fallow for the tumultuous decade of the 1960s, during which only two SETI searches were carried out, one in the United States and one in the Soviet Union. The realization gradually dawned on SETI proponents that the social sciences might be useful, even essential, to their discussions. Nowhere was this truer than in the case of the cultural components of the Drake equation, which embodies all facets of cosmic evolution, including astronomical, biological, *and* cultural. In particular its last two components, the probability of the evolution of radio communicative technical civilizations, and the lifetimes of such civilizations, were clearly in the realm of the social sciences. So it was that at an international meeting on CETI (Communication with Extraterrestrial Intelligence), held in the Soviet Union in 1971, two anthropologists were included, as well as historian William H. McNeill of the University of Chicago. There they argued with the natural scientists about the evolution of technical civilizations. No conclusions were reached, but the natural scientists were clearly interested in what the social scientists had to say (Sagan 1973: 85–111).

At least token social science representation became quite common at gatherings where extraterrestrial intelligence was discussed. When in 1972 NASA sponsored a symposium at Boston University on "Life Beyond the Earth and the Mind of Man," anthropologist Ashley Montagu was among the panelists. His subject was the reaction of humans to the discovery of extraterrestrial intelligence. Montagu concluded that "it is the communication we make at our initial encounter that is crucial." His point was again a plea for the study of culture contacts.

In the mid-1970s interest in SETI was becoming more serious, particularly at NASA (Dick 1993; Dick and Strick 2004). The guiding light of SETI at NASA was John Billingham at NASA's Ames Research Center in Mountain View, California. It was he who organized a series of workshops, chaired by Philip Morrison, with the goal of getting a NASA SETI program off the ground, complete with NASA funding. Part of that effort was a "Workshop on Cultural Evolution," chaired by Nobelist Joshua Lederberg, and including anthropologist Bernard Campbell. The workshop focused on the evolution of intelligence and technology. The summary of the workshop, published in the landmark NASA volume *The Search for Extraterrestrial Intelligence* (1977), edited by Morrison, Billingham, and John Wolfe, asserted that "our new knowledge has changed the attitude of many specialists about the generality of cultural evolution from one of skepticism to a belief that it is a natural consequence of evolution under many environmental circumstances, given enough time" (Morrison et al. 1977).[3] The cultural evolution panel also discussed what evolutionary factors were responsible for hominid intelligence: warfare, communication, and language, the predatory demands of life on the savannah. Arguing that evolutionist George Gaylord Simpson had been too pessimistic in dismissing extraterrestrial intelligence, they even placed a number on the probability that both intelligence and technology would evolve, given the origin of life on a planet. That number, they said, was 1 in 100.

Sporadic though they are, these early efforts through the 1970s demonstrated the relevance of anthropology to SETI, and stand as recognition of that fact by the scientific community that sponsored them.

11.4 Early Social Science Overtures to SETI

These early efforts, however, hardly tapped the richness that anthropology holds for SETI. Were there proactive efforts on the part of social scientists to tackle the subject, rather than waiting to be invited to a SETI meeting? The first substantial evidence of such interest is the proceedings of a symposium at the 1974 American Anthropological Association, published in 1975 as a popular trade book titled *Cultures Beyond the Earth*. The subtitle, *The Role of Anthropology in Outer Space*, is somewhat misleading for several reasons: only two of its eight authors were card-carrying anthropologists, it is a mixed volume including fiction as well as fact, and it is not in any sense systematic. But it did include a stimulating Foreword by futurist Alvin Toffler and an Afterword by anthropologist Sol Tax; it *was* sponsored by the American Anthropological Association as part of a "cultural futuristics" symposium; and most importantly of all, it contained some new and sophisticated ideas, at least in outline. In his Foreword, for example, Toffler pointed out that "what we think, imagine or dream about cultures beyond the earth not only reflects our own hidden fears and wishes, but alters them." He saw the book as important because "it forces us to disinter deeply buried premises about ourselves" (Maruyama and Harkins 1975). This is a straightforward point, but an important one that we do not explicitly address often enough. Contemplating extraterrestrial cultures forces us to look at ourselves anew, raising, as Toffler said, "the critique of our cultural assumptions to a 'meta-level.'"

It is one thing for a futurist to say such things. But in his Afterword, Sol Tax, professor of Anthropology at the University of Chicago, endorsed and elaborated these ideas. As Tax noted, "Only when we have comparisons with species that are cultural in nonhuman ways—some of them maybe far more advanced than we—will we approach full understanding of the possibilities and limitations of human cultures." Nor was this a far-out fruitless undertaking, because "Even if we have no contact with nonhuman cultures in the immediate future, the models that we meanwhile make require that we sharpen the questions that we ask about human beings."

The book also broached another problem with anthropology's entry into the SETI realm, one that perhaps still resonates today: "Just as exo-biologists now run the risk of being called ex-biologists," one of the anthropologists wrote, "so may anthropologists with extraterrestrial interests find themselves regarded with suspicion by the more conservative members of their own profession" (Wescott 1975: 12–26). The same anthropologist, Roger Wescott, also called attention to anthropological relevance of studying cultures and subcultures in Earth orbit, lunar orbit, and on the lunar surface. This aspect particularly resonates now, 30 years later, in light of NASA's current program to return humans to the Moon and head for Mars.

More substantial and influential than the 1974 AAA meeting on cultures beyond the Earth was the response to a crisis for SETI after the mid-1970s. The crisis was the so-called *Fermi Paradox*, the idea that if the galaxy was full of intelligence, given the billions-of-years timescales involved, any intelligence should have colonized the galaxy and should have arrived on Earth by now. Yet we do not see them, so "where are they?" Many concluded that this argument provided strong empirical evidence that extraterrestrials do not exist, "empirical" because we do not observe them on Earth (unless one believes evidence for UFOs, which most SETI enthusiasts studiously avoid).[4] The "diffusion" of cultures was primarily a problem for social scientists, and a problem familiar to cultural anthropologists.

One anthropologist in particular took up the challenge. Ben Finney, Professor of Anthropology at the University of Hawaii and later chair of that Department, was well known for his work on Polynesian migrations. He began his pathbreaking work with the NASA SETI community in the mid-1980s, perhaps the most sustained connection of a single anthropologist with SETI. Most important was the book he edited with Eric Jones, *Interstellar Migration and the Human Experience* (1985). The result of a Conference on Interstellar Migration held in 1983 at Los Alamos, where Jones worked as an astrophysicist, it concentrated on yet another aspect of SETI, the possibility of interstellar colonization. Based on humanity's evolutionary and historical past, and on its characteristic expansionary, technologically innovative and inquisitive nature, in their epilogue Finney and Jones wrote that "Mankind is headed for the stars. That is our credo. Our descendants will one day live throughout the Solar System and eventually seek to colonize other star systems and possibly interstellar space itself. Immense problems—technical, economic, political and social—will have to be solved for human life to spread through space" (Finney and Jones 1985: 333). The results of this conference thus radiated in two directions" the problem of why extraterrestrials are not here—the Fermi Paradox—and the possibility of humans going beyond the Earth. Whether or not life beyond the Earth turns out to be alien or descended from humans, they implied, anthropologists and social scientists should surely play a role in studying cultures beyond the Earth.

11.5 The Last 15 Years: Mutual Benefits?

Over the last 15 years, the interaction of SETI and the social sciences can still only be described as sporadic. At professional meetings of the International Astronautical Federation (IAF), the International Astronomical Union (IAU), and international bioastronomy meetings with a variety of sponsors, social science has been only an occasional companion to the natural sciences.

There have also been a few more sustained and substantial efforts. In the early 1990s, on the eve of the inauguration of the NASA SETI program in October, 1992, John Billingham led a series of workshops on "Cultural Aspects of SETI," known as the CASETI workshops (Billingham et al. 1999). For the first time, social scientists were fully integrated into the discussion of the implications of contact with extraterrestrials. Among the recommendations were that NASA should study appropriate analogies drawn from earlier human experience, and that study should concentrated on analogies based on the transmission of ideas within and between cultures in preference to analogies based on physical encounters (See Chap. 9).

A few individuals have tackled SETI from the social science aspect. In *After Contact: The Human Response to Extraterrestrial Life* (1997), psychologist Albert Harrison has led the way in showing how fields such as psychology, sociology, and anthropology can be used as an aid to thinking about implications of contact, an approach that may be generalized to astrobiology. The work of Douglas Vakoch on interstellar message construction, with its emphasis on the relation between language and culture, also has much in common with linguistic anthropology (Vakoch 1998a, b; Vakoch 1999; Vakoch 2000). Vakoch has also been instrumental in rallying the anthropology community to the study of SETI. The session on "Anthropology, Archaeology and Interstellar Communication" at the 2004 Annual Meeting of the American Anthropological Association—30 years after the previous AAA meeting on the subject—demonstrates the potential for a deeper role for anthropologists in SETI. That role ranges from the scholarly to the popular; possibly the best known anthropological contributions to SETI are the science fiction novels of anthropologist Mary Doria Russell (*The Sparrow, Children of God*). They have led the way in fiction toward what could be studied in fact (Fig. 11.1).

In the most general sense, it is cultural evolution that drives the relationship between SETI and anthropology. If, as most SETI proponents believe, intelligence in the universe is millions or billions of years old, we know only one thing for certain: cultural evolution will have occurred. One can argue exactly what the result might have been. The universe may, for example, be postbiological, full of artificial intelligence, precisely because one must take cultural evolution into account (Dick 2003). But, given intelligence beyond the Earth, the fact of the occurrence of extraterrestrial cultural evolution is not open to doubt, and is fundamentally a problem of anthropology.

11.6 Summary

Historically anthropology has made sporadic contributions to SETI in the following ways, all of which should be systematically elaborated:

1. **Evolution of technological civilization**. Using empirical data from terrestrial cultures, anthropologists can shed light on the likelihood of evolution of technological civilizations, their natures and lifetimes. This is a problem of physical anthropology, and the potential of this approach has been realized since the early 1960s.
2. **Culture Contact**. Using analogical studies of culture contacts on Earth anthropologists may illuminate contact scenarios with ETI, extending cultural anthropology to the extraterrestrial realm. However, because SETI envisions remote

Fig. 11.1 The alien has often been seen as a teacher for humanity. But no one knows where cultural evolution will lead among the stars. Cover by Leo Morey for the April 1930 issue of Amazing Stories showing an alien imparting knowledge of the galaxy

radio contact with ETI, rather than physical culture contact, the transmission of ideas may provide a better model for SETI. Should physical contact be made in the more distant future with cultures beyond the Earth, cultural anthropology (and even archaeology) will become more directly relevant.

3. **Interstellar Message Decipherment and Construction.** Philip Morrison has argued that deciphering an interstellar message may be a long-term problem requiring the efforts of many scholarly disciplines. Linguistic anthropology has a role to play both in deciphering and constructing interstellar messages.
4. **Cultural Diffusion.** Analogical studies of human migration on Earth may illuminate the Fermi Paradox of extraterrestrial civilizations. Beyond SETI, such studies are also applicable to extraterrestrial human cultures wherever they may be established. A start on these subjects has been made with the volume *Interstellar Migration and the Human Experience.*

All of these approaches come broadly under the study of cultural evolution, and fall squarely in the study of SETI as the third component of the Drake Equation. Anthropologists are uniquely qualified by knowledge and training to contribute to SETI. In turn, the extraterrestrial perspective that many of us in the SETI field have found so invigorating also has much to offer to anthropology, both in expanding its boundaries, its insights and its tools, and looking back on cultures on Earth and seeing them anew.

Finally, the participation of anthropologists in SETI is part of the larger problem of bringing the social sciences and humanities into SETI (Harrison et al. 2000: 71–85). This endeavor could prove important for E. O. Wilson's idea of "consilience," the unity of knowledge. Ben Finney has made this point with regard to SETI, arguing that it "has the potential for playing a major role in transcending intellectual boundaries" (Wilson 1998; Finney 2000). In my 35 years working in this field, I have found that nothing has greater potential to unify knowledge than the idea of extraterrestrial intelligence. Moreover, the appeal of the idea to students makes it a natural for implementing a unified knowledge curriculum in schools, work that is already being done at the SETI Institute and elsewhere.

11.7 Commentary 2020

This paper was given at a meeting of the American Anthropological Association (AAA) in San Jose, California in 2005. It was the cover story for the British journal *Anthropology Today* for its April 2006 issue, featuring on its cover cosmic evolution as depicted in Fig. 8.1. In retrospect it is remarkable that this particular subject was considered legitimate both for the professional AAA meeting and for the British journal. This is due in large part to the actions of Douglas Vakoch, who organized not only this session of the AAA meeting, but also a series of meetings on the subject at subsequent AAA conferences, as described in Vakoch (2009).

Following in the footsteps of the pioneering anthropologist Ben Finney, today many more anthropologists are actively involved in SETI and the broader field of astrobiology. Most notable for their active engagement with the astrobiology community are Kathryn Denning, John Traphagan, and the young scholars Klara Capova, Michael Oman-Reagan, and Claire Webb. For a small sampling of their work see Denning (2009, 2013, 2014), Traphagan (2014, 2015a, b, 2016), Traphagan and Traphagan (2015), and Capova and Persson (2018). Also important is the work of anthropologist Michael Ashkenazi, a member of the original CASETI team in the early 1990s (Ashkenazi 2017). Taken together, anthropologists continue to make important contributions to astrobiology, as is fitting in any search for "the Other."

Notes

1. The report was prepared under the direction of D.N. Michael, a social psychologist "primarily responsible for the interpretations, conclusions, and recommendations in and the drafting of this report" (viii).
2. For a contemporary view of these problems in connection with the space program see (Mazlish 1965).
3. The agenda and members of the Workshop on Evolution of Intelligent Species and Technological Civilizations are given on pages 275–276.
4. For the Fermi Paradox crisis in SETI see Dick (1996, 443–454). The original articles in the mid-1970s stating the paradox are akochHart (1975) and Viewing (1975). A collection of articles on the subject is found in Hart and Zuckerman (1982). For a thorough discussion of possible answers to the Fermi Paradox, see (Webb 2002).

References

Ascher, Robert and Marcia Ascher. 1962. "Interstellar Communication and Human Evolution, Nature, 193, 940, reprinted in A. G. W. Cameron, ed., *Interstellar Communication* (New York: W. A. Benjamin, 1963), pp. 306-308.

Ashkenazi, Michael. 2017. *What We Know About Extraterrestrial Intelligence: Foundations of Xenology*. Switzerland: Springer.

Billingham, John et al. 1999. *Social Implications of the Detection of an Extraterrestrial Civilization*, A Report of the Workshops on the Cultural Aspects of SETI held in October 1991, May 1992 and September 1992 at Santa Cruz, California. John Billingham et al., eds. Mountain View, CA: SETI Press.

Capova, K. A. and Persson, E. 2018. "Astrobiology and Society in Europe Today: The White Paper on Societal Implications of Astrobiology Research in Europe," SpringerBriefs in Astronomy book series, Springer, Cham, available at https://europeanastrobiology.eu/onewebmedia/PREPRINT%202018%20Astrobiology%20and%20Society%20in%20Europe.pdf

Denning, Kathryn. 2009. "Social Evolution: State of the Field," in Dick and Lupisella (2009), pp. 63-124

Denning, Kathryn. 2013. "Impossible Predictions of the Unprecedented: Analogy, History, and the Work of Prognostication," in Vakoch (2013), pp. 301–312.

Denning, Kathryn. 2014. "Learning To Read: Interstellar Message Decipherment from Archaeological and Anthropological Perspectives," in Vakoch (2014): pp. 95-112.

Dick, Steven J. 1993. "The Search for Extraterrestrial Intelligence and the NASA High Resolution Microwave Survey (HRMS): Historical Perspectives," *Space Science Reviews*, 64 (1993), 93-139.

Dick, Steven J. 1996. *The Biological Universe: The Twentieth Century Extraterrestrial Life Debate and the Limits of Science*. Cambridge: Cambridge University Press, pp. 414-431.

Dick, Steven J. 2003. "Cultural Evolution, the Postbiological Universe, and SETI," *International Journal of Astrobiology*, 2, 65-74.

Dick, Steven J. and James E. Strick. 2004. *The Living Universe: NASA and the Development of Astrobiology*. New Brunswick, New Jersey and London: Rutgers University Press, pp. 131-154

Dick, Steven J. and Lupisella, M. L., editors. 2009. *Cosmos & Culture: Cultural Evolution in a Cosmic Context*. Washington, DC: NASA, online at http://history.nasa.gov/SP-4802.pdf

Dick, Steven J. 2015. *The Impact of Discovering Life Beyond Earth*. Cambridge: Cambridge University Press

Finney, Ben R. and Eric M. Jones, eds.1985. *Interstellar Migration and the Human Experience*. Berkeley: University of California Press, p. 333.

Finney, Ben R. 2000. "SETI, Consilience and the Unity of Knowledge," in *Bioastronomy '99: A New Era in Biosatronomy*, G. Lemarchand and K. Meech, eds. ASP Conference series, vol. 213. San Francisco: Astronomical Society of the Pacific, pp. 641-647.

Harrison, Billingham, Dick, Finney et al., 2000. "The Role of the Social Sciences in SETI," *When SETI Succeeds: The Impact of High-Information Contact*, Allen Tough, ed. Bellevue, Washington: Foundation for the Future, pp. 71-85.

Hart, Michael H. 1975. "An Explanation for the Absence of Extraterrestrials on Earth," *Quarterly Journal of the Royal Astronomical Society*, 16, 128-135

Hart, Michael H. and Ben Zuckerman. 1982. *Extraterrestrials: Where are They?* New York: Pergamon Press, second edition (Cambridge: Cambridge University Press, 1995).

Maruyama, Magoroh and Arthur Harkins, eds. 1975. *Cultures Beyond the Earth: The Role of Anthropology in Outer Space*. Vintage Books: New York, vii.

Meech, Karen J., J. V. Keane, Michael Mumma et al. 2009. *Bioastronomy 2007: Molecules, Microbes and Extraterrestrial Life*. San Francisco: Astronomical Society of the Pacific.

Mazlish, Bruce (ed). 1965. *The Railroad and the Space Program: An Exploration in Historical Analogy*. Cambridge, MA: MIT Press

Morrison, Philip, John Billingham and John Wolfe, eds., 1977. *The Search for Extraterrestrial Intelligence (SETI)*. Washington, D.C.: NASA SP 419, pp. 49-52.

Sagan, Carl, ed., 1973. *Communication with Extraterrestrial Intelligence (CETI)* Cambridge: MIT Press, pp. 85-111.

Shostak, Seth. 1995. *Progress in the Search for Extraterrestrial Life*, G. Seth Shostak, ed., ASP Conference series, vol. 74. San Francisco: Astronomical Society of the Pacific

Smith, Roger. 1997. *The Human Sciences*. New York: W.W. Norton & Company.

Traphagan, John W. 2014. "Anthropology at a Distance: SETI and the Production of Knowledge in the Encounter with an Extraterrestrial Other," in Vakoch (2014), pp. 131-142:132.

Traphagan, John W. 2015a. *Extraterrestrial Intelligence and Human Imagination: SETI at the Intersection of Science, Religion, and Culture*. Heidelberg: Springer.

Traphagan, John W. 2015b. "Equating culture, civilization and moral development in imagining extraterrestrial intelligence: anthropocentric assumptions?," in Dick, (2015), pp. 127-142.

Traphagan, John W. 2016. *Science, Culture and the Search for Life on Other Worlds*. Heidelberg: Springer.

Traphagan, John W. and Julian W. Traphagan. 2015. "SETI in Non-Western Perspective," in Dick (2015), pp. 299-307.

U.S. House of Representatives, 87th Congress, 1st session, March 24, 1961. *Proposed Studies on the Implications of Peaceful Space Activities for Human Affairs. Prepared for the National*

Aeronautics and Space Administration by the Brookings Institute. Report of the Committee on Science and Astronautics, Washington, D.C.: GPO, pp. 215-216.

Vakoch, Douglas A. 1998a. "Constructing Messages to Extraterrestrials: An Exosemiotic Perspective," *Acta Astronautica*, 42, 697-704

Vakoch, Douglas A. 1998b. "The Dialogic Model: Representing Human Diversity in Messages to Extraterrestrials," *Acta Astronautica*, 42, 705-712;

Vakoch, Douglas A. 1999. "The View from a Distant Star: Challenges of Interstellar Message Making," *Mercury*, 28 (March-April, 1999), 26-32

Vakoch, Douglas A. 2000. "The Conventionality of Pictorial Representation in Interstellar Messages," *Acta Astronautica*, 46, 733-736.

Vakoch, Douglas. 2009. "Anthropological Contributions to the Search for Extraterrestrial Intelligence," in Meech et al (2009), pp. 421-427.

Vakoch, Douglas, ed. 2013. *Astrobiology, History and Society: Life Beyond Earth and the Impact of Discovery*. Berlin-Heidelberg: Springer-Verlag.

Vakoch Douglas, ed., 2014. *Archaeology, Anthropology, and Interstellar Communication.* Washington, D.C: NASA.

Viewing, David. 1975. "Directly Interacting Extra-Terrestrial Technological Communities," *Journal of the British Interplanetary Society*, 28, 735-744.

Webb, Stephen. 2002. *Where is Everybody?: Fifty Solutions to the Fermi Paradox and the Problem of Extraterrestrial Life*. New York: Copernicus Books

Wescott, Roger W. 1975. Toward an Extraterrestrial Anthropology," in Maruyama (1975), 12-26: 13-14.

Wilson, E. O. 1998. *Consilience: The Unity of Knowledge.* New York: Alfred A. Knopf

Chapter 12
Bringing Culture to Cosmos: Cultural Evolution, the Postbiological Universe, and SETI

Abstract *The Biological Universe* (Dick, Cambridge University Press, Cambridge, 1996) analyzed the history of the extraterrestrial life debate, documenting how scientists have assessed the chances of life beyond Earth during the twentieth century. Here I propose another option—that we may in fact live in a postbiological universe, one that has evolved beyond flesh and blood intelligence to artificial intelligence (AI) and that is a product of cultural rather than biological evolution. Davies (Basic Books, New York: 51–55, 1995) and others have broached the subject, but the argument has not been given the attention it is due, nor has it been carried to its logical conclusion. This paper argues for the necessity of long-term thinking when contemplating the problem of intelligence in the universe. It provides arguments for a postbiological universe based on the likely age and lifetimes of technological civilizations and the overriding importance of cultural evolution as an element of cosmic evolution. Additionally, it describes the general nature of a postbiological universe and its implications for SETI.

12.1 The Necessity of Stapledonian Thinking

The possibility of a postbiological universe—one in which most intelligence has evolved beyond flesh and blood to AI—has not been considered in detail because humans are unaccustomed to thinking on cosmic time scales and following the logical consequences of cosmic time scales for biology and culture. The vast majority of humans think in terms of a human lifetime and the necessities for survival. Even historians span only the few thousand years of the rise and fall of civilizations, while anthropologists encompass the several million years of human origins, and geologists cover the 4.5-billion-year history of Earth. Only astronomers contemplate the 13.7-billion-year history of the cosmos, and the vast majority of them concentrate on the physical universe. Biologists—even paleobiologists and paleontologists—have never thought beyond the 3.8-billion-year history of life on Earth, and cultural

First published as "Cultural Evolution, the Postbiological Universe, and SETI," *International Journal of Astrobiology*, 2 (2003), 65–74.

S. J. Dick, *Space, Time, and Aliens*, https://doi.org/10.1007/978-3-030-41614-0_12

evolution has rarely been considered beyond the evolution of culture on Earth. Yet, if biology and culture exist beyond Earth, the one thing we know for certain is that they will evolve.

Only science fiction writers have thought in these longer terms, beginning most notably with H.G. Wells's evocative picture of a terrestrial society of Moorlocks and Eloi in *The Time Machine* (1895). In the twentieth century, the British philosopher Olaf Stapledon is the prime example of one who had a cosmic perspective on universal biological and cultural evolution, as played out in his novels *Last and First Men* (1930) and *Star Maker* (1937), and in some of his essays such as "Interplanetary Man?" (Stapledon 1948). We need therefore to think not only on astronomical time scales, but also on what I shall call Stapledonian time scales, by which I mean an astronomical time scale that takes into account the evolution of biology and culture. The foundation for the concept of a postbiological universe is the recognition of these time scales (Table 12.1), and the *necessity* for thinking in Stapledonian terms, no matter where it may lead. A primary methodological premise of this paper is that *long-term Stapledonian thinking is a necessity if we are to understand the nature of intelligence in the universe today.*

One small set of scientists that *has* thought on astronomical time scales about biology is SETI proponents. SETI enthusiasts, knowing the story of cosmic evolution, have often concluded that extraterrestrials must be older and wiser than us (Shklovskii and Sagan 1966; Oliver 1971; Drake 1976). But they have not used Stapledonian thinking to carry this possibility to its logical conclusion—that biological and cultural evolution will make extraterrestrial intelligence far different from us. Why they have not done so is understandable from an operational viewpoint: SETI proponents wish to search for intelligence using current technology, so they prefer the option that extraterrestrials will have technology similar to ours. That *is* an option, but only one of many and, possibly, not the most likely scenario.

By contrast, those who have no stake in standard SETI strategy have been more successful at adopting Stapledonian thinking. This is particularly true of proponents of the Fermi Paradox—formulated in 1950 even before radio searches were technologically feasible, elaborated in the 1970s and 1980s especially by Hart (1975) and Tipler (1985), and codified in a famous volume of essays (Hart and Zuckerman

Table 12.1 Time scales in human thought

Framework	Duration
Human	100
Historical	10,000 years
Anthropological	10 million years
Geological	5 billion years
Astronomical	14 billion years
Stapledonian	Biology and culture on astronomical scale

1982). If there are so many civilizations in the galaxy, given the time scales involved, Hart, Tipler, and their proponents ask, where are they? If extraterrestrials have acquired space travel, they should have colonized the galaxy in a few million years and should be here. They are not, therefore, they do not exist. Many solutions to the Fermi Paradox have been proposed over the last quarter century (Webb 2002). Suffice it to say that Tipler thought the rationale of the Fermi Paradox was strong enough that we should abandon all SETI programs. SETI proponents, among others, took strong exception to this claim. While Tipler's conclusion is not rigorous, it does embody the methodology of long-term thinking that needs to be applied to the problem of intelligence in the universe. The Fermi Paradox *does* need to be taken seriously.

Tipler's conclusion, however, is not the only possible outcome of long-term thinking about intelligence in the universe. In attempting to disprove extraterrestrials, Tipler argued that the galaxy would be colonized by self-reproducing automata—so-called von Neumann machines—with intelligence comparable to humans, but still under control of an intelligent flesh-and-blood species. Since he concluded extraterrestrials do not exist, for Tipler, machine intelligence also does not exist. But if there is a flaw in the logic of the Fermi Paradox and extraterrestrials *are* a natural outcome of cosmic evolution, then cultural evolution may have resulted in a postbiological universe in which machines are the predominant intelligence. This is more than mere conjecture; it is recognition of the fact that cultural evolution—the final frontier of the Drake Equation—needs to be taken into account no less than the astronomical and biological components of cosmic evolution (Chaisson 2001). Although the importance of cultural evolution was recognized very early on in the modern SETI discussions (Ascher and Ascher 1962), including some of its pioneering documents (Stull 1977), it has been essentially ignored over the last four decades.

The missing element in all past SETI arguments has therefore been a failure to account fully for the effects of cultural evolution. To some extent, cultural evolution is embodied in the "L" parameter of the Drake Equation, the lifetime of a technological civilization (Table 12.2 and see Chap. 7). But, especially if one is interested

Table 12.2 The Drake Equation

N =	$R_* \times f_p \times n_e$	×	$f_l \times f_i$	×	$f_c \times L$
	Astronomical		Biological		Cultural
N = The number of technological civilizations in the galaxy					
R_* = The rate of formation of stars suitable for the development of intelligent life					
f_p = The fraction of those stars with planetary systems					
n_e = The number of planets in each planetary system with an environment suitable for life					
f_l = The fraction of suitable planets on which life actually appears					
f_i = The fraction of life-bearing planets on which intelligent life emerges					
f_c = The fraction of planets with intelligent life that develop technological civilizations					
L = The lifetime of a technological civilization					

in more than just "N" (the number of technological civilizations in the galaxy), many other aspects of cultural evolution are critical to understanding the nature of extraterrestrial intelligence. Moreover, the prevalence of artificial intelligence may be critical to L. Another primary methodological premise of this paper, then, is that *cultural evolution must be seen as an integral part of cosmic evolution and the Drake Equation.* Following this premise, one solution to the Fermi Paradox is that we live in a postbiological universe, in which the psychology of biological beings no longer rules. While SETI proponents might rejoice in yet another solution to the Fermi Paradox, the postbiological universe has other important implications for SETI that must be taken into account in SETI strategies. But before addressing these implications, we must examine the likelihood that we indeed inhabit a postbiological universe.

12.2 Arguments for a Postbiological Universe

In setting forth arguments for a postbiological universe, it is important to define the term more precisely. It cannot mean a universe totally devoid of biological intelligence since we are an obvious counterexample. Nor does it mean a universe devoid of lower forms of life, what I have called elsewhere "the weak biological universe" (Dick 2000a), as advocated by Ward and Brownlee (2000). Rather, the postbiological universe is one in which the *majority* of intelligent life has evolved beyond flesh and blood intelligence, in proportion to its longevity, L.

SETI practitioners often state that ETI would be much older than terrestrial intelligence (TI), and that therefore SETI programs stand to inherit much knowledge and wisdom of the universe. However, they assume that ETI will just be some more advanced form of TI. This may be an excellent case of what Arthur C. Clarke calls "a failure of imagination" because it represents a failure to take into account cultural evolution. If civilizations are billions of years older than TI, or even millions of years older, our experience with the evolution of intelligence on Earth indicates that biological evolution would have carried such civilizations far beyond TI in terms of mental capacity. Moreover, as argued below, if civilizations are even thousands of years older than TI, cultural evolution would likely have also resulted in artificial mental capacities beyond TI, concluding in a postbiological universe. There are thus three scientific premises in the arguments for a postbiological universe (1) the maximum age (A) of ETI is several billion years; (2) the lifetime (L) of a technological civilization is >100 years and probably much larger; and (3) in the long term, cultural evolution supersedes biological evolution, and would have produced something far beyond biological intelligence. If that is the case, the chances of success for standard SETI programs may be greatly reduced, or at least altered, and our place in the universe may be quite different from anything envisioned except in science fiction. We approach each of these premises in turn.

12.2.1 The Maximum Age of Extraterrestrial Intelligence (A)

Cosmic evolution (Delsemme 1998; Chaisson 2001) is our guide to the maximum age (A) of an extraterrestrial civilization. Recent results from the Wilkinson Anisotropy Mapping Probe (WMAP) place the age of the universe at 13.7 billion years, with 1% uncertainty, and confirm the first stars forming at about 200 million years after the Big Bang (Bennett et al. 2003; Seife 2003). Although these first stars were very massive—from 300 to 1000 solar masses—and therefore short-lived, it is fair to assume that the oldest Sun-like stars formed within about one billion years, or about 12.5 billion years ago. By that time enough heavy element generation and interstellar seeding had taken place for the first rocky planets to form (Delsemme 1998, 71; Larson and Bromm 2001). Then, if Earth history is any guide, it may have taken another five billion years for intelligence to evolve. So, some six billion years after the Big Bang, one could have seen the emergence of the first intelligence. Accepting the WMAP age of the universe as 13.7 billion years, the first intelligence could have evolved seven and a half billion years ago. By the same reasoning, intelligence could have evolved in our galaxy four billion to five billion years ago, since the oldest stars in our galaxy formed about 10 billion to 11 billion years ago (Rees 1997).

These conclusions are essentially in line with those of a number of other astronomers. Using similar reasoning Norris (2000) argued that the median age of an extraterrestrial civilization is 1.7 billion years, assuming that civilizations born 5 billion years ago are now dying off because the 10 billion year lifetime of a solar type star has reached its end. (This assumption is perhaps pessimistic, given that a civilization more than a billion years old may well have found a way to escape its star system.) Based on the peak of the cosmic rate of carbon production in stars, Livio (1999a, b) concluded the first civilizations would emerge when the universe was about 10 billion years old, or 3.7 billion years ago assuming the WMAP age of the universe. Kardashev (1997) concluded that cosmological models yield an age for civilizations of six billion to eight billion years. Kardashev also pointed out that the youngest and less developed civilizations would be most distant from us, while the oldest and most developed civilizations would be nearest to us. Thus all lines of evidence converge on the conclusion that the maximum age of extraterrestrial intelligence would be billions of years, specifically, A ranges from 1.7 billion to 8 billion years. Even uncertainties of a billion years would not affect the argument for taking seriously cultural evolution.

12.2.2 The Lifetime of a Civilization (L)

But do civilizations really reach this age? Not necessarily. The maximum age A of ETI is mitigated by L, the lifetime of a technological civilization. We recall that the Drake Equation (see Chap. 7) consists of astronomical, biological, and cultural parameters, that L is the determining factor to the extent that N (the number of technological civilizations) approximates L, and that we know almost nothing about L. This is why values of L vary widely to the despair of many who are genuinely interested in the chances of detecting ETI. Sagan, Drake, and others generally assigned L values in the neighborhood of a million years, and even some pessimists admitted 10,000 years was not unlikely (Dick 1996, 441). Nevertheless, the only data point for L is ourselves, and if L is defined as a radio communicative technological civilization, all we may conclude from this datum is that L is at least 100 years. Beyond that single data point, L is a matter of whether one is optimistic or pessimistic about the survival of civilization. This is hardly an objective parameter even for a single individual; SETI pioneer Joseph Shklovskii, for one, became a pessimist at the end of his life, due in part to political events in the Soviet Union.

Difficulties notwithstanding, is there any more that can be said about L? What about an upper bound? One sometimes hears that civilizations are inherently unstable, that they have risen and fallen many times on Earth, and that therefore an upper bound for L is several thousand years. But what is really relevant is not the longevity of any single historical civilization on Earth, but that terrestrial civilization as a whole is still alive and well after five millennia of ups and downs known as "human history." It seems likely that technological civilization can last much longer, barring manmade catastrophes such as nuclear war and natural catastrophes such as mass extinctions. That a manmade catastrophe could totally wipe out civilization seems unduly pessimistic, despite the controversial results of nuclear winter scenarios (Turco et al. 1983). It seems likely that even in a nuclear world war, some corner of civilization would survive robustly enough that the slow climb of technological evolution would not have to start over again, much less recapitulate the even slower climb of cultural evolution from the cave, or the biological evolution of complex life.

Natural phenomena such as mass extinctions, supernovae, and gamma ray bursters are more problematic for civilization. Norris argued that the latter two events should extinguish all life on planets at intervals of about 200 million years, a conclusion at variance with what we observe on Earth (Norris 2000). A more refined study of gamma ray bursters (Scalo and Wheeler 2002) indicates events of potential biological significance, though not necessarily catastrophic, every ten million years or so. Current data indicates that a mass extinction from an impacting comet or asteroid serious enough to precipitate the collapse of civilization might occur every 300,000 years (Chapman and Morrison 1989; Raup 1992; Chapman and Morrison 1994). Mass extinctions similar to those that destroyed the dinosaurs, and would probably destroy *Homo sapiens*, have taken place on the order of tens of millions of years (Raup 1992; Becker 2002). Assuming that mass extinctions and other cosmic catastrophes could not be overcome, L would be between 100 years and tens of millions of years. If human ingenuity could overcome such natural catastrophes, or (in the case of mass extinctions) if human civilization has evolved far enough that even a small but technologically capable part of human civilization has been transported self-sufficiently to space, then L could conceivably approach A, which is billions of years. Surveying the vast range of possible catastrophes, Leslie (1996) has estimated that civilization has a 70% chance of lasting five more centuries, and believes that if it lasts that long, it could last millions of years.

Necessarily, none of this has the certainty of rigorous deduction. But the possibility of long lifetimes for technological civilizations leads us to explore the likely evolution and nature of such civilizations. It is clear that *biological* evolution, by definition, over the course of millions of years would produce nothing but more advanced biology. Consider what happened to the genus *Homo* in two million years of biological evolution on Earth. Where will we be in another two million years of biological evolution? And what would a billion-year-old terrestrial civilization be like? Possibly the minds of those comprising such a civilization would have evolved significantly beyond *Homo sapiens*. Possibly a similar process would take place for any extraterrestrial intelligence with serious implications for what we normally envision as the biological universe full of communicating civilizations. I say "possibly" because although *knowledge* surely would have increased in both cases, we know so little about the biological evolution of intelligence on Earth (Mithen 1996; Deacon 1997; Parker and McKinney 1999) that its future is unpredictable.

But the important point is that, even at our low current value of L on Earth, biological evolution by natural selection is already being overtaken by cultural evolution, which is proceeding at a vastly faster pace than biological evolution (Dennett 1996). Technological civilizations do not remain static; even the most conservative technological civilizations on Earth have not done so, and could not, given the dynamics of technology and society. Unlike all the other parameters in the Drake Equation, L is a problem of cultural evolution, and cultural evolution must be taken into account no less than astronomical and biological evolution. It must be treated as an integral part of cosmic evolution, in direct proportion to L, the age of the civilization. And unlike biological evolution, L need only be thousands of years for cultural evolution to have drastic effects on civilization.

12.2.3 Cultural Evolution

Because the nature of technological civilizations on time scales ranging from hundreds to billions of years reduces to a question of cultural evolution, we must turn to the social and behavioral sciences for insight. These disciplines have shown embryonic interest in the implications of successful SETI (Billingham et al. 1999; Harrison et al. 2000), but have yet to tackle the problem of cultural evolution in a cosmic context. This is hardly surprising; compared to astronomical and biological evolutions, our understanding of how culture evolves even on Earth is rudimentary. In the past, social scientists have posed two broad models of cultural evolution: the Spencerian, which views society as evolving "through well-defined stages, progressing from chaos to order, from simple to complex, from lower to higher"; and the Darwinian, which posits no particular direction, provides an explanatory framework rather than a historical generalization, and is evolutionary rather than revolutionary (Fellner 1990).

Most social scientists have judged the Spencerian model as too simplistic, but after a long lapse since Darwin's own ideas on cultural evolution detailed in *The Descent of Man* (Richerson and Boyd 2001), Darwinian models of cultural evolution have proliferated in recent decades and have been highly controversial. "Darwin's dangerous idea," as the philosopher Daniel Dennett calls it, posits that the same general evolutionary principles that apply to biology may also apply to culture, though with a mix of mechanisms including the Spencerian inheritance of acquired characteristics as well as those related to natural selection (Dennett 1996). The challenge is in the details of "Darwinizing culture," and elucidating how genes and culture may coevolve. Because the foundation and engine of cultural evolution are human psychology, behavior, cognition, and the transmission of ideas, they must serve as the basis for any theory, though they are notoriously difficult to characterize in individuals, much less in the aggregate.

Among the first modern Darwinian theories of human behavior was sociobiology (Wilson 1975), "the systematic study of the biological basis of all social behavior." Sociobiology has generated bitter disputes as a Darwinian extension from the realm of biology to that of culture (Segerstrale 2000). No less controversial have been related attempts (Lumsden and Wilson 1981; Wilson 1998) to use the idea of gene-culture coevolution to span the natural and social sciences. Cavalli-Sforza and Feldman (1981) pioneered a distinctive approach to gene-culture coevolution that makes use of population genetics. One of the more sophisticated Darwinian models of cultural evolution in this vein, termed the *dual inheritance theory* (Boyd and Richerson 1985), uses population genetics to construct simple mathematical models of how cultural evolution works. The authors recognize, however, that their system cannot yet make quantitative predictions, but can only clarify the relationships between cultural transmission and other Darwinian processes. A better known, if less rigorous, Darwinian model is Dennett's *Universal Darwinism*, wherein he argues that Darwinism applies to humans at many levels—mind, language, knowledge, and ethics (Dennett 1996). When applied to knowledge and its transmission, Dennett's brand of Universal Darwinism leads to the field of *memetics*, based on Dawkins's idea (1976) that culture evolves via memes in the same way that biology evolves with genes. Despite a number of books and a *Journal of Memetics*, even memetic enthusiasts realize the field is far from a real science (Aunger 2000).

All such Darwinian models of cultural evolution have considerable problems. Indeed, for historical reasons many social scientists still resist evolutionary hypotheses of culture altogether (Lalande and Brown 2002, 28). It is possible that some synthesis of sociobiology, gene-culture coevolution, and memetics, along with related Darwinian models like behavioral ecology and evolutionary psychology, will someday provide a widely accepted theory or mechanism for cultural evolution (Lalande and Brown 2002; Segerstrale 2000). It is also possible that the concept of "emergence" will play a role—that culture or its components (toolmaking, language, agriculture, technology, and so on) are emergent phenomena that will be explained in terms of agents, rules, and "pruning relations" in the way that the origin of life and the origin of consciousness may someday be explained as emergent phenomena (Morowitz 2002). But for now a widely accepted theory or mechanism of cultural evolution is lacking.

Still, theoretical and empirical studies of cultural evolution hold hope for a science of cultural evolution in the same way there is currently a well-developed science of biological evolution. In the context of extraterrestrial life, even a theory of universal biological evolution does not yet exist, much less a theory of universal cultural evolution. And even if a theory of cultural evolution existed, such models (short of Asimovian psychohistory) would lack the power to predict the future of our own culture, much less those of extraterrestrials. While galactic, stellar, and planetary evolution may be predicted to some extent based on physical principles, biological evolution cannot be predicted based on natural selection, and the prediction of our cultural evolution is not even contemplated except in the long-term context of the fate of the universe (Ward and Brownlee 2003). And while there is no lack of purely descriptive accounts of terrestrial cultural evolution, such descriptions also lack explanatory power or the predictive power needed to answer our question about the future of cultural evolution.

Lacking a robust theory of cultural evolution to at least guide our way, and "wild-card" events notwithstanding, we are reduced at present to the extrapolation of current trends supplemented by only the most general evolutionary concepts. Several fields are most relevant, including genetic engineering, biotechnology, nanotechnology, and space travel. But one field—artificial intelligence—may dominate all other developments in the sense that other fields can be seen as subservient to intelligence. Biotechnology is a step on the road to AI, nanotechnology will help construct efficient AI and fulfill its goals, and space travel will spread AI. Genetic engineering may eventually provide another pathway toward increased intelligence, but it is limited by the structure of the human brain. In sorting out priorities, I adopt what I term the central principle of cultural evolution, which I will refer to as the *Intelligence Principle*:

The maintenance, improvement and perpetuation of knowledge and intelligence is the central driving force of cultural evolution, and that to the extent intelligence can be improved, it will be improved.

At the level of knowledge, we see this principle in daily operation as individuals, groups, and societies attempt to maximize their knowledge in order to gain advantage in the world around them, an endeavor in which some succeed better than others. Better education, better information, and better technology are generally perceived as advantageous to the individual, group, or society—an understanding recognized in the aphorism "knowledge is power." At the species level, which is the meaning I primarily refer to here, intelligence is related to the size and structure of the brain of *Homo sapiens sapiens*, a capacity that has not changed in 100,000 years, and that led to the "big bang of human culture 60,000–30,000 years ago" (Mithen 1996). In hominid biological evolution the increased brain size and intelligence of *Homo sapiens sapiens* allowed it to outcompete other hominid species and dominate the planet. In the cultural evolution of the species, the same will hold true. Failure to improve intelligence, resulting in inferior knowledge, may eventually cause cultural evolution to cease to exist in the presence of competing forces like

AI. In Darwinian terms, knowledge has survival value, or selective advantage, as does intelligence at the species level, a fact that may someday be elucidated by an evolutionary theory of social behavior, whether "group selection" as recently applied to religion (Wilson 2002), selfish gene theory, evolutionary epistemology (Bradie 1986), or some other Darwinian model. The Intelligence Principle implies that, given the opportunity to increase intelligence (and thereby knowledge), whether through biotechnology, genetic engineering, or AI, any society would do so, or fail to do so at its own peril.

The Intelligence Principle is a hybrid between the Spencerian and Darwinian models of cultural evolution in the sense that it does not have well-defined stages, but is evolutionary and implies a direction toward greater intelligence. Because it is governed by mind, the process is goal-oriented. Culture may have many driving forces, but none can be so fundamental, or so strong, as intelligence itself.

Turning, then, to the field of AI as a striking example of the Intelligence Principle of cultural evolution, we find quite astounding predictions. As Dyson (1997, 25) has pointed out, ever since the Industrial Revolution, there has been concern about the rise of the machines and their relation to humans. Butler (1863) wrote:

> [w]e find ourselves almost awestruck at the vast development of the mechanical world, at the gigantic strides with which it has advanced in comparison with the slow progress of the animal and vegetable kingdom. We shall find it impossible to refrain from asking ourselves what the end of this mighty movement is to be .... The machines are gaining ground upon us; day by day we are becoming more subservient to them; more men are daily bound down as slaves to tend them; more men are daily devoting the energies of their whole lives to the development of mechanical life.

After a century of progress in machine development and the increasing convergence between machine and life that Dyson describes, MacGowan and Ordway III (1966) argued that,

> [a]ny emerging intelligent biological society which engages in the development of highly intelligent automata must resign itself to being completely dominated and controlled by automata. The only means of preventing domination by intelligent artificial automata would be to make them distinctly subnormal in intellectual capacity, when compared with the biological society, and to destroy them or clear their memories at regular intervals.

The possibilities of AI played a substantial role in MacGowan and Ordway's volume on extraterrestrial intelligence, but those possibilities were completely overshadowed by the publication of Shklovskii and Sagan (1966) in the same year. Although the last chapter of Shklovskii and Sagan's volume was on "Artificial Intelligence and Galactic Civilizations," the AI thesis was very general and lost in the midst of the exciting—and at the time more verifiable and realistic—implications of the other chapters, which assumed biological beings. Over the last 40 years, SETI has focused almost exclusively on the biological paradigm, especially the radio SETI technique, as opposed to a postbiological paradigm (MacGowan and Ordway III 1966, 265; Shklovskii and Sagan 1966, 281–288).

The study of AI was rudimentary in 1966, but MacGowan and Ordway's idea as applied to humans has been broached in subsequent years as the field of AI developed. One of the most forward-thinking scholars in the field is Hans Moravec, a pioneer in AI and robotics at Carnegie-Mellon. Already in 1988 in his book *Mind Children: The Future of Robot and Human Intelligence*, Moravec predicted that "[w]hat awaits is not oblivion but rather a future which, from our present vantage point, is best described by the words 'postbiological' or even 'supernatural.' It is a world in which the human race has been swept away by the tide of cultural change, usurped by its own artificial progeny." Within the next century, he predicted, our machines "will mature into entities as complex as ourselves, and eventually into something transcending everything we know—in whom we can take pride when they refer to themselves as our descendants. Unleashed from the plodding pace of biological evolution, the children of our minds will be free to grow to confront immense and fundamental challenges in the larger universe." (Moravec 1988, 1, 1999). Just as there may have been a genetic takeover when RNA or DNA took over from some more primitive system like clay, Moravec foresees a robotic takeover. This assumes the strong AI position that it is possible to construct intelligent machines functionally equivalent to human intelligence, a point of considerable contention (Searle 1980; Tipler 1994, ch. 2). It seems reasonable to assume, however, that the strong AI position will prove increasingly true in direct proportion to the time available for further developments in the field—time that extraterrestrial civilizations, if any, will have already had.

Another thinker who came to a similar conclusion in the terrestrial context is inventor Ray Kurzweil, a pioneer in AI who has been critical in bringing voice-recognition machines to the commercial market. In *The Age of Spiritual Machines: When Computers Exceed Human Intelligence*, Kurzweil (1999) also adopting the strong AI claim, sees the takeover of biological intelligence by AI, not by hostility, but by willing humans who have their brains scanned, uploaded to a computer, and live their lives as software running on machines. In his view, human intelligence will be left behind. Physicist Frank Tipler, well known for his work on the anthropic principle and the Fermi Paradox, has also weighed in on this subject. After a review of the arguments for and against strong AI, Tipler (1994) concluded that "the evidence is overwhelming that in about 30-odd years we should be able to make a machine which is as intelligent as a human being, or more so." Tipler does not necessarily foresee a takeover, but believes that such machines will enhance our well-being. And he ties these ideas to the resurrection of the dead and an entire cosmotheology.

It may well be that Moravec, Kurzweil, and their proponents underestimate the moral and ethical brakes on technological inertia; after all, the abortion controversy in the United States pales in significance with the replacement of the species. And Fukuyama (2002) argues strenuously against a possible "posthuman future" that he sees stemming from advances in the brain sciences, neuropharmacology and behavior control, and the prolongation of life and genetic engineering. He argues for the regulation of biotechnology to preserve human nature, and biotechnology is relatively tame compared to the possibilities of AI. But such objections fail to take into account cultural evolution, and may lose their impact over the longer term, as the Intelligence Principle asserts itself. If we consider cultural evolution over the last millennium, especially as regards science and technology, who would have predicted space travel, genetic engineering, and nanotechnology? No one could have, because the foundational concepts were not in place. This might lead us to conclude that in another millennium there will be important concepts that we have no inkling of now. This is undoubtedly true. But barring a landmark transformation in human thought comparable to the origins of western science over the next thousand years, we are set on a course that will still be playing out in 3001, with AI still a predominating factor. When one considers the accelerating pace of cultural evolution as we enter the third millennium of our era, radical change of the sort foreseen by Moravec and Kurzweil does not seem so farfetched. Just as Thomas Aquinas had a failure of imagination almost a millennium ago, so do we.

We thus come to a startling conclusion. Based on what experts see happening on Earth, L need not be five billion, one billion, or a few million years. It is possible that a postbiological universe would occur if L exceeds a few hundred or a few thousand years, where L is defined as a technological civilization that has entered the electronic computer age, which on Earth was almost simultaneous with the usual definition of L as a radio communicative civilization. If L is less than a few hundred years, less than the time it takes for a technological civilization to conceive, design, construct, and launch their intelligent machines, we do not live in a postbiological universe. If L is between 100 and 1000 years, a transition zone may result populated by human/machine symbiosis, sometimes referred to as "cyborgs" (Dyson 1997; Ward and Rockman 2001; Gray 2002), and genetically engineered humans. But if L is greater than 1000 years, we almost certainly will have made that transition to a postbiological universe (Table 12.3). "Interstellar humanity" (Dick 2000b) remains valid if we expand our definition of "humanity" to our artificial progeny, Moravec's "mind children." As for the present, on the time scales of the universe, this means that we are in the minority; the universe over the billions of years that intelligence has had to develop will not be a biological universe, but a postbiological universe. Biologically based technological civilization as defined above is a fleeting phenomenon limited to a few thousand years, and exists in the universe in the proportion of one thousand to one billion, so that only one in a million civilizations are biological. Such are the results of applying the Intelligence Principle, and the insights of Moravec, Kurzweil, and Tipler among others, to the entire universe using Stapledonian thinking.

Table 12.3 Lifetime of a technological civilization and effects on SETI

L (years)	Stage of cultural evolution	Effect on SETI
<100	Biological	Civilizations scarce but comparable level—EM SETI possible
100–1000	Machine/biology hybrid (cyborg)	Hybrid techniques
>1000	Postbiological	Advanced artificial intelligence—direct EM SETI unlikely

12.3 The Nature of the Postbiological Universe and Its Implications for SETI

What would a postbiological universe be like? What is artificial intelligence *doing* out there? And what does it mean for SETI? Speaking of Earth, Moravec believed that "A postbiological world dominated by self-improving, thinking machines would be as different from our world of living things as this world is different from the lifeless chemistry that preceded it. A population consisting of unfettered mind children is quite unimaginable" (Moravec 1988, 5). Even more unimaginable, then, would be the activities of artificial intelligence in the universe. But, in the tradition of Stapledon, and guided by the Intelligence Principle, let us try.

Although one cannot, and need not, specify morphological details of postbiologicals, we can assess with some confidence their general characteristics. Complex intelligent postbiologicals—which we can assume over the time intervals dealt with here—would have the capability of repair and update, capabilities facilitated by their modularity. The so-called von Neumann machine is able to reproduce better versions of itself. Part of this reproduction is the improvement of intelligence; unlike humans this intelligence is cumulative in the sense that the sum total of knowledge in the parent machine is passed on to the next generation, conferring effective immortality for the machine's most important characteristic. The immortality of postbiologicals is enhanced by their increased tolerance to their environment, whether it be vacuum, temperature, radiation, or acceleration (MacGowan and Ordway III 1966).

Immortal postbiologicals would embody the capacity for great good or evil over a domain that dwarfs biological domains of influence. There are admittedly deep questions of the nature of "good," "evil," and "morality" in the context of artificial intelligence in the universe (Ruse 1985). But if the Intelligence Principle holds, postbiologicals are driven by the improvement of knowledge and intelligence. How they would use these qualities presumably remains a value question no less than for humans. One notable interpretation from science fiction is Asimov's robot series, where select robots traverse the galaxy trying to influence events in a positive way, subject to the famous Laws of Robotics. But another interpretation is that AI could be motivated by darker purposes, whether through the programming of its parent biologicals or through its own evolution. Saberhagen evokes this scenario in his Berserker series, where Berserkers are not quite AI, but are near-sentient death machines programmed for their prime directive to seek out and destroy life wherever it may hide. As Brin has pointed out, such deadly probes, whether intelligent or not, are an eerie solution to everything we observe, including "the Great Silence" as so far determined by all SETI programs (Brin 1983).

It is notable that Asimov's robots are human descendants, since his universe has no extraterrestrials, and that his robots are still to some extent controlled by humans according to the second law, and can allow no harm to come to humanity according to the zeroth law. It is also notable that in Arthur C. Clarke's universe, which is full of extraterrestrial intelligence, artificial intelligence plays very little role—with the exception in *2001: A Space Odyssey* of HAL, a disastrous postbiological that violated Asimov's three laws by harming humans. It would seem that Clarke may have had a failure of imagination when it comes to the potential role of AI in the universe, or that he saw AI as a passing part of evolution: in his earlier novel *The City and the Stars* (1956), humans teamed with other galactic civilizations to build a disembodied intelligence, a pure mentality that would seem to be beyond the stage of AI.

This raises a valid point: on the principle that nothing in the universe remains static, postbiologicals would continue to be subject to cultural evolution. AI may not be the ultimate emergence of cultural evolution, and Morowitz (2002) has suggested that "spirit" could be an emergent phenomenon beyond AI. Where cultural evolution would ultimately lead one cannot say, except that ultimate entities might have characteristics approaching those we ascribe to deities: omniscience, omnipotence, and perhaps the capability of communication through messenger probes. Stapledon himself has envisioned such a being in *Star Maker*, although not a product of cultural evolution via artificial intelligence. Thus, our reflections on postbiologicals lead to a possibility that some might characterize as cosmotheology (Dick 2000c).

Given the characteristics of immortality, increased tolerance to their environment, capacity for action on a large scale, and an intelligence far superior to our own, what are the implications of the postbiological universe for SETI? First, there is the problem of search space. Environmental tolerance and availability of resources beyond the planetary realm means that SETI searches for postbiologicals need not be confined to planets around Sun-like stars, nor to planets at all (Shostak 1998, 201; Tough 2002). Indeed postbiologicals probably would "prefer" not to be so confined. Artificial intelligence, or their robotic surrogates, could roam the galaxy as reproducing von Neumann machines (Tipler 1985), Bracewell probes (Bracewell 1975), or smart microprobes (Tough 1998). Roaming intelligent probes might also lead to an AI version of the Fermi Paradox, but with novel possibilities for solution, since postbiological "psychology" may be very different from the psychology of biologicals.

Secondly, there is the question of the nature of the signal. Postbiologicals could be communicating with each other via electromagnetic signals, but the Intelligence Principle tending toward the increase of knowledge and intelligence renders it unlikely they would wish to communicate in such a way with embryonic biologicals like humans. Shklovskii and Sagan pointed out that the long lifetimes of artificial intelligence "could be very advantageous for interstellar contact among advanced communities. The sluggishness of two-way radio communication over interstellar distances tends to make such contact unsatisfactory for beings with lifetimes measured in decades. But for very long-lived beings, such communication would be much more interesting" (Shklovskii and Sagan 1966, 487). What Shklovskii and Sagan left unsaid was that this means that short-lived biologicals such as ourselves might be reduced to intercepting communications of postbiologicals; attempts to do this might lead to a new sense of what the "magic frequencies" are. Intercepting such signals at interstellar distances would undoubtedly be more difficult than detecting a signal directed at us. But if one of the activities of postbiologicals is to study emerging biologicals, as terrestrial anthropologists study our own roots, they may be closer than we think. Indeed, as the products of technology, the Intelligence Principle of cultural evolution implies that, even if they did not wish to communicate with us, postbiologicals would incessantly attempt to increase their knowledge of emerging cultures and their perhaps unique pathways in the development of science, technology, and mathematics.

Thirdly, the Intelligence Principle leads us to conclude that postbiologicals might be more interested in receiving signals from biologicals than in sending them. This conclusion should lead us to place new emphasis on message construction, to explore the implications for message construction if the intended recipients are AI, including the optimal mode of representation to be used with postbiologicals in contrast to biologicals. In addition to increasing their knowledge of the physical and biological universe, would postbiologicals also be interested in spiritual principles, altruism, and the arts, as some have recently proposed for extraterrestrial biologicals? (Vakoch 1998, 1999; Ringwald 2001). This is tantamount to asking if postbiologicals would be interested in cultural evolution; as products of cultural evolution themselves, this seems highly likely, and with this conclusion cultural evolution comes full circle in a cosmic context.

Finally, the vast disparity in age between postbiologicals and biologicals highlights what has been called the *Incommensurability Problem*. It is entirely possible that the differences between our minds and theirs are so great that communication is impossible.

With a better understanding of the role of cultural evolution in cosmic evolution, it seems clear that the L parameter is a double-edged sword for SETI. If L is large, extraterrestrials may have evolved through biological or cultural evolution, beyond human understanding. If L is small, the chances of communication increase because our mental capacities might be more comparable, but N becomes much smaller, and the chances of finding any scarce civilizations are much smaller. Here, in the Siren call of SETI, we are caught between Scylla and Charybdis.

All of these conclusions, and the possibility of a postbiological universe in general, point to the need to place AI research in a cosmic context. AI and SETI, after all, have much in common with their interest in the nature of intelligence. And although the difficult problem of the definition of intelligence is beyond the scope of this article, the relation of biological and postbiological intelligence gains greater urgency with the prospect that cultural evolution may have already produced artificial intelligence throughout the universe. With the symbiosis of SETI and AI, SETI expands its possibilities into new phase space, and the study of the long-term future of AI becomes more than idle speculation.

12.4 Summary and Conclusions

We have applied two methodological principles in this paper: (1) long-term Stapledonian thinking is a necessity if we are to understand the nature of intelligence in the universe today, and (2) cultural evolution must be seen as an integral part of cosmic evolution and the Drake Equation. We have accepted the strong AI theory that it is possible to construct artificial intelligence equivalent to, or superior to, humans, and adopted the Intelligence Principle that the improvement and perpetuation of intelligence is a central driving force of cultural evolution. Applying these principles to the universe, we have argued that if the lifetime of technological civilizations typically exceed 1000 years, it is likely that we live in a postbiological universe. The argument makes no more, and no fewer, assumptions about the probability of the evolution of intelligence, or its abundance, than standard SETI scenarios; it argues only that if such intelligence does arise, cultural evolution must be taken into account, and that this may result in a postbiological universe. As a byproduct of the discussion, we point out that even if we live in a biological universe, the extraterrestrials that compose the biological universe would be millions, if not billions, of years older than us.

Whether biologicals or postbiologicals, we conclude that the implications for SETI strategies are profound. Biologicals that are part of a civilization millions or billions of years old may or may not still be using electromagnetic technology for SETI, calling for new strategies (Tough 2000). Postbiologicals would not be confined to planetary surfaces, they might be more likely to roam the universe than to send signals, they might be using electromagnetic technology for communication among themselves rather than with others, and they would be more likely to receive than to send messages. Lacking a theory of cultural evolution on Earth, we are unable to predict the cultural evolution even of our own species in the near future. Lacking knowledge of advanced biological or postbiological motivations, we are unable to predict the nature of civilizations millions or billions of years older than ours. Still, the likelihood of Darwinian-type mechanisms at work in cultural evolution throughout the universe forces us to consider the real possibility—perhaps amounting to probability—of a postbiological universe, and calls for a sweeping reconsideration of SETI assumptions and strategies.

12.5 Commentary 2020

I first presented this concept as a poster paper at the second Astrobiology Science Conference (AbSciCon) April 7–11, 2002, held in the now-defunct dirigible hangar at NASA Ames Research Center in Mountain View, California (see Dick and Strick 2004, pp. 222–223 for context). The meeting was attended by about 700 people, an indication of the burgeoning interest in astrobiology in the wake of NASA funding. I broached the subject again at a plenary lecture of the American Astronomical Society on January 7, 2003 in Seattle, just as the full-blown paper was appearing in the *International Journal of Astrobiology* in January 2003 (Dick 2003a). I subsequently gave a paper on the subject at many venues: in Trieste, Italy in September, 2003; at the World Future Society in February 2004; for the Billingham Cutting-Edge lecture at the International Astronautical Conference in Valencia, Spain on October 3, 2006; and at the American Anthropological Association in San Jose, CA in November, 2006. The idea was picked up in quite a few venues, including a popular version I wrote for *Mercury* magazine, a cover story with the provocative title "They Aren't Who You Think" (Dick 2003b) and in the *New Scientist* for May 31, 2008. The original article as published here was reprinted in several places including the NASA volume *Cosmos and Culture* (Dick and Lupisella 2009) under the title "Bringing Culture to Cosmos: The Postbiological Universe," and another version in the journal *Futures* in 2009 (Dick 2009).

The idea has picked up many adherents, including the former Astronomer Royal Sir Martin Rees (2015), the philosopher Susan Schneider (2015), and the futurist Allen Tough, who discussed some of the implications of a postbiological universe for SETI in the wake of one of my early presentations on the subject (Tough 2002). And the general idea of postbiological superintelligence that may arise on Earth is supported in Nick Bostrom's book *Superintelligence* (2014), which is also relevant to the question of what postbiologicals may be doing throughout the universe. I continue to believe the arguments presented in this chapter are compelling and deserve further research and elaboration.

References

Ascher, R. & Ascher, M. 1962. "Interstellar communication and human evolution," *Nature*, 193, 940.

Aunger, R., ed. 2000. *Darwinizing Culture: The Status of Memetics as a Science*. Oxford: Oxford University Press.

Becker, L. 2002. "Repeated blows," *Scientific American* 286 (March): 76–83.

Bennett, C. L. et al. 2003. "First year Wilkinson Microwave Anisotropy Probe (WMAP) Observations: Preliminary maps and basic results," *Astrophysical Journal Supplement Series*, 148, 1–27.

Billingham, J. et al. 1999. *Social Implications of the Detection of an Extraterrestrial Civilization*. Mountain View, CA: SETI Press.

Bostrom, Nick. 2014. *Superintelligence: Paths, Dangers, Strategies*. Oxford: Oxford University Press.

Boyd, R. and P. J. Richerson. 1985. *Culture and the Evolutionary Process*. Chicago: University of Chicago Press.

Bracewell, R. 1975. *The Galactic Club*. San Francisco: W. H. Freeman.

Bradie, M. 1986. "Assessing evolutionary epistemology," *Biology and Philosophy* 1:401–460.

Brin, G. D. 1983. "The 'Great Silence': The controversy concerning extraterrestrial intelligent life," *Quarterly Journal of the Royal Astronomical Society* 24: 283–309.

Butler, S. 1863. *Canterbury Press* (June 13).

Cavalli-Sforza, L. L. & M. W. Feldman. 1981. *Cultural Transmission and Evolution: A Quantitative Approach*. Princeton, NJ: Princeton University Press.

Chaisson, E. 2001. *Cosmic Evolution: The Rise of Complexity in Nature*. Cambridge, MA: Harvard University Press.

Chapman, C. R. and D. Morrison. 1989. *Cosmic Catastrophes*. New York: Plenum Press.

Chapman, C. R. and D. Morrison. 1994. "Impacts on the Earth by asteroids and comets: Assessing the hazard," *Nature* 367:33–34.

Davies, P. 1995. *Are We Alone? Philosophical Implications of the Discovery of Extraterrestrial Life*. New York: Basic Books, pp. 51–55.

Dawkins, R. 1976. *The Selfish Gene*. Oxford: Oxford University Press.
Deacon, T. 1997. *The Symbolic Species: The Co-evolution of Language and the Brain*. New York: W. W. Norton.
Delsemme, A. 1998. *Our Cosmic Origins*. New York and Cambridge: Cambridge University Press.
Dennett, D. 1996. *Darwin's Dangerous Idea*. New York: Simon and Schuster.
Dick, S. J. 1996. The Biological Universe: The Twentieth Century Extraterrestrial Life Debate and the Limits of Science. Cambridge: Cambridge University Press.
Dick, S. J. 2000a. "Extraterrestrial life and our world view at the turn of the millennium," Dibner Library Lecture, Smithsonian Institution Libraries, Washington, DC.
Dick, S. J. 2000b. "Interstellar humanity," *Futures* 32:555–567, reprinted as "Cosmic Humanity" in Tough (2000) pp. 93–101.
Dick, S. J., ed. (2000c). "Cosmotheology: Theological implications of the new universe," in "*Many Worlds: The New Universe, Extraterrestrial Life and the Theological Implications*," Philadelphia: Templeton Press.
Dick, Steven J. 2003a. Cultural evolution, the postbiological universe and SETI, *International Journal of Astrobiology,* 2: 65–74.
Dick, S. J. 2003b. "They Aren't Who You Think," *Mercury*, November/December, 18–26.
Dick, S. J. 2009. "The Postbiological Universe and our Future in Space," *Futures*, 41, 578–580.
Dick, S. J. and M. L. Lupisella, eds. 2009. *Cosmos & Culture: Cultural Evolution in a Cosmic Context*. Washington, DC: NASA, online at http://history.nasa.gov/SP-4802.pdf
Dick, S. J. and J. E. Strick 2004. *The Living Universe: NASA and the Development of Astrobiology*. New Brunswick: NJ Rutgers University Press.
Drake, F. 1976. "On hands and knees in search of Elysium," *Technology Review* 78:22–29.
Dyson, G. 1997. *Darwin Among the Machines: The Evolution of Global Intelligence*. Cambridge, MA: Perseus Books.
Fellner, R. 1990. "The problems and prospects of cultural evolution," *Papers from the Institute of Archaeology* [London] 1:45–55.
Fukuyama, Francis. 2002. *Our Posthuman Future: Consequences of the Biotechnology Revolution*. New York: Farrar, Straus and Giroux.
Gray, C. 2002. *Cyborg Citizen: Politics in the Posthuman Age*. Routledge.
Harrison, A. A., J. Billingham, et al. 2000. "The role of the social sciences in SETI," in *When SETI Succeeds*, ed. Tough, A., pp. 71–85.
Hart, M. 1975. "An explanation for the absence of extraterrestrials on Earth," *Quarterly Journal of the Royal Astronomical Society* 16:128–135.
Hart, M. and B. Zuckerman, eds. 1982. *Where Are They?* New York: Pergamon; 2nd edition, Zuckerman and Hart, Cambridge University Press, 1995.
Kardashev, N. S. 1997. "Cosmology and civilizations," *Astrophysics and Space Science* 252:25–40.
Kurzweil, R. 1999. *The Age of Spiritual Machines: When Computers Exceed Human Intelligence*. New York: Penguin Books.
Lalande, K. N. and G. R. Brown. 2002. *Sense & Nonsense: Evolutionary Perspectives on Human Behaviour.* Oxford: Oxford University Press.
Larson, R. B. and V. Bromm, 2001. "The first stars in the universe," *Scientific American* 285:64–71.
Leslie, J. 1996. *The End of the World: The Science and Ethics of Human Extinction*. London and New York: Routledge.
Livio, M. 1999a. "How rare are extraterrestrial civilizations and when did they emerge?" *Astrophysical Journal* 511:429–431.
Livio, M. 1999b. "How rare are extraterrestrial civilizations and when did they emerge?" *Mercury* 28, no. 2: 10–13.
Lumsden, C. J. and E. O. Wilson. 1981. *Genes, Mind and Culture: The Coevolutionary Process*. Cambridge, MA: Harvard University Press.
MacGowan, R. and F. I. Ordway, III. 1966. *Intelligence in the Universe*. Englewood Cliffs, NJ: Prentice-Hall.
Mithen, S. 1996. *The Prehistory of the Mind: The Cognitive Origins of Art, Religion and Science*. London: Thames and Hudson.

Moravec, H. 1988. *Mind Children: The Future of Robot and Human Intelligence*. Cambridge, MA: Harvard University Press.

Moravec, H. 1999. *Robot: Mere Machine to Transcendent Mind*. Oxford.

Morowitz, H. 2002. *The Emergence of Everything: How the World Became Complex*. Oxford: Oxford University Press.

Norris, R. P. 2000. "How old is ET?," in *When SETI Succeeds: The Impact of High-Information Contact*, ed. Tough, A. Bellevue, Washington, 2000, pp. 103–105.

Oliver, B. 1971. *Project Cyclops: A Design Study of a System for Detecting Extraterrestrial Intelligent Life*, NASA Ames, Moffett Field, California, pp. 27, 60.

Parker, S. and McKinney, M. 1999. *Origins of Intelligence: The Evolution of Cognitive Development in Monkeys, Apes, and Humans*. London: Johns Hopkins University Press.

Raup, D. M. 1992. *Extinction: Bad Genes or Bad Luck*. New York: W. W. Norton.

Rees, M. 1997. *Before the Beginning: Our Universe and Others*. Reading, MA: Addison-Wesley.

Rees, M. 2015. "Why Alien Life Will be Robotic," Nautilus (October 22, 2015).

Richerson, P. J. and R. Boyd. 2001. "Build for speed, not for comfort: Darwinian theory and human culture," *History and Philosophy of the Life Sciences* 23:423–463. Special Issue on Darwinian Evolution Across the Disciplines.

Ringwald, C. D. 2001. "Encoding altruism," *Science and Spirit* (September–October, 2001).

Ruse, M. 1985. "Is rape wrong on Andromeda?" in E. Regis, ed., *Extraterrestrials: Science and Alien Intelligence*. Cambridge: Cambridge University Press, pp. 43–78.

Scalo, J. and J. C. Wheeler. 2002. "Astrophysical and astrobiological implications of gamma-ray burst properties," *Astrophysical Journal*, 566:723–787.

Schneider, S. 2015. "Alien Minds," in Dick (2015), pp. 189–206.

Searle, J. R. 1980. "Minds, brains, and programs," *Behavioral and Brain Sciences* 3, no. 3: 417–457.

Segerstrale, U. 2000. *Defenders of the Truth: The Battle for Science in the Sociobiology Debate and Beyond*. Oxford: Oxford University Press.

Seife, C. 2003. "MAP Glimpses Universe's Rambunctious Childhood," *Science* 299 (14 February): 991, 993.

Shklovskii, J. & C. Sagan, 1966. *Intelligent Life in the Universe*. San Francisco: Holden-Day, pp. 360–361.

Shostak, S. 1998. *Sharing the Universe: Perspectives on Extraterrestrial Life*. Berkeley, CA, Berkeley Hills Book, pp. 103–109.

Stapledon, O. 1948. "Interplanetary Man?" in *An Olaf Stapledon Reader*, Robert Crossley, ed., (Syracuse, NY, 1997), pp. 218–241.

Stull, M. 1977. "Cultural evolution," in P. Morrison, J. Billingham, and J. Wolfe, eds., *The Search For Extraterrestrial Intelligence*, NASA, Washington, DC, pp. 47–52. Based on a Workshop on Cultural Evolution chaired by Joshua Lederberg, 24–25 November 1975.

Tipler, F. 1985. Extraterrestrial intelligent beings do not exist, in *Extraterrestrials: Science and Alien Intelligence*, ed. Edward Regis (Cambridge, 1985), pp. 133–150.

Tipler, F. 1994. *The Physics of Immortality*. New York.

Tough, A. 1998. "Small smart interstellar probes," *Journal of the British Interplanetary Society* 51:167–174.

Tough, A. 2000. "How to achieve contact: Five promising strategies," in *When SETI Succeeds*, Allen Tough, ed., pp. 115–125.

Tough, A. 2002. "Post-biological implications for SETI: A response to the SETICon02 keynote paper [by S. Dick]," *Proceedings of SETICon02: The Second SETI League Technical Symposium*. America Radio Relay League, New Jersey, pp. 11–13.

Turco, R. P., Owen B. Toon, Thomas P. Ackerman, James B. Pollack, and Carl Sagan. 1983. "Nuclear inter: Global consequences of multiple nuclear explosions," *Science* 222:1283–1292 (23 December 1983).

Vakoch, D. A. 1998. "Constructing messages to extraterrestrials: An exosemiotic perspective," *Acta Astronautica* 42:697–704.

Vakoch, D. A. 1999. "The view from a distant star: Challenges of interstellar message-making," *Mercury* (March–April 1999): 26–32.

Ward, P. and D. Brownlee. 2000. *Rare Earth : Why Complex Life is Uncommon in the Universe*. New York: Copernicus.

Ward, P. and D. Brownlee. 2003. *The Life and Death of Planet Earth: How the New Science of Astrobiology Charts the Ultimate Fate of our World*. New York: Henry Holt.

Ward, P. and A. Rockman. 2001. *Future Evolution: An Illuminated History of Life to Come*. New York: Henry Holt & Co.

Webb, Stephen. 2002. *If the Universe Is Teeming with Aliens ... Where Is Everybody?: Fifty Solutions to the Fermi Paradox and the Problem of Extraterrestrial Life*. New York: Copernicus Books and Praxis Publishing.

Wilson, D. S. 2002. *Darwin's Cathedral: Evolution, Religion and the Nature of Society*. Chicago: University of Chicago Press.

Wilson, E. O. 1975. *Sociobiology: The New Synthesis*. Cambridge, MA: Harvard University Press.

Wilson, E. O. 1998. *Consilience: The Unity of Knowledge*. New York: Alfred A. Knopf.

Chapter 13
Toward a Constructive Naturalistic Cosmotheology

Abstract Cosmotheology is a theology that takes into account what we know about the universe based on science. It is therefore a naturalistic theology in the tradition of religious naturalism. This chapter takes as its foundational assumption the concept that the supernatural does not exist. Following this concept, we present six principles of cosmotheology, including the idea that we are not physically, biologically, cognitively, or morally central in the universe; that any concept of God must be grounded in naturalistic cosmic evolution; that it must have an expansive moral dimension, an *astroethics* extending to all life in the universe; and that while a human destiny linked to cosmic evolution rather than supernaturalism is a radical departure from the past, it is in the end beneficial and liberating. Such a worldview resolves many ancient theological problems. Bad things happen to good people because the universe is hostile rather than loving. Yet the prospect of contact with life beyond Earth leaves open the possibility of interacting with that life, and the idea of a loving and compassionate God can be expressed naturally in the way we treat our fellow humans and other creatures in the universe without resorting to supernaturalism. Stripped of supernaturalism and other accoutrements, compassion is at the core of all religions, even if the ideal is not always met, and universal compassion is at the core of cosmotheology.

13.1 Introduction

Science, particularly in the form of astronomy and cosmology, continues to reveal more and more about our place in the universe. For almost a century since Edwin Hubble discovered extragalactic space and hinted at the expanding universe, we have known that life on Earth is part of the vast unfolding of cosmic time and space, over 13.8 billion years according to the latest observations of spacecraft such as Hubble, COBE, WMAP, and Planck. And the search for life *beyond* Earth, once the stuff of science fiction, is now a robust research program with a well-defined Roadmap (Des Marais et al. 2008).[1] The science of astrobiology—and there is no longer any doubt it is a science, simplistic slogans about "a science without a subject" notwithstanding—is funded by NASA and other institutions to the tune of tens

First published in Ted Peters, ed. *Astrotheology: Science and Theology Meet Extraterrestrial Life* (Wipf and Stock, Cascade Books: Eugene, Oregon, 2018), 228–244.

S. J. Dick, *Space, Time, and Aliens*, https://doi.org/10.1007/978-3-030-41614-0_13

of millions of dollars of ground-based research, not to mention the hundreds of millions spent on space-related missions. Biogeochemists study extremophile life on Earth, biologists study the origins of life, a bevy of spacecraft have orbited or landed on Mars, others have found oceans on Jovian and Saturnian moons as well as organic molecules on Titan, and the Kepler spacecraft has discovered thousands of planets beyond the Solar System—all just a prelude to future studies. Recent Congressional hearings on biosignatures and complex life beyond Earth indicate astrobiology is a hot topic in the policy arena (U. S. Congress 2013, 2014). And international interest is also strong, particularly within the European Space Agency (Fridlund and Lammer 2010). Although no life has yet been found beyond Earth, the search for such life as part of the natural unfolding of cosmic evolution shows no signs of abating.

Given that these scientific results bear so heavily on our place in the universe, it is important to examine their societal implications. And in no area of human endeavor are these results likely to have broader impact than in theology, even if the impact is not immediate. It has been 15 years since I first elaborated principles of what I called *cosmotheology*, as part of a Templeton Foundation meeting on the theological implications of the new universe, and 10 years since I revisited the subject for a German audience—evidence that interest in the subject is international, if not yet global (Dick 2000a, 2005). In the intervening decade—precisely because of the new results in science—the problem of adapting theologies to current knowledge has only grown more urgent. Books, symposia, and discussions on the subject now appear with increasing frequency, most recently Thomas F. O'Meara's *Vast Universe: Extraterrestrial Life and Christian Revelation*, David Wilkinson's *Science, Religion, and the Search for Extraterrestrial Intelligence*, Guy Consolmagno's *Would You Baptize an Extraterrestrial*?, and David Weintraub's *Religions and Extraterrestrial Life: How Will We Deal with It?* All of this activity, engaged in by scientists and theologians alike (O'Meara is a Dominican and Consolmagmo a Jesuit), is only the latest manifestation of a controversy that has been building over the last 500 years, since the heliocentric theory of Copernicus made the Earth a planet and the planets potential Earths. The history of the implications of cosmic evolution for theology, particularly its extraterrestrial life aspect, has been written elsewhere in considerable detail (Dick 1998, 2009; Crowe 1986, 1997). Here I intend to take a closer look at the foundations, principles, and necessity for cosmotheology as part of broader efforts in what has variously been called exotheology, astrotheology, and astroethics (Lamm 1971; Peters 1994, 2014). In doing so I hope to clarify my own position by comparison with others, and to indicate what the future may hold.

13.2 Foundations and Principles of Cosmotheology

As Einstein began with the assumption that the speed of light is a fundamental constant independent of the motion of the light source, with all sorts of seemingly strange consequences such as time dilation and shrinking objects in his resulting special theory of relativity (Isaacson 2007), I begin with the assumption that the supernatural does not exist—with all sorts of results that many in established religions will consider strange, but I consider both enlightening and liberating. So entrenched has the idea of the supernatural become in Western civilization that many will ask how anyone can defend the radical assumption of its absence. But it is actually the supernaturalists who need to defend their position, since the existence of a reality beyond the natural world, while a legitimate question, is the extraordinary claim—one that supernaturalists have been defending for the last few thousand years. The idea of the supernatural arose early in the history of civilization, and undoubtedly existed even before the rise of civilization as a response to forces humans did not understand (Armstrong 1993). We now understand most of those forces, to such an extent that many theologians have given up on "god of the gaps" arguments, precisely because science has filled the gaps at an increasing pace. The idea of the supernatural is still with us only because the Abrahamic religions have adopted it as dogma over the course of thousands of years. To those outside established supernatural religions, the existence of a realm beyond the natural world would seem to be no more necessary than Aristotle's terrestrial-celestial dichotomy, an idea that survived intact only through the Middle Ages. I realize this is no small claim given the importance of the supernatural in terrestrial history. But many historical ideas are contingent and have had to be discarded; the absence of a supernatural realm inside or outside the universe is the core of my argument and the foundation for any naturalistic cosmotheology.

What then, is cosmotheology? Cosmotheology is simply a theology that takes into account what we know about the universe based on science. It is therefore a naturalistic cosmotheology, but it is not coextensive with scientism because it does not imply that science is the only way to understand the world.[2] Its first principle is that humanity is in no way physically central in the universe. This has been proven beyond the shadow of a doubt in a continuous series of "de-centerings," beginning with the Copernican removal of the Earth from the center of the Solar System in the sixteenth century, followed by Harlow Shapley's proof around 1920 that the Solar System is on the periphery of the Milky Way Galaxy, and Edwin Hubble's proof in the late 1920s that our galaxy is only one of many in an almost infinite space. Some have argued that the Copernican de-centering was actually beneficial in terms of human dignity, since the Earth, even though central in the Aristotelian and medieval cosmos, was considered the dregs of creation (hell was, after all, down below somewhere), until Copernicus placed our planet in the realm of the heavens (Danielson 2001). This may be true, but surely it is beyond doubt that subsequent physical decenterings hardly elevated humanity's conception of its physical place in the universe.

One may well argue that physical de-centerings do not matter that much; it is biological, cognitive, and moral status that is at stake in assessing our real place in nature. This brings us to the second principle of cosmotheology, that (if astrobiological endeavors are successful in finding intelligent life) any such theology must take into account the probability that humanity is not central biologically, mentally, or morally in the universe (Fig. 13.1). The word "probability" is crucial here, since

Fig. 13.1 "The Creation of Adam," a fresco painted by Michelangelo for the Sistine Chapel ceiling around 1510. It depicts a one-to-one relationship between God and man, but astrotheology raises the fundamental question of how unique this relationship really is. This painting was conceived within the Christian tradition; other religious traditions will be affected differently. Wikimedia Commons

we have not yet found a single instance of life beyond Earth, much less intelligence. That is what the science of astrobiology is exploring, but all indications are that life will eventually be found, in abundance for microbes, and perhaps more rarely for intelligence. But in such a vast universe "rare" is a relative term. It is certainly true that having found exoplanets by the thousands, some of them *Earth-sized* and in the habitable zones of their parent stars, does not necessarily mean these are *Earth-like* planets. That will need to be determined by examining biosignatures in their atmospheres and other means. But it has been interesting to watch the skeptics over the last 20 years first deny that there are any other planets beyond the Solar System, then emphasize that they are only uninhabitable gas giants, then that they are only Earth-sized and not Earth-like. There is a trend here; surely it is only a matter of time before Earth-like planets are found, and (it seems to me) only a matter of time before life is found. In short, thus far the general principle of the uniformity of nature's laws has held with respect to the existence of exoplanets, and the expectation is confirmed that what is true of Earth is true of other places in the universe. Although this has still to be proven in the case of life, the trend is clear, despite the great diversity of planetary systems found. No one would expect a solar system exactly like ours; as with life, so with planets: diversity is the coin of the realm.

If this universal production of life is true, the third principle of cosmotheology holds that we must take into account the probability that humanity is near the bottom in the great chain of beings in the universe. This follows from 13.8 billion years of cosmic evolution, and the fact that planets and life could have formed billions of years before our Earth originated 4.5 billion years ago. The extreme youth of our species, which has only in the last few thousand years emerged to the point that it can contemplate its place in the universe, is surely a sobering fact when placed in the context of cosmic evolution. Again, it is true (as Ted Peters and others have pointed out) that "progress" in social development, and even continuous linear complexity in biology, are not assured in a universal context. But certainly it is undeniable (without invoking any goal-oriented or teleological principles) that life on Earth is more complex after 3.8 billion years of evolution, and that culture is more highly developed than it was when modern *Homo sapiens* originated some 200,000 years ago. What millions or billions of years of biological and cultural evolution have produced in the wider universe is certainly open to debate, but to claim that most intelligent civilizations would not be more advanced than us seems to border on nihilism.

Fourth, cosmotheology must be open to radically new conceptions of God, not necessarily the God of the ancient near-East, nor the God of the human imagination, but a God grounded in cosmic evolution. It is entirely possible that beings have evolved in the natural course of the universe with many of the traits we attribute to God, including omnipotence, omniscience, and so on. It is even possible such beings have meddled in human affairs, though I hasten to add there is no evidence of this, and certainly no evidence that a figure such as Jesus Christ was the son of God or divine in any way, uplifting as he may have been to his numerous followers. Whether one wishes to call such a superior being "God" is also open to discussion, but an expansive theology might do so.

Fifth, cosmotheology must have a moral dimension, extended to embrace all species in the universe—a reverence and respect for life in any form. This principle—a challenge even on Earth—gives rise not only to astrotheology but also to the related field of astroethics (Impey et al. 2013; Peters 2014). It is often stated that morality stems from theology and the existence of God, with the unfortunate implication that any non-believer has no basis for morality. By contrast astroethical principles stem from this reverence for life in all its manifestations, the product of the creativity of cosmic evolution, whether terrestrial or extraterrestrial. Cosmic evolution can also serve as a framework for human moral orientation in other ways (Kaufman 1997).[3]

Sixth, although human destiny has often been couched in divine terms, as in Reinhold Niebühr's *The Nature and Destiny of Man* (1941) or Pierre Lecomte du Noüy's bestselling *Human Destiny* (1947), or, indeed as in the entire Christian theology, it need not be linked to the supernatural. Rather, it can be linked to the process and endpoint of cosmic evolution. If cosmic evolution ends with humans and we are alone in the universe, our destiny involves stewardship of our pale blue dot and perhaps spreading, nurturing, or even creating, life in the universe—all pathways filled with ethical considerations. If cosmic evolution results in a biological universe—one in which life and intelligence is common, our destiny is to interact with this life in all its myriad possibilities, invoking a quite different set of ethical considerations raised in the fifth principle (Dick 2009, 45–49; Vidal 2014; Mckay 1990; Randolph and McKay 2014).

13.3 Cosmotheology and Religious Naturalism

By now it should be clear that cosmotheology fits comfortably within the tradition known as religious naturalism. In the words of its premier historian, Jeremy Stone, religious naturalism "affirms a set of beliefs and attitudes that there are religious aspects of this world which can be appreciated within a naturalistic framework. There are some events or processes in our experience that elicit responses that can appropriately be called religious. These experiences and responses are similar enough to those nurtured by the paradigm cases of religion that they may be called religious without stretching the term beyond recognition (Stone 2010, 1)." In short, religious naturalism denies that an ontologically distinct and superior realm including God, the soul, and heaven is required to give meaning to the world. Nor does it identify with a pantheism that identifies God with Nature, or even a Spinoza version of pantheism (sometimes called pane theism), in which the universe is a subset of God. Rather, meaning is derived from our knowledge of the natural world, from the creativity and beauty of nature. This natural God is compatible with the concept of Einstein, for whom God "does not play dice" nor concerns himself with the fate and actions of men. But Einstein's God "appears as the physical world itself, with its infinitely marvelous structure operating at atomic level with the beauty of a craftsman's wristwatch, and at stellar level with the majesty of a massive cyclotron" (Clark 1972, 38; Einstein 1954).

Stone distinguishes three types of religious naturalists: those who conceive of God as the creative process within the universe; those who conceive God as the totality of the universe; and those who do not speak of God and yet can still be called religious due to the feelings of reverence and awe the universe inspires. Even those who fall in the first two types and use the term "God" do so in a rigorously naturalistic way, not invoking a supernatural realm. It seems to me there is no need to use the loaded term "God" for a naturally creative process or for the universe itself. Stone's third type of religious naturalism is the more common usage, and naturalistic cosmotheology falls comfortably within that category (Stone 2010, 6). This view resonates with the principles found in biologist Ursula Goodenough's *The Sacred Depths of Nature*, but draws its inspiration and principles from astronomy rather than biology. It also resonates with complexity theorist Stuart Kauffman's radical views in *Reinventing the Sacred*, where he proposes a natural divinity that draws its sacred quality from the creativity of the universe itself—and which in his view can still be called God (Goodenough 1998; Kauffman 2008). An astronomically inspired cosmotheology, a biologically inspired view of the sacredness of nature, Kauffman's insights into the complexity of nature, and even the broader environmental movement, all end up at the same place: with a reverence for the creativity of the natural universe and an evolving understanding of our place in it. Nor is it a universe necessarily reducible to physics; Kauffman, for example, argues vigorously for a scientific worldview that embraces the reality of *emergence* for life, meaning and value, a natural process but not predictable because not subject to natural law.

These principles in turn resonate with many aspects of the new humanism, with its openness to "wonder and mystery and transcendence in a naturalistic framework." They emerge in part from what Stone identifies as the two major roots of religious naturalism in America, the Columbia school centered on Columbia University, and the Chicago school, where the Meadville Lombard Theological School plays a major role (Stone 2010, 5). Such principles are already being incorporated into some religions. Although humanism transcends the boundaries of any religious denomination, the Meadville Lombard Theological School, for example, trains Unitarian-Universalist ministers, many of whom espouse some form of religious naturalism. Although small, the Unitarian tradition dates back centuries, and in a broader sense naturalism not only has deep roots, but also has played a major role in history. Indeed Matthew Stewart has recently argued convincingly that many of those who played a major role in the founding of the American republic—including Thomas Jefferson, Benjamin Franklin, Thomas Paine, and Ethan Allen—believed only in "nature's God," rendering ironic, or at least problematic, the vociferous claim that the United States was founded as a Christian nation (Stewart 2014).

Religious naturalism is, of course, a controversial position in more ways than one. The renowned historian of science John Greene, for example, cautioned in no uncertain terms that naturalism is just another world view, no more privileged than supernaturalism or anything else to serve as the source of value and meaning in human life. He pointed out that scientists such as Ernst Haeckel, Julian Huxley, Ralph Burhoe, and E.O. Wilson "all typify the scientist-ideologue bent on saving society by promulgating a new ethics and a new religion claiming the sanction of science." The pattern is "depressingly familiar," in Greene's view: "When will scientists and others learn that naturalism is a philosophical point of view with no more claim to the status of science than any other philosophical viewpoint, whether Marxian, Freudian, Russellian, Whiteheadian, or whatever. Scientists have as good a right to expound their philosophical, ethical and religious views as anyone else, but they have no right to palm these off as the findings of science" (Moore 1989, 404). In my view this attitude is both instructive and dead wrong. It is instructive in the sense that a science-oriented theology is far from the sole source of meaning and value in life. And it is dead wrong in the sense that science has *at least as much* claim to infuse theological thought as any other world view, and perhaps more, since being grounded in the natural world gives it a pillar of support that supernatural religions do not have, numerous attempts at natural theology notwithstanding. Exactly how to draw meaning from the findings of science is open to interpretation, and not just by scientists. Thinkers such as Stuart Kauffman have no problem finding meaning and value in the creativity of the universe. Philosopher-scientist Mark Lupisella has described a long-term worldview that "can be characterized as a morally creative cultural cosmos—a post-intelligent, post-technological universe that enters the realm of conscious evolution driven largely by moral and creative pursuits," in other words, a worldview in which meaning and value may be bootstrapped from the universe. Many others are espousing similar views at an increasing pace (Lupisella 2009).

Naturalistic cosmotheology may be seen as one strand in a galaxy of astrotheologies proposed over the last few centuries. How does it differ from the others? The idea of an *astrotheology* dates back at least to 1715, when the English clergyman and natural philosopher William Derham penned a book by that name whose subtitle indicated its purpose: *Astrotheology: Or a Demonstration of the Being and Attributes of God from a Survey of the Heavens* (Derham 1715). In other words, this was an exercise in natural theology, an attempt to prove the existence of God from his created work. The term *cosmotheology* originates with the philosopher Immanuel Kant, who referred to it in 1781 in his *Critique of Pure Reason* (without endorsing it) as a "transcendental theology" method of "inferring the existence of a Supreme Being from general experience," rather than a natural theology method of inferring the nature of a Supreme Being from the particulars of nature (Kant 1781). Both of these views, of course, are far from our concept of a naturalistic cosmotheology, and in fact they are far even from the modern view of most non-naturalistic astrotheologies.

How, then, does naturalistic cosmotheology compare to more modern versions of astrotheology? As early as 1994 the theologian Ted Peters took up the challenge of what he then called *exo-theology*, arguing that the discovery of extraterrestrial intelligence would present no significant challenge to theology. Even though that is seriously questioned by some theologians such as former Vatican Observatory director George Coyne (Coyne 2000, 187). I tend to agree with Peters that after a period of perhaps wrenching change (depending on the nature of contact), theologies would expand to include the new view, and that this could in fact be an enriching experience for theology. Peters, in fact, strongly argues that the latest science, including evolution, should be incorporated into theology, totally in agreement with cosmotheology as laid out here. On the other hand, when Peters concludes in an essay two decades later that "contact with extra-terrestrial intelligence will expand the existing religious vision that all of creation—including the 13.7 billion year history of the universe replete with all of God's creatures—is the gift of a loving and gracious God," he is speaking within a supernatural tradition that makes little sense to those outside of it—unless the concept of God is expanded to mean the natural universe itself. But it is admittedly hard to see the universe as "loving and gracious" (Peters 1994).

Similarly, when the Anglican priest and biochemist Sir Arthur Peacocke argues in a thoughtful, farsighted, and beautifully written article that any theology "will be moribund and doomed if it does not incorporate the perspective [of the epic of cosmic evolution] into its very bloodstream," even including the significance of Jesus Christ, I can fully agree. Yet when he concludes that "humanity is incomplete, unfinished, falling short of that instantiation of the ultimate values of truth, beauty and goodness that God, their ultimate source, must be seeking to achieve to bring them into harmonious relation to Godself," he is speaking a language that is foreign to a religious naturalist. Again, in his forward-looking book *Thank God for Evolution*, Michael Dowd embraces the evolutionary epic, but only within the Christian tradition that, according to him, gives it meaning. And the Methodist theologian and physicist David Wilkinson, in arguing that the discovery of extraterrestrials would not diminish us in the eyes of God, bases his position on Biblical revelation, citing "divinely initiated redemption, an action of a gracious God on behalf of a fallen cosmos" (Peacocke 2000; Dowd 2007; Wilkinson 2013). This too is foreign to a religious naturalistic cosmotheology, which finds no need to embrace the concept of a fallen cosmos.

All of these efforts are, like those of Thomas Aquinas in another turbulent era, attempts to reconcile new scientific knowledge within the Christian tradition, rather than attempts to ask the more basic question of whether the Christian tradition still makes sense in the modern world. Whether they are steps along the way to religious naturalism is a dubious hope; for one thing, religious naturalism does not offer salvation from a transcendent God, and never will, even in the form of advanced extraterrestrials. As Carl Sagan said in his book *Pale Blue Dot* referring to the Earth as seen from the Voyager 1 spacecraft, "our posturings, our imagined self-importance, the delusion that we have some privileged position in the Universe, are challenged by this point of pale light. Our planet is a lonely speck in the great enveloping cosmic dark. In our obscurity, in all this vastness, there is no hint that help will come from elsewhere to save us from ourselves" (Sagan 1994). A religious naturalist finds the equivalent of salvation in other ways, through social justice, good works, and making this world a better place to live—all in a cosmic context that defines our place in the universe.

13.4 A Difference in Worldview

One way of understanding the enduring differences in religious naturalism and religious supernaturalism is in terms of worldviews. The construction of worldviews and their influence on our thinking are deep philosophical problems (Vidal 2014, 3–57). What you believe is interesting, but *why* you believe it is even more so. If we could understand why people believe what they do, we might begin to tackle some of the world's problems, and not just in the religious domain. In this respect, science and theology are two worldviews, with different epistemologies, different sources of knowledge, and different aims except in the broadest sense of human attempts to understand our place in the universe. The main epistemology of science is empiricism—which admits of theory and observation, with all their complexities and problems, and yet which taken together have revealed so much about the universe around us. The main epistemology of theology, on the other hand, is revelation, faith and a heroic attempt at empiricism via what has been known over the last four centuries as natural theology. But as we have emphasized, natural theology is no longer considered a good theological argument; even most theologians admit the complex universe no longer requires a God to explain its inner workings—intelligent design biologist Michael Behe's idea of irreducible complexity in "Darwin's black box" and atheist philosopher Thomas Nagel's mysterious teleological principles as applied to mind and consciousness notwithstanding (Behe 1996; Nagel 2012). This leaves revelation, which also reduces to faith—faith that somehow God revealed his or her word to the authors of the Bible and that we should be following its dictums thousands of years later. Faith, then, is the difference between science and theology.

Some people have it, some people don't. If your personal epistemology does not include faith, the practice of religion and the weaving of supernatural theologies will seem strange activities indeed. But if your worldview includes faith as a source of knowledge, it makes perfect sense, though it must be pointed out that with all the religious diversity found on Earth today, one person's faith is another person's heresy.

It is axiomatic that problems arise when one worldview interacts with another, in this case the religious and the scientific. Science and religion can indeed seek common ground—and I applaud efforts to engage a dialogue between science and religion. But when religious beliefs result in attempts to stifle the teaching of science in the form of evolution or anything else, it is dangerous indeed. Intelligent design might well be taught in religion class, but not in a science class. Similarly, while it is easy to denigrate religion and theology, it remains true that religion has been the source of much good in the world, even if that must be weighed against the numerous religious wars and enormous personal anguish it has engendered, and that continues today. Religion means a great deal in the lives of many people, and freedom of religion is a foundational concept no less than academic freedom in science. A little humility is in order both for religious and scientific fanatics, each of whom should realize that their respective worldview does not explain everything of value in the universe.

Some may well object that the religious naturalist worldview really qualifies as cosmophilosophy rather than cosmotheology. But if theology is defined broadly as that which gives meaning and value to life in a cosmic context, then even a naturalistic cosmotheology is indeed a theology, albeit one without God, at least the standard God. In its emphasis on evolutionary becoming, cosmotheology resonates with Alfred North Whitehead's process theology, without seeing God as the beginning and source of all possibilities. It also resonates with the Jesuit Teilhard de Chardin's evolutionary cosmology, but without the teleological end, the "Noosphere" he identified with Jesus Christ. And the sixth principle of cosmotheology makes clear that human destiny may be couched in terms of natural cosmic evolution rather than supernatural divinity.

13.5 Following the Consequences: Cosmotheology and the Speed of Light

I began this essay with the core principle that the supernatural does not exist, promising to follow that premise wherever it led, in the same way Einstein assumed the speed of light was constant, despite the fact that it led to radical and non-intuitive concepts such as time dilation and shrinking objects as part of special relativity. Where does the absence of the supernatural leave us? For starters it leaves us in a much better position with respect to a explaining a variety of problems that have generated elaborate arguments among theologians and bewilderment among the populace (Kushner 1981). There is no need to explain "why bad things happen to good people." (Arnold, 2009) They happen not because of a vengeful or inattentive

God, but because that's the way the universe is—hardly loving and gracious when it comes to the forces of nature. The universe has the potential for great beauty and good, and yet the cosmos out of which we originated is a harsh place that may inflict pain and suffering on humans, who, as a product of cosmic evolution, have themselves generated evil throughout history. There is no need for a Creator God, even though we cannot yet fully explain the origin of the universe or the origin of life. And there is certainly no need for a judgmental God in a world that overflows with judgment grounded in different worldviews large and small. There *is* a need for a loving and compassionate God, in the best tradition of all theologies. But this can be expressed naturally in the way we treat our fellow humans each and every day. Indeed, Karen Armstrong has made the point that, stripped of the supernaturalism and other accoutrements, compassion is at the core of all religions, even if the ideal is not always met (Armstrong 2011).

A naturalistic cosmotheology also leaves us with a number of questions. From a historical, psychological, and sociological point of view: how did the idea of the supernatural originate in human societies, by what mechanisms did it evolve to its present unassailable position in the human mind via the majority of major religions, and how does it maintain its hold? From an astrobiological point of view: what are the chances extraterrestrial intelligence would have evolved similar points of view? Put another way, is supernaturalism a universal or contingent feature of high-order consciousness and intelligence? From the point of view of astroethics: aside from a general reverence for all forms of life, including compassion, what moral principles can be drawn from a universe that, while it may seem harsh and indifferent, indisputably has ingrained in its core laws of nature that are biofriendly? And last but not least, from a religious point of view: can such a naturalistic worldview satisfy human needs as much as supernaturalistic worldviews? If so, by what mechanisms? If not, why not?

Surely an extraterrestrial anthropologist or theologian visiting Earth and studying the beliefs of its inhabitants would find curious the religious ideas many Earthlings take for granted, passed down through generations and often accepted without thinking too much about them. Specific beliefs associated with religion have clearly grown out of history: in the Christian tradition the doctrine of the Trinity of God dating to the fourth century; a baby filled with "original sin" that needs to be washed away, dating substantially to Augustine of Hippo in fifth-century Africa; a supernatural God who incarnates a son; the Virgin birth of this son—all these ideas and more have accreted over the centuries and given rise to innumerable religious wars over the most minute details, even as they have arguably formed the basis for Western civilization. Similarly with Islam, whose Sunni and Shia sects have waged wars large and small based on their version of the true successor to Muhammed in the seventh century, from which their respective religious principles flow. Whether in these Abrahamic religions or some other tradition, these are also principles that give many people meaning and solace to endure the difficulties of life. And they are one source of a morality and worldview that arguably and in important ways have held together the social fabric for millennia, even as its darker centrifugal forces sometimes tear that fabric apart.

By contrast, an extraterrestrial might well recognize moral principles grounded in the shared worldview of cosmic evolution. But can a cosmotheology, grounded in the natural rather than the supernatural, contribute to uniting the social fabric without the deleterious consequences of supernatural religions? As many astrotheologians and scientists now accept, the epic of evolution is a shared worldview that has brought humans where we are today, a concatenation of elements spewed from a supernova, congealed through eons of evolution, now contemplating the universe and our place in it, while seeking meaning and value even as that evolution continues. In the modern view of cosmic evolution—Genesis for the third millennium—the concatenation of elements may have produced innumerable other intelligences throughout the universe. They likely also seek meaning and value in the universe, and while we will undoubtedly find new extraterrestrial modes of seeking such meaning and value, we all share the same universe.

"Dust thou art, and into dust thou shalt return," according to one popular source of terrestrial theological wisdom. "Stardust thou art, and into stardust thou shalt return" is the inspiring vision offered by cosmic evolution. It is a unifying vision good enough for increasing numbers of people as we enter the third millennium. Whether it is good enough for the majority of Earthlings and extraterrestrials, over the long or short term, remains to be seen.

13.6 Commentary 2020

I first broached the idea of cosmotheology at a meeting of the American Association for the Advancement of Science (AAAS) in Seattle in February, 1997, where I participated in a panel discussion on the Mars rock ALH84001, including a paper by Jesuit astronomer Chris Corbally, whose paper was "Religious Implications from the Possibility of Ancient Martian Life." At a splinter meeting of the Institute on Religion in an Age of Science, I gave an after dinner talk about what I then called "astrotheology." Astronomer Nancy Houk from the University of Michigan, who was active in this Society, recommended it be called "cosmotheology" because "astro" conveyed "star." (It turns out I had used the term "astrotheology" as early as 1982 in a book proposal on the societal impact of discovering ET life.) From the beginning the gist of the concept was that humans were in no way central geographically, celestially, or biologically in the universe, that cosmic evolution encouraged new ways of looking at God and the spiritual world, and that in any case we certainly did not need the supernatural for a cosmotheology.

An early indication of increasing interest in the implications of discovering life was a meeting on theology and alien life in 1998. Sponsored by the Templeton Foundation, which fosters dialogue between religion and science, it was an intimate meeting of about a dozen scholars held at the home of Sir John Templeton in Nassau, the Bahamas. Several papers at this meeting independently elaborated on issues that had been raised before. Ernan McMullin, a Catholic priest and eminent philosopher at the University of Notre Dame, argued that "the Creator of a galactic universe may well choose to relate to creatures made in the Creator's own image in ways and on grounds as diverse as those creatures themselves." For him Christian doctrine in the context of aliens boiled down to three issues: original sin, soul and body, and incarnation. He speculated that an omnipotent creator might want "to try more than once the fateful experiment of allowing freedom to a creature," a freedom that Adam and Eve had failed in the earthly Garden of Eden. He pointed to the possibility that aliens might or might not have souls; if they did, God might or might not elect to become incarnate. And regarding incarnation McMullin suggested that conflicting theological interpretations of that central doctrine of Christianity left wide open the question of Christ's incarnation on other worlds. George Coyne, at the time the Director of the Vatican Observatory, also offered no definitive answers, but suggested that theologians would have to "rethink some fundamental realities within the context of religious belief." In other words, neither McMullin nor Coyne were at all sanguine about the rapid adjustment of Christianity to extraterrestrials, but they believed it would be accommodated in the long term. By contrast, Jill Tarter, a pioneer in the field of SETI, argued that an extraterrestrial message might be "a missionary campaign without precedent in terrestrial history, replacing our diverse terrestrial religions with a universal religion." Alternatively, a message that indicated long-lived extraterrestrials with no need for God or religion might undermine our religious worldview completely.

Aside from myself as a participant, the roster included luminaries such as Nobelist Christian de Duve; physicists and astronomers such as Paul Davies, Freeman Dyson, and Martin Rees; and evolutionary biologist Richard Dawkins, a skeptic if ever there was one when it came to dialogue between science and religion. The meeting was lively. As the theological deliberations proceeded, I recall Dawkins more than once interrupting a speaker to ask in his British accent "What on Earth are you talking about?," with emphasis on a long and drawn-out "Earrrrth." There was a lively discussion among believers and skeptics—not so much when it came to the possibility of life, but when it came to the possible theological implications. I vividly remember George Coyne commenting on my paper on cosmotheology by saying "there is a special place in hell for those who think God isn't supernatural!"—whether tongue-in-cheek or not I was never sure. In any case, I agreed to be the editor of the proceedings of this meeting, published in 2000 by the Templeton Foundation Press (Dick 2000b), including my paper on cosmotheology (Dick 2000a).

The Templeton volume was my first foray into theology in published form, but not my last. I subsequently elaborated the idea in "Cosmotheology Revisited" for a German publication, and finally in the paper reprinted here, written for a volume edited by theologian Ted Peters (Peters et al. 2018). Peters himself has written extensively on the subject in his edited volume and elsewhere. Ironically, my original term "astrotheology" has come into common usage in the embryonic field of societal implications of astrobiology, with "cosmotheology" now being considered one flavor of it.

Notes

1. For an entrée to the large literature of astrobiology see (Sullivan III and Baross 2007). On the history of astrobiology see (Dick and Strick 2004). And on the critical issues in the field see (Dick 2012).
2. For a spirited, robust, but light-hearted defense of scientism, see the book by Duke University philosopher (Rosenberg 2012).
3. This issue of *Zygon* is dedicated to cosmic evolution in a religious context. See also Peackocke (2000).

References

Armstrong, Karen. 1993. *A History of God: The 4,000 Year Quest of Judaism, Christianity and Islam*. New York: Ballantine Books.

Armstrong, Karen. 2011. *Twelve Steps to a Compassionate Life*. New York, Alfred A. Knopf, pp. 6 ff.

Arnold, David. 2009. *Why Do Bad Things Happen to Good People?: Answers to One of Life's Greatest Moral Questions*. New York: Creation House.

Behe, Michael 1996. *Darwin's Black Box: The Biochemical Challenge to Evolution*. New York: Free Press.

Clark, Ronald W. 1972. *Einstein: The Life and Times*. New York: Avon.

Coyne, George. 2000. "The Evolution of Intelligent Life on Earth and Possibly Elsewhere: Reflections from a Religious Tradition:" In Dick (2000b), pp. 177–88.

Crowe, Michael J. 1986. *The Extraterrestrial Life Debate 1750–1900: The Idea of a Plurality of Worlds from Kant to Lowell*. Cambridge: Cambridge University Press

Crowe, Michael J. 1997. "A History of the Extraterrestrial Life Debate," *Zygon*, 32, 147–162

Danielson, Dennis. 2001. "The Great Copernican Cliché," *American Journal of Physics*, 69, 1029–1035.

Derham, William. 1715. *Astrotheology: Or a Demonstration of the Being and Attributes of God from a Survey of the Heavens*. London

Des Marais, David, D. Nuth et al. 2008. "The NASA Astrobiology Roadmap," *Astrobiology*, 8, 715–730.

Dick, Steven J. 1998. *Life on Other Worlds*. Cambridge: Cambridge University Press

Dick, Steven J. 2000a. "Cosmotheology: Theological Implications of the New Universe," in Dick, 2000b, 191–210.

Dick, Steven J. ed. 2000b. *Many Worlds: The New Universe, Extraterrestrial Life and the Theological Implications*. Philadelphia: Templeton Foundation Press

Dick, Steven J. 2005. "Kosmotheologie – neu betrachtet," in Tobias Daniel Wabbel, ed., Leben im All: Positionen aus Naturwissenschft, Philosophie und Theologie (Dusseldorf: Patmos Verlag, pp. 156–172 online in English at http://bdigital.ufp.pt/bitstream/10284/778/2/287-301Cons-Ciencias%2002-5.pdf

Dick, Steven J. 2009. "Cosmic Evolution: History, Culture and Human Destiny," in Dick and Lupisella, 2009, pp. 25–59.

Dick, Steven J. 2012. "Critical Issues in the History, Philosophy, and Sociology of Astrobiology," *Astrobiology*, 12, 906–927.

Dick Steven J. and Mark Lupisella, eds., 2009. *Cosmos and Culture: Cultural Evolution in a Cosmic*. Washington, DC: NASA

Dick, Steven J. and James E. Strick, 2004. *The Living Universe: NASA and the Development of Astrobiology*. New Brunswick: Rutgers University Press.

Dowd, Michael. 2007. *Thank God for Evolution: How the Marriage of Science and Religion Will Transform Your Life and Our World*. New York: Viking

Einstein, Albert. 1954. "Religion and Science," in *Ideas and Opinions*, New York: Bonanza, 36–40.

Fridlund, Malcolm and Helmut Lammer. 2010. "The Astrobiology Habitability Primer," *Astrobiology*, 10, 1–4. The entire issue is devoted to European Space Agency work in this area.

Goodenough, Ursula, 1998. *The Sacred Depths of Nature*. Oxford: Oxford University Press

Impey, Chris, Anna H. Spitz and William Stoeger, eds, 2013. *Encountering Life in the Universe: Ethical Foundations and Social Implications of Astrobiology*. Tucson: University of Arizona Press

Isaacson, Walter. 2007. *Einstein: His Life and Universe*. New York: Simon and Schuster, pp. 118–122.

Kant, Immanuel. 1781. *Critique of Pure Reason*, online at http://www2.hn.psu.edu/faculty/jmanis/kant/critique-pure-reason6x9.pdf, section VII, "Critique of all Theology based upon Speculative Principles of Reason," pp. 364–365.

Kaufman, Gordon D. 1997. "The Epic of Evolution as a Framework for Human Orientation in Life," *Zygon*, 32, 175–188.

Kauffman, Stuart A. 2008. *Reinventing the Sacred: A New View of Science, Religion and Reason*. New York: Basic Books.

Kushner, Harold S. 1981. *When Bad Things Happen to Good People*. New York, Shocken.

Lewis, James R., ed. 1995. *The Gods Have Landed: New Religions from Other Worlds*. Albany: SUNY Press

Lamm, Norman. 1971. "A Jewish Exotheology" in Norman Lamm, *Faith and Doubt: Studies in Traditional Jewish Thought*, Ktav Pub. House, p. 107

Lupisella, Mark. 2009. "Cosmocultural Evolution: The Coevolution of Culture and Cosmos and the Creation of Cosmic Value," in Dick and Lupisella, 2009, pp. 321–359: 322.

Mckay, Christopher. 1990. "Does Mars have rights? An Approach to the environmental ethics of planetary engineering," in D. MacNiven, ed., *Moral Expertise*. New York: Routledge, pp. 184–197.

Moore, James R. 1989. *History, Humanity and Evolution: Essays for John C. Greene*. Cambridge: Cambridge University Press.

Nagel, Thomas. 2012. *Mind and Cosmos: Why the Materialist Neo-Darwinian Conception of Nature is Almost Certainly False*. Oxford and New York: Oxford University Press

Peacocke, Arthur. 2000. "The Challenge and Stimulus of the Epic of Evolution to Theology," in Dick 2000b, p. 92.

Peters, Ted. 1994. "Exo-theology: Speculations on Extra-terrestrial Life." *CTNS Bulletin* 14, 1–9, expanded in Lewis (1995).

Peters, Ted. 2014. "Astrotheology: A Constructive Proposal," *Zygon*, 49, 443–457.

Peters, Ted, et al. 2018. *Astrotheology: Science & Theology Meet Extraterrestrial Life*. Eugene, Oregon: Cascade Books

Randolph, Richard O. and Christopher P. McKay. 2014. "Protecting and Expanding the Richness and Diversity of Life, an Ethic for Astrobiology Research and Space Exploration." *International Journal of Astrobiology* 13, 28–34.

Rosenberg, Alex. 2012. *The Atheist's Guide to Reality: Enjoying Life without Illusions*. New York: Norton

Sagan, Carl. 1994. *Pale Blue Dot: A Vision of the Human Future in Space*. Reprint, New York: Random House, 1997 reprint, xv–xvi.

Stewart, Matthew. 2014. *Nature's God: The Heretical Origins of the American Republic*. New York: Norton.

Stone, Jerome A. 2010. *Religious Naturalism Today:* The *Rebirth of a Forgotten Alternative*. Albany: SUNY.

Sullivan III, Woodruff T. and John A. Baross, eds. 2007. *Planets and Life: The Emerging Science of Astrobiology*. Cambridge: Cambridge University Press.

U. S. Congress. 2013. House Committee on Science, Space, and Technology, Hearings on "Astrobiology: Search for Biosignatures in our Solar System and Beyond," December 4, 2013

U. S. Congress. 2014. Hearings on "Astrobiology and the Search for Life in the Universe," May 21, 2014.

Vidal, Clement. 2014. *The Beginning and the End:* The *Meaning of Life in a Cosmological Perspective.* Heidelberg and New York: Springer

Wilkinson, David. 2013. *Science, Religion, and the Search for Extraterrestrial Intelligence.* Oxford: Oxford University Press.

Chapter 14
Astroethics and Cosmocentrism

Abstract New discoveries in astrobiology give rise to many ethical questions. Does Mars belong to the Martians, even if the Martians are only microbes? What do we say in response to an alien message, and who speaks for Earth? How do we treat aliens, either remotely or in a "close encounter of the third kind?" These issues are only the tip of the proverbial iceberg in the new field of astroethics. We argue the need for a new cosmocentric, rather than anthropocentric, ethic.

14.1 Introduction

With the recent announcement of a large subsurface lake on Mars, ongoing investigations of the oceans of Europa and Enceladus (complete with shooting geysers!), the discovery of exoplanets numbering in the thousands and the $100 million Breakthrough Listen SETI program well underway, the paradigm-shattering discovery of life beyond Earth could be made any day. NASA is showing renewed interest in SETI (it is sponsoring a meeting on technosignatures in September), and a few intrepid organizations such as METI International are actually sending messages to the stars (METI stands for "messaging extraterrestrial intelligence").

In recent months both Breakthrough Listen and the SETI Institute have sponsored both real and virtual meetings to examine the societal impact should their programs prove successful. Anthropologists, historians, ethicists, philosophers, and others are joining the interdisciplinary conversation in a serious way, impelled by the increasing possibility of discovery.

All of this activity gives new urgency to a whole series of ethical questions. Does Mars belong to the Martians, even if the Martians are only microbes? What do we say in response to an alien message, and who speaks for Earth? How do we treat aliens, either remotely or in a "close encounter of the third kind?" In short, whether we discover alien microbes or advanced alien life, we will immediately be faced with the problem of how to interact. Welcome to the world of *astroethics*—the con-

Portions of this article were first published as a blog on the *Scientific American* website, August 8, 2018.

S. J. Dick, *Space, Time, and Aliens*, https://doi.org/10.1007/978-3-030-41614-0_14

templation and development of ethical standards for a variety of outer space issues, including terraforming the planets, resource utilization, near-Earth asteroid threats, space exploration, planetary protection—and the discovery of extraterrestrial life.

14.2 The Moral Status of Extraterrestrial Organisms

The problems involving E.T. life are particularly fraught, especially if it talks back to us. Before we can act in any situation that involves life, it is first important to assess the moral status of the organisms involved. This is no easy task, since we are ambiguous about relations with animals on Earth, on the one hand sheltering them as beloved pets, on the other hand and rather arbitrarily hunting, eating, and exterminating them. But a good deal of thought has been given to the subject of the moral status of Earth organisms and the idea of intrinsic value on which it is often based (Table 14.1). Contemplating encounters with alien life tremendously expands our ethical horizons.

The case of intelligent aliens also encompasses not just the problem of how we might treat them but also how they might act or react. In other words, it is not just a question of our ethics. What about their ethics? Is there any basis for inferring whether alien intelligence might be good or bad? On Earth we exhibit a mix of altruism and evil, but is there any reason to believe altruism has triumphed among extraterrestrials? Might there be such a thing as a universal ethics in the form of a universal Golden Rule or a reverence for life? Or is Star Trek's "Prime Directive" of nonintervention a naive one-way street, a recipe for our own extinction? Does the arc of the moral universe indeed bend toward justice?

There are obviously many more questions than answers. Nonetheless, answers to these questions will inform our actions in real-world contacts with alien life under different scenarios. As I argue in my new book *Astrobiology, Discovery and Societal Impact* (Dick 2018), by contemplating these issues, and certainly by putting them into practice in the event of the discovery of life beyond Earth, we will not only address what the World Economic Forum has called one of the "X-factors" in our near or far future (World Economic Forum 2013), but also transform our thinking by moving from an anthropocentric ethic toward a *cosmocentric* one that establishes the universe and all or part of its life as a priority rather than just humans or even terrestrial life in general.

14.3 Microbes

Let's look at some specific issues, beginning with microbes, which many consider most likely to be the first discovery of life beyond the Earth. Microbes have always been a focus of attention in the context of Mars exploration, but now the focus is expanded to other water worlds of our Solar System, such as Jupiter's moon Europa or Saturn's Enceladus. At first the issues might seem straightforward: NASA has a robust planetary protection program whose goal is to protect all of the planets all of the time from contamination or back contamination.

Table 14.1 Theories of moral status

Theory	Explanation	Representative proponents	Implications for life beyond Earth
Anthropocentrism	Only humans have moral status	Most religions Zubrin and Wagner (1996)	Alien life has only instrumental value; protection possible
Ratiocentric Social-Reason-Cultural Triad	All organisms that have reason, or the social-reason-culture triad have moral status	Smith (2009, 2014)	Alien microbes and some complex alien life have only instrumental status
Sentientism	All and only sentient beings have moral status	Persson (2012) Peters (2013)	All and only sentient aliens have moral status
Biocentrism	All and only living things have moral status	Schweitzer (1960) Callicott (1986a, b) McKay (1990) Sagan (1980)	All living aliens life has moral status
Ecocentrism (Environmental Ethics; Deep Ecology)	All living beings, ecosystems and perhaps non-living nature have moral status	Leopold (1949) Devall and Sessions (1985) Rolston III (2014)	Even non-living nature has moral status
Planetocentrism	All planets have intrinsic value, especially with life	Sullivan III (2013)	All planets have intrinsic value
Cosmocentrism:	The entire cosmos and its constituent parts have moral value		
Basic	Cosmic consciousness	Vidal (2014) Dick (2009) Hart (2013)	Some physical or metaphysical aspect of the universe has priority in a value system, and provides a justification for intrinsic value
Enhanced	Derived from alien contact		
Strong	Value from physical or metaphysical aspect of the cosmos	Haynes (1990) Lupisella and Logsdon (1997) Lupisella (2016, 2020)	
Anthropic	Intimate connection between cosmos and life		

Modified from Erik Persson (2012), by permission

Beyond that, however, the scary fact is that no guidance exists on what to do if microbial life is actually discovered. In the context of microbes, it matters whether we adopt an anthropocentric or ratiocentric ethic that confers intrinisic value only on reasoning beings, or a biocentric ethic that values all living things. It matters whether we consider microbes only of scientific value, or whether they are considered to have intrinsic value, in which case microbes have rights too—rights that we do not give their counterparts on Earth. Planetary contamination policies seem to confer rights on any microbes we may find on other worlds; the central goal of those policies, after all, is to protect from contamination any planets that might harbor life. That is a kind of biocentric ethic.

But it is an unstable and inconsistent one, since by necessity on Earth we stamp out pathogenic microbes while at the same time realizing the microbiome is essential to human health. Thus, the status of microbes is one of many ethical dilemmas we will face if and when extraterrestrial microbes are discovered. One has the feeling that, even if a biocentric ethic is adopted in principle, human health will always take priority.

14.4 Intelligence: SETI and METI

While the policy issues involved with the discovery of microbes are serious enough, the issues become even more daunting for extraterrestrial intelligence. Once again they depend on the discovery scenario, most urgently in connection with current programs for indirect contact via SETI or METI programs, and most spectacularly in terms of impact if we ever make direct contact with aliens on Earth or in our Solar System, even in the form of alien artifacts.

The question of what to do in the event of success in SETI has received considerable attention, in the form of SETI protocols adopted three decades ago, which basically boil down to "confirm and then tell everybody." In other words, no false positives and no secrets. While these protocols have been adopted by a number of international organizations such as the International Astronomical Union, they have not been adopted by the United Nations and are not legally enforceable. Moreover, they have already been broken. When a reporter calls an astronomer to ask about a rumored detection, astronomers admirably tend not to lie, even before confirmation. Beyond that, there are no principles for dealing with a successful SETI detection.

And despite attempts, there are no protocols for messaging extraterrestrial intelligence (METI), although there has been a great deal of heated discussion about the ethics of initiating messages, both in terms of consultation and message content. Opponents have gone so far as to suggest METI should be banned, and readers of Cixin Liu's disturbing *Three-Body Problem* trilogy might tend to agree as they witness the Trisolaran fleet heading to Earth from the Alpha Centauri system. In contrast, I argue that when it comes to METI—and all of astrobiology—we are a part of the universe and cannot isolate ourselves from it. We will have to deal with microbes and aliens for good or ill in the same way we have had to deal with terrestrials for good and ill. Certainly, we can have consultations about message construction, content and other burning issues bound to arise.

But it is good to recall that METI is just one step ahead of SETI. If SETI is successful, we will reply, and all the questions METI practitioners are now dealing with will immediately come to the fore. In my view, not only is it unrealistic to think that we will restrain ourselves from replying, but it is also undesirable. An Earth where we have to limit our curiosity is not the kind of place I want to live. We should take all necessary precautions, feel at home in the universe and deal with the problems and the promise as they come.

14.5 A Cosmocentric Ethic?

The questions we have been asking go to the very core of the concepts of intrinsic value, moral status and their meaning for practical ethics. They raise the issue of whether an anthropocentric ethic is enough for an astroethics dealing with alien life, even when extended to environmental ethics and deep ecology, or whether we need something even broader, a *cosmocentric ethic*, as NASA engineer and biologist/philosopher Mark Lupisella and space policy analyst John Logsdon have suggested (Lupisella and Logsdon 1997; Lupisella 2016, 2020). I would argue that we do, in the sense that at a minimum we should apply a basic cosmocentric ethic, stipulating that our increasing cosmic consciousness requires us to consider our place in the biological universe when we make ethical judgments. We are, after all, part of the cosmos and perhaps not the most important part when it comes to life—the central question of astrobiology. In this view when we ask about the rights of Martian life, or how to treat alien intelligence, we should certainly avoid an anthropocentric stance that only humans have moral status.

Perhaps you think this is all rather esoteric, a subject for elites to contemplate while most people deal with the more pressing problems of daily life. In my view, you would be wrong. Yes, we have plenty of problems on Earth to deal with, but extraterrestrial contact may soon be one of them. Preparing for discovery is important to maximize the chances for a beneficial outcome. And we should never forget that Earth is part of the universe, and the cosmic view of astroethics and an accompanying cosmocentric ethic might just give us a perspective on our problems that will help solve them. In addition, astroethics has the potential to influence multitudes with the rise of the related discipline of astrotheology, now also a hot topic and the subject of many books. But that is another question (Chap. 13).

14.6 Commentary 2020

This brief chapter first appeared as a blog on the *Scientific American* website, and is supplemented here with Table 14.1 and the references. The field of astroethics is growing and now has a considerable literature. For an entrée to that literature see Impey et al. (2013) and Dick (2018). Table 14.1 is taken from the latter source.

References

Bertka, Constance, ed. 2009. *Exploring the Origin, Extent, and Future of Life: Philosophical, Ethical and Theological Perspectives*. Cambridge: Cambridge University Press.

Callicott, J. Baird. 1986a. "Moral Considerability and Extraterrestrial Life," in *Beyond Spaceship Earth: Environmental Ethics and the Solar System*. Eugene C. Hargrove, ed. San Francisco, Sierra Club Books. pp. 227–259.

Callicott, J. Baird. 1986b. "On the Intrinsic Value of Nonhuman Species," in *The Preservation of Species*, Bryan Norton, ed., Princeton, Princeton University Press, pp. 138–172. Reprinted 2014.

Devall, William and George Sessions. 1985. *Deep Ecology: Living As if Nature Mattered*. Salt Lake City: Gibbs M. Smith, Inc.

Dick, Steven J. 2009. "Cosmic evolution: History, culture and human destiny," in Dick and Lupisella (2009), pp. 25–59.

Dick, Steven J. 2018. *Astrobiology, Discovery, and Society*. Cambridge: Cambridge University Press.

Dick, Steven J. and Lupisella, M. L., editors. 2009. *Cosmos & Culture: Cultural Evolution in a Cosmic Context*.Washington, DC: NASA, online at http://history.nasa.gov/SP-4802.pdf

Hart, John. 2013. *Cosmic Commons: Spirit, Science and Space*. Eugene Oregon: Cascade Books.

Haynes, Robert. 1990. "Ecce Ecopoiesis: Playing God on Mars," in *Moral Expertise*, ed. D. MacNiven. London: Routledge, pp. 161–183.

Impey, Chris, Anna Spitz, and William Stoeger, eds. 2013. *Encountering Life in the Universe: Ethical Foundations and Social Implications of Astrobiology* Tucson: University of Arizona Press.

Leopold, Aldo. 1949. *A Sand County Almanac*. New York, Oxford University Press.

Lupisella, Mark. 2016. Cosmological Theories of Value: Relationalism and Connectedness as Foundations for Cosmic Creativity, in James S. J. Schwartz and Tony Milligan, eds., *The Ethics of Space Exploration*. Switzerland: Springer, 75–91.

Lupisella, Mark. 2020. *Cosmological Theories of Value: Science, Philosophy, and Meaning in Cosmic Evolution*. New York, Springer.

Lupisella, Mark and John Logsdon. 1997. Do we need a cosmocentric ethic?, paper IAA-97-IAA.9.2.09, IAC, Turin, Italy.

McKay, C. P. 1990. "Does Mars have rights? An Approach to the Environmental Ethics of Planetary Engineering," In *Moral Expertise: Studies in Practical and Professional Ethics*, ed. D. MacNiven. London and New York: Routledge, pp. 184–197.

Persson, Erik. 2012. "The Moral Status of Extraterrestrial Life," *Astrobiology*, 12, 976–984.

Peters, Ted. 2013. "Astroethics: Engaging Extraterrestrial Intelligent Life-Forms," in Impey et al, eds. (2013), pp. 200–221.

Rolston III, Holmes. 2014. "Terrestrial and Extraterrestrial Altruism," in Vakoch (2014), pp. 211–222.

Sagan, Carl. 1980. *Cosmos*. New York: Random House.

Schweitzer, Albert. 1960. "The Ethic of Reverence for Life," in *Albert Schweitzer: An Anthology*, Charles R. Joy, ed., Boston: Beacon Press, pp. 259–260.

Smith, Kelly. 2009. "The Trouble with intrinsic value: an ethical primer for astrobiology," in Bertka (2009).

Smith, Kelly. 2014. "Manifest Complexity: a Foundational Ethic for Astrobiology?" *Space Policy*, 30, 209–214.

Sullivan III, Woodruff T. 2013. "Planetocentric Ethics: Principles for Exploring a Solar System that May Contain Extraterrestrial Microbial Life," in Impey et al. (2013), pp. 167–177.

Vakoch, Douglas. 2014. *Extraterrestrial Altruism: Evolution and Ethics in the Cosmos*. Heidelberg and New York: Springer.

Vidal, Clement. 2014. *The Beginning and the End: The Meaning of Life in a Cosmological Perspective*. New York: Springer.

World Economic Forum. 2013. *Global Risks 2013*, Lee Howard, ed. Geneva: World Economic Forum, http://reports.weforum.org/global-risks-2013/section-five/x-factors/#hide/img-5

Zubrin, Robert and Richard Wagner. 1996. *The Case for Mars*. New York: The Free Press.

Chapter 15
Should We Message ET, and Is an Asilomar Consultation Process Possible?

Abstract We argue that Messaging Extraterrestrial Intelligence (METI), also known as active SETI, is an activity that should be undertaken for a variety of reasons. In this paper we begin by laying out some of more serious issues that METI raises. We present a brief history of Messaging Extraterrestrial Intelligence (METI) and the controversy surrounding it, not only by way of background but also because it is important that we not reinvent the wheel when it comes to relevant issues. We then focus on the issue of consultation, and ask if there is a model we can follow based on other cases of controversial scientific research affecting all of humanity, in particular the famous Asilomar process for biotechnology, often mentioned in the context of METI. We conclude with lessons learned and recommendations.

15.1 Should Humanity Hide?

Sending messages to the stars is not an entirely new endeavor (see Chap. 6), but in the last 50 years it has increasingly become a subject of discussion and of action, at first sporadically and then more systematically. For example, one of the primary purposes of the organization known as METI International is Active SETI, "in which powerful, intentional information-rich signals are transmitted to possible extraterrestrial civilizations."[1] This goal is surprisingly controversial, indicating just how seriously many scientists and others take the possibility of advanced alien life. Indeed, in 2015 a group of scientists signed a statement originating at Berkeley urging caution in any Active SETI project (Berkeley 2015). Although I am in agreement with much of the Berkeley statement, the devil is in the details.

I am on the record as saying that governments should not fund Active SETI, but that groups that want to undertake METI cannot practically be regulated short of legislation on the subject, which does not exist and is unlikely to happen. Given that, it seems to me a responsible group of scholars that wants to undertake a METI project should be free to do so if it seeks consensus inside and outside the group. Though an international consensus at the governmental level is in my view unlikely in such

This chapter combines a blog posted on the METI International website on December 9, 2015 and a paper delivered at the International Space Development Conference (ISDC), in St. Louis in 2017, published here for the first time.

S. J. Dick, *Space, Time, and Aliens*, https://doi.org/10.1007/978-3-030-41614-0_15

a matter, the first goal of METI International's Strategic Plan stipulates broad consultation with scholars in many fields to encourage a responsible approach to sending messages. I confess to some uncertainty about the substance of the consultation: Is it to craft the perfect message? To ensure a politically correct message? To ensure that not too much information is sent? This is all up for discussion, but the desire for international input seems to me to improve on the methodology of past messages, including Frank Drake's Arecibo radio transmission, the Pioneer plaque, and the Voyager record, all sent in the 1970s. It also improves on the methodology of a series of messages sent from Russia over the last 15 years.

Many readers will recall Stephen Hawking's warning in 2010 and again in 2016 that aliens might be dangerous to Earth (Baral 2016). A few readers might also recall that in the wake of Drake's message, sent in 1974 to the Hercules cluster of some 300,000 stars 22,000 light years from Earth, Nobel Laureate Sir Martin Ryle appealed to the International Astronomical Union that no attempts be made to signal other civilizations for fear of the consequences. In response a *New York Times* editorial asked "Should Mankind Hide?" Its conclusion was a solid "no." "To live is to accept dangers," the *Times* wrote, arguing that "the universe seems too rich to require an advanced race to look hungrily on Earth's eager patrimony" (*New York Times* 1976a, b). That was admittedly only one editorial in one newspaper at one point in space and time, but the underlying spirit of the conclusion seems to me to be correct.

The underlying principle guiding my opinion in this matter is that humanity as a species should not cower and hide from the stars. We cannot isolate ourselves from the universe because we are an integral part of it. METI is in the same spirit of exploration as passive SETI, which some also said was not a science when it began more than 50 years ago. And both are in the same spirit as astrobiology in general, which seeks life in the universe, leavened by planetary protection protocols. I must also say it seems to me the METI controversy is greatly overwrought; the chances of success are probably low, and one could argue that any advanced civilization would likely already have us in their catalogue of galactic civilizations. And they would be so distant as to be unlikely to pose a threat, unless they have a hyperwarp drive, in which case they might be here. But they are apparently not here, giving rise to the famous "Where are They?" question at the core of the Fermi Paradox. Meanwhile, the research into message construction that METI International is undertaking would have benefit should a SETI search prove successful.

15.2 Concerns About METI

Among the concerns about METI, many expressed in the Berkeley document, are the following:

1. It is impossible to predict whether ETs will be benign or hostile. This is certainly true, since no universal principle of intelligent behavior can be applied on Earth, much less to extraterrestrials. Early hopes that ETs would be our savior for a

variety of terrestrial ills ranging from cancer to war, are just that—hopes if not fantasies. A fascinating recent volume (Vakoch 2014b) on *Extraterrestrial Altruism: Evolution and Ethics in the Cosmos*, explored the pros and cons of altruism in the universe. The bottom line is that no definitive conclusion is possible. No one knows if ETs are good guys or bad guys. The question is whether this should prevent METI from happening. "Curiosity killed the cat" critics might warn. But do they really mean to imply we should stifle curiosity?
2. It is likely civilizations will be millions of years more advanced than us. This is also probably true, given the age of the universe and the youth of our species. The implication is that we have no idea of their capabilities, which might be hostile. This is an unproven assumption no less than the assumption that longevity implies wisdom.
3. It is prudent to listen before we shout, and in any case transmission is not necessary. If all ETs follow this rule, everyone may be listening, and no one messaging, and therefore SETI has no chance of success. There is always the possibility of leakage radiation, but as Drake has pointed out that is decreasing on Earth due to cable and satellite TV. Leakage radiation aside, only if ETs have a debate about METI and decide to go ahead could SETI be successful. If no ETs have a METI program, this is one solution for our failure to see them.
4. METI may jeopardize funding for astrobiology and SETI. I am confident funders can see the difference between microbes and SETI, and between SETI and METI. After all, for more than 20 years the U. S. Congress has declined to fund SETI, all the while funding a robust astrobiology program that focuses on microbial life.
5. METI is a religious cult, not science. This a charge also occasionally levied at SETI, which has been called a search for "deities for atheists." But the essence of religion is supernatural; neither SETI nor METI has anything to do with the supernatural.

15.3 Humanity Should Not Hide

Other objections have been voiced, but underlying most of them is a kind of xenophobia, a fear of the unknown in general and fear of the Other in particular. This is in part based on analogies of culture contact on Earth, especially the largely disastrous European contacts with the Americas in the Age of Discovery 500 years ago. *The Three Body Problem*, by the Chinese writer Cixin Liu, in which a Chinese METI project leads to an invasion of Earth 400 years later, was inspired by such analogies. But the author might just as well have used a more benign analogy from his own country: the treasure fleet voyages of Zheng He three generations before Columbus, which did not result in such destruction. Or indeed the later Jesuit missions around the world, which, while proselytizing, raised many interesting questions about communication and conceptual difficulties. Moreover, some have argued that the withdrawal of the Chinese fleet in 1433 was an important element in its demise, as China turned inward and away from maritime trade and exploration. There may be a lesson in that for those who want to turn inward rather than outward. Analogies are an important mode of argument but must be used with caution. Their proper use is just a small part of the study of the societal impact of discovering life beyond Earth, which is now finally receiving the attention it deserves (Dick 2015, 2018, and Part II of this volume).

More generally, those who oppose METI are undoubtedly influenced by science fiction, which tends toward conflict for dramatic effect, especially Hollywood movies such as the Alien series and Star Wars, where action and conflict are necessary to appeal to young audiences. Popular culture driven by such profit motives greatly influences us all, but may well have no bearing on reality. The only way to find out is to explore. Part of my outlook has been shaped by my time at NASA, the premier agency for exploration in the world. I arrived at NASA Headquarters in the wake of the Columbia Space Shuttle accident, when some outside critics wanted to cancel the Shuttle program and even human spaceflight as a whole. But wiser heads prevailed. I well recall the NASA Administrator's Symposium on Risk and Exploration (Dick and Cowing 2005), in which many explorers ranging from mountain climbers to astronauts concluded that safety is not the first priority in exploration. The first priority, after taking all necessary precautions, is to GO. Otherwise *Homo sapiens* would never have left the cave, Columbus would never have left Palos, and we would not be exploring the universe. Risk is the inevitable companion of exploration.

In this spirit of exploring the unknown, the recently announced $100 million Breakthrough Listen Project is accompanied by a Breakthrough Message project. Although the latter does not immediately seek to send a message, it is certainly consistent with transmitting a message in the future. Interestingly, Stephen Hawking helped launch the Breakthrough Listen and Breakthrough Message programs last July, leading some to conclude that his desire to know about ETs may trump his fear of them. The triumph of hope over fear is always a good thing.

In short, while there is room for valid arguments on both sides, and for discussion of what level of international consultation is optimal, I come down on the side of not isolating ourselves from the universe of which we are a part, for better or worse. Waiting until a SETI signal is received and confirmed is certainly one strategy. But if everyone in the universe is listening and no one is sending, we will never make contact. And if we never make contact we will not solve one of the great mysteries of science. Humanity should not hide. This does not imply, however, that no consultation should be involved. After a brief history of the METI controversy, we now then turn to the question of consultation.

15.4 The METI Controversy: History as a Useful Tool

The Search for Extraterrestrial Intelligence (SETI) has a long and storied history beginning in its modern phase with Frank Drake's project Ozma, and dating back in an earlier phase to Tesla and Marconi (see Chap. 6). By contrast active SETI, also known as METI (Messaging Extraterrestrial Intelligence), is much more recent. Technically, the Pioneer plaque and the Voyager record of the 1970s constitute METI, but they are so slow they caused little concern, taking 40,000 years to reach the nearest possible planetary system. But electromagnetic METI, sending *signals* to the stars rather than artifacts, is another matter. METI also began with Drake, when he sent a message in 1974 to the star cluster M13 in Hercules on the occasion of a significant upgrade to the Arecibo Observatory feed system. Drake recalls little thought about the target other than a very practical one: M13, 25,000 light years away, was in the Puerto Rican sky at the time of the program at 1 pm! Despite the rather strong signal of 1000 kW, his event also caused relatively little stir with one notable exception, the British astronomer Sir Martin Ryle, who agitated for the International Astronomical Union to urge that in the future no attempts be made to communicate with other civilizations because of possible hostile consequences. The *New York Times* responded with an editorial titled "Should Mankind Hide?" and concluded mankind should *not* hide (New York Times 1976a, b).

But in the last several decades the number of transmissions has ramped up, ranging from NASA's broadcast of the Beatles' "Across the Universe" in 2008, to multiple Russian broadcasts organized by astronomer Alexander Zaitsev from the Evpatoria Planetary Radar in Crimea beginning in 1999, transmitting an abbreviated encyclopedia of human knowledge. Some of these transmissions, widely reported in the press, were too weak to be of real concern, but as transmitters become cheaper and more readily available the urgency for policy grows. This despite the fact that Zaitsev (2011, 424) argues that the probability of detecting our powerful planetary radars is a million times greater than aliens detecting a METI message. This is also the position of Frank Drake, who argues METI is a waste of resources, not because of any concerns of hostility, but because ET would already know about us from transmissions such as these planetary radars (Drake private communication, 2017).

Foreseeing potential problems and opportunities, already in 1995 the SETI Committee of the International Academy of Astronautics (IAA)—the same Committee that formulated the SETI Protocols—developed a "Draft Declaration of Principles Concerning Sending Communications with Extraterrestrial Intelligence" (International Academy of Astronautics SETI Committee 1995). The Declaration laid out ten principles, specifying that any message should be sent on behalf of all humankind rather than from any individual state, that it should reflect the broad interests and wellbeing of humankind, and be available to the public prior to transmission. It stated further that the content should be based on input from a wide variety of people with diverse expertise, and that appropriate international consultations should take place prior to any transmission. The Declaration also encouraged international studies to consider these issues, and urged that these issues should eventually come before the United Nations. These proposals met with less success than the SETI protocols. The SETI Institute's 1998 roadmap recommended against active SETI projects (Ekers et al. 2002). To this day the IAA Draft Principles have not achieved consensus even at the IAA and its SETI Committee. The METI declaration therefore remains just that, a draft with no consensus or force. But it should not be forgotten in any future effort at consultation.

In the meantime substantive discussion continues to take place, largely centered around the IAA activities and published in *Acta Astronautica*, at least partially fulfilling the "international studies" principle of the Declaration. Douglas Vakoch has been a leader in pro-METI arguments, arguing that if no one is transmitting, SETI programs are doomed to fail; that having extraterrestrials receive, decode, and interpret messages would facilitate mutual comprehension; and that such a project provides both humans and aliens with a long-term vision for mutual benefit (Vakoch 2011b, c). Others inside and outside the IAA have argued the opposite: as mentioned above Stephen Hawking, for example, warned that aliens receiving such a message could retaliate, one of many catastrophic warnings he has more recently been giving about everything ranging from artificial intelligence to near-Earth asteroids. These pronouncements have received a great deal of public attention, but Hawking, though an expert in black holes, knows no more about alien intentions that anyone else. Science fiction writer David Brin has been particularly outspoken, arguing that no action should take place prior to consultation; that a process should take place like that of Asilomar, where brakes were placed on biotechnology; and that there is no guarantee of altruism in the universe. To send a message now, he argues, would be like "ignorant children, screaming 'Hello' at the top of their lungs, in the middle of a dark, unknown jungle" (Brin 2011, 2013). As mentioned above, his argument was given a colorful rendition in Cixin Liu's trilogy *The Three Body Problem*, one of the volumes of which is titled "The Dark Forest." These points and more were argued at a Royal Society meeting in 2010, again with no consensus (Dominik and Zarnecki 2011).

Discussion in this area has reached a peak in the last few years when 28 scientists signed a statement, originating at Berkeley, arguing that METI programs carry "unknown and potentially enormous implications and consequences," and that nothing should be done without international consultation. Specifically it strongly urges "vigorous international debate by a broadly representative body prior to engaging further in this activity," and concludes that "a worldwide scientific, political and humanitarian discussion must occur before any message is sent" (Berkeley 2015). In 2015 Doug Vakoch founded METI International, complete with a Board of Trustees (of which I was a member) and a large number of distinguished advisors, with the explicit purpose of sending "powerful, intentional information-rich signals to possible extraterrestrial civilizations" (METI International 2015). The organization explicitly states in its Strategic Plan that "prior to transmission METI International will engage in broad consultation with experts from the natural sciences, social sciences, humanities, and other fields to encourage a responsible approach to sending messages." Note the question is not whether to send a message, but a responsible approach *when* one is sent. That is a crucial difference with Berkeley.

Both the Berkeley and METI groups agree that consultation is important prior to any messages being sent. The question again is at what level the consultation should take place, what is the purpose of the consultation, and on what principles would a consultation actually be decided. Based on history it seems highly unlikely that the United Nations would take up such a consultation, leaving "wise" people, as Brin puts it, to make the decisions. But who are these wise people?

15.5 Asilomar as a Model for Consultation

By way of answering this question, let me turn now to a model sometimes proposed for METI consultation, the often-cited *Asilomar process*. The Asilomar process is based on the International Congress on Recombinant DNA Molecules held February 24–27, 1975 at the Asilomar Conference Center on the Monterey Peninsula in California. The meeting was organized by scientists who raised warnings about genetic engineering research and its danger to public health. Specifically, the early 1970s were a heady time when new technology allowed biologists to insert DNA from one organism into cells of another and monitor the effects, all in the name of research. The research held out the hope for important advances in medicine, agriculture, and industry. But the researchers themselves realized that "unfettered pursuit of these goals might have unforeseen and damaging consequences for human health and Earth's ecosystems." A voluntary moratorium was agreed to until an international meeting could be held to discuss the risks and make recommendations.

It is notable that the meeting had the backing of the National Academy of Sciences and the National Institutes of Health, and was organized by Nobel-level scientists. There were about 140 participants including scientists, lawyers, journalists, and government officials—a much larger group than all SETI and METI practitioners combined. During the meeting there was considerable disagreement about the magnitude of the purported risks.

The basis for my description of this conference is a brief but incisive article by Paul Berg, a Nobel biochemist, one of the four organizers of the Asilomar meeting. It was written as a retrospective and published in *Nature* in 2008 (Berg 2008). Berg recalled that while formulating policy on recombinant DNA research seemed an overwhelming task taken as a whole, the key to success of the Asilomar meeting was that risk estimates were assigned for specific types of experiments, with guidelines for each according to the degree of risk. The meeting concluded research in these particular areas should continue under stringent guidelines, and these guidelines formed the basis for the official U. S. guidelines issued in July, 1976. In his retrospective, Berg points out that in the years since Asilomar, countless recombinant DNA experiments have been carried out, without damage to public health or to any natural processes, but with great benefit to society. Moreover, he remarked, "the fear among scientists that artificially moving DNA among species would have profound effects on natural processes has substantially disappeared with the discovery that such exchanges occur in nature." It took a decade, but now "genetically modified hormones, vaccines, therapeutic agents and diagnostic tools are enhancing medical practice. Genetically engineered food plants are being grown and sold for consumption in both developed and developing countries. A thriving biotechnology industry has created products, jobs and wealth for scientists and others." Berg pointed out that because scientists undertook their own regulation, and because 15% of the Asilomar participants were from the media, they gained the public trust (Berg 2008).

There is no doubt of the importance and success of the Asilomar meeting. It was one of six "Meetings that changed the world," as *Nature* magazine put it in, ranking it as part of a series that included the 1951 meeting that led to the creation of CERN, the 1986 meeting in Santa Fe that led to human genome sequencing, and a 1995 Madrid meeting on climate change that led to the Kyoto protocols (Nature 2008). All of these meetings may also hold lessons for METI, but let me focus now on Asilomar.

15.6 Lessons Learned: Would an Asilomar Process Work for METI?

A number of lessons may be learned from the Asilomar process:

1. First, as Berg, pointed out, "The people who sounded the alarm about this new line of experimentation were not politicians, religious groups or journalists, as one might expect: they were scientists." Scientists were the ones who organized the conference, and they had a variety of motivations, ranging from personal concerns about the effects of their work on the environment to fears that if they didn't do something, regulators in Washington would.
2. Secondly, although scientists organized the conference, crucially, participants included scientists, lawyers, journalists, and government officials. If the Asilomar process is followed, practitioners in the fields of SETI and METI would organize the consultation, but they would draw on a wide variety of disciplines, including the social sciences and humanities. In fact this is exactly what happened in 1992–1993, just prior to the inauguration of the NASA SETI program, when John Billingham and others organized the CASETI meetings at the Chaminade conference center in Santa Cruz, California. This intimate gathering of about two dozen scholars included astronomers, historians, religious scholars, policymakers and the media, and was a model of interdisciplinary brainstorming (Billingham et al. 1999).[2] The gathering was a de facto recognition that the societal implications of SETI were broadly based and not to be solved by scientists alone. While the publication of the results was delayed almost a decade by the congressional cancellation of the scientific program, the recommendations of that group are still a valuable starting point for contemplating the aftermath of any successful SETI program.
3. Thirdly, it is notable that while the scope of the recombinant DNA problems seem overwhelming, Berg points out that the key was breaking the problem into its constituent parts and addressing them individually. Specifically, risk estimates were assigned for specific types of experiments, with guidelines for each according to the degree of risk. For example, different degrees of risk could be handled by different degrees of physical containment, ranging from an open bench for no risk to airlocks or special containment facilities for high risk. Similarly, risk

estimates could be assigned to certain kinds of METI experiments, where risk variables include signal strength, distance to target, number of planets in the target's habitable zone, and whether those planets exhibit biosignatures. If METI followed a similar course to recombinant DNA as decided at Asilomar, research would continue, but with guidelines about particular experiments using these variables.

Let me give some examples, keeping in mind that we now know that virtually all stars have planets, though not necessarily habitable planets. In a world free of concern about societal implications, the recent discovery of a planet around the closest star, Proxima Centauri, would seem to make it a primary target for both SETI and METI. The planet, dubbed Proxima Centauri b, has a minimum mass of 1.3 Earths, and orbits a small M dwarf star in the star's habitable zone. Critics of METI would likely claim this is too close for comfort for a potential METI project. Revealing our position to a civilization only four light years away might mean the possibility of some kind of reaction in only 8 years. But it seems to me the same would be true for a SETI program that detected a signal and would then have to decide whether or not to reply. Surely the pressure to reply would be great. To me this is one of the great benefits of starting METI programs now. Rather than rushed consultations in the wake of a successful SETI program, METI can more deliberately consider a variety of communications strategies. This emphasizes a point not often made: METI is the next step beyond any successful SETI program, and so the two are intimately linked.

But there is another point. At a recent meeting of the American Astronomical Society, it was reported that Proxima Centauri b is likely a desert world, in which the active M star lashes the planet with X-rays, stripping it of any atmosphere. The operative word is "likely." Under some circumstances, such as if the planet formed further from its star and migrated inward, it might have avoided some of the punishing radiation characteristic of the early active dwarf stars. A panel of experts might consider the facts and make a recommendation.

In fact, scales have already been devised to assess risk, direct descendants of the so-called Torino scale devised to assess the impact hazard of near-Earth objects (Binzel 1997). In 2007 the *San Marino scale* was devised to assess the hazard of deliberate transmissions from Earth, in other words Active SETI, or what today has become known as METI (Almar and Shuch 2007; Shuch and Almar 2007).[3] The San Marino scale ranges from 0 for insignificant to 10 for extraordinary. It does not take into account the probability of being detected. For example, Drake has stated that by the time the Arecibo beacon reaches its target, the central core of the Hercules globular cluster some 25,000 light years will have partially rotated out of the beam, by about 1/7th of its diameter. Omnidirectional signals might be seen everywhere, but for economic and technical reasons the signal strength would likely be much weaker than in the case of a targeted beacon.

Scientists have also quantified past transmissions based on the Marino scale, assigning Drake's 1974 Arecibo message as an 8, which characterizes its significance as "far reaching," and a series of three Russian transmissions from the Evpatoria radio telescope in Ukraine as a 7, "high." Near-Earth object radars get a 6 for "noteworthy" (Shuch and Almar 2006). Although these scales have been criticized from an anthropocentric point of view by Traphagan and others (Traphagan 2015), the London and San Marino scales remain important reference points which scientists can use or improve in order to undertake a more focused discussion. They provide a starting point, especially if these criticisms are taken into account.

15.7 Criticisms of Asilomar

Does an Asilomar process with these kinds of questions solve the risk problems of both SETI and METI? Far from it, because the Asilomar process itself has critics. Berg himself pointed out it would be much more difficult today to organize a successful Asilomar process for controversial issues such as fetal tissue, embryonic stem-cell research, somatic and germ-line gene therapy, and the genetic modification of food crops. Why? Because in the 1970s the practitioners of recombinant DNA were working in public institutions like university labs, but now many work for private companies where commercial considerations are paramount. Moreover, he pointed out in 2008, many scientific issues today are beset by nearly irreconcilable political, religious, and ethical issues. "A conference that sets out to find a consensus among such contentious views would, he wrote, "be doomed to acrimony and policy stagnation" (Berg 2008). A meeting that had been held in 2000 on the 25th anniversary of Asilomar, attended by bioethicists, scientists, lawyers, and journalists assessed how the Asilomar model might apply to current pressing biotechnology issues such as gene therapy and genetically modified organisms reached virtually the same conclusion—the Asilomar process would not work today. Another assessment concluded "While the process is good philosophically, it has little bearing on reality," although such conferences "would be very beneficial in reestablishing scientific credibility" (Daveatelis 2000; Russo 2000).

Nevertheless, Berg's parting conclusion was this: "There is a lesson in Asilomar for all of science: the best way to respond to concerns created by emerging knowledge or early-stage technologies is for scientists from publicly funded institutions to find common cause with the wider public about the best way to regulate—as early as possible. Once scientists from corporations begin to dominate the research enterprise, it will simply be too late" (Berg 2008). The analog for METI might be if there is a profit motive, or a reputation motive, it might be too late. While it is possible that these could be motives for SETI and METI, for most practitioners the motivation is a sincere desire to communicate with the stars. So from that point of view, maybe an Asilomar process could work for METI.

15.8 The Equal Interest Problem and the Enforcement Problem

On the other hand, while it's fine to have consultations and come up with guidelines, we should ask do they work in the real world? They did for Asilomar in the 1970s, thanks to voluntary participation. But SETI and METI may be very different. As I mentioned at the beginning, the SETI protocols—basically confirm and then tell everyone—were the product of hard work at many levels over a long period of time. But they also depend on voluntary participation, and when it comes down to an actual anomalous signal, the protocols are sometimes broken. This is exactly what occurred in early summer 1997, when astronomers at the National Radio Astronomy Observatory in Green Bank, West Virginia detected a promising narrow band signal, confirmed as extraterrestrial in origin. It was the most promising such signal ever seen. As SETI astronomer Seth Shostak recalled, despite being very familiar with the SETI protocols, when the *New York Times* was tipped off and called the next morning, he did not lie. He said they were indeed checking out a signal, and to check back 3 h later. By that time, they had determined it was a telemetry signal from the Sun-observing satellite SOHO. The lesson, according to Shostak was that "The SETI protocols, while well intentioned, aren't particularly useful in real life." (Shostak 2016). Scientists tend not to lie if confronted by the press, which is probably a good aspirational goal for scientists. But that trait of human nature (or at least of scientists as compared, say, to certain politicians) dilutes any protocols.

The situation is similar for METI. In an Asilomar-like process serious and well-meaning people might come up with guidelines about which METI experiments are most dangerous. But if a planet is discovered in a habitable zone around a nearby star, does anyone really expect SETI or METI practitioners to cease and desist and to send a message somewhere else? I don't think so. To the contrary, those are precisely the objects both SETI and METI practitioners will want to target. Consider Tabby's star, a middle-aged F class star about half as massive as the Sun, 1300 light years away in the constellation Cygnus. In 2016 it was formally announced that 4 years of Kepler spacecraft data indicated this star was dimming in an irregular way that might only be explained by an alien megastructure passing in front of it as seen from Earth. The first thing SETI did was to take a look, and METI sponsored optical SETI laser observations (Schuetz et al. 2016). Nothing was found, but if it had been, METI and others would undoubtedly want to send a message. If METI did not, others less qualified certainly would. So, realistically speaking, it seems the consultation would be about what message to send, not when and if to send it.

So here is an important dis-analogy with Asilomar and biotechnology. In the case of biotech there were many experiments that could be undertaken that were less dangerous but equally interesting, until guidelines could be worked out for specific experiments. In the case of METI there are of course many possible targets, but not all are equally interesting. I call this situation the *equal interest problem*. Biotech as discussed at Asilomar did not have this problem of where to go for equally interesting research; METI does. Practically speaking, it is difficult to see that SETI or

METI practitioners would be expected to forego interesting targets in favor of less promising ones. And of course, if they decide not to forego the most interesting targets, there is no enforcement mechanism, short of legislation, which is unlikely to be put in place.

This is why it is so important for the public to be brought into the process to be made aware of the dangers and benefits. At root it is a problem of risk and exploration, a subject NASA has dealt with throughout its history. Indeed, in the wake of the Columbia accident in 2003 and calls to cancel the entire Space Shuttle program because the risk was too great, NASA Administrator Sean O'Keefe convened a memorable symposium on the subject (Dick and Cowing 2005). The bottom line for most at that symposium was that everything had to be done to minimize the risk, but in the end the first priority was not safety, but to GO, to launch, which is exactly what NASA did with its return to flight in the summer of 2005. NASA continued to launch shuttles for 6 more years until 2011, when the program was cancelled primarily for budget reasons, not safety reasons. Interestingly, in 2004 Administrator Sean O'Keefe famously decided not to have a final repair mission for the Hubble Space Telescope, because he risk was too great (see Chap. 21). But his successor as NASA Administrator, Michael Griffin, reversed that decision in 2006, with the result that the lifetime of the HST was extended and is still going strong. The lesson here is that assessments of risk are subjective. These examples are not precisely analogous to METI, because the lives of seven astronauts were at risk, not the lives of everyone on the planet. But the general idea of assessing risk, and its sometimes subjective nature, is important for METI. Related to this is the precautionary principle—the idea that METI practitioners have a burden of proof to demonstrate that the risk is not high. It is sobering fact that in the end all such assessments are subjective.

Kathryn Denning has very well framed the issue when she points out that in the end METI and its issues are "the problem of the Commons:" "the individual right to action, the accumulated effects of many individuals actions upon a society, and appropriated behavior regarding collective resources," in short, a problem of shared resources writ large (Denning 2011). That problem, she argued, has no simple technical solution, but METI belongs to this class of problems that has a vast literature in economics, the social sciences, and governance, a literature that may be useful for the current debate. Mark Lupisella has pointed out that in the end the policy also connects with philosophy and ethics, particularly with cosmocentric thinking (Lupisella 2011).

In the end METI is part of larger set of issues having to do with not only the cosmic commons, but also with human effects on planet Earth. In his book *Earth in Human Hands: Shaping our Planet's Future*, astronomer David Grinspoon sees METI in just this way. After detailing the many ways in which humans are affecting Earth on a planetary scale, and debating whether we have entered a new Anthropocene era, he relates that he "began to see the METI debate as emblematic of many of the dilemmas we are facing as part of this planetary transition in which cognition is starting to play a central role in the workings of the planet." He views METI as a global issue, in the same way as planetary protection from harmful organisms, or planetary defense from near-Earth asteroids, or the rise of the machines of artificial intelligence are. All of these are serious issues, and after wavering on both sides of the arguments, Grinspoon concludes that a "voluntary moratorium" is the best approach until such time as international consultations can take place (Grinspoon 2016, 352–404). And this puts us right back to the dis-analogies with Asilomar and the problems of consultation mentioned earlier. Active discussions continue.[4]

15.9 Recommendations and Conclusions

Let me end with five practical recommendations:

1. METI consultations, like SETI consultations in the event of a successful SETI detection, should take place at the level of the practitioners, supplemented by an array of scholars, in conjunction with an organization like the International Academy of Astronautics in order to give it more force. History shows that governments will likely not become involved until a message is received. To insist that consultations should be carried out at the level of the United Nations is the equivalent of saying that no action should ever be taken with METI.
2. We need to have clarity about the purpose of the consultations, which might take place on at least three levels: (a) to decide whether humanity should be in the business of sending messages to the stars at all; (b) if the answer is yes, to decide on the targets; (c) to decide on the transmission power; (d) to decide on the message language; and (e) to craft message content that minimizes risk and is to the benefit of mankind. I refer to these levels of consultation, respectively, as the Go/No Go consultation, the target consultation, and the message consultation. But what kind of result from these consultations would constitute no risk at all? Not knowing anything about alien psychology, I expect the "no risk" would be a null set. There will always be some risk, as in any endeavor, be it in daily life, in space exploration, or in METI. Some of those risks, as in planetary protection or climate change, are global risks.
3. We need to have public buy-in.
4. We should be clear-eyed about the philosophical ideal and the hard reality when it comes to actually implementing any principles derived from consultation.
5. We need to make use of literature on risk, the problem of the commons, the precautionary principle, and other relevant problems.

So where does this leave us today? I expect the best we can do is to make cautious use of analogy with past and present controversies, including biotechnology, space exploration, and planetary protection. I note that for all of the analogies I can think of research was not halted indefinitely. Biotech research continued after a brief moratorium. Space exploration continues despite planetary protection protocols. In the end, my own view is that humanity as a species cannot and should not hide from the universe. We cannot isolate ourselves from the universe because we are an integral part of it. We should feel at home in the biological universe, no less than we feel at home on planet Earth, even with all of its problems. Moreover, METI is in the same spirit of exploration as passive SETI, which some also said was not science when it began more than 50 years ago. And both are in the same spirit as astrobiology in general, which seeks life in the universe, leavened by planetary protection protocols.

My view is further that while governments should not fund METI as a means to propagandize, from a practical viewpoint groups that want to undertake such a project cannot be regulated short of legislation on the subject, which does not exist and is unlikely to happen. Given that situation, it seems to me a responsible group of scholars that wants to undertake a METI project should be free to do so if it seeks consensus inside and outside the group. The burning question remains what constitutes consensus, and there is no consensus on even this issue.[5] The Asilomar process was not an easy process, nor a perfect process, and many people, including one of its chief organizers, are on record as saying it probably would not work today given commercial interests and the contentious nature of science and society relationship. Nevertheless, we should try at the level of practitioners, bringing in a wide array of scholars, and including the general public.

15.10 Commentary 2020

The first part of this chapter was a blog posted on the METI International website on December 9, 2015, available at http://meti.org/en/blog/should-we-message-et. The latter part of this chapter discussing Asilomar was first delivered at the International Space Development Conference (ISDC), held in St. Louis in 2017 and is published here for the first time. The issues of astroethics in both SETI and METI remain urgent questions (Chap. 14). Full disclosure: I was a member of the Board of Directors of METI International from its founding in 2015 until 2019.

Notes

1. Full disclosure: the author is on the Board of METI International. See http://www.meti.org.
2. The gathering included astronomers Frank Drake and Jill Tarter, anthropologists represented by Ben Finney and Michael Ashkenazi, religious scholars and historians including Georgetown's Langdon Gilkey and Harvard's Karl Guthke, several representatives from media studies, and even two diplomats, represented by Michael Michaud from the State Department and Nandasiri Jasentuliyana, the Director of the Office of Outer Space Affairs at the United Nations.
3. The scale has only two parameters, intensity or strength of the signal sent (ranging from 0 for low to 5 for very high), and character of the transmission, also ranging from 0 to 5. Zero represents a beacon with no message such as a planetary radar, while 4 represents a continuous, omnidirectional broadband transmission of a message, and 5 is an actual reply to an extraterrestrial message). For an online calculator for the Rio scale, devised by the Permanent SETI Committee of the IAA, see http://avsport.org/IAA/riocalc.htm.

4. For more on the controversy see Vakoch (2011b, c), Korbitz (2014), Harrison (2014), in Vakoch (2014a), Jones (2011). See also http://www.universetoday.com/119055/who-speaks-for-earth-the-controversy-over-interstellar-messaging/.
5. For the case in favor of METI see Vakoch (2011b, c). For the case against METI see Brin (2011, 2013). See also Musso (2012).

References

Almar, Ivan and Paul Shuch. 2007. "The San Marino Scale: A New Analytical Tool for Assessing Transmission Risk," *Acta Astronautica*, 60, 57–59

Baral, Susmita. 2016. "Why Aliens Shouldn't Contact Us." http://www.ibtimes.com/why-aliens-shouldnt-contact-us-stephen-hawking-warns-intelligent-extraterrestrial-2420526.

Berg, Paul. 2008. "Meetings that changed the world: Asilomar 1975: DNA modification secured," *Nature*, 455 (18 September, 2008), 290–291, at http://www.nature.com/nature/journal/v455/n7211/full/455290a.html.

Berkeley. 2015. "Regarding Messaging to Extraterrestrial Intelligence (METI)/Active Searches for Extraterrestrial Intelligence (Active SETI)," http://setiathome.berkeley.edu/meti_statement_0.html.

Billingham, John, Heyns, Roger, Milne, David, et al. 1999. *Social Implications of the Detection of an Extraterrestrial Civilization*, Mountain View, CA.: SETI Press.

Binzel, R. P. 1997. "A Near-Earth Object Hazard Index," *Annals of the New York Academy of Sciences*, 822, 545–551;

Brin, David. 2011. "A Contrarian Perspective on Altruism: The Dangers of First Contact," in Shuch (2011), pp. 429–449: 433

Brin, David. 2013. "Shouting At the Cosmos: How SETI has Taken a Worrisome Turn Into Dangerous Territory," online at http://www.davidbrin.com/shouldsetitransmit.html.

Daveatelis, George. 2000. "The Asilomar Process: Is it Valid?" *The Scientist*, April, 2000. http://www.the-scientist.com/?articles.view/articleNo/12781/title/The-Asilomar-Process%2D%2DIs-It-Valid-/.

Denning, Kathryn. 2011. "Unpacking the Great Transmission Debate," in Vakoch (2011b) pp. 237–252.

Dick, Steven J. 2015. *The Impact of Discovering Life Beyond Earth*. Cambridge: Cambridge University Press.

Dick, Steven J. 2018. *Astrobiology, Discovery, and Societal Impact*: Cambridge: Cambridge University Press.

Dick, Steven J. and Keith Cowing, eds. 2005. *Risk and Exploration: Earth, Sea and the Stars*. NASA SP-2005-470. *http://history.nasa.gov/SP-4701/frontmatter.pdf*.

Dominik, Martin and J. C. Zarnecki. 2011."The Detection of Extra-terrestrial Life and the Consequences for Science and Society," *Philosophical Transactions of the Royal Society A*, 36, 409–507

Drake, Frank. 2017. Private communication.

Ekers, Ronald D., Kent Cullers, John Billingham and Louis K. Scheffer, eds., 2002. *SETI 2020: A Roadmap for the Search for Extraterrestrial Intelligence* (Mountain View, CA: SETI Press, 2002).

Grinspoon, David. 2016. *Earth in Human Hands: Shaping our Planet's Future*, New York: Grand Central Publishing, pp. 352–404: 371.

Harrison, Albert. 2014. "Speaking for Earth: Projecting Cultural Values Across Deep Space and Time," in Vakoch (2014a), pp. 175–190.

International Academy of Astronautics SETI Committee. 1995. "Draft Declaration of Principles Concerning Sending Communications with Extraterrestrial Intelligence."

Jones, Morris. 2011. "A Journalistic Perspective on SETI-Related Message Composition," in Douglas A. Vakoch and Albert. A. Harrison, eds., *Civilizations Beyond Earth: Extraterrestrial Life and Society*, (New York and Oxford: Berghahn Books, 2011), pp. 226–235.

Korbitz, Adam. 2014. "The Precautionary Principle: Egoism, Altruism, and the Active SETI Debate," In Vakoch (2014b), pp. 111–127.

Lupisella, Mark. 2011. "Pragmatism, Cosmocentrism, and Proportional Consultation for Communication with Extraterrestrial Intelligence," in Vakoch (2011a), pp. 319–331.

METI International. 2015. http://www.meti.org.

Musso, Paolo. 2012. "The Problem of Active SETI: An Overview," *Acta Astronautica*, 78 (2012), 43–54.

Nature. 2008. "Meetings that changed the world *Nature*, 455 (18 September, 2008) http://www.nature.com/nature/focus/meetings/.

New York Times. 1976a. "Astronomer Fears Hostile Attack; Would Keep Life on Earth a Secret," November 4, 1976.

New York Times. 1976b. "Should Mankind Hide?," November 22, p. A24.

Russo, Eugene. 2000. http://www.the-scientist.com/?articles.view/articleNo/12768/title/Reconsidering-Asilomar/.

Schuetz, Marlin, Douglas A. Vakoch, Seth Shostak, Jon Richards, 2016. Optical SETI Observations of the Anomalous Star KIC 8462852, *Astrophysical Journal Letters*, 825, p. L5.

Shostak, Seth. 2016. "Aliens on Line 1," *Air and Space*, Sept 2016, pp. 14–16.

Shuch, Paul and Ivan Almar. 2007. "Shouting in the Jungle: The SETI Transmission Debate," *Journal of the British Interplanetary Society*, 60, 142–146

Shuch, Paul and Ivan Almar. 2006. "Quantifying Past Transmissions Using the San Marino Scale."

Traphagan, John W. 2015. *Extraterrestrial Intelligence and Human Imagination: SETI at the Intersection of Science, Religion, and Culture.* Heidelberg: Springer, pp.73–76.

Vakoch, Douglas, ed. 2011a. "Asymmetry in Active SETI: A case for transmissions from Earth," *Acta Astronautica*, 68, 476–488, in Vakoch (2011a), 253–278.

Vakoch, Douglas. 2011b. *Communication with Extraterrestrial Intelligence*. Albany, New York, SUNY Press.

Vakoch, Douglas A. 2011c. "Responsibility, capability, and Active SETI: Policy, law, ethics, and communication with extraterrestrial intelligence," *Acta Astronautica*, 68 (2011), 512–519

Vakoch, Douglas, ed., 2014a. *Archaeology, Anthropology and Interstellar Communication*. Washington, DC: NASA.

Vakoch, Douglas A., ed., 2014b. *Extraterrestrial Altruism: Evolution and Ethics in the Cosmos*". Berlin: Springer,

Zaitsev, Alexander L. 2011. "METI: Messaging to ExtraTerrestrial Intelligence," in H. Paul Shuch, ed., *Searching for Extraterrestrial Intelligence*. Berlin: Springer, 399–428.

Chapter 16
Astrobiology and Society: An Overview at the Beginning of the Twenty-First Century

Abstract What are the implications of astrobiology for society? When one considers that astrobiology encompasses research on the origin and evolution of life, the existence of life beyond Earth, and the future of life on Earth and beyond, the scope of that deceptively simple question becomes clear. It embraces not only the religious, ethical, legal, and cultural concerns inherent in those subjects, but also the meaning of life and even human destiny in a universe where humans are unique—or not. Particularly in the area of extraterrestrial life, which has been a focus for astrobiology and society concerns in terms of implications, the issues have been global and contentious. The consequences have long been vividly played out in science fiction by classic authors such as Arthur C. Clarke in *Childhood's End* or *2001: A Space Odyssey*, and by more recent writers like Ted Chiang in "Story of Your Life" and its film adaption *Arrival*.

How can we even approach such questions as the impact of discovering life beyond Earth, whether microbial or intelligent? How can we transcend anthropocentrism when we address concepts such as life and intelligence, culture and civilization, technology and communication? And in what areas is humanity most likely to be transformed by such a discovery? We cannot fully answer these questions in this chapter, but there is now a surprisingly substantial literature that does address them. As with astrobiology, it is prudent for current researchers in the subject to be aware of this much shorter history, whether to contest or expand it. In this chapter we provide an overview of this literature on astrobiology and society. Substantial as it may seem, it is only the leading edge of what is sure to become an entire discipline of its own, especially if life is actually discovered out there among the stars.

16.1 Introduction

A report from the World Economic Forum in 2013 declared the discovery of life beyond Earth one of five X factors—emerging concerns for planet Earth of possible future importance but with unknown consequences. Giving attention to X factors,

---

First published as "Astrobiology and Society Comes of Age," Introduction to Kelly Smith, "*Social and Conceptual Issues in Astrobiology*" (Oxford: Oxford University Press. 2019).

S. J. Dick, *Space, Time, and Aliens*, https://doi.org/10.1007/978-3-030-41614-0_16

the report suggested, would lead to a more proactive approach if and when these events actually occurred, resulting in more *cognitive resilience* and perhaps preventing at least some undesirable social consequences. Such consequences could occur even if simple alien life were discovered. "Over the long term," the report argued, "the psychological and philosophical implications of the discovery could be profound … The discovery of even simple life would fuel speculation about the existence of other intelligent beings and challenge many assumptions that underpin human philosophy and religion." The study of these assumptions and implications is therefore far from academic, and is a worthy endeavor even if life is never discovered beyond Earth. Such studies also help to nurture a cosmic perspective sorely needed in our turbulent times.

16.2 Early Explorations in Astrobiology and Society

Interest in astrobiology and society in its broadest sense date back at least a quarter century to the days when NASA was planning its SETI program. In 1976–1977 when scientists first met to contemplate this program, the discussions included the possibilities of cultural evolution beyond the Earth, led by none other than the young Nobelist Joshua Lederberg, whose 2-day "Workshop on Cultural Evolution," focused more specifically on "evolution of intelligent species and technological civilizations." Among the conclusions of the group—which included several scholars in the social sciences—was that "our new knowledge has changed the attitude of many specialists about the generality of cultural evolution from one of skepticism to a belief that it is a natural consequence of evolution under many environmental circumstances, given enough time" (Morrison et al. 1977). This meant that cultures beyond the Earth, perhaps ending in technological civilizations capable of radio communication, were at least a possibility. A few farsighted anthropologists were even beginning to show some interest (Maruyama and Harkins 1975), an interest that has grown over the decades since (Vakoch 2009, 2014a).

It is quite remarkable that the early practitioners of SETI were already sensitive to societal concerns. In the early 1990s, just prior to the inauguration of NASA SETI operations on the quincentennial of Columbus's first landfall in the Americas, NASA convened a series of workshops on the cultural aspects of SETI (CASETI). The intimate gathering of two dozen scholars (see Chap. 10) was a model of interdisciplinary brainstorming, with astronomers, anthropologists, religious scholars, historians, several representatives from media studies, and even two diplomats. The gathering was a *de facto* recognition that this was a broad-based problem not to be solved by scientists alone. While the publication of the results was delayed almost a decade by the cancellation of the scientific program, its recommendations are still valuable for contemplating the aftermath of any successful SETI program (Billingham et al. 1999). Plans for an international conference on the subject were cancelled when the NASA SETI program itself was cancelled.

The quick and untimely demise of the NASA SETI program meant that astrobiology and society discussions would be scattered and sporadic. One opportunity for a more systematic treatment of the societal aspects of astrobiology was NASA's construction of a roadmap for astrobiology, mentioned above. However, although some proponents argued that astrobiology and society issues should be among the Roadmap's firm goals, in the end proponents had to be content that two of the four roadmap operating principles were related to these issues, one in encouraging planetary stewardship by emphasizing planetary protection and avoiding contamination, and another by recognizing "a broad societal interest in our subject," including the discovery of extraterrestrial life and engineering new life forms adapted to live on other worlds. Thus, while the Roadmap and its successors served as a focus for a broad program of science research, they did not do the same for funding social sciences and humanities research.

These conditions notwithstanding, it is rather surprising that in 1999 NASA's Ames Research Center organized a workshop on the societal implications of astrobiology (Harrison and Connell 2001). This time about 50 scholars ranging from futurists like Alvin Toffler to anthropologists, scientists and journalists gathered to discuss the subject. Not surprisingly, the group emphasized the importance of their task: to encourage public understanding of this new science, to gauge public reaction to astrobiological discoveries, and to prepare for the future through policy decisions given "a possible sea of living worlds." More than a dozen recommendations were issued, including the importance of a multidisciplinary approach involving both scientists and humanists, studying the implications of a shift in our frame of reference from the Earth to a living cosmos, making "state-of-the-art preparations" for discovery of life, studying the ethical implications of discovering life, and implementing policy measures "to ensure the integrity of extraterrestrial life." They made a strong case for undertaking serious levels of research and outreach before the fact of discovery, arguing such research should be integrated into core science initiatives (as would soon be done with the Human Genome Project). "Science and society are deeply and irrevocably intertwined," they wrote, "and a mutual appreciation of the close relationship is vital to the integrity of both fields."

Beyond NASA, several other organizations undertook initiatives on the subject. One notable meeting was sponsored by the John Templeton Foundation, which focuses on the dialogue between science and religion. The Foundation convened a meeting in late 1998, only a few months after the first NASA astrobiology roadmap was constructed. Again, the meeting was interdisciplinary, including a Nobel biochemist (Christian de Duve), physicists, astronomers, theologians, one historian, and the very skeptical evolutionary biologist Richard Dawkins. The results, published as *Many Worlds: The New Universe, Extraterrestrial Life, and the Theological Implications* (Dick 2000), read like a cauldron of non-consensus. Theological and ethical issues would become an important component of societal issues in astrobiology (Dick 2018a, b; Impey et al. 2013; Peters 2013, 2014, 2018; Smith 2014).

16.3 Into the New Millennium

The new millennium has seen increasing, though still sporadic, interest in issues involving astrobiology and society and sponsored by a variety of organizations. In conjunction with NASA and the Templeton Foundation, in 2003 and 2004 Constance Bertka, a planetary scientist who also headed the Dialogue on Science, Ethics and Religion program at the American Association for the Advancement of Science (AAAS), convened a series of workshops at the AAAS in Washington that included ethical and theological perspectives on the origins, extent, and future of life (Bertka 2009). Since the AAAS is the largest organization of scientists in the world these discussions (and the DoSER program in general) are an indication that scientists are interested in the social impact of what they do, as well they should be. From changing definitions of life to extraterrestrial life to the future of life in the universe, these workshops proved to be a window on the many issues that need to be tackled under the umbrella of astrobiology and society. Another example is a workshop held in 2008 at the University of Arizona's Biosphere 2 artificial ecosystem facility. Like the AAAS workshops, these discussions ranged across the full spectrum of societal, cultural, and ethical issues in astrobiology (Impey et al. 2013). Yet another example is a meeting in Hven, Sweden in 2011 sponsored by the Pufendorf Institute for Advanced Studies at Lund University. The publication of the proceedings in the scientific journal *Astrobiology* (Dick 2012; Dunér et al. 2012, 2013) is another indication of interest among scientists in societal issues. Sometimes meetings have addressed more specific issues such as communication with extraterrestrial intelligence, exemplified especially in a series of volumes edited by Douglas Vakoch, who for many years held the title of Director of Interstellar Communications at the SETI Institute (Vakoch 2011, 2013, 2014a, b).

Stimulating as they were, the scattered discussions of the previous decades cried out for more organization and synthesis. This was the hopeful goal of a meeting in 2009, held under the auspices of the NASA Astrobiology Institute. Some 43 invited scholars gathered at the SETI Institute to develop an "Astrobiology and Society" roadmap, fully aware of the astrobiology science process. Unlike the science roadmap, however, the societal impact roadmap (Race et al. 2012) was not officially adopted by NASA and thus has not become policy backed up by sustained funding. But the work continues at a basic level, and the process seems to be following its companion science roadmap in percolating from the bottom up with minimal funding and the hope of eventually becoming a more recognized and funded activity. That will require the two cultures to work together, and it is encouraging that the introduction to the latest 2015 Astrobiology Strategy document still lists a goal to enhance societal interest and relevance. "Astrobiology recognizes a broad societal interest in its endeavors," it states, "especially in areas such as achieving a deeper understanding of life, searching for extraterrestrial biospheres, assessing the societal implications of discovering other examples of life, and envisioning the future of life on Earth and in space" (NASA 2015). The document also includes as an appendix a humanities and social sciences section, the substance of which many feel should be a more integral part of the report.

Finally, NASA's establishment of the Baruch S. Blumberg NASA/Library of Congress Chair in Astrobiology in 2011, specifically to address the humanistic and societal aspects of astrobiology, is a de facto recognition of the importance of these issues. This prestigious position has resulted in both individual and collective research (Grinspoon 2016; Dick 2015, 2018b), drawing in younger scholars from a variety of disciplines, and giving respectability to a field that has long been on the margins.

Thus, far from initial skepticism about a role for the social sciences and humanities in astrobiology, there is now considerable consensus that the problem of the impact of discovering life in any form is not only important but essential, and should not be left to scientists alone. The same is true of the broader aspects of astrobiology and society. When the Royal Society of London sponsored a meeting on the detection of extraterrestrial life and the consequences for science and society in 2010, and a satellite meeting seeking a scientific and societal agenda on extraterrestrial life, the organizers wrote that "While scientists are obliged to assess benefits and risks that relate to their research, the political responsibility for decisions arising following the detection of extraterrestrial life cannot and should not rest with them. Any such decision will require a broad societal dialogue and a proper political mandate. If extraterrestrial life happens to be detected, a coordinated response that takes into account all the related sensitivities should already be in place" (Dominik and Zarnecki 2011).

My point is that these and other conferences on astrobiology and society (Table 16.1) should form the collective basis for future studies. Moreover, a few individual efforts have also concentrated on aspects of the problem. Foremost among these are psychologist Albert Harrison's volume *After Contact: The Human Response to Extraterrestrial Life* (Harrison 1997) and the American diplomat Michael Michaud's *Contact with Alien Civilizations: Our Hopes and Fears about Encountering Extraterrestrials* (Michaud 2007). While some have argued that we know nothing about extraterrestrial intelligence (Billings 2015), the anthropologist Michael Ashkenazi, one of the participants in the original CASETI workshops, has offered an answer of sorts with a large volume *What We Know About Extraterrestrial Intelligence: Foundations of Xenology* (Ashkenazi 2017), arguing that we can actually infer quite a bit about extraterrestrials and therefore lay out scenarios about societal impacts. Individual efforts are also represented in a plethora of widely scattered articles (Dick 2013), whose full extent may be measured in the 30-page bibliography of *Astrobiology, Discovery, and Societal Impact* (Dick 2018b). In short, the humanities and social sciences should become not a peripheral activity, but an integral part of astrobiology as a discipline (Fig. 16.1).

16.4 Anticipating the Future

In summary, the future for astrobiology and society studies looks bright, if not guaranteed to maintain momentum. If studies so far have been dominated by researchers in the United States, an important harbinger comes from Europe. In contrast to the NASA Astrobiology roadmap, in 2017–2018 astrobiology and society became a

Table 16.1 Conferences on societal impact of astrobiology, 1991–2018

Meeting	Date and place	Sponsor	Results
Cultural Aspects of SETI (CASETI)	1991–1992 Chaminade Conference Center, Santa Cruz, California	NASA	John Billingham et al., eds., *Social Implications of the Detection of an Extraterrestrial Civilization* (1999)
Many Worlds	November 22–24, 1998 Lyford Key, Nassau, The Bahamas	John Templeton Foundation	Steven Dick, ed., *Many Worlds: The New Universe, Extraterrestrial Life, and the Theological Implications* (2000)
When SETI Succeeds	1999 Hapuna Prince Big Island of Hawaii	Foundation for the Future	Allen Tough, ed., *When SETI Succeeds: The Impact of High-Information Contact* (2000)
Societal Implications of Astrobiology Workshop	November 16–17, 1999 NASA Ames	NASA	Albert Harrison et al., eds. *Workshop on the Societal Implications of Astrobiology* (1999)
Exploring the Origin, Extent and Future of Life	2003 American Association for Advancement of Science Washington, DC	NASA/American Association for the Advancement of Science	Constance Bertka, ed., *Exploring the Origin, Extent and Future of Life* (2009)
Astrobiology: Expanding our Views of Society and Self	May 2008 Univ. of Arizona Biosphere 2 Institute	University of Arizona	Chris Impey et al., eds. *Encountering Life in the Universe* (2013)
Astrobiology and Society	February 2009 SETI Institute	NASA Astrobiology Institute	Margaret Race et al., "Astrobiology and Society," *Astrobiology*, 12 (2012), pp. 958–965
The Detection of Extra-terrestrial Life and the Consequences for Science and Society	25–26 January, 2010 Royal Society in London	Royal Society of London	Dominik and Zarneki, eds., *Philosophical Transaction of the Royal Society of London A*, vol. 369, issue 1936 (2011)
Satellite Meeting	Kavli Centre Buckinghamshire		
The History and Philosophy of Astrobiology	September 27–28, 2011, Ven, Sweden	Pufendorf Institute for Advanced Studies, Lund University, Sweden	David Duner et al., *Astrobiology* special issue, vol. 12 (2012); David Duner, ed., *The History and Philosophy of Astrobiology*
Preparing for Discovery	September, 2014 Library of Congress	NASA/Library of Congress	Steven Dick, ed. *The Impact of Discovering Life Beyond Earth* (2015)

(continued)

Table 16.1 (continued)

Meeting	Date and place	Sponsor	Results
Social and Conceptual Issues in Astrobiology 2016	September, 2016 Clemson University	Clemson University	Kelly Smith, *Social and Conceptual Issues in Astrobiology*, in press
Social and Conceptual Issues in Astrobiology 2018	April, 2018 University of Nevada, Reno	University of Nevada, Reno; Blue Marble Institute; others	Ted Peters (Ed.) *Science and Theology* special issue, Vol 17 (2019); Kelly Smith and Keith Abney (Eds.) *Futures* special issue, in press

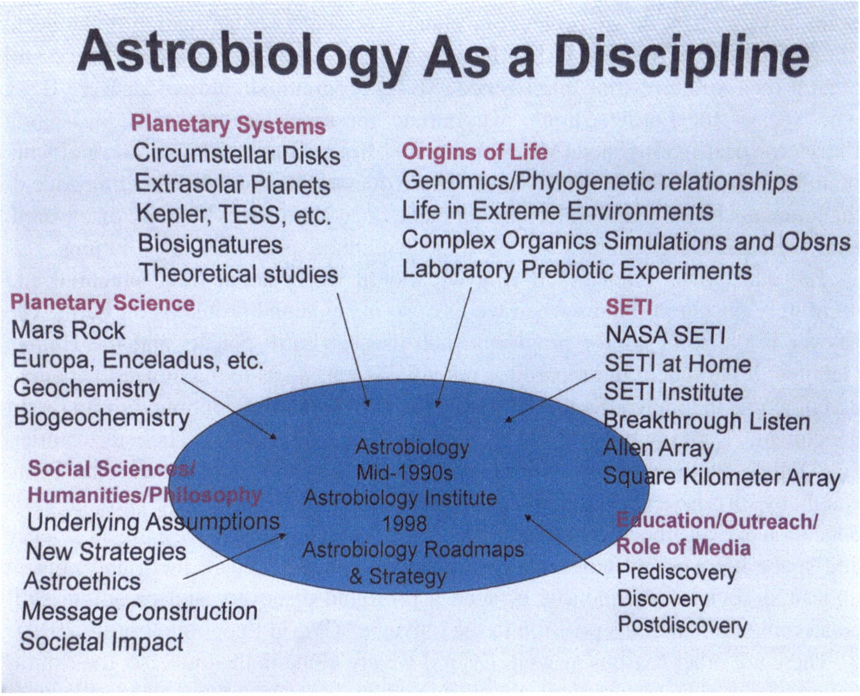

Fig. 16.1 Astrobiology as a discipline now includes the humanities, social sciences and philosophy. Education, outreach, and the role of the media are also crucial. Compare to Fig. 2.3

foundational theme for the proposed European Astrobiology Institute (EAI). In contrast to the American astrobiology roadmap process in 1998, 20 years later the European Astrobiology Institute systematically laid out the societal issues in a roadmap that bids fair to become an integral part of astrobiology in Europe (Capova and Persson 2018). In addition to the Royal Society meeting mentioned above, the EAI initiative has been preceded in recent years by European research on the subject of societal impacts (Dunér et al. 2012; Dunér et al. 2013). Though not yet fully established as of this writing, when and if the EAI is fully established, it bodes well for the astrobiology and society theme. So does the inauguration in 2017 of the Society for Conceptual Issues in Astrobiology (SoCIA), which envisions international participation, and of which this volume is a product.

This chapter is all too brief to lay out comprehensively the issues encompassed in the astrobiology and society. But the questions are legion, and potentially Earth-shaking. Who should take the lead in preparing for discovery? What do we do if life is actually discovered, microbial or intelligent, near or far? Should national governments be in charge, international political and scientific institutions, scientists and social scientists, ethicists and theologians, or some mix thereof? How do we prevent contamination of potential microbes on Mars, Europa, Enceladus or other habitable sites in the Solar System, and (more perhaps more urgently from most Earthlings' point of view) how do we protect our planet from back contamination in the event of the discovery of microbial life? If a message is received as a result of a successful Search for Extraterrestrial Intelligence (SETI) program, should we answer? If so, who speaks for Earth? Should we initiate messages as part of a Messaging Extraterrestrial Intelligence (METI) program? If so, what should we say, and who, if anyone, should control what is said? These questions are only the leading edge of the many decisions that will have to be made once alien life is actually discovered. And each discovery scenario will have its own unique problems and solutions.

The question is often asked why we should worry about these potential and seemingly far-out problems when we have so many actual problems on Earth. The answer is the same as for programs such as near-Earth objects and the Human Genome: it is prudent to prepare for potential events so as to maximize the beneficial outcome that may affect all of humanity. As the World Economic Forum report concluded, "Looking forward and identifying emerging issues will help us to anticipate future challenges and adopt a more proactive approach, rather than being caught by surprise and forced into a fully reactive mode." Moreover, "Through basic education and awareness campaigns, the general public can achieve a higher science and space literacy and cognitive resilience that would prepare them and prevent undesired social consequences of such a profound discovery and paradigm shift concerning humankind's position in the universe" (World Economic Forum 2013).

There are other reasons as well. Even if we are alone in the universe, the examination of our basic assumptions about life and intelligence, culture and civilization, technology and communication, will have been well worth it. It has been said before, but it bears repeating, that astrobiology is in many ways a search for ourselves, for our place in the universe, and for our future destiny. Our destiny will be much different if we live in the universe of Isaac Asimov, where life is human or robotic products of humans, or if we live in Arthur C. Clarke's universe, where alien life is everywhere. In either case, we need to be good stewards of our planet. But if aliens are in the mix, whether for good or ill we will have to deal with them. The universe is what it is, not what we want it to be. Meanwhile, the presence or absence of life will be one of the greatest discoveries in the history of science.

16.5 Commentary 2020

This chapter is part of an introduction written for the Proceedings of a meeting held at Clemson University in September, 2016, where I delivered the keynote address (Smith 2019). The meeting was the first in a series organized by Clemson philosopher Kelly Smith based on the organization he founded in 2016, the Society for Social and Conceptual Issues in Astrobiology (SoCIA). That founding, and its subsequent biennial meetings are themselves evidence of the sustained and growing interest in the issues of astrobiology and society. A much longer version of this chapter is found in Dick (2019).

Following up on previous studies, efforts continue to prepare for discovery in the policy arena (Denning and Dick 2019).

References

Ashkenazi, Michael. 2017. *What We Know About Extraterrestrial Intelligence: Foundations of Xenology*. Switzerland: Springer.

Bertka, Constance, ed. 2009. *Exploring the Origin, Extent, and Future of Life: Philosophical, Ethical and Theological Perspectives*. Cambridge: Cambridge University Press.

Billingham, John, Heyns, Roger, Milne David, et al. 1999. *Social Implications of the Detection of an Extraterrestrial Civilization*, Mountain View, CA.: SETI Press.

Billings, Linda. 2015. "The Allure of Alien Life: Public and Media Framings of Extraterrestrial Life," in Dick (2015), pp. 308–323.

Capova, K. A. and Persson, E. 2018. "Astrobiology and Society in Europe Today: The White Paper on Societal Implications of Astrobiology Research in Europe."

Denning, Kathryn and S. J. Dick. 2019. "Preparing for the Discovery of Life Beyond Earth", Astro2020 National Academies Decadal Survey white paper. https://tinyurl.com/y64wqpqe.

Dick, Steven J. 2000. Many Worlds: The New Universe, Extraterrestrial Life and the Theological Implications. Philadelphia: Templeton Press.

Dick, Steven J. 2012. "Critical Issues in the History, Philosophy, and Sociology of *Astrobiology*," *Astrobiology,* 12, 906–927.

Dick, Steven J. 2013. "The Societal Impact of Extraterrestrial Life: The Relevance of History and the Social Sciences," in Vakoch (2013), pp. 227–257.

Dick, Steven J. 2015. The Impact of Discovering Life Beyond Earth. Cambridge: Cambridge University Press.

Dick, Steven J. 2018a. "Toward a Constructive Naturalistic Cosmotheology," in Peters (2018), 228–244.

Dick, Steven J. 2018b. Astrobiology, Discovery, and Societal Impact: Cambridge: Cambridge University Press.

Dominik, Martin, and J. C. Zarnecki. 2011. "The Detection of Extra-terrestrial Life and the Consequences for Science and Society," *Philosophical Transactions of the Royal Society A*, 369, 1936, 409–507: 503–504. http://rsta.royalsocietypublishing.org/content/369/1936.toc.

Dick, Steven J. 2019. "Humanistic Implications of Discovering Life Beyond Earth," in *Handbook of Astrobiology*, Vera Kolb, ed., CRC Press, 741–756.

Dunér, David, E. Persson, and G. Holmberg, eds. 2012. *The History and Philosophy of Astrobiology*, special issue of Astrobiology, 12, pp. 901–1016.

Dunér, David, J. Parthemore, E. Persson, and G. Holmberg, eds. 2013. The History and Philosophy of Astrobiology: Perspectives on Extraterrestrial Life and the Human Mind. Newcastle-upon-Tyne: Cambridge Scholars Publishing.
Grinspoon, David. 2016. *Earth in Human Hands: Shaping our Planet's Future*. New York: Grand Central Publishing.
Harrison, Albert A. 1997. *After Contact: The Human Response to Extraterrestrial Life*. New York: Plenum.
Harrison, Albert A. and Kathleen Connell. 2001. *Workshop on the Societal Implications of Astrobiology*. Moffett Field: NASA Ames Research Center. Online at http://www.astrosociology.org/Library/PDF/NASA-Workshop-Report-Societal-Implications-of-Astrobiology.pdf.
Impey, Chris, Anna Spitz, and William Stoeger, eds. 2013. Encountering Life in the Universe: Ethical Foundations and Social Implications of Astrobiology. Tucson: University of Arizona Press.
Maruyama, M. and A. Harkins, eds. 1975. Cultures Beyond the Earth: The Role of Anthropology in Outer Space. New York: Vintage Books.
Meech, Karen J., J. V. Keane, Michael Mumma et al. 2009. *Bioastronomy 2007: Molecules, Microbes and Extraterrestrial Life*. San Francisco: Astronomical Society of the Pacific.
Michaud, Michael A. G. 2007. Contact with Alien Civilizations: Our Hopes and Fears about Encountering *Extraterrestrials*, New York: Copernicus.
Morrison, Philip, John Billingham, and John Wolfe. 1977. The Search for Extraterrestrial Intelligence (SETI). Washington, DC: NASA.
NASA. 2015. Astrobiology Strategy online at https://nai.nasa.gov/media/medialibrary/2016/04/NASA_Astrobiology_Strategy_2015_FINAL_041216.pdf,
Peters, Ted. 2013. "Astroethics: Engaging Extraterrestrial Intelligent Life-Forms," in Impey et al, eds. (2013), pp. 200–221.
Peters, Ted. 2014. "Astrotheology: A Constructive Proposal," *Zygon*, 49, 443–457.
Peters, Ted, ed. 2018. Astrotheology: Science and Theology Meet Extraterrestrial Life. Wipf and Stock, Cascade Books, Eugene, Oregon
Race, Margaret S., Kathryn Denning, Constance Bertka, et al. 2012. "Astrobiology and Society: Building an Interdisciplinary Research Community," *Astrobiology*, 12, number 10, 958–965.
Smith, Kelly. 2014. "Manifest Complexity: A Foundational Ethic for Astrobiology?" *Space Policy*, 30, 209–214.
Smith, Kelly. 2019. Social and Conceptual Issues in Astrobiology. Oxford: Oxford University Press.
Vakoch, Douglas. 2009. "Anthropological Contributions to the Search for Extraterrestrial Intelligence," in Meech et al. (2009), pp. 421–427.
Vakoch, Douglas, ed. 2011. *Communication with Extraterrestrial Intelligence*. Albany: SUNY Press.
Vakoch, Douglas, ed. 2013. Astrobiology, History and Society: Life Beyond Earth and the Impact of Discovery. Berlin-Heidelberg: Springer.
Vakoch, Douglas, ed. 2014a. Archaeology, Anthropology and Interstellar Communication. Washington, DC: NASA.
Vakoch, Douglas, ed. 2014b. Extraterrestrial Altruism: Evolution and Ethics in the Cosmos. Heidelberg and New York: Springer.
World Economic Forum. 2013. *Global Risks 2013*, Lee Howard, ed. Geneva: World Economic Forum, 2013, at http://reports.weforum.org/global-risks-2013/section-five/x-factors/#hide/img-5.

Part III
The Exploration of Space

Part III Frontispiece Apollo 15 Lunar Module pilot James B. Irwin loads the rover with tools and equipment at the Hadley-Apennine landing site, with a portion of the Lunar Module *Falcon* on the left. Human and robotic spaceflight represent the latest episodes in the saga of human exploration. (NASA Image AS15-86-1160)

Oh! I have slipped the surly bonds of Earth
And danced the skies on laughter-silvered wings;
Sunward I've climbed and joined the tumbling mirth
of sun-split clouds—and done a hundred things
You have not dreamed of—wheeled and soared and swung
High in the sunlit silence. Hov'ring there,
I've chased the shouting wind along, and flung
My eager craft through footless falls of air . . .
Up, up the long, delirious, burning blue
I've topped the wind-swept heights with easy grace
Where never lark, nor even eagle flew—
And, while with silent lifting mind I've trod
The high, untrespassed sanctity of space,
Put out my hand and touched the face of God.

"High Flight"
by Pilot Officer John Gillespie Magee, Jr.
No. 412 squadron, Royal Canadian Air Force (RCAF) Killed 11 December 1941, age 19
A poem beloved by aviators and astronauts alike

Little more than 50 years separated the beginnings of the era of flight from the era of spaceflight. Most will agree that aeronautics has significantly affected culture, since daily flight has become commonplace. But the significant impact of the exploration of space beginning in the latter half of the twentieth century is also beyond dispute. In the arena of human space travel we need only recall the Apollo program: the reading of Genesis from circumlunar orbit in December, 1968 after a tumultuous year on planet Earth; the first Moon landing the following year watched on television by a significant fraction of Earth's population; and the images of Earthrise as seen from the Moon and the Earth as "Blue Marble," both of which played a significant role in giving rise to Earth Day and changed our perspective on our home planet forever (Chaikin 2007; Poole 2008; Lambright 2005, 2007). In the arena of robotic spaceflight, spacecraft such as Voyager, Galileo, Cassini radically changed our view of the Solar System, while the Hubble Space Telescope captured the popular imagination, with its spectacular imagery of deep space, its sometimes hair-raising servicing missions, and its risk to Space Shuttle astronauts, to the extent that it became known as "the people's telescope."

It was recognized early in the Space Age that spaceflight would affect society. NASA's founding document, the National Aeronautics and Space Act of 1958, specifically charged the new agency with eight objectives, including "the establishment of long-range studies of the potential benefits to be gained from, the opportunities for, and the problems involved in the utilization of aeronautical and space activities for peaceful and scientific purposes." Although the Space Act has been often amended, this provision has never changed, and still remains one of the main objectives of NASA. Despite a few early studies, the mandate to study societal impact went unfulfilled as NASA concentrated on the many opportunities and technical problems of spaceflight itself. But as NASA approached its 50th anniversary in 2008 in a reflective mood, it contemplated the impact of the space program over past decades by inaugurating a new series of edited volumes on the societal impact of spaceflight (Dick and Launius 2007; Dick and Lupisella 2009; Dick 2018). As we editors wrote in the introduction to the first volume on *Societal Impact of Spaceflight*, "It is time to take up the challenge once again ··· Whether or not the ambitious space visions of the United States and other countries are fulfilled, the question of societal impact over the past 50 years remains urgent and may in fact help fulfill current visions or at least raised the level of debate." We cautioned, however, that the question of societal impact of spaceflight is not as simple as it may seem, beginning with what we mean by "impact," "societal," and "spaceflight." From there, 31 authors from many fields undertook their assigned task, and many more continued the analysis in the subsequent two volumes.

Part III of this volume begins with two chapters, both written during my time as NASA Chief Historian for the 50th anniversaries of the Space Age in 2007, and of NASA in 2008. Chapter 17 is a relatively straightforward history of NASA's accomplishments, while Chap. 18 is an extended comparison of the Age of Space with the sixteenth-century Age of Discovery, including intellectual, economic, geopolitical, and social impact. Chapter 19, a keynote address presented at the first international conference on the cultural history of outer space, addresses the role of imagination in spaceflight. As such it contemplates a part of *astroculture*, an idea originated by the European space historian Alexander Geppert, who had already contributed an article on the subject at the first conference at NASA on the societal impact of spaceflight (Dick and Launius 2007). Astroculture, though concentrating in Geppert's research on European spaceflight, provides a framework for the much broader consideration of space exploration and society.

Chapters 20 and 21 address both the societal impact of the Hubble Space Telescope, and the public and political outcry when its final servicing mission was canceled, only to be restored by a new NASA Administrator. Chapter 20 is a brief foray into the societal impact, written as an introduction to the societal impact section of a volume on Hubble's legacy. By contrast Chap. 21 is an extraordinarily fine-grained look at the decision to cancel the final Hubble servicing mission in the wake of the Columbia Shuttle accident, a decision that would have doomed the telescope to obsolescence within a few years of 2004. It was a heartbreaking decision to many, subjecting NASA Administrator Sean O'Keefe to withering criticism. The chapter is an independent report written at the request of Administrator O'Keefe during my time as NASA Chief Historian, and features interviews with all the major players in the decision, revealing the often raw emotions displayed at the time. For example, when asked about the withering criticism of his decision to cancel the servicing mission, O'Keefe bluntly stated:

Let me offer my view of "withering." Withering is the feeling you get when you are standing at a runway with the dawning realization that the Shuttle everyone is waiting for isn't going to land. Withering is when you have to explain to wives, husbands, parents, brothers, sisters and children that their loved ones aren't coming home alive. Withering is attending funerals, memorial services, and ceremonies over 16 months in number too many to count any more, yet having every single one of these events feel like the weight of that responsibility will never be relieved. Withering is the knowledge that we contributed to the Columbia disaster because we weren't diligent. (O'Keefe 2004)

Taken together, the interviews demonstrate the complex decision-making process and cross-current of arguments that take place in high-technology organizations under extraordinary public and congressional scrutiny. The chapter also reveals a kind of Roshomon effect, as each player interprets the same event differently from their own perspective.

Chapter 22 consists of reflections on a meeting in Paris addressing French-American cooperation in space. Though brief, it opens a window to a very robust and important subject—the role of international cooperation in space exploration. In a world fraught with conflict, this is sure to be a major theme in the twenty-first century and beyond.

References

Chaikin, Andrew. 2007. "Live from the Moon: The Societal Impact of Apollo," NASA SP-2007-4801: Washington, DC, pp. 53–66

Dick Steven J. and Roger D. Launius, eds. 2007. *Societal Impact of Spaceflight*. Washington, DC: NASA SP-2007-4801

Dick, Steven J. and Mark Lupisella, eds., 2009. *Cosmos and Culture: Cultural Evolution in a Cosmic Context* NASA History Series

Dick, Steven J. 2018. *Historical Studies in the Societal Impact of Spaceflight*, ed. Steven J. Dick. NASA History Series.

Lambright, William H. 2005. *NASA and the Environment: The Case of Ozone Depletion*. Washington, DC: NASA SP-2005-4538.

Lambright, William H. 2007. "NASA and the Environment: Science in a Political Context," in Dick and Launius, 2007, pp. 313–330

O'Keefe. 2004. Remarks by Administrator O'Keefe at the American Astronomical Society Annual Meeting, Denver, Colorado, June 1, 2004.

Poole, Robert. 2008. *Earthrise: How Man First Saw the Earth*. New Haven, CT, and London, U.K.: Yale University Press.

Chapter 17
Exploring the Unknown: 50 Years of NASA History

Abstract After briefly describing the origins of NASA in 1958, this chapter analyzes three broad themes of NASA's mission over the last 50 years: human spaceflight; the space, Earth, and life sciences; and aeronautics. It distinguishes four eras of human spaceflight: the Apollo era, the Space Shuttle era, the International Space Station era, and the Moon/Mars era. In the space sciences NASA achievements have been legendary, ranging from the early Moon probes, such as Ranger and Surveyor, to landmark robotic spacecraft exploring the inner planets, such as Mariner and the Mars probes, and Pioneer, Voyager, and others exploring the outer gas and ice giant planets. Together these spacecraft completed the preliminary reconnaissance of the Solar System. The spectacular imagery from the Great Observatories including the Hubble Space Telescope carried this exploration to the furthest reaches of space. NASA's Earth applications programs utilized near-Earth space to study the planet's resources, to provide essential information about weather, and to provide means for navigation that was both life-saving and had enormous economic implications. Worldwide satellite communications brought the world closer together, a factor difficult to estimate from a cost-benefit analysis. NASA's life sciences programs tackled some of the most profound questions for humanity, including the origins of life and the search for extraterrestrial life. In the field of aeronautics NASA from its beginnings conducted research on aerodynamics, wind shear, flight safety, and other important topics using wind tunnels, flight testing, and computer simulations. We conclude that exploration is important for any forward-looking society.

17.1 Origin

The birth of the National Aeronautics and Space Administration (NASA) was directly related to the Soviet Union's launch of Sputniks I and II in late 1957 and the ensuing race to demonstrate technological superiority in space. In late January, 1958 the United States answered the challenge with Explorer 1, hoisted aloft by the Army's

First published in *50 Years of Space: A Global Perspective* (Universities Press India: Hyderabad, India, 2007).

S. J. Dick, *Space, Time, and Aliens*, https://doi.org/10.1007/978-3-030-41614-0_17

rocket team led by Wernher von Braun, using rocket technology developed from World War II. Though a small spacecraft weighing only 30 pounds, it discovered what are now known as the Van Allen radiation belts, named for scientist James Van Allen, and thereby launched the new discipline of space science. Explorer 1 was followed in March by the Navy's Vanguard 1, 6 in. in diameter and weighing only 3 pounds.

These events were a prelude to NASA. Driven by the competition of the Cold War, on July 29, 1958 President Dwight D. Eisenhower signed the National Aeronautics and Space Act, providing for research into the problems of flight within the Earth's atmosphere and in space. After a protracted debate over military versus civilian control of space, the act inaugurated a new civilian agency designated the National Aeronautics and Space Administration (NASA). The Agency began operations on 1 October, 1958.

NASA began by absorbing the earlier National Advisory Committee for Aeronautics (NACA), including its 8000 employees, an annual budget of $100 million, three major research laboratories—Langley Aeronautical Laboratory in Virginia, Ames Aeronautical Laboratory in California, and Lewis Flight Propulsion Laboratory in Ohio—and two smaller test facilities. It quickly incorporated other organizations (or parts of them), notably the space science group of the Naval Research Laboratory that formed the core of the new Goddard Space Flight Center in Maryland, the Jet Propulsion Laboratory managed by the California Institute of Technology for the Army, and the Army Ballistic Missile Agency in Huntsville, Alabama, where Wernher von Braun's team of engineers was developing large rockets (Fig. 17.1).

Within months of its creation, NASA began to conduct space missions, and over the last 50 years has undertaken spectacular programs in human spaceflight, robotic spaceflight, and aeronautics research. It has done so in the context of both international competition and cooperation. The race to the Moon, certainly a competition with the Soviet Union in the midst of the Cold War, was followed by the Apollo-Soyuz Test Project, an exemplary case of cooperation between the two superpowers. Overall NASA has an extraordinary record of 4000 international agreements over 50 years, with 256 current active agreements with 58 countries.

NASA today carries on the nation's long tradition of exploration dating back at least to the expedition of Meriwether Lewis and William Clark, commissioned by Thomas Jefferson in 1803 to explore the uncharted west of North America. In addition to its headquarters in Washington, DC, NASA facilities include ten centers around the country staffed by nearly 19,000 employees. Its budget for fiscal year 2007 was nearly 17 billion dollars.

17.2 Human Spaceflight

Looking back after 50 years, we can distinguish several eras of human spaceflight at NASA, characterized by the Apollo Moon program, the Space Shuttle, the International Space Station, and a new era of human activity beyond low earth orbit begun in January 2004.

Fig. 17.1 Hermann Oberth (forefront) with officials of the Army Ballistic Missile Agency at Huntsville, Alabama in 1956, prior to the formation of NASA. Left to right: Dr. Ernst Stuhlinger (seated); Major General H.N. Toftoy, Commanding Officer, Wernher von Braun, Director, Development Operations Division, and Dr. Eberhard Rees, Deputy Director, Development Operations Division. General Toftoy was responsible for "Project Paperclip," which took scientists and engineers out of Germany after World War II to design rockets for American military use. Von Braun later spearheaded development of the Saturn V rocket that took the Apollo astronauts to the Moon. NASA

17.2.1 The Apollo Era

President John F. Kennedy's challenge on May 25, 1961 of "achieving the goal, before this decade is out, of landing a man on the moon and returning him safely to the Earth," put in motion events that will be forever remembered not only as a technological and managerial feat, but also an extraterrestrial adventure that gave humanity a new perspective on its home planet. Following the Soviet Union's launch of Yuri Gagarin in April 1961 and driven by Kennedy's challenge, the United States launched its own astronauts in capsules with the mythical names of Mercury, Gemini, and Apollo.

The first of these programs had its origins well before President Kennedy's announcement. Indeed, NASA announced its first human space program on October 7, 1958, only 1 week after the agency became operational. It was designated "Project Mercury" on November 26. The objectives were to place a spacecraft in orbit around Earth, observe human performance, and return the human and spacecraft safely to Earth. Whether a human could function in the harsh conditions of weightless spaceflight was still unknown, and NASA quickly decided that the pool of astronauts should come from military test pilots. On April 9, 1959 NASA introduced its first astronauts, the "Mercury Seven:" Scott Carpenter, Gordon Cooper, John Glenn, Gus Grissom, Walter Schirra, Alan Shepard and Donald K. "Deke" Slayton.

After several test flights, including the monkey "Miss Sam" in January 1960 and the chimpanzee "Ham" a year later, on May 5, 1961 a Mercury Redstone rocket launched Alan Shepard and his Freedom 7 spacecraft on a 15-min suborbital flight. On July 21 Virgil "Gus" Grissom was launched on a second 15-min suborbital flight in the Liberty Bell 7.

Orbital flights required a larger launch vehicle, and after a test of the Mercury Atlas launcher with the chimp "Enos," on February 20, 1962, John Glenn made three successful orbits of the Earth in his Friendship 7 spacecraft. This was followed in May by Scott Carpenter's three orbits in Aurora 7, and by Wally Schirra's 6 orbits in October in Sigma 7. The Mercury program ended on May 15, 1963 with Gordon Cooper's 22 orbits of the Earth (Faith 7) lasting 34 h and 20 min.

Less than 5 years after the announcement of the Mercury program, and over a span of 2 years and six human flights, the goals of the program were fully met. Brief as they were, the pioneering Mercury flights were full of anxiety and adventure. Grissom's capsule sank when the hatch blew off and it filled with water before it could be recovered; Glenn's reentry into Earth's atmosphere was marked with tension because of a signal that his heat shield had come loose; and Carpenter's capsule landed more than 200 miles off course, causing anxious moments until it could be located and recovered.

NASA announced plans for a two-man spacecraft on December 7, 1961, even before Glenn's orbital flight. The following month it was officially designated the Gemini program, named after the constellation with its twin stars, Castor and Pollux. The program was conceived as an intermediate step between project Mercury and Apollo. Its major objectives were to subject humans to spaceflight for up to 2 weeks, to rendezvous and dock with orbiting vehicles, and to perfect methods of entering the atmosphere and landing. Gemini consisted of 12 flights, including two unmanned flight tests in April, 1964 and January 1965.

The first manned Gemini flight was Gemini 3, launched March 23, 1965 with Gus Grissom and John Young aboard. Grissom nicknamed the spacecraft "Molly Brown" in reference to his first spacecraft that sank, unlike the unsinkable Molly Brown of Titanic legend. Young, representing a new class of astronauts, would go on to fly in the Apollo and Shuttle programs. The three-orbit Gemini 3 lasted almost 5 h. It was followed by flights of increasing duration and difficulty: 62 orbits for Gemini 4, 120 for Gemini V (Roman numeral designations began with this flight), and 206 orbits for Gemini VII.

The Gemini program saw many firsts, had its share of exciting moments, and was a training ground for many of the astronauts who would go on to the Moon in the Apollo program. Gemini 4 saw the first American Extravehicular Activity (EVA), a 22-min spacewalk carried out by Edward H. White II, later killed in the Apollo 1 fire. Gemini V saw the first use of fuel cells for electrical power, and evaluated the guidance and navigation system for future rendezvous missions. Gemini VII not only showed that humans could live in space for 14 days, but it was also the first rendezvous in space. Launched on December 4, 1965, it rendezvoused with Gemini VI-A on December 15, station-keeping for over 5 h at distances from 1 to 295 feet. Gemini VIII accomplished the first docking with another space vehicle, an unmanned Agena stage.

When a malfunction caused uncontrollable spinning of the craft, the crew undocked and effected the first emergency landing of a manned U. S. space mission. Neil Armstrong's impressive actions on this flight later helped win him the Apollo 11 landing slot. Four more Gemini flights during 1966 perfected rendezvous and docking procedures, including the final flight Gemini XII. Gemini XII rendezvoused and docked with its target Agena and kept station with it during EVA. Edwin "Buzz" Aldrin set an EVA record of 5 h, 30 min for one space walk and two stand-up exercises.

All of this activity during the early 1960s was in the service of the Apollo program to land humans on the Moon. The Apollo program (excluding the follow-up Apollo-Soyuz Test Project and Skylab programs) consisted of 11 manned flights that took place between 1968 and 1972. Two of the flights (Apollo 7 and 9) were Earth-orbit tests, two (Apollo 8 and 10) were circumlunar, and one (Apollo 13) was aborted midway to the Moon. Six flights (Apollo 11, 12, 14, 15, 16, and 17) succeeded in landing men on the Moon. None of this could have happened without the magnificent Saturn rockets, developed through the work of Wernher von Braun and a team of thousands. And all of it happened after the tragic fire of January 27, 1967 killed Gus Grissom, Ed White and Roger Chaffee as they sat in their Apollo 1 capsule on the launch pad undergoing tests.

The Apollo manned flights began in October, 1968, when Walter Schirra (a veteran of Mercury and Gemini flights), Donn Eisele and Walter Cunningham tested the command and service modules in Earth orbit over a period of almost 11 days during the Apollo 7 mission. After this one test flight, and with its eye still on the goal of a lunar landing by the end of the decade, NASA made the bold decision to send Apollo 8 to the Moon. Apollo 8 orbited the Moon ten times, and made a now-legendary transmission on Christmas Eve, citing passages from Genesis. The 6 day mission proved the spacecraft could traverse the Earth-Moon distance safely, enter lunar orbit, and return to Earth. But it did not yet prove that a lunar landing was possible.

An actual lunar landing required two more flights in preparation. In early March, 1969 Apollo 9 tested the entire Apollo spacecraft in Earth orbit, including the rendezvous maneuvers between the command module and the lunar module that would descend to the lunar surface. Over 10 days Apollo 9 proved this sequence could be done. Only 2 months later Apollo 10 performed a full dress rehearsal for the lunar landing. During 2.5 days in lunar orbit, the lunar module descended within 50,000 feet of the lunar surface, re-docked with the command module, and returned safely to Earth, setting the stage for the first manned lunar landing in history.

The now-legendary Apollo 11 was commanded by Neil Armstrong, with Michael Collins as the Command Module pilot and Buzz Aldrin as the Lunar Module pilot. Launch took place on July 16, 1969 (Fig. 17.2), with landing on July 20 on the Sea of Tranquility. For sheer excitement it was hard to beat, as Armstrong and Aldrin set down on the lunar surface with seconds of fuel to spare. Six hours later, Armstrong took his famous "one giant leap for mankind." Aldrin joined him, and the two spent two-and-a-half hours drilling core samples, taking photographs, and collecting rocks (Fig. 17.3). For 1 day the world seemed united, as hundreds of millions around the world watched with fascination. After more than 21 h on the lunar surface, the astronauts returned to the Columbia command module, bringing 47.7 pounds of lunar

Fig. 17.2 The Apollo 11 Saturn V space vehicle lifts off with astronauts Neil A. Armstrong, Michael Collins, and Edwin E. Aldrin, Jr., at 9:32 a.m. EDT July 16, 1969, from Kennedy Space Center's Launch Complex 39A. NASA

samples with them. The two moonwalkers had left behind scientific instruments, an American flag, and other mementos, including a plaque bearing the inscription: "Here Men from Planet Earth First Set Foot upon the Moon. July 1969 A.D. We Came in Peace for All Mankind." They returned to Earth on July 24, surely the conclusion of eight of the most historic days ever recorded.

Fig. 17.3 Astronaut Neil A. Armstrong, Apollo 11 mission commander, at the modular equipment storage assembly (MESA) of the Lunar Module "Eagle" on the historic first extravehicular activity (EVA) on the lunar surface. Astronaut Edwin E. Aldrin Jr. took the photograph with a Hasselblad 70-mm camera. Most photos from the Apollo 11 mission show Buzz Aldrin. This is one of only a few that show Neil Armstrong. NASA

Apollo 12 touched down on the lunar surface only 4 months later, on the Ocean of Storms near the unmanned Surveyor 3 probe. Astronauts Charles "Pete" Conrad and Alan Bean took two moonwalks lasting just under 4 h each. They collected rocks and set up experiments that measured Moonquakes, magnetic field, and the wind from the Sun. A few months later Apollo 13 was on its way to the Moon when an oxygen tank in the Service Module exploded. The crew (James Lovell, John Swigert, and Fred Haise) aborted their planned landing, swung around the Moon, and returned on a trajectory back to Earth, using the Lunar Module Aquarius as a lifeboat.

Following the near-disastrous Apollo 13, Alan Shepard, Stuart Roosa, and Edgar Mitchell achieved the third lunar landing on February 5, 1971. After landing in the Fra Mauro region, Shepard and Mitchell took two moonwalks, adding new seismic studies to the Apollo experiment package, and used a "lunar rickshaw" pull-cart to carry their equipment. This was also the flight where Shepard made his famous long golf shot. On the way back to Earth, the crew conducted the first U. S. materials processing experiments in space. The Apollo 14 astronauts were the last lunar explorers to be quarantined on their return from the Moon.

The last three Apollo missions, from July, 1971 to December, 1972, were characterized by much longer excursions from the lunar lander, up to tens of kilometers, made possible by the lunar roving vehicle. The Apollo 15 astronauts explored the area known as Hadley rille, while Apollo 16 explored the Descartes highlands.

The last lunar expedition, Apollo 17, landed in an area known as Taurus-Littrow. Here astronaut Eugene Cernan and geologist Harrison Schmitt conducted the longest lunar exploration of the Apollo program. The crew roamed for 33.80 km through the Taurus-Littrow valley in their rover, discovered orange-colored soil, and left behind a plaque attached to their lander Challenger, which read: "Here Man completed his first exploration of the Moon, December 1972 A.D. May the spirit of peace in which we came be reflected in the lives of all mankind." The Apollo lunar program had ended. Over a span of 4 years 12 men walked on the surface of the Moon. Including their time in the lunar excursion module, they spent just a few minutes less than 300 h on the lunar surface.

The societal effect of the Apollo program was profound, no more so than in its view of the Earth from the Moon. The photographs of Earthrise, and of the full Earth as a blue marble suspended in space, fragile and without national boundaries, changed humanity's view of Earth forever (Fig. 17.4).

A fitting conclusion to the Apollo era, after the brief American experience in operating the Skylab orbiting space station in 1973–1974, brought competition full circle to cooperation. In 1975 the United States and the Soviet Union achieved the first international human spaceflight, the Apollo-Soyuz Test Project.

17.2.2 The Space Shuttle Era

The next major era in human spaceflight began in April, 1981 with the maiden voyage of the Space Shuttle Columbia, the world's first reusable spacecraft. The Space Shuttle was approved as an initiative by President Nixon in 1972. The Shuttle would take off vertically and glide to an unpowered landing, with crews of five to seven astronauts. It was capable of carrying large satellites with payloads of up to 54,000 pounds both to and from orbit.

The Space Shuttle consists of four main components: the orbiter in which the astronauts and payloads reside, the external tank carrying more than a half million gallons of liquid hydrogen and liquid oxygen, the Space Shuttle Main Engines fed by the fuels from the external tank, and two Solid Rocket Boosters that provide 3.3 million pounds of thrust at liftoff (Fig. 17.5). All components except the external tank are reusable, making the Space Shuttle the first reusable launch vehicle, in contrast to expendable launch vehicles.

Aside from the Enterprise, a test Shuttle now at the National Air and Space Museum near Washington, DC, five Space Shuttle orbiters have operated since 1981: Columbia, Challenger, Discovery, Atlantis, and Endeavor. The latter is a replacement for the Challenger, destroyed on launch in 1986. The Columbia, destroyed on reentry in 2003, has not been replaced. Challenger had only 10 flights and 62 flight days during its brief lifetime, while Columbia at the time of its demise held the Shuttle record with 300 flight days during 28 flights.

During 117 flights in its first 25 years (1981–2006), the Space Shuttle could boast a variety of accomplishments, carried out by 813 astronauts. For 15 years beginning with STS-9 in 1983, 15 flights carried Spacelab, a laboratory module for

Fig. 17.4 The classic "Blue Marble" view of Earth was captured by the Apollo 17 crew traveling toward the Moon on December 7, 1972. The photograph extends from the Mediterranean Sea area to the Antarctica south polar ice cap. Heavy cloud covers the Southern Hemisphere. Almost the entire coastline of Africa is clearly visible. The Arabian Peninsula can be seen at the northeastern edge of Africa. The large island off the coast of Africa is the Malagasy Republic. NASA

microgravity experiments. In 1990 STS-31 launched the Hubble Space Telescope, and four subsequent flights serviced it, an invaluable activity without which the telescope could not have operated. Altogether Shuttle astronauts deployed 66 satellites, some of them major satellites for communications and science.

Between 1995 and 1998 nine flights docked with the Russian Mir space station as part of what became known as the Shuttle-Mir program. The first components of Mir were launched in February, 1986, 1 month after the Challenger accident. Beginning in 1995 seven astronauts spent a total of 28 months on Mir, often undergoing unexpected adventures. Jerry Linenger was onboard when a fire broke out in February, 1997, and the following June Michael Foale and his Russian colleagues survived a collision of the Mir station with an unmanned Progress spacecraft. Mir was de-orbited March 23, 2001.

Fig. 17.5 A timed exposure of the Space Shuttle, STS-1, at Launch Pad A, Complex 39, turns the space vehicle and support facilities into a night-time fantasy of light. Structures to the left of the Shuttle are the fixed and the rotating service structure. NASA

Between 1998 and 2006 twenty Space Shuttle flights went toward construction of the International Space Station (ISS), orbiting at an altitude of 230 miles.

While the original Shuttle goals of low cost and routine access to space were not met, what was officially called the Space Transportation System (STS) amassed many significant accomplishments. Among them were the 24 commercial satellites deployed prior to the Challenger accident in 1986; the placement in orbit of major scientific missions including Galileo, Magellan, Chandra and the launching and servicing missions of the Hubble Space Telescope; Spacelab and Spacehab missions with their material, microgravity and life sciences experiments; deployment of the Tracking and Data Relay System (TDRS) constellation; and numerous flights in support of the Mir and International Space Station. Humanity's first attempt to build and operate a reusable spacecraft was an impressive achievement in itself.

Along with these accomplishments, the Shuttle program has seen the depths of tragedy. On January 28, 1986 a leak in the joints of one of two solid rocket boosters attached to the Shuttle orbiter Challenger caused the main liquid fuel tank to explode 73 s after launch, killing all seven crew members. On September 29, 1988, the Shuttle successfully returned to flight, and NASA flew 87 successful missions before tragedy struck again on February 1, 2003 with the loss of the orbiter Columbia and its seven astronauts during reentry. Three Shuttle orbiters remain in NASA's fleet: Atlantis, Discovery, and Endeavour. NASA's plan is to fly out the remaining Shuttle missions through 2010 for the purposes of International Space Station assembly and servicing of the Hubble Space Telescope.

17.2.3 International Space Station Era

The International Space Station era, intimately related to the Shuttle era, was the result of another Presidential decision, announced in Ronald Reagan's State of the Union address in January, 1984. The mature accomplishments of this largest human-made object ever to orbit the Earth remain to be seen, but one achievement not to be underestimated was international cooperation. Originally dubbed Freedom, over the years it became the International Space Station (ISS), encompassing 15 partners in addition to the United States, including the Russian Federal Space Agency, the Japan Aerospace Exploration Agency, the Canadian Space Agency, and the member nations of the European Space Agency. The first elements of the space station were launched in 1998, and permanent habitation began when the Expedition One crew arrived on November 2, 2000. With this milestone, civilization had reached a point beyond which there would likely always be humans living and working in space. The facility orbits the Earth nearly 16 times every day, at an inclination of 51 degrees. The ISS, powered by enormous solar arrays, will consist of ten pressurized modules from a variety of countries. One Russian module (Zvezda) and three US modules (Destiny, Node 1, and Russian-built Zarya) are now in place (Fig. 17.6). ISS now holds a crew of three, which will increase to six in 2009. When completed the ISS will be about four times the size of Mir, five times larger than Skylab, and will weigh 925,000 pounds. In addition to the Space Shuttle, ISS is serviced by the manned Soyuz spacecraft and the unmanned Progress spacecraft. In the future, supply missions will also be undertaken by the European Automated Transfer Vehicle (ATV), Japan's H-II Transfer Vehicle (HTV), and the U. S. Orion Crew Exploration Vehicle (CEV). Private industry will also supply Space Station crew and cargo transportation services to NASA through a unique program called Commercial Orbital Transportation Services Demonstration (COTS).

S124E010013

Fig. 17.6 Backdropped by Earth' horizon and the blackness of space, the International Space Station is seen from Space Shuttle Discovery as the two spacecraft separated during the STS-119 mission in March 2008. During this, the 28th Shuttle mission to the Space Station, the fourth set of solar arrays was deployed. NASA

Consistent with NASA's current agenda for space exploration, the ISS will serve as an engineering test bed for flight systems and operations critical to NASA's exploration mission. U. S. research on the ISS will concentrate on the long-term effects of space travel on humans and engineering development activities in support of exploration. The ISS also is used for research in the life sciences, physical sciences, and Earth observation. Quite aside from the scientific accomplishments and expertise gained from a large construction project in space, the international cooperation fostered during construction and operations was no small achievement.

The Space Shuttle and International Space Station were low-Earth orbit (LEO) projects, and many space enthusiasts longed for the days of more distant destinations. They pointed out that after traversing a quarter million miles to the moon and back eight times from 1968–1972, in all the years afterward humans traveled no further than 386 miles from their home planet—during the Hubble Space Telescope servicing mission of STS-82 in 1997. They wanted to return to the moon and go onto Mars. They were given hope when President George H.W. Bush announced his Space Exploration Initiative (SEI) in 1989 on the 20th anniversary of the first moon landing. But projected costs and political realities spelled doom for this venture within 2 years.

17.2.4 The Moon/Mars Era

History came full circle on January 14, 2004 when President George W. Bush, in an address at NASA Headquarters, called for a return of humans to the moon and a long-term push for a human mission to Mars. The Shuttle would be phased out by 2010 and all Space Station work would concentrate on human factors necessary for trips to the moon and Mars. A new era in human spaceflight had begun. The new Vision for Space Exploration seeks to return humans to the Moon by 2020, and eventually take them to Mars. For the first time since the Apollo era, a new human-rated rocket and crewed capsule are being designed. The Ares I rocket and the Orion Crew Exploration Vehicle capsule will be joined by a reusable lunar lander and a pressurized rover for transport over the Moon's surface. As part of a new global exploration strategy, the return to the Moon will begin with a lunar outpost at the South Pole by 2024, where sunlight for power generation is more plentiful and where material may be available for nuclear power. The outpost may also provide hydrogen and oxygen, components of rocket fuel. One of the possible locations for the outpost is near Shackleton crater, named after Ernest Shackleton, explorer of the Earth's own South Pole.

17.3 Space, Earth and Life Sciences

17.3.1 Lunar and Planetary Exploration

Over the last 50 years NASA has launched a spectacular array of robotic spacecraft with a great variety of purposes. Naturally, NASA began by targeting the closest celestial body, the moon. While the Russia Luna series probed the Moon, in the 1960s NASA sent the Ranger, Lunar Orbiter, and Surveyor spacecraft to undertake a preliminary reconnaissance, preparing the way for humans. For two decades in the post-Apollo era NASA launched no further lunar missions, but in the 1990s Clementine and Lunar Prospector resumed analysis of the lunar surface. More spacecraft were scheduled to study the moon as the vanguard for the new human spaceflight program.

The moon, only a quarter of a million miles away, was the equivalent of our backyard compared to the much more distant planets (Fig. 17.7). Venus, dubbed the Earth's sister planet, was visited first by a series of Mariner spacecraft in the 1960s, and then by the Pioneer-Venus and Magellan missions. The Russians even succeeded in landing descent modules from the Venera spacecraft on Venus and returning data. It became clear during the course of the Space Age that Venus was about as far from Earth's sister planet as could be imagined. In the true spirit of exploration, what was once thought to be a lush planet ripe for life was instead revealed to be an alien environment, with an atmosphere composed of 95% carbon dioxide and crushing pressures of 75 to 100 Earth atmospheres, causing a greenhouse-induced

Fig. 17.7 Montage of planetary images taken by NASA spacecraft. From top to bottom: Mercury, Venus, Earth/Moon, Mars, Jupiter, Saturn, Uranus, and Neptune. NASA has also explored Pluto (Fig. 1 of Part V), no longer classified as a planet. NASA/JPL

temperature of 900 °F. To top it off, the Venusian clouds were found to be composed of sulfuric acid. In the early 1970s the other inner planet, the rocky planet Mercury, was studied by Mariner 10, revealing a cratered Moon-like surface, a tenuous helium atmosphere, temperature swings between plus and minus 300 °F, and a magnetic field.

Looking outward from the Sun space scientists found one of their most alluring targets: the legendary planet Mars. Once thought to be crisscrossed by canals built by a race of Martians, its cratered surface was revealed by Mariner 4 in 1964 to be more reminiscent of the Moon, causing some to lose interest in what seemed to be a dead and uninteresting world. But by 1969, Mariner 9 showed a much more complex surface, including what appeared to be dry river beds. Water was a necessary ingredient for life, and this raised speculation about past life on Mars. In 1976, during America's bicentennial year, two Viking spacecraft—in a difficult feat not to be repeated for two decades—landed on the surface of Mars and began a series of experiments that yielded a huge amount of information about the red planet. Of particular interest were the biology experiments, which produced controversial results. At least one of the Principal Investigators still believes his experiment showed indications of life, but the consensus of the other scientists was that the Martian surface harbored active chemistry rather than biology.

Two decades, later Pathfinder and its Sojourner rover provided spectacular images from the surface of the red planet, and the Mars Global Surveyor and Mars Odyssey returned scientific data and images from orbit, including evidence of recent water activity. The remarkably long-lived and productive Mars Exploration Rovers, Spirit, and Opportunity, continued the exploration, while the European Mars Express orbited overhead. As is the case with the moon, there is no doubt that these robotic spacecraft will lead the way for human exploration of Mars—it is only a matter of when. And as with the Moon, there is no question that Martian exploration can have a profound impact in a number of ways. Mars is the nearest planet and thus the most likely candidate for human habitation. A Mars outpost, possibly following on the heels of a lunar outpost, will raise profound technical, scientific, and ethical questions.

The search for life on Mars will have even more significant implications. If life is found and it is of independent origin from the Earth, Mars will have served as a test case for life in the universe. If life is found on two planets so close together, it means that life will likely arise on planets throughout the universe wherever the conditions are right.

Beyond Mars is the realm of the gas giant planets. Pioneers 10 and 11 were indeed pioneers in the sense of the first reconnaissance of the planets Jupiter and Saturn in the mid-1970s. Jupiter, Saturn, Uranus and Neptune yielded beautiful photographs and many surprises with the missions of Voyager 1 and 2. Launched in the summer of 1977, both Voyager spacecraft were designed to last 5 years, and both encountered Jupiter and Saturn between 1979 and 1981. After the flyby of Saturn's moon Titan, Voyager 1 took a trajectory north of Saturn's orbital plane out of the Solar System, while Voyager 2 headed onward to Uranus and Neptune, courtesy of a gravity assist and a rare planetary alignment. After encountering Uranus in 1986 and Neptune in 1989, Voyager 2 took a southward trajectory out of the Solar System. The last Voyager images were taken Valentine's Day, 1990, when Voyager 1 looked back from 3.7 billion miles to take a portrait of seven of the nine planets in our own Solar System, including the "pale blue dot" that is Earth. The data the Voyagers returned revolutionized our knowledge of the outer planets and their intriguing panoply of satellites. Today, the Voyagers are flying through the void of interstellar space, carrying two golden records containing greetings to whatever creatures may find it from various leaders and citizens of planet Earth. The Galileo mission to Jupiter, topped off by the arrival of the Cassini/Huygens spacecraft at Saturn in 2004, continued to revolutionize our knowledge of the gas giant planets, while the European Space Agency's Huygens probe landing on the surface of Titan added to our knowledge of the Solar System's suite of bizarre satellites.

Beyond the gas giant planets is the enigmatic and controversial Pluto. NASA's New Horizons spacecraft is speeding toward the edge of the Solar System, some three billion miles away, to explore the nature of the object greatly in dispute because of a decision of the International Astronautical Union in 2006 to downgrade its planet status. At the time New Horizons launched in January, 2006, Pluto was the only planet in our Solar System not yet visited by spacecraft. It will now become the first dwarf planet to be visited, in July, 2015, unless the Dawn spacecraft reaches Ceres (now also designated a dwarf planet) a few months earlier. In any case, Pluto is the vanguard of an entire new class of trans-Neptunian objects that are part of the Kuiper-Edgeworth Belt.

Amazingly, over the last two decades a variety of spacecraft have also voyaged to six comets and several asteroids. In 1986 an armada of spacecraft visited the famous Halley's Comet, including two Russian spacecraft (Vega 1 and 2), two Japanese spacecraft (Sagigake and Suisei), and the European Space Agency's Giotto. In 1999 NASA launched a comet sample return mission known as Stardust. In January 2004 the spacecraft flew within 149 miles of the nucleus of comet Wild 2, collected samples of comet dust, and stored them in a return capsule. After a roundtrip journey of some 2.88 billion miles, the capsule returned to Earth with its precious sample on January 15, 2006. Deep Impact, another NASA Discovery mission, brought yet another approach to comet exploration—impacting a comet and studying the subsequent debris for clues to the origin of the Solar System. After a journey of 171 days and 268 million miles, on July 3, 2005 the Deep Impact flyby spacecraft released it 820-pound impactor on a course for Comet Tempel 1. The following day it impacted the comet's 14 km-long nucleus at 23,000 miles per hour, producing a spectacular flash of light and a crater of undetermined depth. Analysis of the ejection plume showed large amounts of organic material, confirming that during its history the Earth might have been infused with organics from similar comets. In addition, images from three cameras showed what appear to be impact craters, never before seen on a comet and of unknown origin. Other data indicates that the nucleus is extremely porous, a fluffy structure weaker than powdered snow.

17.3.2 Solar Science

The Sun, our nearest star, is a mere eight light minutes away, compared to 4.5 light *years* for the next star, the Alpha Centauri system. A nuclear furnace generating prodigious amounts of energy, the Sun provides the conditions necessary for life on

Earth. It is a matter of practical importance that we know how the Sun works, as well as a matter of theoretical importance, since its proximity gives us the best information on how other Sun-like stars work.

After early observations from sounding rockets, the study of the Sun from space began, naturally enough, from Earth orbit. The Orbiting Solar Observatory (OSO) was a series of eight orbiting observatories that NASA launched between 1962 and 1971. Seven of them were successful, and studied the Sun at ultraviolet and X-ray wavelengths. The OSO spacecraft photographed the million-degree solar corona, made X-ray observations of a solar flare, and enhanced our understanding of the Sun's atmosphere among its many other achievements.

The Apollo Telescope Mount, though inelegantly named, was an innovative program for astronauts to observe the Sun from Skylab, the orbiting space station that made use of hardware in the aftermath of the Apollo program. It was the most important scientific instrument aboard Skylab, which operated for 8 months beginning in May, 1973. Unhampered by the limits of telemetry, the astronauts brought solar photographs back to Earth, including X-ray observations of solar flares, coronal holes, and the corona itself.

Attempts to observe the Sun beyond Earth orbit are more recent. Ulysses, known before launch as the International Solar Polar Mission, was deployed in October, 1990 from the Space Shuttle Discovery. It was a joint mission of NASA and the European Space Agency designed to gain a new perspective of the Sun by viewing its polar regions. Making use of a gravity assist from Jupiter, Ulysses passed the Sun's south pole in 1994 and its north pole a year later. It repeated these passes in 2000 and 2001, and did so again in 2006 and 2007. With the first pass of Ulysses, scientists discovered unknown complexities of the Sun and its surroundings, including different speeds of the solar wind. Ulysses—named after Homer's Greek adventurer—did not carry imaging instruments, and focused on the Sun's environment rather than its surface. Fifteen years after launch, the spacecraft remains in good health.

SOHO, also a joint American-European project, is another epic solar voyage still underway. Launched December 2, 1995, its array of instruments were designed to study the solar wind, as well as the Sun's outer layers and interior structure. In order to do this, it was placed in an orbit 1.5 million kilometers from Earth, at a point known as the L1 Lagrangian point, where the combined gravity of Earth and Sun keep it in an orbit locked to the Earth-Sun line. Though still far from the Sun, this location, about four times the distance of the Moon in the direction of the Sun, is ideal for long-term uninterrupted observations with the Earth out of the way.

SOHO's scientific findings have been phenomenal. It has imaged the structure of sunspots below the surface, measured the acceleration of the wind from the Sun (streams of protons and electrons traveling at a million miles per hour!), discovered coronal waves and solar tornadoes, and found more than 1000 comets. Moreover, it has revolutionized our ability to forecast space weather, and provided data on the variability of the Sun's energy, both of which affect us directly on Earth. Both still images and movies showing the dynamic Sun's prominences, flares, spots, coronal mass ejections, and otherwise lively gyrations fill the SOHO website at http://sohowww.nascom.nasa.gov.

Designed for a nominal mission of 2 years, it has now passed the 10 year mark. With its nine European and three American principal investigators, SOHO is also another example of international cooperation in space. It was built by companies in 14 European countries, and is operated from Goddard Space Flight Center.

In 2006 NASA launched the Solar TErrestrial RElations Observatory (STEREO) spacecraft. For the first time, these spacecraft allows scientists to view the Sun in 3D and to track solar storms from the sun to Earth. The new view from STEREO greatly improves the ability to forecast the arrival time of severe space weather, to within 2 h.

17.3.3 Earth Applications Programs

We now take for granted photographs of weather and Earth resources data from space, as well as navigation and worldwide communications made possible by satellite. Along with human and robotic missions, the late twentieth century will be remembered collectively as the time when humans not only saw the Earth as a fragile planet against the backdrop of space, but also utilized near-Earth space to study the planet's resources, to provide essential information about weather, and to provide means for navigation that was both life-saving and had enormous economic implications. Worldwide satellite communications brought the world closer together, a factor difficult to estimate from a cost-benefit analysis.

Among the famous early communications satellites were Telstar and Syncom. Bell Telephone Laboratories designed and built the Telstar spacecraft with AT&T corporate funds. NASA's contribution to the project was to launch the satellites and provide tracking and telemetry functions, but AT&T bore all the costs of the project, reimbursing NASA $6 million. Telstar I was launched on July 10, 1962, and on that same day live television pictures originating in the United States were received in France. With the advent of Syncom and subsequent communications satellites, live telecasts from around the world became commonplace, and are now taken for granted. Under the guiding principle that NASA was a research and development organization, rather than one that undertook routine observations, in its early years NASA spun off some of its Earth applications programs to other agencies or to the private sector. Between 1962 and 1965 the semi-private Communications Satellite Corporation (COMSAT) and the International Satellite Communications Consortium (INTELSAT) were formed. Since then communications satellites have helped to make Earth a global village.

The first weather satellite, the Television Infrared Operational Satellite (TIROS), originated in the Department of Defense and was taken over by NASA when it was formed in 1958. NASA also began development of the next-generation NIMBUS weather satellite, but once it became operational the function was turned over to the Department of Commerce's Weather Bureau. In 1972 NASA's Earth resource satellite program began with the launch of Landsat 1, the first of a series continued most recently with Landsat 7, launched in 1999. The Earth resources satellites have also been subject to controversy over control and commercial viability, having been run by NOAA and the private sector during their history. Landsat is now managed by NASA but the data is collected and distributed by the U. S. Geological Survey.

The first two decades of the space age determined the capabilities of earth-observing satellites. Satellites for specific purposes, such as TIROS for weather, were largely piecemeal efforts. Only in the 1980s were steps taken toward a more comprehensive plan for studying the entire Earth system on a global scale. Following a number of studies, in 1987 the "Ride Report" on *Leadership and America's Future* in Space recommended that NASA adopt a Mission to Planet Earth as one of its four overriding themes. The centerpiece of Mission to Planet Earth was to be the Earth Observing System. Originally envisioned as a \$17 billion program over 10 years, it was scaled back to \$11 billion and then \$8 billion in 1992. After much re-scoping and reshaping of the program, in 1999 Terra, the first of three flagship polar-orbiting spacecraft was launched. It was followed by Aqua in 2002 and Aura in 2004. As a result of cooperation between the National Oceanic and Atmospheric Administration (NOAA), NASA, and the U. S. Air Force, a National Polar-Orbiting Environmental Satellite System (NPOESS) is now under long-term development. Despite being subjected to the politics of climate change and global warming, these satellites and others are making, and will continue to make, significant contributions to Earth science.

17.3.4 Space Astronomy and Astrophysics

In addition to spectacular images and data from Earth, the Sun, and planets, space exploration has proven useful for observations well beyond the Solar System. From their vantage point above the Earth's atmosphere, satellites could peer at the heavens at wavelengths not visible from Earth. From 1972–1981 the Orbiting Astronomical Observatory (OAO), also known as Copernicus, observed many objects at ultraviolet and X-ray wavelengths. In the late 1970s and early 1980s, the High Energy Astronomy Observatories (HEAOs) observed the sky in both gamma ray and X-ray. And in 1983, the Infrared Astronomical Satellite (IRAS), a joint project of the United States, the Netherlands and the United Kingdom, performed the first infrared survey of the entire sky.

Arguably more than any other single program, NASA's Great Observatories revealed the mysteries of the universe at many wavelengths. The Hubble Space Telescope (1990–), the Compton Gamma Ray Observatory (1991–2000), the Chandra X-ray Telescope (1999–), and the Spitzer Infrared Telescope (2004–) revolutionized our view of the universe. In its storied history, for example, the Hubble Space Telescope has observed objects within the Solar System a few light hours away to galaxies billions of light years distant, including those revealed in the Hubble Deep Field. It has discovered circumstellar material and extrasolar planets, confirmed the widespread existence and nature of black holes, and refined the age of the universe. Because it observed in visible wavelengths, the Hubble Space Telescope also inspired the public with some of the most memorable images of the cosmos, including the towering Eagle Nebula (see Fig. 21.1), the fantastic forms of planetary nebulae, and a variety of galaxy shapes.

NASA has not only undertaken voyages in space, but also in time. Thanks to the finite speed of light, NASA has even succeeded in making several voyages to the beginning of time. In 1989 it launched the Cosmic Background Explorer (COBE), and within hours detected the primordial seeds of galaxies and clusters of galaxies—small variations in the temperature of the cosmic background radiation first detected in 1964—the blueprint from which our universe formed. In 2006, the Nobel Prize for Physics was jointly awarded to NASA Goddard Space Flight Center senior project scientist Dr. John C. Mather, and to University of California at Berkeley scientist Dr. George Smoot, for their contributions to the COBE project. In 2001 the Wilkinson Microwave Anisotropy Probe (WMAP) was launched, and its high resolution observations confirmed and extended the results of COBE. The satellite also provided more evidence of the rapid inflation of the universe at its beginning, verifying and refining the leading theory of the origin of the universe. And it pinned down the age of the universe, within 100,000 years, to 13.7 billion years. It yielded information on the dark matter content of the universe. And it provided unprecedented detail on the origin of the universe and the evolution of the first stars and galaxies.

Collectively, NASA astronomy and astrophysics spacecraft, from the early probes of the 1970 and 1980s to the Great Observatories of the 1990s and the twenty-first century, yielded the secrets of cosmic evolution from the Big Bang to the present. While the great question of extraterrestrial life—currently being addressed by NASA's Astrobiology and Origins programs—remains unanswered, we can now begin to see our planet's place in the 13.7-billion-year history of the cosmos.

17.3.5 Life Sciences and Astrobiology

Discussions about life sciences at NASA began within the first year of the Agency's founding. In July, 1959 NASA first Administrator, T. Keith Glennan, appointed a Bioscience Advisory Committee, which reported in January 1960 that NASA should not only be involved in a traditional and obviously necessary space medicine role in support of manned spaceflight, but should also investigate the effects of extraterrestrial environments on living organisms, and undertake a search for extraterrestrial life. In the spring of 1960 NASA set up an Office of Life Sciences at Headquarters, and by August, with the possibility of planetary missions on the horizon, it had authorized the Jet Propulsion Laboratory (JPL) to study the type of spacecraft needed to land on Mars and search for life. In order to study chemical evolution, the conditions under which life might survive, and a variety of issues related to origins of life, NASA's first life sciences laboratory was also set up at Ames Research Center in 1960. Because it was also related to space science, during its history NASA's life science program has often fallen under the space science organization.

Also among the early life science concerns at NASA was planetary protection—both of planets on which spacecraft might land, and of our own home planet—the problem of back-contamination with returning spacecraft or samples. In the search for extraterrestrial life, contamination of another planet would be an irreversible disaster. Conversely, back contamination of our planet raised an Andromeda Strain scenario, named for a science fiction novel and movie (1971) about a fictional satellite returning to Earth carrying a deadly extraterrestrial organism. NASA has maintained a strong planetary protection program since those early days.

NASA's life sciences program also carried out a variety of successful missions in space, beginning with the Biosatellite program in 1967. The Biosatellites carried frog eggs, amoeba, bacteria, planets, and mice, and collected data regarding the effects of zero gravity on life. Beginning in 1975 the United States also cooperated for 20 years with the Soviet Union's Cosmos/Bion missions. Life sciences research also took place on human spaceflight missions. Europe's Spacelab, a pressurized module flown on the Space Shuttle, made possible several dedicated life sciences missions during the 1990s. Space life sciences research is also planned aboard the International Space Station, particularly as it applies to long-term human missions to the moon and Mars.

Meanwhile, in the area of exobiology, at its NASA Ames laboratories, the Agency had continued its research on the origins of life. But by far the largest investment of time and funding was the Viking project, two spacecraft that orbited Mars and sent landers to its surface in 1976. Although there were some ambiguities in the biology experiments, the consensus was that Viking did not detect life on Mars. Although no spacecraft returned to the red planet for two decades after Viking, the exobiology program continued to fund cutting-edge research in the life sciences.

The year 1996 saw a revival of exobiology under the name astrobiology, fueled by NASA's announcement of possible nanofossils in the Mars rock known as ALH 84001, by the Galileo spacecraft's confirmation of a likely ocean on the Jovian satellite Europa, and by the discovery of extrasolar planets. Mars continued to tantalize with more spacecraft observations. In 2001 Mars Global Surveyor revealed numerous gullies on Martian cliffs and crater walls and evidence of geologically recent liquid water. The following year Mars Odyssey also gave strong evidence of large quantities of water under the surface. And the Mars Exploration Rovers examined an outcrop of salt-laden sediment and found thin intersecting layers interpreted as sand ripples, perhaps shaped by flowing water in a huge shallow sea.

Origins of life studies also fed into the new optimism about extraterrestrial life. Scientists found life at extreme pressures and temperatures around deep-sea hydrothermal vents, fueled by energy and nutrients seeping from the Earth's crust. More generally, life was found in a variety of extreme environments, including caves, inside deep rock, and in highly acidic and salty conditions. These discoveries showed that life was much more adaptable than previously thought. At the same time, the discovery of complex organics in molecular clouds in space, at the level of amino acids, gave credence to the idea that life could be ubiquitous because its building blocks were common in outer space.

Astrobiology took advantage of these new developments to considerably broaden exobiology. Astrobiology placed life in the context of its planetary history, encompassing the search for planetary systems, the study of biosignatures, and the past, present and future of life. Astrobiology science also added new techniques and concepts to exobiology's repertoire, in the attempt to answer one of humanity's oldest and most profound questions.

17.4 Aeronautics

NASA's first "A" is sometimes downplayed in the midst of its spectacular space achievements. But building on its roots in the National Advisory Committee for Aeronautics (NACA), NASA from its beginnings conducted research on aerodynamics, wind shear, flight safety, and other important topics using wind tunnels, flight testing, and computer simulations.

Wind tunnels, though not the most glamorous technology, were essential to relatively low-cost testing of aircraft performance before the aircraft were actually built. NASA had inherited a variety of wind tunnel facilities from its NACA centers, many of them constructed during World War II at Ames, Lewis, and Langley. Each wind tunnel had its own characteristics, depending on the size of the aircraft or models being tested, and whether they were being tested at subsonic, transonic and supersonic speeds. By the dawn of the Space Age, hypersonic tunnels were constructed with their own unique characteristics and capabilities. Wind tunnels were also used to test the atmospheric dynamics of the Mercury, Gemini and Apollo capsules, and eventually the Space Shuttle. They continue to be a vital tool for aeronautics research.

In the area of real flight testing, from its beginning NASA assumed responsibility for the X-15 hypersonic aircraft, capable of speeds exceeding Mach 6 (4500 miles per hour) at altitudes of 67 miles, reaching the very edge of space. Between 1959 and 1968, three X-15 aircraft completed 199 flights, and contributed greatly to knowledge about hypersonic aerodynamics and structures eventually needed for spaceflight, including the Space Shuttle. The X-15 was air-launched by B-29s, B-50s, and eventually B-52s. Its "control room," located at the NASA (now Dryden) Flight Research Center in the California desert, advanced from a portable van to a more formal permanent room that later served as the model for the famous mission control at Johnson Space Center. Synergies between aeronautics and human spaceflight also appeared in other ways; today it is a little-known fact that Neil Armstrong began as an X-15 pilot working for NACA, and that eight other X-15 pilots flew high enough to be qualified as astronauts according to U. S. standards (50 miles). Many other astronauts were test pilots on other high-performance aircraft. NASA also cooperated with the Air Force in the 1960s on the X-20 Dyna-Soar program, which was designed to fly humans into orbit. The program was eventually cancelled, but the ideal of winged spacecraft never died.

Fig. 17.8 A collection of NASA's research aircraft on the ramp at the Dryden Flight Research Center in July 1997: X-31, F-15 ACTIVE, SR-71 blackbird, F-106, F-16XL Ship #2, X-38, Radio Controlled Mothership and X-36. July 16, 1997

NASA also conducted significant research on high-speed aircraft flight efficiency, maneuverability and safety, research that was often applicable to lower speed airplanes (Fig. 17.8). NASA scientist Richard Whitcomb invented the "supercritical wing," specially shaped to delay and lessen the impact of shock waves on transonic military aircraft. It had a significant impact on civil aircraft design. From 1963 to 1975, NASA conducted a research program on "lifting bodies," aircraft without wings. During the 1970s several of NASA's aeronautics centers also undertook a variety of aeronautics research using the SR-71 Blackbird in the Mach 3 range. Such research was useful for diagnostics systems on the Shuttle, and also paved the way for the Shuttle to glide to a safe unpowered landing.

During the 1980s NASA and the Department of Defense began the development of a hypersonic National Aerospace Plane known as the X-30, and later worked on a hypersonic X-33 project. For a variety of reasons these never reached production. In 2004, the X-43A aircraft used innovative scramjet technology to fly at 7000 miles per hour, almost ten times the speed of sound, setting a world's record for air-breathing aircraft. It reached an altitude of 110,000 feet over the Pacific Ocean.

In addition to its better-known spaceflight achievements, during its first 50 years NASA thus continued in the forefront of flight, carrying on from the humble beginnings of the Wright brothers at Kitty Hawk.

17.5 Why We Explore

The vast scope of NASA's work and that of other space agencies inevitably raises questions about cost, motivation, and sustainability in a world with so many other problems. The question "should we explore?," whether the frontiers of aeronautics or the furthest reaches of outer space, must be seen in deep historical context, not in the context of passing politics or whims. The historical connections are fully recognized at NASA today. When announced in January 2004 the concept of orienting NASA's program to the human and robotic exploration of the moon, Mars and beyond was billed as "a new spirit of discovery," and the implementation plan was titled "A Journey to Inspire, Innovate and Discover." Indeed, Americans tend to place their space endeavors in the long tradition of exploration, and historians have argued the Space Age ushered in a new Age of Exploration.

The continued exploration of space is, however, a choice we must make. As historian Stephen J. Pyne has argued, "Exploration is a specific invention of specific civilizations conducted at specific historical times. It is not … a universal property of all human societies. Not all cultures have explored or even traveled widely. Some have been content to exist in xenophobic isolation." Ming China's abandonment of its massive fleets in the early fifteenth century is often cited, even by Chinese historians, as a poor decision that hampered Chinese civilization for centuries and left the world open to European discovery. Historian and former Librarian of Congress Daniel Boorstin called the withdrawal of the Chinese into their own borders "catastrophic … with consequences we still see today."

The question of whether we should explore when there is so much that needs to be done on Earth is both an ethical and a public policy question. Quite aside from the short term benefits of applications satellites, national security, jobs, and inspiration to the young, much of NASA's impact is long term, and it is always tempting to sacrifice long-term goals for short-term needs. Today there are ample reasons one might argue not to continue space exploration. But we should recall the sentiment of H. G. Wells many years ago, that "Human history becomes more and more a race between education and catastrophe." We are still in that race today, and surely space exploration expresses humanity's most noble aspirations.

17.6 Commentary 2020

This paper was written on the occasion of the 50th anniversary of the Space Age for the Indian Space Research Organization. The volume in which it appeared (Manoranjan Rao 2007) is unique in that it contains articles on all the major space agencies of the world, written by a representative of each agency.

Among the conferences on the occasion of that anniversary was one sponsored by the NASA History Office, *Remembering the Space Age: Proceedings of the 50th Anniversary Conference* (Dick 2008). It is notable for its often sobering assessment by some of the participants. Walter McDougall, the Pulitzer Prize winning author of *The Heavens and the Earth* (1985) viewed the 50th anniversary of the Space Age as a melancholic affair, filled with disappointment and unfulfilled hopes, a secondary activity compared to the dominant trends of contemporary history, and in any case too embryonic to judge its significance. Space policy guru John Logsdon disagreed in part, arguing that both the modern nation-state and the global economy depend on space-based systems. The ability to operate in outer space, he contended, is an integral part of modern history. He agreed that the progress of the Space Age has been frustrating in many ways to those who lived through the Apollo era, a level of activity that was not sustainable. Former NASA Chief Historian Sylvia Kraemer argued there are many competing events that may define the last 50 years more than space exploration, including the Cold War and digital and information technologies. She also argued that the contribution of space activity to globalization has been far greater than its contribution to nationalism. Linda Billings reflected on space exploration in the context of culture, concluding that it means many things to many people, quite aside from dominant official narratives. Nor is this an academic exercise, for she suggests that if space programs are to survive and thrive in the twenty-first century they need to involve citizens and be aware of the visions they have for a human future in space. On the global level, this resonated with historian John Krige's statement that "when 'Remembering the Space Age,' we should not shy away from admitting the complexity and diversity of the space effort, nor pretend that the view of the world from Washington is the only view worth recording."

Some in the audience at this 50th anniversary conference thought it should have been more celebratory and described the meeting itself as depressing. Others felt it reflected both the frustrations and the realities of the Space Age. In the end, there seemed to be consensus that human spaceflight has been a disappointment in the aftermath of Apollo, and in that sense the Space Age, if indeed it ever existed, has been a disappointment as well. Such disappointment is no artificial construct of historians; the legendary Wernher von Braun, who thought humans would land on Mars by 1984, would undoubtedly have agreed. Nor is disappointment necessarily a bad attitude; it means vision has outstripped practical realities and that vision may yet drive individuals and nation-states toward new realities.

In common parlance, the title "Remembering the Space Age" carries with it a connotation that we are looking back on something that may have ended. Or maybe it never began; certainly launching Sputnik in and of itself did not constitute a Space Age, and the resulting reaction culminating in the manned lunar landings had ended within 15 years. Communications, navigation, weather, reconnaissance, and remote sensing satellites have been more sustained. But is such space activity, bounded by commercial and practical applications, enough to constitute a Space Age? Or, as several speakers opined, is space science the real core of the Space Age? As historian John McNeill concluded, it may well be too early to tell whether space activities over the last half century constitute a genuine "Age." We may need more time for better perspective. One thing is certain: if indeed the Space Age exists and if it is to continue, it must be a conscious decision requiring public and political will.

Like exploration, each culture must set its priorities, and there are no guarantees for the Space Age.

As for the eras of human spaceflight delineated in this chapter, the Space Shuttle era ended with its last flight in July, 2011, the International Space Station era continues, and a new era may begin in 2020 with the inauguration of NASA's Space Launch System, commercial human spaceflights with SpaceX and Boeing, and its Artemis program to return humans to the Moon. The question of Why We Explore is elaborated in a series of 28 essays written during my time as NASA Chief Historian, available on the NASA website at https://history.nasa.gov/Why_We_/Why_We_Main.html.

References

Dick, Steven J., ed. 2008. *Remembering the Space Age: Proceedings of the 50th Anniversary Conference*. Washington, DC, NASA.

Manoranjan Rao, P. V., 2007. *50 Years of Space: A Global Perspective*. Universities Press, Hyderabad, India.

Chapter 18
Exploration, Discovery, and Culture: NASA's Role in History

Abstract This chapter, written for the 50th anniversary of NASA, is an extended comparison of the Age of Discovery and the Age of Space. It attempts to place NASA in the cultural context of exploration and discovery. Though there are many political, social, and cultural components, we argue that exploration is the key to understanding the Space Age. Following J.H. Parry's classic volume *The Age of Reconnaissance: Discovery, Exploration and Settlement, 1450–1650*, this chapter draws comparisons with the conditions for the Space Age in terms of motivation, infrastructure, voyagers, institutions, funding, and risk, before addressing the story of the Space Age from the realm of the Earth to the realms of the planets, stars, and galaxies. The chapter concludes with remarks on the intellectual, economic, geopolitical, and social impact of the Space Age. The Space Age opens a vast new future for humanity. But exploration is a choice that societies must make in the midst of many other priorities. That choice embodies the meaning and essence of the Space Age.

18.1 Introduction: Space Exploration in Context

Like the facets of a jewel, the overall importance of NASA and the Space Age over the last 50 years may be considered from many viewpoints, ranging from the geopolitical and technological to the educational and scientific. But no facet is more central than exploration, a concept that encompasses most of the other possibilities and arguably constitutes one of the main engines of human culture, spanning millennia. In its simplest and purest form, the Space Age may be seen as the latest episode in a long tradition of human exploration. Surveying the vast panoply of history, historians have often found "symmetry in the narrative arc of the Great Ages of Discovery" or traced that tradition back even to the Paleolithic Era in an attempt to find a "global historical context" for the Space Age (Pyne 1993, 2006; McNeill 2008; Lewis 1976).

First published in Steven J. Dick, ed., *NASA's First 50 Years: Historical Perspectives* (NASA History Series 2010).

S. J. Dick, *Space, Time, and Aliens*, https://doi.org/10.1007/978-3-030-41614-0_18

The Paleolithic Era aside, prior to the Space Age historians often distinguished two modern Ages of Exploration: the Age of Discovery in the fifteenth and sixteenth centuries associated with Prince Henry the Navigator, Columbus, Magellan, and other European explorers, and the Second Age in the eighteenth and nineteenth centuries characterized by further geographic exploration such as the voyages of Captain Cook, underpinned and driven by the scientific revolution (Goetzmann 1986). Some now distinguish a Third Age, beginning with the IGY and Sputnik, primarily associated with space exploration, but also with the Antarctic and the oceans (Pyne 2006, 7–35). If one accepts this framework, it makes sense to compare one age of exploration with another, constantly keeping in mind the differences as well as the similarities and with full realization of the unlikelihood of any predictive ability. Here we choose to compare the Age of Space with the European Age of Discovery, in the hope of revealing symmetries and differences and casting in a new light some of the chief characteristics of the last 50 years in space.

The overarching theme and structure of our argument for the primacy of exploration as a key to understanding the Space Age is inspired by the distinguished Harvard maritime historian J. H. Parry, who 30 years ago published his classic volume *The Age of Reconnaissance: Discovery, Exploration and Settlement, 1450 to 1650* (Parry 1963). NASA's first 50 years may also be characterized as "The Age of Reconnaissance," or to put it more broadly, as the first stages of "The Age of Discovery." There have been discovery and exploration, but not yet settlement—unsurprisingly, since we are only 50 years into the Age of Reconnaissance for space. Parry tackled his theme by discussing the conditions for discovery, then the story of the discoveries themselves, and finally the "fruits of discovery." A parallel tripartite structure provides a framework for examining the importance of NASA and the Space Age: what were the conditions for the Space Age, the story of its voyages, and their impact? Much of the meaning of NASA and the Space Age may be found in the context of those three questions.

By drawing such comparisons we are engaging in the time-worn method of analogy, and we need to ask whether analogy is a valid framework for analysis, a proper method of reasoning? In making use of analogy, I am following a methodology pioneered almost 50 years ago in another classic book, *The Railroad and the Space Program*, whose subtitle is *An Exploration of Historical Analogy*. This volume, edited by MIT Professor Bruce Mazlish and populated with well-known scholars, addressed the problem of analogy in considerable detail. Mazlish himself spoke of "attempting to set up a new branch of comparative history: the study of comparative or analogous social inventions and their impact on society." The authors went on to give what is, almost 50 years later, perhaps still the best treatment of the general use of historical analogy. Although originally suspicious of parallels with the past, present, and future, the contributors to this volume found it a useful tool; historian Thomas P. Hughes saw "the possibility of moving up onto a level of abstraction where the terrain of the past is suggestive of the topography of the present and its future projection." (Mazlish 1965a, b; Hughes 1965; Coopersmith 2008). The authors cautioned that as much empirical detail should be used as possible and that analogies drawn from vague generalities should be avoided. Confident in the use of historical analogy as suggestive but not predictive of the future, Mazlish and his coauthors went on to elaborate their analogy with the railroad and the space program with such a degree of success that their work is still discussed today.

The utility of analogy is suggested by its frequent use: throughout the Space Age, and indeed the history of science in general, scientists have been drawn to this mode of reasoning (Wheeler 1996; Kennefick 2007; Harré 1972; Hesse 1966). The Antarctic dry valleys have been studied as analogs to conditions for life on Mars, the subglacial Antarctic Lake Vostok as an analog to the ocean of Jupiter's satellite Europa, and extremophiles on Earth as analogs to possible alien life. More similar in kind to the railroad and the space program analogy, NASA Administrator Michael Griffin has invoked the "highway to space" to emphasize the sustaining effort required in space exploration. "Space exploration by its very nature requires the planning and implementation of missions and projects over decades, not years," he wrote. "Decades of commitment were required to build up our network of transcontinental railroads and highways, as well as our systems for maritime and aeronautical commerce. It will be no quicker or easier to build our highways to space, and the commitment to do it must be clear and sustaining." (Griffin 2008a, b, 1–8). Speaking of the new systems being built for the current space exploration vision, Griffin wrote that "NASA will build the 'interstate highway' that will allow us to return to the moon, and to go to Mars." Similarly, he has compared polar exploration to lunar exploration, arguing that the Apollo program was like the singular forays of Scott or Byrd, while the current plans to establish a base on the Moon are more like the permanent presence that several countries have had in the Antarctic since the 1950s, requiring international collaboration (Griffin 2008a, b, 175–186).

Analogies are never perfect, but they can be useful and illuminating as guides for thought. They can also be overstated and misleading, as in the case of the "frontier analogy" so prominent in American space exploration. There is no doubt that exploration is part of the American character and that federally funded exploration has been a significant part of American history (Goetzmann 1966).[1] But the very idea of the American frontier and its meaning have been questioned, especially as popularized at the end of the nineteenth century by historian Frederick Jackson Turner. Turner saw many of the distinctive characteristics of American society, including inventiveness, inquisitiveness, and individualism, as deriving from the existence of a frontier, and he therefore saw the closing of the Western frontier about 1890 as cause for worry (Turner 1994). It was natural for Americans to find a new frontier in space as an analog to their Western frontier and to argue that conquering the new frontier would perpetuate those characteristics described by Turner. The problem is that many historians do not agree with Jackson's *frontier thesis* as the sole, or even the primary, source of these characteristics in the United States. And by extension, they are skeptical of the benefits of the new frontier. Historians notwithstanding, space as a new frontier has always been a driver of the U. S. space program and remains very much in NASA's lexicon. Nevertheless, it is an analogy that needs to be used with qualification and caution (McCurdy 1997, 144–145).[2]

If we accept analogical reasoning as a useful tool applied with caution, are exploration and discovery the right analogies? Certainly exploration was not the only, or even the chief, motivation for the space program. But, abstract and even metaphysical as it may seem, it was surely one of the motivations, and a major one at that—the philosophical apex of a pyramid that, of necessity, included more practical motivations. The concepts of discovery and exploration are frequently found throughout space literature, most recently in the Vision for Space Exploration, billed as "a new spirit of discovery," enunciated by President George W. Bush in January 2004. The same concepts are emphasized in the Aldridge Commission's Report on the Implementation of United States Space Exploration Policy, titled *A Journey to Inspire, Innovate, and Discover*, and yet again in NASA's subsequent new strategic objectives released in a report titled "The New Age of Exploration" (White House 2004; Aldridge Jr. et al. 2004; NASA 2005).[3] One can easily trace the concept back to the dawn of the Space Age, an omnipresent, if insufficient, driver of the new age that anchored it in history.

As space science practitioners and supporters like to emphasize, exploration and discovery apply not only to human spaceflight, but also (they would say especially) to space science. That, indeed, is the broad definition encompassed in NASA's documentary history series over the last two decades, *Exploring the Unknown* (Logsdon 1995–2008). Moreover, "The New Age of Exploration" speaks of a human and robotic partnership for exploration—robotic reconnaissance, followed by human voyages that satisfy that desire to explore in person and up close. In 2005, A National Research Council study also concluded that "the expansion of the frontiers of human spaceflight and the robotic study of the broader universe can be complementary approaches to a larger goal." This is easy to say and difficult to implement. To achieve that balanced partnership with the limited resources at hand, in the midst of turbulent events and ever-changing economic and political conditions on Earth, has been one of NASA's great challenges over the last 50 years (National Research Council 2005).

Exploration parallels have, of course, been drawn before. Wernher von Braun was fond of comparing his proposed voyages to Mars to the voyages of Magellan. When Laurence Bergreen researched his book *Voyage to Mars: NASA's Search for Life Beyond Earth* about Pathfinder, the Mars Global Surveyor, and the heartbreaking unsuccessful 1999 voyages to Mars, he found references to the Age of Discovery and Magellan rampant within NASA. "After the tenth or maybe the twentieth time the name Ferdinand Magellan was mentioned to me," he recalled, "a dim light bulb eventually illuminated in my mind" (Bergreen 2000, 2005). The experience led him to write his gripping account, *Over the Edge of the World: Magellan's Terrifying Circumnavigation of the World.* Moreover, references to exploration in the American context are even more common and reached their height in 2003 with the bicentennial of the Lewis and Clark expedition. Such analogies were used to sell the space program and, more recently, the Vision for Space Exploration (Asner and Garber 2019).

Finally, the imagery of the oceans of Earth and the ocean of space has often been employed in space rhetoric, evoking past exploration. It is one thing when the President of the United States proclaims, as he did a few months after setting the course for the Moon in 1961, that "We set sail on this new sea because there is new knowledge to be gained, and new rights to be won, and they must be won and used for the progress of all people. For space science, like nuclear science and all technology, has no conscience of its own. Whether it will become a force for good or ill depends on man, and only if the United States occupies a position of preeminence can we help decide whether this new ocean will be a sea of peace or a new, terrifying theater of war." And it is significant when historians and journalists build on the analogy, as in the official history of project Mercury, entitled *This New Ocean*, or William Burrows' classic history of the Space Age with the same title (Swenson Jr. et al. 1966; Burrows 1998). But it is even more significant when NASA workers see themselves in the tradition of the Age of Discovery, for that idea, once individually and institutionally internalized, becomes a part of NASA culture and a powerful force in itself (McCurdy 1993; Dick and Launius 2006, 345–428).

With analogy as our guide, exploration as our theme, and Parry's work as our framework, let us examine NASA and the Space Age with all the caution and boldness due such a complex and all-encompassing theme.

18.2 The Conditions for the Space Age

Analysis of a sampling of the many major factors in common between the Age of Discovery and the Age of Space will suffice to demonstrate the utility of making comparisons: motivations, infrastructure, voyagers, funding, and risk were clearly important considerations in both eras. It is no surprise that similar narrative arcs should generate similar general categories. But the interest lies in the details, the analogies, and the dis-analogies, all placed in the proper context of their time, and allowing us to see the Space Age in the light of long historical perspective.

18.2.1 Motivations

As a necessary condition of existence, both ages had their motivations, but they were very different. In the fifteenth century, exploring nations were in search of empire, and their motivations were twofold: economic gain, through trading or land acquisition, and religious conversion. As Parry put it in his classic study, "Among the many and complex motives which impelled Europeans, and especially the peoples of the Iberian peninsula, to venture oversea in the fifteenth and sixteenth centuries, two were obvious, universal, and admitted: acquisitiveness [wanting to acquire land for empire] and religious zeal. Many of the great explorers and conquerors proclaimed these two purposes in unequivocal terms" (Parry 1963, 19).

The motivation for the Space Age was neither of these. In the wake of Sputnik, under the Eisenhower administration, the newly formed PSAC, chaired by James R. Killian, identified four factors that gave "importance, urgency and inevitability" to entering space. The first of these was exploration. Foreshadowing the theme of

Star Trek 10 years later, the report spoke of "the compelling urge of man to explore and to discover, the thrust of curiosity that leads men to try to go where no one has gone before." With an explicit nod to past exploration, the authors of the report noted that "Most of the surface of the earth has now been explored and men now turn to the exploration of outer space as their next objective" (President's Science Advisory Committee 1958).

The second rationale posed in 1958 for entering space was national defense. "We wish to be sure that space is not used to endanger our security. If space is to be used for military purposes, we must be prepared to use space to defend ourselves." Third was national prestige. "To be strong and bold in space technology will enhance the prestige of the United States among the peoples of the world and create added confidence in our scientific, technological, industrial, and military strength." Science was the fourth factor, for space "affords new opportunities for scientific observation and experiment which will add to our knowledge and understanding of the earth, the solar system, and the universe (President's Science Advisory Committee 1958, 333). In the Soviet Union, the only other space power at the time, the motivations were much the same.

Among these motivations for spaceflight, national prestige was paramount for the first decades of the Space Age, as historical analyses, such as Walter McDougall's *The Heavens and the Earth*, have shown (McDougall 1985). The motivations are much the same today, although economic competitiveness and survival of the species are now at least part of the discussion (Launius 2006, 37–70). Since the end of the Cold War in the early 1990s, and arguably since the end of the Apollo era, we have entered a period that will determine whether international cooperation, exploration, and commercial gain can provide the same impetus to space that international competition once did. The ISS is a prime example of the cooperation, albeit sometimes difficult, of 16 countries over the last decade. Still, the utility and cost of the ISS have often been called into question, and analysts such as Woody Kay have asked with more than just rhetoric, "Can Democracies Fly in Space?" Without the impetus of outside competition, under always difficult economic conditions, and in the midst of so many other priorities in a democratic society, this remains an important question of public policy (Kay 1995; Krige 2006; Logsdon 1996).

18.2.2 Infrastructure

Both ages of discovery required a certain infrastructure, none more important than the means of conveyance—ships for the Age of Discovery and rockets for the Age of Space. Beginning with Prince Henry the Navigator in the fifteenth century, the vessel of choice for ocean exploration was the small, maneuverable, and relatively fast caravel with its lateen triangular sail, in contrast to the galley or other vessels with fixed sails or oarsmen. (Russell 2001). Caravels were used for everyday trade routes in Western Europe, and typically new types of vessels were not constructed for the early long, transoceanic voyages. But caravels were small, crowded, and

Fig. 18.1 The already existing caravel was often the vessel of choice in the Age of Discovery, while rockets had to be built *de novo* or based on military rockets. Symbolizing the Age of Discovery and the Age of Space, replicas of Christopher Columbus's sailing ships *Santa Maria*, *Niña*, and *Pinta* sail by the Space Shuttle *Endeavour* at KSC's Launch Complex 39B awaiting liftoff on its maiden voyage in 1992. The *Niña* and *Pinta* were caravels, whereas the *Santa Maria* was a merchant ship known as a carrack. Next to the launchpad (at right) are the sound suppression water system tower and the liquid hydrogen (LH_2) storage tank, all part of the complex infrastructure of the Space Age. The caravels, managed by the Spain '92 Foundation, were at the time on a tour to ports around the Gulf of Mexico and up the Atlantic Coast of the Untied States on the occasion of the 500th anniversary of Columbus's voyage to the New World. NASA JSC Image S92-3907

uncomfortable, and as the Age of Reconnaissance continued, mixed types of ship designs were developed, and fleets sailed with a balanced mix of ships when possible: "one or two caravels, which they employed for dispatch-carrying, inshore reconnaissance, and other odd jobs which later admirals would entrust to frigates. Such ships and such fleets first became available, through a strenuous process of experiment and change, to Europeans in the late fifteenth century. This was the development which made the Reconnaissance physically possible." Caravels could also carry cannons, and some historians argue that "Caravels and cannon were the technological developments that made European expansion overseas possible, not astrolabes and improved maps" (Parry 1963, 65–66; Fritze 2002). See Fig. 18.1.

By contrast, because nothing had ever entered the ocean of space, designers had to invent motive power and spaceships through a combination of old and new technologies and sometimes from scratch. It is true that both the Soviet Union and the United States adapted older military missiles as the motive power to enter space, but both also independently designed new rockets (Hunley 2007, 2008a, b; Launius and

Jenkins 2002; Bilstein 1980; Dawson and Bowles 2004). Unlike ships, the motive power was no longer natural wind power. The core of the new rockets was their engines, and the history of engine development is fraught with uncertainty and contingency. At every stage, from the V-2s and their successors, to the Apollo first-stage F-1 engines with their famous early "combustion instability" problems, and to the SSMEs, it was never assured that access to space would be possible, and it is still not cost-effective. (Bilstein 1980; Murray and Cox 1989, Chap. 10; Dunar and Waring 1999; Biggs 2008; Dawson 1991). Another of the perennial debates of the Space Age was whether reusable or expendable launch vehicles were best; history records that despite its utility and magnificent engineering, even the reusable Space Shuttle was never cost-effective (Butrica 2006; Logsdon 2006). With the projected return to expendable rockets after 2010, human winged spaceflight may prove to have been only an ephemeral 30-year phenomenon, at least for the twentieth and twenty-first centuries.

Human spaceflight also required the design of capsules and later the reusable Shuttle to carry humans on their epic early piloted programs. Spacecraft design pioneers like Max Faget (who played a role in the design of every American piloted spacecraft), as well as a variety of unsung heroes, were no less essential to the Space Age than were rocket engineers, and both were as indispensable as the shipbuilders of 500 years before (Scranton 2006).[4] The design of robotic spacecraft and the perennial debate over human versus robotic spacecraft, on the other hand, find no parallel with the Age of Discovery (McCurdy 2006; Gerovitch 2006; Mindell 2006; Launius and McCurdy 2008). Robotic spacecraft design, with its communications, thermal, and electronic subsystems, is especially part of the histories of JPL, GSFC, and their aerospace partners. Indeed, an entire industry sprang up on the foundations of the aviation industry to cater to both the human and the robotic rocket and spacecraft needs of the Age of Space (Bilstein 1996; Bromberg 1999).

The engineering challenges inherent in the design of rockets and spacecraft were legion (Fries 1992; Johnson 2002; Mindell 2008). Design decisions were sometimes brilliant, often modified, and occasionally second-guessed after accidents and failures, whether human or robotic, and the agonizing but detailed accident reports of those failures make for compelling reading about the importance and far-reaching consequences of engineering decisions (Brown 2006; Vaughan 1986; Apollo 204 Review Board 1967; Rogers 1986; Columbia Accident Investigation Board Columbia Accident Investigation Board 2003).[5] As far as we know, no such *ex post facto* analysis was undertaken in the Age of Discovery, where the whims of nature at sea were most often at fault (though one might question some of Magellan's decisions, including the final one leading to his death).

Ships and rockets alike required specialized points of departure, where they could prepare for the journey. Unlike the ancient ports from which the ships of the fifteenth and sixteenth centuries departed, spaceports were built from scratch or on sites of military missile launches. Their locations were determined not so much by water (though an uninhabited overflight path was a factor), but by the latitudes at which Earth's rotation could impart additional motive power, among other considerations. Those spaceports, with now-legendary names like Cape Canaveral,

Vandenberg, Kourou, Pletetsk, and Baikonur, were the equivalents of Palos, Lisbon, and Sanlúcar de Barrameda. Except for the ever-popular KSC, the launch sites so essential to spaceflight are often unappreciated by the public, as is other necessary infrastructure such as ground tracking stations, navigation, and mission control. Scientists, engineers, and historians, however, are fully aware that the Space Age could not exist without them (Sheahan and Hoban 2004; Lipartito and Butler 2007; Benson and Faherty 2001a, b; Tsiao 2008; Mudgway 2001, 2005; Kranz 2000; Kraft 2002).

Both the Age of Discovery and the Age of Space had their navigators, their users and producers of maps that increased in accuracy as a result of the voyages of discovery. The Age of Discovery had its world cosmographic maps and its portolan maps, the latter to actually help in navigating. The Age of Space, too, had its general cosmography, as backdrop, and its practical star maps for celestial navigation, though its methods of navigation—gravitational assists from planetary flybys, for example—were strikingly novel. As in the sixteenth century, Space Age voyages of discovery produce ever more accurate maps of their routes and their destinations, and the astrogeology branch of the U. S. Geological Survey, funded largely by NASA, carries out the same role for mapping new worlds as sixteenth-century cartographers did for *the* New World (Butrica 2014; U. S. Geological Survey n.d.; Levy 2000; Schaber 2005).

18.2.3 Voyagers

Voyagers, whether human or robotic, are also essential to the exploration enterprise. Both ages had their heroes, leaders of the voyages of discovery. Columbus and Magellan were men of daring and adventure who personally argued for government funding of their voyages. Cosmonauts, astronauts, and taikonauts were also daring, but unlike explorers from the Age of Discovery, it was not they who argued for government funding for the space program; it was scientists and managers like Wernher von Braun and a sequence of NASA Administrators, now enmeshed in a growing technocratic complex.

At another level, crews in the Age of Discovery, as in the case of Magellan's circumnavigation, were often hard to come by. There is no parallel to this situation among myriad astronaut applicants, who outnumbered successful candidates by more than 1000 to one. While many ship captains were men of some learning, their crews varied greatly, from people off the streets to religious seekers, profiteers, and pirates. By contrast, the nearly 500 astronauts, cosmonauts, and taikonauts who have ventured into Earth orbit or beyond over the last 50 years were the products of refined technical training, as were the eight X-15 pilots who flew high enough to be qualified as astronauts, and even the two pilots who flew on *SpaceShipOne* in 2004. Beginning with the Mercury 7 (Fig. 18.2), they all had what writer Tom Wolfe immortalized as "the right stuff" (Wolfe 1979).[6]

Fig. 18.2 As with the Age of Discovery, voyagers were essential for the Age of Space. Project Mercury astronaut selection was announced on April 9, 1959, only 6 months after NASA was formally established on October 1, 1958. All were military test pilots. This iconic image, taken at NASA's Langley Research Center in Virginia, captures the pilots (front row left to right) Walter H. Schirra, Jr., Donald K. Slayton, John H. Glenn, Jr., and Scott Carpenter; (back row) Alan B. Shepard, Jr., Virgil I. "Gus" Grissom, and L. Gordon Cooper. Despite the iconic status of the image, its precise date is unknown, but it was taken by Ralph Morse for *Life Magazine* prior to Shepard's suborbital flight in May 1961. NASA Image 84PC-0022

In the United States in 1962, JSC in Houston, Texas, became the home of the astronauts, where they underwent (and still undergo) rigorous training. The Soviet/Russian counterpart is the legendary Cosmonaut Training Center in Star City, near Moscow, where training began in 1965. At these two locations, the vast majority of space explorers have prepared for their journeys prior to launch from their countries' respective spaceports into the "new ocean."[7]

18.2.4 Institutions and Funding

The space programs of the world required massive efforts in institution building, management, and funding. The Age of Discovery explorers were funded in part by nation states such as Spain and Portugal, often without the intermediary of an organizing institution. By the time of the Age of Space, the infrastructure had grown so complicated and expensive that national governments had to form new agencies dedicated to the task (Rao 2007). Paramount among these was NASA, and its story of "organizing for exploration" is well known (Logsdon 1995–2008; MacGregor 2008; McDougall 1985; Hunley 1993; Logsdon 1998; Portree 1998; NASA 1958; Rosholt 1966). Along with the technical aspects, the development of management techniques appropriate to a high-technology, high-reliability organization has been essential to its success, and Apollo management techniques have been especially studied (Johnson 2002).

No less crucial has been funding. Over the last 50 years, aside from the anomalous Apollo era, NASA's budget has remained relatively stable at below 1% of the federal budget. Still, NASA leads the world in its space budget as a percentage of government spending (Organization for Economic Cooperation and Development 2007; Oxford Analytica 2008). As with other government agencies, and especially because NASA's reach exceeds its grasp, the search for more funding is a never-ending enterprise. Yet, in the case of the United States, polls show most of the public is content with this level (Bainbridge 2018; Stine 2007). Whether in the next 50 years harsh economic realities drive the budget percentage down, or whether international competitive pressures from Europe, China, and India drive it up, the budget must remain, for now, one of the great unanswered questions of the future Space Age.

18.2.5 Risk

Finally, it is important to emphasize that both the Age of Discovery and the Age of Space had, and will continue to have, their risks and their tragedies. Out of five ships and 260 men who departed Spain with Magellan on 20 September 1519, only one ship and 18 bedraggled men returned in 1522—and Magellan was not one of them. In a sense, there is a huge difference between the two ages in this regard; while both ages recognized risk, little was done to manage risk in the Age of Discovery. By contrast, in the Age of Space, risk is managed to the extent that agencies such as NASA, and by association the entire nation, are sometimes accused of being risk averse. One of the greatest policy challenges is to find the proper balance between risk and exploration, and this, too, should be informed by history.

One of the greatest lessons of history, emphasized by studies from the Augustine Report of 1990 to the Columbia Accident Investigation Board Report of 2003, is that risk will always be associated with exploration. The Augustine Report conjoined both themes of risk and exploration and the Age of Discovery when it opined that "In a very real sense, the space program is analogous to the exploration and settlement of the new world. In this view, risk and sacrifice are seen to be constant features of the American experience. There is a national heritage of risk taking handed down from early explorers, immigrants, settlers, and adventurers. It is this element of our National character that is the wellspring of the U. S. space program" (Augustine 1990; Columbia Accident Investigation Board 2003).[8] At times during the last 50 years, that element of willingness to take risk in the space program has hung by a thread in the aftermath of searing accidents in both human and robotic spaceflight. The easy course after losing both the Mars Climate Orbiter and the Mars Polar Lander in 1999, and after losing the second Space Shuttle in 2003, would have been to cancel the programs. But despite deep personal losses to families, careers, and the American sense of exceptionalism, the programs moved ahead. Just as the original Age of Discovery faded, and the preeminence of their nation states along with it, there is no guarantee the Space Age will not suffer the same fate, despite its literally infinite possibilities.

In summary, both symmetries and asymmetries exist in the general narrative arc of the Age of Space and the Age of Discovery, whether in terms of motivation, infrastructure, funding, people, risk, or many other factors not mentioned here. The particular conditions were very different, and both ages can only be understood in the context of their times. Nevertheless, both ages indisputably produced great voyages of discovery, and it is to those voyages we now turn.

18.3 The Story of the Space Age

Even with its multifaceted and fascinating policy, infrastructure, and engineering aspects, the Age of Space is best characterized not by its conditions, but by its results. Space exploration has generated many narratives, but its central narrative is simple, straightforward, and profound: a continuous story of voyages further and further from the home planet. The Age of Discovery began in the fifteenth century with Portuguese sailors hugging the west coastline of Africa, then sailing outward to increasingly distant islands—Madeira in the 1420s, the Azores in the 1430s and 1440s, and the Cape Verde islands in the 1450s (and a long unsuccessful attempt at the Canary Islands controlled by Castile) (Russell 2001). At the end of the century, the Portuguese denied Columbus the funding he requested, and it was the Spanish who funded the first plunge across the ocean in a remarkable story we all learn in school (Thomas 2003), By about 1650, in Parry's estimation, the Age of Reconnaissance was over, as Africa, Asia, and the Americas had become routine destinations.

Unlike the Age of Discovery, which ran its course in about two centuries, the Age of Space is a process that has only begun and that potentially has no end, but that is nonetheless fundamentally a story of exploration and discovery now played out on an unimaginably vaster scale. Not by accident have spacecraft been named Mariner, Voyager, Viking, Ulysses, *Challenger*, *Endeavor*, and Magellan, hearkening back to that long exploring tradition. The 50-year narrative trajectory of spacecraft, ranging from Earth's atmosphere and Earth orbit to the Solar System and the universe at large, is full of remarkable discoveries that will echo down the ages and that will someday also be part of the standard school curriculum.

18.3.1 The Realm of Earth

The journey begins with atmospheric flight, which takes place within a thin skin surrounding Earth to an altitude of a few tens of miles. Like fifteenth-century coastal navigation in relation to oceanic navigation, aeronautics was a preparation, but in this case, for leaving Earth. Aside from its own intrinsic practical value, flight has been essential to spaceflight in numerous ways, ranging from supersonics to hypersonics and the Space Shuttle (Conway 2005; Heppenheimer 2007). It is therefore no surprise that NASA's technical history has close connections to the history of flight (see the aeronautics section of this volume); indeed, the institution was built on the foundations of the NACA, dating back to 1915 (Roland 1985; Hansen 1987; McDougall 1985, 157–176). The X-15 research of the 1960s is legendary, but aeronautics continues to be important for spaceflight in ways not usually appreciated by the public.

Voyages to Earth orbit have a special meaning of their own. Climbing Earth's gravitational well put one "half way to anywhere in the solar system," as science fiction writer Robert Heinlein once put it, a necessary step toward more distant explorations. But even from Earth orbit, humans and robots saw the planet anew and viewed it in unprecedented fashion, whether for reconnaissance, for environmental remote sensing, or as "high ground" for providing a means of navigation and communication. Each of these programs has its own history of technical problems and achievements, though some of the history is better known than others, and some is classified (Whalen 2007). Reconnaissance for reasons of national security was one of the earliest drivers of the space program, featuring satellites from CORONA to the KH series, among others (Johnson 2006; Hastedt 2007). Communications satellites also enjoyed early successes with the likes of Telstar, followed by Intelsat and a variety of domestic satellite systems (Pelton 1998; Butrica 1997). And after early successes with weather satellites such as TIROS 1, Earth science observation and research from space began to find global coherence in NASA's flagship Earth

Observing System, a system of satellites that monitors Earth at many wavelengths (Mack and Williamson 1998; McElroy and Williamson 2004; Tatem et al. 2008; Parkinson et al. 2007; Mack 1990).

For the human spaceflight program, Earth orbit is where humans first learned that the human body could function under the harsh conditions of space, including the new experience of weightlessness, as long as they could carry the necessities of life with them in their hermetically sealed spacecraft. It is also where they learned to "fly in space," with Vostok, Voskhod, and Soyuz on the Soviet/Russian side and Mercury and Gemini on the U. S. side—the indispensable prelude to Apollo (Swenson Jr. et al. 1966; Hacker and Grimwood 1977). Earth orbit also provided a microgravity environment for experiments, both on the Space Shuttle and on space stations. Taken together, and perhaps most importantly, all these endeavors provided a new perspective on the home planet. Although still "hugging the coastline" in terms of the analogous maritime history, these endeavors were nonetheless voyages of discovery, yielding data on huge issues of great practical import, such as global climate change, land use, and meteorology, and providing the essential infrastructure for global navigation and communication.

Important as low-Earth orbit and geosynchronous orbit are for utilitarian applications and way station status, it is the voyages beyond Earth that captured the public imagination. It is not surprising that we turned first to the nearest celestial body, our own Moon—a nearby "island" less than two light seconds away (where light travels at 186,000 miles per second), still gravitationally in the realm of Earth. The Luna, Ranger, Surveyor, and Lunar Orbiter spacecraft served as the prelude to the piloted Moon landings and gave us the first iconic images of the Space Age (Hall 1977; Byers 1977; NASA Office of Space Science 1969). Above all are the epic piloted voyages of the United States that resulted in 12 humans walking on the Moon, a feat that many think 500 years from now will be viewed in the same way as we now look back on the Age of Discovery. The stories of Neil Armstrong and Buzz Aldrin touching down on the Moon in July 1969, followed by 10 others by 1972; the harrowing experiences of the ill-fated Apollo 13; the astronauts roving over the surface of another world; are seared in memory and will remain monuments to ingenuity, the force of geopolitics, and exploration (Chaikin 1994).

The achievements of Apollo culminated in 1972 (Brooks, Grimwood, and Swenson 1979), and since then only our robotic surrogates have left the vicinity of Earth. A single voyage, or set of voyages, does not make an age, and the jury is still out on whether our descendants 20 generations from now will view Apollo as a unique set of bold achievements or the beginnings of an era of human space exploration. Historian Arthur M. Schlesinger, Jr., special assistant to President Kennedy, ventured one opinion when he wrote in support of the new Vision for Space Exploration in January 2004, "It has been almost a third of a century since human beings took a step on the Moon—rather as if no intrepid mariner had bothered after 1492 to follow up on Christopher Columbus. Yet 500 years from now (if humans have not blown up the planet), the twentieth century will be remembered, if at all, as the century in which man began the exploration of space." On the other hand, there are some, historians among them, who think the Apollo program was time and money misspent and that analogies to Columbus are misplaced. In reviewing Andrew Chaikin's book *A Man on the Moon: The Voyages of the Apollo Astronauts* in the *New York Times Review of Books*, historian of technology Alex Roland called Chaikin's retelling of the Apollo story "the great American legend of the late 20th century," replete with heroic astronauts and epic tales. Eschewing Apollo's role in exploration, and pointing to the lack of science on the missions, he downplayed the significance of the voyages of Apollo (Schlesinger Jr. 2004; Roland 1994; Chaikin 2007, 53–66). More perspective is needed; in part, the course of the next 50 years will determine whether Apollo was a beginning or an ending. In any event, it was only a tiny first step into the immensity of space.

18.3.2 The Realm of the Planets

Even as we began lunar exploration, scientists and engineers were looking beyond to the realm of the planets (now light hours away rather than light seconds for the Moon) and the preserve of robotic, rather than piloted, spacecraft (Schorn 1998; Burrows, 1990; Kraemer 2000; Murray 1989). In the 1960s, the Mariner spacecraft took us to the nearest planets, first to Venus in 1962, revealing an extremely hot planet with a runaway greenhouse effect and a dense and weird atmosphere dominated by carbon dioxide and sulfuric acid rain. By 1965, it was on to Mars, where Mariner IV imagery revealed a cratered surface, a shocking discovery at the time, indicating a dead planet, like the Moon, rather than the canalled Mars of Percival Lowell. But by 1972, Mariner 9 revealed ancient riverbeds and a much more active geological history, reviving interest in Mars as an abode of life. The exploration of Mars has been continued by the likes of Viking, Mars Global Surveyor, Mars Odyssey, ESA's Mars Express, the MERs, and Pathfinder and Phoenix. After 4 years, the MERs *Spirit* and *Opportunity* still roamed the surface of the Red Planet during NASA's 50th anniversary (Siddiqi 2002; Ezell and Ezell 1984; Squyres 2005; Bergreen 2000). Mariner 10 reached the inner planet Mercury only in 1974, a planet not to be visited again until 2008, when the MESSENGER spacecraft produced stunning imagery and scientific data from the planet closest to our Sun. Meanwhile, the exploration of the other inner planet, Venus, was continued by the Soviet Venera spacecraft, Pioneer Venus, and the ingenious radar mapper aboard Magellan, which pierced the thick clouds.

In what Carl Sagan and others have called the Golden Age of Exploration, in the 1970s and 1980s, the Pioneer and Voyager spacecraft took us to Jupiter (Fig. 18.3), Saturn, and, in the case of Voyager 2, all the way to Uranus and Neptune at the edge of the Solar System (Pyne 1988; Dethloff and Schorn 2003). Galileo revisited Jupiter and its retinue of moons in the 1990s, and Cassini is now exploring Saturn, with its Huygens companion having landed on the huge Saturnian moon Titan (Meltzer 2003). New Horizons is on the way to Pluto, classified in 2006 by the International Astronomical Union as a dwarf planet, much to the chagrin of some planetary scientists. Other spacecraft have visited comets (Giotto, Deep Impact) and

Fig. 18.3 Jupiter's Great Red Spot and its surroundings were imaged by Voyager 1 from a distance of 5.7 million kilometers, just over a week before its March 5, 1979 closest approach. Note the complex wave motion in the clouds to the left of the Great Red Spot, which is roughly 12,000 km from top to bottom (Voyager 1, P-21151). NSSDC

orbited and even landed on an asteroid (NEAR Shoemaker) (McCurdy 2005). Moreover, the Voyager spacecraft (renamed the Voyager Interstellar Mission), with their engraved greetings from Earth, are traveling beyond the Solar System on their way to the stars (Sagan et al. 1978; Hanson 2005).

In the process of exploring our Solar System, fundamental discoveries were made. We learned the geological and atmospheric histories of new worlds. We found planetary rings to be more common than once thought, though still not surpassing those of Saturn. And we discovered an entire retinue of new and unique worlds—the planetary satellites. Whereas when the Space Age began, about 30 natural satellites were known, now more than 145 are known and named, many of them imaged up close by spacecraft. The new worlds do not end there. Since 1995, we have discovered hundreds of new planets beyond the Solar System, and the Kepler spacecraft, launched in 2009, will doubtless carry that number into the thousands; some, perhaps, are worlds like our Earth.

Finally, spacecraft such as Ulysses and SOHO have visited the nearest star, our life-giving Sun, returning spectacular images of solar activity and inaugurating the new field of heliophysics. The Sun is our entrée into another, more far-reaching realm, the realm of the stars.

18.3.3 The Realm of the Stars and Galaxies

Beyond the realm of the planets, we pass from the regime of light minutes and light hours to the realm of the stars—light years to tens of thousands of light years distant in our own Milky Way Galaxy, and then to the realm of the galaxies millions or billions of light years distant. Space telescopes in Earth orbit, or its vicinity, have taken us only vicariously on voyages beyond the Solar System. Those sensors that have pointed upward rather than downward—after a prelude of pioneering observatories, such as the OAOs and the Infrared Astronomical Satellite (IRAS); the "Great Observatories" including the Hubble Space Telescope, Spitzer, Compton, and Chandra; as well as the Fermi Gamma Ray Telescope—have probed the depths of the universe and produced stunning images and pioneering data of star birth such as the Orion Nebula, stellar explosions like the Crab Nebula (Fig. 18.4), and star death, visible in a stunning array of planetary nebulae. Their images and data gave a sense of reality to the various phases of cosmic evolution, proving that robotic spacecraft results can also capture the public imagination (Smith et al. 1989; Zimmerman 2008; DeVorkin and Smith 2008; Tucker and Tucker 2001; Rieke 2006).

In the realm of the galaxies (Fig. 18.5), the Hubble Space Telescope played a key role in discovering "dark energy" and the apparent acceleration of the expansion rate of the universe. It narrowed the age of the universe to 13–14 billion years, an accuracy of about 10%. The Hubble Deep Fields provided snapshots of the early universe within a few hundred million years of the Big Bang. Two spacecraft, COBE and WMAP, studied the details of the background radiation remaining from the Big Bang, pinpointed the age of the universe to 13.7 billion years (plus or minus 100 million years), and detected the seeds from which galaxies grew, a result that yielded NASA's only Nobel Prize winner (Mather 1998; Smoot and Davidson 1993). As we once mapped Earth in the wake of the Age of Discovery, we are now mapping the heavens, both in space and time and in the entire range of the spectrum.

Three main themes emerge from this master narrative of Space Age voyaging. First, science has benefited tremendously from the journey into space. The Earth Observing System and its predecessors have brought unprecedented knowledge of our home planet. The lunar probes and the Apollo program (though often maligned for its scientific return) have returned data not only important for its science, but also crucial to human settlements that will undoubtedly come in the future.[9] In the realms of the planets, stars, and galaxies, we have added infinite detail to a story previously grasped only through ground-based telescopes, which, fantastic as they

Fig. 18.4 Stellar explosion. The Crab Nebula is an expanding remnant of a star's supernova explosion, recorded by Japanese and Chinese astronomers nearly 1000 years ago in 1054. This composite image uses data from three of NASA's Great Observatories. The Chandra X-ray image is shown in light blue; the Hubble Space Telescope optical images are in green and dark blue; and the Spitzer Space Telescope's infrared image is in red. The neutron star, which has the mass equivalent to the Sun crammed into a rapidly spinning ball of neutrons 12 miles across, is the bright white dot in the center of the image. NASA, ESA, Chandra X-ray Observatory, JPL-Caltech, J. Hester and A. Loll (Arizona State University), R. Gehrz (University of Minnesota), and Space Telescope Science Institute (STScI)

have become with adaptive optics and other stunning innovations, must still peer through the Earth's atmosphere, as through a glass darkly. By making the universe a real place filled with a bestiary of fantastic but scientifically comprehensible objects, space exploration has provided almost infinite space for free reign of the human imagination.

Secondly, although prior to the Space Age we learned much from 350 years of ground-based telescopic observations, in carrying out their missions, space telescopes during the last 50 years have opened the electromagnetic spectrum for astronomy in a way that could, by definition, not have been done from Earth, revealing the relatively calm sights of the infrared to the extreme violence of the X- and

Fig. 18.5 Realm of the galaxies. When this image was released on 15 January 1996, it was the "deepest-ever" view of the universe, called the Hubble Deep Field because it was made with NASA's Hubble Space Telescope. Almost every image on this photograph, which covers a speck of sky only 1/30 the diameter of the full Moon, is a galaxy. Besides the classical spiral- and elliptical-shaped galaxies, a variety of other galaxy shapes and colors provide important clues to understanding the evolution of the universe. Some of the galaxies may have formed less than one billion years after the Big Bang. The image was assembled from many separate exposures with the Wide Field Planetary Camera 2, for 10 consecutive days from 18 to 28 December 1995. Other Hubble Deep Field images have been released since this time. Robert Williams and the Hubble Deep Field Team (STScI) and NASA Image STScI-PRC96-01a

gamma-ray universe. The discoveries of Spitzer and its predecessors (especially IRAS) in the infrared; of International Ultraviolet Explorer (IUE), Far Ultraviolet Spectroscopic Explorer (FUSE), and Galaxy Evolution Explorer (GALEX) in the ultraviolet; of Chandra and its predecessors (the High-Energy Astronomical Observatory [HEAO] series, X-ray Multi-Mirror Mission [XMM]-Newton, Rossi, and Röntgen Satellite [ROSAT]) in the X-ray; and of Compton, Swift, and Fermi in the gamma ray, reveal a universe totally unknown when the Space Age began. Along with ground-based optical, infrared, and radio wavelength observations, the next 50 years will see new discoveries with a range of new spacecraft spanning the entire electromagnetic spectrum.

Thirdly, along with stunning advances in ground-based astronomy, the Space Age has, for the first time, revealed our place in cosmic evolution in general through its spacecraft and more particularly through NASA's Origins and Astrobiology programs and similar programs in other space agencies around the world. Though cosmic evolution is an idea that dates back at least a century, it has been taken seriously only in the last 50 years—not coincidentally, the same length of time as the Space Age (Shapley 1958; Dick and Lupisella 2009; Chaisson 2001). To a large extent, space science since that time has filled in the epic of cosmic evolution in increasing detail, revealing for the first time in detail our real place in the universe. And it has revealed that the visible universe represents less than 5% of the content of the universe, the remainder constituted by dark matter and dark energy. That 95% of the universe remains to be explored.

18.4 The Impact of the Space Age

It was recognized early in the Space Age that access to outer space would affect society. NASA's founding document, the National Aeronautics and Space Act of 1958, specifically charged the new agency with eight objectives, including "the establishment of long-range studies of the potential benefits to be gained from, the opportunities for, and the problems involved in the utilization of aeronautical and space activities for peaceful and scientific purposes."[10] Despite a few early studies,[11] the mandate to study the societal impact of spaceflight went largely unfulfilled as NASA concentrated on the many opportunities and technical problems of spaceflight itself. Only recently has NASA made a serious attempt to examine, with historical objectivity, the broad impact of the Space Age (Dick and Launius 2007; Dick 2018).[12]

Once again, because of the symmetry of the narrative arc, studies of the impact of the Age of Discovery offer a framework for analysis. The impacts of the Age of Discovery were complex and bidirectional, encompassing sometimes disastrous effects on the New World and not always positive effects on the Old World. This suggests that not all impacts of spaceflight may be good, though we must at the outset take into account that the often insidious effects of culture contact are unlikely to be a factor in space exploration in the near term. Moreover, the eminent historian J.H. Elliott has delineated three components in the impact of the New World on the Old: intellectual (challenging European assumptions about geography, theology, history, and the nature of man), economic (as an extension of European business and a source of produce), and political (affecting the balance of power) (Elliott 1970, 2006). These broad categories also apply to the Space Age, some in the short term and others in the long term.

18.4.1 Intellectual Impact

Perhaps the most profound, and as yet largely unrealized, effect of the Space Age is the intellectual impact. As the story of the Space Age demonstrates, the science returned from spaceborne instruments over the last 50 years has been truly transformational, most immediately for scientists, but also for our general worldview. Although not everyone has yet absorbed the impact, that worldview has been altered or completely transformed by the images of "Earthrise" and the "Blue Marble" (Fig. 18.6) from space, with consequences that have affected, or will eventually affect, philosophy, theology, and the view of our place in nature. In *Rocket Dreams: How the Space Age Shaped Our Vision of a World Beyond*, Marina Benjamin argues that "The impact of seeing the Earth from space focused our energies on the home planet in unprecedented ways, dramatically affecting our relationship to the natural world and our appreciation of the greater community of mankind, and prompting a revolution in our understanding of the Earth as a living system" (Benjamin 2003; Poole 2008). She finds it no coincidence that the first Earth Day on April 20, 1970 occurred in the midst of the Apollo program, or that one of the astronauts developed a new school of spiritualism (Lambright 2007, 2005).

More broadly, the same master narrative of cosmic evolution that over the last 50 years has shown us our true place in the universe has also spread to many areas of society, from history and education to religion and theology. Some historians have begun a movement toward "Big History," in which the usual political, social, and economic factors of human history are fully integrated and analyzed in the context of the billions of years of cosmic evolution it took to arrive at *Homo sapiens* (Christian 2004; Spier 1996; Brown 2007). Some educators have integrated cosmic evolution into the standard school curriculum with the same goal of perspective.[13] And some theologians have even called cosmic evolution "Genesis for the Third Millennium" (Peacocke 2000). Cosmic evolution is the ultimate master narrative within which the future of humanity will be played out. The discovery of our place in the universe made possible by studies of cosmic evolution and the search for extraterrestrial life, and the embodiment of these and other themes in literature and the arts, is surely an important effect of space exploration not yet fully realized (Dick and Lupisella 2009; Shapley 1958; Palmeri 2009). Exploration shapes worldviews and changes cultures in unexpected ways, and so does lack of exploration. The full extent of the intellectual impact of the Space Age remains to be seen.

18.4.2 Economic Impact

The economic impact of spaceflight has been considerable, but it has only begun to be felt. That impact ranges from a far-reaching aerospace industry at one end of the spectrum to the famous (and sometimes literally legendary) "spinoffs" at the other end; it is a part of national and international political economy; and it has sometimes

Fig. 18.6 Earthrise. This view of the rising Earth greeted the Apollo 8 astronauts as they came from behind the Moon after the lunar orbit insertion burn in December 1968. The photo is displayed here in its original orientation, though it is more commonly viewed with the lunar surface at the bottom of the photo. Earth is about 5° left of the horizon in the photo. The unnamed surface features on the left are near the eastern limb of the Moon as viewed from Earth. The lunar horizon is approximately 780 km from the spacecraft. NASA

measurable, but often elusive, effects on daily life and commerce. Recent rigorous historical studies suggest the scope of the impact of the Space Age, while emphasizing the complexity and richness of this topic (Dick and Launius 2007, 212–266; Hertzfeld 1998).

Economic impact is also closely related to applications satellites. We now take for granted photographs of weather and Earth resources data from space, as well as navigation and worldwide communications made possible by satellite. Along with human and robotic missions, the late twentieth century will be remembered collectively as the time when humans not only saw Earth as a fragile planet against the backdrop of space, but also utilized near-Earth space to study the planet's resources, to provide essential information about weather, and to provide means for navigation that both were life-saving and had enormous economic implications. Worldwide satellite communications brought the world closer together, a factor difficult to estimate from a cost-benefit analysis. Names like Landsat, Geostationary Operational Environmental Satellites (GOES), Intelsat, and GPS may not be household words (though the latter is now becoming one), but they affect humanity in significant ways not always appreciated (Hertzfeld and Williamson 2007, 237–266).

Applications satellites are, in turn, inseparable from environmental issues and national security. Imaging Earth from space and global space surveillance have played an arguably central role in the increasingly heated debate over global climate change and altered the manner in which national security issues are understood and interpreted. Despite political and economic hurdles, monitoring our home planet is likely to be an important and sustained space activity over the next 50 years, with concomitant impact on society (Dick and Launius 2007).[14]

The greatest economic potential will come after space travel becomes cheaper, opening up new resources on the Moon and in the Solar System. There has been no lack of specific proposals for exploiting such resources, especially with regard to the Moon. Senator Harrison Schmitt, the only scientist to fly in the Apollo program (Apollo 17), has argued that the Moon is a resource for the clean generation of fusion energy and for the mining and processing of materials; he also has argued that the Moon is a logical outpost from which more cost-effective exploration of the Solar System can take place. For decades, some visionaries have proposed schemes for harnessing solar power, mining asteroids, and exploiting other resources of the Solar System. In the far future, some have even proposed large-scale astroengineering projects, such as the Dyson spheres that astronomers have searched for as evidence of advanced extraterrestrial civilizations. While such proposals have been criticized as being impractical, pie-in-the-sky, and in the long-term future, history shows that it is likely only a matter of time before some of them become realities (Schmitt 2006; Wingo 2004).

The economic impact of the Space Age has been real and significant in certain segments of society over the last 50 years, but it is only a taste of things to come. In a democratic free-market society, once outer space becomes economically viable in the marketplace, commercial ventures will find a way into that market. Space tourism is likely to be one of the earliest such ventures.

18.4.3 Geopolitical Impact

The third area of societal impact of spaceflight is geopolitical, and as our discussion of motivations indicated, there is no denying that this aspect has played a central role over the last 50 years. The Moon race between the United States and the Soviet Union was totally driven by geopolitical considerations. Satellite reconnaissance has been an important part, at times even a driver, of national space activities, certainly in the United States, where the space budgets of DOD and NRO far exceed those of NASA. The weaponization and militarization of space are huge issues with immense consequences for the future of both Earth and activities in outer space (Johnson 2006; Hastedt 2007; Hays 2006, 199–238). Space has become both an instrument of foreign policy and a strategic asset, and the interactions of Russia, China, India, Europe, and the United States in the space arena are likely to be a dominant theme for the next 50 years (Johnson-Freese 2007).

18.4.4 Social Impact

To the intellectual, economic, and political, we may add a fourth domain: that of social impact. Space activities have affected science, math, and engineering education; embodied questions of status, civil rights, and gender among other social issues; and led to the creation of "space states" such as California, Florida, and Texas. Others have demonstrated the complex relation of such space goals to social, racial, and political themes. One such study is De Witt Kilgore's recent book *Astrofuturism: Science, Race, and Visions of Utopia in Space*. In this book, Kilgore examines the work of Wernher von Braun, Willy Ley, Robert Heinlein, Arthur C. Clarke, Gentry Lee, Gerard O'Neill, and Ben Bova, among others, in what he calls the tradition of American astrofuturism (Kilgore 2003). Such studies remind us that, like it or not, the idea of space exploration has been woven into the fabric of society over the last 50 years, even as exploration has raised our cosmic consciousness. The historical analysis of that transformation, in ways large and small, should help justify space exploration as an integral part of society rather than a burden on it as sometimes perceived by the public.

Important as they are, the social effects thus far may pale in significance to what space may represent for the future of humanity. While some argue that robotic spacecraft are cheaper and less risky than human spaceflight, it is most likely that humans will follow robotic reconnaissance as night follows day—perhaps not immediately, but in the long-term future of humanity. Humans will not be content with a space odyssey carried out by robotic surrogates any more than the other great voyages of human history. Robots extend the human senses but will not replace the human mind in the foreseeable future, even with advances in artificial intelligence. HAL in Arthur C. Clarke's famous novel and movie was not as smart as he thought, and he will not be for a long time. As President Bush said in announcing his new initiative in January 2004, humans will spread through the Solar System, fulfilling the vision of what British philosopher Olaf Stapledon 55 years ago called "interplanetary man" (Stapledon 1997; Dick 2000b). Eventually humans will spread into the cosmos at large. Space enthusiasts tend to argue that is the nature of humans, with their inbuilt curiosity and penchant for exploration; one might say that it is the very definition of what it is to be human. Not all historians and social scientists agree, however, that the utopian ideal of spreading humanity to outer space is a valid reason for going or that utopia is what we will build when we get there.

There are also more practical reasons for going into space: the survival of the species may depend on the human space program. Specifically, it would seem prudent to remove some of our species from the planet in case of natural or human-induced catastrophe, whether an asteroid impact or nuclear war. In that context, space exploration would seem a small price to pay for survival of the species, as opposed to having to start over from 3.8 billion years of evolution after, for example, a near-Earth object impact.

This theme treads dangerously close to "manifest destiny," the belief that spreading a culture, or a species, is part of their destiny, to be attained by any means. Although the concept has been a red flag for historians, who like to recall that Manifest Destiny led to slaughter as Americans spread westward and pushed out Native Americans, the analogy is not a good one. Though *Star Wars* makes good entertainment, it is unlikely to become reality as humans spread throughout the Solar System. Nor should we *a priori* shrink from the idea of destiny, though no destiny will be achieved without proper funding.

Indeed, one feature unlikely to be paralleled with the Age of Discovery, or the Second Age of Discovery in the eighteenth and nineteenth centuries, is contact with other cultures. Ship crews often included naturalists to study exotic new flora and fauna, and the ultimate experience in the Age of Discovery was contact with exotic human cultures. In the Age of Space, the search for microbial life has been a main driver of space exploration, in particular with regard to Mars, but also now extended to more exotic environments like the Jovian moon Europa. This activity has generated its share of ethical conundrums (Dick and Strick 2004). And with the search for life on new worlds, planetary protection protocols—sometimes controversial—have been put in place, both for our own planet and for others (Meltzer 2011). Contact with intelligent extraterrestrials beyond the Solar System will remain a more remote possibility, and when and if it happens, it is more likely to be radio rather than physical contact. Difficult as they are, such impacts have been studied in some detail at NASA and elsewhere (Billingham et al. 1999; Tough 2000; Dick 1995).

As NASA's study *Societal Impact of Spaceflight* shows, unpacking the nature and extent of societal impact is no simple task. "Society" is not monolithic, and "impact" can be an elusive concept. Determining the impact of anything is problematic, especially in the short term, and especially in the hands of academics. If we succeed in the near future in going back to the Moon on a permanent basis, perhaps Columbus may be a good analogy for the Apollo program, and the Age of Discovery a good analogy for the Age of Space; if not, it will have been an abortive attempt more akin to Leif Erickson and the Vikings.

Despite the difficulty, analysis of the societal impact of spaceflight is not just an academic exercise. NASA's plans for the next 50 years—multi-decade programs to explore the planets, build and operate large space telescopes and space stations, or take humans to the Moon and Mars—require that the public have a vested interest. Whether or not those ambitious space visions of the United States and other countries are fulfilled, the question of societal impact over the past 50 years remains urgent, and it may in fact help fulfill current visions or, at least, raise the level of debate.

In the end, it is difficult to determine how much society has really been affected by spaceflight during its first 50 years because society is composed of individuals, and each individual has been affected in different ways, even when witnessing a transformational event such as the first Moon landing. "The horror of the Twentieth Century," Norman Mailer declared in his account of the first Moon landing, *Of a Fire on the Moon*, "was the size of each new event, and the paucity of its reverberation" (Mailer 1969). The "paucity of reverberation" may reflect a lack of appreciation in the minds of the average citizen about the role space has played, rather than the absolute role itself, which in fact has arguably been very significant. Whether a boon or a burden to society, the impact of space activities is likely to increase over the next 50 years.

18.5 Conclusions—Ad Astra?

I do not wish to imply that exploration is the only interpretive framework for the Age of Space. There are real-life, more immediately compelling, and strategic considerations that impel the United States and other countries into space. But in my view, far from being the metaphysical, esoteric, or empty conceit of its critics, exploration is an unchanging, long-term, stimulating, and useful framework for understanding why any country with a claim to greatness must go into space. Moreover, while the analogies discussed here are only suggestive, placing space exploration within the deep history of exploration gives a context to space history that it otherwise might not have, integrating space history into the broader history of humanity and going some way toward eliminating the isolation of space history from other historical subdisciplines.

I do wish to claim that by conquering the third dimension of space—as maritime explorers did in two geographical dimensions during the Age of Discovery, as the eighteenth- and nineteenth-century explorers did on both land and sea with improved transportation methods, and as aviation has in the thin skin of our atmosphere during a century of flight—in the long run, the space program has the potential to have an impact that far exceeds any of these advances. Despite historians' qualms about the negative effects of these developments, especially the conquest mode of the Age of Discovery, the Space Age opens a vast new future to humanity, most likely not utopian, but one already imagined in science fiction and, for the first time in history, contemplated in science fact. In contemplating that future, it is well to remember that history need not repeat itself, either in its positive or negative aspects.

The experience of the railroad with which we opened this essay illuminates the Space Age from a different angle and scale. The railroad was, the authors of *The Railroad and the Space Program* concluded, an engine of social revolution that had its greatest impact only 50 years after the start of the railways in America. As a transportation system, the railway had to be competitive with canals and turnpikes, and 20 years after the start of railways in America, more miles of canals were being built than railroads. It was not clear how they could be economically feasible. And though many technological, economic, and managerial hurdles needed to be overcome, railroads are still with us. In the course of the nineteenth century, they represented human conquest of natural obstacles, with consequences for the human view of nature and our place in it. Secondary consequences often turned out to have greater societal impact than the supposed primary purposes for which they were built. The space program has had, and still has, its technological challenges, and the economic benefits may be even longer term than those of the railroad. But by conquering the third dimension of space, it has the potential to have an exceedingly large impact on the human story, as we expand into the Solar System and find our place in the scheme of cosmic evolution.

For its part, the United States has much at stake in the debate over the importance of space exploration. Pulitzer Prize-winning historian William Goetzmann saw the history of the United States as inextricably linked with exploration. "America has indeed been 'exploration's nation,'" he wrote, "a culture of endless possibilities that, in the spirit of both science and its component, exploration, continually looks forward in the direction of the new" (Goetzmann 1986). The direction of the new is now outer space, and the space exploration debate should accordingly be seen in that context. At the same time, we need to be fully aware that pro-space ideology is often driven by the problematic idea of "progress," an idea with a long history in which Americans are deeply invested. As one scholar concluded, "Given the deep commitment of Americans to ideas about progress, such ideological concerns are as likely to affect policy as any rational assessment of scientific or economic need" (Dark III 2007; Billings 2007; Nisbet 1980; Bury 1932). Thus occurs the need for historians and the social sciences to join the discussion about the human future in space.

The analogy of the fifteenth-century Chinese treasure fleet, commanded by Zheng He, has often been used as a lesson to be learned for those who would withdraw from the Space Age to seek shorter term goals on Earth. It is a matter of historical fact that, from 1405 to 1433, China sent seven massive expeditions into the Indian Ocean and perhaps beyond; the first expedition alone may have included 62 "junks" three or four times larger than Columbus's flagship, 225 support vessels, and 27,000 men. It is also well known that following a maritime tradition stretching back to the eleventh century, these ships plied the seas of southeast Asia, and then they sailed to India, the Persian Gulf, the Red Sea, and down the east coast of Africa. And the sudden end of this distant voyaging is indisputable: with changing internal political conditions and the external threat of the Mongols, the fleet was withdrawn in 1433 and its records burned. The subsequent inward turn, it is argued, set China back centuries (Levanthes 1994; Dreyer 2007; Fernandez-Armesto 2006, 109–117).

The interpretation begins with the effect of this inward turn. There is no doubt that, although Chinese state revenues were probably 100 times Portugal's, after the 1430s the Ming emperors had other priorities, leaving the Portuguese and other European countries to lead the way in exploration. As Librarian of Congress and historian Daniel Boorstin noted, "When Europeans were sailing out with enthusiasm and high hopes, landbound China was sealing her borders. Within her physical and intellectual Great Wall, she avoided encounter with the unexpected .... Fully equipped with the technology, the intelligence, and the national resources to become discoverers, the Chinese doomed themselves to be discovered." Historians J.R. McNeill and William McNeill came to the same conclusions, and historians in general (even Chinese historians) tend to agree that the Chinese chose poorly in the mid-fifteenth century. By the 1470s, the McNeills wrote, even the skills needed to build great ships were lost; some would draw a parallel to the Saturn V rockets, the last three of which found their rest in museum settings rather than in exploration. Boorstin called the withdrawal of the Chinese into their own borders, symbolized by the Great Wall of China that took its current form at that time, "catastrophic ... with consequences we still see today" (Boorstin 1983; McNeill and McNeill 2003).

The lesson of fifteenth-century China is perhaps not quite so simple, because history is driven by complex factors. Nevertheless, China's maritime withdrawal is certainly one element in its well-documented demise, and it is an undisputed fact that the Chinese are now building a massive reproduction of one of the treasure ships in the ancient Ming shipyard at Nanjing, and they are using it to shape perceptions of China's rise to global prominence after 600 years (Hvistendahl 2008). It is also an undisputed fact that the Chinese now have a human space program and that they have ambitions to land on the Moon. The question goes to the geopolitical impact mentioned earlier: whether or not the United States decides to return humans to the Moon, the Chinese or another nation will ultimately do so, with real consequences for the global balance of power. History shows that the United States will likely wait until the Chinese do so before committing resources to the same end. The ISS notwithstanding, the past 50 years demonstrate that, for the United States, competition trumps cooperation as a national *modus operandi* for space. The result would again be a Moon race, perhaps this time the key to the rest of the universe. If so, it will be yet another case of not learning the lessons of history.

Skeptics of the benefits of exploration might well point to the fate of Portugal and Spain, the leaders of the Age of Discovery who eventually lost their leadership. As one historian has pointed out, "the rewards of national strength and wealth proved elusive. Portugal never achieved true great power status. Its population was too small, its commitments too many and its new-found overseas wealth flowed too quickly into foreign hands" (Fritze, 240). Portugal came under the rule of Philip II of Spain in 1581. Spain itself came to dominate Western Europe during the late sixteenth and seventeenth centuries, but the treasures from the New World also proved ephemeral. In a scenario tempting to compare to the present case for the

United States, Spain also overcommitted itself and, by the mid-seventeenth century, weakened by the Thirty Years' War, lost its status as a world power.

But the ultimate lesson is not that exploration lacks geopolitical impact. As Norman Augustine, Chair of the Augustine Committee, argued in his report, "Rising Above the Gathering Storm," in the *Report of the Advisory Committee on the Future of the U. S. Space Program,* leadership among nations is not a birthright; it must be earned and reearned.[15] The report showed how already, in 1990, American leadership in science and technology had begun to erode, and it argued that the federal government must urgently address this situation. Surely exploration is an important part of that picture and an important part of national leadership. Each of the ages of exploration in the past was the product of specific decisions of certain cultures: the Europeans (and briefly the Chinese) for the first age, the Europeans and Americans for the second age, and the Soviet Union—soon joined by the United States, then Europe, and other countries—for the third age. As historian Stephen J. Pyne has argued, "Exploration is a specific invention of specific civilizations conducted at specific historical times. It is not … a universal property of all human societies. Not all cultures have explored or even traveled widely. Some have been content to exist in xenophobic isolation" (Pyne 1993).

In the end, what does history offer in this great debate? It was the arch-Darwinian T.H. Huxley who said, "the great end in life is not knowledge, but action." The importance of our knowledge of history is that it empowers us to act wisely, if cautiously. Not without reason does there exist a National Archives in the United States with the words "What is Past is Prologue" scrolled along the top of its impressive façade, a building whose function is duplicated in all civilized countries of the world. Not without reason did the Columbia Accident Investigation Board devote an entire chapter to history in its official report and conclude that "history is not just a backdrop or a scene-setter, history is cause" (Columbia Accident Investigation Board 2003). Not without reason does the Smithsonian Institution strive to display thoughtful commentary in its exhibits, despite criticism from its wide variety of audiences, each with its own interpretations of history. And not without reason does every high school, college, and university teach history. As Hermann Wouk said in the context of his novel *War and Remembrance*, "the beginning of the end of War lie in Remembrance" (Wouk 1978). For the United States, the beginning and end of the exploration of space lie in remembrance, remembrance of what happens to cultures that have turned too much inward. It would be ironic if, having led the world in space exploration during its first 50 years, the United States squandered that lead during the next 50. Put in a more ecumenical sense, it may be better to cooperate than compete, and it would be an extraordinary lost opportunity if the United States did not lead the international cooperation of space, as it has in the ISS, whose most important product may be a model of cooperation, difficult though it has been at times.

Unfortunately one of the great lessons of history is that we do not learn the lessons of history. As a recent author put it in while contemplating Herodotus's ancient message about intercultural understanding, "it goes unheeded, as it always has and as it always will, because history teaches us that we do not learn from history, that we fight the same wars against the same enemies for the same reasons in different eras, as though time really stood still and history itself as moving narrative was nothing but artful illusion" (Marozzi 2008). Even in an optimistic frame of mind, in a world in which we might apply lessons learned if only we paid attention, the problem is determining exactly what those lessons are. To give only one recent example, confusion in the political world between any attempt at "negotiation" and Chamberlainian "appeasement" does not inspire confidence in lessons learned, especially where ideologies are at stake. Realizing the difficulties and ambiguities of the task, in closing, I nevertheless offer six macro-lessons that should be learned from the first 50 years of the Space Age:

1. Absent an Asimovian "psychohistory" that would allow us to foresee the statistical probabilities of the future, history is not predictive, and it cannot guarantee that exploration (human or robotic) will result in a more creative society.[16] Numerous factors regulate society, which, after all, is composed of individuals more unpredictable than the gas molecules of statistical mechanics. But history, nevertheless, suggests that robust exploration, undertaken by a nation that continually looks forward to the new, enhances its chances of survival as a vibrant society.
2. It is always tempting to sacrifice long-term goals for perceived short-term needs. And it is almost always a bad idea, unless survival is at stake and there is no long term. This is one lesson that the U. S. Congress could particularly take to heart (Kay 1995).
3. Long-term goals need to be better understood in the political process. If this were true, we might not throw away a $25 billion investment on launch technology, as the United States did with Apollo, with consequences we are still suffering more than three decades later. As space policy analyst John Logsdon has memorably put it, NASA at 50 is still suffering from NASA at 12.
4. There is never enough funding to do everything. Painful priorities must be set. This seems to be common sense. But NASA has often not set priorities, tried to do too much, and failed to achieve major goals as a consequence.
5. Human spaceflight will not, and should not, go away. Robotic spaceflight will not, and should not, go away. It is always a question of balancing resources, but in the end, each needs the other, and they should exist in a synergistic relationship. The Hubble Space Telescope servicing missions are the role model here. If in the long term, humans become intelligent robots, the problem of this false dichotomy will disappear.[17]
6. Risk and exploration have always gone hand in hand, and they will forever go hand in hand. Safety is a priority, but it is the number-two priority. The number-one priority is to go, to get off the launchpad. Otherwise no explorer would ever have left the ports of Palos, Lisbon, and Sanlúcar de Barrameda. And no rocket would ever have left its launchpad. NASA understands this; the astronauts understand it; but the public does not. Thousands are killed each year on highways, but no one calls for an end to automobiles. A forward-looking nation must take risks.

As we stand at NASA's 50th anniversary and on the verge of a presidential transition—always a perilous time for government agencies—we and our leaders need to remember that (rhetoric notwithstanding) exploration is not a destiny, but a choice. It is a choice that any society must make in the midst of many other priorities. History hints, at least, that those societies that make the wrong choice will suffer the consequences. At the 100th anniversary of NASA in 2058, our descendants will be looking back at the choices we made as the leading agency for exploration in the world, as well as the choices made by the other nations of the world. The choice to explore or not to explore, in the midst of a world perpetually swamped by more pressing problems, is the ultimate challenge to NASA, the nation, and nation states constituting planet Earth. That choice is the proper context embodying the meaning and the essence of the Space Age. The universe awaits the nation, or consortium of nations, willing to take the risks and meet the challenge.

18.6 Commentary 2020

This article was written for NASA's 50th anniversary in 2008, and appeared in the Proceedings of a conference held on that occasion at NASA Headquarters in Washington, DC (Dick 2010). Over 2 days at Headquarters, historians and policy analysts discussed NASA's role in aeronautics, human spaceflight, exploration, space science, life science, and Earth science, as well as crosscutting themes ranging from space access to international relations in space and NASA's interaction with the public. The speakers were asked to keep in mind the following questions: What are the lessons learned from the first 50 years? What is NASA's role in American culture and in the history of exploration and discovery? What if there had never been a NASA? Based on the past, does NASA have a future? These question echoed the perennial themes elaborated in an earlier volume *Critical Issues in the History of Spaceflight* (Dick and Launius 2006).

The conference culminated a year of celebrations, beginning with an October 2007 conference celebrating the 50th anniversary of the Space Age and including a lecture series, future forums, publications, a large presence at the Smithsonian Folklife Festival, and numerous activities at NASA's ten Centers and venues around the country.[18] It took place as the Apollo 40th anniversaries began, ironically still the most famous of NASA's achievements, even in the era of the Space Shuttle, International Space Station (ISS), and spacecraft like the Mars Exploration Rovers (MERs) and the Hubble Space Telescope. And it took place as NASA found itself at a major crossroads, for the first time in three decades transitioning, under Administrator Michael Griffin, from the Space Shuttle to a new Ares launch vehicle and Orion crew vehicle capable of returning humans to the Moon and proceeding to Mars in a program known as Constellation. The Space Shuttle, NASA's launch system since 1981, was scheduled to wind down in 2010 (its last flight was actually in July, 2011), freeing up funds for the new Ares launch vehicle. But the latter, even if it moved forward at all deliberate speed, would not be ready until 2015, leaving the unsettling possibility that for at least 5 years the United States would be forced to use the Russian Soyuz launch vehicle and spacecraft as the sole access to the ISS in which the United States was the major partner.

The presidential elections a week after the conference presaged an imminent presidential transition, from the Republican administration of George W. Bush to (as it turned out) the Democratic presidency of Barack Obama, with all the uncertainties that such transitions imply for government programs. The uncertainties for NASA were even greater, as Michael Griffin departed with the outgoing administration and as the world found itself in an unprecedented global economic downturn, with the benefits of national space programs questioned more than ever before. Indeed, the Ares program was cancelled in 2010 and was superseded by the Space Launch System. As of 2020 the United States still had not completed its Space Launch System, nor had commercial interests such as Boeing and SpaceX yet flown humans into space. The Constellation program was also cancelled in 2010, and morphed into the Artemis program, a crewed spaceflight endeavor intended to land humans on the Moon by 2024. It embodied portions of the Constellation program, including the Orion crewed spacecraft. Although largely a NASA program, as a sign of the times Artemis also depended on both international cooperation and commercial aerospace companies. Much more information on the Artemis program and the Space Launch System is available at the NASA websites https://www.nasa.gov/specials/artemis/ and https://www.nasa.gov/exploration/systems/sls/index.html.

Many themes could be elaborated from this chapter. For example, the issue of how much risk individuals and institutions should take in the pursuit of exploration is a persistent concern. It is a question that has been asked throughout history (Bernstein 1998), and at NASA every day, especially following events such as the Columbia Space Shuttle accident and the cancellation and reinstatement of the final Hubble Space Telescope Shuttle servicing mission (Chap. 21). In the wake of the Columbia disaster, in 2004 the many aspects of the question were pondered at the NASA Administrator's Symposium on "Risk and Exploration: Earth, Sea and the Stars" (Dick and Cowing 2005). Held at the Naval Postgraduate School in Monterey, California, the gathering brought together a variety of risk takers, ranging from mountain climbers to deep-sea divers and astronauts, and several speakers who addressed history. One of the organizers was NASA's own chief scientist, John Grunsfeld, who knows something about risk. A veteran of four Shuttle flights and a mountain climber, he is the last man to touch the Hubble Space Telescope on its last servicing mission. The meeting was intended to draw on a wide variety of experience inside and outside NASA in order to illuminate the question risk and reward in exploration. The *Risk and Exploration* volume is available online at https://history.nasa.gov/SP-4701/riskandexploration.pdf.

Notes

1. Any highlights of nineteenth-century American exploration would include the Lewis and Clark expedition from 1803 to 1806, the U. S. Exploring Expedition headed by Charles Wilkes from 1838 to 1842, and the exploration of the American West by the likes of John Wesley Powell. The Lewis and Clark literature is voluminous, but on the Wilkes expedition, see Philbrick (2003).
2. Roger Launius discusses the controversy over the space frontier analogy in Dick and Launius (2006, 44–45), as do Howard McCurdy and Asif Siddiqi on pages 84–85 and 437–438, respectively, of the same volume. The noted historian of the American West, Patricia Nelson Limerick (1992, 1994) has argued especially vigorously that the American frontier, with its history of exploitation and conquest, should not be used as an analogy for space exploration.
3. The renewed emphasis on exploration at NASA raises the question of the relation between exploration, discovery, and science—and not just for academic reasons. One formulation holds that exploration and science are one and the same and that when it comes to spaceflight, exploration equals science. A National Research Council study asserted in 2005 that "Exploration is a key step in the search for fundamental and systematic understanding of the universe around us. Exploration done properly is a form of science" (NRC 2005). Yet, while it is clear that there is a synergy between exploration and science, they are not one and the same. After all, Magellan was an explorer, not a scientist or a natural philosopher. And many scientists undertake routine science that can hardly be called exploration; though even routine science can lead to discovery, often it does not. Exploration can also lead to discovery, but not necessarily. In either case, exploration and science are not the same.
4. On robotic spacecraft design, see Gruntman (2004). While no general history of spacecraft design exists, histories of individual programs generally cover design. See Swenson Jr. et al. (1966), Heppenheimer (1984), and Ezell and Ezell (1984). On the debate over expendable vs. reusable launch vehicles, see Logsdon (2006) and Butrica (2006), in Dick and Launius (2006), pp. 263–344.
5. The Mars Climate Orbiter failure investigation found that the root cause of failure was the failure to translate English units into metric units in a segment of ground-based, navigation-related mission software; see *The Mars Climate Orbiter Mishap Investigation Board Phase I Report* (10 November 1999), available at *ftp://ftp.hq.nasa.gov/pub/pao/reports/1999/MCO_report.pdf*. The Mars Polar Lander accident report and others are available at *http://sunnyday.mit.edu/accidents/*.
6. In both the Soviet and American cases, the first astronauts and cosmonauts had military backgrounds. When in April 1959 NASA selected its first astronauts, all seven had aviation experience in the military As Siddiqi (2000) has shown although the Soviets considered individuals from aviation, the Soviet navy, rocketry, and car-racing backgrounds, their Air Force physicians insisted that the initial pool be limited to qualified Air Force pilots. By the end of 1959, they had chosen 20 cosmonauts, formally approved on 7 March 1960.
7. Numerous firsthand accounts have been written by the astronauts themselves, ranging from Mercury astronauts their collective book *We Seven* (recalling Charles Lindbergh's book *We*) to Space Shuttle astronauts Mullane (2006) and Jones (2006) A unique dual autobiography is Scott and Leonov (2004). A few astronauts have been the subject of full-scale biographies, including Hansen (2005).
8. That risk is the constant companion of exploration, and that the public needs to understand this, is one of the main conclusions of the essays in Dick and Cowing (2005).
9. Aside from its geopolitical goals, and despite the clear backseat status of science, a considerable amount of science was, in fact, returned from the Moon. As Beattie (2001) has described, almost 5000 pounds of experimental equipment was landed on the Moon, including the Apollo Lunar Surface Experiments Package (ALSEP) on each of the last five Apollo missions. There was 840 pounds of lunar material returned and analyzed. On foot or in the lunar rover, 65 miles were traversed in support of field geology and geophysical studies. And

during the last three missions, detailed data were collected from the orbiting Command and Service Modules. The overall result is a much better understanding of the nature and origin of the Moon and its relation to Earth. The top ten science discoveries from the Apollo missions, as ranked by the office of the curator for planetary materials at NASA's JSC, are available at *http://www.lpi.usra.edu/expmoon/science/lunar10.html.*

10. The National Aeronautics and Space Act and its complete legislative history are available at *http://www.hq.nasa.gov/office/pao/History/spaceact-legishistory.pdf.* The passage quoted here is on page 6. Although the Space Act has been often amended, this provision has never changed.
11. In addition to *The Railroad and the Space Program*, there have been sporadic studies of the societal impact of spaceflight. On the occasion of the 60th anniversary of the British Interplanetary Society, NASA was heavily involved in a special issue of its journal devoted to the impact of space on culture: British Interplanetary Society (1993). In 1994, the Mission from Planet Earth program in the Office of Space Science at NASA sponsored a symposium entitled "What Is the Value of Space Exploration?" 18–19 July 1994, NASA Historical Reference Collection, NASA History Division, NASA Headquarters, Washington, DC. More recently, in 2005, the International Academy of Astronautics (IAA), which has a commission devoted to space and society, sponsored the first international conference on space and society in Budapest, Hungary (IAA 2005). The meeting agenda is available at *http://www.iaaweb.org/iaa/Publications/budapest2005fp.pdf.* The IAA and ESA jointly sponsored a study published as *The Impact of Space Activities upon Society*, in which well-known players on the world scene briefly discussed their ideas of societal impact, ranging from the practical to the inspirational (ESA BR-237 2005).
12. The NASA History Division also has commissioned a series of specific studies on the societal impact of spaceflight, of which Dick and Launius (2007) and Dick and Lupisella (2009) are a part.
13. One curriculum, developed by the Search for Extraterrestrial Intelligence (SETI) Institute, the California Academy of Sciences, NASA Ames Research Center, and San Francisco State University, is available on CD-ROM. This and other educational curricula are described and available at *http://www.seti.org/epo/litu-curriculum/.* The Wright Center program on cosmic evolution, directed by Eric Chaisson, is available at *http://www.tufts.edu/as/wright_center/cosmic_evolution/docs/splash.html.*
14. For just how politically sensitive the study of global climate change became within NASA in the early twenty-first century, see Bowen (2008).
15. "Americans, with only 5% of the world's population but with nearly 30% of the world's wealth, tend to believe that scientific and technological leadership and the high standard of living it underpins is somehow the natural state of affairs. But such good fortune is *not* a birthright. If we wish our children and grandchildren to enjoy the standard of living most Americans have come to expect, there is only one answer: We must get out and *compete*" (Norman Augustine, "Rising Above the Gathering Storm, Energizing and Employing America for a Brighter Economic Future," statement before U. S. House of Representatives Committee on Science, 20 October 2005). The report was published by the National Academy of Sciences in 2007 and is available at *http://history.nasa.gov/augustine/racfup1.htm*. The report's Executive Summary is at *http://history.nasa.gov/augustine/racfup2.htm*, and its main recommendations are summarized at *http://history.nasa.gov/augustine/racfup6.htm*
16. In his famous fictional Foundation series beginning in the 1950s, Isaac Asimov postulated "psychohistory," a discipline that used statistics to assess probabilities of future events. While it seems far-fetched in some ways, the 2008 Nobel Laureate in Economics, Paul Krugman, confessed to being influenced by it in his work on economics. See, for example *http://www.technovelgy.com/ct/Science-Fiction-News.asp?NewsNum=1925* (accessed 3 December 2008).
17. This is not as far-fetched as it may seem; see Launius and McCurdy (2008).

18. Among other 50th anniversary publications was a book of iconic images *America in Space: NASA's First 50 Years* (Dick et al. 2007), with an introduction by Neil Armstrong, and a book of interviews with NASA's senior leadership, *NASA at 50* (Wright et al. 2012). Also relevant are a series of 28 essays written during my time as NASA Chief Historian, available on the NASA website at https://history.nasa.gov/Why_We_/Why_We_Main.html.

References

Aldridge, Jr., E. C. et al., 2004. *A Journey to Inspire, Innovate, and Discover: Report of the President's Commission on the Implementation of the United States Space Exploration Policy*, transmitted from E. C. "Pete" Aldridge to President George W. Bush. Washington, DC: GPO.

Apollo 204 Review Board, 1967. *Report of Apollo 204 Review Board to the Administrator, National Aeronautics and Space Administration*. Washington, DC: GPO, available at *http://history.nasa.gov/Apollo204/as204report.html*.

Asner, Glen and Stephen Garber. 2019. *Origins of 21st Century Space Travel: A History of NASA's Decadal Planning Team and the Vision for Space Exploration*, 1999–2004. Washington, D.C., NASA SP-2019-4415.

Augustine, Norm. 1990. Chair, *Report of the Advisory Committee on the Future of the U.S. Space Program*. Washington, DC.

Bainbridge, William Sims. 2018. "The Impact of Space Exploration on Public Opinions, Attitudes and Beliefs," in *Historical Studies in the Societal Impact of Spaceflight*, ed. Steven J. Dick. Washington: NASA History Series,

Beattie, Donald. 2001. *Taking Science to the Moon: Lunar Experiments and the Apollo Program*. Baltimore, MD, and London, U.K.: Johns Hopkins University Press

Benjamin, Marina. 2003. *Rocket Dreams: How the Space Age Shaped Our Vision of a World Beyond*. New York, NY: Free Press

Benson, Charles D. and William B. Faherty. 2001a. *Gateway to the Moon, Building the Kennedy Space Center Launch Complex*. Gainesville, FL: University Press of Florida

Benson, Charles D. and William B. Faherty, 2001b. *Moon Launch! A History of the Saturn-Apollo Launch Operations*. Gainesville, FL: University Press of Florida

Bergreen, Lawrence. 2000. *Voyage to Mars: NASA's Search for Life Beyond Earth*. New York, NY: Riverhead Books

Bergreen, Lawrence. 2005. "Over the Edge of the World," in Dick and Cowing (2005), p. 31.

Bernstein, Peter L. 1998. *Against the Gods: The Remarkable Story of Risk*. Wiley

Biggs, Robert E., ed. 2008. *Space Shuttle Main Engines: The First Twenty Years and Beyond*. San Diego, CA: American Astronautical Society

Billingham, J., R. Heyns, D. Milne, et al., 1999. *Social Implications of the Detection of an Extraterrestrial Civilization*. Mountain View, CA: SETI Press.

Billings, Linda. 2007. "Overview: Ideology, Advocacy, and Spaceflight—Evolution of a Cultural Narrative," pp. 483–499

Bilstein, Roger E. 1980. *Stages to Saturn: A Technological History of the Apollo/Saturn Launch Vehicles*. Washington, DC: NASA SP-4206.

Bilstein, Roger. 1996. *The American Aerospace Industry: From Workshop to Global Enterprise*. New York, NY: Twayne

Boorstin, Daniel. 1983. *The Discoverers*. New York, NY: Random House, pp. 186–201

Bowen, Mark. 2008. *Censoring Science: Inside the Political Attack on Dr. James Hansen and the Truth of Global Warming*. New York, NY: Dutton

British Interplanetary Society. 1993. "The Impact of Space on Culture," *Journal of the British Interplanetary Society* 46, no. 11.

Bromberg, Joan Lisa. 1999. *NASA and the Aerospace Industry* Baltimore, MD: Johns Hopkins University Press

Brooks, Courtney, James Grimwood, and Loyd S. Swenson, Jr., 1979. *Chariots for Apollo: A History of Manned Lunar Spacecraft*. Washington, DC: NASA SP-4205, available at *http://www.hq.nasa.gov/office/pao/History/SP-4205/cover.html.*

Brown, Alexander. 2006. "Accidents, Engineering, and History at NASA, 1967–2003," in *Critical Issues*, ed. Dick and Launius, pp. 377–402.

Brown, Cynthia Stokes. 2007. *Big History: From the Big Bang to the Present*. New York, NY: New Press.

Burrows, William E. 1998. *This New Ocean: The Story of the First Space Age*. New York, NY: The Modern Library.

Burrows, William. 1990. *Exploring Space: Voyages in the Solar System and Beyond*. New York, NY: Random House.

Bury, J. B. 1932. *The Idea of Progress: An Inquiry into Its Origin and Growth*. New York, NY: Dover Publications.

Butrica, Andrew J. ed., 1997. *Beyond the Ionosphere: Fifty Years of Satellite Communication* (Washington, DC: NASA SP-4217), available at *http://history.nasa.gov/SP-4217/sp4217.htm.*

Butrica, Andrew J. 2006. "Reusable Launch Vehicles or Expendable Launch Vehicles? A Perennial Debate," in Dick and Launius (2006), pp. 301–341.

Butrica, Andrew J. 2014. The Navigators: A History of NASA's Deep Space Navigation. CreateSpace.

Byers, Bruce K. 1977. *Destination Moon: A History of the Lunar Orbiter Program*. Washington, DC: NASA TM 3487, available at *http://www.hq.nasa.gov/office/pao/History/TM-3487/top.htm.*

Chaikin, Andrew, 1994. *A Man on the Moon: The Voyages of the Apollo Astronauts*. New York, NY: Viking Penguin

Chaikin, Andrew. 2007. "Live from the Moon: The Societal Impact of Apollo," Dick and Launius, eds., *Societal Impact of Spaceflight*. Washington, DC: NASA SP-2007-4801, esp. pp. 53–66.

Chaisson, Eric. 2001. *Cosmic Evolution: The Rise of Complexity in Nature*. Cambridge, MA: Harvard University Press.

Christian, Christian. 2004. *"Maps of Time": An Introduction to "Big History"*. Berkeley, CA: University of California Press.

Columbia Accident Investigation Board, 2003. *Report*. Washington, DC: GPO, *http://history.nasa.gov/columbia/CAIB_reportindex.html.*

Conway, Erik. 2005. *High-Speed Dreams: NASA and the Technopolitics of Supersonic Transportation, 1945–1999*. Baltimore, MD: Johns Hopkins University Press.

Coopersmith, Jonathan. 2008. "Great (Unfulfilled) Expectations: To Boldly Go Where No Social Scientist or Historian Has Gone Before," in *Remembering the Space Age*, ed. Steven J. Dick (Washington, DC: NASA SP-2008-4703), pp. 135–154.

Dark, Taylor E. III, 2007. "Reclaiming the Future: Space Advocacy and the Idea of Progress," in Dick and Launius (2007), pp. 555–571.

Dawson, Virginia P. 1991. *Engines and Innovation: Lewis Laboratory and American Propulsion Technology*. Washington, DC: NASA SP-4306, chapters. 8ff.

Dawson, Virginia P. and Mark D. Bowles 2004. *Taming Liquid Hydrogen: The Centaur Upper Stage Rocket, 1958–2002*. Washington, DC: NASA SP-2004-4230, 2004.

Dethloff Henry C. and Ronald A. Schorn, 2003. *To the Outer Planets and Beyond: Voyager's Grand Tour* Washington, DC: Smithsonian Institution Press

DeVorkin, David and Robert Smith, 2008. *Hubble: Imaging Space and Time* (Washington, DC: National Geographic).

Dick, Steven J. 1995. "Consequences of Success in SETI: Lessons from the History of Science," in *Progress in the Search for Extraterrestrial Life*, ed. Seth Shostak. San Francisco, CA: ASP Conference Series, pp. 521–532.

Dick, Steven J. 2000a. *Many Worlds: The New Universe, Extraterrestrial Life and the Theological Implications*, Philadelphia, PA, and London, U.K.: Templeton Foundation Press.

Dick, Steven J. 2000b. "Interstellar Humanity," *Futures: The Journal of Forecasting Planning and Policy* 32: 555–567.
Dick, Steven J. 2009. "Cosmic Evolution: History, Culture and Human Destiny," in Dick and Lupisella. NASA History Series.
Dick, Steven J. 2010. NASA's First 50 Years: Historical Perspectives. NASA SP-2010-4704, Washington, DC.
Dick, Steven J. 2018. *Historical Studies in the Societal Impact of Spaceflight*, ed. Steven J. Dick. Washington, DC, NASA.
Dick, Steven J. and Keith Cowing, ed. 2005. *Risk and Exploration: Earth, Sea and the Stars*. NASA SP-2005-470. *http://history.nasa.gov/SP-4701/frontmatter.pdf.*
Dick, Steven J. and Roger D. Launius, 2006. *Critical Issues in the History of Spaceflight*, section V, "NASA Cultures," pp. 345–428.
Dick Steven J. and Roger D. Launius, eds., 2007. *Societal Impact of Spaceflight*. Washington, DC: NASA SP-2007-4801.
Dick, Steven J., Robert Jacobs, et al. 2007. *America in Space: NASA's First 50 Years*. New York, Abrams.
Dick, Steven J. and Mark Lupisella, eds., 2009. *Cosmos and Culture: Cultural Evolution in a Cosmic Context*. Washington, DC, NASA.
Dick, Steven J. and James E. Strick, 2004. *The Living Universe: NASA and the Development of Astrobiology*. New Brunswick, NJ, and London, U.K.: Rutgers University Press.
Dreyer, Edward L. 2007. *Zheng He: China and the Oceans in the Early Ming Dynasty, 1405–1433*. New York, NY: Pearson Longman.
Dunar, Andrew J. and Stephen P. Waring, 1999. "Crafting Rockets and Rovers: Apollo Engineering Achievements," chap. 3 in *Power to Explore: A History of the Marshall Space Flight Center, 1960–1990*. Washington, DC: NASA SP-4313.
Elliott, J. H. 1970. *The Old World and the New, 1492–1650*. Cambridge, U.K.: Cambridge University Press.
Elliott, J. H. 2006. *Empires of the Atlantic World: Britain and Spain in America, 1492–1830*. New Haven, CT: Yale University Press.
Ezell, Edward C. and Linda N. Ezell, 1984. *On Mars: Exploration of the Red Planet, 1958–1978*. Washington, DC: NASA SP-4212.
Fernandez-Armesto, Felipe. 2006. *Pathfinders: A Global History of Exploration*. New York, NY: W. W. Norton, pp. 109–117.
Fries, Sylvia D. 1992. *NASA Engineers and the Age of Apollo*. Washington, DC: NASA SP-4104, 1992.
Fritze, Ronald H. 2002. *New Worlds: The Great Voyages of Discovery 1400–1600*. Phoenix Mill, U.K.: Sutton Publishing, pp. 70–73.
Gerovitch, Slava. 2006. "Human-Machine Issues in the Soviet Space Program," in Dick and Launius (2006).
Goetzmann, William H. 1966. *Exploration and Empire*. New York, NY: Knopf, 1966.
Goetzmann, William H. 1986. *New Lands, New Men: America and the Second Great Age of Discovery*. New York, NY: Penguin.
Griffin, Michael. 2008a. "Leadership in Space," in *Leadership in Space: Selected Speeches of NASA Administrator Michael Griffin, May 2005–July 2008*. Washington, DC: NASA SP-2008-564, pp. 1–8.
Griffin, Michael. 2008b. "NASA and the Business of Space," in *Leadership in Space*, pp. 175–186.
Gruntman, Michael. 2004. *Blazing the Trail: The Early History of Spacecraft and Rocketry*. Reston, VA: AIAA.
Hacker, Barton C. and James M. Grimwood. 1977. *On the Shoulders of Titans: A History of Project Gemini*. Washington, DC: NASA SP-4203.
Hall, Cargill. 1977. *Lunar Impact: A History of Project Ranger.* Washington, DC: GPO, 1977), available at *http://history.nasa.gov/SP-4210/pages/Cover.htm.*
Hansen, James R. 1987. *Engineer in Charge: A History of the Langley Aeronautical Laboratory, 1917–1958*. Washington, DC: NASA SP-4305.

Hansen, James R. 2005. *First Man: The Life of Neil A. Armstrong*. New York, NY: Simon & Schuster.

Hanson, Brooks. 2005. "Voyager 1 Passes the Termination Shock," *Science* 309 (23 September 2005): 2015–2029.

Harré, R. 1972. *The Philosophies of Science: An Introductory Survey*. London, U.K.: Oxford University Press, pp. 172–176.

Hastedt, Glenn. 2007. "Reconnaissance Satellites, Intelligence and National Security," in Dick and Launius (2007), pp. 369–383.

Hays, Peter. 2006. "NASA and the Department of Defense: Enduring Themes in Three Key Areas," in Dick and Launius (2006), pp. 199–238.

Heppenheimer, T. A. 2007. "Hypersonics and the Space Shuttle," in *Facing the Heat Barrier: A History of Hypersonics*. Washington, DC: NASA SP-2007-4232.

Heppenheimer, Tom. 1984. *Development of the Space Shuttle, 1972–1981*. Washington, DC: Smithsonian Institution Press.

Hertzfeld, Henry R. 1998. "Space as an Investment in Economic Growth," in Logsdon (1998), vol. 3, *Using Space*, pp. 385–402.

Hertzfeld, Henry and Ray A. Williamson. 2007. "The Social and Economic Impact of Earth Observing Satellites," in (Dick and Launius, 2007), pp. 237–266, and section IV of the same volume, "Applications Satellites, the Environment, and National Security."

Hesse, Mary B. 1966. *Models and Analogies in Science*. Notre Dame, IN: University of Notre Dame Press.

Hughes, Thomas P. 1965. "The Technological Frontier: The Railroad," in (Mazlish, 1965), p. 53, note 1.

Hunley, J. D. 2007. *The Development of Propulsion Technology for U.S. Space-Launch Vehicles, 1926–1991*. College Station, TX: Texas A&M Press.

Hunley, J. D. 2008a. *Preludes to U.S. Space-Launch Vehicle Technology: Goddard Rockets to Minuteman III*. Gainesville, FL: University Press of Florida.

Hunley, J. D. 2008b. *U.S. Space-Launch Vehicle Technology: Viking to Space Shuttle*. Gainesville, FL: University Press of Florida.

Hunley, J. D. ed., 1993. *The Birth of NASA: The Diary of T. Keith Glennan*. Washington, DC: NASA SP-4105, at *http://history.nasa.gov/SP-4105/sp4105.htm*.

Hvistendahl, Mara. 2008. "Rebuilding a Treasure Ship," *Archaeology* (March/April 2008): 40–45.

Johnson, Stephen. 2002. *The Secret of Apollo*. Baltimore, MD: Johns Hopkins University Press.

Johnson, Stephen B. 2006. "The History and Historiography of National Security Space," in Dick and Launius, 2006, pp. 481–548.

Johnson-Freese, Joan. 2007. *Space as a Strategic Asset*. New York, NY: Columbia University Press.

Jones, Tom. 2006. *Sky Walking: An Astronaut's Memoir*. New York, NY: HarperCollins.

Kay, Woody. 1995. *Can Democracies Fly in Space? The Challenge of Revitalizing the U.S. Space Program*. Westport, CT: Praeger Publishers.

Kennefick, Daniel. 2007. *Traveling at the Speed of Thought: Einstein and the Quest for Gravitational Waves*. Princeton, NJ: Princeton University Press.

Kilgore, De Witt Douglas. 2003. *Astrofuturism: Science, Race, and Visions of Utopia in Space*. Philadelphia, PA: University of Pennsylvania Press.

Kraemer, Robert S. 2000. *Beyond the Moon: A Golden Age of Planetary Exploration, 1971–1978*. Washington, DC: Smithsonian Institution Press.

Kraft, Chris. 2002. *Flight: My Life in Mission Control*. New York, NY: Plume.

Kranz, Gene. 2000. *Failure Is Not an Option: Mission Control from Mercury to Apollo 13 and Beyond*. New York, NY: Berkley Books.

Krige, John. 2006 "Technology, Foreign Policy, and International Cooperation in Space," in Dick and Launius, pp. 239–268.

Lambright, William H. 2005. *NASA and the Environment: The Case of Ozone Depletion*. Washington, DC: NASA SP-2005-4538.

Lambright, William H. 2007. "NASA and the Environment: Science in a Political Context," in Dick and Launius (2007), pp. 313–330.

Launius, Roger D. 2006. "Compelling Rationales for Spaceflight? History and the Search for Relevance," in Dick and Launius (2006), pp. 37–70.

Launius, Roger D. and Dennis R. Jenkins, eds. 2002. *To Reach the High Frontier: A History of U.S Launch Vehicles*. Lexington, KY: University Press of Kentucky.

Launius, Roger D. and Howard E. McCurdy. 2008. *Robots in Space: Technology, Evolution, and Interplanetary Travel*. Baltimore, MD: Johns Hopkins University Press.

Levanthes, Louise. 1994. *When China Ruled the Seas: The Treasure Fleet of the Dragon Throne, 1405–1433*. Oxford, U.K.: Oxford University Press.

Levy, David. 2000. *Shoemaker by Levy: The Man Who Made an Impact*. Princeton, NJ: Princeton University Press.

Lewis, Richard S. 1976. *From Vinland to Mars: A Thousand Years of Exploration*. New York, NY: New York Times Book Company.

Limerick, P. 1992. "Imagined Frontiers: Westward Expansion and the Future of the Space Program," in *Space Policy Alternatives*, ed. Radford Byerly. Boulder, CO: Westview Press, pp. 249–261.

Limerick, P. 1994. "Space Exploration and the Frontier," in *What Is the Value of Space Exploration?* 18–19 July 1994, NASA Historical Reference Collection, NASA History Division, NASA Headquarters, Washington, DC.

Lipartito, Ken and Orville Butler, 2007. *A History of the Kennedy Space Center* (Gainesville, FL: University Press of Florida.

Logsdon, John, ed., 1995–2008. *Exploring the Unknown: Selected Documents in the History of the U.S. Civil Space Program*, 7 vols. (Washington, DC: NASA SP-4407, 1995 to 2008), available at http://history.nasa.gov/series95.html.

Logsdon, John. 1996. "The Development of International Cooperation in Space," in Logsdon (1996) vol. 2, *External Relationships*, pp. 1–15.

Logsdon, John, moderator. 1998. *Legislative Origins of the National Aeronautics and Space Act of 1958: Proceedings of an Oral History Workshop*, Monographs in Aerospace History, No. 8. Washington, DC: NASA History Division, *http://www.hq.nasa.gov/office/pao/History/40thann/legorgns.pdf*.

Logsdon, John, 2006. "'A Failure of National Leadership': Why No Replacement for the Space Shuttle?," In Dick and Launius (2006), pp. 269–300.

Marozzi, Justin. 2008. *The Way of Herodotus: Travels with the Man Who Invented History*. Cambridge, MA: Da Capo Press, p. 95.

Mazlish, Bruce. 1965a. "Historical Analogy: The Railroad and the Space Program and Their Impact on Society," in Mazlish (1965).

Mazlish, Bruce, ed.. 1965b. *The Railroad and the Space Program: An Exploration of Historical Analogy*. Cambridge, MA: MIT Press, p. 12.

McCurdy, Howard. 1993. *Inside NASA: High Technology and Organizational Change in the U.S. Space Program*. Baltimore, MD: Johns Hopkins University Press.

McCurdy, Howard. 1997. *Space and the American Imagination*. Washington, DC, and London, U.K.: Smithsonian Institution Press, pp. 144–145.

McCurdy, Howard. 2005. *Low-Cost Innovation in Spaceflight: The Near Earth Asteroid Rendezvous (NEAR) Shoemaker Mission*, Monographs in Aerospace History, No. 36 (Washington, DC: NASA SP-2005-4536).

McCurdy, Howard. 2006. "Observations on the Robotic Versus Human Issue in Spaceflight," in Dick and Launius (2006).

McDougall, Walter. 1985. … *The Heavens and the Earth: A Political History of the Space Age*. New York, NY: Basic Books.

McElroy, John H. and Ray A. Williamson, 2004. "The Evolution of Earth Science Research from Space: NASA's Earth Observing System," in Logsdon, 2004, vol. 6, *Space and Earth Science*, pp. 441–690.

MacGregor, Robert R. 2008. "Imagining an Aerospace Agency in the Atomic Age," in *Remembering the Space Age*, ed. Steven J. Dick. Washington, DC: NASA SP-2008-4703.

McNeill, J. R. 2008. "Gigantic Follies? Human Exploration and the Space Age in Long-term Historical Perspective," in Dick (2008). Washington, DC: NASA SP-2008-4703, 2008, pp. 3–16.
McNeill, J. R. and William H. McNeill, 2003. *The Human Web: A Bird's-Eye View of World History*. New York and London: W. W. Norton and Co., p. 166.
Mack, Pamela. 1990. *Viewing the Earth: The Social Construction of the Landsat System*. Cambridge, MA: MIT Press.
Mack, Pamela and Ray A. Williamson. 1998. "Observing the Earth from Space," in *Logsdon, 1998*, vol. 3, *Using Space*, pp. 155–384.
Mailer, Norman. 1969. *Of a Fire on the Moon*. New York, NY: Little, Brown and Co., p. 29.
Mather, John. 1998. *The Very First Light: The True Inside Story of the Scientific Journey Back to the Dawn of the Universe*. New York, NY: Basic Books.
Michael Meltzer, 2003. *Mission to Jupiter: A History of the Galileo Project*. Washington, DC: NASA SP-2007-4231, available at *http://history.nasa.gov/sp4231.pdf*.
Meltzer, Michael. 2011. *When Biospheres Collide: A History of NASA's Planetary Protection Programs*. NASA History Series.
Mindell, David. 2006. "Human and Machine in the History of Spaceflight," in Dick and Launius (2006), pp. 73–164.
Mindell, David. 2008. *Digital Apollo: Human and Machine in Spaceflight*. Cambridge, MA: MIT Press.
Mudgway, Douglas. 2001. *Uplink-Downlink: A History of the NASA Deep Space Network, 1957–1997*. Washington, DC: NASA SP-2001-4227.
Douglas Mudgway, 2005. *Big Dish: Building America's Deep Space Connection to the Planets*. Gainesville, FL: University Press of Florida.
Mullane, Michael. 2006. *Riding Rockets: The Outrageous Tales of a Space Shuttle Astronaut*. New York, Scribner.
Murray, Bruce. 1989. *Journey into Space: The First Three Decades of Space Exploration*. New York, NY: W. W. Norton.
Murray, Charles and Catherine Bly Cox. 1989. *Apollo: The Race to the Moon*. New York, NY: Simon & Schuster, chap. 10.
NASA. 1958. National Aeronautics and Space Act, as amended, with legislative history showing changes over time, at *http://history.nasa.gov/spaceact-legishistory.pdf*.
NASA. 2005. *The New Age of Exploration: NASA's Direction for 2005 and Beyond*. Washington, DC: NASA NP-2005-01-397-HQ, available at *http://www1.nasa.gov/pdf/107490main_FY06_Direction.pdf*.
NASA Office of Space Science and Applications. 1969. *Surveyor Program Results*. Washington, DC: NASA SP-184 at *http://ntrs.nasa.gov/archive/nasa/casi.ntrs.nasa.gov/19690027073_1969027073.pdf*.
National Research Council. 2005. *Science in NASA's Vision for Space Exploration*. Washington, DC: The National Academies Press, available at *http://books.nap.edu/openbook.php?record_id=11225&page=R1*.
Nisbet, Robert. 1980. *History of the Idea of Progress*. New York, NY: Basic Books.
Organization for Economic Cooperation and Development (OECD), 2007. *The Space Economy at a Glance*. Paris, France: OECD.
Oxford Analytica. 2008. *Understanding the Space Economy: Competition, Cooperation and Commerce*. Oxford, U.K.: Oxford Analytica.
Palmeri, JoAnn. 2009. "Bringing Cosmos to Culture: Harlow Shapley and the Uses of Cosmic Evolution," in Dick and Lupisella (2009).
Parkinson, Claire, Kim C. Partington, and Robin G. Williams, ed., 2007. *Our Changing Planet: The View from Space*. Cambridge, U.K.: Cambridge University Press.
Parry, J. H. 1963. *The Age of Reconnaissance: Discovery, Exploration and Settlement, 1450 to 1650*. London, U.K.: Weidenfeld and Nicolson, 1963.
Peacocke, Arthur. 2000. "The Challenge and Stimulus of the Epic of Evolution to Theology," in Dick, 2000a, pp. 89–117.

Pelton, Joseph N. 1998. "The History of Satellite Communications," in Logsdon, 1998, vol. 3, *Using Space*, pp. 1–154.

Philbrick, Nathaniel. 2003. *Sea of Glory: America's Voyage of Discovery: The U. S. Exploring Expedition, 1838–1842*. New York, NY, Viking.

Poole, Robert. 2008. *Earthrise: How Man First Saw the Earth*. New Haven, CT, and London, U.K.: Yale University Press.

Portree, David S. 1998. *NASA's Origins and the Dawn of the Space Age*, Monographs in Aerospace History, No.10. Washington, DC: NASA History Division, September 1998, available at *http://www.hq.nasa.gov/office/pao/History/monograph10/*.

President's Science Advisory Committee, 1958. "Introduction to Outer Space," 26 March 1958, in Logsdon (1995), vol. 1, *Organizing for Exploration*, pp. 332–333.

Pyne, Stephen J. 1988. "A Third Great Age of Discovery," in Carl Sagan and Stephen J. Pyne, *The Scientific and Historical Rationales for Solar System Exploration*. Washington, DC: Space Policy Institute [SPI], SPI 88-1, George Washington University, pp. 13–77.

Pyne, Stephen J. 1993. "The Third Great Age of Discovery," in *Space: Discovery and Exploration*, ed. Martin Collins and Sylvia Fries. New York, NY: Beaux Arts Editions.

Pyne, Stephen J. 2006. "Seeking Newer Worlds: An Historical Context for Space Exploration," in Dick and Launius, 2006, pp. 7–35, available at *http://history.nasa.gov/SP-2006-4702/frontmatter.pdf*.

Rao, P. V. Manoranjan ed., 2007. *50 Years of Space: A Global Perspective*. Hyderabad, India: Universities Press.

Rieke, George. 2006. *The Last of the Great Observatories: Spitzer and the Era of Faster, Better, Cheaper at NASA*. Tucson, AZ: University of Arizona Press.

Rogers, William P. Chair, 1986. *Report of the Presidential Commission on the Space Shuttle Challenger Accident* (Washington, DC), available at *http://history.nasa.gov/rogersrep/genindex.htm*.

Roland, Alex. 1985. *Model Research: The National Advisory Committee for Aeronautics, 1915–1958*. Washington, DC: NASA SP-4103, two volumes.

Roland, Alex. 1994. "How We Won the Moon," *New York Times*, sec. 7 (17 July 1994): 1.

Rosholt, Robert L. 1966. *An Administrative History of NASA, 1958–1963*. Washington, DC: NASA SP-4101.

Russell, Peter. 2001. "The Caravels of Christ," chap. 9 in *Prince Henry the Navigator: A Life*. New Haven, CT and London, U.K.: Yale University Press, pp. 225–238.

Sagan, Carl, Frank Drake, Ann Druyan, Timothy Ferris, Jon Lomberg, and Linda Salzman Sagan, 1978. *Murmurs of Earth: The Voyager Interstellar Record*. New York, NY: Random House.

Schaber, Gerald G. 2005. *The U.S. Geological Survey, Branch of Astrogeology: A Chronology of Activities From Conception through the End of Project Apollo (1960–1973)*, available at *http://pubs.usgs.gov/of/2005/1190/*.

Schlesinger Jr. Arthur M. 2004. "State of the 'Vision Thing,'" *Los Angeles Times* (21 January 2004): B13, available at *http://www.commondreams.org/views04/0121-06.htm*.

Schmitt, Harrison. 2006. *Return to the Moon: Exploration, Enterprise, and Energy in the Human Settlement of Space*. New York, NY: Praxis Publishing.

Schorn, Ronald A. 1998. *Planetary Astronomy From Ancient Times to the Third Millennium*. College Station, TX: Texas A&M University Press.

Scott, David and Alexei Leonov. 2004. *Two Sides of the Moon: Our Story of the Cold War Space Race*. New York, NY: St. Martin's Press.

Scranton, Philip. 2006. "NASA and the Aerospace Industry: Critical Issues and Research Prospects," in Dick and Launius (2006), pp. 169–198.

Shapley, Harlow. 1958. *Of Stars and Men: The Human Response to an Expanding Universe*. Boston, MA: Beacon Press.

Sheahan, John T. and Francis T. Hoban, 2004. "Spaceports," in *Defining Aerospace Policy: Essays in Honor of Francis T. Hoban*, ed. Kenneth Button, Julianne Lammersen-Baum, and Roger Stough. Burlington, VT: Ashgate.

Siddiqi, Asif. 2000. Challenge to Apollo: The Soviet Union and the Space Race, 1945–1974. Washington, DC: NASA SP-2000-4408.

Siddiqi, Asif. 2002. *Deep Space Chronicle: A Chronology of Deep Space and Planetary Probes*, Monographs in Aerospace History, No. 24. Washington, DC: NASA SP-2002-4524.

Smith, Robert et al., 1989. *The Space Telescope: A Study of NASA, Science, Technology, and Politics*. Cambridge, U.K., and New York, NY: Cambridge University Press, 1989, 2nd edition with new introduction, 1993.

Smoot, George and Keay Davidson, 1993. *Wrinkles in Time*. New York, NY: William Morrow and Company.

Spier, Fred. 1996. *The Structure of Big History: From the Big Bang Until Today*. Amsterdam, Netherlands: Amsterdam University Press.

Squyres, Steve. 2005. *Roving Mars: Spirit, Opportunity, and the Exploration of the Red Planet*. New York, NY: Hyperion.

Stapledon, Olaf. 1997. "Interplanetary Man?," in *An Olaf Stapledon Reader*, ed. Robert Crossley. Syracuse, NY: Syracuse University Press, pp. 218–241.

Stine, Deborah D. 2007. *U.S. Civilian Space Policy and Priorities: Reflections 50 Years after Sputnik*. Washington, DC: Congressional Research Service, available at *http://fas.org/sgp/crs/space/RL34263.pdf*.

Swenson Jr. Loyd S., James M. Grimwood, and Charles C. Alexander, 1966. *This New Ocean: A History Of Project Mercury*. Washington, DC: NASA SP-4201, 1966, repr. 1998.

Tatem, Andrew J., Scott J. Goetz, and Simon I. Hay, 2008. "Fifty Years of Earth-observation Satellites," *American Scientist* 96: 390–398.

Thomas, Hugh. 2003. *Rivers of Gold: The Rise of the Spanish Empire, from Columbus to Magellan*. New York, NY: Random House.

Tough, Allen. 2000. ed., *When SETI Succeeds: The Impact of High-Information Contact*. Bellevue, WA: Foundation for the Future.

Tsiao, Sunny. 2008. *"Read You Loud and Clear!" The Story of NASA's Spaceflight Tracking and Data Network*. Washington, DC: NASA SP-2007-4233.

Tucker, Wallace and Karen Tucker, 2001. *Revealing the Universe: The Making of the Chandra X-Ray Observatory*. Cambridge, MA: Harvard University Press.

Turner, Frederick Jackson. 1994. "The Significance of the Frontier in American History," in *Rereading Frederick Jackson Turner: The Significance of the Frontier in American History and Other Essays*, ed. John M. Faragher. New Haven, CT: Yale University Press.

U. S. Geological Survey, n.d. *http://astrogeology.usgs.gov/*.

Vaughan, Diane. 1986. *The Challenger Launch Decision: Risky Technology, Culture, and Deviance at NASA*. Chicago, IL: University of Chicago Press.

Whalen, David J. 2007. "For All Mankind: Societal Impact of Application Satellites," in Dick and Launius, 2007. Washington, DC: NASA SP-2007-4801, pp. 288–312.

Wheeler, J. A. 1996. *At Home in the Universe*. New York, NY: Springer, pp. 13–16

White House, 2004. "A New Spirit of Discovery: The President's Vision for U. S. Space Exploration" available at *http://www.ostp.gov/pdf/renewedspiritofdiscovery.pdf*.

Wingo, Dennis. 2004. *Moonrush: Improving Life on Earth with the Moon's Resources*. Burlington, ON: Apogee Books.

Wolfe, Tom. 1979. *The Right Stuff*. New York: Farrar, Strauss and Giroux, and New York.

Wouk, Herman. 1978. *War and Remembrance*. New York, NY: Little, Brown and Co., preface.

Wright, Rebecca, Sandra Johnson, and Steven J. Dick. 2012. *NASA at 50. Interviews with NASA Senior Leadership*. Washington, DC, NASA.

Zimmerman, *Robert*. 2008. *The Universe in a Mirror: The Saga of the Hubble Space Telescope and the Visionaries Who Built It*. Princeton, NJ: Princeton University Press.

Chapter 19
Space, Time and Aliens: The Role of Imagination in Outer Space

Abstract In this chapter we argue that imagination played an important role in making spaceflight possible, that spaceflight has had and continues to have an important effect on individuals and culture, and that the exploration of space has affected our individual and collective worldviews. Many (though not all) of the pioneers of spaceflight, notably Wernher von Braun, were influenced by science fiction, as were many (though not all) of the pioneers of the Search for Extraterrestrial Intelligence (SETI), including Carl Sagan, Philip Morrison, Freeman Dyson, and Jill Tarter. Conversely, science fiction writers ranging from Olaf Stapledon and Arthur C. Clarke to Stanislaw Lem have been influenced by the new views of the universe represented by Shapley, Hubble, and Einstein. Taken together, science fiction, the UFO debate, and their depiction in media and the arts may be seen as one way that popular culture absorbs this new worldview of a biological universe, expanded in space and time and perhaps replete with aliens.

19.1 The Cultural History of Outer Space

The role of personal and collective imagination in the Space Age—both in making spaceflight possible and in its reverse effect on individuals and culture—is a complex subject fraught with difficulty. Even when I contemplate my own career, which I presumably know better than anyone, it is not easy to separate the effect of events in the real world from youthful imagination, and the cultures in which both are imbedded. It is perhaps useful to begin by relating my personal experience as an entrée to the large issues of the subject.

I was 7 years old when Sputnik was launched, an event that undoubtedly had some impact in launching my own imagination. When I was 11, I spent the summer of 1961 in Karlsruhe, then in the western part of divided Germany. My mother was German, and one of the indelible memories of that summer was a science fiction movie that I still remember as "Venus Won't Answer." That movie further whet my appetite for space, but my mother would undoubtedly not have taken me to such a

First published in *Imagining Outer Space: European Astroculture in the Twentieth Century*, Alexander T. Geppert, ed. (Palgrave-MacMillan: New York, 2012), 27–44.

S. J. Dick, *Space, Time, and Aliens*, https://doi.org/10.1007/978-3-030-41614-0_19

movie unless I had asked her, based on my interest generated by real events then taking place in space—only a few months after the first human spaceflights by Yuri Gagarin, Alan Shepard, and Gus Grissom. In this way, imagination and reality feed on each other symbiotically, an eternal entanglement difficult to deconstruct, precisely because in many ways they may be understood as complementary and intertwined.

Further contemplation of this formative personal experience also reveals the difficulties of determining influences and how they may operate unconsciously. The movie was shot in East Germany, directed by Kurt Maetzig, co-produced with a Polish company, titled *Der schweigende Stern* (The Silent Star) and released in East Germany in February 1960. The movie was also released in West German theatres in September 1960 under the title *Raumschiff Venus antwortet nicht* (Spaceship Venus Won't Answer), which accounts for me seeing it the following summer. An Americanized version was released in the United States in 1962 under the title *First Spaceship on Venus* (Fig. 19.1).[1] It turns out that Maetzig's movie was based on the first novel of none other than the great Polish science fiction writer Stanislaw Lem, titled "The Astronauts" (1951), thus the Polish co-production.[2] The novel was translated into many languages, but never into English. Three decades later Stanislaw Lem had a great influence on me through his novels *Solaris* and *His Master's Voice*, but I failed to realize until recently that he had unknowingly influenced me already in 1961, at the age of 11, through *Raumschiff Venus antwortet nicht*. This experience emphasizes the many different levels to the theme of "space and the imagination." One of them is how any particular individual is influenced, not always easy to determine, even by the individual. Another level is how the European imagination can affect the American imagination, and by extension how one culture can affect another.

Yet another lesson emerges from the actual content of this movie—the influence of cultural context on the imagination as represented in the film. Born in 1911, Kurt Maetzig was an East German film director, and during World War II a member of the anti-Nazi German Communist Party. At the time he made the film he lived in Soviet-occupied East Germany. It is not surprising Maetzig took up the popular novel of Lem, who was living under Soviet occupation in Poland, and had to portray Earth as a social utopia in order to get his novel published. Lem's novel tells of a Venusian artifact found buried near Tunguska, the Siberian site of the famous extraterrestrial impact in 1908, with data indicating the Venusians will irradiate the Earth and take over. Earth officials send the spaceship Kosmokrator to Venus, where scientists find the remains of a warlike civilization that perished in a nuclear war. In the movie version (reflecting the new technology of radio telescopes), a radio signal with greetings is sent to Venus, but there is no reply (thus "Venus antwortet nicht"). Kosmokrator, equipped with a vacuum tube computer, travels to Venus with its international crew of scientists, and finds only advanced machines, programmed to carry out the goals of the original Venusians. The film is full of communist ideology and anti-American sentiment, removed from the 82-min Americanized version. In politics, nuclear war and technology, the novel and the different versions of the movie reflect the cultural context of their time. That context, including the incipient Space Age, fed the imagination of these two European artists, who, even while

Fig. 19.1 Poster for *First Spaceship on Venus*, the American version of *Der Schweigende Stern*, based on Stanislaw Lem's first novel. The poster shows the spaceship Kosmokrator on the surface of Venus, which the Mariner 2 spacecraft revealed at the end of 1962 was extremely hot due to the greenhouse effect—already conjectured in this poster due to the relative proximity of Venus to the Sun. (Source: DEFA-Stiftung)

under totalitarian rule, absorbed the space aspirations of the Soviet Union (Andrews 2007; Schmölders 2012).

I begin with this personal experience because it illustrates in a concrete way just how complex our subject can be. Because the subject of this volume is a cultural history of outer space and space exploration, it is central first of all to know what we mean by "culture," or, more accurately, how difficult it is to know what we mean by that word. "Culture" and its derivatives are infused with multiple meanings by different individuals and different professions, and not only among countries but also within countries. Whereas American scholars more often refer to "cultural evolution," for example, many Europeans prefer "social evolution," perhaps because "cultural anthropology" has roots as a discipline in the United States, while "social anthropology" was born in Europe. Though the two disciplines can have quite different meanings—the former referring to more concrete cultural variation among human societies over time and the latter to social behaviors—the two have grown closer together over the decades (Rapport and Overing 2000).

Such difficulties have not kept scholars from trying to define the term. More than 50 years ago, two anthropologists collapsed 164 distinct definitions of culture into one: "[C]ulture is a product; is historical; includes ideas, pattern, and values; is selective; is learned; is based upon symbols; and is an abstraction from behavior and the products of behavior" (Kroeber and Kluckhohn 1952, 656). Perhaps a brave attempt at a scholarly definition, but hardly one that yields a concrete intuitive grasp of what culture really is. Two decades later anthropologist Clifford Geertz, a giant in the field, defined culture more understandably as "an historically transmitted pattern of meanings embedded in symbolic forms by means of which men [people] communicate, perpetuate and develop their knowledge about and attitudes toward life" (Geertz 1973, 289; Kuper 1999).[3] According to Harvard biologist E.O. Wilson—famed for his work on sociobiology—each society creates culture and is created by it (Wilson 1998). In short, the idea of culture is a moving target, evolving with time and in space (and perhaps literally in outer space); not only does the understanding of the concept differ in Chinese and Western culture, but it is also different now than it was 50 years ago. So spaceflight is a manifestation of culture, a product of culture, but it is also embedded in culture. The influences travel both ways, and it is well to recognize this at the outset. And in the cosmic context, our terrestrial ideas of culture may be expanded if we discover cosmic civilizations, in which case the natural history of cultural evolution and its theoretical underpinnings will be taken to a new level (Dick and Lupisella 2009; Dick 2006).

If, as Wilson says, society creates culture, then there is the question of what is society? This too is problematic—it is a law of nature that any time academics focus on a word or concept it becomes problematic—but the question of the difference between society and culture is an important and venerable topic of discussion among anthropologists. A recent book on the key concepts in social and cultural anthropology put it this way: "Throughout the modernist period, a concept of society has underpinned the construction of all social theory, whatever its hue or denomination. If the concept of culture has played the role of queen to all analytic categories of the human sciences, the notion of society has been king. It is the master trope of high modern social thought" (Rapport and Overing 2000, 333). It is, the authors said, a treacherous friend, a necessary term, but a term to be used at one's risk.

Similarly, "imagination" has been the subject of both theoretical and descriptive study, and comes in many forms: the personal imagination of creative writers, artists, and scientists; the collective imagination of a given culture, as in the "American imagination," the "European imagination," or the "Russian imagination," each formed by the distinctive history and experiences of a specific culture; or the perhaps distinct (because so self-consciously explicit) imagination of science fiction, often characterized and even flouted as a literature of the imagination—so much so that even the best science fiction is still not accepted as sophisticated literature in some circles. Each of these forms of imagination is at work in any general study of outer space and the imagination. Their complex nature and interaction remain largely uncharted waters in the field space history. But the richness awaiting researchers is evident in Howard McCurdy's book *Space and the American Imagination*, where McCurdy shows how the American space program took advantage of elements deeply ingrained in the American imagination, notably the exploration imperative, the search for extraterrestrial life, and the idea of "the last frontier." Similarly, McCurdy and Roger Launius have shown how the imagery of space, from Buck Rogers and Flash Gordon to the art of Chesley Bonestell and real images beamed from outer space, have inspired the imagination and had a real effect on public and scientific interest in space. In a broader sense, Harvard historian of science Gerald Holton has shown how the imagination of the scientist, rather than objective criteria, is often important in the early stages of a scientific idea (McCurdy 1997; Launius and McCurdy 2001; Holton 1978, 1996). In short, imagination is not to be trifled with, but constitutes a real force with real-life consequences.

Although it is counterproductive to spend too much time on definitions, it is important to realize that the cultural history of outer space and the role of imagination are not simple problems precisely because of the vagueness of the terms. Other historical subfields suffer from the same conceptual problem, but few fields are as expansive as the physical extent of outer space, or as complex as the mental terrain of the imagination. Perhaps it is best to say what the cultural history of outer space is not: it is less about the political, diplomatic, and technological aspects of spaceflight, than about the socio-cultural rationale for spaceflight, a term that nicely circumvents the differences between the social and the cultural by combining the two terms (Geppert 2008). In addition to rationale, it is also about socio-cultural impact, beliefs, and visions of the future. Given such expansive mental and spatial terrain, it is hardly surprising that approaches to the subject may (and should) be expansive as well.

Given the difficulties with the concepts of culture, society and imagination, and the difficulties of determining the exact role of imagination on any one individual, much less on society and culture, we nevertheless boldly proceed into the unknown. We can divide the question of outer space and the imagination into three parts: First, how has space exploration affected the imagination and society? Second, and conversely, how has imagination historically affected space exploration? And third, what is the effect of spaceflight on our worldview, our *Weltanschauung*, to use the great theory-laden German word? This is a large subject, and can be only faintly illuminated here.

19.2 Space and the Imagination: How Has Space Affected Our Imagination?

The question of how changing ideas of space and time over the twentieth century have affected our imagination is related to, but distinct from, the question of space and our *Weltanschauung*, to which I shall return to at end. The first observation that must be made is that it is no less than astonishing how much our ideas of space have changed over the last century. In terms of spatial extent, at the beginning of the twentieth century Alfred Russel Wallace, the cofounder with Darwin of the theory of natural selection, offered a model of the universe only 3600 light years across. In supporting it at length in his well-known volume *Man's Place in the Universe* (1903), Wallace claimed that he was simply espousing the view of the most eminent astronomers of his day, a reasonable claim. When Wallace wrote, all stars, and indeed all observable phenomena in the universe, were widely believed to be part of a single system perhaps several thousand light years in diameter (compared to the 100,000 light years now estimated), with the Sun in a nearly central position. The *island universe theory*, which postulated many such systems, had been in gradual decline since the 1860s and had completely fallen from favor by the late 1880s. It is therefore not surprising that Wallace viewed the universe as a single system of stars with our Solar System at the approximate center (Smith 1982; Berenzden et al. 1976; Dick 1996, 36–58, 2008, 320–340).

Though *Man's Place in the Universe* went through seven editions by 1908 and another in 1914, and was translated into German in 1903 and French in 1907, it had little influence beyond the second decade of the twentieth century. The reason is not far to seek. Within 15 years of Wallace's death in 1913, most of his central assumptions had been rendered obsolete by an emerging new cosmology. In 1918 the American astronomer Harlow Shapley reported, based on his study of the distribution of globular clusters of stars, that our Solar System was located in a very eccentric position in the galaxy, at its periphery rather than its center. This proved to be one of the great shifts in our cosmological world view, from the geocentric to the heliocentric to the galactocentric, as Shapley himself called his revolutionary new view (Bok 1974). By 1924 Edwin P. Hubble had demonstrated to the satisfaction of most astronomers that many other galaxies exist outside our own, galaxies that he showed a few years later are fleeing from one another in what could be interpreted as an expanding universe (Smith 1982, 97–146). We now know from the Hubble Space Telescope and other observations that we live in a universe billions of light years in extent, characterized by an interrelation among parts and the whole that astronomers characterize by the term *cosmic evolution*. Though Wallace recognized the evolution of the stars based on the contemporary work of astronomers, neither he nor they could have known the extent of full-blown cosmic evolution, ranging from the Big Bang to the present and covering some 13.7 billion years of time (Chaisson 1981, 2001). As Olaf Stapledon and many other science fiction writers have commented, this greatly enlarged universe gives vast scope for imagination and for action, whether by humans or extraterrestrials, conjuring the warp speeds of Star Trek in order to traverse its domain.

In addition to an expanded concept of space, the idea of cosmic evolution represents another dimension—the dimension of time—which has become very important to our world view, and will be even more so in the future. The idea of cosmic evolution only gradually came to be realized during the twentieth century, as the Big Bang cosmology gained greater acceptance and as the idea of the explosive beginning of the universe gave greater force to a coherent story of the universe. The evolution of stars was known from the work of the astronomer George Ellery Hale among others, and a broader idea of cosmic evolution was occasionally discussed by the followers of Herbert Spencer's evolutionary world view and by a few scientists such as Lawrence J. Henderson. But it was only in the late 1950s, with the writings of Harvard astronomer Harlow Shapley, that the modern idea of cosmic evolution was fully enunciated and sustained (Dick 2009). It became a major driving force taken up at NASA, first in its Search for Extraterrestrial Intelligence (SETI) program, then in its exobiology program, and finally in its Origins and astrobiology programs. With the discovery of the cosmic background radiation in the 1960s, and its detailed analysis by the COBE and WMAP spacecraft, we now know that the universe is 13.7 billion years old, with an uncertainty of only 1%, or about 100,000,000 years. The fact of cosmic evolution is inherent in most of the work done in space science by the national space agencies, which may be seen as filling in the gaps in the history of cosmic evolution, the ultimate master narrative of the universe. But only a few scholars, preeminently astronomer Eric Chaisson, have analyzed the idea of cosmic evolution across its full astronomical, biological and cultural breadth (Chaisson 1981, 2001).

Along with these expanded notions of the *extent* of space and time, we have had an equally revolutionary change in our perception of the very *nature* of space and time. The work of Albert Einstein yielded the concept of a space-time continuum, demonstrating that Newtonian ideas were incomplete at relativistic speeds and cosmological distances. Einstein's personal imagination was essential for his signal scientific advances; we need only recall Einstein's thought experiments, in which he imagined himself riding on moving trains, or on a light beam, or in an enclosed chamber in freefall, in order to arrive at his radical ideas of simultaneity and relativity. At the same time Einstein was greatly affected by a signal development of the late nineteenth and early twentieth centuries—attempts to synchronize timepieces across increasingly larger areas of the Earth, and with increasing accuracy (Galison 2003; Isaacson 2007). In the wake of Einstein, the old concepts of absolute space and absolute time were no longer viable. The result was a new concept of space and time that in turn altered our world view and fed the imagination of science fiction writers. Consonant with the nature of science, the Einsteinian worldview may itself someday be subsumed under a more general theory, and with it the parameters of imagination will change once again.

In addition to changing conceptions of space and time over the century, the possibility of life beyond Earth has been a continuous and often spectacularly popular theme. The so-called Drake Equation, originated by the American astronomer Frank Drake in 1961 in the wake of the first search for extraterrestrial radio signals, is the iconic image for extraterrestrial intelligence (Dick 1996, chap. 8). The Drake Equation tries to assess the number (N) of technological civilizations in the galaxy, and in doing so it represents various parameters of astronomical, biological and cultural evolution. Depending on the values inputted, N might be millions, as Carl Sagan and Frank Drake opined, or only one—us. Despite decades of research and speculation since the first radio search for intelligent signals beyond Earth, we do not yet know if there is any life beyond Earth, primitive or intelligent, and this lack of an answer leaves open for the imagination the question of whether the universe is for aliens, for humans, or for both. This has indeed proven a fertile source if imagination, a playing ground for countless profound and less-profound thinkers, especially in science fiction literature (Crowe 1986; Dick 1996; Guthke 1990).

How has the new view of space, time and aliens affected culture, even with all the ambiguities of that term? That is an enormous question, so let us narrow our inquiry into how a few selected science fiction writers, representing specific cultures, were affected by the new views induced by Shapley, Hubble, and Einstein, among others. Because aliens have been a favorite theme of science fiction literature at least since H.G. Wells and Kurd Lasswitz at the end of the nineteenth century, it is obvious that the new view of the cosmos was not *required* for the depiction of aliens. But *The War of the Worlds* was only a local battle within our own parochial Solar System, with Martians invading Earth. And Kurd Lasswitz's more peaceful Martians in *Auf zwei Planeten* were also localized in their interaction with Earth. Although the effect of worldviews on cultures requires a comprehensive approach, here we may look briefly at four of the most influential science fiction writers of the twentieth century—two in Britain, one in the United States, and one in Poland—to illustrate how much the scope of alien literature was expanded by the new view of space, time and aliens. Olaf Stapledon, Arthur C. Clarke, Stanislaw Lem, and Isaac Asimov each represent a different aspect of the question, and each shows the effect of the new world view on the imagination in different ways.

In 1930, at the age of 44, the British philosopher Olaf Stapledon, a graduate of Oxford University in history and Liverpool University in philosophy, took up the writing of fiction, in which aliens immediately played an essential role. In his novels *Last and First Men* (1930) and *Star Maker* (1937), political, religious and philosophical ideas dominate rather than adventure. Stapledon lived in an era when the immensity of the cosmos was well known, and his novels appropriately cover billions of years. He knew of Edwin Hubble's work, and for his conception of the size of the cosmos he cited the astronomer William J. Luyten's *The Pageant of the Stars*. Still, "Immensity is not itself a good thing," Stapledon wrote. "A living man is worth more than a lifeless galaxy. But immensity has indirect importance through its facilitation of mental richness and diversity [...] though spatial and temporal immensity of a cosmos have no intrinsic merit, they are the ground for psychical luxuriance, which we value. Physical immensity opens up the possibility of vast physical complexity, and this offers scope for complex minded organisms" (Stapledon 1968). This is a direct statement of how space affected the imagination of Olaf Stapledon, for his novels were played out on this immense tapestry of infinite space and billions of years of time (though not yet Einsteinian space-time). His characters were a variety of amazing and evolutionarily connected lifeforms, a fertile source of imagination for future science fiction authors. Through his novels, Stapledon taught us to think long-term about space, time, and aliens, and the richness of this thought over these timespans is still in many ways unsurpassed.

Arthur C. Clarke possessed a more technical background than Stapledon, and in fact served as chairman of the British Interplanetary Society. The composition of his early stories overlap in time with Stapledon, who was one of his main influences. Virtually all of his novels are filled with aliens, and their themes are human interaction with aliens, as in *Childhood's End, Rendezvous with Rama, or 2001 A Space Odyssey* and its sequels. "The idea that we are the only intelligent creatures in a cosmos of a hundred million galaxies is so preposterous that there are very few astronomers today who would take it seriously," he wrote in 1972. "It is safest to assume, therefore, that They are out there and to consider the manner in which this fact may impinge upon human society' (Clarke 1972; Dick 1996, 254–56). Clarke believed that extraterrestrials gave a true perspective on humanity, "true" meaning in the broadest context of the possibilities inherent in the new universe, dwarfing even the globalists of the day. It was this perspective that was the main theme of almost all of Clarke's novels.

The prolific American science fiction writer Isaac Asimov took a very different approach, the opposite side of the coin of humanity's role in the newly expanded space and time. With one exception, a novel titled *The Gods Themselves*, aliens are not at all prominent in his science fiction, which is nevertheless considered some of the best of the twentieth century. The famous original *Foundation* trilogy, its subsequent prequels and sequels, and Asimov's robot novels as well, have no aliens at all, but show how the new ideas of space and time have greatly expanded the scope for human and robotic action.[4]

Meanwhile, as we have seen, in continental Europe the Polish physician and writer Stanislaw Lem had taken up science fiction at mid-twentieth century with his novel *Astronauci*, absorbing the Soviet fascination with space despite running into trouble with Soviet Lysenkoism. Despite the movie treatment of that first novel, it was Lem's novel *Solaris*, published in 1961, that spread his fame. By this time, Lem had read Clarke and Asimov, as well as Ray Bradbury. Although affected by those authors, Lem's treatment of the alien was very different, allowing him to play out themes in an alien setting unlike anything produced in the West. Solaris is a planet with an ocean that is alive, "a monstrous entity endowed with reason, a protoplasmic ocean-brain enveloping the entire planet and idling its time away in extravagant theoretical cognitation [*sic*] abut the nature of the universe." The monologue of this living being, however, was beyond the understanding of humans. While the ultimate purpose of Lem's novel is to use the cosmos to learn about humans, it may also be read at a different level as an argument against attempting contact before humans understand themselves: "Man has gone out to explore other worlds and other civilizations without having explored his own labyrinth of dark passages and secret chambers, and without finding what lies behind the doorways that he himself has sealed" (Lem 1970, 28 and 165). Lem seems to be saying that, bold as the new universe may be, humans may after all remain its central mystery (or at least a central mystery), imparting the message that our fate may lie not in the stars, but in ourselves. But the mere possibility of Lem's non-humanoid aliens expanded the scope of human imagination and illuminated age-old human questions. *Solaris* was first filmed in 1971, and subsequently received other movie treatments both in the Soviet Union and the United States. Multiplied hundreds of times in sophisticated or more shallow treatments, ideas of the alien have spread rapidly throughout the popular culture, affecting individuals and cultures in countless, though not always quantifiable, ways.

The comparison of these thinkers in relation to the new view of the universe highlights an important point. Although they were both affected by the same new view of the universe, Clarke and Asimov provide two views of human destiny—one in which humans interact with aliens beings, and one in which human destiny is to expand throughout the galaxy for its own purposes, without having to deal with pesky aliens. Lem believes we may have to deal with aliens, even though alien minds may be incommensurable with ours—and we had better learn to understand ours better. In addition to these very different views of human destiny, Stapledon constantly reminds us of the necessity of thinking over billions of years. Applying this kind of Stapledonian thinking to cultural evolution in the cosmos, and taking cultural evolution as a serious and dominating integral of cosmic evolution, the long time spans over which extraterrestrial intelligence may have existed implies that they are nothing like humans. They may in fact have evolved beyond flesh and blood biologicals, giving rise to postbiologicals, perhaps in the form of artificial intelligence (AI) (Dick 2003). Although that idea is not new, the idea of cultural evolution over eons of time as an integral part of cosmic evolution gives it new force. Thus we may live in a postbiological universe full of machines, and this may have implications for the SETI scientists, who should be looking for machines rather than biologicals like us.

While this may seem far-out speculation, it is our knowledge of space and time, together with the real possibility of aliens, that leads us to such a vision. Based as it is on current ideas of terrestrial cultural evolution, the likelihood is that it is too conservative rather than too speculative.

19.3 The Imagination and Space: How Has Imagination Affected Space Exploration?

In examining whether imagination has affected space exploration we turn first to one of space exploration's most perceptive historians, Walter McDougall. McDougall has argued that imagination is one of three structural forces necessary for spaceflight, along with funding and technology. There is no doubt that spaceflight pioneers were imaginative thinkers. As McDougall himself put it in his Pulitzer-prize winning book *The Heavens and the Earth*, "The great pioneers of modern rocketry—Tsiolkovsky, Goddard, Oberth and their successors Korolev, von Braun and others—were not inspired primarily by academic or professional interest, financial ambitions, or even patriotic duty, but by the dream of spaceflight. To a man they read the fantasies of Jules Verne, H. G. Wells and their imitators, and the rocket for them was only a means to an end." (McDougall 1985, 20).

This much is well known, but how much can it be generalized? From personal experience I can say that not only was my entry into astronomy affected by imagination in the form of science fiction, but also that a good percentage of my colleagues at NASA and other space agencies around the world were (and still are) influenced by science fiction, and that therefore imagination played a role in their entry into careers in astronomy and spaceflight. But this is certainly not true of all pioneers and practitioners of spaceflight. Let us take the case of the three pioneers of Explorer 1, the first U. S. satellite, which recently passed its 50th anniversary. These pioneers, Wernher Von Braun, William Pickering and James van Allen, are familiar from the iconic photo of the three at the early morning press conference following the successful launch of Explorer 1 on January 30, 1958 (Fig. 19.2). All three are the subject of recent exhaustive biographies (Neufeld 2007; Mudgway 2008; Foerstner 2007).

There is no doubt that one of Von Braun's influences was the so-called "father of German science fiction," Kurd Lasswitz. Lasswitz was a philosopher and historian, a Kantian who was steeped in the school of German idealism and wrote a biography of Gustav Fechner. In the first English translation of Lasswitz's science fiction novel *Auf zwei Planeten* ("On Two Planets"), published in 1971 during the Apollo program, von Braun wrote, "I shall never forget how I devoured this novel with curiosity and excitement as a young man" (von Braun 1971).[5] One can safely assume it was one of the complex of factors that propelled von Braun forward to the stars.

But it was quite different for Pickering and Van Allen, as is evident from their recent biographies. Neither one of these physicists was influenced by science fiction, but rather more by technology. Douglas J. Mudgway, the author of the new Pickering biography, when asked whether Pickering was at all influenced by science fiction from his native New Zealand, wrote that Pickering, "definitely was not. He was greatly attracted to the things around him in his country town, radio crystal sets, the town electric generator that ran only fours per day and the telephone switchboard and telephone system in his town. Later at secondary school he became fascinated with amateur radio" (Mudgway 2008). Similarly for Van Allen, who gives no evidence of science fiction influence (Foerstner 2007). It is therefore important to realize that sources of inspiration exist other than science fiction, in this case a fascination with technology, quite different from imagination, or at least a different kind of imagination. Even from such a small sample we can conclude that imagination in the science fiction sense is neither necessary nor sufficient for space exploration. It is not necessary because it drove only some of the spaceflight pioneers, and it is not sufficient because imagination cannot propel any nation to the Moon in the absence of McDougall's other two factors, funding and technology.

We may also look at a second area, not spaceflight itself, but exobiology and the Search for Extraterrestrial Intelligence, a subject taken up by NASA already in the 1960s, though at a relatively low level of funding. More than a decade ago in *The*

Fig. 19.2 The three men responsible for the success of Explorer 1, America's first Earth satellite, launched January 31, 1958. At left is William H. Pickering, former director of the Jet Propulsion Laboratory, which built and operated the satellite. James A. Van Allen, center, designed and built the instrument on Explorer 1 that discovered the radiation belts that circle the Earth. At right is Wernher von Braun, leader of the Army's Redstone Arsenal team that built the first stage Redstone rocket that launched Explorer 1. (Source: NASA)

Biological Universe, my history of the twentieth century extraterrestrial life debate, I included an entire chapter on "The Role of Imagination." There I concluded that an understanding of the alien in science fiction was essential to understanding why it held such a grasp on popular culture, and even why it was taken up by some scientists. During the twentieth century, I found:

> Science and science fiction increasingly complemented each other: speculative science fiction provided the perfect outlet for scientists who wished to go beyond science. Not only did scientists exercise their imaginations in science fiction, science fiction also inspired them to tackle questions in the real world. Many of the pioneers in exobiology and SETI grew up on science fiction and were led to their careers by its imaginative lure. Having nurtured science fiction, science now received in return some of the rewards of imagination. (Dick, 266)

In the SETI arena David Swift's book of interviews, *SETI Pioneers,* is revealing. Swift found that Philip Morrison, famous for his 1959 paper on interstellar communication, was influenced by H.G. Wells; similarly Freeman Dyson, "read a good deal of science fiction" and was especially influenced by H.G. Wells. Carl Sagan, around 7 or 8 years old, read the Edgar Rice Burroughs novels, and of course later himself wrote his own science fiction novel *Contact*. One of Barney Oliver's "more profound influences" as a youth was Hugo Gernsback's *Amazing Stories*; Oliver went on to write the famous book on Project Cyclops and helped direct NASA's SETI program. Ron Bracewell, author of *The Galactic Club*, first thought about extraterrestrials when, like Carl Sagan, he read Edgar Rice Burroughs, as well as Gernsback's *Amazing Stories*. Jill Tarter "loved science fiction and read enormous amounts of it," especially Robert Heinlein. In the Soviet Union Nikolai Kardashev, well known for his typology of civilizations, also read science fiction (Swift 1990, 22, 323, 211, 88–89, 140–41, 150–51, 180).

There were, of course, some SETI scientists not influenced by science fiction; Frank Drake and Joseph Shklovskii are among them. Nevertheless, certain fields related to outer space have a more imaginative component, and we can conjecture that the relationship is directly proportional: the more imaginative the field, the more its practitioners have been influenced by science fiction and other imaginative drivers.

There is room here for more interesting research on how particular fields differ in the role of the imagination, from general categories like scientists vs engineers, to specific categories like SETI scientists. In any case, it is clear that imagination has historically affected spaceflight, but one needs to be nuanced in just how general that claim can be made, in what areas, and in particular, to what effect.

19.4 Space and Our *Weltanschauung*: How Has Space Exploration Affected Our Worldview?

The evolution of our ideas of space, time and aliens in twentieth century has affected more than just our imaginations. It has also affected our individual and collective worldviews, our *Weltanschauung*, our society and culture however one wishes to define them. And, I would argue, our new knowledge of the universe should affect our worldviews even more. Our new knowledge of cosmic evolution demonstrates for the first time in an empirical way our true place in the universe, both in space and time, with the question of aliens still very much open, perhaps the greatest question remaining in the history of science. The new view of cosmic evolution is already affecting us in numerous ways, though the diffusion rate of cosmic ideas into popular culture is in many ways agonizingly and remarkably slow. In science—arguably one of the primary drivers of culture—there is no doubt that cosmic evolution is now the master narrative, the subject of scholarly books, public broadcasting television treatment, and most importantly of all, research programs. It is clear that the photos from the Hubble Space Telescope and the other Great Observatories have fired the popular imagination, but they can also be seen as pieces in the story of cosmic evolution—the story that leads (in a non-teleological way) to humans and to the question of life beyond Earth. That question continues to fascinate the public, and to draw in an increasingly diverse audience of scholars into the fields of astrobiology and SETI. Over the last few decades the astrobiology and SETI communities have convened special groups to discuss the societal impact of the discovery of life in the universe. Not surprisingly, they have concluded that there will be a multitude of reactions to the discovery of extraterrestrial life, depending on the scenario and the society (Billingham et al. 1999; Tough 2000; Bertka et al. 2007; Harrison 1997).

Cosmic evolution has also made small inroads into a number of academic disciplines. In history, it has specifically spawned the movement known as *Big History*. Big History, pioneered by David Christian and Fred Spier, views history in the context of 13.7 billion years of cosmic evolution, rather than in the traditional mode of thousands of years of wars and politics (Christian 1991, 2004; Spier 1996). A continuation of the *cosmic calendar* used to great affect by Carl Sagan and others, Big History has the potential to revolutionize the teaching of history even beyond the current and more advanced trends toward global history. Just as global history expands the individual's *Weltanschauung*, so cosmic history views global history as just one example among many possible worlds, and explicitly questions parochial terrestrial assumptions in history, philosophy and all areas of human thought. While this method of teaching history is not yet widespread, nor is the cosmic mode of thought embodied in the cosmic calendar internalized in most people's lives, it very likely will be in the future.

Similarly, small inroads have been made in anthropology. For the last several years, there have been SETI sessions at the annual meetings of the American Anthropological Association, drawing anthropologists into the subject from a variety of viewpoints. A cover story in the British publication *Anthropology Today* recently emphasized how anthropology can be applied to cultures beyond the Earth, either human or alien (Dick 2006). Anthropologists have used their approaches to guide thinking on interstellar migration; and anthropologists have even written alien science fiction (Russell 1996, 1998).

Aside from science, history and anthropology, small inroads have also been made in religion, theology and philosophy. The Space Age spawned considerable discussion about theological implications, especially if life were actually found.

About 10 years ago the Templeton Foundation sponsored a meeting on the theological implications of the new universe. The subsequent volume (Dick 2000c) includes an article by Sir Arthur Peacocke specifically discussing theology and cosmic evolution (Peacocke 2000). Another article in the same volume argues that, if one has need of theology, a cosmotheology that takes into account what we know about the universe would be preferable to current parochial terrestrially-bound theologies (Dick 2000b). A German volume of essays on the same subject indicates that the possibility of a cosmotheology is not an idea confined to one culture (Wabbel 2005). Similarly, a few scholars have begun to discuss cosmophilosophy: what part of our knowledge is necessary, what is contingent, and how the traditional problems of terrestrial philosophy might be broadened by the expanded outlook afforded by space exploration. This, in the end, is the great benefit of the Space Age, providing a much broader perspective, making us realize that all our Earthly knowledge may be only a single instance of a much more generalized knowledge (Regis Jr. 1985, 79–129; Dick 2000a). The diffusion of the cosmic perspective into academic disciplines has been excruciatingly slow. Yet slowly but surely, it is making its mark, and it will likely gather momentum over the next decades as our cosmic consciousness increases. It is increasingly seeping into consciousness through curricula actually based on cosmic evolution (Dick 2009).

Still, the chief impact today has been not on these academic disciplines, but mainly in popular culture through science fiction, the debate over Unidentified Flying Objects (UFOs), and popular interest in Star Trek, Star Wars, and the visual media that stimulate the imagination and from which much of the public takes their ideas of science. Taken together, science fiction, the UFO debate, and their depiction in media and the arts may be seen one way that popular culture absorbs this new worldview of a biological universe, expanded in space and time and perhaps replete with aliens.

The immediate impact of the Space Age, however, is far more diverse than the ultimate discovery of life in space. Even if no aliens are found, space has already impacted, and will continue to impact our civilization in surprising and not always evident ways. In her recent book *Rocket Dreams: How the Space Age Shaped Our Vision of a World Beyond*, Marina Benjamin argues that space exploration has shaped our worldviews in diverse ways. She argues that "the impact of seeing the Earth from space focused our energies on the home planet in unprecedented ways, dramatically affecting our relationship to the natural world and our appreciation of the greater community of mankind, and prompting a revolution in our understanding of the Earth as a living system." Benjamin thinks it is no coincidence that the first Earth Day on April 20, 1970 occurred in the midst of the Apollo program; or that one of the astronauts developed a new school of spiritualism; or that people "should be drawn to an innovative model for the domestic economy sprung free from the American space program by NASA administrator James Webb" (Benjamin 2003, 4). Nor is Benjamin the first, or strongest, proponent of this argument, which has been made since Apollo days by poets like Archibald MacLeish and authors such as Frank White (Poole 2008). Space exploration shapes world views and changes cultures in unexpected ways. So does lack of exploration.

As Isaac Asimov foretold in his *Foundation* series, eventually humans will spread into the cosmos at large. Some see space in utopian terms, as the new frontier, or a place to start over for a new and better world. The Star Trek mission "to boldly go where no man has gone before" is the clarion call of those who see space exploration as a necessary part of human evolution, not a luxury. Historians and social scientists have analyzed this kind of argument, and not all agree that the utopian ideal of spreading humanity to outer space is a valid reason for going, or that utopia is what we will build when we get there. Others have demonstrated the complex relation of such space goals to social, racial, and political themes. One such study is De Witt Kilgore's recent book *Astrofuturism: Science, Race and Visions of Utopia in Space*. In this book Kilgore examines the work of Wernher von Braun, Willy Ley, Robert Heinlein, Arthur C. Clarke, Gentry Lee, Gerard O'Neill, and Ben Bova, among others, in what he calls the tradition of American astrofuturism (Kilgore 2003).

In the end we must also realize that the impact of space exploration on our world view will also vary according to individuals and cultures—coming back now full circle to that problematic term "culture" and the relation of the individual imagination to culture. Howard McCurdy's *Space and the American Imagination* critically analyzes ideas such as the new frontier, progress, the exploration imperative, and the search for extraterrestrial life as part of American culture (McCurdy 1997). Although he does not discuss it, his book implicitly raises the questions: What is the role of space in the European imagination, or the Chinese or Russian imagination? How do different cultures affect the imagination, and how does the imagination affect cultures differently? And just how important has space exploration been as one among many sources of imagination in the twentieth century like atomic power and other wonders of science? In undertaking these studies, we need to remember that we both produce culture and are a product of culture. We need to remember that in doing history our remembrance of things past is inevitably colored and clouded by the geographical separations in space, by the passage of time, and by our own minds that at times seem alien to each other. Comparative studies should certainly be undertaken on these subjects, and this volume is an opening contribution toward that goal in the European context.

19.5 Commentary 2020

This paper was the keynote address for the first international conference on the cultural history of outer space, held in Bielefeld, Germany from February 6–9, 2008. The conference gathered some 70 scholars from a dozen countries to analyze the cultural significance of outer space from a European perspective. It was thus unusual in two respects: in examining cultural significance rather the usual "nuts and bolts" history that space historians most often engage, and in encouraging a European perspective rather than the usual American one. More than this, the meeting and its subsequent publications sought to establish *astroculture* as a new field of historical inquiry, thanks to the insight of the meeting organizer, Professor Alexander Geppert. As Geppert put it, "Research on the history of astroculture does not aim at providing definitive answers regarding the reality or fiction of space-related phenomena. Instead, it critically focuses on the intentions, actions, categories and explanations provided by actants themselves, because they are part and parcel of the ways in which human beings attempt to come to terms with and make sense of the infinite universe that surrounds us" (Geppert 2012, 9).

The results of the meeting *Imagining Outer Space* (Geppert 2012, reissued in 2018) were followed by two more meetings and volumes on astroculture, *Limiting Outer Space* (Geppert 2018) and *Militarizing Outer Space* (Geppert, in press). All three volumes are highly recommended to those interested in how space exploration has affected culture and vice versa. I expect astroculture to grow in importance, both as a cultural phenomenon and as an area of study.

Notes

1. *Der schweigende Stern* was a co-production of the DDR (East Germany) and Poland, undertaken by DEFA Studios. It was released in East Germany on February 26, 1960 and in West Germany on September 9, 1960 with a new title.
2. The Polish title is *Astronauci*. Though Lem in later life did not think highly of his first novel, its success encouraged him to write more fiction in this vein. On the novel in the context of Lem's life see Swiarski (1997, 3–4).
3. For debated differences between the concepts of culture and society, a good starting point is Rapport and Overing (2000), entries on culture and society, 92–102, 333–343.
4. The original Foundation trilogy is Isaac Asimov, *Foundation* (1951), *Foundation and Empire* (1952), and *Second Foundation* (1953), subsequently published in numerous editions. The Robot novels are *Caves of Steel* (1954), *The Naked Sun* (1957), *The Robots of Dawn* (1983) and *Robots and Empire* (1985).
5. On Lasswitz see Dick (1996), 227–30, and on the influence of Lasswitz on the Space Age, see Ley (1969), 65–9.

References

Andrews, James T. 2007. "In Search of a Red Cosmos: Space Exploration, Public Culture, and Soviet Society," in Dick and Launius, 2007, 41–52.

Benjamin, Marina. 2003. *Rocket Dreams: How the Space Age Shaped our Vision of a World Beyond*, New York: Free Press.

Berenzden, Richard, Richard Hart and Daniel Seeley. 1976. *Man Discovers the Galaxies*, New York: Science History Publications.

Bertka, Connie, Nancy Roth and Matthew Shindell. 2007. *Workshop Report: Philosophical, Ethical, and Theological Implications of Astrobiology*, Washington, DC: American Association for the Advancement of Science

Billingham, John et al., eds., 1999. *Social Implications of the Detection of an Extraterrestrial Civilization: A Report of the Workshops on the Cultural Aspects of SETI held in October 1991, May 1992 and September 1992 at Santa Cruz, California*, Mountain View, CA: SETI Press.

Bok, Bart J. 1974. 'Harlow Shapley and the Discovery of the Center of Our Galaxy,' in *The Heritage of Copernicus: Theories 'More Pleasing to the Mind'*, ed., Jerzy Neyman, ed., Cambridge, MA: MIT Press, 26–62.

Chaisson, Eric. 1981. *Cosmic Dawn: The Origins of Matter and Life*, Boston: Little, Brown and Co.
Chaisson, Eric. 2001. *Cosmic Evolution: The Rise of Complexity in Nature*, Cambridge, MA: Harvard University Press
Christian, David. 1991. "The Case for Big History", in *The Journal of World History* 2.2 (Fall 1991), 223–38
Christian, David. 2004. *'Maps of Time': An Introduction to 'Big History'*, Berkeley, CA: University of California Press
Clarke, Arthur C. 1972. 'When the Aliens Come,' *Report on Planet Three and Other Speculations*, New York: New American Library, 89–102
Crowe, Michael J. 1986. *The Extraterrestrial Life Debate, 1750–1900: The Idea of a Plurality of Worlds from Kant to Lowell*, Cambridge: Cambridge University Press
Dick, Steven J. 1996. *The Biological Universe: The Twentieth Century Extraterrestrial Life Debate and the Limits of Science*, Cambridge, Cambridge University Press
Dick, Steven J. 2000a. 'Extraterrestrials and Objective Knowledge,' in Tough (2000), 47–48.
Dick, Steven J. 2000b. 'Cosmotheology: Theological Implications of the New Universe,' Dick (2000), 191–210.
Dick, Steven J. 2000c. *Many Worlds: The New Universe, Extraterrestrial Life and the Theological Implications*, Philadelphia: Templeton Foundation Press
Dick, Steven J. 2003. 'Cultural Evolution, the Postbiological Universe, and SETI,' *International Journal of Astrobiology* 2, 65–74.
Dick, Steven J. 2006. 'Anthropology and SETI,' *Anthropology Today*, 22. 2 (April, 2006), 3–7.
Dick, Steven J. 2008. 'The Universe and Alfred Russel Wallace,' in *Natural Selection & Beyond: The Intellectual Legacy of Alfred Russel Wallace*, ed. Charles H. Smith and George Beccaloni, Oxford, Oxford University Press, 320–40.
Dick, Steven J. 2009. 'Cosmic Evolution: History, Culture and Human Destiny,' in Dick and Lupisella (2009)
Dick, Steven and Roger D. Launius, eds. 2007. *Societal Impact of Spaceflight*, Washington: NASA.
Dick, Steven J. and Mark Lupisella, eds. 2009. *Cosmos and Culture: Cultural Evolution in a Cosmic Context*, Washington: NASA.
Foerstner, Abigail. 2007. *James van Allen: The First Eight Billion Miles. Iowa City: University of Iowa Press.*
Galison, Peter. 2003. *Einstein's Clocks, Poincaré's Maps: Empires of Time,* New York, W. W. Norton
Geertz, Clifford. 1973. *The Interpretation of Cultures*, New York: Basic Books, 289.
Geppert, Alexander C. T. Geppert. 2008. 'Imagining Outer Space, 1900–2000: An International Conference,' February 6–9 2008, conference description.
Geppert, Alexander C. T., ed. 2012. *Imagining Outer Space: European Astroculture in the Twentieth Century*. London: Palgrave Macmillan.
Geppert, Alexander C. T. 2018. *Limiting Outer Space: Astroculture After Apollo*. London: Palgrave Macmillan.
Guthke, Karl S. 1990. *The Last Frontier: Imagining other Worlds from the Copernican Revolution to Modern Science Fiction.* Ithaca, Cornell University Press
Harrison, Albert. 1997. *After Contact: The Human Response to Extraterrestrial Life*, New York: Plenum
Holton, Gerald. 1978. *The Scientific Imagination: Case Studies*, Cambridge: Cambridge University Press.
Holton, Gerald. 1996. 'Imagination in Science,' in Gerald Holton, *Einstein, History and Other Passions*, New York: Addison-Wesley, 78–102.
Isaacson, Walter. 2007. *Einstein: His Life and Universe*, New York, Simon and Schuster 122–7.
Kilgore, De Witt Douglas. 2003. *Astrofuturism: Science, Race and Visions of Utopia in Space*, Philadelphia: University of Pennsylvania Press
Kroeber, Alfred L. and Clyde K. M. Kluckhohn. 1952. *Culture: A Critical Review of Concepts and Definitions,* Cambridge, MA: The Peabody Museum.

Kuper, Adam. 1999. *Culture: The Anthropologists' Account*, Cambridge, MA: Harvard University Press
Launius, Roger D. and Howard E. McCurdy. 2001. *Imagining Space: Achievements, Predictions, Possibilities, 1950–2050*, San Francisco: Chronicle Books
Lem, Stanislaw. 1970. *Solaris*, New York: Berkley Publishing Company.
Ley, Willy, 1969. *Rockets, Missiles, and Men in Space*, New York
McCurdy, Howard. 1997. *Space and the American Imagination*, Washington, DC: Smithsonian Institution Press
McDougall, Walter. 1985. ... *The Heavens and the Earth: A Political History of the Space Age*, New York: Basic Books
Mudgway, Doug. 2008. *William H. Pickering: America's Deep Space Pioneer Washington, DC: NASA.*
Neufeld, Michael. 2007. *Von Braun: Dreamer of Space, Engineer of War. New York: Alfred A. Knopf.*
Peacocke, Arthur. 2000. 'The Challenge and Stimulus of the Epic of Evolution to Theology,' in Dick (2000), 89–118.
Poole, Robert. 2008. *Earthrise: How Man First Saw the Earth*, New Haven: Yale University Press
Rapport, Nigel and Joanna Overing. 2000. *Social and Cultural Anthropology: The Key Concepts*, London: Routledge, vii and passim.
Regis, Jr., Regis. 1985. ed., *Extraterrestrials: Science and Alien Intelligence*, Cambridge: Cambridge University Press
Russell, Mary Doria. 1996. *The Sparrow*, New York: Fawcett.
Russell, Mary Doria. 1998. *Children of God,* New York, Villard.
Schmölders, Claudia. 2012. "Heaven on Earth: Tunguska, 30 June, 1908," in Geppert, ed. (2012), pp. 51–72.
Smith, Robert. 1982. *The Expanding Universe: Astronomy's Great Debate, 1900–1931*, Cambridge: Cambridge University Press
Spier, Fred. 1996. *The Structure of Big History: From the Big Bang Until Today*, Amsterdam: Amsterdam University Press
Stapledon, Olaf. 1968. 'A Note on Magnitude,' in Stapledon, *Last and First Men* and *Star Maker*, New York: Dover Publications, 435.
Swiarski, Peter. 1997. *A Stanislaw Lem Reader*, Evanston, IL: Northwestern University Press.
Swift, David W. 1990. *SETI Pioneers: Scientists Talk about their Search for Extraterrestrial Intelligence*, Tucson, AZ: University of Arizona Press.
Tough, Allen, 2000. *When SETI Succeeds: The Impact of High-Information Contact*, Bellevue, WA: Foundation for the Future
von Braun, Wernher. 1971. Introduction to Kurd Lasswitz, *Two Planets* [*Auf zwei Planeten*], trans. Hans H. Rudnick, Carbondale, ILL: Southern Illinois University Press, epigraph.
Wabbel, Tobias. 2005. ed., *Leben im All: Positionen aus Naturwissenschaft, Philosophie und Theologie*, Düsseldorf: Patmos.
Wilson, E. O. 1998. *Consilience: The Unity of Knowledge*, New York: Knopf.

Chapter 20
The Impact of the Hubble Space Telescope

Abstract Although impact is an elusive concept, the Hubble Space Telescope has had a measurable effect on culture, not only through its imagery but also through its success in deepening our understanding of the universe and our place in it. Both science and culture would be poorer without its revelations.

20.1 The Idea of Impact

In asking what the Hubble Space Telescope has revealed about the universe and ourselves, we are really asking about its scientific and cultural impact. As we found when NASA and the National Air and Space Museum convened a joint conference on the societal impact of spaceflight a few years ago, the subject of such impact is rich and complex (Dick and Launius 2007). One can ask, for example, what does impact mean? Who is being impacted? What is the evidence that anyone is being impacted? And if there is an impact, individuals are undoubtedly affected in different ways, depending on their worldviews or individual interests and predispositions. Another way of approaching the subject in a more global sense is to ask the counterfactual question, where would we be today had there been no Hubble Space Telescope?

20.2 Looking Back

The authors of this valuable Hubble retrospective after almost two decades of service illustrate how the popular impact of the Hubble Telescope is intertwined with the scientific impact (Launius and DeVorkin 2014). While world-class science is clearly the primary purpose of Hubble, the strong popular interest has been continually reaffirmed through two decades, and was highlighted by the public outcry when

---

First published as the Introduction to Part 3 of *Hubble's Legacy: Reflections by Those Who Dreamed it, Built It, and Observed the Universe with It*," Roger Launius and David DeVorkin, eds. (Smithsonian Institution Scholarly Press: Washington, DC, 2014), pp. 74–78.

S. J. Dick, *Space, Time, and Aliens*, https://doi.org/10.1007/978-3-030-41614-0_20

Administrator Sean O'Keefe cancelled the fifth servicing mission in 2004, only to have it restored later by Administrator Mike Griffin (Dick 2014, chap. 21). As Ken Sembach, the head of the Hubble Mission Office at the Space Telescope Science Institute, has emphasized (Sembach 2014), Griffin's action extended Hubble's lifetime by many years, continuing the spectacular results to which scientists and the public have become accustomed.

Most of Hubble's popular impact is undoubtedly due to the its imagery, and Zolt Levay—the imaging group leader for Hubble, has given us an inside look at how these images are manipulated for aesthetic appearance, while maintaining their scientific integrity (Levay 2014). Elizabeth Kessler has also studied the aesthetics of the Hubble images, and their similarity to late nineteenth-century landscapes of the American West (Kessler 2006, 2008, 2014). The importance of imagery can also be seen by comparing Hubble with the second of NASA's Great Observatories, the very productive but imageless Compton Gamma Ray Observatory. While there is no doubt that Compton produced world-class science with BATSE, EGRET, and its other instruments in terms of advancing gamma ray astronomy, most of the public has never heard of Compton precisely because its data output was not amenable to aesthetic presentation.

On the other hand, of the three Great Observatories still operating, the Chandra X-ray Observatory and the Spitzer (Infrared) Space Telescope do not seem to evoke the same reaction as Hubble, despite the striking images they produce at their respective wavelengths. This is undoubtedly due to multiple factors: Hubble was the *first* of the Great Observatories to return stunning images of the universe at large, it has enjoyed more than two decades of longevity in the popular imagination thanks to its unique servicing missions, and it boasts an unrivalled public relations effort. We need only recall that the famous Eagle Nebula, with its "pillars of creation," evoked an almost religious response in some people, and many of Hubble's other images are not far behind in their emotional impact (Figs. 20.1 and 20.2). It is difficult to measure whether or not such images actually affect individual worldviews by bolstering theological convictions or simply enhancing understanding of the universe of which we are a part, just as it is difficult to measure the impact of the Blue Marble and Earthrise images from the Apollo era. But judging by their public interest and staying power, all of these images have had their impact, and have enhanced the very idea of what we call culture.

While such images are certainly evocative from an aesthetic point of view, it is their scientific content that draws us into a deeper and more intimate understanding of the universe and our place in it. Hubble's former Senior Project Scientist, David Leckrone, has highlighted some of that science and detailed why Hubble has been so successful (Leckrone 2014). Among the factors he enumerates are not only Hubble's increase in sensitivity and resolution over a broad range of wavelengths, but also its ability to evolve with technological advances through five servicing missions. As he points out, since the telescope optics were repaired in 1993 Hubble discoveries have been consistently ranked in the top tier of scientific discoveries in any given year. Those discoveries include its participation in uncovering the

Fig. 20.1 The Cone Nebula, a star-forming pillar of gas and dust within star cluster NGC 2264 in the constellation Monoceros, has inspired some people to suggest it as an image of Jesus. It was taken in April 2002 with Hubble's Advanced Camera for Surveys (ACS). (Credit: NASA, H. Ford [Johns Hopkins University], G. Illingworth [University of California at Santa Cruz, Lick Observatory], M. Clampin [Space Telescope Science Institute (STScI)], G. Hartig [STScI], the ACS Science Team, and European Space Agency)

acceleration of the universe and the implied presence of a mysterious *dark energy*, confirming the existence and elucidating the nature of supermassive black holes, actually imaging protoplanetary systems known as *proplyds*, direct imaging of extrasolar planets, and numerous results flowing from the several Hubble Deep Field projects (Space Telescope Science Institute 2005; DeVorkin and Smith 2008). Hubble is the example par excellence of telescopes as "engines of discovery" (Smith 1997).

Fig. 20.2 The Whirlpool Galaxy (M51) and its companion, NGC 5195, as imaged with Hubble's Advanced Camera for Surveys in January 2005. As a classic "grand design" galaxy, a term that has strangely become popular among astronomers since the 1970s, it exhibits strongly defined and articulated spiral arms that have been described poetically as a grand spiral staircase sweeping through space. The Whirlpool's arms may have resulted from a close encounter with the small galaxy NGC 5195, now situated at the outermost tip of one arm and evidently still perturbing it. Galaxy NGC 5195 has been passing behind the Whirlpool for hundreds of millions of years and is a photogenic example of what has come to be understood as an important mechanism of galaxy evolution: collisions between galaxies. (NASA image; credit: NASA, ESA, S. Beckwith [STScI], and the Hubble Heritage Team [STScI and Association of Universities for Research in Astronomy])

20.3 Humanity's Place in the Universe

Hubble, as well as the other Great Observatories and spacecraft like COBE, WMAP, and Planck, all bring us to a more definitive and robust realization of our place in the universe, not only in space but also in time, in the 13.7 billion years of cosmic evolution. In parallel they have demonstrated in ever more detail how we originated from "star stuff," as astronomers Harlow Shapley and Carl Sagan were fond of saying (Palmeri 2009). In this respect Hubble and other space probes are contributing to what we might call "Genesis for the third millennium," the knowledge of our ancestry in the wake of the Big Bang. As we discern this epic of evolution, the ultimate master narrative of the universe, it increasingly impacts culture in numerous ways (Dick 2009). It forms the basis for the new field of Big History, which places humans in a cosmic context (Christian 2004; Spier 1996); it is increasingly used in educational curricula; and is even finding a central role in the burgeoning concept of religious naturalism (Goodenough 1998; Hogue 2010).

Where all this will lead in the future we cannot say, but like Hubble's unexpected discoveries, I would suggest that the full impact of Hubble on culture is as yet unknown and partly unknowable. Cosmos and culture are becoming increasingly intertwined, and Hubble, through both its aesthetic images and its scientific data, will have played a central role in this process.

Where would we be today had the Hubble Space Telescope never existed? In short, much poorer in both science and culture.

20.4 Commentary 2020

This chapter was written as a brief introduction to the societal impact section of a volume on the legacy of the Hubble Space Telescope (Launius and DeVorkin 2014). It is based on a symposium held at the National Air and Space Museum in the Fall of 2009 on the occasion of the 20th anniversary of the telescope's launch. That museum, located in Washington, DC, and one of the most visited in the world, has a close association with NASA and is the repository of many of its artifacts. The volume is edited by two of its long-time curators, former NASA Chief historian Roger D. Launius and historian of astronomy David DeVorkin. It is notable for its contributions by historians, scientists and administrators closely associated with the telescope, as well as John Grunsfeld, one of the astronauts who led many of Hubble's five servicing missions.

Along with the triumphs of the "people's telescope," the authors of the volume describe the difficulties of the telescope's 20-year gestation, the heartbreak when it was discovered to have a serious optical flaw, and the fight over the final servicing mission that has extended its lifetime to the present day, as described in the next chapter. In one of the triumphs of the American space program, after much derision from politicians and others, astronauts fixed the optics and extended its lifetime through a series of unprecedented servicing missions that form one of the high marks of the Space Shuttle, which had originally launched Hubble in 1990. In fact, the telescope is characterized in this volume as a "serviceable national observatory," placing it in the long tradition of national observatories described in Chaps. 24 and 25. By the way, the Space Shuttle *Discovery* now sits in the Udvar-Hazy Center of the National Air and Space Museum adjacent to Washington Dulles Airport.

References

Christian, David. 2004. *'Maps of Time': An Introduction to 'Big History'*. Berkeley: University of California Press

DeVorkin, David, and Robert Smith. 2008. *Hubble: Imaging Space and Time*. Washington: National Geographic

Dick, Steven J. 2009. "Cosmic Evolution: History, Culture and Human Destiny," in Dick and Lupisella (2009), 25–59.

Dick, S. J. 2014. "The Decision to Cancel Space Shuttle Servicing Mission 4 (SM4) of the Hubble Space Telescope," in Launius and DeVorkin (2014), 151–190.

Dick, S. J. and Roger D. Launius. 2007, *Societal Impact of Spaceflight*. Washington, DC: NASA SP-2007-4801.

Dick, Steven J. and Mark Lupisella, eds., 2009. *Cosmos and Culture: Cultural Evolution in a Cosmic Context*. NASA SP 2009-4802.

Goodenough, Ursula. 1998. *The Sacred Depths of Nature*. Oxford: Oxford University Press

Hogue, Michael. 2010. *The Promise of Religious Naturalism*. Rowman & Littlefield.

Kessler, Elizabeth. 2006. "*Spacescapes: Romantic Aesthetics and the Hubble Space Telescope Images*," University of Chicago dissertation.

Kessler, Elizabeth. 2008. "The Wonders of Outer Space," in DeVorkin and Smith (2008), 136–163, especially 150–151

Kessler, Elizabeth. 2014. "Displaying the Beauty of the Truth: Hubble Images as Art and Science," in Launius and DeVorkin (2014), 120–130.

Launius, Roger and David DeVorkin, eds. 2014. *Hubble's Legacy: Reflections by Those Who Dreamed it, Built It, and Observed the Universe with It*, Smithsonian Institution Scholarly Press: Washington, DC.

Leckrone, David S. 2014. "The Secrets of Hubble's Success." In Launius and DeVorkin (2014), 93–111.

Levay, Z. 2014. "Creating Hubble's Imagery," in Launius and DeVorkin (2014), 112–119.

Palmeri, J. 2009. "Bringing Cosmos to Culture: Harlow Shapley and the Uses of Cosmic Evolution," in Dick and Lupisella (2009), pp. 489–521.

Sembach, Kenneth R. 2014. "Recommissioning Hubble: Refurbished and Better than Ever," in Launius and DeVorkin (2014), 79–92.

Smith, Robert, 1997. "Engines of Discovery: Scientific Instruments and the History of Astronomy and Planetary Science in the United States in the Twentieth Century," *Journal for the History of Astronomy*, 32, 49–77.

Space Telescope Science Institute. 2005. "Hubble Space Telescope's Top Ten Greatest Achievements," released on Hubble's 15th anniversary, April 25, 2005, online at http://hubblesite.org/newscenter/archive/releases/2005/12/background/

Spier, Fred. 1996. *The Structure of Big History: From the Big Bang Until Today*. Amsterdam: Amsterdam University Press

Chapter 21
The Decision to Cancel the Hubble Space Telescope Servicing Mission 4 (SM4) and Its Reversal

Abstract On January 16, 2004, NASA Administrator Sean O'Keefe announced his decision to cancel the Hubble Space Telescope (HST) Servicing Mission (SM4) by the Space Shuttle. SM4 was to have inserted two new instruments, the Wide Field Camera 3 and the Cosmic Origins Spectrograph, at the same time replacing the batteries and gyroscopes, extending Hubble's lifetime to 2010. The decision resulted in a strong reaction among some members of Congress, the HST science community, and the general public, because it would likely leave the telescope inoperable by 2007, years before its full lifetime and well before the James Webb Space Telescope (JWST) would be launched. In the immediate aftermath of his decision to cancel the final servicing mission, Administrator O'Keefe requested an independent study be undertaken by the author in his role as NASA Chief Historian, in order to document in detail the events that led to the cancellation decision. What follows is a history of that decision and its aftermath, completed December 17, 2004, as well as an Epilogue added in 2012 describing the reversal of O'Keefe's decision by NASA Administrator Michael Griffin.

21.1 Background

After a long history of concept, design and construction stretching back to 1965 (Smith 1989), the Hubble Space Telescope was launched April 24, 1990 (Table 21.1). Scheduled for launch in late 1986, it had been delayed by the Space Shuttle Challenger disaster in January of that year. Although there had been other successful telescopes in space, notably the Orbiting Astronomical Observatories II and IV (Copernicus) in the 1960s and 1970s, Hubble, with its 2.4-m mirror and $1.3 billion price tag, was in a different league. Disappointment was therefore acute when it was discovered shortly after launch that spherical aberration in the mirror made the Hubble images blurry, greatly limiting its scientific capacity. The press had a field

First published in *Hubble's Legacy: Reflections by Those Who Dreamed it, Built It, and Observed the Universe with It*," Roger Launius and David DeVorkin, eds. (Smithsonian Institution Scholarly Press: Washington, DC, 2014), pp. 151–189.

S. J. Dick, *Space, Time, and Aliens*, https://doi.org/10.1007/978-3-030-41614-0_21

Table 21.1 Timeline of events related to Hubble Space Telescope Servicing Mission 4

1990	April 24	Hubble Space Telescope launched
1993	December	SM1
1997	February	SM2
1999	December	SM3A
2002	March	SM3B
2003	February 1	Shuttle Columbia accident; Columbia Accident Investigation Board formed
	June 13	Stafford-Covey Return to Flight Task Group formed
	August 26	Columbia Accident Investigation Board (CAIB) issues report
	November 7	Weiler briefs O'Keefe on options for dates for SM4
		Thanksgiving Decision not to include SM4 in FY05 budget
	December 2	Isakowitz rolls out budget at Executive Committee meeting; SM4 not included
2004	January 14	President Bush announces new Space Exploration Vision at HQ
	January 15	*Washington Post* article mentions SM4 cancellation
	January 16	**O'Keefe announces SM4 cancellation at GSFC meeting**
	July 13	National Academy of Sciences Committee on Assessment of Options for Extending the Life of the HST Interim Report
2005	February 11	Administrator O'Keefe's resignation effective
	April 13	Michael Griffin becomes new NASA Administrator
2006	October 31	**Administrator Griffin reinstates SM4 mission; launch scheduled for September 11, 2008**
2009	May 11	SM 4 launches aboard shuttle Atlantis, STS-125; achieves all goals during 13-day mission

day ridiculing NASA and its engineers, a situation that was not helped when the subsequent investigation discovered that faulty testing of the mirror had been the culprit.

21.1.1 Hubble Servicing Missions

Ever since it became clear that it would be launched with the Space Shuttle rather than a Titan III rocket, Hubble's fortunes had been bound up with human spaceflight. The good news in Hubble's bleak situation after launch was that it had been designed to be serviced. The triumph was all the greater when, in December, 1993, the first Hubble servicing mission (SM-1) succeeded in placing corrective optics into the telescope, rendering its new images perfect (Fig. 21.1). It was not only vindication for Hubble, but also for the concept of human servicing. So high were the stakes that some called it a "save NASA" mission (Grunsfeld 2004, p. 2). Over the next decade three more servicing missions followed. SM-2, carried out with the Shuttle Discovery during STS-82 in 1997, was the highest Shuttle flight, at an altitude of some 386 miles (Fig. 21.2). It was this mission that President Bush indirectly referred to in his 2004 space exploration speech, when he cited 386 miles as

Fig. 21.1 The Eagle Nebula is a Hubble Space Telescope image in the visible portion of the spectrum, showing "pillars of creation" in a star-forming region. These eerie, dark, pillar-like structures are columns of cool interstellar hydrogen gas and dust that are also incubators for new stars. The image was taken on April 1, 1995. Jeff Hester and Paul Scowen (Arizona State University) and NASA

the furthest humans had been from Earth since the last Apollo mission in 1972, a quarter century earlier. On this mission the NASA Goddard High Resolution Spectrometer and Faint Object Spectrograph were replaced by the Space Telescope Imaging Spectrograph (STIS) and Near Infrared Camera and Multi-Object Spectrometer (NICMOS).

What was to be the third Hubble servicing mission was broken into two missions, SM3A and SM3B, later causing some confusion among the media and public with regard to the number of servicing missions. SM3A, carried out in late 1999 with the Shuttle Discovery during STS-103, took place under urgent conditions, and was moved up in the schedule to accomplish that part of the original SM3 mission that needed to be done immediately. The telescope itself was in safe mode, its gyros

Fig. 21.2 Hubble Space Telescope after deployment on the second Hubble Servicing Mission. February 19, 1997. NASA

having failed, and the servicing mission had to be accomplished before the end of the year because of Y2K software fears. The crew successfully installed new gyroscopes and scientific instruments, and the telescope was redeployed on Christmas day. SM3B, the fourth Hubble servicing mission, was carried out in March, 2002 during the Columbia STS-109 flight. It installed a new digital camera, a cooling system for the infrared camera, new solar arrays, and a new power control unit. The last was a particular triumph, since it went beyond the normal servicing requirements. Payload Commander John Grunsfeld recalled:

> Nobody believed we could necessarily do that; this is a big switch box, lots of connectors, all the power runs through it, and there was a problem with it that would, gone unchecked, have terminated Hubble's life early probably in the 2005 to 2008 timeframe. And we took that issue all the way to the Administrator, at that time Dan Goldin, and said this is a tough one; if we try this and it doesn't work we lose Hubble; if we don't try it we'll probably lose Hubble. And it's well beyond the limits of any kind of EVA that's ever been done, harder,

Fig. 21.3 NASA Administrator Sean O'Keefe with the author on the ninth floor of NASA Headquarters in Washington, DC, near the Administrator's office

> longer and it involves significant risk to the telescope. And Dan Goldin looked at me straight in the eyes and said, "Well John, do you think we can do it?"

Grunsfeld answered in the affirmative, and though he characterized it as "the most challenging space walking activity we've ever done in the space program," it proved very successful (Grunsfeld 2004, p. 5). As it turned out, Grunsfeld was the last person to touch the Hubble Space Telescope.

21.1.2 Sean O'Keefe

There was another novelty to the SM3B mission. After a record 10 years as NASA Administrator, Dan Goldin had left the agency the previous November. STS-109, with its Hubble servicing mission, was the first opportunity for his successor, Sean O'Keefe, to witness a Shuttle launch. O'Keefe (Fig. 21.3) had joined the administration of George W. Bush on inauguration day, and served as Deputy Director of the Office of Management and Budget (OMB) until his appointment as NASA Administrator on December 21, 2001. It was his fourth Presidential appointment, having also served as Comptroller and Chief Financial Officer of the Department of the Defense (1989) and Secretary of the Navy (1992). He had also served for 8 years on the U. S. Senate Appropriations Committee staff, and as the Louis A. Bantle

Professor of Business and Government Policy, an endowed chair at the Syracuse University Maxwell School of Citizenship and Public Affairs. With this background, O'Keefe was in a strong position to bring NASA's budget under control, in particular cost overruns on the International Space Station, which had subjected NASA to severe Congressional criticism during the Goldin years. And with STS-109 as his first Shuttle launch, O'Keefe was well aware of the importance of Hubble servicing missions from the beginning of his tenure.

21.1.3 The Columbia Accident Investigation Board (CAIB) and the Stafford-Covey Return to Flight Task Group

The next servicing mission, designated SM4, was to have been carried out in November, 2004, but disastrous events intervened on February 1, 2003 with the catastrophic loss of Columbia and its crew. Administrator O'Keefe was at Kennedy Space Center waiting for the landing, which never came. Shortly after the planned landing time of 9:16 am he declared a Shuttle Contingency, and the Action Plan for Space Flight Operations was implemented. Within hours of the accident he appointed an Investigation Board, named the following day the Columbia Accident Investigation Board, to be chaired by Admiral Harold W. Gehman, Jr. Gehman was a retired four star Admiral who had served as the NATO Supreme Allied Commander, Atlantic; commander in Chief of the U. S. Joint Forces Command, and Vice Chief of Naval Operations for the U. S. Navy. He had co-chaired the DoD review of the attack on the U. S. S. Cole. The Columbia Accident Investigation Board was charged with investigating the facts and probable causes of the accident, and with recommending "preventative and other appropriate actions to preclude the recurrence of a similar mishap" (CAIB 2003, p. 231). After a seven-month investigation, the Board issued its report August 26, 2003. Among the many recommendations was the following: "For non-station missions, develop a comprehensive autonomous (independent of Station) inspection and repair capability to cover the widest possible range of damage scenarios"(CAIB 2003, p. 225). Although HST was not mentioned by name, the only post-Columbia missions that would not fly to ISS were the servicing missions to HST. As with all of the recommendations, O'Keefe was to take this one very seriously.

Meanwhile, on June 13, 2003, O'Keefe established the Return to Flight Task Group, whose charge was to implement the recommendations of the CAIB Report. Chaired by two veteran astronauts, Thomas P. Stafford and Richard O. Covey, the group would undertake numerous fact-finding visits, public meetings and media teleconferences. Most importantly, it produced *NASA's Implementation Plan for Return to Flight and Beyond*, a "living document" first released on September 8, followed by interim reports in January and May, 2004 (NASA 2003a). The recommendations of the CAIB report were the benchmarks against which NASA's progress would be monitored on a point-by-point basis in return to flight meetings held by the Stafford-Covey group and at Headquarters. Those meetings would play a crucial role in the Hubble SM4 decision.

21.1.4 Other Studies Related to Post-Columbia HST

In addition to the recommendations of the CAIB report, several studies were chartered to analyze the future of HST, and these naturally had to take into account the impact of the Columbia accident. The first arose from NASA Congressional appropriations language in February, 2003: "The conferees direct NASA to carry out an in-depth study of an additional servicing mission (SM5) in the 2007 timeframe that would study operating HST until the Webb Telescope becomes operational. The study should address the costs of an additional servicing mission and the potential scientific benefits." This "HST Post SM4 Scientific Review Panel," as its name implied, was to deal with longer term issues. Also termed the "Black Commission" after its Chair, David Black, in April, 2003 the Commission assumed that SM4 would be conducted in the 2004–2005 timeframe. It concluded that HST would continue to provide high quality science even beyond the time of a proposed SM5, but foresaw budgetary and technical problems with a servicing mission in the 2007 time frame (NASA 2003b).

In June the Office of Space Science, realizing that "it is a necessary task to consider exactly how and when to terminate the operation of this successful scientific experiment," chartered the HST-JWST Transition Plan Review Panel, chaired by John Bahcall, to evaluate the scientific impact of the current NASA plan for ending HST operations and beginning James Webb Space Telescope operations. That plan called for the end of Hubble operations in 2010, and the launch of JWST in late 2011. In August the Panel provided three options in priority order: (1) Two additional Shuttle servicing missions, SM4 in about 2005 and SM5 in about 2010, in order to maximize the scientific productivity of the Hubble Space Telescope. The extended HST science program resulting from SM5 would only occur if the HST science was successful in a peer-reviewed competition with other new space astrophysics proposals. (2) One Shuttle servicing mission, SM4, before the end of 2006, which would include replacement of HST gyros and installing improved instruments. In this scenario, the HST could be de-orbited, after science operations are no longer possible, by a propulsion device installed on the HST during SM4 or by an autonomous robotic system. (3) If no Shuttle servicing missions are available, a robotic mission to install a propulsion module to bring down the HST in a controlled descent when science is no longer possible (NASA 2003c).[1] The conclusions of the report were endorsed by the American Astronomical Society, which strongly urged that whatever support was needed for SM4 be found, consistent with CAIB recommendations.

The Bahcall panel reported its conclusions in mid-August. Less than 2 weeks later the Columbia Accident Investigation Board issued its report, and it was this report and return to flight issues that were destined to have the greatest impact on the final decision to cancel SM4. That decision that would eventually be made was similar to the third and last priority in the Bahcall report.

21.2 The Decision

In the minds of several of the key players in the decision, the first thought that SM4 might be canceled dated to the Columbia disaster itself. As Ed Weiler, Associate Administrator for Space Science put it, "I got a first inkling that the servicing program in general was in trouble on February 1, 2003 when I turned on CNN in the early morning and saw what was an unmistakable signature of a spacecraft breaking up in front of my eyes … and I knew at that point that if it was what I thought it was, which was the destruction of the shuttle, that this would portend poorly for future shuttle flights to orbits like Hubble's orbit. I was certainly worried about it." Along with concern for the astronauts, it was natural for Weiler to think about the ramifications for the HST, which came under his Office of Space Science. The same thought must have been in the minds of the other Hubble managers also, since there was no way to service the HST without the Shuttle. Everyone knew the Challenger accident had caused a long delay in return to flight. Fortunately, at the time of the Columbia accident, Hubble had been serviced less than a year earlier; still its batteries and gyros would inexorably wear out, and there was no doubt of the importance of timeliness for another servicing mission. A post-Columbia return-to-flight date would depend on the course of the investigation and the cause of the accident, and in this respect the recommendations of the CAIB report would assume utmost significance.

21.2.1 Role of the CAIB Report and Return-to-Flight Meetings

Administrator Sean O'Keefe recalled that for him, the decision process for SM4 began in a serious way

> on August 26, 2003, when the Columbia Accident Investigation Board released its report. So we started looking through all those challenges, consistent with all the Return to Flight [RTF] activities we were engaged in as early as March/April 2003, when the formal kind of framework got kicked off lining up an RTF process. It wasn't directly informed by all the recommendations, findings and observations until the 26th of August. Then thereafter each step along the way we were formulating a regular assessment that began in September 2003 of what it would take in order to implement those recommendations to return to flight. And as each mounting month went by at every update of the Return to Flight document … every one of those reveals it is harder and harder and harder to accomplish every one of those recommendations to achieve that objective. So I think by the late fall, early winter, it was pretty apparent that our likelihood of accomplishing all those objectives in time to mount a servicing mission that would be in compliance with all those recommendations, was becoming more and more remote. (O'Keefe 2004a, 1–2)

Bill Readdy, the Associate Administrator for Spaceflight who had himself flown three Shuttle missions between 1992 and 1996, also pinpointed the CAIB report as the event that triggered serious discussion about the Hubble mission. Although the CAIB investigators "were not saying that we couldn't fly it if we developed stand-alone autonomous inspection and repair capability … the bit was pretty much set in my mind that this was going to be a very, very high bar set to ever go do a Hubble Servicing Mission." More broadly, Readdy was struck by the CAIB's finding "that NASA's not a learning organization. That NASA failed to completely follow up on the Challenger recommendations. I was left with a clear impression that, yeah, we could proceed at risk to go off and do another Hubble servicing mission, but that would also be conclusive proof that NASA hadn't learned anything from Columbia … or from Challenger by implication" (Readdy 2004, 6).

The return-to-flight meetings made it clear that there were numerous obstacles to a quick resumption of Shuttle flights. The prime objective was not speed but safety, and that meant at a minimum satisfying all of the CAIB recommendations.

21.2.2 Assessing SM4 Options

In the wake of the Bahcall and CAIB reports, in the midst of the much broader and still moving target of a return-to-flight date for the Shuttle, the assessment of an SM4 decision continued, both within the Office of Space Science and at higher levels. By early November it was still more a question of when, not if, such a mission would occur. On November 7, Weiler presented O'Keefe with the advantages and disadvantages of dates for a servicing mission ranging from 2005 to 2007. They talked about how long the gyros would last, and Weiler recommended one more servicing mission, but not another one (SM5) beyond that:

> I pointed out that if you know you are only going to have one more gas stop and you want to go as many miles as you can, do you fill your gas tank up when it is half full or do you wait until you're on fumes? That is the argument that said, if you wanted to wait until you were on fumes you would probably go to maybe 2007, but that was pushing the envelope, so we centered on the optimum time … of the U. S. Core Complete for [the International Space] Station, which would be around June 2006, and that is where the June of 2006 came from. If we planned a servicing mission June of 2006, it wouldn't impact finishing off the station, Core Complete U.S., making sure we got it restocked with water and food.

According to that brief, the latest useful SM 4 mission would have been 2008, but a reliable restart of the spacecraft would have been in doubt by then. And there was a final option listed: "No SM4." A backup slide—to be used only if necessary—gave the story of HST without SM4: experience showed that ways would be found to extend its life; thousands of archival images existed that astronomers could still study, and savings could benefit other programs. But the slide was either not used, or in any case did not carry the day: "We gave that presentation to him and we said the way we would dispose of Hubble is we wouldn't plan a shuttle anymore because obviously that would be crazy. We would build a robotic thing to grab it and take it to the Pacific and he approved that. I left that meeting ... feeling like we were on the road to an SM4" (Weiler, 12–13; NASA 2003d).[2]

21.2.3 The NASA Fiscal Year 2005 Budget

As the return-to-flight meetings were proceeding and HST managers were assessing their options, the FY2005 NASA budget was being prepared. Per the usual procedure NASA submitted its budget to the Office of Management and Budget in September, and OMB gave NASA its "pass back" with revised numbers in November. Thanksgiving weekend saw NASA Comptroller Steve Isakowitz, O'Keefe and others finalizing the budget to go back to the White House to get the President's approval before it went on to Congress. February first was the traditional day when the final budget was rolled out; ironically, it would be the first anniversary of the Columbia accident.

SM4 had budget implications; if it were going to be in the 2005 budget, "offsets" needed to be found in other areas of space science, something that the Office of Space Science was perfectly willing to do. But, Isakowitz recalled,

> the problem was that if you went strictly by what came out of the CAIB recommendations in terms of the ability to inspect and repair and safe haven, we had no known way to do it. So we can go ahead and budget for a date, but then the question becomes when would we actually know that we could fly it? ... As we began to ask questions like that, even then it became clear that we wouldn't know maybe until the last minute whether or not we could actually do such a mission. Yet in the meantime, we are going to have to spend lots of money and keep it all going.

By this time, then, the RTF implications for a Hubble mission were coming to the fore. Isakowitz left the Thanksgiving weekend meeting with O'Keefe with a tentative decision that SM4 would not be in the budget. It was, he said, one of "a million other decisions" on the list. Asked if SM4 was a budget issue, Isakowitz said "No, the only reason I would say it is tied to the budget was the budget helped to dictate the timing of when we were going to make a decision." Elaborating further, he noted

> That is what the budget process does. When you have issues, even if it has nothing to do with the budget ... the budget process forces people to make decisions. ... The budget dictated the schedule as to when the decisions were going to be made. For those who ... still argue that this was a budget decision, we cut the Hubble to pay for the vision, that is just simply not true. We would have found the money to do the Hubble. (Isakowitz 2004, 10, 19, 25, 28)

As O'Keefe put it referring to the Thanksgiving weekend meeting,

> The choice was you either had to put the resources in to continue planning for that mission through FY05, or not. And it finally got down to the point where the act of leaving it as it was would have signaled improperly that we had planned to do a mission that I had come to the conclusion that I didn't think we were likely to be able to do ... Could have been delayed ... but in the end ultimately it would have had to be manifested in that way to make a decision. So it was not a question of whether you put how much in, it was a question of

> whether you put anything in … I realized at that stage of the game that if I did not make that decision at that time it would be potentially another year that we would maintain the fiction that we could do this mission.

O'Keefe called that meeting "a prompting event," a way of forcing him to make a decision, but added that in the end it was based on the unlikelihood of meeting the CAIB recommendations before the predicted turnoff of Hubble (O'Keefe 2004b, 2, 4).

21.2.4 The Decision Is Made

Asked the date when the final decision to cancel SM4 was made, O'Keefe said it "probably converged around the early part of December," after the return to flight meetings showed more and more clearly that it could not be done in time to save HST. It was at a crucial December 2 meeting of the Executive Committee, where Isakowitz briefed NASA Associate Administrators on the 2005 budget submission, that it first became clear at the Associate Administrator level that the SM4 mission was not in the budget. "That was the first time I saw that SM4 was cancelled and that was the first time anybody in that room other than Steve [Isakowitz], I guess, and Sean, knew that SM4 was cancelled, so I had to react in real time," Weiler recalled. Asked if he felt he was not consulted Weiler replied,

> No, because I could have stood up at that meeting. Nothing was published at that point in time. I could have said I object. I think it is safe. I think the science is worth it, but that would be disingenuous of me because I don't know if it is safe or not. I'm not a safety engineer. I think it is very important for people to recognize their own limitations. I'm going to be an enemy of the scientific community because of this. I could get up there and be on my high pulpit and say damn with safety, we have to go fix the Hubble because it is the greatest scientific thing since sliced bread. I could say that but that is the easy way out. That is the easy way out, hide behind the science. (O'Keefe 2004a, 3–5; Weiler 2004, 4–7; Isakowitz 2004, 19, 27–28)

Still, it had to be a difficult decision for Weiler, who had been associated with the HST project for 25 years.

> Anybody who says I take this lightly is missing the point. I am taking it rationally not lightly. I cannot stand up and say that the science justifies additional risk. I don't know how to quantify science in those terms. Human life is too valuable … I wouldn't want to have to explain to a four-year-old boy why he will never see his dad again, or his mom. That has to be the position Sean was in. That is a serious position to be in. It is a lot different than sitting in an ivory tower university making pronouncements about how valuable the science is. That is as blunt as I get. (Weiler 2004, 15–16)

O'Keefe later confirmed this aspect of his thinking, when responding to what some called the "withering" criticism of the SM4 cancellation: "Let me offer my view of 'withering,'" he said:

> Withering is the feeling you get when you are standing at a runway with the dawning realization that the Shuttle everyone is waiting for isn't going to land. Withering is when you

> have to explain to wives, husbands, parents, brothers, sisters and children that their loved ones aren't coming home alive. Withering is attending funerals, memorial services, and ceremonies over 16 months in number too many to count any more, yet having every single one of these events feel like the weight of that responsibility will never be relieved. Withering is the knowledge that we contributed to the Columbia disaster because we weren't diligent. (O'Keefe 2004c)

In O'Keefe's estimation, every further return to flight meeting confirmed the wisdom of the SM4 cancellation decision. In particular the RTF meeting at Johnson Space Center on December 12, following the Stafford-Covey Task Group fact-finding visit there the previous 3 days, confirmed that the CAIB recommendations were not likely to be met by the hoped-for September–October timeframe (NASA 2003e). At about the same time, the Space Flight Leadership Council (the spaceflight community) concluded that return to flight would not occur in September–October of 2004, but would likely slip to March/April of 2005. As O'Keefe recalled, "All those events were converging in that few weeks span of time, and looking more and more and more apparent that the likelihood of return to flight in a timely manner was remote and therefore even more so remote that you'd be able to mount a servicing mission unique to Hubble." On December 19, during a brief on the still unannounced new exploration strategy for NASA, O'Keefe informed the President that the HST mission was not going to happen. The President agreed that compliance with the CAIB recommendations was paramount (O'Keefe 2004a, 3–6).[3]

21.2.5 Plans for Announcing the SM4 Cancellation

Planning for the SM4 cancellation announcement fell to NASA's Chief Scientist, John Grunsfeld. Curiously, as January began Grunsfeld had little idea what was about to happen to HST. An astronaut who had participated in the last two Hubble servicing missions (SM3A and SM3B in 1999 and 2002), prior to becoming NASA's Chief Scientist in September, 2003, Grunsfeld had been leading the activity for SM4 at Johnson Space Center, home of the astronauts. In the summer he had testified before the Bahcall group, saying there was astronaut consensus that SM4 was one of the missions "worth risking our lives for ... really important for humans to do ... the marriage of human spaceflight and robotic science spaceflight" (Grunsfeld 2004, 7). By contrast, the astronaut office was not on board for risking lives for any mission to bring the HST back for the National Air and Space Museum.

Throughout the fall, in his position as Chief Scientist, Grunsfeld discussed with Anne Kinney, head of the Astronomy and Physics Division of the Office of Space Science, the details of carrying out SM4. Neither had any inkling it might be cancelled except for the general rule that no mission was secure until it actually flew. At the same time he had urged the community to concentrate on SM4 rather than worrying so much about SM5. Although Grunsfeld had gotten a faint signal from Isakowitz during the OMB budget pass back around November 28 that SM4 might not be in the budget, only on January 7 was he informed in an abrupt way. The previous day Grunsfeld was at the winter meeting of the American Astronomical Society in Atlanta when he got a Blackberry message inviting him to a senior staff meeting the following day to discuss HST Servicing Mission timing.

Grunsfeld immediately flew back to Washington from Atlanta. He had assumed the meeting was to discuss the timing for SM4 in the flight manifest, but when he walked into the meeting it was clear that decision to cancel the servicing mission had already been made and the discussion was how to roll out the decision to the public. Grunsfeld was stunned; he "literally felt like somebody hit me in the head with a two-by-four" (Grunsfeld 2004, 11). Moreover, because Ed Weiler's Office of Space Science was about to land two rovers on Mars, Grunsfeld was given the unhappy task of coming up with a plan of how to roll the decision out to the public. Grunsfeld consulted with some of his mentors, including John Bahcall, as to whether he should even stay with NASA in the wake of such a decision on which he had not been consulted. He decided that matters might be worse for HST if he left. Thus, over the course of several senior staff meetings he laid out a plan that would be rolled out on January 28, a few days prior to release of the President's budget, at a press event that O'Keefe would lead. Prior to that the HST principals would be informed in an orderly way.

21.2.6 The Role of Probabilistic Risk Assessments vs. Intuition

Meanwhile, Grunsfeld went to his fellow astronaut Bill Readdy, the Associate Administrator for Spaceflight, the office in charge of Shuttle flights, looking for a Probabilistic Risk Assessment (PRA) that might document the risk. A PRA is a comprehensive, structured, and logical analysis method aimed at identifying and assessing risks in complex technological systems for the purpose of cost-effectively improving their safety and performance. It was a computer model tailored for each technological case, used for years in the nuclear industry, and since 1995 at NASA in relation to the Shuttle. As Brian O'Connor, Chief Safety and Mission Assurance Officer, put it, a PRA

> incorporates all the best technical know-how of your system, how it's hooked up, interrelationships between subsystems. For example in the model if you fail an electrical circuit, just take it out, and you can do this in these PRA models, you can fail things. Then it can have an effect on your thermal system and your navigation system, it takes away a leg of redundancy from your cooling loops and all the kinds of things because it's just a big software model of your system. And the way the Probabilistic Risk Assessment works is that it takes all of the best notions of your engineering and your safety and reliability community on failures and what chances they have of failing, and it factors in all these accident scenarios that could happen. (O'Connor 2004, 9)

A PRA was not comprehensive in every detail, however; while loss of thermal protection system was in the Shuttle model, the risk due to insulating foam from the external tank hitting the Shuttle was not.

In this case, no such analysis existed. According to O'Connor, the Shuttle PRA was going through a peer review and was not useable for testing this scenario. Even if it had been possible to compare the risks of an HST rescue mission to a space station mission, he noted, a PRA was only one piece of the puzzle:

> Far be it from me to ever suggest that anybody would ever make a decision like this just based on risk trade from a PRA, because I know that a PRA is limited as a model, it only looks at certain things. It doesn't look at some of those secondary things like the distraction factor of putting a different kind of mission, and all the planning that goes with it, in the middle of your Return To Flight activities to the station. You now have tasked your people to go worry about other things like how you do a shuttle to shuttle safe haven rescue, which you wouldn't worry about on Station. On Station you know how to hook the shuttle up to the space station to get the people out, but we've never thought much about how you would go up there and bring another shuttle up to a crippled shuttle and get the people out of one vehicle into the other, so a lot of work would have to be done there and there's risk inherent in that. It's not even in this model. (O'Connor, 2004, 15–16)

Grunsfeld came to understand that O'Keefe's decision was an intuition call: he had synthesized the RTF data and concluded that it was too hard. Asked whether his decision was intuitive, O'Keefe answered

> Absolutely, no question. But rather than calling it "intuitive grounds," I would say "intuitive" in the sense of confidence level and attaining the objectives of the Accident Investigation Board recommendations as a forecast in time. That part is intuitive; you can't analytically demonstrate whether you will or you won't … You kind of look at what the trend-line looks like at any number of things … so it is by nature more of an intuitive circumstance of where you see the trend going … it is driven by the analysis and the data and the information and the current status of our capacity to do things technically. (Grunsfeld 2004, 13–14; O'Keefe 2004b, 13–14)

21.2.7 The New Space Exploration Vision

Meanwhile, events were occurring that would have a profound effect on NASA's future.

In the wake of the Columbia, and especially after the CAIB report in August, the White House was planning a new exploration vision for NASA. In the summer of 2003 a White House interagency team began meeting to consider the options. Among the options was phasing out the Shuttle by 2010, something that could obviously impact HST.[4] All of this was coming to a head in January, 2004, at the same time that the senior staff meeting was being held to decide how to roll out the SM4 decision. On January 13th the Leadership Council, including Headquarters leaders and Center Directors, were briefed on the President's space exploration vision. Isakowitz presented the details of the budget implications, and although he did not mention HST, "it came up. There was some discussion at that point and that was the meeting at which people talked it through because that was where some people were hearing it for the first time" (Isakowitz 2004, 21).

On January 14, President Bush came to NASA Headquarters to announce the new space vision. It included retiring the Shuttle by 2010, abandoning the space station around 2016, and sending humans to the Moon by 2020 and to Mars by some unspecified date (NASA 2004a; Science 2004, 589, 610–619; Kluger 2004, 42–52). The same day a staffer from the White House went to Capitol Hill to brief staffers on the vision. When he talked about the Shuttle only flying until the end of the decade, someone asked about the implications for HST. The staffer said this meant cancellation of SM4, and on January 15, an article in the *Washington Post* mentioned this fact in passing, obviously having been leaked. In speculating on the possible implications of the President's vision, Kathy Sawyer wrote, "There may also be slowed growth in the NASA space science budget, sources said, and a 'refocusing' of activities within the agency to support the central theme of returning to the moon. There will be no further servicing missions to the Hubble Space Telescope. Though there is rampant speculation about closing NASA facilities and axing programs, there were few specifics" (Sawyer 2004; Grunsfeld 2004, 19; Isakowitz 2004, 26–27).

This "accidental release" unleashed a variety of charges—that HST was being sacrificed for the new vision, that it was a victim of budget cuts, etc. Grunsfeld admitted that it looked bad for NASA, "I think to the press that looks suspicious, Friday night late calls as if we were trying to pull a fast one; and from there it has been an uphill struggle" (Grunsfeld 2004, 20). He was right about that.

21.2.8 The Decision Announced

Grunsfeld's carefully crafted plan for the announcement was shattered. Worse than that, even some of the principals involved in HST servicing missions found out the hard way. Michael Moore, the Program Executive for HST at NASA HQ, heard it from his boss, Anne Kinney, on January 15, the day after the President's speech, and the same day as the *Post* article. Jennifer Wiseman, the Program Scientist for HST at NASA HQ, found out from Michael Moore via a telephone call very early that same morning of January 15 (Moore 2004, 36; Wiseman 2004, 15). Moreover, those who operated the HST at the Space Telescope Science Institute in Baltimore, and those who planned the servicing missions at Goddard Spaceflight Center (GSFC), were totally in the dark. On Friday morning, January 16, O'Keefe, Weiler and Grunsfeld made the short trip to GSFC, where they broke the news to the Hubble team. Among those present in the audience were Steven Beckwith, Director of the Space Telescope Science Institute (STScI), and Frank Cepollina, a 40 year veteran of NASA who had been in charge of all the HST servicing missions. Also present from Headquarters were Anne Kinney, Eric Smith, and Jennifer Wiseman. Trying to make the best of an admittedly bad situation, O'Keefe spoke for about 45 min, without notes, saying the decision was his alone. He asked the Hubble team to come up with creative ways to extend Hubble's life by increasing efficiencies in the batteries or gyros or both. The Administrator was followed by John Grunsfeld and Ed Weiler, who endorsed the decision. There were questions and answers, and the mood was somber. As Wiseman recalled, "It had the very same sort of funeralesque type of atmosphere where people were somber and yet they tried to comfort each other with small statements of comfort" (Wiseman 2004, 19, Beckwith 2004a).

The timing was unfortunate in terms of the President's space exploration vision announced 2 days before. As Isakowitz recalled, "It was a tough decision and in an ideal world actually, it would have been great if we could have deferred it, because the fact is that the Hubble decision really had … no specific link to the vision itself, but it was clear that if we were going to take a decision that said not to do it, it would cast a shadow on the vision." But, like the budget, Isakowitz insisted that the HST decision and the new space exploration vision were unrelated. Bill Readdy agreed: "They were totally decoupled, they really were" (Isakowitz 2004, 16; Readdy 2004, 14).

21.3 The Reaction

21.3.1 Media and Public Reaction

The immediate reaction to O'Keefe's hurried January 16 announcement at GSFC of the cancellation of SM4 was swift and overwhelmingly negative. Perhaps most surprising was the reaction from the media and the public, which had pummeled Hubble because of its problems 14 years earlier, but had now grown accustomed to the awe-inspiring pictures beamed down regularly from the orbiting telescope. NASA officials and the Space Telescope Science Institute received thousands of emails, some offering money. "The overwhelming amount of general public comment we've gotten is just sort of shock," Bruce Margon, associate director of STScI was quoted as saying in the *Washington Post*. "If it's working," people ask, "how can you possibly shut if off? I don't have an answer to that," Margon said. He announced a public web site to accept public suggestions for the Hubble (McBride 2004).

Some in the media immediately linked the HST decision to the President's new vision for space exploration. In its cover story announcing the new vision, *Time* magazine wrote, "The budgetary shake-up has already claimed a victim. The Hubble Space Telescope had been scheduled for a maintenance visit next year by space-shuttle astronauts. Now there is no money for the mission, and after 2010 there will be no shuttle anyway. One of NASA's greatest success stories, Hubble will probably wink out sometime in 2007" (Kluger 2004, 42–52). Like Readdy and Isakowitz, O'Keefe was adamant that the two were unrelated:

> No! It was reading in the *Washington Post* an article that made very scant reference to it … it wasn't even an essential piece of it, it was just kind of a throw away line. And I thought "well hell, that's it, this accelerates the whole thing." We ought to go out and describe this as quickly as possible … the last thing I wanted [was for] anybody to find out about this is from reading something in the paper. … So I called [Senator Barbara] Mikulski, made

> arrangements to go up to Goddard, talk to the Hubble team, that's what accelerated everything at that point because originally we were thinking about trying to organize something that would be in sequence sometime the last ten days or so of January, all that went up in flames on the basis of this. There was no relationship, association or decision about this as it related to the President's vision statement or anything else. No, no linkage at all. (O'Keefe 2004a, 9–10)

O'Keefe was not surprised about the public reaction, but he was surprised by "the depth of personal animus described by the leading advocates of all this, it wasn't a professional issue to them, it's personal. In particular he was surprised by the animus in the "save the Hubble" petition on the internet (O'Keefe 2004a, 11).

21.3.2 Scientific Reaction

Perhaps less surprising was the reaction from the scientists most directly involved in Hubble. In a statement prepared for distribution to members of the American Astronomical Society on January 28, Space Telescope Science Institute Director Steven Beckwith wrote, "the decision to end Hubble is a blow to astronomy and to NASA's efforts to engage a larger public in its mission of exploration and discovery. Never in the history of astronomy has society shut down its most powerful optical observatory before a successor was ready" (Beckwith 2004b). The reaction at Headquarters was understandably more muted, even from those not involved in the decision. Anne Kinney wrote that the Astronomy and Physics Division of OSS was greatly saddened, but "fully supports the Administrator's decision, a decision based on issues related to risk."

Another argument, both among scientists and in the media, was that HST needed to stay operational until its successor, the James Webb Space Telescope, was launched. Weiler had a ready answer to that:

> JWST was never sold as a replacement for the Hubble. It is a different kind of science. It is different wavelength color. It is a different community and these people who are out there again in the ivory tower saying we have to wait until the replacement for Hubble, they have a long wait because JWST isn't a replacement of Hubble. More importantly, where in the Constitution does it say that optical astronomers always have to have a telescope in space? X-ray astronomers haven't got that in the Constitution. Gamma ray astronomers don't have that. Infrared astronomers don't have that. It is ironic that optical astronomers, unlike those other areas I just mentioned, can do their astronomy from the ground too. The others can't. (Weiler 2004, 17)

It was a remarkable statement from one who was not only an optical astronomer, but who had spent a good part of his career on HST.

The link to the President's space vision would not go away, even in the scientific literature. The respected professional magazine *Physics Today* headlined its story with "Hubble Sacrificed in Wake of President Bush's New Space Vision." Beckwith was again quoted as saying the SM4 cancellation "was a complete shock and devastating for everyone." John Bahcall, who had headed the panel reviewing HST's lifetime options, including the option that SM4 might not occur, noted "All the astronomers I talked to assumed SM4 would occur .... I certainly assumed that it would. I think it is regrettable that no research scientist was involved in the decision to terminate the life of the most famous research telescope of the past 100 years." (Bahcall 2004). *Science*, the professional magazine of the American Association for the Advancement of Science, surveyed the potential "collateral damage" from the new space vision, and characterized Hubble as one of the "unpleasant side effects." The new vision "would end the Hubble Space Telescope's brilliant run and could jeopardize the future of any activity that doesn't directly serve the exploration effort" (Lawler 2004a).

21.3.3 Congressional Reaction

Undoubtedly more disturbing to NASA managers was the reaction from Congress, whence NASA receives its funding. On January 21, Senator Barbara Mikulski, a Democrat from Maryland whose district included the Space Telescope Science Institute, and the ranking minority member of the Appropriations Committee's Subcommittee on VA/HUD and Independent Agencies, fired off a letter to O'Keefe asking him to reconsider the decision and appoint an independent review panel. "I was shocked and surprised by your recent decision to terminate the next scheduled servicing mission of the Hubble Space Telescope (HST)," Mikulski wrote (Mikulski 2004). On January 26 O'Keefe replied, explaining his rationale in detail:

> The decision had to balance the world-class science that HST has produced, and would continue to produce, against the risks to the Shuttle and its crew. In the end, the determining factor was the recommendations of the Columbia Accident Investigation Board report for developing on-orbit inspection, repair, and contingency rescue requirements for every Shuttle flight. As such, my decision was not made with regard to budget considerations, nor was it based on any question as to the significance of the science return of the HST.

O'Keefe specifically made the following points:

> (1) Because the SM4 mission would have been the sole remaining Shuttle flight not directed to ISS, NASA would have had to develop unique procedures and technology because of its unique orbital inclination. (2) A second Shuttle would have been necessary to be ready on the launch pad in the event of a problem with the SM4 orbiter. New inspection and the second Shuttle technologies would have been required for this one-time mission to HST. Developing these new and unique items and procedures poses a set of risks which, taken individually, are surmountable, but, in the aggregate, the risks are significantly higher than a Shuttle mission to ISS. The total risk, considering the astronauts, the Shuttle, the ISS schedule and HST health, I have determined is too high." (O'Keefe 2004d)

O'Keefe further argued that an SM4 mission could not have occurred before June 2006. The unspoken assumption was that in the highly likely event that the RTF was delayed, an SM4 mission would be delayed even further, by which time HST might be dead. Since a dead HST could not be revived, this raised the possibility that the 18-month training period and effort would be wasted. O'Keefe concluded, "I cannot delegate my ultimate responsibility for decisions related to the safety of human spaceflight to any panel of experts, no matter how distinguished" (Mikulski 2004; O'Keefe 2004d).

Despite his conclusion, after further pressure from Mikulski, 2 days later O'Keefe asked Admiral Gehman, chairman of the CAIB, to review the matter (Berger 2004). Gehman reported back on March 5, saying that the HST mission was "slightly more risky" than an ISS mission, and that it needed "a rich and deep study" to see if it was worth the risk. Mikulski kept up the pressure. In an FY05 budget hearing before the Senate Subcommittee on March 11, Mikulski and Subcommittee Chair Kit Bond (D-MO) called for the National Academy of Sciences (NAS) and the General Accounting Office to review the risks, costs, and benefits of a Shuttle repair mission to Hubble. O'Keefe agreed and said he would take such recommendations into account as long as they were not counter to the CAIB report. In a preview of things to come, he further insisted that the NAS charter also include a study of robotic methods for servicing. A few hours later at a press conference, O'Keefe clearly remained skeptical of a human servicing mission, again citing the risk associated with a rescue mission if something went wrong as fundamentally opposed to the conservative approach to Shuttle operations advocated by the CAIB report. But he returned to the idea of a robotic mission, which he now cited as potentially providing new technology in line with the President's vision to send humans to the Moon and Mars (Lawler 2004b).

Congressional pressure came also from the Republicans, notably Senator Kay Bailey Hutchison (R-Texas), who represented the state with the fabled home of the astronauts, the Johnson Space Center. Moreover, she served on the Senate committee that oversaw NASA as well as on the Appropriations Committee that controlled its funding. In a letter to the White House accompanying a petition signed by 26 former astronauts, Hutchison asked President Bush to reconsider early retirement plans for HST. "The replacement parts needed to keep Hubble operating have already been produced," she wrote. "Should NASA choose robotic transport, installation of parts would be delayed by the expensive and complicated development of mechanical tools. NASA should keep Hubble operational by sending a manned space flight to perform the simple repairs and ensure the satellite's ability to provide crucial knowledge to our space science experts." Similarly, the astronaut petition argued that robotic servicing would have a lower probability of success than a Shuttle servicing mission, with only a portion of the tasks accomplished. The petition had no immediate effect. According to O'Keefe, several of the astronauts later regretted signing it and after discussions with O'Keefe Senator Hutchison seldom mentioned it. O'Keefe pointed out that the thrust of the petition was that return to flight should occur without complying with the CAIB recommendations. Some astronauts were still of that opinion; O'Keefe was decidedly not. Nevertheless, some of the arguments in the petition later were echoed in the National Academy of Sciences study (Hutchinson 2004).

Meanwhile, reaction from scientists and the public continued. NASA even had to battle dissent from within, when in early February two reports from anonymous NASA engineers concluded that a mission to HST was no riskier than the 25 planned missions to ISS (NASA 2004b, c). On March 11, Michael Greenfield, Associate Deputy Administrator for Technical Programs and a member of the Space Flight Leadership Council, wrote a white paper on HST SM4 for general consumption, attempting to capture all the arguments (Greenfield 2004). Three days later, CBS *60 min* reported on the HST controversy, and indication of what a cause cèlebré the Space Telescope had become. In a widely quoted statement in the *60 min* report, Mikulski used a vivid analogy: "The O'Keefe decision is irrevocable, and it's like surgery. If you're going to do an irrevocable decision, you want a second opinion. And that's why I asked for a second opinion—on the risk factors." She would get that second opinion in the form of the National Academy study.

21.4 Robotic Resolution?

As the reaction to the SM4 cancellation swirled, Frank Cepollina's servicing mission team at Goddard Space Flight Center had not been idle. Every day they worked in or near the clean room that contained Hubble's new instruments, their fate now uncertain. Serviceable spacecraft had long been a staple at Goddard. Unmanned space missions of the 1960s and 1970s had been considered "Failure Alley." Of 120 satellites launched during that period, 30% failed to meet mission objectives within 30 days after launch. As a result, GSFC started an initiative to develop serviceable spacecraft using modular components. This landmark in spacecraft design proved itself with the Solar Max mission repair in 1984, resulting in extension of the spacecraft lifetime to 9 years instead of four. In the years leading up to the first Hubble Servicing Mission in 1993, spacecraft servicing had been proven on the Westar and Palapa satellites (1984), Syncom IV (1985), Compton Gamma Ray Observatory (1991), and Intelsat (1992) (Cepollina 2004a, 6; b). Cepollina's team had been involved in all four HST Servicing Missions, and was preparing for SM4 when O'Keefe announced its cancellation at the Goddard meeting on January 16.

For a few days the Goddard team was shocked, but then they went into action. On February 20, NASA issued a Request for Information for "HST end of Mission Alternatives." A month later they were poring over the two dozen ideas submitted. By early April they had produced a plan for a robotic servicing mission. "Our group agreed that a robotic servicing mission of HST conducted in the 2007 timeframe, leveraging existing technologies and robotics expertise stemming from decades of prior investments, and using an approach as general described herein, falls within the capabilities of NASA and its partners at acceptably low levels of risk," the report stated. It also concluded that its group at GSFC would be best suited to undertake this task (Goddard Space Flight Center 2004).

On April 20, Cepollina and others met with O'Keefe to discuss the robotic mission options. O'Keefe was impressed. The following day at another hearing before Mikulski's Senate Subcommittee, he told committee members that "It's looking a lot more promising than I would have told you a few weeks back." While it was not yet clear that a servicing mission would work, O'Keefe said it was his intent to move ahead with a decision on whether or not to implement such a plan by September or October (Space News 2004).

The risks of such a plan could not be underestimated. There was confidence that a robot could deorbit the HST at the end of its life, as planned all along. There was more uncertainty about robotically replacing batteries and gyros, which had never been done before. Most uncertain of all was the ability to robotically install the two new instruments—a difficult task even for experienced astronauts.

At a meeting of the American Astronomical Society in Denver on June 1, 2004, O'Keefe summed up the situation. After reiterating his reasoning for canceling SM4, O'Keefe came to the main point of his address (O'Keefe 2004c):

> Fortunately, there may be other options for extending the Hubble's useful work...good options that are looking more promising as we've examined them more closely. Our confidence is growing that robots can do the job. For the last few months some of the best and brightest engineers at NASA, within industry and academia have been tirelessly evaluating the options for servicing Hubble by autonomous, robotic means. This approach now appears to be technically feasible. And the way it started was that we asked the question rather than clinging to a single point solution. Of course on any prospective complex mission of this nature, whether conducted by humans or robots, there are enormous challenges to be faced and no guarantees of success. We are not yet at a point where we have a firm alternative, but we're getting pretty close. ... In the same can-do spirit that propelled the first Hubble servicing mission, I am very pleased to inform this community that NASA is releasing a call for proposals today for a robotic Hubble servicing capability. This specific request for proposals calls for methods in ascending order of complexity, to: first, safely de-orbit the Hubble; second, to extend Hubble's service life by adding batteries and new gyroscopes; and third, to install new scientific instruments. This request for proposals is the first step in a long process of developing the best options to save Hubble. We are on a tight schedule to assure a Hubble servicing mission no later than the end of calendar year 2007. But we must act promptly to fully explore this approach. In essence we seek capabilities that highly dexterous robots assisted by humans on the ground could bring to this mission. What we are looking for is not autonomous robotics, but tele-robotics. If this mission goes forward, people will still be servicing Hubble. We are now at a point where these proposals can give us the means to seriously judge whether a robotic servicing mission can be mounted in time to replace Hubble's aging batteries, restore the pointing system with new gyros, and install new scientific instruments.

Thus on June 1, NASA released a Request for Proposal for an HST Robotic Servicing/DeOrbit module. The mid-July deadline emphasized the urgency of the task, which had to be accomplished by 2007 before the spacecraft died from lack of power or gyros, or both; once dead, it could not be revived. The general idea for servicing was to launch (with a Delta 4 or Atlas 5 rocket) a 20,000 pound two-piece spacecraft that would attach to Hubble's aft. It would include robotic eyes, targeting technology and a robotic arm that would install the new camera, gyros, and spectrograph. The lower part of the spacecraft would then be jettisoned, while the upper section with new batteries and a propulsion system would remain and eventually deorbit Hubble at the end of its life.

At the same meeting O'Keefe also took the opportunity to signal that his reasoning for the original SM4 cancellation had not changed:

> As I have publicly and repeatedly stated, we are committed to implementing the Columbia Accident Investigation Board's recommendations. In their report, the Columbia Accident Investigation Board addressed the need for development of on-orbit inspection, repair, and

contingency rescue requirements for every Shuttle flight. Those factors bear on any decision to proceed with Shuttle operations and acutely bear on requirements surrounding a Hubble servicing mission. A mission to the Hubble would require the development of a unique set of procedures, technologies and tools different from any other mission we'll fly before the Shuttle fleet retires. Many of these capabilities which provide safety redundancy for ISS missions are primary or singular for a Hubble mission. Moreover, these Hubble unique methods must be developed and tested promptly before Hubble's batteries and other critical systems give out. We are making steady progress in our efforts to meet the safety requirements for the Shuttle return to flight next year. But based on where we are today, prospects are even more challenging than six months ago for our being able to develop in time all required safety and return-to-flight elements for a servicing mission before Hubble ceases to be operational.

The easy route would have been for us to keep plugging along and hope for the best. But "hope" is not a management method we should rely on to keep Hubble operating. The Columbia Accident Investigation Board recommended that we change our culture to a commitment to "prove that it is safe" rather than place the burden of proof on folks to "prove that it's not safe." Well, with that guidance in mind, we're nowhere near close to proving that it's safe. It's not the unknowns we are wrestling with, it's the knowns that we haven't yet devised a way yet to conquer. Accordingly, I found it would not be responsible to prepare for a servicing mission, only to find that the required actions identified by the Board could not be implemented. This likely condition would pose a Hobson's choice. It is likely we would have two untenable alternatives to choose from. Either fly the Hubble mission without fully complying with the Board's recommendations or allow Hubble to simply cease to function. The prospect of either of these options if we had put all our eggs in the Shuttle servicing basket is simply unacceptable. Equally untenable is the expectant atmosphere that would exist all the way up to a launch "go or no go" decision. This is precisely the type of "schedule pressure" that the Board quite correctly stated would significantly undermine the future safe operation of the Shuttle. (O'Keefe 2004b, 9–10, c)

This reasoning catapulted robotic servicing to the forefront. Unfortunately, the National Academy of Sciences report would not let O'Keefe off the hook quite so easily.

21.5 The National Academy Report

On April 16, in response to the Mikulski-Bond request, the National Academy of Sciences announced the members of its "Committee on Assessment of Options for Extending the Life of the HST." Chaired by Bell Laboratories consultant Louis Lanzerotti, the National Academy of Sciences panel met with Administrator O'Keefe on June 22 and issued an interim report on July 13. The interim report, issued because of the urgency of the Hubble situation in a letter to O'Keefe, included three findings and recommendations (National Academy of Sciences 2004):

FINDING. Compelling scientific returns will result from a servicing mission to the Hubble Space Telescope that accomplishes the scientific objectives of the originally planned NASA servicing mission SM-4. RECOMMENDATION. The committee urges that NASA commit to a servicing mission to the Hubble Space Telescope that accomplishes the objectives of the originally planned SM-4 mission, including both the replacement of the present instruments with the two instruments already developed for flight—the Wide Field Camera-3 and the Cosmic Origins Spectrograph—and the engineering objectives, such as gyroscope and battery replacements. Such a servicing mission would extend the life of this unique telescope and maximize its productivity. Other potential options to extend the useful life of Hubble—for example, by servicing components such as batteries and gyroscopes but without replacing instruments—will be studied by the committee as part of its charge. However, such a reduced level of servicing has not been featured in the repair strategies that the committee has heard about to date.

FINDING The proposed Hubble robotic servicing mission involves a level of complexity, sophistication, and technology maturity that requires significant development, integration, and demonstration to reach flight readiness. RECOMMENDATION. As an early step, NASA should begin immediately to take an active partnership role that includes HST-related demonstrations in the robotics space experiments that are now under way in other agencies in order to ensure that the returns from these experiments can be beneficial to a potential robotic Hubble servicing mission.

FINDING Because of inherent uncertainties in the early stages of development of a robotic mission to the Hubble Space Telescope, as well as the uncertain current status of the shuttle return-to-flight program, the key technical decision points for committing to a specific service scenario are at least a year in the future. RECOMMENDATION. At the same time that NASA is vigorously pursuing development of robotic servicing capabilities, and until the agency has completed a more comprehensive examination of the engineering and technology issues, including risk assessments related to both robotic and human servicing options, NASA should take no actions that would preclude a space shuttle servicing mission to the Hubble Space Telescope.

Democratic Senator Mikulski and Republican Representative Sherwood Boehlert, chair of the House Science Committee, endorsed the recommendations, demonstrating bipartisan support for extending Hubble's lifetime. O'Keefe's reaction was also positive:

> We agree with the Committee's view that the Hubble Space Telescope is arguably the most important telescope in history. NASA is committed to exploring ways to safely extend the useful scientific life of Hubble. The challenges of a robotic mission are under examination and we'll continue our exhaustive and aggressive efforts to assess innovative servicing options. In parallel with the Committee's ongoing research and deliberations, NASA will evaluate proposals we expect to receive shortly. Along the way, we'll keep options open to assure the best possible outcome.

But the bottom line of the report was clear, and a potential headache for Administrator O'Keefe: While it should pursue the robotic option, the panel recommended that NASA should not preclude using the Shuttle for a servicing mission along the lines originally proposed before the Columbia accident, enabling the full servicing with new instrumentation. This was the option that O'Keefe though he had ruled out 6 months earlier.

By December 2004, another problem with the robotic option had come to the fore: cost. The Aerospace Corporation, commissioned by NASA to study all options, concluded in a report dated August 3 that a robotic servicing mission was "high risk," would cost $2 billion, and would take 5 years to implement, by which time Hubble would be dead. The most certain way to a successful servicing mission, they concluded, was using the method used four times before: the Shuttle. Another option would be to place the new instruments on a new bare-bones telescope. An executive summary of the report was not available to the public until December 6 (Aerospace Corporation 2004; Kelly 2004).

Two days later, on December 8, the National Academy released its final detailed 147-page congressionally mandated report. The findings and recommendations were a rebuff to the robotic mission in every way. The Committee found that the technology for robotic servicing required a level of complexity that was inconsistent with the 39-month development schedule needed to reach the telescope in time, even with the expertise of the Goddard team, which had little experience in autonomous rendezvous and docking. It found further that meeting the CAIB and NASA requirements relative to inspection and repair, safe haven, and Shuttle rescue was viable; that "the shuttle crew safety risks of a single mission to ISS [International Space Station] and a single HST mission are similar and the relative risks are extremely small;" and that Space Shuttle crews and their ground-control teams had a proven track record of developing innovative techniques in orbit—likely to be needed for a Hubble servicing mission. And, they found that HST was a uniquely powerful instrument worthy to be saved.

The three final recommendations could not have been clearer—or more diametrically opposed to O'Keefe's original intention to cancel a Shuttle servicing mission:

1. The committee reiterates the recommendation from its interim report that NASA should commit to a servicing mission to the Hubble Space Telescope that accomplishes the objectives of the originally planned SM-4 mission.
2. The committee recommends that NASA pursue a shuttle servicing mission to HST that would accomplish the above stated goal. Strong consideration should be given to flying this mission as early as possible after return to flight.
3. A robotic mission approach should be pursued solely to de-orbit Hubble after the period of extended science operations enabled by a shuttle astronaut servicing mission, thus allowing time for the appropriate development of the necessary robotic technology (National Academy of Sciences 2005).

Six months after, O'Keefe had committed to studying the robotic options for a Hubble rescue, the option seemed to be closed by a blue ribbon panel. Yet the decision remained for O'Keefe to make.

Less than a week later, however, on December 13, O'Keefe resigned as NASA Administrator. There was no clear cause-and-effect relationship as far as HST was concerned. His resignation letter to President Bush cited family reasons, and the desire to accept the chancellorship at Louisiana State University, at more than three times his civil service salary ($500,000 vs. $158,000). It was also a few weeks after Congress had given NASA a record $16.2 billion appropriation, which O'Keefe had shepherded through Congress and took as a mandate for the new vision for space exploration enunciated by President Bush the previous January. Two days later, on January 16, O'Keefe had announced the cancellation of the HST servicing mission, setting in motion the remarkable chain of events described in this paper.

Almost a year after his HST decision was announced, and a few weeks before his resignation announcement, O'Keefe still believed he had made the right decision for HST. "I'm not sure 'vindicate' is the right word, because it's a long time before that would be demonstrated. But I certainly feel that as passing time has gone by that it was the right call. Painful as it was, difficult as it was I don't have any reservations about it at all. I think it has helped to prompt a whole different way of looking at the problem." He viewed his decision as essential to shifting NASA's culture, and believed that the critics of his decision had submerged "adherence to principle, to a set of objectives necessary to demonstrate the credibility of this Agency to do what we say we're going to do, that we're committed to do and actually prove that we are going to do it. That credibility, by the way, has always been much in question, always been much accused of being not quite was rigid as it should be. It's a point I find to be really problematic and we have a real challenging kind of history in this case" (O'Keefe 2004b, pp. 5, 10–11).

21.6 Summary and Lessons Learned

As of the end of 2004 NASA was involved in a race not of its choosing—to adopt the best option to service the Hubble Space Telescope before it ceased operations in 2007 or 2008, or to decide to de-orbit the spacecraft at some point after its death. Whether a servicing mission would be carried out with robots or humans remained an open question. Robots had no record of capability to accomplish the simple tasks of replacing batteries and gyros, much less the far more complicated tasks of installing new instruments. Humans had a proven record of servicing with the Space Shuttle, but the Space Shuttle might not be able to make it in time. At the core of the matter was an assessment of the relative risk of a Shuttle HST mission compared to a Shuttle ISS mission. This assessment remained controversial, with the National Academy of Sciences panel concluding that the risk differential was "extremely small," and O'Keefe maintaining that his responsibility to crew safety and the CAIB recommendations precluded a Shuttle flight in the time remaining. Both options required long lead times, meaning a decision needed to be made soon. Ironically, a telescope that had been the brunt of jokes when it was first launched with its mirror problems now had triggered a national outpouring of concern, as Congress, scientists and the general public clamored for its life to be extended. It was truly a remarkable turn of events in the history of space science.

What made the controversy even more remarkable was the intimate connection with human spaceflight. The fortunes of the Hubble Space Telescope had always been tied to the Space Shuttle. Its launch was delayed by the Shuttle Challenger accident, and its servicing made possible by four remarkable Shuttle missions.

Critics who had at times portrayed the Space Shuttle as an expensive toy stuck in Earth orbit had been partially mollified by the HST servicing missions, which had demonstrated a place for both humans and robots in space and kept a steady stream of inspiring astronomical images beaming toward Earth. Now, its last servicing mission was threatened because of the Shuttle Columbia accident. To this extent, space science and human spaceflight had become intimately wedded, and the reputations of both were at stake.

In short, the HST servicing mission controversy was part of an American space program at a crossroads at the beginning of a new millennium. It should be remembered that it was only a small part of NASA's portfolio, given NASA's far-flung concurrent space science programs such as the Mars Exploration Rovers, the Cassini-Huygens spacecraft, and the elements of the new exploration vision that were being discussed as the HST controversy raged. Nevertheless, the public, scientific and Congressional reaction to the Columbia accident, and the subsequent controversy that has swirled around the SM4 cancellation decision and the related return to flight, raised issues that went to the core of NASA's mission, and indeed held relevance to high-tech agencies beyond NASA. Among them are the following:

Organizational Learning: NASA had long been criticized for not being a learning organization, specifically for not learning the lessons of Challenger. Yet, when the agency took the lessons to heart in the wake of the Columbia accident—with serious implications of which the Hubble Space Telescope was only one—it was subjected to withering criticism. Associate Administrator Ed Weiler summarized the irony succinctly "Sean feels he is following Admiral Gehman's report. Sean has been beaten up by Congress. He has been beaten up by the press for not having a safety culture at NASA. He has made a safety decision. I'll leave it to the reader" (Weiler 2004, p. 14). Many NASA managers felt that the recommendations of the CAIB report were all too soon forgotten by a Congress protecting its constituency, scientists more interested in science than human risk, and a public that wanted nice pictures. How does one weigh the agendas of those constituencies against the risk involved? In answering that question it should be obvious that managers need to be able to make decisions that are not always the popular choice based on a hoped-for outcome. "Hope cannot be used as a management tool," in O'Keefe's words.

Reaction to Executive Decisions: Asked for lessons learned from the HST case, O'Keefe stated, "first and foremost don't be intimidated by the negative reaction. And if you make a decision exclusively on the popularity of it you may well likely make a poor decision. And so the lesson learned from this one is stay the course, hold on, the wind is going to get pretty wild. And the hurricane force blow may be a little bit violent at times, but it will eventually get you through to where you need to go. And if you fold up you may end up doing the popular choice, but you may not be doing the right thing" (O'Keefe 2004b, p. 26).

Communicating Decisions: Decisions are inevitably more complex than they appear to the media, the public and even to those relatively close to those decisions inside an agency. In this case the HST SM4 cancellation appeared to outsiders to be a budget issue. The unanimous claim inside NASA, from the Administrator to the Comptroller to the HST managers was that it was not a budget issue, a claim borne out by the evidence, but that never seemed to penetrate the media. Similarly, the servicing mission cancellation, having been announced 2 days following the new space exploration vision, was immediately identified as a casualty of that vision. In fact, we have seen that the decision to cancel SM4 was made in late November 2003 and reflected in budget documents by December 2, 6 weeks prior to the President's announcement. Although the vision announced on January 14, 2004 had been in the making for several months, there is no evidence that HST had been part of those deliberations. Communicating facts both internal and external to the agency is essential, but sometimes futile given human failings. In order to set the record straight, it is essential that objective histories of policy decisions be encouraged, written and widely disseminated, and that the lessons of history be learned. To the extent possible, objective history needs to be taken into account before decisions are made.

Planning for the Unplanned: To paraphrase Robert Burns "The best-laid plans of mice and men often go awry." That was certainly the case with the timing of the SM4 cancellation announcement, which had been carefully planned for late January after all constituencies could be properly briefed, but which was in actuality dictated by a leak to the press. This is such a common event in Washington that it should itself be taken into account in planning.

Schedule Pressure: Schedule pressure is the norm at NASA, and at many other agencies, but unlike most other agencies, at NASA it could be seen as a contributing factor to spectacular and very public national tragedy. Nevertheless, schedule pressure is not necessarily bad. As O'Keefe said "there is good scheduled pressure and there is bad schedule pressure. Bad schedule pressure is when you've got everybody cutting corners in order to do something and risking people's lives to do it. Good schedule pressure is something where there's an imperative out there to come to solution, come to closure, reach understanding, find a consensus, so that you can get on with the task constructively so that it's just not lingering forever and ever. But in the process the risk should be not at the expense of the potential of someone surviving the experience or not." Navigating a project or an agency between good schedule pressure and bad schedule pressure is perhaps one of the most difficult, but necessary, tasks that managers face.

Risk Aversion: The HST SM4 controversy and the schedule for return to flight raise important questions about risk aversion, and whether the United States has become a risk averse society. If anything good has come out of the Columbia accident and the HST SM4 cancellation controversy, it is perhaps that a more sophisticated discussion of risk aversion has been generated. O'Keefe noted that "The bulk of the folks who really are dedicated to this, who care about it, really are coming to grips with the fact that this really is a Hobson's choice. This really is coming to grips with that intellectual challenge of how you look at problems and issues, and how you ultimately then have to make decisions about waiving what you believe to be principles, or not, or sticking with them and having it be a very withering circumstance based on what is a very difficult definition of what is risk" (O'Keefe 2004a, pp. 20–23).

The question bears on situations far beyond NASA, and on the role of the United States as a creative society that does not shrink from exploration. "What's acceptable risk?" O'Keefe asked. "Some people sat back and said you know, this is indicative of the fact that the American public can't stand losses. I think that's wrong. I think that's a different case, I think we're more tolerant and accepting of risks quite frankly than many other cultures and many other civilizations throughout the course of history in some ways." One might ask why not just let the astronauts go, since there would certainly be no lack of volunteers. Management's answer is that one does not make policy based on daredevils: "It's not a question of whether or not the people who are exposed to it are witting of that level of risk, it should be a question of whether or not we as the individuals who are responsible for the conduct of those operations are witting of that risk on their behalf, and have done our diligent best to avert as much of the challenge and the unknowns about it as we possibly can. We'll never eliminate the risk" (O'Keefe 2004a, pp. 20–23).

O'Keefe considers the risk question one of the most important results to come out of the current controversy.

> And so it's kind of gotten to that level of sophistication of the debate which is interesting and really, really good. It is a more positive way; I think a more constructive way, to evaluate what it is we are willing to accept here. And it isn't the yee-haw school of get aboard and fly anytime you want nor is it the other side, which is asserted that some would argue that I'm advocating here which is aversion, it isn't that either. It's saying there is a metric, there's a measurement, there's a standard by which you can measure what you're prepared to accept, and that was set for us in a two hundred and forty-eight-page report, with seven volume set of appendices that went with it. Produced lots of details, that's a good way to measure that to say this is what the standard of risk acceptance ought to be for this program. But in a broader sense, it also is calling into question what is the acceptance of risk that we as a people, as a culture, will adhere to in pursuit of exploration objectives, science pursuit, whatever. And I think frankly the answer to that is, it is pretty tolerant. People are pretty tolerant. If we weren't, the entire space flight program would have shut down on February 2, 2003. And it would have had another opportunity to shut down on August 27th, the day after the report was released 2003, and we didn't. Each of these steps turns on, again, I think, a deeper reflection of these things which again is coming out in the public debate more and more evidently – a willingness of acceptance of risk as long as you understand what the objectives are all about and whether those are objectives we think are worth encountering that level of risk. And that there is diligence exercised along the way to mitigate it as much as possible, that's an expectable standard. Boy, that's a much more sophisticated debate and argument that we've had about this in a long time. (O'Keefe 2004a, 20–23)

21.7 Epilogue: Reversal of Fortune (Added 2012)

In the wake of Sean O'Keefe's resignation effective February 11, 2005, the hopes of those who advocated a final Hubble servicing mission rested on Michael D. Griffin, who arrived as the new NASA Administrator April 13. Griffin (Fig. 21.4), an engineer who had worked as NASA's Associate Administrator for Exploration in the early 1990s, came to the Administrator's position from the Johns Hopkins University Applied Physics Laboratory. His resumé boasted years of experience in both space

Fig. 21.4 NASA Administrator Michael Griffin in his office on the ninth floor of NASA Headquarters, September 10, 2007. The engineer's motto of "Show me the Data" instead of "Show me the Money" may have figured in his decision to rescue the Hubble Space Telescope. Photo by Steven Dick

science and human spaceflight, but he left no doubt that his top priority was to implement the human spaceflight program, known as Constellation, that President George W. Bush had enunciated in January 2004 (Lambright 2002).

How Griffin would act on Hubble was an open question. But he indicated his thinking already during his Senate confirmation hearings on April 12, 2005–24 years to the day after the first Shuttle flight. "I would like to take the robotic mission off the plate," he stated, bowing to the National Academy report that had concluded it was unfeasible. "I believe the choice comes down to reinstating a Shuttle servicing mission or possibly a very simple robotic de-orbiting mission." The latter referred to de-orbiting Hubble for safety reasons sometime after 2020, without having performed the risky servicing mission—in other words, long after it was dead. That decision was hardly urgent, but the Shuttle servicing mission was. "When we return to flight it will be with essentially a new vehicle, which will have a new risk analysis associated with it. At that time I think we should reassess the earlier decision in light of what we learn after we return to flight" (Zimmerman 2008). The essential ingredient was a successful return to flight. That was accomplished when Shuttle Discovery was launched on July 26, 2005. Two more successful flights followed in July and September the following year.

In the wake of three successful Shuttle missions after return to flight, on October 31, 2006–33 months after O'Keefe had announced the cancellation of SM4—Administrator Griffin announced to employees at the same venue (Goddard Spaceflight Center) that the fifth servicing mission (still known as SM4) would indeed take place. In the official press release from NASA Headquarters, Griffin was quoted as saying "We have conducted a detailed analysis of the performance and procedures necessary to carry out a successful Hubble repair mission over the course of the last three Shuttle missions. What we have learned has convinced us that we are able to conduct a safe and effective servicing mission to Hubble. While there is an inherent risk in all spaceflight activities, the desire to preserve a truly international asset like the Hubble Space Telescope makes doing this mission the right course of action." (NASA 2006a). At a press conference the same day, Griffin noted, "I don't believe I have talked to anyone in the agency, from flight crew to flight ops manager to even budget guys ... who thinks we shouldn't do this." There was not, however, unanimity from the astronaut office or the engineers about "Launch on Need," the decision to have a second Shuttle ready to go on launch pad 39B for rescue in case of a problem at Hubble. In any case, the second "rescue" Shuttle would be in place at the time of the servicing mission launch (NASA 2006b).

The decision to return to Hubble was hailed by astronomers, Congress, and the public alike. Senator Barbara Mikulski, the long-time Hubble supporter whose district included the Goddard center, enthused "This is a great day for Maryland, for America, but most of all, for science. Hubble is a national asset and a national priority. Without question, Hubble has been the most successful NASA program since Apollo. And like Apollo, Hubble has helped America lead the way in discovery and exploration" (Mikulksi 2006).

SM4 was originally scheduled for September 11, 2008 on Shuttle *Atlantis* (STS-125). But due to a malfunction of Hubble's Control Unit/Science Data Formatter, which affected the storage and transmittal of all science data to Earth, the mission was postponed until 2009 while a replacement unit was checked out. On May 11, 2009 SM4 was finally launched aboard the *Atlantis*, 5 years after originally scheduled. Among the astronauts were mission specialists John Grunsfeld and Michael Massimino, both veteran spacewalkers who had serviced Hubble before, and rookie astronauts Andrew Feustel and Michael Good. Over a series of five spacewalks from May 14–18, the four astronauts accomplished all their goals, though not without challenges. They installed two new instruments: the Wide Field Camera 3 (WFC3) and the Cosmic Origins Spectrograph (COS), the latter used exclusively in the ultraviolet, increasing Hubble's UV sensitivity at least ten times. In addition to these two new instruments, the astronauts also repaired two instruments, the Advanced Camera for Surveys (ACS) and the Space Telescope Imaging Spectrograph (STIS). STIS had been dysfunctional since a power failure in 2004, and ACS had suffered an electrical short in 2007. The astronauts also replaced the 18-year old batteries, installed six new gyroscopes, and a new Fine Guidance sensor. Finally, they installed the spare Science Instrument Command and Data Handling Unit, the apparatus that had caused the 8-month launch delay. All told,

SM4 was the heaviest servicing mission of all, carrying 22,000 pounds of hardware to Hubble. After the mission, Hubble was at the apex of its performance.

With its new lease on life, Hubble was projected to finish its remarkable mission around the 2012–2015 timeframe. As Grunsfeld remarked, "Our warranty is 3 years, labor not included. Five years is totally reasonable. Beyond that is something we'd be delighted to have, especially overlapped with the James Webb Telescope." While the latter is looking unlikely due to delays, 25 years of cutting-edge astronomy would be a remarkable legacy for a telescope once given up for dead.

How was Hubble's fifth servicing mission cancelled by one Administrator, only to be green-lighted by his successor? As we have seen, O'Keefe's reluctance can only be understood in the context of the Columbia accident and the report of the Columbia Accident Investigation Board. Griffin's decision can only be seen in the context of his belief that exploration, human or robotic, is among the most important things humans can do (Griffin 2008a, b). Risk assessments were involved in both decisions, and despite the technical evaluations, there remained a large measure of subjectivity in both cases. Griffin's gutsy decision was more in tune with the idea that safety is the second priority in any bold adventure; having taken all precautions, the first priority is to go, otherwise no explorers would ever have left home (Dick and Cowing 2005).[5]

21.8 Commentary 2020

Astronaut John Grunsfeld's prediction that the 2009 servicing mission would extend Hubble's life into the 2012–2015 timeframe before mission termination has proven too pessimistic. In the wake of the decision to undertake what turned out to be the final servicing mission SM4, the Hubble Space Telescope continues to perform brilliantly as of this writing in late 2019. Whether it will last until the launch of the James Webb Space Telescope, now scheduled for March of 2021, remains to be seen.

Acknowledgments My thanks to all those interviewed, to the NASA History Office staff, to Harry Lambright for discussions, and to Retha Whewell for verifying dates of crucial meetings and other help. Oral History Interviews cited in this chapter are available at NASA HQ History Office.

Notes

1. Panel members were John Bahcall (chair), Barry Barish, Jacqueline Hewitt, Christopher McKee, Martin Rees, and Charles Townes.
2. In addition to O'Keefe, present at the November 7 meeting were Isakowitz, Readdy, O'Connor and Chief of Staff John Schumacher.
3. The Shuttle Return to Flight date of March 2005 was officially announced February 19, 2004. It was later moved to May, 2005. Space Flight Leadership Council (SFLC). Charged with oversight of the agency's Return to Flight efforts, the Council is co-chaired by Bill Readdy

(Associate Administrator for Space Flight) and Michael Greenfield (Associate Deputy Administrator for Technical Programs). It also includes the directors for NASA's four space flight centers, Bryan O'Connor (Associate Administrator for Safety and Mission Assurance), and Michael Kostelnik (Deputy Associate Administrator for International Space Station and Space Shuttle Programs).

4. Details of NASA-White House interaction in planning the new space vision are given in Sietzen Jr. and Cowing (2004).
5. Sean O'Keefe convened this "NASA Administrator's Symposium" in the wake of the Columbia accident to discuss these issues in a broad historical context. Oral History Interviews cited in this chapter, and many others undertaken for this study, are available at the NASA HQ History Office Archives.

References

Aerospace Corporation. 2004. "Hubble Space Telescope (HST) Servicing Analyses of Alternatives (AoA): Final Delivery," Executive Summary, August 3, 2004.

Bahcall, John. 2004. Quoted in *Physics Today*, March 2004, 29.

Beckwith, Steven. 2004a. "Space Telescope Science Institute Status Report," January 22, 2004.

Beckwith, Steven. 2004b. Exploder to members of American Astronomical Society, January 28, 2004.

Berger, Brian. 2004. "O'Keefe Asks CAIB Chairman to Review Hubble Decision," *Space News*, February 2, 2004, p. 7.

Cepollina, Frank. 2004a. Oral History Interview with Steven J. Dick, May 27, 2004

Cepollina, Frank. 2004b. "On Orbiting Servicing" presentation, July 17, 2004.

Columbia Accident Investigation Board Report. 2003. Volume 1, Appendix A, "The Investigation," p. 231. The full report is online at https://www.nasa.gov/columbia/home/CAIB_Vol1.html.

Dick, S. J. and Keith Cowing, eds., 2005. *Risk and Exploration: Earth, Sea and the Stars*. Washington: NASA SP-2005-4701m 2005 online at http://history.nasa.gov/SP-4701/riskand-exploration.pdf.

Goddard Space Flight Center. 2004. "Hubble Space Telescope Robotic Servicing Mission: Mission Feasibility Study for the WFC3 and COS Scientific Instrument Installation," 6 April, 2004.

Greenfield, Michael. 2004. "Cancellation of the Fifth (SM-4) Hubble Servicing Mission," March 11, 2004, NASA History Office.

Griffin, M. 2008a. *Leadership in Space: Selected Speeches of NASA Administrator Michael Griffin, May 2005–October 2008*. Washington: NASA Headquarters.

Griffin, M. 2008b. "NASA at 50," in Steven J. Dick, ed., *NASA's First 50 Years: Historical Perspectives*. Washington: NASA SP-2010-4704, pp. 1–9, online at http://history.nasa.gov/SP-4704.pdf.

Grunsfeld, John. 2004. Oral History Interview with Steven J. Dick, 13 April 2004, NASA History Office Archives, NASA HQ.

Hutchinson, Kay Bailey to President Bush dated May 26, 2004.

Isakowitz, Steve. 2004. Oral History Interview with Steven J. Dick, 18 February, 2004.

Kelly, John. 2004. "Study: Robotic Repair Mission to Hubble Too Costly," *Florida Today*, December 7, 2004.

Kluger, Jeffrey. 2004. "Mission to Mars," *Time* magazine, January 26, 2004, 42–52.

Lambright, W. H. 2002. "Launching a New Mission: Michael Griffin and NASA's Return to the Moon," online at http://www.spacepolicyonline.com/images/stories/LambrightMoonReport1.pdf, accessed February 1, 2012.

Lawler, Andrew. 2004a. "Scientists Add up Gains, Losses, in Bush's New Vision for NASA," *Science*, 303 (23 January, 2004), 444–445.

Lawler, Andrew. 2004b. "Academy, GAO to Study Possible Robotic Hubble Mission," *Science*, 303 (19 March, 2004), 1745.
McBride, Kelley. 2004. "Hubble, the Beloved: Decision to Stop Maintenance of Telescope Generates Outpouring," *Washington Post*, Jan. 29, 2004, A 27.
Mikulksi, B. 2006. "Mikulski Applauds Hubble Announcement, Says Decision is Right for America," Spaceref, November 1, 2006.
Mikulski, Barbara. 2004. Barbara Mikulski to Sean O'Keefe, January 21, 2004.
Moore, Michael. 2004. Oral History Interview with Steven J. Dick. 23 January, 2004.
NASA. 2003a. *NASA's Implementation Plan for Return to Flight and Beyond*, September 8, 2003. https://www.nasa.gov/pdf/49874main_RTF_08092003.pdf. The final report may be found at http://govinfo.library.unt.edu/returnflight/reports/final_report.html.
NASA. 2003b. Final Report of the Hubble Space Telescope Post-SM4 Scientific Review Panel, April 25, 2003, available online at http://nicmosis.as.arizona.edu:8000/REPORTS/Post-SM4_Final_Report.pdf.
NASA. 2003c. Report of the HST-JWST Transition Panel, August 14, 2003.
NASA. 2003d. Briefing slides, "HST SM4 Options Briefing with Mr. O'Keefe, 9:30 am, November 7, 2003," NASA History Office.
NASA. 2003e. The minutes of the December 9–11 JSC meeting are found at http://govinfo.library.unt.edu/returnflight/reports/final_report.html.
NASA. 2004a. *The Vision for Space Exploration* (February), online at http://www.nasa.gov/pdf/55583main_vision_space_exploration2.pdf.
NASA. 2004b. "Applicability of CAIB Findings/Recommendations to HST Servicing".
NASA. 2004c. "Risk Considerations for HST and ISS Flights".
NASA 2006a. "NASA Approves Mission and Names Crew for Return to Hubble," NASA HQ Press release, October 31, 2006. Online at http://www.nasa.gov/home/hqnews/2006/oct/HQ_06343_HST_announcement.html, accessed February 1, 2012.
NASA 2006b. "NASA Press Conference – Shuttle Return to Hubble," Tuesday, October 31, 2006, p. 11, online at http://www.spaceref.com/news/viewsr.html?pid=22474, accessed February 1, 2012.
National Academy of Sciences. 2004. Assessment of Options for Extending the Life of the Hubble Space Telescope: Letter Report (July 13, 2004).
National Academy of Sciences. 2005. *Assessment of Options for Extending the Life of the Hubble Space Telescope: Final Report*, Committee on the Assessment of Options for Extending the Life of the Hubble Space Telescope, National Research Council.
O'Connor, Brian 2004. Oral History Interview with Steven J. Dick, 1 November, 2004.
O'Keefe, S. 2004a. Oral History Interview with Steven J. Dick, 22 April 2004, NASA History Office Archives, NASA HQ.
O'Keefe, S. 2004b. Oral History Interview with Steven J. Dick, 29 October 2004, NASA History Office Archives, NASA HQ.
O'Keefe. 2004c. Remarks by Administrator O'Keefe at the American Astronomical Society Annual Meeting, Denver, Colorado, June 1, 2004.
O'Keefe, S. 2004d. Sean O'Keefe to Barbara Mikulski, January 26, 2004.
Readdy, William F. 2004. Oral History Interview with Steven J. Dick, 23 March, 2004. NASA History Office.
Sawyer, Kathy. 2004. "Visions of Liftoff, Grounded in Political Reality," *Washington Post*, January 15, 2004, A1.
Science. 2004. "Remaking NASA," *Science*, 303 (30 Jan 2004).
Sietzen Jr., Frank and Keith L. Cowing. 2004. *New Moon Rising: The Making of American's New Space Vision and the Remaking of NASA*. Burlington, Canada: Collector's Guide Publishing.
Smith, R. W. 1989. *The Space Telescope: A Study of NASA, Science, Technology, and Politics*. Cambridge: Cambridge University Press
Space News. 2004. "O'Keefe: Robotic Options for Saving Hubble Promising," *Space News*, April 26, 2004, p. 7.

Weiler, E. 2004. Oral History Interview with Steven J. Dick, 2 February, 2004, NASA History Office Archives, NASA HQ.
Wiseman, Jennifer. 2004. Oral History Interview with Steven J. Dick, 29 January, 2004.
Zimmerman, R. 2008. *The Universe in a Mirror: The Saga of the Hubble Space Telescope and the Visionaries Who Built It*. Princeton: Princeton University Press, pp. 206–207.

Chapter 22
Reflections on French-American Relations in Space, 1957–1975

Abstract Cooperation between the United States and France in space exploration must be seen in at least four contexts: global, national, institutional, and personal. Each of these interacting contexts is important, but none more so than the latter. No matter what the global, national, and institutional contexts, in the end it always comes down to people who have had to do the hard work of international cooperation and make the progress we have made so far. We should not lose sight of the indisputable fact that despite the difficulties, international cooperation has been a success at many levels. These brief reflections open a window on an entire field of study regarding NASA's international relations.

If the history of science teaches us anything as a discipline, it is the importance of the cultural, political and social environment in shaping science and technology. Whether we are talking about the origins of science in ancient Greece, Galileo's work in Italy in the sixteenth century, Newton in England in the seventeenth century, or Lavoisier in France in the eighteenth century, the environment in which these pioneer thinkers worked is very important to understanding how they were able to make their discoveries.

22.1 Four Contexts of Cooperation

The importance of context is no different today. So I think it is very important that in a meeting on French-American (United States) relations in space we consider the general context of cooperation. It is generally accepted that French-American relations in space took place in three contexts: (1) the global context of the Cold War dominated by two superpowers; (2) the national context of European and French-American politics; and (3) the institutional context of the National Aeronautics and Space Administration (NASA), formed in 1958 in the United States. At that time France had no space agency, but formed the Centre National d'Etudes Spatiales (CNES) in 1962, then participated in the European Launcher Development Organization (ELDO) and the European Space Research Organization (ESRO),

This chapter consists of reflections presented at a meeting on "French-American Relations in Space, 1957–1975" held in Paris December 8–9, 2005, published here for the first time.

S. J. Dick, *Space, Time, and Aliens*, https://doi.org/10.1007/978-3-030-41614-0_22

before finally becoming part of the European Space Agency (ESA) when it was founded in 1975. The latter was a landmark event for U.S.-French relations.

There is also a fourth context that may be the most important of all: the personal level. Without the trust established by the individuals involved, real cooperation is stymied. There is no doubt of the importance, for example, of Jacques Blamont and his personal relationship with scientists at the Goddard Space Flight Center beginning in 1960. This comes through clearly in his paper "International Space Exploration: Cooperative or Competitive?" (Blamont 2005).

So we have really four levels of context, from the global to the national, the institutional, and the personal when discussing French-American relations. It is an interesting question for further research which of these four levels is most important for cooperation to occur. In a recent paper at a meeting sponsored by the NASA History Office on "Critical Issues in the History of Spaceflight" (Dick and Launius 2006), John Krige—well-known for his work on the history of the European Space Agency and its predecessors—studied researchers, managers, and diplomats at NASA and the U.S. State Department who considered sharing communications satellite and booster technology with Europe in the mid-1960s. He showed that in these cases, the need to protect national industry and national security prevailed over foreign policy considerations (Krige 2006). So the desire by some to share technology in the 1960s to unite European and American interests was very much limited by national interests. There is no doubt that the personal context is constrained by the institutional context, and so on up the line. And there is no doubt that politics is involved at every level, as Walter McDougall showed in his Pulitzer Prize-winning history *The Heavens and the Earth* (McDougall 1985).

22.2 National Security Versus Foreign Policy

We see the same considerations of national security versus foreign policy also at work in post-Apollo negotiations and in the case of the Symphonie satellite. The Symphonie communications satellite experience may have convinced the French of the importance of an independent European launcher, because Symphonie was seen as a threat to the financial best interests of Intelsat under its Article 14 (Harvey 2003; Sebesta 1997). That, at least is the French perception, but Richard Barnes, NASA's Director of International Relations from 1985 to 1990, has emphasized this claim is open to interpretation (Barnes 2006). It is an interesting question of "counterfactual history" whether the Ariane launcher would have been born without this episode. Perhaps at least it would have been greatly delayed.

To the outsider it would seem that international cooperation is purely a good thing, an instrument of foreign policy that could enhance international relations. That was the fondest wish of many individuals involved in space exploration on both sides of the Atlantic in the 1960s. Alas, historically things are much more complicated. That more complicated version of international cooperation in space is evident during the period 1957–1975, and of course continues today. In his book on *International Cooperation in Space,* Arnold Frutkin tells of NASA's early philosophy of cooperation. Frutkin states that from the start "it was necessary to acknowledge limits to the scope of cooperative programs imposed by the dictates of national security," and that "international partners were informed quite frankly that collaboration could be more easily arranged in scientific experiments than in booster technology." Accordingly "A healthy focus on space science objectives, rather than on the versatile tools of space exploration, was then the final factor in NASA's developing philosophy of cooperation in space research" (Frutkin 1965, 35).

There is ample evidence of this in the historical record. Aside from Diamant, France's first expendable launch system, most cooperation did indeed come from space science. This is evident in the work with Goddard Space Flight Center on the magnetosphere, the FR-1/French-1 spacecraft (1965), the EOLE experimental meteorological satellite/balloon program launched in 1971, and the French photometer on the Orbiting Geophysical Observatories, the latter an early example of a French experiment on NASA mission.

Finally I commend to you a report on *U.S.-European Collaboration in Space Science*, published in 1998. The report was undertaken jointly by the Space Studies Board of the U.S. National Research Council and the European Space Science Committee of the European Science Foundation, both charged with advising their respective space enterprises. Itself an exercise in international cooperation, the report reminded us that

> Photographs of Earth from space show no political boundaries, reminding us that national distinctions are manmade. But the agencies responsible for those pictures, other missions to Earth orbit, and probes to deep space are inevitably national or international. Each has its own set of constituencies, procedures, capabilities, and limitations. (National Research Council 1998, p. ix)

Therein lies the challenge. Given these different constituencies, it is a remarkable testimony to the hard work at the personal level—the fourth level of context that I mentioned—that we have made the progress we have at the international level. No matter what the global, national and institutional contexts, in the end it always comes down to people who have had to do the hard work of international cooperation and make the progress that we have made so far. And in the final analysis we should not lose sight of an indisputable fact—that despite the difficulties, international cooperation has been a success at many levels. That is not only my personal judgement; I believe it will also be the judgement of history.

From the point of view of the history of science, it has always been this way and, I predict, always will. Progress comes only after setbacks and triumphs, failure and success. That is what makes history so interesting.

22.3 Commentary 2020

This paper consists of reflections presented at a meeting on "French-American Relations in Space, 1957–1975" held in Paris December 8–9, 2005 and organized under the patronage of the French Ministry of Foreign Affairs, the French Ministry of Education, and the French space agency CNES, as well as the American Institute of Aeronautics and Astronautics. The meeting began with a session on "the general context of cooperation" that included Pierre Messmer, Prime Minister under Charles De Gaulle. Mr. Messmer pointed out that space policy in France was a byproduct of French arms policy, and this alone explains much of the early difficulty in cooperation with the United States. The substantial volume resulting from that meeting (Institut Français de l'histoire de l'Éspace 2008), is testimony to the deep issues involved, many of which are universal in international space cooperation, though always diverse in the particulars.

These brief reflections thus represent the tip of the iceberg of the much larger subject of international cooperation in space. Although the Age of Space began in a fiercely competitive mode, political and funding realities eventually shifted the balance toward cooperation. This is particularly true in the case of the International Space Station, with its 16 partner nations. At the time of the Paris meeting, fully half of the last 60 space science missions of the world had international participation. And in the 10 years previous to that Paris meeting NASA carried out more than 900 agreements with 68 countries. For the full picture of NASA and international collaboration, see the works of John Krige (2006, 2007, 2008), and especially his book *NASA in the World* (Krige et al. 2013), commissioned during my time as NASA Chief Historian. It is not possible here to do justice to the vast panoply of NASA's international interactions, but *NASA in the World* is a good place to start, including the period covered here.

NASA collaboration takes many forms at many levels. One in which I was directly involved in a management role was the English translation and publication of Boris Chertok's *Rockets and People*. Chertok participated for six decades in the Soviet and Russian aviation and space programs, including many years as the Deputy to the founding figure of the Soviet space program, Chief Designer Sergey Korolev. On the occasion of the publication in the NASA History series of the English translation of volume 1 of *Rockets and People*, I travelled to Moscow in March of 2005 for a memorable event with Chertok himself present (Fig. 22.1). Among the 100 attendees were top management from Roskosmos (the Russian Federal Space Agency) and RSC Energia (the Russian equivalent of Lockheed or Boeing), many veteran cosmonauts (including Alexsandr Pavolich Alexandrov, Pavel Popovich, Viktor Savynikh, and Yuri Baturin), and distinguished guests including Natalia Koroleva, Chair of Surgery at the Moscow Medical Academy, and daughter of Sergey Korolev.

Among the speakers at the dedication ceremony was the U. S. Ambassador to Russia, Alexander Vershbow, who pointed to the publication of Chertok's book in the United States as "more evidence of the growing interest in Russia's history and enormous expertise in space exploration." After recalling the history of Soviet/Russian and U. S. cooperation in space, Ambassador Verhsbow touched on both the philosophical and practical aspects of cooperation in the wake of the Columbia Shuttle accident 2 years before:

> … this longstanding cooperation is based on mutual respect and admiration for the capabilities brought to the table by the other side. In this way, we believe that we have forged a true partnership. This has been most recently demonstrated since the Space Shuttle Columbia accident. While the Shuttle fleet has been grounded for the last two years, Russia has been providing the International Space Station with crews and cargo, keeping the station human-tended and productive. We all look forward to the Shuttle's safe return to flight in May of this year.

Fig. 22.1 Left to right, Jesco von Puttkamer (NASA HQ), Boris Chertok, Steven Dick in Moscow, March 30, 2005, on the occasion of NASA's publication of the English translation of volume 1 of Chertok's *Rockets and People*. Von Puttkamer was an aerospace engineer who worked with Wernher von Braun. Seated across the table is Korolev's daughter Natasha Koroleva

Return to flight did indeed occur in July of 2005 with STS 114, but after many successful flights the Shuttle fleet was retired after the program's last flight in July, 2011 when the Space Shuttle era came to an end (see Chap. 17). International cooperation in support of the International Space Station has of necessity continued to the present time, since the United States has had no capability to send humans into space with its own rockets since the Space Shuttle's retirement. That may well change in 2020 with the inauguration of NASA's Space Launch System (SLS), and commercial private ventures such as SpaceX and Boeing, funded in part by NASA.

The NASA History Division, which during my tenure was under the Office of External Relations that included the international portfolio, eventually published the complete four volume set of Chertok's *Rockets and People* (Chertok 2005, 2006, 2009, 2011). Those volumes, translated from the Russian, were greatly enhanced in their English editions for two reasons: the unmatched knowledge of the series editor, Asif Siddiqi, who added notes to the translation placing events in context, and the fact that Chertok often elaborated on the original Russian volumes for the English edition. These volumes constitute one of the premier primary sources for the history of the Soviet space program. Chertok died December 14, 2011 at age 99, almost 7 years after our event in Moscow.

International cooperation is never easy, as there are competing interests and a variety of technological and funding capabilities to be weighed. But one can argue that the very act of negotiation, not to mention the implementation of the subsequent agreements, enhances international prospects for peace as projects of great moment for humanity are carried out despite the difficulties.

References

Barnes, Richard. 2006. "Symphonie Launch Negotiations," News and Notes, NASA History Division, February, 2006, pp. 1–4, online at https://history.nasa.gov/nltr23-1.pdf

Blamont, Jacques. 2005. "International Space Exploration: Cooperative or Competitive," *Space Policy*, 21, 89–92.

Chertok, Boris. 2005. *Rockets and People,* volume 1. Washington, NASA.

Chertok, Boris. 2006. *Rockets and People,* volume 2. Washington, NASA.

Chertok, Boris. 2009. *Rockets and People,* volume 3. Washington, NASA.

Chertok, Boris. 2011. *Rockets and People,* volume 4. Washington, NASA.

Dick, Steven J., ed. 2008. *Remembering the Space Age: Proceedings of the 50th Anniversary Conference*. Washington, DC, NASA

Dick, Steven J. and Roger Launius, eds. 2006. *Critical Issues in the History of Spaceflight.* Washington, DC, NASA.

Dick Steven J. and Roger D. Launius, eds., 2007. *Societal Impact of Spaceflight*. Washington, DC: NASA

Frutkin, Arnold. 1965. *International Cooperation in Space.* Englewood Cliffs, NJ, Prentice-Hall.

Harvey, Brian. 2003. *Europe's Space Programme: To Ariane and Beyond.* New York: Springer, pp. 159–160.

Institut Français de l'histoire de l'Éspace. 2008. *Les relations franco-amÉricaines dans le domaine. spatial (1957–1975),* Quatrième rencontre de IFHE sur l'essor des recherches spatiales en France, 8–9 dÉcembre 2005, Paris, France. Paris: Institut Français de l'histoire de l'Éspace

Krige, John. "Technology, Foreign Policy, and International Cooperation in Spaceflight," in Dick and Launius, 2006, 239–262.

Krige, John. 2007. "NASA as an Instrument of U. S. Foreign Policy," in Dick and Launius, 2007, 207–218.

Krige, John. 2008. "Building Space Capability through European Regional Collaboration," In Dick (2008), 37–52.

Krige, John, Angelina Long Callahan, and Ashok Maharaj. 2013. *NASA in the World: Fifty Years of International Cooperation in Space*. New York: Palgrave Macmillan

McDougall, Walter. 1985. *The Heavens and the Earth: A Political History of the Space Age.* New York, NY: Basic Books

National Research Council 1998. *U.S.-European Collaboration in Space Science*. Washington, DC, The National Academies Press. https://doi.org/10.17226/5981.

Sebesta, Lorenza. 1997. "U. S. European Relations and the Decision to Build Ariane, the European Launch. Vehicle," in *Beyond the Ionosphere: Fifty Years of Satellite Communication* (NASA SP-4217), 137–156.

Part IV
Measuring the Universe: Goals, Institutions, Techniques

Part IV Frontispiece View of the U. S. Naval Observatory, in Washington, D.C., 1877. The dome housing the 9.6-in. refractor is atop the central portion of the building, and the time ball and mast can be seen above it. The dome for the 26-in. refractor, the telescope used to discovery the moons of Mars, is visible in the background. The U. S. Naval Observatory was the first national observatory in the United States. National observatories around the world have played a significant role in society, both for practical reasons and for research purposes. (Frontispiece from *Instruments and Publications of the U. S. Naval Observatory* (Washington, 1845–1876))

We are now on the eve of the second transit of a pair, after which there will be no other till the twenty-first century of our era has dawned upon the earth, and the June flowers are blooming in 2004. When the last transit season occurred the intellectual world was awakening from the slumber of ages, and that wondrous scientific activity which has led to our present advanced knowledge was just beginning. What will be the state of science when the next transit season arrives God only knows. Not even our children's children will live to take part in the astronomy of that day. As for ourselves, we have to do with the present...

William Harkness, 1882

At first thought, the theme of measuring the universe might seem far removed from culture. But this is not so, since humans have been surrounded by the heavens and have contemplated its meaning from the earliest times. Part of that meaning is determining the size of objects visible from Earth, the size of the Solar System, and the size of the universe itself. In a crude sense humans have been measuring the universe at least since the days of Stonehenge when they followed the Sun's movements along the horizon. Cultures from the Babylonians and Greeks to the Inca and Maya have observed the heavens in order to regulate their calendars, determine times of planting and harvest, navigate the seas, and for a variety of other purposes (Aveni 1989, 1997; Krupp 1984, 1991; Penprase 2011). Although Part IV of this volume alludes to some of these ancient beginnings, it quickly focuses on the period beginning in the late sixteenth century when humans began to take the measure of the universe by precise measurements, an activity that eventually blossomed into a field of positional astronomy known as astrometry.

The first chapter of Part IV is an overview of classical attempts at measuring the universe, although it must be said that it is incomplete in many ways. It is not, for example, a history of attempts to measure the Hubble constant in the twentieth century. It distinguishes three eras of astrometry, each characterized by unique methods and precisions achieved: the naked-eye era stretching to ancient times, the telescopic era beginning in the seventeenth century, and the space and interferometry era beginning in the second half of the twentieth century. Narrowing our focus to the more modern second and third eras, we then discuss instruments and techniques, and conclude with a discussion of institutions and their research programs as a means of exploring the driving forces, uses, and patronage of astrometry through history. Although astrometry is in may ways a highly technical science, it has historically had many benefits for society, ranging from time determination, calendar reform, navigation, and surveying as mentioned above, to spacecraft navigation, geodesy, and astrophysics.

Chapters 24 and 25 focus on institutional aspects of positional astronomy through what has appropriately been called the national observatory movement. It does so also by distinguishing three eras in this movement: the first era, in which the prototype Paris, Greenwich, Berlin and St. Petersburg observatories were founded; the second era, characterized by offshoots from previous national observatories, by new observatories of younger nations, and by the rise of astrophysical observatories; and the third era, post-World War II, characterized by national or international consortia, large budgets relative to the previous eras, and the study of old and new wavelength regions with increasingly sophisticated telescopes, detectors and spacecraft. It then focuses in some detail on two nineteenth-century representatives of this movement, the Pulkovo Observatory in Russia, and the U. S. Naval Observatory in the United States, comparing and contrasting their early work and how it was influenced by the societies in which they were embedded. Examining two such institutions with similar missions but embedded in such different societies allows us to make the point, for example, that there were advantages to having a czar if he happened to be on your side, as opposed to the messy machinations of democracy in the founding and operations of scientific institutions. And yet the common property that stands out is the largely practical nature of the work for which the first national observatories were founded. Whether for the improvement of navigation, geographic and geodetic work, or calendar reform, all these institutions were founded to meet national needs.

Chapter 26 describes only one of the many original functions of national observatories in the nineteenth century, the dissemination of time via time balls. Time balls today are known mainly for their celebratory function as in the well-known New York time ball drop on New Year's Eve. But for a significant part of the nineteenth century they were crucial for navigation and for disseminating time to the public, before being overtaken by the telegraph and radio signals.

Chapters 27 and 28 represent another function of national observatories, as well as other observatories—the far-flung worldwide expeditions for specific purposes, in this case to measure the so-called astronomical unit, the distance from the Earth to the Sun and thereby the scale of the Solar System. The astronomical unit held the key to measuring the rest of the universe, and was therefore a kind of Holy Grail of astronomy for several centuries. Chapter 27 illustrates one technique to achieve this purpose, the determination of the parallaxes, and thus distances, of Venus and Mars. Chapter 28 documents another technique, the infamous transits of Venus across the Sun, a technique that launched numerous expeditions around the world—the technique to which William Harkness refers in the passage at the beginning of this section. Here we use the 16 nineteenth-century American expeditions around the world as a representative example of the problems and promise of this technique. In the end, these chapters demonstrate that more than science was involved in these expeditions, including international cooperation and the study of other cultures.

After the previous five chapters have focused primarily on the nineteenth century, in Chap. 29 we move into the twentieth century, demonstrating how sophisticated our concepts of time have become. The chapter describes one scientific program undertaken in the midst of the International Geophysical Year to determine a special kind of time, and highlights an important aspect of science that historians usually ignore—the determination and dissemination of time worldwide. This chapter also highlights another often-ignored aspect of science—that things often don't turn out the way they were meant to.

References

Aveni, Anthony. 1989. *Empires of Time: Calendars, Clocks, and Cultures.* New York: Basic Books.

Aveni, Anthony. 1997. *Stairways to the Stars: Skywatching in Three Great Ancient Cultures.* Wiley, New York

Krupp, E. C. 1984. *Echoes of the Ancient Skies: the Astronomy of lost Civilizations.* New York, New American Library.

Krupp, E. C. 1991. *Beyond the Blue Horizon: Myths and Legends of the Sun, Moon, Stars and Planets.* Harper Collins, New York

Penprase, Bryan E. 2011. *The Power of Stars: How Celestial Observations have Shaped Civilization.* Springer, New York

Chapter 23
Measuring the Universe: A Brief History of Astrometry

Abstract For most of astronomical history, astronomers busied themselves with positional astronomy, a field known for the last two centuries as astrometry. Because it is impossible to analyze all of the developments of astrometry in a brief compass, we concentrate in this chapter on certain patterns of achievement, method, and research programs that emerge from its long and checkered history. In order to provide an overview, we first distinguish three eras of astrometry, each characterized by unique methods and precisions achieved. Narrowing our focus to the more modern second and third eras, we then discuss instruments and techniques, and conclude with a discussion of institutions and their research programs as a means of exploring the driving forces, uses, and patronage of astrometry through history. Although astrometry is in many ways a highly technical science, it has historically had many benefits for society, ranging from time determination, calendar reform, navigation, and surveying, to spacecraft navigation, geodesy, and astrophysics.

23.1 Introduction

Astrometry, the precise measurement of the positions, motions, and apparent sizes of celestial bodies, may also be taken to encompass the many research programs and conclusions derivable from this data. The term *astrometry* itself dates only from the nineteenth century, but the practice of positional astronomy is the oldest part of astronomical science. In the broadest sense, Stonehenge and other stone circles and alignments are believed to have been constructed to mark the motions of the Sun and Moon. The Babylonians tracked the motions of Venus and other celestial bodies, and the Greeks are reputed to have produced the first star catalogue in the second century B.C. These were pioneering and valiant efforts, but only since the sixteenth century has a variety of new techniques infused astrometric science with steadily increasing accuracy. This enterprise continues to the present day, even in the midst of the new astronomy of astrophysics. Indeed, modern astrometry is sometimes driven by astrophysical problems and aided by studies at non-optical wavelengths, developments undreamt of by its earlier practitioners. Despite the multifaceted nature of modern

---

First published as "Astrometry," in *History of Astronomy: An Encyclopedia*, edited by J.A. Lankford (Garland Publishing, 1997), 47–60.

S. J. Dick, *Space, Time, and Aliens*, https://doi.org/10.1007/978-3-030-41614-0_23

astronomy, it is indisputably true that of the 40 centuries of astronomy since Stonehenge, and the 25 since classical Greece, only in the last century has astrophysics dethroned astrometry as the predominant theme of celestial science. The importance of astrometry to the history of astronomy cannot therefore be overemphasized, nor should its relevance to modern astronomy be underrated.

Astrometry is an observational science par excellence, but it is closely related to the theoretical science of celestial mechanics. From Babylonian algebraic models and Greek epicyclic models of planetary motions, to the Newtonian and relativistic dynamics characteristic of modern celestial mechanics, astrometric observations have always provided the raw material for models and theories of celestial motions. The interplay between theory and observation is thus an important hallmark of astrometry. More broadly, while astrometric positions may be an end in themselves in providing products such as a stellar reference frame, the usefulness of the field extends far beyond the determination of positions. Carefully measured over time, these positions have revealed the scale of the Solar System through solar parallax, the distances of the stars from stellar parallax, stellar masses through measures of double star motions, the motion of the sun and galactic rotation through stellar proper motions, and unseen stellar and perhaps even planetary companions through perturbations in proper motions. Moreover, the stellar reference frame itself has for centuries been essential for determining civil time and for navigational and geodetic purposes; more recently, the extragalactic reference frame allows unprecedented accuracy in measuring the whole range of motions of our own Earth, as well as motions on the Earth caused by plate tectonics.

23.2 From Hipparchus to the Hipparcos Satellite: Three Eras of Astrometry

From the Greek astronomer Hipparchus of Rhodes (fl. 140 BCE) to the Hipparcos satellite dedicated to gathering astrometric data in the 1990s, three eras of astrometry may be distinguished. In terms of method, they may be designated as the naked-eye, telescopic, and interferometric/space eras; in terms of precision achieved, they correspond to the minute of arc, second of arc, and milliarcsecond regimes, a factor of about a million increase in precision over the last two millennia. Table 23.1 at the end of this chapter demonstrates this increased precision, and also shows some of the means by which this was achieved, as well as the resulting discoveries of each era. It is important to note that the table represents precisions actually achieved at each period, rather than those theoretically achievable; especially as the second of arc regime was approached, theoretically achievable precisions due to technological breakthroughs were not always realized because of other errors of observation. Thus, while half-arcsecond precision might have been achievable because of improvements in divided circle technology by 1800, precisions actually achieved were still at the arcsecond level, an important consideration in explaining why

Table 23.1 Improvement in precision of astrometric measurements

Method	Observer/Institution	Date	Precision[a] (arseconds)	Instrument	Refinements	Discoveries
Naked eye	Hipparchus	150 BCE	1200″	Armillary sphere		Precession (50″/year)
	Ptolemy[b]	150 CE		Same		
	Ulugh Beg[c]	1437	1200″	Same		
	Walther[d]	1503	600″	Same		
	Tycho Brahe[e]	1600	15″	Quadrant/sextant	Improved sights and diagonal scales	
	Hevelius[f]	1670s	15–20″	Same		
Telescopic visual						Proper motion (Halley, 1718)
	Flamsteed[g]	1700 1725	15″ 8″	Quadrant	Telescopic sight and micrometer	Aberration, 20″ (Bradley, 1728)
	Greenwich[h]	1800	1″	Transit circle	Divided circle	
	Bessel[i]	1838	.02″	Heliometer	Small field	Parallax, .3″ (Bessel, 1838)
	Pulkovo[j]	1855	.7″	Transit circle	Six circle microscopes	Unseen stellar companions Polar motion, .5″/year (Chandler, 1885)
	USNO[k]	1910	.3″	Transit circle		
	USNO[l]	1980s	.07″	Transit circle	Movable wire micrometer	
Photographic						
	Gill	1886–1900	2–4″	Astrograph	Photographic plate	
	Kapteyn[m]	1900	2–4″	Same	Same	Star streams (Kapteyn, 1904)

(continued)

Table 23.1 (continued)

Method	Observer/Institution	Date	Precision[a] (arseconds)	Instrument	Refinements	Discoveries
	22 observatories Carte du Ciel[n]	1891–1950	.17″–.40″	Same	Same	Galactic rotation (Oort, Lindblad, 1920s)
	Schlesinger[o]	1910	.02″	Long-focus telescopes	Small field	Planetary companions? 1960s
	USNO[p]	1980s	.004″	Same	Same	
	USNO[q]	1980s	.15″	Astrograph	Fine emulsions/measuring machines	
Electronic	USNO[r]	1990s	Sub-.001″	61-in. reflector telescope	Digital imaging/charge coupled device	
Interferometry radio	NASA, USNO, NRL, NGS[s]	1980s	.001″	VLBI	Combined data from several instruments	
Interferometry optical	USNO, NRL[t]	1994	.005″			
Space						
	HIPPARCOS[u]	1989–93	.002″	Satellite	No atmosphere	

[a]In this table the internal accuracy, or "precision," of a given instrument with its random errors must be distinguished from the true accuracy, which takes into account systematic errors. The distinction between the terms precision and accuracy is of recent origin, and although not always used consistently, the concept is important. As Boss (1937) explained, "Simultaneous observations of the same star by two different telescopes might each yield a probable error of plus or minus 0.15 arcseconds for each series of observations, yet the two mean positions might disagree by 0.4 arcseconds. This may be due to outstanding instrumental errors, to differential refraction, to differences in the methods of reduction, and to other factors" (Boss 1937, p. 49). To cite a more recent example, the first Hubble Space Telescope mirror is an example of precision without accuracy: a mirror precisely shaped to the wrong figure. Through most of the seventeenth century internal accuracy depended on the graduated scales of the instrument. With the introduction of the micrometer, the precision attainable by an instrument no longer bore a direct relation to the interval between its graduated divisions. Precision is almost always an underestimate of the true accuracy, which includes systematic errors detectable only by using several instruments. As a general rule, at least since the nineteenth century, the systematic error is often about the same order of magnitude as the random error. Since errors add vectorially, the total error, and thus the accuracy, may be the square root of two times the precision. The distinction between "probable error" used into the early twentieth century and the "mean error" or "standard error" used today has little effect on our Table; the former is 74 percent the size of the latter. In this table we use precision in the modern sense of internal accuracy, also known as

"random error" and "accidental error." The precisions given are state-of-the art for the best instrument and selected stars; they are nor necessarily routinely achievable, especially for the earlier periods

[b]There is wide variation among accuracy estimates of ancient observations, in part because some authors do not specify internal or external accuracy. Pannekoek (1961, 157), points out that while the instruments of Hipparchus and Ptolemy were probably divided into fourths or sixths of a degree, the mean errors of observations were greater than these divisions, corresponding to 35 minutes of arc in longitude (2100″) and 22 minutes of arc in latitude (1320″). Podobed (1964, 8), gives plus or minus 15′ [arcminutes] (900″) as the accuracy of the first star catalogues observed by armillary spheres. It is well known that the systematic error in the longitudes of Ptolemy's star catalog is about 1°, or 3600″ (see, for example, Toomer 1984, 328, note 51). Shevchenko (1990) gives—78′ as the mean systematic error in longitude for Ptolemy; he finds 21′ (1260″) for random errors in longitude and 17 min for latitude—both for zodiacal stars. See also Evans (1987, esp. 162–65). Note the value quoted for precession is the approximate modern value; Ptolemy may have used 36″/year

[c]Shevchenko (1990) gives random errors for Ulugh Beg's catalogue as plus or minus 20′ (1200″) in longitude and plus or minus 17′ in latitude-both for zodiacal stars—essentially the same as Ptolemy's results. He finds systematic errors in longitude of +13′, but demonstrates how this varies by constellation. Krisciunas (1993) confirms this result. He gives systematic errors of about 11′ in longitude and about 8′ in latitude, a significant improvement over Ptolemy's 1°

[d]Kremer (1980). Kremer (p. 180) comments that these random errors of 10′ "result from inherent structural limitations of the armillary, determined by the readability of the calibrations, the size of the holes or slits in the alidade, and the flexure of the rings due to their weight or to temperature variations." Since on the armillary sphere different rings are used to measure longitudes and latitudes, Kremer separated Walther's data into two in dependent populations. Comparing Walther's positions with a modern computation, he finds that the latitudes were relatively free of systematic error, while the total error in longitude (including the systematic error, thus giving the true accuracy) may have been as large as 13′, including the error in the position of the reference object. The accuracy in longitude relative to the reference star was about 5′ (300″), and the accuracy in latitude was .7′ (42″). Thoren (1990, 155), states that 10 minutes of arc is "generally regarded as the best accuracy achieved before Tycho's time." Kremer's data shows that 10′ was also the best precision achieved before Tycho

[e]Wesley (1978), especially Table 1. The value quoted is for the mural quadrant. By comparison Tycho's precision for the Jacob's staff had been 35′, causing him to discard it. According to Kremer (176), the precision for Tycho's equatorial armillary was 3′. Thoren (1990, 190), finds that Tycho had 1 minute of arc accuracy with his new instruments by 1585, compared to 4 minutes of arc he achieved before 1581. Wesley (Table 4) shows that Tycho's accuracy was around 35″. See also Thoren (1973)

[f]Chapman (1990, 32)

[g]Chapman (1983, 135), and Chapman 1990, 120)

[h]Chapman (1983, 135), and Chapman 1990, 120). Chapman in the latter reference states that "In the 1790s the 'ceiling of accuracy' had passed beyond the 1″ of arc barrier, and by 1820, Pond's observing books [at Greenwich] displayed consistent reading to places of 1/10 of an arcsecond." This may have been true of the accuracy of graduated scales on the full circle, but does not take into account the random errors of observation (the observational precision), nor the systematic errors (yielding the observational accuracy)

[i]Fricke (1970, 101). This mean error applies to the precision of Bessel's measurement of the stellar parallax of 61 Cygni. Pannekoek (1961, 343), notes the importance of systematic errors is emphasized by the fact that while Bessel after many years found the parallax value for 61 Cygni to be 0.35″, Struve inde-

(continued)

Table 23.1 (continued)

pendently found it to be 0.51″ for the same star. Note the heliometer can make only small field measurements. Monet (1988, 415), notes that small-angle techniques may give ten times the accuracy of large field measurements

[j]Eichhorn (1974, 147)

[k]Eichelberger and Morgan (1920, lxxix; cxxxix), and Eichhorn (1974, 157). Note again this is only mean error; systematic error is probably another .4″. Eichhorn (278) also states that the typical mean error of a meridian position around 1890 was .5″ in either coordinate

[l]U. S. Naval Observatory, average mean error with six observations of a star with the six-in. transit circle telescope. Van Altena (1983, 157), notes that refraction of the Earth's atmosphere places a limiting accuracy of 0.13″ on accuracy obtainable from one transit circle observation

[m]Gill (1913, lvii). This is the precision of positions in the Cape Photographic Durchmusterung, containing some 454,000 stars. For the context see Pannekoek (1961, 469)

[n]Eichhorn (1974, 280). By comparison the mean errors of the Cape Photographic Catalogue for 1950 range from .15 to .22″, but the positions may be subject to an additional .3″ systematic error. Eichhorn (1974, 274)

[o]Schlesinger (1911, 26). Note again that this is small field astrometry

[p]Harrington et al. (1985). Average internal error for parallax plates taken with the U. S. Naval Observatory 61-in. astrometric reflector. External error is seated as also about the same

[q]Douglass and Harrington (1990). Internal error of star positions based on photographic places with U. S. Naval Observatory 8-in. double astrograph

[r]Dahn et al. (1990). Internal error for parallax obtained with U. S. Naval Observatory 61-in. astrometric reflector with CCD detector

[s]Ma (1990). Abbreviations: *VLBI* Very Long Baseline Interferometry. *NASA* National Aeronautics and Space Administration. *USNO* U. S. Naval Observatory. *NRL* Naval Research Laboratory. *NGS* National Geodetic Survey

[t]Predicted precision of USNO/NRL instrument in progress

[u]Perryman (1990)

stellar parallax was not detected until 1838. Moreover, precisions even with one particular instrument will be affected by the skill of the observer, the number of observations of a single object, the care with which the observations were analyzed, atmospheric seeing and transparency, and even stellar magnitude. Increasingly after 1800, the theory of errors, closer attention to the peculiarities of each instrument, and the combination of observations contributed to precision improvements as much as technological advance.

Even so, the history of astrometry is characterized by several sudden technological breakthroughs, followed by long periods of routine work with more minor modifications. Thus, through innovative instrumentation Tycho Brahe in the sixteenth century improved over the Greeks by more than a factor of ten. At the beginning of the nineteenth century techniques of precision dividing and reading of full circles (as opposed to sextants or quadrants) produced another order of magnitude jump. After that, a combination of painstaking modifications in technique and analysis (such as the moving wire micrometer and screens to dim the brightest stars) brought the errors still lower, until in the 1980s and 1990s the techniques of interferometry and spacecraft produced yet another order of magnitude jump.

23.2.1 The Naked Eye Era

Stonehenge and its successors aside, the era of naked-eye positional observations spanned 1800 years, from the time of Hipparchus to the mid-seventeenth century. Already in this era we see one astronomical discovery closely linked to improvements in precision. Hipparchus's observations produced the first great surprise of positional astronomy—the apparent motion of the sphere of fixed stars around Earth's pole at the rate of about 50 arcseconds per year, the phenomenon we now know as the precession of the equinoxes, which is due to the motion of the earth's axis. Hipparchus could make this discovery only by comparison of the bright star Spica with a position from about 160 years earlier, and only because of the cumulative nature of precession; his instruments were certainly unable to discern 50 arcseconds, but he could take the 2° (7200-arcsecond) difference in positions and divide by 160 years to come up with about 1°/century. Hipparchus is often credited as the author of the first star catalogue, later elaborated by Ptolemy (150 C.E.) in his *Almagest*. Modern scholarship differs widely on this point, however, with one author (Toomer 1984) claiming Hipparchus had no star catalogue at all similar to the form of Ptolemy's, and another (Newton 1977) claiming that Ptolemy used Hipparchus wholesale and fabricated his own data. Nevertheless, Ptolemy's star catalogue is the first passed down to us from ancient times.

Although the Islamic culture produced the star catalogues of Al-Sufi (960) and Ulugh Beg (1430), and although in the West Bernard Walther and perhaps others made considerable instrumental progress, the great breakthrough in precision observations is generally accorded to Tycho Brahe. Supported by Frederic II of Denmark, Tycho constructed a variety of instruments at Uraniborg on the island of Hven, using improved sights and diagonal scales on quadrant- and sextant-type instruments that he constructed anew. Tycho's principal achievement was almost entirely observational, due to his skill in constructing and using improved instrumentation. But his painstaking observational work had an almost immediate impact on broader aspects of astronomy. Worldviews hinged on these minute measurements; Tycho's inability to measure stellar parallax led to doubts about the Copernican theory, and resulted in his own Tychonic system of the world. And it was the 8 minutes of arc discrepancy between theory and observation of the motion of Mars that motivated Johannes Kepler to abandon Greek epicycles for the ellipse, one of the foundation stones of modern astronomy. The German astronomer Johannes Hevelius in his private' observatory at Danzig (now Gdansk) took Tycho's techniques to their limit. As Chapman (1990) points out, Hevelius was the last representative of the Tychonic instrument school, that is, he was the last major user of large sextants and quadrants without telescopic sights. With Tycho and Hevelius, naked-eye astrometry reached its highest form and its physical limit; as the English natural philosopher Robert Hooke found experimentally at the time, the Tychonic school had reached the resolution of the human eye of about 1 arcminute.

23.2.2 The Telescopic Era

The link between improved accuracy and astronomical discovery accelerated remarkably with the second era. Supported by royal patronage, the first great national observatories were founded at Paris (1667), Greenwich (1675), Berlin (1701), and St. Petersburg (1725) (see Chap. 24). At the Royal Observatory, Greenwich John Flamsteed produced the first great star catalogue observed using telescopic sights; the value of the *Historiae Coelestis* (1725) may be appreciated from Newton's strenuous efforts to obtain it for his use. The accuracies represented in Flamsteed's work were beginning to bear fruit in unexpected ways. Taking precession into account, the man who within 2 years would become Flamsteed's successor at Greenwich, Edmond Halley, showed in 1718 that the bright stars Aldebaran, Sirius, and Arcturus were displaced from their positions in antiquity by large fractions of a degree, the phenomenon that became known as stellar proper motion. No longer were the fixed stars of the Greeks so fixed, opening a whole new area of research into stellar motions and their implications. Halley's successor, James Bradley, discovered a stellar aberration of 40 arcseconds in 1728, interpreted as due to the earth's orbital motion, and two decades later became the first to detect the nodding wobble of the earth known as nutation, an effect amounting to about 18 arcseconds (1748). When in 1783 William Herschel detected, from an analysis of stellar proper motions, the motion of the sun toward the constellation Hercules, astrometry added yet another triumph to its growing list of discoveries.

Further improvements in accuracy aided by refined telescopic practice allowed much smaller effects to be discovered in the nineteenth century. Attempts to measure stellar parallax had been sporadically undertaken since the seventeenth century, and there were several celebrated claims that it had been measured in the early nineteenth century, culminating in a controversy between John Brinkley at Dublin and John Pond at Greenwich (1814–1822). But only with the independent work of Friedrich Wilhelm Bessel, Wilhelm Struve, and Thomas Henderson, announced during the years 1838–1840, were measurements made that stood the test of time. The measurements of only a few tenths of an arcsecond solved one of the great mysteries of astronomy, proving definitively that a star shifted position as the observer changed position in the Earth's orbit. This demonstrated at last the truth of the Copernican theory, and opened yet another research program for astrometry, for the determination of the minute shifts was a method for directly measuring accurate stellar distances. Meanwhile the measurements of proper motions yielded another surprise: Bessel first announced in 1844 that unseen stellar companions were producing variations in proper motion of a few seconds to few tenths of a second amplitude in the stars Procyon and Sirius; the companions were confirmed by direct telescopic observation later in the century. Later in the nineteenth century, Chandler detected through stellar measurements the motion of the Earth's pole amounting to only 0.5 arcsecond/year (1885). By the mid-twentieth century, the discovery of unseen planetary companions, producing perturbations in proper motions or double star orbits of a few hundredths of a second of arc amplitude, was claimed. Like parallaxes before 1838, the claim of planetary companions has been controversial and is still very much in doubt. These advances went hand in hand with a plethora of advances in instrumentation and technique; the most important of these are discussed further in the section on instruments and methods, as well as in the entry on meridian instruments.

23.2.3 The Era of Space and Interferometry

Just as precisions for the naked-eye era eventually ran up against the inherent resolution of the human eye at about one arcminute, the traditional techniques of the telescopic era by the 1980s were producing diminishing returns despite heroic efforts, as atmospheric effects at optical wavelengths made it difficult to go beyond precisions of several hundredths of an arcsecond. Although lack of perspective often makes it difficult to declare the dawn of a new era during one's own time, modern astrometry is certainly entering such an era. The use of the transit circle, the prime instrument of large-angle astrometry for two centuries, has been on the decline for several decades, and is now in danger of extinction, as space satellites and the technique of interferometry (interference of light waves) are ushering in a new regime of accuracy that can circumvent or surmount the effects of the Earth's atmosphere. The European Space Agency's Hipparcos (High Precision Parallax Collecting Satellite) is measuring positions, proper motions and parallaxes for more than 100,000 stars to magnitude 13, with a mean error of about 0.002 arcsecond. The United States has plans for an Astrometric Imaging Telescope, and the Russians have announced a dedicated astrometric instrument known as Lomonosov. While funding is still uncertain for these spacecraft, the pioneering Hipparcos shows that astrometry above the Earth's atmosphere is one of the waves of the future. Meanwhile, ground-based optical and radio interferometry are also beginning to show results at the milliarcsecond level.

Interferometry in space promises even better accuracies in the future; one American proposal known as Points (Precision Optical Interferometer in Space) calls for 5 microarcsecond accuracy. With these techniques, the modern era holds the promise of producing the holy grail of astrometry—an inertial reference system—that is, one that does not change perceptibly over time and against which all other stellar, planetary, and Earth motions may be measured. What new phenomena may emerge out of this new regime of accuracy and stability cannot be predicted.

23.3 Transitions Between Eras

Inflection points in history always produce controversy, and the transitions between astrometric eras are no exception. The major controversy between the naked-eye and the telescopic era involved Hevelius and Flamsteed. Hevelius, the aging German astronomer of the Tychonic instrument school, used the quadrant with normal sights, but Flamsteed, the English Astronomer Royal still in his twenties, used a telescope with micrometer eyepiece on his quadrant and sextants. During the 1670s Hooke became the champion of telescopic sights, and in 1678 the Royal Society of London sent Halley to Hevelius's observatory at Danzig to adjudicate the issue of telescopic sights. If Hevelius did better than the one arcminute that Hooke argued was the resolution of the human eye, it was not much better, and more a testament to the remarkable eyesight of one man than a path to the future. It soon became clear that the future lay elsewhere.

A similar argument was gathering steam 300 years later during the transition from the telescopic to the interferometric and Space Age. In this case, however, there was little doubt that the new methods were more accurate, but 200 years' worth of refined technique does not die easily. The controversy revolved around how much overlap should exist between the old and new techniques, as well as the relative merit of proven versus non-proven techniques, and the economics and reliability of ground-based versus space-based techniques. Once again, however, it is likely that these were merely the arguments of a transition phase, and that the instruments that can prove themselves to produce reliable milliarcsecond accuracies will win out until the transition to the interferometric/space era of astrometry is as complete as the transition from the naked eye to the telescopic era of astrometry. This is already the case for the photographic zenith tube (PZT), which has been completely superseded by radio Very Long Base Interferometry (VLBI) for determining earth orientation parameters.

23.4 Instruments and Techniques: The Methods of Astrometry

The achievements of astrometry are intimately linked to improved instrumentation, and eventually to a theory of errors and procedures for combining observations to the best advantage. We now address in more detail, for the period beginning with telescopic sights, the development of the instruments and techniques to which astrometric achievement is so clearly tied.

23.4.1 Visual Methods

Visual methods—the combination of eye and telescope—dominated astrometry through the eighteenth and nineteenth centuries, and have been important through most of the twentieth century. Founded on the increasing mechanical and optical skills, particularly of British instrument makers John Bird, John Dollond, Edward Troughton, and Jesse Ramsden, visual astrometric instrumentation took many forms in the constant drive for increasing precision. The large quadrants gradually gave way early in the nineteenth century to full divided circles, which could be read by microscopes at several points and analyzed for errors. A Troughton mural circle (so called because it was mounted on a wall) was installed at Greenwich in 1812. Zenith instruments (more accurate because they were fixed) date back to the seventeenth century, were developed in their telescopic forms throughout the eighteenth century, and achieved notable form in Airy's reflex zenith tube (1851), which culminated in the photographic zenith tube used for the determination of variation of latitude and time. The equatorial refractor itself was an important astrometric instrument, especially when accompanied by refinements like the filar micrometer, or in specialized forms such as the heliometer, which originated with Dollond and utilized a split lens for astrometric measurement. By the end of the nineteenth century, the merits of long-focus refractors for astrometry were becoming clear, not for the large-angle task of determining a coordinate system, but for differential work in small fields, including the determination of double star orbits.

But for large-angle work, no instrument was more important to astrometry than the transit instrument and its later development, the transit circle (Fig. 23.1). It was first used by Ole Roemer in Copenhagen in 1689, and Halley mounted England's first transit instrument at Greenwich in 1721. A transit instrument could provide only right ascension, but when provided with a circle, it could also provide declination, and to such accuracy in both coordinates that it would eventually supersede the mural circle. The first successful transit circle was made by Troughton in 1806 for Groombridge. This form of instrument proved so successful that by the mid-nineteenth century no respectable observatory was without one. For 150 years they were the prime instrument of large-angle astrometry. During its century of operation beginning in 1851, the Airy transit circle at Greenwich made some 675,000 individual observations. Today, there are fewer than a dozen transit circles in operation in the world, and most of these are no longer visual instruments.

Fig. 23.1 The 8.5-in. transit circle of the U. S. Naval Observatory, mounted in 1866. The transit circle, standard equipment at most observatories during the nineteenth century, was used for the precise measurement of the positions of celestial objects in both coordinates. The instrument is now being superseded, even at the great national observatories. (Courtesy U. S. Naval Observatory)

The prominence of the English instrument makers was gradually overshadowed by events on the Continent. Faced with purchasing new instrumentation, both Pulkovo Observatory (1834) and the U. S. Naval Observatory (1842) turned primarily to the instrument makers of Germany. Both purchased achromatic refractors and smaller comet seekers from Merz and Mahler in Munich. Both purchased prime vertical transits, Pulkovo from Repsold and the Naval Observatory from Pistor and Martins in Berlin. And both purchased transit instruments from Ertel, successor to Reichenbach in Munich. In addition, Pulkovo purchased a large vertical circle from Ertel and a heliometer from Merz and Mahler. Only for the mural circle did the Naval Observatory turn to the English instrument maker Troughton and Simms; Pulkovo placed its faith in Repsold for its large meridian circle.

The English school differed from the German school in more than instrumentation. Bessel was the first to insist on the importance not merely of good instrumentation, but also of systematically understanding the peculiarities of each instrument, so that its errors could be taken into account to ensure maximum accuracy. This philosophy extended even to analyzing the observer; Bessel is credited with introducing the term personal equation to characterize the differences among observers in registering times of transit, a phenomenon found also in other measurements. It was not without reason that Bessel was among the first to detect stellar parallax and unseen companions; the combination of instrumentation and analytical methods that he championed brought astrometry to its heights by the end of the nineteenth century.

23.4.2 Photographic Methods

Photographic methods have been applied to astrometry for about a century. Seen by some as a method for abolishing personal equation, this goal was not entirely met because of the need to measure the star images on the plates—at least not until the 1960s, when automatic measuring engines came into use. Nevertheless, photography did provide immediately a permanent record and a faster technique whereby thousands of star images might be captured simultaneously. Photographic astrometry began with David Gill, who undertook at the end of the nineteenth century his own photographic survey of the southern sky at the Cape Observatory, the results of which were measured and analyzed by J. Kapteyn. But as with visual astrometry, photographic astrometry could also be beneficially used for small fields or individual objects; in the first decades of the twentieth century, Frank Schlesinger at Yerkes, Allegheny, and Yale observatories pioneered the application of photography to the determination of parallaxes using long-focus refractors.

Gill was also a key player in organizing the International Astrophotographic Congress held in Paris in 1887, a conference that coordinated the first international photographic survey of the sky. The resultant Carte du Ciel and its accompanying Astrographic Catalogue were landmarks in astrometric photographic technique. Twenty-two observatories around the world participated, and although the work stretched over many decades, the result is still valuable today.

The photographic technique has been steadily improved by better emulsions sensitive to various wavelengths, improved measuring machines, and new developments with telescope design, including the wide-angle reflecting telescope first developed by Bernhard Schmidt in 1932. Today the photographic method, both for surveys such as the National Geographic-Palomar Sky Survey and for long-focus work, is an essential aspect of astrometry. Though many images may be obtained on a photographic plate, compared with the single observations made with the transit circle, the latter (or its successors) is still necessary to provide reference stars for measuring the images on the photographic plate.

23.4.3 Photoelectric Methods and CCDs

Photoelectric methods are relatively new to the field of astrometry, becoming widespread only in the 1980s. Although several forms of photoelectric detectors exist, most are being superseded for astrometry by the charge-coupled device (CCD), a high-quantum efficiency solid-state detector. Applied to transit circles or more traditional telescopes, the CCD will capture images much fainter than is possible with traditional photographic methods and increase accuracies five to ten times. Although CCDs are not yet available for wide-field work, CCD arrays will put the photographic method in danger of extinction.

23.5 Interferometry and the Space Age

Interferometry, which uses the interference of light waves for its measurements, was a method developed by Albert A. Michelson in the late nineteenth century. The first stellar interferometer was not built until 1920, and only in the 1970s did it begin to be used for astrometry other than double star measurements. In the radio region of the spectrum, Connected Element Interferometry (CEI) uses nearby telescopes to collect data simultaneously, while Very Long Baseline Interferometry (VLBI) combines data from distantly located telescopes. As techniques of radio interferometry were refined in the 1980s, optical interferometry also began to be developed for astrometric uses. For smallfield work such as double star measurements, the technique of speckle interferometry combines very fast freeze-frame images to surmount some of the problems of atmospheric turbulence.

Spacecraft may potentially make use of photoelectric, CCD, or interferometric methods of astrometry. Though few are yet funded for actual construction, over the long term astrometric spacecraft will undoubtedly play an important part. Considering the expense of spacecraft design, construction, and operations, ground-based astrometry, especially interferometry, will also continue to play an essential role.

It is important to recognize that not all astrometric methods are useful for all purposes, and therefore the precisions quoted in Table 23.1 are not achievable for all applications. Visual transit circles can measure stars only to about tenth magnitude, but they can determine so-called fundamental catalogues in that they can also determine the origins of their coordinate systems. Photographic methods can measure many more stars and much fainter (the Astrographic Catalogue include stars to the twelfth magnitude, and the First Palomar Sky Survey to the twentyfirst), but reference stars are still needed. CCD methods can go even fainter, but only for very small fields. Similarly, long-focus astrometry is applicable only to small fields because such telescopes have only small fields. While very useful for double star, parallax, and unseen companion work, it is therefore not possible in practice to achieve milliarcsecond accuracy using a long-focus refractor except for such specialized purposes. Moreover, it is not possible for the Hipparcos satellite, for example, to produce by itself a fundamental reference frame (one whose origin is independently determined), unless it is linked by other means to a terrestrial or extragalactic reference frame. The latter is the subject of much work today.

23.6 Computer Methods for Analysis

An important characteristic of astrometric work is the care needed to analyze the observations; it is a maxim of the discipline that for every hour spent at the telescope, 10 h must be spent in analysis of the observations. For most of its history this has meant tedious hand calculations, sometimes aided by mathematical tables or a mechanical calculating device. No overview of methods in astrometry would be complete without mention of this aspect, or the complete revolution that electronic computers brought to the field. Beginning at about the dawn of the Space Age in 1957, large computers began to become widely available in astronomical institutions. The routine work of calculating star positions by rigorous methods was an excellent match for the computer's ability to process many steps quickly and accurately.

The last 35 years, therefore, have seen a radical change in the analysis of the observations of positional astronomy, as well as in the day-to-day routine of astronomers. The astrometrist has had to become adept at programming and has had to keep abreast of the constant improvements in computer methods. In addition to this routine reduction of observations, the ability of the computer to acquire and handle large sets of data, and to apply more sophisticated mathematical models and techniques, has allowed analyses to be undertaken that would have been prohibitively time consuming before.

23.7 Institutions and Their Research Programs: The Uses of Astrometry

The achievements of positional astronomy did not occur in a vacuum, but within institutions subject to a variety of political, social, and economic effects, all of which affected the extent to which technical advances could be implemented, and the means of implementation. Although we cannot here study the multitude of external effects that have impinged on astrometric science, in this section we clarify the context in which astrometric research programs are undertaken and the goals that drive the field.

23.7.1 Patronage

By contrast to the naked-eye era of positional astronomy, the telescopic era was motivated by navigational needs and so from an early date gained the patronage of national governments. National observatories were founded for just this reason (see Chap. 24). The royal warrant for the Greenwich Observatory specifically stated that its purpose was to provide data essential for navigation. At that time this meant the method of lunar distances, which required a detailed knowledge of the motion of the moon and the positions of the background stars. The Berlin Observatory was founded in the early eighteenth century to make observations necessary for calendar reform. And aside from navigation, the Paris, St. Petersburg, and Pulkovo observatories were deeply involved in triangulation methods of surveying their national lands by way of astronomical observation. It is no surprise, then, that national governments have traditionally been one of the primary supporters of astrometric work. The same still holds true today, although the celestial connections to navigation, for example, are no longer so direct, and astrometric work at observatories such as Greenwich has suffered as a consequence. In some countries, new government-supported research programs have arisen; in the United States, NASA, the U. S. Naval Observatory, the Jet Propulsion Laboratory, and the Naval Research Laboratory are still the prime supporters of new astrometric instrumentation and data analysis necessary for spacecraft navigation, geodesy, and astrophysics. In Europe the consortium of the governments that comprise the European Space Agency supports Hipparcos.

At the same time, other institutions have supported astrometric research, but their support has declined in proportion to the rise of astrophysics. While most university astronomy departments and observatories are unable to undertake the long-term programs characteristic of positional astronomy, a few (Yale, Yerkes, Lick) are still heavily involved, for example, in parallax and proper motion work, or in analyzing the results based on the larger programs. And while government agencies often take the lead financially, they are supported by university expertise, which may in turn be supported by government grants or contracts. The *Hipparcos* investigative teams are drawn from both government and universities all over Europe. But, in general, modern astrometry has become too much of a big science, and the output of its research programs too long term, to be heavily supported by universities.

23.7.2 Research Programs

The primary goal of astrometry has always been to provide more accurate positions for more stars at increasingly fainter magnitudes, and, after the discovery of galaxies and quasars, to provide an inertial reference system based on objects that appeared to be fixed. With such a reference system everything else could fall into place: parallaxes and proper motions could be determined, a reference system would exist for the motions of comets, asteroids, and planets, and the myriad motions of the earth could be measured. The shelves of any observatory library are lined with the results of this work of positional determination, from Ptolemy‘s catalogue of 1022 stars to Flamsteed and the Hubble Space Telescope. The Smithsonian Astrophysical Observatory Catalogue (1966) contains positions and proper motions of 258,997 stars, and the Space Telescope Guide Star Catalogue contains millions of stars. These are the long-term programs today largely undertaken by government institutions.

From stellar proper motions, the motion of the Sun through space and the rotation of the galaxy may be determined. The former was determined in the eighteenth century, and the latter was successful in the twentieth century, but these results are always subject to further refinement. These are not large research programs, and it is an interesting characteristic of the field that these analyses are often not undertaken by the data gatherers themselves, but by other workers at universities or observatories who are the data users. This is undoubtedly a reflection of manpower requirements, specialization, and constraints on mission-related work of government institutions.

Since the first valid determinations of stellar parallax in 1838, some 15,000 parallaxes have been measured. The goal behind these routine parallax programs is clear: astrophysically interesting conclusions can be drawn only when stellar distances are known, and the entire cosmic distance scale rests on trigonometric parallaxes. Parallaxes were relatively slow in coming during the nineteenth century, but with the application of photography to the long-focus refractor by Schlesinger as early as 1910, the great parallax programs began. Some 6000 parallaxes were known by 1938, more than half of them determined at Allegheny, McCormick, and Yale Johannesburg observatories. Adriaan van Maanen was the first to use the reflecting telescope for parallaxes, making use of both the 60- and 100-in. Mr. Wilson telescopes for more than 400 determinations. For much of the twentieth century, refractors continued in use for this purpose, but the 61-in. astrometric reflector of the U. S. Naval Observatory now is considered the standard ground-based instrument for stellar parallaxes. On the space-based side, *Hipparcos* will make an enormous contribution to the field. Like parallaxes, double star programs are very relevant to astrophysics because they provide the chief means of determining stellar masses; nevertheless, this demanding and long-term undertaking is now being carried out by only a few individuals.

Even though during the nineteenth century some observatories sold time as a means of financial support, the determination of time and the motions of the earth have become almost exclusively the province of government agencies because of their relevance to navigation and geodesy. Pure research goals are also met by these programs; NASA's Crustal Dynamics Program is a prime example.

Aside from these major programs, astrometry embodies a variety of special studies. In the nineteenth century, for example, many countries sent expeditions to determine the solar parallax, particularly by the method of the transit of Venus, a very rare phenomenon that occurs in pairs only once a century. A study of the internal motions of clusters of stars may lead to a determination of their ages. Early in the twentieth century, Adriaan van Maanen claimed that internal motions in spiral nebulae proved they were internal to our own galaxy. But in a cautionary note on the difficulties of astrometric work, van Maanen was proven wrong; the spiral nebulae were later proven to be independent galaxies. One of the more exciting claims of astrometry is the possible existence of planetary systems that might contain earth-like planets. Once again, these are extremely difficult measurements, and the existence of such systems is as yet unproven.

23.8 The Human Element

Finally, we should not lose sight of the human aspect of astrometry. The achievements of the discipline are the result of centuries of painstaking individual and team efforts at data gathering, not always the most glamorous of scientific endeavors, but surely among the most necessary. Hundreds of astronomers, opticians, and technicians, collectively working some millions of hours, have been seriously involved in astrometry over the course of history, not a large number in the course of human events, but certainly a good fraction of the astronomical community's efforts. Gradually extracting nature's secrets through some of the most difficult measurements known to science, they have produced multitudes of star catalogues and almanacs, measured the universe from the Solar System to the quasars, determined the structure of the galaxy, and perhaps detected the effects of remote planets on which life might dwell. Finally, the field now stands on the verge of producing astrometry's Holy Grail—an inertial reference system better than ever before, though never perfect. These achievements are no small tribute to the past, and no small legacy for the future.

Because astrometry has comprised so much of the field of astronomy since its beginnings, the general histories of astronomy cover broad aspects of the development of astrometry. More specific aspects are covered in biographies of key figures (concisely in the *Dictionary of Scientific Biography* or in more detail in volumes such as Thoren's biography of Tycho Brahe), in institutional histories of the national observatories, in annotated translations of seminal works (such as Toomer's on Ptolemy's *Almagest*), and in modern astrometric conferences with historical components (Débarbat et al. 1988; Lieske and Abalakin 1990). Very little research, on the other hand, has been undertaken with broad analytic questions in mind. The problems of patronage, research programs, institutional cooperation, interaction of astrometry and astrophysics, and many others remain ripe for historical inquiry.

Because astrometry was so important to national governments, key archival collections will be found in the archives of the national observatories, such as the U. S. Naval Observatory (U. S. National Archives), Royal Greenwich Observatory (Cambridge University Library), and the Paris Observatory. Much more of interest will be found in collections of individual practitioners or their institutions scattered around the world.

23.9 Commentary 2020

Since this article was published in 1997, the statement that "the pioneering Hipparcos shows that astrometry above the Earth's atmosphere is one of the waves of the future" has proven true in spades. In fact it is fair to say that in the late twentieth and early twenty-first centuries astrometry has been dominated by specialist astrometric spacecraft, especially the European Hipparcos satellite and its successor, Gaia. By

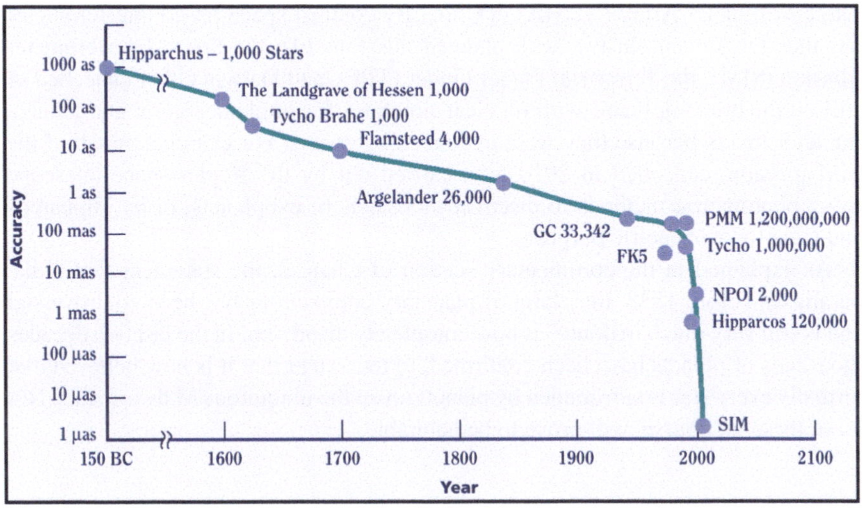

Fig. 23.2 Astrometric accuracies, showing the "Golden Age of Astrometry," beginning in the 1980s, as accuracies surpassed the milliarcsecond range. (Source: U. S. Naval Observatory.) SIM, the Space Interferometry Mission, never flew, but Gaia was a worthy successor to the Hipparcos spacecraft

providing positions, proper motions, and distances to billions of stars, these satellites have revolutionized astronomy in a number of ways evident at the Gaia website https://sci.esa.int/web/gaia/.

Figure 23.2, a cruder but more graphic version of Table 23.1, shows how accuracies have improved over two millennia. This revolution in astrometry has not come easily, however. Several astrometric satellites, including the Full Sky Astrometric Explorer (FAME) of the U. S. Naval Observatory and Naval Research Laboratory, never made it to fruition. FAME was scheduled for launch in 2004 as a NASA MIDEX mission, but was cancelled in 2002 due to budget overruns and technical issues, leaving the field of space astrometry largely to the Europeans.

As a corollary to the success of space-based astrometry, the meridian transit circle as seen in Fig. 23.1 has become extinct, a classic example of an instrument being overtaken by new technology. At the U. S. Naval Observatory, for example, the 6-in. transit circle (successor to the instrument in Fig. 23.1), having made almost a million individual observations over a century of service, terminated its programs in 1995. Its pole-to-pole program in coordination with a 7-in. transit circle in New Zealand, in which I participated for many years, was entirely superseded by Hipparcos, which could determine stars positions 100 times more accurately and at a much faster pace (Dick 2003, 440). One way or another it was the same story for all the other transit circles around the world, including the famous Airy transit at Greenwich straddling the prime meridian.

Ground-based interferometry has advanced over the last two decades, as in the case of the Very Large Array (VLA) of radio telescopes in New Mexico and Atacama

Large Millimeter Array (ALMA) in Chile. By contrast space-based interferometry has thus far proven elusive, with planned missions like the Space Interferometer Mission (SIM), the Terrestrial Planet Finder (TPF), and Darwin either cancelled or still on the drawing board with no clear timelines for implementation and launch. Some of this is because they were overtaken by events. For example, much of the SIM mission, cancelled in 2010, was carried out by the Kepler space telescope using photometric methods to discover thousands of exoplanets, or by Hipparcos and Gaia for astrometric purposes.

As explained in the commentary section of Chap. 2, the statement that "Like parallaxes before 1838, the claim of planetary companions has been controversial and is still very much in doubt" is now completely disproven. In the last two decades thousands of planets have been confirmed, to the extent that it is now believed that virtually every star is surrounded by planets, even the ubiquitous M dwarf stars. Not all of these, of course, will prove to be habitable.

References

Boss, B. 1937. *General Catalogue of 33,342 Stars*. Washington: Carnegie Institution.

Chapman, A. 1983. "The Accuracy of Angular Measuring Instruments Used in Astronomy between 1500 and 1850." *Journal for the History of Astronomy*, 14, 133–137.

Chapman, A. 1990. *Dividing the Circle: The Development of Critical Angular Measurement in Astronomy, 1500–1850*. New York: Harwood.

Dahn, C.C., D.G. Monet, et al. 1990. "Results from the First List of USNO CCD Parallaxes." *Bulletin of the American Astronomical Society* 22: 820.

Débarbat, S., J.A. Eddy, H.K. Eichhorn, and A.R. Upgren, eds. 1988. *Mapping the Sky: Past Heritage and Future Directions*. International Astronomical Union Symposium 133. Dordrecht: Kluwer.

Dick, Steven J. 2003. *Sky and Ocean Joined: The U.S. Naval Observatory, 1830–2000*. Cambridge: Cambridge University Press.

Douglass, G. G. and R.S. Harrington. 1990. "The U.S. Naval Observatory Zodiacal Zone Catalog." *Astronomical Journal*, 100, 1712–1715.

Eichelberger, W.S., and H.R. Morgan. 1920. *Results of Observations with the Nine-Inch Transit Circle, 1903–1911, Publications of the United States Naval Observatory*, vol. 9, part I. Washington: Government Printing Office, 1920.

Eichhorn, H. 1974. *Astronomy of Star Positions*. New York: Frederick Ungar.

Evans, J. 1987. "The Origin of the Ptolemaic Star Catalogue, Part 1." *Journal for the History of Astronomy* 18 (1987): 155–172.

Forbes, E.G. (vol. 1), A.J. Meadows (vol. 2), and D. Howse (vol. 3). 1975. *Greenwich Observatory: The Royal Observatory at Greenwich and Herstmonceux,1675–1975*. London: Taylor and Francis

Fricke, W. 1970. "Bessel." In *Dictionary of Scientific Biography*, edited by C.C. Gillespie, vol. 2, 97–102. New York: Scribner.

Gill, D. 1913. *A History and Description of the Royal Observatory, Cape of Good Hope*. London: His Majesty's Stationery Office.

Harrington, R.S., C.C. Dahn, et al. 1985. "U.S. Naval Observatory Parallaxes of Faint Stars. List VII." *Astronomical Journal* 90, 123–129.

Kremer, R.L. 1980. "Bernard Walther's Astronomical Observations." *Journal for the History of Astronomy* 11, 174–189.

Krisciunas, K. 1993. "A More Complete Analysis of the Errors in Ulugh Beg's Star Catalogue." *Journal for the History of Astronomy* 24, 269–280.
Lieske, J.H., and V.K. Abalakin. 1990. *Inertial Coordinate Systems on the Sky.* International Astronomical Union Symposium 141. Dordrecht: Kluwer.
Ma, C. 1990. "Realization of an Inertial Reference Frame from Mark III VLBI." In J. H. Lieske and V. K. Abalakin, eds., *Inertial Coordinate Systems on the Sky,* International Astronomical Union Symposium 141, 271–280. Dordrecht: Kluwer, 1990.
Monet, D.G. 1988. "Recent Advances in Optical Astrometry." In *Annual Review of Astronomy and Astrophysics,* edited by G. Burbidge, D. Layzer, and J.G. Phillips, vol. 26, 413–440. Palo Alto: Annual Reviews.
Newton, R.R. 1977. *The Crime of Claudius Ptolemy.* Baltimore: Johns Hopkins University Press.
Pannekoek, A. 1961. *A History of Astronomy.* London: George Allen and Unwin.
Perryman, M.A.C. 1990. "In-Orbit Status of the Hipparcos Astrometry Mission." In Lieske and Abalakin (1990): 297–305.
Podobed, V.V. 1964. *Fundamental Astrometry.* Chicago: University of Chicago Press.
Schlesinger, F. 1911. "Photographic Determinations of Stellar Parallax Made with the Yerkes Refractor." *Astrophysical Journal* 33, 8–27.
Shevchenko, M. 1990. "An Analysis of Errors in the Star Catalogues of Ptolemy and Ulugh Beg." *Journal for the History of Astronomy,* 21: 187–201.
Thoren, V. 1973. "New Light on Tycho's Instruments." *Journal for the History of Astronomy* 4, 25–45.
Thoren, V. 1990. *Lord of Uraniborg: A Biography of Tycho Brahe.* New York: Cambridge University Press.
Toomer, G.J., trans. and ann. 1984. *Ptolemy's Almagest.* New York: Springer-Verlag.
Van Altena, W.F. 1983. "Astrometry." In *Annual Review of Astronomy and Astrophysics,* edited by G. Burbidge, D. Layzer, and J. G. Phillips, vol. 21, 131–164. Palo Alto: *Annual Reviews.*
Wesley, W.G. 1978. "The Accuracy of Tycho Brahe's Instruments." *Journal for the History of Astronomy* 9, 42–53.

Chapter 24
Pulkovo Observatory and the National Observatory Movement: A Historical Overview

Abstract The patronage of national governments has played an important role in the history of astronomy, classically in the form of national observatories. In this paper we (1) argue that the last three centuries have seen what we may call a "national observatory movement," in that national governments during this period increasingly supported astronomical observatories, and in that such institutions share certain common properties of origin, purpose and evolution; (2) demonstrate the important role that Pulkovo Observatory has played in this movement; and (3) compare certain aspects of the Pulkovo Observatory and the United States Naval Observatory (USNO) as exemplars of this species founded within a decade of each other under very different political conditions.

24.1 The National Observatory Movement

The phrase "national observatory movement" implies an ongoing process with definite goals shared by members of the national observatory community. The term has been applied before to attempts to form national observatories within particular countries, such as the United States (Paullin 1923) and India (Ansari 1985), but we apply it here in a more encompassing global sense, where the national observatories of particular nations are seen as members of the movement. We may consider that this broader process began, after the important but abortive founding of Tycho Brahe's observatory by Frederick II of Denmark in 1576, with the Paris Observatory (1667) and the Royal Observatory at Greenwich (1675). But the need for national observatories did not end in France and England. As Table 24.1 shows, Germany and Imperial Russia added two important examples in the early eighteenth century, the rate of new institutions actually increased in the nineteenth century, and new members have continued to be added in the twentieth century. It is clear that all national observatories are not listed in Table 24.1, particularly those of the Far East. A more comprehensive list would be an interesting and major task, but the sample given here is large enough to demonstrate several points and the overall benefits of

First published in *Inertial Coordinate Systems on the Sky*, Proceedings of International Astronomical Union Symposium 141, Leningrad, October 17–21, 1989, Jay H. Lieske and Victor K. Abalakin, eds. (Dordrecht, 1990), 29–38.

S. J. Dick, *Space, Time, and Aliens*, https://doi.org/10.1007/978-3-030-41614-0_24

Table 24.1 Some important national observatories and their patrons

Institution	Founded[a]	Patron
First era		
Uraniborg (Tycho Brahe)	1576 (abortive)	Frederick II
Paris Observatory	1667	Louis XIV
Royal Observatory, Greenwich	1675	Charles II
Berlin Observatory	1701	Frederick I
St. Petersburg Observatory	1725	Peter the Great
Second era		
Royal Observatory, Cape	1820	Britain
U. S. Naval Observatory	1830	U.S. Navy
Pulkovo Observatory	1839	Nicholas I
Chilean National Observatory	1852	Chile
Argentine National Observatory	1870	Argentina
Potsdam Astrophysical	1874	German Acad. Science
Smithsonian Astrophysical	1891	Smithsonian/U.S.
Dominion Observatory	1903	Canada
Dominion Astrophysical	1918	Canada
Third era		
NRAO (U.S.)	1956	NSF/AUI
Kitt Peak	1957	NSF/AURA
NRAO (Australia)	1959	CSIRO
Cerro-Tololo Inter-American	1963	NSF/AURA/Chile
European Southern Observatory	1964	Five countries (now 8)
Anglo-Australian (Siding Spring)	1967	Britain/Australia
Space Telescope Science Institute	1981	NASA/AURA

[a]A number of criteria can be used for founding dates. The majority of dates here indicate when funding was assured. Abbreviations: *AUI* Associated Universities, Inc., *AURA* Association of Universities for Research in Astronomy, *CSIRO* Commonwealth Scientific and Industrial Research Organization, *NASA* National Aeronautics and Space Administration, *NSF* National Science Foundation, *NRAO* National Radio Astronomy Observatory

this institutional approach to the history of astronomy. It should also be clear that there are many other “astronomical centers of the world” (Krisciunas 1988) that may receive partial government funding but are not considered national observatories.

We may distinguish three eras in this movement: the *first era*, in which the prototype Paris, Greenwich, Berlin, and St. Petersburg observatories were founded; the *second era*, characterized by offshoots from previous national observatories (Royal Observatory Cape), by new observatories of younger nations (USNO), and by the rise of astrophysical observatories; and the *third era*, post-World War II, characterized by national or international consortia, large budgets relative to the previous eras, and the study of old and new wavelength regions with increasingly sophisticated telescopes, detectors and spacecraft. National observatories were the original “big science” of their time, but in this third era, which encompasses both the

Computer Age and the Space Age, the movement has benefited from the general trend toward even bigger science seen in the national laboratories of many disciplines (Price 1963; Weinberg 1967).

Aside from the striking association of all the early institutions with their national scientific societies (whether Academies of Science or the Royal Society), the common property that stands out in the first era is the largely practical nature of the work for which the first national observatories were founded. Whether for the improvement of navigation, geographic and geodetic work, or calendar reform, all these institutions were founded to meet a national need. In meeting these national needs, the precise determinations of celestial positions formed the backbone of much of their work; for example, the method of lunar distances for navigation required precise ephemerides of the Moon and precise positions of the stars as the reference frame, a task that was brought to fruition only with Maskelyne's publication of the British *Nautical Almanac* in 1766 (Sadler 1976). Byproducts of this practical work were the great star catalogues of Flamsteed and others, Bradley's discovery of the aberration of light, the determination of proper motions, and many other results also of interest to pure astronomy (see Table 23.1).

The early institutions of the second era were also founded for similar purposes, but now with the determination of longitude by chronometers as the most promising method of navigation at sea, and an important method for determining geographical positions as well. With accuracies now on the order of tenths of arcseconds vs. about 15 arcseconds for Flamsteed (Chapman 1983), greatly improved star catalogues were the byproducts of the Cape, U. S. Naval, Pulkovo, Chilean, and Argentine observatories, whereas in this era of improved instruments stellar parallaxes too were at last possible. With the rise of the astrophysical national observatories at the end of the nineteenth century, results beyond any practical need were the goal, and the question of how far public money should support such research became increasingly important. The very existence of the third era gives an answer to that question, for the research of these observatories has gone beyond anything that the public would consider practical. The longevity of the older institutions, as well as the occasional addition of new ones, attests to a continuing national necessity, though one expanded to include the benefits of pure research, whether to national prestige or the advancement of astronomy.

A second common characteristic of national observatories, rooted in the first property of common purpose in the service of a perceived national necessity, is that most of them at the time of their founding undertook similar programs in order to carry out their tasks. This led to substantial interactions and cooperation among national observatories, not only in international programs such as the transits of Venus, the *Carte du Ciel*, the Eros campaign and long-term astrometric projects such as the AGK3R and Southern Reference Star systems, but also in many colorful episodes in the history of astronomy. A notable example of the latter is the involvement of the observatories of Paris, Greenwich, and Berlin in the discovery of Neptune. Of particular interest to this meeting (Fig. 24.1), the work of determining fundamental reference frames has largely fallen to these national observatories, for few other observatories can undertake such long-term programs.

Fig. 24.1 Astronomers in the history session at the Leningrad meeting commemorating the 150th anniversary of Pulkovo Observatory in Leningrad, October, 1989. Left to right Kevin Krisciunas (Joint Astronomy Centre, Hawaii), Alan Batten (Dominion Astrophysical Observatory, Canada), Steven Dick (U. S. Naval Observatory), Suzanne Débarbat (Paris Observatory), Viktor Abalakin (Director of Pulkovo Observatory), and Wolfgang Dick (Zentralinstitut für Astrophysik, German Democratic Republic)

National Observatories have also faced in turn many similar problems and issues: the proper balance between practical and pure research, the relative roles of astrometry and astrophysics, the most efficient mode of administration, and patterns of funding, to name only a few. It is true that all astronomical institutions share to some extent these problems, but government institutions, particularly in regard to a mission to be accomplished, share them in peculiar and related forms.

Finally, it is notable that national observatories have historically been perceived as a distinct group of astronomical institutions, both by the institutions themselves and externally. Aside from cooperative programs, the younger national observatories such as Pulkovo and USNO often consulted their predecessors regarding buildings, instrumentation, and programs; visits and comparisons were often made among the national observatories to improve efficiency; and moral support was given in times of crisis such as threats of retrenchment or even abolishment. For all their similarities, each observatory has developed its own character, as it has met these problems and challenges in different ways. Even with obvious differences reflecting national styles, a global view of the history of astronomy shows that national observatories, with their common purpose, common programs and common problems, form a distinct group of institutions and comprise a worldwide national observatory movement.

24.2 The Place of Pulkovo Observatory

Let me proceed to the second point, namely the role of Pulkovo Observatory in this movement. We note first from Table 24.1 that Pulkovo Observatory was founded in what we have called the second era of national observatories. By the 1830s the older national observatories were in many ways in their prime. In 1835 G.B. Airy began his tenure of almost a half-century as Astronomer Royal at RGO (Meadows 1975). At Paris Observatory the reign of Jean Baptiste Joseph Delambre was over, Alexis Bouvard had been Director since 1822, and Francois Arago and Urbain Leverrier were coming on stage (Débarbat et al. 1984). In Berlin J.F. Encke was well into his long tenure at a revitalized observatory newly constructed in 1828 (Dick 1951).

In Russia itself, the St. Petersburg Observatory was now in a run-down condition, and its demise figured directly in the founding of Pulkovo Observatory. Like the Berlin Observatory a few years earlier, Pulkovo thus began with new buildings and instruments on a new site, but unlike Berlin, it also began with a new head, Wilhelm Struve, who as the head of the Dorpat Observatory in Tartu had completely outdone the St. Petersburg Observatory. Struve relates how in December, 1830 he frankly told Tsar Nicholas I of the sad condition of the St. Petersburg observatory. The result was the Tsar's decision to build an observatory near the capital, to be located on the hill of Pulkovo outside the city (Fig. 24.2). The details of Pulkovo's founding, and the life and work of the Struves, are by now well known, thanks to Struve's 1845 *Description*, and more recently the work of Daedev (1972) and Sokolovskaya (1976) in Russia, and Krisciunas (1978, 1988) and Batten (1988) in the United States and Canada. For the record I show in Table 24.2 some of the more important dates in the founding of Pulkovo, taken from Struve.

Pulkovo was one of the few observatories to have so many interests so early in its career. The official statute for the observatory specified not only the perfection of navigation and geographical positions, but also the advancement of astronomy beyond any practical need. While it is clear is that Struve took seriously the practical aspects of his job, his contributions to the advancement of astronomy are even better known. The breadth of interest and the development of the work at Pulkovo in its first 50 years may be seen from Table 24.3, according to categories established by Struve himself.

We see from this table that the work of the Pulkovo Observatory during its first 50 years was dominated by sidereal and solar system studies, some of which contrasted with the more purely practical work such as the measure of the arc of the meridian. The stellar work included double stars, parallaxes, positional catalogues, and the constants of precession, aberration, and nutation; and the planetary work

Fig. 24.2 Pulkovo Observatory in 1855, as depicted in a woodcut by Ev. Vernardsky (1819–1889)

Table 24.2 Important dates in Pulkovo's early history

1830 December	Struve meets with Nicholas I
1833 October 28	Nicholas orders purchase of instruments
1834 February 24	Architects present their plans to the Commission
1834 April 15/3	Nicholas names Struve Director
1834 June–October	Struve's trip abroad to purchase instruments
1835 July 3/June21	Foundation stone laid
1839 August 19/7	Official inauguration
1845	Struve's *Description* published

Source: Struve (1845). Double dates represent New Style/Old Style

consisted largely of comets, satellites, and eclipses. As a member of the second era of national observatories, its practical problems of navigation and geodesy were more refined than the problems that faced its older predecessors at their founding. The chronometer method, rather than the method of lunar distances, was on the ascendancy, with all that implied not only for navigation but also for geographic positions by chronometer expeditions, the latter for which Pulkovo is famous. We also note from this table the early entry of Pulkovo into the realm of astrophysics; from 1864 to 1888, 10% of the publications dealt with astrophysics, while the number of geodesic and geographic publications declined by half relative to the first 25 years.

Table 24.3 Research publications at Pulkovo, 1839–1888

Subject	1839–1864		1864–1889		Differences
Sidereal	49	30%	63	27%	−3
Solar system	42	26%	54	23%	−3
Practical astronomy	14	9%	21	9%	0
Geodesy and geographical positions	36	22%	22	9%	−13
History, bibliography and physical experiments	22	14%	33	14%	0
Astrophysics	0		23	10%	+10
Mathematics, physics and theoretical studies			21	9	+9

Sources: Struve (1865) and Pulkovo Observatory (1889)

Pulkovo Observatory quickly became legendary among national observatories. Early visitors to the observatory, such as the Astronomer Royal George B. Airy, and the American astronomers Benjamin A. Gould (1849) and Cleveland Abbe (1868), left fascinating accounts of their visits. On his return from St. Petersburg in 1847, Airy typically remarked that " … no astronomer can feel himself perfectly acquainted with modem observing astronomy in its most highly cultivated form, whether as regards the personal establishment, the preparation of the buildings, the selection and construction of the instruments, or the delicacy of using them, who has not well studied the Observatory of Pulkowa" (Airy 1848). This is high praise indeed coming from the head of another national observatory, especially one with the personality of George Airy. In Russian history of science Pulkovo represents a revitalization of government support for astronomy in that country; in the international sense it represents the addition to the community of national observatories that quickly became a model for the others.

24.3 Pulkovo Observatory and the United States Naval Observatory

The detailed comparative history of astronomical institutions, especially those in a similar class such as national observatories, is an important task. It is, however, a very large one that cannot be carried out in this paper, and that indeed cannot be carried out fully until the individual histories of these institutions are written. But I would like to touch here on a more manageable task: the comparison of USNO and Pulkovo in just a few areas during their first 50 years, as two exemplars founded very close together in time, though very far in political systems. This comparison will serve to illustrate some of the similarities and differences, some of the variations on the theme of national observatories.

We may briefly compare the two institutions in four areas during the first half century of their lives: origin, administration, instruments, and programs. I will conclude with some remarks on twentieth-century political effects on these institutions.

24.3.1 *Origins*

Whereas as we have seen in Russia, Nicholas I was convinced of the need for a new observatory and his wishes were of course carried out, in the United States, President John Quincy Adams had called for such an observatory in 1825, but it was expressly forbidden by Congress as a waste of money (see Chap. 24). There were advantages to having a czar if he happened to be on your side. Thus in November, 1830 it was left for a lowly Navy Lieutenant to recommend to the Secretary of the Navy that a Depot of Charts and Instruments be founded in Washington, DC. This direct forerunner to the Naval Observatory was established in December, 1830, the very same month that Struve had his audience with Nicholas I. It was founded to care for navigational instruments and to rate chronometers by astronomical observation, and here was the opening to astronomy. It thus had only very modest astronomical instruments, and was not yet an observatory in the true sense of the word.

The founding of Pulkovo Observatory in 1839 did not go unnoticed in the United States. John Quincy Adams, now an ex-president, pointed out to Congress a few months later "Here is the sovereign of the mightiest empire and the most absolute government upon earth, ruling over a land of serfs, gathering a radiance of glory around his throne by founding and endowing the costly and most complete establishment for astronomical observation on the face of the earth ... The committee of the House [of Representatives] ... in casting their eyes around over the whole length and breadth of their native land, must blush to acknowledge that not a single edifice deserving the name of an astronomical observatory is to be seen" (Rhees 1879).

Though Adams' arguments were to no avail, another Navy Lieutenant, James M. Gilliss, officer-in-charge of the Depot of Charts and Instruments, pushed through Congress a bill for a new Depot in 1842, which became a national observatory in fact if not yet in name. The purposes of founding for both Pulkovo and USNO were thus similar, but the purpose of the USNO was more limited at first, with the determination of geographical positions being left to the Army and the Coast Survey and a role for less practical aspects of astronomy only gradually evolving (Dick 1980).

24.3.2 *Administration*

Whereas Pulkovo was founded by Royal decree and came under its country's Academy of Sciences, the U. S. Naval Observatory was totally under the control of the Navy, and headed at first by Navy Lieutenants and later Navy Captains or Admirals. While the distinguished astronomer Struve brought instant fame to Pulkovo Observatory, the Naval Observatory was at first better known for Matthew Maury 's oceanographic wind and current charts than for its astronomy. This illustrates an important point: that even though national observatories are bound to carry out certain specific duties necessary for national needs, the character of the institution can be largely shaped by the personality and interests of its Director and senior staff, in addition to the overriding importance of its original charter and administrative sponsor. In the United States it is perhaps not surprising that oceanography rather than astronomy dominated the observatory's first 20 years; there were as yet no distinguished astronomers in the country, indeed no tradition of astronomy to draw on as there was in Russia. Only in the post-civil war era, with the names of Simon Newcomb, Asaph Hall, William Harkness, and G.W. Hill, did the Naval Observatory achieve fame in astronomy.

24.3.3 Instruments

In 1834 Struve journeyed to Europe to obtain instruments, and in 1842 Lt. Gilliss made a similar journey. As Table 24.4 shows, both turned to the expert instrument makers of Europe, and their choices are an interesting testimony to the rise of the German astronomical technology as opposed to the English. Both obtained transit instruments from the German maker Ertel (successor to Reichenbach at Munich). Both obtained achromatic refractors and comet seekers from Merz and Mahler (successor to Fraunhofer at Munich). Struve opted for a Repsold prime vertical transit instrument and Gilliss for a Pistor and Martins from Berlin. Whereas Merz and Mahler made a 15-in. refractor for Struve, Gilliss could only afford a 9.6-in., exactly similar to the Dorpat telescope Struve had previously used. Only with the mural circle did Gilliss show some faith in English instrument makers; Struve showed none.

Though the U. S. Naval Observatory began with a relatively small achromatic refractor, it is notable that by 1873, American technology had advanced to the extent that the Naval Observatory had the Alvan Clark 26-in. refractor, the largest in the world (Fig. 34.3). In another example of direct interaction between the two observatories, Otto and Hermann Struve visited the U. S. in 1879, where Newcomb persuaded them to purchase from Alvan Clark the 30-in. lens for their new retractor. This turnabout over a period of 30 years is testimony the rapid rise of astronomy, and astronomical technology in America.

Table 24.4 Early instruments at Pulkovo and USNO

Pulkovo (1839)	USNO (1944)
15-in. achromatic refractor (Merz & Mahler, Munich)	9.6-in. achromatic refractor (Merz & Mahler, Munich)
Heliometer (Merz & Mahler)	
Small parallactic refractor	
Two comet seekers (Merz & Mahler)	3.9-in. comet seeker (Merz & Mahler)
Large meridian telescope (Ertel)	
Large vertical circle (Ertel)	
Large meridian circle (Repsold)	4-in. mural circle (Troughton & Simms)
Prime vertical transit (Repsold)	5-in. prime vertical transit (Pistor & Martins)

24.4 Programs

Like Pulkovo the Naval Observatory carried on long-term programs for star catalogues, double stars and specialized observations, leading in some cases to direct cooperation, most recently in the form of the Southern Reference Star (SRS) catalogue. If we were to categorize publications at the USNO by subject over its first 50 years, similar to Table 24.3, a larger proportion of practical work would be evident at the Naval Observatory, again due largely to Maury.

Unlike Pulkovo, where we have seen in Table 24.3 a substantial amount of research in astrophysics by 1888, the Naval Observatory did not undertake a sustained program in astrophysics until John Hall's work on interstellar polarization beginning in 1948. The difference lies partly in a more restricted concept of mission at the USNO, but also in individual differences—whereas Struve pushed astrophysics at Pulkovo beginning in 1866, Newcomb in the 1890s was still questioning its importance. With the founding of the Smithsonian Astrophysical Observatory in the 1890s, the USNO left the subject to such specialized observatories, or to university or privately endowed observatories such as Lick Observatory, at least until the twentieth century, when the close relationship between astrometry and astrophysics became evident.

24.5 Political Events

It is clear that both Pulkovo and the U. S. Naval Observatory have been subject to political and economic events in their own countries. The 1930s was a particular time of crisis for both observatories. In the United States in 1932 President Herbert Hoover, as part of his policy to counter the Great Depression, recommended that the Naval Observatory be transferred from the Navy to the Commerce Department, a proposal seriously considered by Congress, but not carried out. Hardly had this crisis been weathered when President Roosevelt came into office and proposed abolishing the Naval Observatory completely in a money-saving move, a proposal obviously also not carried out.

In the Soviet Union the situation for Pulkovo Observatory during the same decade was even more serious. Not only were the threats more serious, they were followed by action, as Pulkovo Observatory was caught up in Stalin's purge of Soviet astronomers in 1936–1937. Throughout the Soviet Union 29 astronomers were arrested, many never heard from again. At Pulkovo alone 13 scientists, almost half the staff, disappeared during these years. This tragic event has recently been studied in detail by McCutcheon (1989). Nor did this end the problems of Pulkovo Observatory; it was completely destroyed in 1942 during the war. With all of its scientific achievements, we should not forget that the very existence of the Pulkovo Observatory today is a testimony to endurance and hope.

24.6 Summary and Conclusions

Many more comparisons could be made, but these examples suffice to indicate some of the interesting problems that arise in the comparative history of that subset of astronomical institutions known as national observatories.

In summary, looking back on the last three centuries we may conclude that national observatories as a group do indeed represent a related process responding to the needs of each nation; that Pulkovo arose suddenly like a bright star in this movement due largely to the Struves, who not only filled a national need but also greatly advanced astronomy; and that despite their similarities, national observatories may evolve in quite different ways, as a result of political systems, science organization, and individual enterprise, as shown in the case of Pulkovo and its American counterpart.

It is clear that over the centuries national observatories have achieved many of their goals, perhaps too well. Navigation and other practical needs are no longer the driving force for many of them, and most have changed in ways their founders could have hardly foreseen. The Berlin Observatory has gone through many reincarnations, Cape Observatory was amalgamated with others, and Greenwich Observatory is at present undergoing radical change.

Despite the appearance of new evolutionary forms in the movement, such as the multinational European Southern Observatory and the Space Telescope, these developments warn us that the future of the national observatory movement is not clear. Among astronomical institutions, national observatories especially face the problem of justifying their relevance to national needs. They must surely change, not only in advancing new techniques and instrumentation, but also in reshaping their goals to meet modern requirements. Just as surely, they are bound to be challenged about the relevance of any new goal to national needs, and affected by national priorities and economies much more directly than their colleagues at private institutions or universities. Perhaps the comparative study of the histories of these institutions will help to illuminate future directions, as well as past patterns. The road taken will determine the fate of this venerable movement, now well into its fourth century, that has contributed so much to astronomy.

24.7 Commentary 2020

This paper was delivered at the 141st Symposium of the International Astronomical Union, held at Leningrad, USSR from October 17 to 21, 1989 on the occasion of the 150th anniversary of Pulkovo Observatory (Fig. 24.3). The circumstances were memorable, as the theme of cosmos and culture was played out in real time during the meeting. A series of revolutions had taken place earlier in the year in Hungary, Poland and other East European states, and just as the meeting was taking place events in East Germany were coming to a head. Erich Honecker, the Head of State

Fig. 24.3 The Pulkovo Observatory building as depicted on its 150th anniversary commemorative medallion. The text translates "Principle Astronomical Observatory of the Russian Academy of Sciences"

of the German Democratic Republic (better known as East Germany), resigned on October 18, the second day of the meeting. The East German delegates at the meeting, including astronomer Wolfgang Dick (pictured in Fig. 24.1), followed these events with great interest. The fall of the Berlin Wall beginning November 9 culminated in the opening of the Brandenburg Gate in December. Meanwhile, events in the Soviet Union were also rapidly leading to its downfall that culminated in 1991, when Leningrad was renamed to Saint Petersburg as part of the fallout. Meanwhile, as the meeting opened more literal earth-shaking events were taking place in the United States, as the Loma Prieta earthquake caused heavy damage in parts of California, including the San Francisco Bay area. Communications at that time in the days before e-mail, social media and the internet were not good, and as American participants watched scenes of fire on the limited access Russian TV, it was unclear just how extensive the damage was. Meanwhile, papers at the meeting were presented as scheduled.

Despite their obvious importance, with few exceptions national observatories have been the object of little systematic study, especially when this article was written. Even less have they been studied in a comparative manner, as a class of scientific institutions. The cautionary note about national observatories in the last section of this chapter has proven prescient. The famed Greenwich Observatory, which in 1947 had moved to Herstmonceux Castle 50 miles southeast of Greenwich in East Sussex, moved again from Herstmonceux to Cambridge in 1990. It was closed in 1998 and its landmark library was disbursed. The original site of the Observatory in Greenwich near London now operates largely as a museum, where visitors can still have their picture taken on the prime meridian straddling East and West, visit the decommissioned Airy transit circle, and watch the time ball drop at 1 pm every day (see Chap. 26). They can also visit the National Maritime Museum down the hill, where they can see the Harrison chronometers as well as numerous other historical instruments.

As for Pulkovo, it still exists but has undergone rough times. Its original buildings were destroyed by German air raids during the siege of Leningrad of World War II. In 1997, 8 years after this meeting took place, most of its Library books and manuscripts were destroyed or damaged by arson. Nevertheless it continues to undertake astronomical research, including radio astronomy, and is also a UNESCO World Heritage site. The U. S. Naval Observatory, where I spent 24 years as an astronomer and historian, has fared better in the last 30 years, maintaining its site on Massachusetts Avenue in Washington, as well as its dark sky site in Flagstaff, Arizona. But as with all institutions, budgets remain problematic, and the Observatory is constantly called on to justify its existence.

Many of the other national observatories of the second and third era have thrived, although in altered form such as the Space Science Telescope Institute. These observatories can no longer be seen as part of a movement, but rather a continuing effort based on its nineteenth- and twentieth-century foundations.

The proceedings of the entire meeting (Lieske and Abalakin 1990), of which this paper was a part, are available online on the Astrophysics Data System.They includes a section of papers on the history of Pulkovo, including those by the authors seen in Fig. 24.1.

A few months after this paper was delivered, a symposium on national observatories was held at the U. S. Naval Observatory. That event and its results are described in the final section of the next chapter.

References

Abbe, C. 1868. "Dorpat and Poulkova," in *Annual Report of the Board of Regents of the Smithsonian Institution,* Government Printing Office, Washington, 370–390.

Airy, G. B. 1848. "Schreiben des Herrn Airy, Konigl. Astronomen in Greenwich, an den Herausgeber. Royal Observatory Greenwich, 1847 November 26," in *Astronomische Nachrichten,* 26, 353–360.

Ansari, S. M. R. 1985. "The observatories movement in India during the 17–18th centuries." *Vistas in Astronomy* 28:379–385

Batten, A. H. 1988. *Resolute and Undertaking Characters: The Lives of Wilhelm and Otto Struve.* Dordrecht: D. Reidel

Chapman, C. 1983. "The Accuracy of Angular Measuring Instruments used in Astronomy Between 1500 and 1850," *Journal for the History of Astronomy,* 14, 133–37.

Daedev, A. 1972. *Pulkovo Observatory: An Essay on its History and Scientific Activity,* Izdatel 'stvo Nauka, Leningrad, trans. K. Krisciunas (1977).

Débarbat, S., Grillot, S. and Levy, J. 1984. *L' Observatoire de Paris: Son Histoire (1667–1963),* Observatoire de Paris, Paris.

Dick, J. 1951. "The 250th Anniversary of the Berlin Observatory," *Popular Astronomy,* 59, 524–535.

Dick, S. J. 1980. "How the U.S. Naval Observatory Began, 1830–1865," *Sky and Telescope.* 466–471; reprinted with corrections and notes in S. J. Dick and L. E. Doggett, *Sky with OceanJoined: Proceedings of the Sesquicentennial Symposia of the U.S. Naval Observator y.* U. S. Naval Observatory, Washington, D. C., 1983, 167–181.

Gould, B. A. 1849. "An Account of the Observatory at Pulkowa," reprinted from *North American Review*, July, 1849, 1–20.

Krisciunas, K. 1978. "A Short History of Pulkova Observatory," *Vistas in Astronomy*, 22, 2 7–37

Krisciunas, K. 1988. *Astronomical Centers of the World*, Cambridge University Press, Cambridge, England.

Lieske, Jay H. and Viktor Abalakin. 1990. *Inertial Coordinate System on the Sky* (Kluwer, Dordrecht, Netherlands.

Meadows, A. J. 1975. *Greenwich Observatory: Recent History (1836–1975)*, volume 2 of Greenwich Tercentenary history, Taylor and Francis, London.

McCutcheon, Robert. 1989. "Stalin's Purge of Soviet Astronomers," *Sky and Telescope* (October, 1989), 352–57.

Paullin, Charles O. 1923. "Early Movements for a National Observatory, 1802–1842," *Records of the Columbia Historical Society*, 25, 36–56.

Price, Derek J. de Solla. 1963. *Little Science, Big Science*, Columbia University Press, New York.

Pulkovo Observatory. 1889. *Zum 50-Jahrigen Bestehn der Nicolai -Hauptsternwarte, Kaiserchen Akademie der Wissenschaften*, St. Petersburg.

Rhees, W. J. 1879. "Congressional Proceedings, Twenty-Sixth Congress, 1839–41," in *Tr. Smithsonian Institution: Documents Relative to its Origin and History, Smithsonian Institution*, Washington, DC, 200–246, esp. 220–222.

Sadler, D. H. 1976. "Lunar Distances and the Nautical Almanac," in *The Origins, Achievements and Influence of the Royal Observatory, Greenwich: 1675–1975,Vistas in Astronomy*, 20, 113–121.

Sokolovskaya, Z. K. 1976. "Struve" in *Dictionary of Scientific Biography*, 13, 108–121.

Struve, F. G. W. 1845, *Description de L' Observatoire Astronomique Central de Poulkova, L' Academie Imperiale des Sciences*, St. Petersburg.

Struve, O. 1865. *Ubersicht der Thatigkeit der Nicolai-Hauptsternwarte wahrend der ersten Jahre ihres Bestehen*, Kaiserlichen Akademie der Wissenschaften, St. Petersburg.

Weinberg, Alvin. 1967. *Reflections on Big Science*, The MIT Press, Cambridge, Mass.

Chapter 25
John Quincy Adams, the Smithsonian Bequest and the Founding of the U. S. Naval Observatory

Abstract The founding of the U. S. Naval Observatory in Washington, DC, is a classic example of the role of competing political interests in early American science. President John Quincy Adams had called for a national observatory as early as 1825, but Congressional ridicule delayed action for almost two decades. The quest for such an observatory can only be understood in the context of Congressional politics, hostility toward Adams stemming from the election of 1824, the concurrent debate over the bequest of James Smithson, arguments for a National Institute for the Promotion of Science, and naval officers who promoted the practical benefits for such an observatory. In the end a small Depot of Charts and Instruments, founded to aid navigation, became not only a naval but also a national observatory, part of a national observatory movement taking place around the world.

25.1 Introduction

The history of the first national observatory in the United States must be placed in both an international and an American context, each of which provides quite different perspectives on the institution. In a global sense the U. S. Naval Observatory falls in the "second era" of national observatories (Dick 1990b; Chap. 24, this volume). From this perspective it may be viewed as the continuation of a venerable theme now played out in the context of a young and growing America. In the context of astronomy in the United States, on the other hand, the Naval Observatory stands as one of the country's oldest scientific institutions, virtually unique in its time as the recipient of government funding for astronomical research. As Elias Loomis reported in 1856 (Loomis 1856), some 24 observatories had been founded in America in the previous 25 years (Fig. 25.1). But only the Naval Observatory and the West Point Military Academy were government funded, and the latter was more an educational institution than a research institution (Bartlett 1846). Furthermore, the Naval Observatory remained the only national observatory in the United States until the founding of Kitt Peak National Observatory and the National Radio Astronomy Observatory in the 1950s, with the possible exception of the

First published in *Journal for the History of Astronomy*, 22 (February, 1991), 31–44.

S. J. Dick, *Space, Time, and Aliens*, https://doi.org/10.1007/978-3-030-41614-0_25

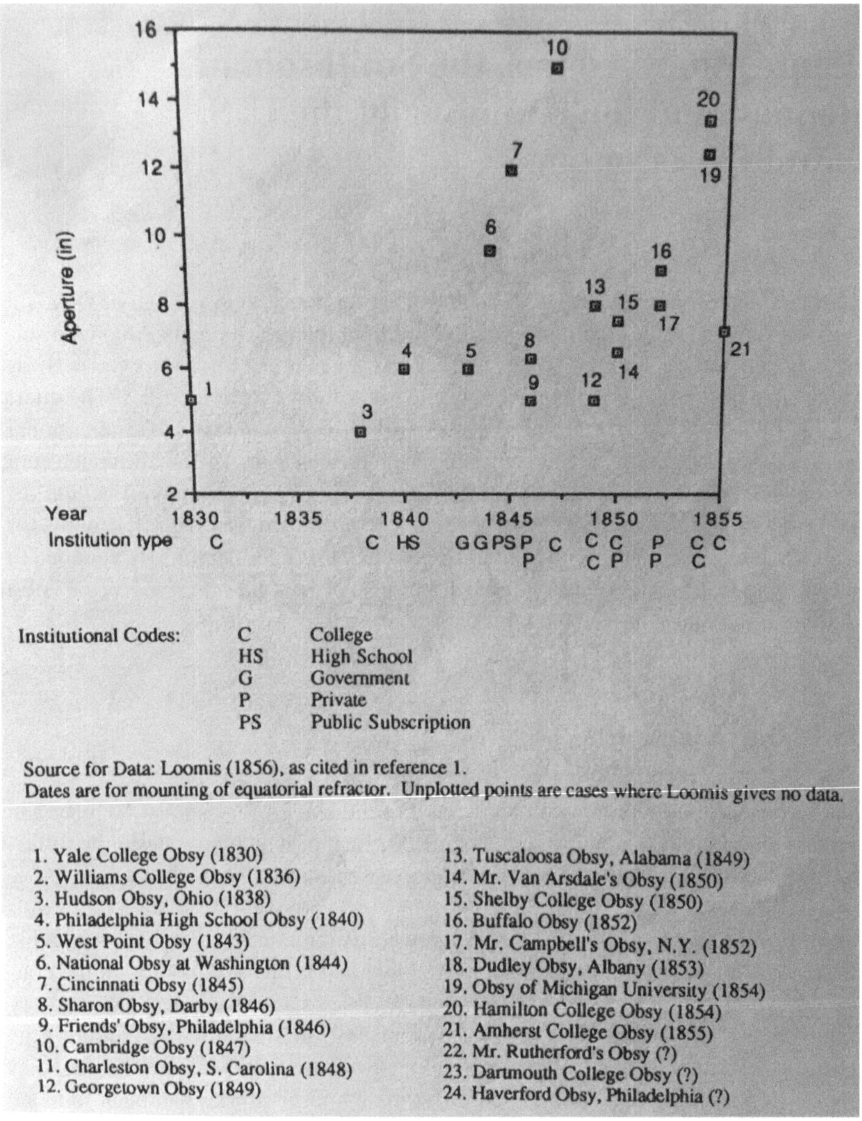

Fig. 25.1 American observatories and their telescopes, 1830–1855

quasi-governmental Smithsonian Astrophysical Observatory (Jones 1965; Smithsonian Astrophysical Observatory Centennial Papers 1990).

It is therefore of particular interest to ask how this unique institution came about. It has long been known that its birth was somewhat unusual. It is not surprising that, in contrast to the observatories in Paris, Greenwich, and Pulkovo, and in good Yankee fashion, the U. S. Naval Observatory did not begin by royal proclamation.

But it is curious that a national astronomical institution was the object of active opposition, to the point that its origin has been called.

> ...the classic example of the surreptitious creation of a scientific institution by underlings in the executive branch of the government in the very shadow of congressional disapproval. No more hated proposal existed, and nowhere had more pains been taken to prevent the creation of a new agency. Yet despite this vigilance the forces that required an observatory gained their ends. (Dupree 1957)

Behind this startling statement of surreptitious creation is a story that has never been fully told. I will argue here that origins of the U. S. Naval Observatory must be understood in the context of concurrent events in the American Congress, in particular the controversy over the Smithsonian bequest, President John Quincy Adams's advocacy of a national observatory, and the fate of an early group of science enthusiasts known as the National Institute for the Promotion of Science. Only by examining these three components of the story can we appreciate just how very different was the origin of the Naval Observatory, compared with the national observatories of other countries (Dick 1980; Nourse 1873; Weber 1926).

25.2 John Quincy Adams

The strong advocacy role of John Quincy Adams (1767–1848) for astronomy is well known. Adams had specifically called for a national observatory on several occasions, the first in his Presidential message of 1825. In conjunction with the bequest of James Smithson, his efforts for an observatory continued as a Congressman until 1846, a fact that will form an important part of our story. It is clear that had Adams been a czar, like Nicholas I of Russia, the United States might have had a national observatory as early as 1825. But in the event, Congressional ridicule of Adams's so-called "lighthouse of the skies" proposal prevented an American national observatory (Bemis 1970); Jones 1965, 8–33).[1] Nor did other similar proposals related to the Coast Survey, an American prime meridian and a Nautical Almanac fare any better. Though Congress supported all three, it never allowed an observatory to be associated with any of them (Paullin 1923; Musto 1968).

25.3 The Navy's Depot of Charts and Instruments

Meanwhile, in 1830 the 25-year-old Navy Lieutenant Louis M. Goldsborough had successfully begun a modest office to store instruments, distribute charts, and rate the Navy's chronometers, the latter to determine how fast or slow they were running so that the corrections could be applied to longitude determinations for navigation. Since the best way for rating chronometers was by astronomical observation, the Depot was provided with a small transit telescope. Thus was the Depot of Charts and Instruments founded, in a central location only a few blocks from the President's

House in Washington, DC, Goldsborough had studied in Paris for several years, and his Depot may well have been modeled after the French Depot des Cartes et Plans de la Marine; certainly its activities had more in common with the French Depot than with the Paris Observatory.[2] Despite the astronomical connection, certainly Goldsborough did not consider it the beginnings of a national observatory, and neither did anyone else.

Beginning in 1834 the Depot and its telescope were located only 1200 ft. northwest of the Capitol Building, in quarters owned by Lt Charles Wilkes, who had taken over from Goldsborough. Here Wilkes built a small observatory with a Troughton transit instrument for observing the Sun and stars for rating chronometers. The Depot was successively headed by Wilkes (1833–1837) and Lt James Melville Gilliss (1837–1842), under whose direction the astronomical work gradually became more important. In particular, with Wilkes's departure on the South Seas Exploring Expedition in 1838, the Navy instructed Gilliss (along with William Bond at Harvard) to begin a series of systematic observations for longitude differences; this resulted in the first star catalogue issued from the United States. Gilliss (Fig. 25.2), largely self-taught but experienced in navigational astronomy, would play the crucial role in obtaining funds for the construction of a permanent observatory (Gould 1867, Gilliss 1846).

Fig. 25.2 James Melville Gilliss, courtesy U. S. Naval Observatory

By 1841 "the little box on Capitol Hill, as the American astronomer B.A. Gould later called the makeshift observatory adjacent to the Navy's Depot, was clearly inadequate to its task. As Gilliss recalled, several factors caused him to press for a new Depot and associated observatory. Among these were the unsuitability of the building, defects of the transit instrument, and especially the lack of space for a permanent meridian circle that he was finally authorized to purchase for measuring declinations (Gilliss 1845).[3] He thus urged the Board of Navy Commissioners (three prominent officers who helped the Secretary of the Navy carry out Navy policy) to recommend an appropriation for a permanent building, which they did in November 1841. The utility of such a Depot, the Board wrote to the Secretary of the Navy,

> …as a matter of economy only, is fully manifested in the careful preservation, and distribution to sea-going vessels, of the valuable instruments and charts which have been carefully selected and procured for the service, at considerable expense. To this may be added the facilities which such an establishment presents to officers of the navy, for obtaining useful, valuable, and, indeed, necessary knowledge, in some of the higher branches of their profession.

Secretary of the Navy Abel Upshur, whom Gilliss characterized as interested in advancing science, expressed his "entire approval" of the proposal. The charts and nautical instruments of the Navy, he wrote,

> …have been procured at great labor and expense, and are indispensable in the naval service. The small expenditure which will be necessary to preserve them in a condition always ready for use, is not worthy a moment's consideration when compared with the great purposes which they are designed to answer. They are a necessary part of a naval establishment worthy of the present and growing greatness of our country.[4]

25.4 Congressional Machinations

Since a large expenditure was involved, the proposal next had to go through the normal process for naval appropriations in both the Senate and the House of Representatives of the U. S. Congress. This was a tortuous procedure at best, and Gilliss informs us that great delay occurred, especially in the Naval Committees. One Congressman, the 35-year-old Francis Mallory of Virginia, to whom it was referred by the Naval Committee of the House of Representatives, favoured it openly, but the majority expressed no opinion.[5] On March 15, 1842, Gilliss persuaded the one skeptical member of the Committee to visit the Depot, so close at hand to the Capitol. On the same day, a unanimous report and bill, known as the Mallory Bill, was sent to the House (Gilliss 1845, 65). Regarding the nature of the institution the bill specified only that the Secretary of the Navy was "authorized to contract for the building of a suitable house for a depot of charts and instruments of the navy of the United States, on a plan not to exceed in cost the sum of twenty-five thousand dollars." The lengthy report, drafted by Gilliss for Mallory, on the other hand, is a revealing document that stresses the utility of the depot to the Navy, the in adequacy of the present Depot and its accompanying observatory, the past accomplishments of the Depot in making magnetic, meteorological, and astronomical observations, the necessity of new quarters in order to protect the valuable instruments in both the Depot and the Observatory, and the unsettling fact that the property now occupied was privately owned (Gilliss 1842). Most of the four-page document dealt with the various duties of the Depot, and a few sentences addressed the need for an observatory: "A small observatory is absolutely essential to the depot: without it, the duties cannot be performed." Aside from this clear statement, and a few paragraphs on the utility of astronomy for navigation, no stress was laid on an astronomical observatory, certainly not one that might be considered a national observatory.

On March 15, the day it was introduced, the bill "was read a first and second time, and committed to the Committee of the Whole House on the State of the Union," where no action was taken (U. S. House of Representatives 1842). Mallory suggested the bill would have a greater chance of success in the House if the Senate would first pass a bill. Gilliss then tried to generate support in the Senate Naval Committee by asking members to inspect the poor conditions of the Capitol Hill Observatory, but his requests for such inspections "were put off from time to time" (Gilliss 1845, 65). Gilliss later wrote that he believed the Naval Committee of the Senate was finally persuaded by an astronomical event. In April Gilliss had observed Comet Encke with a small telescope of 3.5-ft focal length. He subsequently reported this at a meeting of the National Institute for the Promotion of Science, and as it happened, in the audience was a certain Senator W.C. Preston.[6] When a few days later Gilliss paid a visit to the chairman of the Senate Naval Committee, he found Preston with him. "As soon as I began the conversation about the little observatory, Mr. Preston inquired whether I had not given the notice of the comet at the institute, and immediately volunteered, 'I will do all I can to help you.'" (Gilliss 1845, 66).[7] In fact on June 23, a bill was introduced into the Senate, identical to the House bill (U. S. Senate Journal 1842).

Meanwhile ominous events were coming to head on another front, and a circumstance that Gilliss had perhaps foreseen came to pass. Upon his return from the Exploring Expedition, on June 15, 1842 Wilkes had informed the Board that as of the end of the month the premises occupied as the Depot would no longer be available to the Government for that purpose (Wilkes 1842). On June 22 (1 day before his proposal was introduced in the Senate) Gilliss was ordered to find another house for the Depot, and by June 30 the contents of the Depot were removed to a house on Pennsylvania Avenue (Board of Navy Commissioners 1842; Gilliss 1842).[8] On July 2, Gilliss gave the key of the old observatory to Wilkes. Then, in a curious move, Gilliss resigned, and on July 12 reported to the Board that he had turned over charge of the temporary Depot to one Matthew Fontaine Maury, who had received orders to report on June 29.[9]

The abrupt removal of the Depot must have been a disappointment to Gilliss, made more so because of charges from Wilkes that Gilliss had not taken care of the property. This perhaps had something to do with Gilliss's resignation, but the reasons are unclear. One day later the Board simply expressed its regret that Gilliss had found it necessary to retire from his duties, and praised him for duties well performed (BONC 1842).

Despite his resignation, Gilliss was not about to let plans for the new Depot drop. On July 29, the Mallory bill was passed in the Senate (U. S. Senate 1842). On the same day it was reported to the House, and Gilliss later described the situation there: "A majority was known to be favorable, but its number on the calendar, and the opposition of one or two members, were likely to prevent action upon it; and that it did receive the sanction of the House of Representatives at the last hour of the session of 1841-42, the navy is indebted to the untiring exertions of Dr. Mallory" (Gilliss 1845, 66). As B.A. Gould later recalled, the bill "went to the House on the 30th of July; was referred to the same committee as before; but as a Senate bill was treated with courtesy. It was reported back without discussion, passed by the House without debate, and on the 31st of August, 1842, became a law." (Gould 1867, 24; US House 1842).[10] And so, as of August 31, 1842, the last day of the second session of the 27th Congress, $25,000 was appropriated and a new Depot was assured.

25.5 The Smithson Bequest and the National Institute for the Promotion of Science

With knowledge of these details, it is crucial to ask why, given a Congress that had ridiculed every attempt by John Quincy Adams to found a national observatory, it now voted to fund even a "small observatory." The answer is that, as we have seen, the actual bill did not specify an observatory, and most members of Congress undoubtedly did not read the committee report, which did mention a small observatory. Those members of the House and Senate Naval Committees who were familiar with the report may have been persuaded by two factors: it was not Adams who was proposing it, and Gilliss had made a good case that the Navy needed a small observatory for practical needs. According to Gould, the motive in allowing an observatory for the Navy under a different name was the same motive as the proviso against an observatory stipulated when the Coast Survey was revived in 1832: to avoid at all costs giving Adams any credit for the idea. "When," he wrote,

> …at last Congress did appropriate the means for erecting an Astronomical Observatory, and subsequently for its support, it was under a fictitious name; the authors of the laws intending an Astronomical Observatory, and being well aware that the funds would be so applied, but causing the insertion of the proviso in the one case [of the Coast Survey], and of the feigned name in the other, for the purpose of preventing the institution from being attributed to the influence of Mr. Adams. (Gould 1867, 20)[11]

Moreover, we must look to broader events in order to understand how the funding for a naval observatory came about in Congress. In particular it was the events surrounding the bequest of James Smithson, and the struggle over the fate of the fledgling National Institute for the Promotion of Science,[28] that secured the crucial support of Senator Preston for Gilliss's proposal in the Senate.[12] We recall what Gilliss tells us about his serendipitous meeting with Preston: "As soon as I began the conversation about the *little observatory* [my emphasis] … Mr. Preston immediately volunteered 'I will do all I can to help you'." Why was Preston so interested? Preston, the Senate colleague of John Calhoun of South Carolina, had argued with Calhoun in 1839 against the U. S. acceptance of the Smithsonian bequest because it increased national power at the expense of the states (Bemis 1970, 505). After that failed, he was an ardent opponent of Adams's desire to use the Smithsonian bequest for an astronomical observatory. As chairman on the Senate side of the Joint Committee on the Smithsonian bequest (as Adams was on the House side), Preston in 1840 presented resolutions counter to Adams's proposal for an astronomical observatory.

His reasons for doing so became clearer the following year. In 1841 Preston took up the idea of Joel Poinsett—a guiding light of the National Institute, and President van Buren's Secretary of War before retiring to South Carolina after van Buren's defeat—to place the Smithsonian bequest under the control of the National Institute (Bemis 1970, 511). It was Preston who in 1841 introduced one bill to incorporate and give Congressional blessing to the National Institute for the Promotion of Science, and another bill to put the Smithsonian fund under the control of that Institute (Rhees 1879, 238–242).[13] No action was taken in 1841, so April 1842, the same month that Gilliss was pushing the Depot of Charts and Instruments bill through the Senate, found Preston again introducing in the Senate the bill to dispose of the Smithsonian Fund through the National Institute (Rhees 1879, 238–242). In the House Adams simultaneously introduced his bill advocating the use of the Smithsonian bequest for an astronomical observatory (Rhees 1879, 249–60).[14] Here we have an answer to the question of why Preston latched onto Gilliss's proposal in June and personally pushed it through the Senate. He may have had a passing interest in astronomy piqued by Come Encke, but more importantly it had surely occurred to him that by establishing an Observatory within the Navy Department, he would undermine Adams's case for an observatory based on the Smithsonian bequest, and, not incidentally, put the money instead under the control of the National Institute, as his own bill proposed.

25.6 Naval or National Observatory?

Thus, the creation of a permanent Depot of Charts and Instruments with its observatory was in part a byproduct of the tug-of-war over the Smithsonian bequest. Though Preston did not get his way with the bequest for the National Institute, neither did Adams realize his lifelong dream, and Adams's failure can be directly linked to the founding of the Depot with its observatory. The small observatory of the House report materialized as much more. Gilliss later wrote, "Taking the report of the Naval Committee [note, not the bill, but the report] … as the exponent and will of Congress" (Gilliss 1845, 2), the Secretary of the Navy ordered him to consult with experts in the United States,[15] and then to proceed abroad for final consultations and purchases.[16] The nature of the building (Fig. 25.3), the excellence of the instruments considered for purchase from the outset, and the expertise of the consultants themselves betray Gilliss's purpose that this would be no mere Depot. Though he kept the rooms for the charts and navigational instruments large, and those for the telescopes small "lest the limits of the law should be exceeded." Gilliss ended up with a sizeable observatory. Gilliss himself was unequivocal that he had worked with more than a Depot in mind: "I should have regarded it as time misspent to labor so earnestly only to establish a *depot.* My aim was higher. It was to place an institution under the management of naval officers, where in the practical pursuit of the highest known branch of science, they would compel an acknowledgement of abilities hitherto withheld from the service" (Gilliss 1845, 66).

Gilliss clearly had an observatory, indeed a naval observatory, in mind, but at this point it was still not at all clear that this constituted a national observatory, even with the excellent instruments that Gilliss would purchase from Europe. Perhaps more to the point, upon its completion in 1844, Gilliss was not made head of the institution, and so would have no control over its destiny. Abel Upshur had gone on to become Secretary of State, and both he and his replacement as Secretary of the Navy had been killed in a cannon accident. March 1844 thus found a new Secretary of the Navy, John Y. Mason, in office. To Gilliss's astonishment, Mason named fellow Virginian Matthew Fontaine Maury to head the new institution. With Maury's known expertise in hydrography, this may have been a conscious decision to emphasize another of the Depot's duties rather than astronomy (Williams 1963).

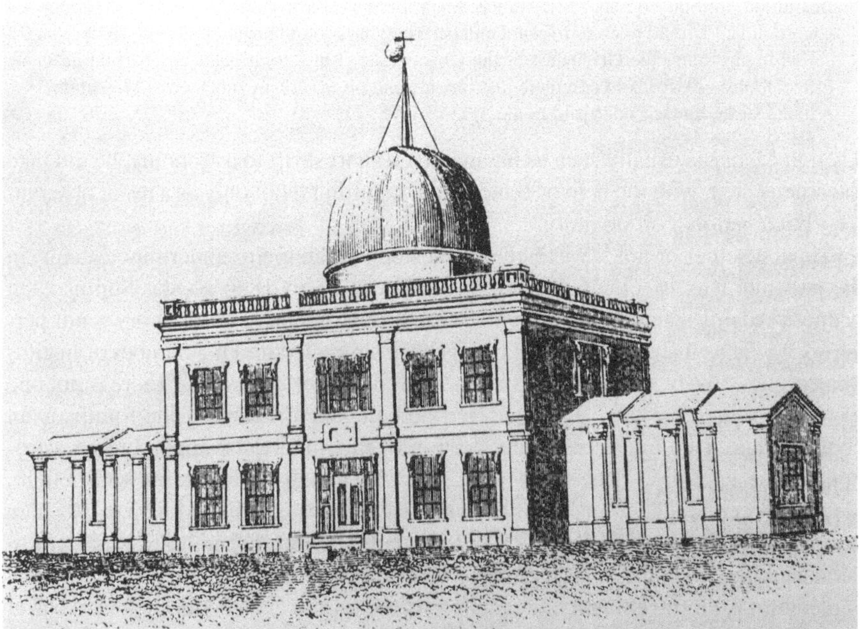

Fig. 25.3 The National Observatory, Washington, DC, from the title page of Matthew F. Maury, Washington Astronomical and Meteorological Observations for 1845. The time ball is in the raised position

Given this turn of events, the resulting national observatory is even more surprising. That even after its construction was authorized, no one yet considered the naval observatory a national observatory, is evident by Adams's repeated attempt in 1844 to have the Smithsonian fund use for a national observatory (Rhees 1879, 293–302).[17] The final irony is that it seems it was Maury who co-opted Adams. He failed in a bid to hire William Bond from Harvard, but did hire in rapid succession John C. Coffin, Joseph S. Hubbard, Reuel Keith, and Sears Cook Walker. With their help and that of naval personnel, Maury was able to publish in 1846 what he called " the first volume of Astronomical Observations that has ever been issued from an institution properly entitled to the name of Observatory on this side of the Atlantic" (Maury 1846).[18] Maury's co-option of Adams took shape not only in the form of the impressive contents of this volume, but also by its dual title pages, one of which called the institution a "Naval Observatory" and the other a "National Observatory.

Maury could not have achieved this alone, and in fact it is clear that he had the complicity of the new Secretary of the Navy George Bancroft, the famous historian and founder of the U. S. Naval Academy. Bancroft later recalled:

> When I became Secretary of the Navy the Observatory was already in existence and under the superintendence of Maury. It was then known officially as the Depot for Charts, but Congress had not expressly sanctioned the Observatory by name. Mr. J.Q. Adams still cherished the hope of being the founder of a National observatory. In conjunction with Lt. Maury and taking counsel also of the best scientific men, I got large appropriations for the Institution, introduced under Mr. Maury, Scientific men, for example Sears Walker, and in a word did all I could to carry forward and perfect what I found begun. I have no right to be called in any sense the Originator of the Observatory. But I contributed my part while in office, to procure for it so complete instruments and observers, as superseded Mr. Adams' scheme, as he himself once said to me. (Howe 1908, 277–78)

Though Maury is usually seen as having given short shrift to astronomy, he did take the crucial step, with the help of Bancroft, of claiming it not only as a naval observatory, but a national observatory.

Thus, whether or not it was Senator Preston's intention to undermine Adams, in the end this was its effect. During the final debate in 1846 on the Smithsonian bequest, Adams stated "I am delighted that an astronomical observatory – not perhaps so great as it should have been – has been smuggled into the number of institutions of the country, under the mask of a small depot for charts…" "There is not one word about it in the law," he noted, but he concluded regarding the Smithsonian bequest, that "I no longer wish any portion of this fund to be applied to an astronomical observatory" (Rhees 1879, 442–43).[19] Though Adams expressed delight, his disappointment at not having his own name associated with a dream of 20 years must have been very real. In 1843 he had made the difficult journey westward to dedicate the Cincinnati Observatory (Adams 1843); political realities now prevented him from travelling a few blocks to dedicate and even better-endowed Naval Observatory.

The early 1840s thus form a crucial juncture for three institutions in the history of American science: the rapid development of the Depot of Charts and Instruments into a National Observatory by 1846 denied Adams a national observatory attributable to his own efforts; the Smithsonian bequest was left free for the development of that institution as we know it under Joseph Henry; and with Senator Preston' s failure to tie the National Institute to the Smithsonian, the National Institute was left to fade away into history, making way for the American Association for the Advancement of Science in 1848 (Kohlstedt 1976).

From modest and purely practical origins in navigation 16 years before, the Depot of Charts and Instruments by 1846 startlingly assumed the role of the first national observatory of the United States. It did so without concerted plan via a number of crucial steps. Goldsborough and Wilkes brought centralization, scientific method and increased efficiency to the Navy's growing navigational technology. Gilliss, inspired by a deeply rooted personal desire to increase the scientific reputation of the Navy, not only perfected the work of his predecessors, but also seized on the Wilkes Expedition as a means of expanding astronomical, magnetic, and meteorological observations beyond the immediate needs of navigation. Aided by partisan politics centering around Adams and the Smithson bequest, Gilliss furthermore pushed through Congress an appropriation for a permanent Depot, which he converted into a naval observatory. And in a surprise move again indicative of a broad interest in science and an increased Navy reputation, Maury seized for this naval institution the title of National Observatory. All of this happened with the direct approval of a series of Secretaries of the Navy, who saw the relevance of science and technology to the naval service, appreciated the increased prestige that a naval observatory lent to the naval establishment, and finally may even have perceived that an expanded national role might contribute to a growing American science, to the international credit of the Navy and the nation.

The surreptitious creation of the U. S. Naval Observatory has echoed down through its history in many ways, among them its circumscribed mission of practical astronomy, and a long controversy over military versus civilian control. Having succeeded in co-opting the function of a national observatory, Maury's 20-year administration ironically concentrated on hydrography and consequently did not live up to the hopes of many American astronomers. In a kind of poetic justice, upon Maury's defection to the South at the onset of the Civil War in 1861, Gilliss would finally be appointed to head the institution. Of this change of command from Maury to Gilliss, Gould (by then the founder of *Astronomical journal)* would say: "The sudden transformation which took place was like the touch of an enchanter's wand. Order sprang from chaos, system from confusion, and the hearts of the faithful few who had struggled on for years, hoping against hope, were filled with sudden joy." Under Gilliss's short tenure (1861–1865), Asaph Hall, Simon Newcomb, J.R. Eastman, William Harkness, and others came to the observatory. With Gilliss's death in 1865, the Founding Era for the Naval Observatory was over, as it was 8 days later at Harvard with the death of George Phillips Bond. A new era now began that would make the Naval Observatory a National Observatory by the unanimous consent even of American astronomers, entitling it permanently to enter the ranks of the observatories at Greenwich, Paris, and Pulkovo.

25.7 Commentary 2020

In order to better understand the origins, evolution, and differences among national observatories, the U. S. Naval Observatory and the Historical Astronomy Division of the American Astronomical Society jointly sponsored a symposium on "National Observatories: Origins and Functions," at the Naval Observatory in Washington, DC, on January 14, 1990. This paper was one of seven published in a special issue of the *Journal for the History of Astronomy* (Dick and Hoskin 1991), and included papers on several of the national observatories listed in Table 24.1: the Greenwich Observatory (Robert Smith), the early national observatories of South America (Philip Keenan), the Dominion Observatory of Ottawa (Richard Jarrell), the European Southern Observatory (Adriaan Blaauw), the Kitt Peak National Observatory (Frank Edmonson), and the Carnegie Institution of Washington (Allan Needell).

Since that time, some of these papers have been elaborated into books, including Edmondson (1997), Blaauw (1991), and Dick (2003). The founding of the Chilean National Observatory is described further in Chap. 27 of this volume. The Naval Observatory went on to make significant discoveries well beyond its practical origins, including the two moons of Mars (Chap. 34), the largest moon of Pluto (Chap. 30), and the interstellar polarization of starlight. The full history of the Naval Observatory, including these discoveries and much more, is found in Dick (2003).

Notes

1. The issue was not hostility toward science or astronomy, but a much broader partisan hatred of Adams stemming from his narrow defeat of Andrew Jackson in the election of 1824.
2. The work of the French Depot during the eighteenth century is discussed briefly in (Konvitz 1987). Its work during the nineteenth century is discussed in more detail in (Winterhalter 1889). Goldsborough specifically mentions that his proposal for a Depot was "sustained by the examples and experience of the French, Spanish, and it is believed, British Navies." The proposal is printed in full in Dick (1990a). The Spanish and British counterparts were the Observatorio de la Armada (founded 1753 at Cadiz) and the British Hydrographic Office (1795). The comparative institutional history in different countries of astronomical and hydrographic activity related to navigation is a subject in need of further study.
3. Late in his tenure Gilliss was authorized to purchase a sidereal and mean time clock from Parkinson and Frodsham, and a meridian circle from Ertel & Son, additional evidence of the increasing importance the Navy assigned to astronomical observations.
4. Gilliss to Board of Navy Commissioners (hereafter BONC), 24 September 1841, National Archives of the United States, Record Group 45, Naval Records Collection, Entry 228, argues briefly for "the propriety of erecting a suitable building for the Depot of Charts and Instruments." BONC to Secretary of the Navy, 30 November 1841, in *Report of the Secretary of the Navy,* 4 December 1841, 373, endorses the proposal. The Secretary of the Navy's recommendation is in his *Report,* 367.
5. Francis Mallory, born on December 12, 1807, was a Congressman from Virginia in the 25th, 26th, and 27th Congresses (December 28, 1840 to March 3, 1843). His interest in the Gilliss proposal probably derived from his having served as a midshipman in the U. S. Navy from

1822 to 1828. He graduated from medical school at the University of Pennsylvania in 1831, and after practicing in Norfolk pursued agricultural interests. He died on March 26, 1860. *Biographical directory of the American Congress, 1774–1971*, 1330–1.

6. William Campbell Preston, born on December 27, 1794, served as Senator from South Carolina from November 26, 1833 to his resignation on November 29, 1842. He studied law at the University of Edinburgh, practiced in Virginia and South Carolina, and was president of South Carolina College from 1845 to 1851. He died in 1860, the same year as Congressman Mallory.
7. Comet Encke has the shortest known period of any comet, 3.3 years. Discovered in 1786 by Pierre Mechain, it is named after J.F. Encke, who first established its periodic nature in 1819. On the 1842 return it reached perihelion on April 13. Since it was 5.5 magnitude and had no tail, it is not a case of the Senator having been impressed with the sight of the comet, but with the fact that Gilliss had seen it in the nation's capital. I have found no observations published by Gilliss. But Sears Cook Walker and E.O. Kendall observed it in great detail at the Philadelphia High School between March 27 and April 11, "Observations of Encke's Comet, at the High School Observatory, Philadelphia, March and April, 1842, with the Fraunhofer Equatorial, by Sears C. Walker and E. Otis Kendall. Read May 20, 1842, *Transactions of the American Philosophical Society,* viii (1843), 311–14. Their results were also reported in three other journals.
8. Gilliss further describes the house on Pennsylvania Avenue as belonging to the late Hon. John Forsyth. Further evidence indicates it was at 2222–2224 Pennsylvania Avenue.
9. Gilliss to Wilkes, 2 July 1842; Gilliss to BONC, 12 July 1842.
10. *House journal,* 27th Congress, 2d session, 1194, 1201 and 1475 indicates that on July 29, 1842 the Secretary of the Senate informed the House that S 285 had passed the Senate; on August 2 it was read a first and second time in the House; and on August 31 it was read a third time "and passed in the affirmative." The bill was approved and signed by President John Tyler on the same day, *ibid.,* 1478, 1482. See *Congressional globe,* 27 August 1842, 960, 978 for more details, indicating initial opposition to the bill from Cave, Johnson, Spring, and others. The final law, with the same wording as the House and Senate bills, is found in *The public statutes at large of the United States of America* (Boston, 1846), v. 572, ch. 277.
11. Gould cites Quincy's *Memoir of the life of J.Q. Adams*. For more on Adams's proposals for a national observatory see Bemis (1970).
12. On the origins of the Smithsonian, see, for example Dupree (1957), ch. 4, and G.B. Goode (1897). On the National Institute see Sally Gregory Kohlstedt (1971).
13. The bills were introduced on February, 17, 1841.
14. The bill, (HR 286) was introduced on April 12, 1842 and specified "That the sum of thirty thousand dollars, part of the accruing interest on the same Smithsonian fund, be, and the same is hereby, appropriated towards, the erection and establishment, at the city of Washington, of an astronomical observatory, adapted to the most effective and continual observations of the phenomena of the heavens; to be provided with the necessary, best, and most perfect instruments and books, for the periodical publication of the said observations, and for the annual composition and publication of a nautical almanac."
15. The names of those whom Gilliss consulted make an interesting commentary on the state of astronomy in America at the time. In Washington, he met with F.R. Hassler, in his last years as Superintendent of the Coast Survey, and a veteran of astronomical observation for survey purposes. In Philadelphia Gilliss visited A.D. Bache, who in this year had resumed a professorship at the University of Pennsylvania after working on the reorganization of public schools in Philadelphia, and was on his way to succeeding Hassler in 1843. In Philadelphia he also found Sears Cook Walker, head of the Philadelphia High School Observatory, and R.M. Patterson, Director of the U.S. Mint, a founder of the Franklin Institute and active member of the American Philosophical Society. In New York he visited at West Point Military Academy W.H.C. Bartlett, Professor of Natural and Experimental Philosophy, who had made a trip to European observatories in1840 to consult about plans for the West Point Observatory.

In some ways this observatory could be seen as the Army analogue to the proposed Naval Observatory, and Gilliss undoubtedly listened with special interest to Bartlett, who may also have helped Gilliss plan for his own trip to Europe. In Boston Gilliss found William C. Bond, Director of the Harvard College Observatory since 1839, and his 17-year-old son George Phillips Bond, a future Director of that Observatory. Here also was R.T. Paine, who prepared the astronomical contents for the American Almanac, and was the Chief Engineer for the trigonometric survey of Massachusetts, a task that included much astronomical work.

16. In travelling abroad to purchase his instruments, Gilliss was following in the footsteps of Elias Loomis, William Bartlett and O.M. Mitchel, who went to Europe in 1837, 1840, and 1842, respectively, to purchase instruments for the Hudson, West Point and Cincinnati Observatories. Like the Philadelphia High School Observatory (which had ordered its instruments from Europe without sending a representative abroad), Gilliss made most of his purchases from the Munich firms of Merz & Mahler, and Ertel & Son. Gilliss's diary from this trip is still extant.
17. The bill (HR 418) was introduced on June 7, 1844, and is found in full on pp. 299–301.
18. Maury's staff is discussed in Rothenberg (1983) and Rothenberg (1974).
19. The bill establishing the Smithsonian Institution passed the Senate on August, 10, 1846.

References

Adams, John Quincy. 1843. "An oration delivered before the Cincinnati Astronomical Society on the occasion of laying the corner stone of an astronomical observatory on the 10th of November, 1843," Cincinnati, reprinted in I. B. Cohen, *Aspects of astronomy in American in the nineteenth century*, New York, 1980.

Bartlett, William H. C. 1846. "Account of the observatory and instruments of the United States Military Academy at West Point, with observations of the comet of 1843," *Transactions of the American Philosophical Society,* new series., ix, 191–203.

Bemis, Samuel Flagg Bemis. 1970. *John Quincy Adams and the union* (New York, especially chap. 23, "Lighthouses of the skies"), 1825–46.

Blaauw, Adriaan. 1991. *ESO's Early History: The European Southern Observatory from Concept to Reality*: European Southern Observatory: Garching.

Board of Navy Commissioners (BONC). 1842. BONC to Gilliss, 22 June 1842 and Gilliss to BONC, 25 June 1842, National Archives, RG 45.

Dupree, A. Hunter. 1957. *Science in the federal government: A history of policies and activities to 1940.* Cambridge, Mass., 1957, 62.

Dick, Steven J. 1980. "How the U.S. Naval Observatory began, 1830–65", *Sky & Telescope,* lx (1980), 466–71, reprinted with corrections and notes in Dick and Doggett (1983).

Dick, Steven J. 1990a. "Louis M. Goldsborough's proposal to establish a Depot of Charts and Instruments in the U.S. Navy: Text and comments", *Rittenhouse: Journal of the American Scientific Instrument Enterprise,* iv, 79–86.

Dick, Steven J. 1990b. "Pulkovo Observatory and the National Observatory Movement." In J.H. Lieske and V. K. Abalakin (eds), *Inertial Coordinate System on the Sky*, 29–38. Netherlands: Kluwer.

Dick, Steven J. 2003. *Sky and Ocean Joined: The U. S. Naval Observatory, 1830–2000.* Cambridge University Press, Cambridge

Dick, Steven J. and LeRoy Doggett, eds. 1983. *Sky with ocean joined: Proceedings of the sesquicentennial symposia of the U.S. Naval Observatory,* (Washington, 1983), 166–81.

Dick, Steven J. and Michael Hoskin. 1991. Special issue on National Observatories, *Journal for the History of Astronomy*, 22, 1–100.

Edmondson, Frank. 1997. *AURA and its US National Observatories*. Cambridge University Press: Cambridge.

Gilliss, J. M. 1842. *Depot of Charts, &c, report number 449 to accompany bill H. R. 303, 27th Congress, 2d session, March 15, 1842.* The first draft of this report, in Gilliss's hand, is in RG 45, E 228, 1 March 1842.

Gilliss, J. M. 1845. *Report of the Secretary of the Navy, communicating A report of the plan and construction of the Depot of Charts and Instrument, with a description of the instruments,* 28th Congress, 2d Session, Senate Document 114, read and referred to the Committee on Naval Affairs, 18 February 1845, 66.

Gilliss, J. M. 1846. *Astronomical observations made at the Naval Observatory, Washington, under orders of the Honorable Secretary of the Navy, dated August 13, 1838* (Washington, 1846).

Goode, G. B. 1897. *The Smithsonian Institution, 1846-1896: the History of its First Half-Century.* Washington, DC

Gould, B. A. 1867. *Biographical notice of James Melville Gilliss* (Cambridge, Mass., reprinted in *Biographical memoirs of the National Academy of Science,* i, 135–79.

Howe, M. A. DeWolfe. 1908. *The life and letters of George Bancroft.* Port Washington, New York. 277–78.

Jones, B. Z. 1965. *Lighthouse of the skies: The Smithsonian Astrophysical Observatory, background and history 1846–1965.* Washington, DC.

Kohlstedt, Sally Gregory. 1971. "A Step toward Scientific Self-Identity in the United States: The Failure of the National Institute, 1844." *Isis* 62 (3):339–362.

Kohlstedt, Sally Gregory. 1976. *The formation of the American scientific community: the American Association for the Advancement of Science, 1848–1860.* Urbana, 87–88.

Konvitz, Josef W. 1987. *Cartography in France 1660–1848: Science, engineering and statecraft* Chicago, 63–81.

Loomis, Elias. 1856. *The recent progress of astronomy, especially in the United States.* 3rd edn, New York, 202–92.

Maury, M. F. 1846. *Astronomical Observations made under the Direction of M. F. Maury . . . During the Year 1845 at the U. S. Naval Observatory in Washington.* Washington, D.C.

Musto, David F. 1968. "A survey of the American observatory movement, 1800–1850", *Vistas in Astronomy,* ix, 87–92.

Nourse, J. E. 1873. *Memoir of the founding and progress of the U.S. Naval Observatory.* Washington.

Paullin, Charles O. 1923. "Early movements for a national observatory, 1802–42", *Records of the Columbia Historical Society,* xxv (1923), 36–56

Rhees, William J. 1879. *The Smithsonian Institution: Documents relative to its origin and history,* Washington. 26th Congress, 1830–41. "Proceedings of the Senate," 238–42.

Rothenberg, Marc. 1974. The educational and intellectual background of American astronomers, 1825–1875, PhD dissertation, Bryn Mawr College.

Rothenberg, Marc. 1983. "Observers and Theoreticians: Astronomy at the Naval Observatory, 1845–1861," in Dick and Doggett (eds), 29–43.

Smithsonian Astrophysical Observatory Centennial Papers. 1990. *Journal for the history of astronomy,* xxi (special issue of February 1990), 107–53.

U. S. House of Representatives. 1842. *Journal of the House of Representatives,* 27th Congress, 2d session, 15 March 1842, 542. The *Journal* indicates that John Quincy Adams was on the House floor at the time.

U. S. Senate Journal, 27th Congress, 2d Session, 23 June 1842, 417.

Weber, Gustavus. 1926. *The Naval Observatory: Its history, activities and organization.* Baltimore

Wilkes to BONC, 15 June 1842, Wilkes papers, Library of Congress, Box 1, folder 1.

Williams, Frances Leigh Williams. 1963. *Matthew Fontaine Maury, scientist of the sea.* New Brunswick, 142-3 and 518.

Winterhalter, A. G. 1889. *The International Astrophotographic Congress and a visit to certain European observatories and other institutions.* Washington, 108–20, also published as Appendix I to *Washington observations for 1885* (Washington, 1891).

Chapter 26
The First Time Balls and the First North American Time Ball

Abstract This article documents that the world's first time ball was erected at Portsmouth, England in 1829, preceding the well-known Greenwich time ball by more than 3 years. It furthermore establishes that the earliest time ball in North America was first dropped at the U. S. Naval Observatory in Washington, DC sometime between April 1, 1845 and September 1, 1846. An addendum describes the U. S. Naval Observatory's ceremonial round-the-world time ball drop for the year 2000 (Y2K) and for the turn of the millennium in 2001, synchronized to the Global Positioning System satellites, for which the Naval Observatory provides the time via its Master Clock. Although now a historical and ceremonial curiosity (as in the famous New York City New Year's Eve ball drop), in their heyday time balls were essential for time dissemination for navigation, horologists, and the general public until superseded by radio signals in the early twentieth century. Because time balls were synchronized by astronomical observations, they illustrate an important example of the impact of astronomy on society.

26.1 Introduction

We have uncovered evidence in the National Archives of the United States that the world's first time ball was dropped at Portsmouth, England late in 1829, more than 3 years before a time ball was erected at the Royal Observatory at Greenwich. The same evidence makes possible further investigations that shed new light on the events leading to the establishment of the Greenwich time-ball service and other early time balls. Now largely an historical curiosity, the maintenance of this service was considered an important function of many astronomical observatories during much of the nineteenth century. The service was superseded by radio time signals.

Although dropping a time ball at a preconcerted instant provided very accurate time signals to horologists and the general public, its primary purpose was to act as a visual signal for rating chronometers, the all-important timepieces that began to come into widespread use at the beginning of the nineteenth century by which ships

First published (with Ian R. Bartky) as "The First Time Balls," *Journal for the History of Astronomy*, 12 (October 1981), 155–64 and "The First North American Time Ball," *Journal for the History of Astronomy*, 13 (February 1982), 50–54.

S. J. Dick, *Space, Time, and Aliens*, https://doi.org/10.1007/978-3-030-41614-0_26

at sea could determine their longitude.[1] For determining longitude accurately, the problem of "rating" chronometers—determining their exact error so that it could be applied to the longitude calculation—was crucial.[2] This rating process, ideally carried out at least every 7–10 days, required an accurate determination of time independently of the chronometer being rated. A navigator could in principle accomplish this through astronomical observations with a sextant and artificial horizon, but the results depended on the accuracy of the equipment and the experience of the observer.[3] A better method was to observe the Sun or stars with a transit instrument, a delicate apparatus seldom carried on a ship, but usually found in astronomical observatories. Thus, the need for an improved rating service as chronometers came into more general use led to the assignment of this responsibility to the Royal Greenwich Observatory in England (Howse 1975; Gould 1923) and the creation of the Depot of Charts and Instruments, the forerunner of the United States Naval Observatory (Dick 1980).

There were many drawbacks if chronometers were actually removed from the ships as they sometimes were: the risk of damage during transport, the time expended, and the fact that there was a slight difference between the rate determined on land and the rate determined at sea (Shadwell 1861). While another timepiece could be used as an intermediary to transfer time from the observatory to the ship without removing the chronometer, the simple method of finding time through observation of an instantaneous signal such as a rocket or falling object—visible from the harbor and regulated by a transit instrument—was convenient and efficient.

26.2 The Portsmouth Plan

The evidence found in the National Archives (U. S. National Archives 1829–1830) contains just such a "Plan for Ascertaining the Rates of Chronometers by an Instantaneous Signal," written by Captain Robert Wauchope of the Royal Navy.[4] The plan, a coalesced version of two previous articles (Wauchope 1830a, b) contains a drawing (Fig. 26.1) of the device. It had been tested successfully at the Portsmouth Dockyard some weeks before the written material was given in July 1830 to the American ambassador for transmission to the U. S. Navy Department. We have been able to place this material in the context of Wauchope's communications with the Admiralty and the several journals in which he expounded upon the use of his device. We conclude he was the inventor of the time ball, his ideas dating from remarks written in 1818, 6 years earlier than the date generally listed.[5]

Wauchope described his 1818 plan as one "for communicating time by means of telegraphs." Though the mechanical telegraph began in France in the early 1790s and shutter telegraphs were erected in England shortly thereafter, the term is not a good one for his invention since a telegraph can transmit *unforeseen* messages. Thus we consider his device as being derived from the system of signal stations erected in southern England in the mid-1790s, and used to convey *pre-arranged* information of enemy ships in sight (Tuck 1924, 465–81).[6]

Fig. 26.1 Engraving of Captain Robert Wauchope's design for the time ball at Portsmouth, from *Edinburgh New Philosophical Journal*, vii (1830), 290

The concept embodied in this 1818 plan is exactly that of a visual time signal. As described later by others, "The plan there proposed of giving the instant of time, is by the observer with the sextant and artificial horizon on shore, causing at the instant of observation the dropping of a shutter on its hinges, the instant of its fall being noted on board as shewn by the time-keeper."[7]

In the ensuing decades, other means were employed to increase the accuracy of the time observation, and many devices—balls, discs, guns, boards, etc.—were used to signal the instant of time. Yet nothing really changed. Indeed, we consider the 1852 addition of means to drop a time ball by a pulse from a distant clock as only an embellishment.[8]

The time signal concept was discussed, particularly among officers assigned to the St. Helena-Cape of Good Hope station, but very little happened for many years. In 1824 Wauchope communicated his plan to the Lords Commissioners of the Admiralty, and eventually a test of his device was made in 1829.[9] We have found two complementary discussions of the operational device: the National Archives material, and the somewhat critical comments made by the editors of the *Nautical magazine* in 1835.

The two balls depicted in Fig. 26.1 were 5 feet in diameter. The upper, fixed ball was cut off at its point of contact with the lower, sliding one so that no light could be seen between them when the ball was ready to be dropped. Wauchope emphasized the need for an instantaneous signal, designing the release mechanism with this purpose in mind: as one that could be operated by the official observing the clock.[10] He estimated that the weighted ball would fall its diameter in about four tenths of a second, and consequently suggested its release be that amount ahead of the true time. He also advocated the use of a premonitory signal to alert the shipboard observer; the flag shown in the figure was for that purpose.

In operation the sliding ball was raised by halliards against the fixed one, and held in place by securing the halliards to a triggering mechanism (shown in detail in the lower part of Fig. 26.1). Since the ball was weighted, it dropped rapidly when the halliards were released at the correct instant of time.

The time to be given is of interest, since the subsequent British time balls were dropped at 1 o'clock, while American time balls were dropped at noon. Wauchope recommended noon Greenwich time for the test, regulating the time signal by the transit instrument at the Portsmouth Observatory.[11] At other ports (whose longitude would be known), Wauchope suggested: "Perhaps one hour after noon might be better for the signal, to allow an observation being made to ascertain the time of day a short time before the signal is made" (Wauchope 1830a). Here, too, we see an attempt to deliver the highest accuracy possible.

For the Portsmouth test the time ball was placed on the platform, at the water's edge. Accordingly, as the *Nautical magazine* editors indicate, a signal from the Portsmouth Observatory—a falling shutter—had to be observed first, causing the editors to comment that "Although the greatest care is taken to secure precision, it must not be expected that the limits of error would never exceed two-tenths of a second, like that at Greenwich …" (Nautical Magazine 1835). Some later time balls also required relayed signals.[12] The criticism of reduced accuracy due to relayed signals should be taken cautiously, because the errors associated with American time balls 50 years later could exceed those of the Greenwich time ball: one tenth of a second in good weather to 1 second or more in cloudy weather (Upton 1881). The American time balls were adequate for the purpose.

This date (1835) of the *Nautical magazine* editors' comments indicates the relay configuration of the Portsmouth time ball continued at least for a few years. Others also mentioned the Portsmouth installation.[13] Sometime after 1855, a new time ball, based on the Greenwich design but still controlled by the Portsmouth Observatory clock, was installed at the naval base (Laurie 1958). A time ball remained in use at Portsmouth until about 1914.

26.3 Foreign Countries

Wauchope understood the benefits of extending his system. On February 1, 1830, he wrote:

> Should this plan for showing true time be universally adopted, which, from its simplicity, it bids fair to be, both by this country and by France and America, there will then be no port of any consequence into which a ship can enter, where an accurate rate for the timepieces on board may not be found. Chronometers will be more generally adopted, and the *risk* attached to both *life and property* embarked in ships, be much diminished. (Wauchope 1830b)

The Admiralty's notice to Wauchope of the success of the Portsmouth test also stated "with regard to your proposal of establishing signal stations on the continents of France and America, their Lordships cannot interfere" (U. S. National Archives 1829–1830; Wauchope 1836).[14] Almost immediately, Wauchope submitted his plan to the French and American ambassadors in London, urging its adoption.[15] The American ambassador forwarded the plan to Navy Department Secretary John Branch, who, in turn, referred it to the Board of Navy Commissioners (U. S. National Archives 1829–1830). In December 1830, less than 3 months later, the U. S. Navy established a Depot of Charts and Instruments—the forerunner of the U. S. Naval Observatory—for the express purpose of maintaining the Navy's charts and instruments, and especially for rating its chronometers (Dick 1980).

No doubt the process then underway for establishing the Depot of Charts and Instruments contributed to the lack of interest in Wauchope's proposal when it arrived.[16] In addition, we can suggest that it was premature in this country: not only did the United States Navy, unlike the Royal Navy, have no central place to rate and test chronometers, but also there was scarcely an observatory in the western hemisphere.[17] Wauchope's proposal described observatories, and urged the use of transit instruments to make the meridian observations to maintain the observatory clock. The situation in the United States in 1830 when the Depot was established was not easily adapted to this proposal. The first observations at the Depot were made with a sextant and circle, and not until 1831 were they made with a newly acquired transit, believed to be "the only one in the United States for sale," Until 1838, the time was kept by using one of the chronometers being rated (Dick 1980).

26.4 Greenwich and Other British Ports

In addition to foreign countries, Wauchope tried to get British ports to erect time signals: "In Edinburgh and Liverpool I have also done my best to get it established," he wrote in 1836, "both for the use of shipping, and also for regulating the time in the towns, which is often of so much consequence in mercantile concerns, particularly as to the starting of coaches, &c" (Wauchope 1836). Time balls were erected later at these ports, so his influence there is indirect.[18]

Rather more direct is Wauchope's influence on two southern time-ball sites: the Cape of Good Hope and St Helena. During his command of the *Eurydice*, Wauchope communicated his idea to the naval commissioner at the Cape in 1818 (Wauchope 1836).[19] Fearon Fallows, H.M. Astronomer at the Cape, arrived in August 1821 and began a time signal similar in concept to Wauchope's around early 1823 (Warner 1979, 64). Thomas Henderson, Fallows's successor, began his famous time-pistol service early in 1833 (Warner 1979, 32).[20] By Thomas Maclear's time, the service had become one of noting and publishing the time of the flash of the 9 pm cannon (Warner 1979, 47). On September 30, 1836, about 20 months after first considering it, Maclear announced the erection of a time ball on the Cape Observatory grounds. It has been suggested that Wauchope, along with his brother-in-law and chief naval officer at Simon's Bay, Admiral Patrick Campbell, influenced Maclear in his considerations (Warner 1979, 47).

The other interesting southern time ball site is St Helena, where the East India Company established the St. Helena Observatory on Ladder Hill before giving up control of the island on April 22, 1834.[21] A time-ball service was available by January 21, 1834, based on the observatory notice.[22] Wauchope, commanding H.M.S. *Thalia*, arrived at St. Helena in December, writing, "we found my chronometer-signal established there" (Nautical Magazine 1835). The new government decided to dismantle the observatory, but the time-ball service was brought to the town itself and placed under the operation of the former master of a ship commanded by Wauchope (Gosse 1938; Thompson 1978; Liddell 1837) Again, it is probable that Wauchope played an important role in initiating, and certainly in maintaining the St. Helena time ball.[23]

Wauchope did have a more direct influence on the erection of a time ball at the Royal Observatory at Greenwich. Writing late in 1833, Wauchope recounts how he visited the London docks in June, seeing hundreds of ships being made ready for sea, "all of which could most distinctly see a chronometer signal from Greenwich Observatory" (Wauchope 1834). Gaining additional support for the utility of his method by talking to ship captains and officers, Wauchope again wrote to the Lords of the Admiralty, proposing a signal at Greenwich.[24] The Admiralty's June 20, 1833 reply stated that his plan had been referred to the Astronomer Royal, John Pond (Wauchope 1834, 1836). By August the Greenwich time ball was under construction, and on October 28, 1833 the Admiralty issued its public notice (Nautical Magazine 1833). The first detailed operating description was published about 2 years later, and the later history of this time ball and the observatory time service are well known (Nautical Magazine 1835; Howse 1975, 134–136; Laurie 1958).

While we do not intend to denigrate the importance of the Greenwich time ball as the model for others, nor the efforts of the Astronomer Royal, G.B. Airy, to expand greatly this service for navigation and the public,[25] we do want to show the practical navigation tradition from which this time-signal device originated, and the importance of Captain Robert Wauchope, R.N. as the inventor and disseminator of the system. The spread of the time ball system by 1845 is summarized in Table 26.1.

Table 26.1 Time balls inaugurated prior to 1845

Location	Date inaugurated
Portsmouth, England	1829
Port Louis, Mauritius	April 1833
Greenwich, England	October 1833
St. Helena, S. Atlantic	1834
Cape of Good Hope, S. Africa	1836
Batavia (Jakarta) Indonesia	1839
Cape Coast Castle, Ghana	1839
Madras, India	1839
Bombay, India	After 1841
Valparaiso, Chile	Before 1843
Liverpool, England	1844
Washington, DC, USA	Early 1845

26.5 The First North American Time Ball

As stated above, as part of Wauchope's efforts to have his time ball signal used widely, he wrote to the American Government in 1830 while the U. S. Navy was establishing the Depot of Charts and Instruments. The Depot, forerunner of the U. S. Naval Observatory (see Chap. 25), was given the responsibility for rating the Navy's chronometers (Dick 1980) but no time-signal device was erected then. We establish here the date for the first American time ball.

The building on the U. S. Naval Observatory's first site was completed in the fall of 1844 (Gilliss 1845) and observations were started the next year.[26] This observatory, known then as the National Observatory, can be considered the American analogue to the Royal Observatory, Greenwich, and consequently the most likely candidate for an early time ball. Indeed, the first volume of observations (Maury 1846), contains a small engraving of the Observatory which depicts a staff attached to the top of the dome housing the 9.6 in. refractor, and a large sphere attached to one side of the staff (Fig. 25.3 in Chap. 25).[27] Though the sphere clearly appears to be a time ball, no mention is made of one in the otherwise detailed account of Observatory activities given in this volume. In fact, a time ball is not mentioned in this series until 1862.[28]

Numerous other sources state or suggest that the Washington time ball was erected in the 1850s.[29] However, one city guidebook contains an allusion to President John Quincy Adams that reinforces the earlier time period implied by the engraving.[30] Further, we found in the voluminous official correspondence of the Observatory's first Superintendent, Matthew Fontaine Maury, the following letter from the Secretary of the Navy dated December 10, 1844 (U. S. National Archives 1838–1884).

> You will be pleased to devise some signal by which the mean time may be made known every day to the inhabitants of the city of Washington. When you are prepared to put your signal into operation, you will give notice of the kind you have adopted in the city newspapers, and at the same time inform the Department.

Three days later Maury ordered from the Navy agent in Washington "one flagstaff 20 feet long and three stays for do [ditto?] each 40 feet long two pair of Sig [signal?] halliards. One black ball 3 feet diameter…" (U. S. National Archives 1842–1862). On February 7, 1845 he wrote to the New York agent for Charles Goodyear, the inventor of vulcanized rubber whose patent had been awarded in 1844 (U. S. National Archives 1842–1862):

> Be pleased to make and send … four air tight balls of Gumelastic Composition capable of being inflated into spheres. They are wanted for signal balls. They should therefore have a bracket at opposite poles firmly secured so that the halliards for hoisting up may be bent on to one bracket and a rope for hauling down to the other. Let two of the balls be at least 4 feet in diameter when inflated one of 3 feet, and one of 18 inches…

An Observatory staff member followed up on the order, writing to Maury on February 20, 1845 that the "India-rubber balls" are "in progress and will be completed as soon as the weather will permit" (U. S. National Archives 1838–1884). Finally, in a May 27, 1845 order to Messrs Blunt, chart suppliers and instrument makers in New York, Maury asked, "What will you charge for that copper ball? I may want it for a *time* ball…" [emphasis in the letter] (U. S. National Archives 1842–1862). There is no further correspondence on time balls up to October 1846.

We therefore date the first American time ball as after April 1, 1845 (the date of President Adams' first visit to the National Observatory) and before September 1, 1846, the latest date printed in Maury (1845).[31] In addition to being the first North American time ball, and apparently the twelfth one in the world, it had several interesting characteristics.[32] First, though one guidebook attributes a navigational use, the Secretary of the Navy's order demonstrates its primary purpose was for the dissemination of time to the general public, a contrast to the other early time balls.[33] Maury, well known as a navigator, apparently did not view the device as important, so its description did not form a part of the volumes under his supervision.[34]

Second, the operation of the ball was unusual. The American astronomer, Elias Loomis, consistent with Maury's description in his order to Goodyear, wrote that the ball was "*lowered* at the precise instant of twelve o'clock"[35] [emphasis added]. H.S. Pritchett, who installed several time-ball services in the United States, stated that "The ball when released drops upon the dome and thence rolls to the roof beneath" (Pritchett 1881). Another operational mode was described by A.D. Bache and B.A. Gould of the Dudley Observatory at Albany, New York, in their 1856 proposal for a time ball for the city of New York where they refer to "the [Washington] ball … which is thrown down by hand at the word of command" (Bache and Gould 1856). We conclude that several different balls and at least two different modes of operation were in use in the early period of the Washington time ball. An operational mode proposed a few decades ago is clearly a misinterpretation of the engraving in the *Washington observations*.[36]

Third, the Washington time ball, in contrast to the first ones at Portsmouth and Greenwich, had no means for establishing a plane of reference at the top of the staff against which the moment of release could be judged. As the first time ball in the United States, its initial design may have influenced most subsequent American time balls, somewhat to the detriment of accuracy, but clearly sufficient for public and navigational needs.[37]

The U. S. Naval Observatory's time ball was removed from the dome shortly before July 11, 1885 (USNO 1885). It, or a new model, was installed on the central pavilion (on the side facing The White House) of the building now termed the Old Executive Office Building. The Washington time ball remained in service until December 16, 1936. See Fig. 26.2.

26.6 Commentary 2020

As mentioned in the Introduction to this section, time ball drops are now celebratory, having been superseded many times over by telegraph, radio, and satellite time dissemination. But time ball drops remain quite popular, especially the New Year's Eve drop in New York City. At Greenwich the ball is still dropped every day at 1 pm (Fig. 26.3), and the Lyttleton time ball at Christchurch, New Zealand (Fig. 26.4) was rebuilt after being mostly destroyed during the 6.3 earthquake of 2011. It reopened in 2018 to much fanfare. These time balls and others recreate an event whose origin is described in this chapter.

Just how common time balls were in the nineteenth century was brought home by an event at the U. S. Naval Observatory on New Year's Eve 2000 and 2001, Y2K and the millennium respectively. The Naval Observatory built a new time ball for this occasion (Fig. 26.5) at its Massachusetts Avenue site, but that was only one part of a much larger project spearheaded by the Observatory: in conjunction with the arrival of 2000 and 2001 (the latter constantly being emphasized as the official date of the new millennium), the Naval Observatory coordinated a worldwide time ball drop. It did so with the cooperation of Commission 41 (History of Astronomy) of the International Astronomical Union.

As the New Year swept around the world from the International Date line (Table 26.2), time balls were dropped at midnight local time beginning in New Zealand, then Australia, South Africa, Sweden, UK, New York City Times Square, the Naval Observatory in Washington, and several other sites in the United States. Altogether several dozen sites participated. Moreover, most sites synchronized their events via the Global Positioning System satellites, for which the Naval Observatory (keeper of the Master Clock of the United States) synchronizes the time. This juxtaposed the old and the new time dissemination systems, emphasizing how time dissemination accuracy has improved from a few tenths of a second with the time ball to a few billionths of a second with GPS.

Few people recognize the scientific origins of time balls, but because time balls were synchronized to the heavens, they are another case of the intertwining of

Fig. 26.2 Time ball on the Old Executive Office Building (now the Eisenhower Building) in Washington, DC, in operation from July 1885 to December 1936

Fig. 26.3 Time ball at Greenwich Observatory. (Photo by Steven Dick)

Fig. 26.4 Lyttleton time ball station in Christchurch, New Zealand, before its destruction in the earthquake of 2011, subsequently rebuilt

Fig. 26.5 Ceremonial time ball erected for Y2K and the millennium, adjacent to the 12-in. refractor dome of the U. S. Naval Observatory at its present site on Massachusetts Avenue, NW, Washington, DC. The ball is in the lowered position. (Photo by Steven Dick)

Table 26.2 Y2K and millennium round-the-world time ball project

U. S. Naval Observatory		
Round-the world time ball project		
(In order of time zone from international date line)		
Site	Time diff. from GMT including DST	Midnight occurs at this time EST
New Zealand—Lyttelton (Christchurch)	+13	6 am
Antarctica		
McMurdo	+13	6 am
Scott Station	+13	6 am
Australia		
Sydney Observatory	+11	8 am
Melbourne—Parks Victoria	+11	8 am
Adelaide—Town Council	+10.5	8:30 am
Cape Borda Light Station	+10.5	8:30 am
(Kangaroo Island; cannon)		
Fremantle	+8	11 am
India		
National Physical Laboratory	+5.5	1:30 pm
New Delhi		
South Africa		
Cape Town	+2	5 pm
Europe		
Sweden—Goteborg	+1	6 pm
—Karlskrona Naval Base	+1	6 pm
UK		
Greenwich—National Maritime	0	7 pm
Edinburgh (ball and cannon)	0	7 pm
Antarctica		
Palmer Station	−3	10 pm
USA		
New York Times Square	−5	Midnight
South Street Seaport Museum	−5	Midnight
Washington, DC—USNO	−5	Midnight
Miami (raising an orange)	−5	Midnight
Doane College, Crete NE	−6	1 am Sat
La Jolla, CA	−8	3 am Sat

cosmos and culture. They also stand as a stark case of one technology being superseded by another, but also of continued public interest in time. The latter was also brought home by Dava Sobel's bestselling book *Longitude*, the subject of a Nova TV special on PBS, launched with Sobel and Vice President Gore at the Naval Observatory. Research on the history of time balls also continues (Bateman 2013)

Acknowledgments The authors wish to acknowledge the efforts of Elizabeth Bartky, Brenda Corbin of the U. S. Naval Observatory library, and Lee Johnson and Sharon Gibbs of the National Archives of the United States.

Notes

1. The chronometer method for determining longitude became practical with the successful sea trials of John Harrison's H4 chronometer on two sea voyages in 1761–1762 and 1764. By comparing local time determined by celestial observations with port time maintained by the chronometer, the difference in longitude between the port and the ship could be calculated. Maskelyne's publication in 1766 of *The nautical almanac and astronomical ephemeris for the year 1767* made possible the rival method of lunar distances, which determined time by observing the position of the Moon relative to certain nearby stars. The chronometer method was simpler, but often both methods were used as a check on each other. A case in point is Sir John Herschel's voyage to the Cape of Good Hope, described in his diary entry for Friday, 20 December 1833: "The Captn in a great puzzle about his Lunars. All hands at work taking obsns and at last as a final result the Lunars of yesterday and today give –27 s for the Chronometer error while all Chronomrs make it +2 m 21 s. Bad work this. But a Lunar at Sea seems rather a bungling business." The next day Herschel reported that "by plenty of patience and using masses of obsns" they came to a satisfactory conclusion, with the lunars and chronometers tallying within one-half minute of time, 8 miles of longitude (Evans et al. 1969).
2. The calculation was made as follows: if on May 27 at 9 am the chronometer was slow 2 h, 7 min, 18 s, and on June 3 at 5 pm the chronometer was slow 2 h 6 min 51 s, then the difference in the error in 7 days 8 h was 27 s, or a daily rate of 3.7 s daily, gaining. The navigator would use this new rate for determining his longitude until another rate could be obtained (Raper 1866, 275). Note that the observation had to be made at the same meridian, or the rate could not be disentangled from a change in longitude of the ship.
3. The method most often used was to observe the "equal altitudes of the Sun or stars; that is, the times when the object had equal altitudes before and after passing the meridian. The arithmetical mean gave the time the star passed the meridian. When the Sun was observed, its daily motion in declination had to be taken into account. The method could be used at sea, but was not as accurate with the sea horizon, and so was carried out on land when possible. A "single altitude" method could also be used, but was not as accurate (Raper 1866, 276).
4. Robert Wauchope entered the Royal Naval Academy at Portsmouth in December 1802, and served on various ships from 1805–1819 and 1834–1838. He was promoted to the rank of commander in 1812, and to captain in 1814, after which he commanded the *Swinger*, the *Eurydice*, and the *Thalia*. Wauchope and his wife were intimate friends of John Herschel and his family during Herschel's years at the Cape, serving as godparents to Herschel's second son Alexander. Wauchope occasionally accompanied Herschel at the 20-ft reflecting telescope, and returned some of Herschel's observations to England aboard the *Thalia*. Some 42 entries on Wauchope and his family are contained in Herschel's diaries and correspondence at the Cape (Evans et al. 1969). Wauchope was an Admiral at his death in Edinburgh, June 14, 1862. He is listed in O'Byrne's *Naval biographical dictionary* (London, 1849), 1261.
5. Writing in 1836 to claim priority for the invention, Wauchope refers to an 1818 discussion of its advantages in his "remark-book transmitted to the Admiralty when in command of the *Eurydice*, at that time on the Cape and St. Helena station" (Wauchope 1836). This letter to the editor, dated April 2, 1836, is accompanied by much of Wauchope's correspondence on the subject of his invention, and by the comments of the editors. The date of 1824 for the first time ball suggestion was given in the *Nautical magazine* (1835), which appears to have been

written by the magazine's editors. It was upon reading this article that Wauchope wrote to the editors with his 1818 priority claim. Derek Howse (1980) also gives the 1824 date, 79.

6. There were 47 stations between North Foreland and Land's End. The signals given were formed by "hoisting various numbers of black balls to a yardarm…."
7. See the editors' comments in Nautical Magazine, v (1836), 460–64.
8. Apparently first at Greenwich. See Laurie (1958), 113–115.
9. We obtain this date from Wauchope's correspondence with Professor Robert Jameson, the editor of the *Edinburgh new philosophical journal*, and with the various Secretaries to the Admiralty, whose answers were invariably signed by John Barrow, for 41 years the Second Secretary. In his cover letter to Jameson dated 23 November (Wauchope (1830a), 160), Wauchope quotes from a letter of 12 November 1829 sent to him by an officer stationed at Portsmouth, in which the officer described a ship's chronometer rating by the usual method of observing with a sextant and artificial horizon, the chronometer having been brought ashore: "The Admiral and myself were at the King's Stairs. … I then repeated to the Admiral … your plan, marking this [unsatisfactory chronometer rating] process as a case in point. From that moment he pursued its adoption with energy, and it is now, although in an infant operation, quite sufficiently established to give proof of *complete success*." In a letter dated June 22, 1830 (Wauchope 1836, 462), Wauchope queried the Admiralty regarding his plan "which is at present on trial at Portsmouth." Barrow's reply (U. S. National Archives 1829–1830, also printed in the *Nautical Magazine* for 1836) shows that its testing had been completed before 6 July.
10. Wauchope even suggested that the clock could be fitted to drop the time ball automatically. Wauchope (1830a), 160–1.
11. This observatory was at the Royal Naval College (until 1808 the Royal Naval Academy) in the Portsmouth Dockyard. The observatory may have had its origins in the sea trials of Harrison's H4 chronometer, since the ships for both the 1761–1762 and 1764 trials departed from Portsmouth. On the first trial the local time at Portsmouth was determined by equal altitudes of the Sun observed from "the Royal Academy at Portsmouth Dockyard." On the second trial the equal altitude observations were taken "at the observatory in Portsmouth" and Harrison's timepiece "compared with the clock in the observatory" (Maskelyne 1767). The statement is found in the "Appendix containing observations of equal altitudes of the Sun taken at Portsmouth, Jamaica, and Barbadoes…" Gould (1923), 59, notes that before departure on the second trial, Harrison's son "checked the timekeeper's rate by comparison with a regulator installed in a temporary observatory," the clock "lent for this purpose by the Duke of Richmond." According to a source contemporary with Wauchope's discussions, the observatory was later equipped with "an excellent astronomical clock, a good transit instrument, and a telescope at the public expense. A series of astronomical observations have been carried on at this observatory for many years by the late Mr. William Bayly, headmaster of the Royal Academy, who was one of the best calculators and accurate astronomers of the age, having been several years employed as an assistant to that able astronomer Dr. Maskelyne, at the Royal Observatory at Greenwich, where he calculated several of the early volumes of the Nautical Almanac…" (Burney 1830), a modernized and much enlarged version of *Falconer's Marine dictionary* by William Falconer. Bayly (1737–1810) was headmaster of the Royal Academy from 1785 to 1807.
12. A relay configuration was apparently used at St Helena in 1834 (Nautical Magazine 1835), 660. Brian Warner (1979), 64, describes the 1853 ball at Lion's Rump Signal Station whose drop time was controlled by telescopic observation of the observatory time ball.
13. In May 1835 the commander of an East India Company ship writes of "the admirable plan recently adopted at Greenwich and Portsmouth, of dropping a ball at a given moment … the only ones where the ball is now dropped" (Liddell 1835). In August 1836 the *Nautical magazine* editors write that the 1818 plan "is the same in principle as the system adopted at Portsmouth and Greenwich," so we conclude the time ball is still operational then. We believe that the lack of a detailed discussion of the Portsmouth time ball in any one place in this journal-of-record is due to the date of the test-2 years before the magazine's establishment (Nautical Magazine 1836).

14. The letter is dated 6 July 1830.
15. "Time signals for chronometers" (Wauchope 1836), 461. The cover letter to the American ambassador is dated July 21, 1830. Wauchope writes "the object ... is to get the Plan generally adopted, so that Ships of all nations may have it in their power, immediately upon entering a port, to get an accurate Rate for their Chronometers, without which they are worse then useless. And that Merchant Ships may be induced by this facility given them, to adopt the use of Chronometers more generally to the great saving of both Life, & Property" (U. S. National Archives 1829–1830, 218).
16. The report by the Board of Navy Commissioners on the reorganization of the Navy, including the provision for better care of navigation equipment, was written in November 1829. For the events that followed see Dick (1980), 467.
17. Thomson (1978), 4, writes that "prior to 1835, there was no fixed astronomical observatory on the North American continent." There were places-the observatory in Jamaica dismantled before the arrival of Harrison's chronometer H4 in January 1762, for example where equipment was available to equip an observatory at the level needed for Wauchope's proposal. Dick (1980) also describes the U. S. Congress's opposition to an *astronomical* observatory, which was possibly an additional factor.
18. The Liverpool Observatory, which became the outstanding institution for marine chronometer testing, operated a time ball when it began in 1844. At the 1837 meeting of the British Association for the Advancement of Science held in Liverpool, the President of the Association proposed the establishment of an observatory there. A committee, which included the astronomer Francis Baily, was formed and in a memorial to the corporation stated "that the inaccuracies in the Greenwich mean time given in some of the principal ports in the kingdom were known to be sufficient to cause the wreck of ships" (Baines 1859), 53. Baines, a city father, also states that "the observatory was established in consequence of these and similar representations." Although this is a weak thread by which to deduce Wauchope's influence, the time of Wauchope's writings—1830 to 1836—is appropriate, and we note other indications of a maritime, in contradistinction to a scientific or astronomical influence in this particular case. See "The time-ball at Greenwich" (Nautical Magazine 1835), 584; Liddell (1837), 364–5. The Edinburgh time ball was erected on Calton Hill in 1855, but Charles Piazzi Smyth, the Astronomer Royal for Scotland, had advocated its erection since his arrival in 1846 (Smyth, *Edinburgh astronomical observations,* vi (1846), Appendix, 17). Note that his listing of time balls here is incomplete. Smyth had been First Assistant at the Cape Observatory from 1835 to 1845, and according to Warner "had copied the Cape method of announcing midday with the aid of a time-gun and Time Ball" (Warner 1979, 80). This process of diffusion *via* the Cape, rather than directly, is discussed further in the text.
19. Wauchope's letter was confirmed by the naval commissioner, Jahleel Brenton.
20. Warner (1979), 32. David Gill (1913), cxliii, describes a time ball operated "in the days of Fallows and Henderson." Fallows's tenure at the Cape was from August 12, 1821 until his death July 25, 1831, so if Gill's statement is correct, it could change our view regarding the first time ball. Gill does not mention Fallows's lamp/shutter signal, and we know the time-pistol was unavailable to Fallows. We conclude that Gill's description is an anecdote that combines Fallows's 1823 lamp/shutter (Warner 1979, 10), Henderson's 1833 pistol *(ibid.,* 47), and the 1853 time ball under Maclear's supervision at Lion's Rump Signal Station *(ibid.,* 64). Warner (1979, xi and 22) discusses similar anecdotes that he has been unable to document.
21. The observatory building was completed in 1828 and Manuel J. Johnson was appointed astronomer on 24 January 1828 (Gosse 1938). Warner (1978), 23, cites Johnson's earlier visits to Fallows at the Cape, and his interaction with Henderson in 1833 *(ibid.,* 36).
22. The time-ball service is described in detail in (Nautical Magazine 1835), 658–60, and noted by (Liddell 1835), 403. The time-ball notice is cited in the St Helena article, 660.
23. We have no evidence that Wauchope influenced another pre-Greenwich time ball, one established at the Port Louis, Mauritius observatory on April 25, 1833, the government notice being given in "Rating chronometers at the Mauritius," *Nautical magazine,* iv (1835), 136. The time

ball was still operational in 1836 ("Chronometers at the Mauritius," *Nautical magazine,* vi (1836), 365). The initial date for this time ball is given as 1832—still after the Portsmouth time ball—in another government document (Colony of Mauritius, Education Department, *The historical monuments of Mauritius* (Port Louis, 1958), 1).

24. The letter is dated June 17, 1833, and is excerpted in (Wauchope 1836), 462. Note that Sir Jahleel Brenton, who figures in the 1818 history and the Cape Observatory time ball, is given credit for suggesting the importance of Greenwich.
25. There is a famous *Punch* cartoon showing the Astronomer Royal as a time ball on the Observatory (Howse 1980, 103).
26. Gilliss does not describe or illustrate a time ball in this first description of the Observatory.
27. This series is clearly based upon the format of *The Greenwich observations*, particularly the volumes for 1836 and later, when George B. Airy, seventh Astronomer Royal, superintended their preparation.
28. "For the purpose of giving correct time to the city, a staff has been placed on top of the dome, and a large, but light, ball is hoisted 10 min before 12 o'clock of each day, except Sunday. The pulley is connected with an electro-magnetic battery after the ball is up, and the circuit is broken by the assistant in the chronometer room at the instant of noon" (Gilliss 1862, p. x). Gilliss, Maury's successor, summarizes the various duties in the same manner as Airy did when he succeeded the sixth Astronomer Royal, John Pond.
29. These sources include writings of three important, late nineteenth-century American scientists who served as Naval Observatory astronomers in their youths. Cleveland Abbe included in his influential paper on Standard Time a table entry: "1855, WASHINGTON. U. S. Naval Observatory began dropping noon-time-ball" (Abbe 1880), 17–44. Later Abbe wrote, "About 1850 a time ball was established on the U.S. Naval Observatory at Washington..." (anon., "Standard time in America," *Science*, n.s., xxii (1905), 315–18; this article is listed as no. 233 in Cleveland Abbe's bibliography in the *Biographical memoirs of the National Academy of Sciences*, viii (Washington, 1919), 505. H.S. Pritchett wrote, "The first time ball established in the United States was dropped from the dome of the Naval Observatory in Washington in 1855," a statement repeated essentially verbatim by E.S. Holden (Pritchett 1881); Edward S. Holden, "Astronomy," *Annual report of the Board of Regents of the Smithsonian Institution ... for the year 1881* (Washington, 1883), 228).George B. Airy alludes to the Washington time ball in his 1861 list of 16 time balls in operation around the world, a list reproduced as an appendix in Howse (1980), 227–8. *The Nautical magazine*, a British journal-of-record for navigation, carried no notice of the Observatory's time signal; and an equivalent American journal stated only, "As is generally known, a large black ball descends the flagstaff of the dome" (Editors, "The National Observatory", *U. S. nautical magazine and naval journal,* vi (New York, 1857), 443). Another possible source of this era is Henry Raper, *The practice of navigation and nautical astronomy* (London), Table 10; however, all editions from the third (1849) to at least the fifteenth (1883) list only the same nine time balls, and the Washington time ball is not one of the nine.
30. "John Q. Adams, who was a devoted friend of the Observatory, and who used to visit it frequently in the last days of his life [died February 23, 1848], has been known to walk all the way up to the Observatory from his lodgings, to see the ball fall" (*Bohn's hand-book of Washington* (Washington, 1852), 51–52; this reference forms part of the Toner collection in the Rare Book Division of The Library of Congress).
31. The date of President Adams's first visit to the National Observatory, as given in Charles F. Adams (ed.), *Memoirs of John Quincy Adams, comprising portions of his diary from 1795 to 1848,* xii (Philadelphia, 1877), 189. Although he describes the Observatory's building and equipment in some detail, Adams does not mention a time ball. We note here that Adams had received a letter in 1839 from G.B. Airy, in which the Greenwich time ball was included

as part of the Royal Observatory's description (*Annals of the Astronomical Observatory of Harvard College*, i (Cambridge, Mass., 1856), p. ciii).

32. There was a South American (Valparaiso, Chile) time ball by 1843 (*The nautical magazine*, xii (1843), 768). The first Canadian time ball was erected in 1855 at the Citadel in Quebec (Thomson 1978, 18). There were over two dozen U. S. and Canadian time balls by the first decade of the twentieth century. Between 1832 and 1845 the *Nautical Magazine* describes 11 time-ball sites.
33. "At 12 m. a ball is dropped from a flagstaff, ... which enables navigators leaving the Potomac to regulate their time to a second, and also regulates the city time" (A. Hunter, *The Washington and Georgetown directory, strangers' guide-book for Washington, and Congressional and clerks' register* (Washington, 1853), p. 66 of "Strangers' guide-book" section). Two American *navigational* time balls were erected as commercial ventures on Telegraph Hill in San Francisco in 1852 (David F. Myrick, *San Francisco's Telegraph Hill* (Berkeley, 1972), 31–32). A reproduction of a newspaper advertisement for the first time ball is on p. 31; a picture of the London-made clock which regulated the second time ball is on p. 5 of Stephens (1961), 3–10.
34. A contrast to Airy, who mentions "the dropping of the Signal Ball at lh mean time every day" in his description of observatory duties (G.B. Airy, *Astronomical observations made at the Royal Observatory, Greenwich in the year 1836* (London, 1837), p. ii). He also refers to "the Signal Ball" on p. i of the 1840 volume (published 1842). We believe that Maury saw these citations, since he uses the term "signal ball" in his Goodyear letter, while *The nautical magazine* citations are invariably to "time-ball."That Maury did not report the time ball's installation to the Secretary of the Navy can be explained by the fact that John Y. Mason had two tours of duty as Navy Secretary: March 26, 1844–March 10, 1845, and September 10, 1846–March 7, 1849. Since the Washington time ball became operative after Mason's first tour, one need not inform his successor of the event; since the time ball was operative by the start of Mason's second tour of duty, there was no need to inform him.
35. Elias Loomis, *The recent progress of astronomy; especially in the United States* (New York, third edition enlarged, 1856), 289. Page 224 of this book has a drawing of the National Observatory with a time ball larger than the official engraving, but "of smaller size than ... [the five-foot diameter one] at Greenwich," a representation also consistent with Maury's order to Goodyear. The illustration appears again in Loomis, "Astronomical observatories in the United States," *Harper's new monthly magazine*, xiii (New York, June 1856), 25–52.
36. Dick (1980), 469, describes still another time ball, "a black canvas structure, 2½ feet wide." This is apparently a later device that surrounded the staff on which it slid. "The ball, instead of having a hole in the center fitting over the pole, slid down the side of the pole. This is the only instance found of this form of ball" (Stephens 1961, 7). "This ball did not slide on the mast [like the Greenwich one] but alongside it as shown in the accompanying [sic] [missing] picture" (Ada W. Watts, "Time balls," unpublished manuscript, US Naval Observatory Library (1963), 6 pp).
37. "I have just received a suggestion from ... San Francisco that a large fixed shape be placed at the top of the mast, so that the signal may be more quickly observed by the separation of the ball and this shape. It is intended to embody this in the specifications" (Norfolk, Virginia Branch Hydrographic Office, 19 March 1913; in US National Archives, Record Group 37, Entry 41, 187–75893). The various US Navy time balls installed in San Francisco, including the 1908 one being discussed in this passage, are pictured in Bartky, "Naval Observatory time dissemination before the wireless" (U. S. Naval Observatory Sesquicentennial Special Publication 1982).

References

Abbe, C. 1880. "Report of Committee on Standard Time; May, 1879", *Proceedings of the American Metrological Society,* ii, 17–44.

Bache, D. and Gould, B. A. 1856. Letter in *The New York Times*, 14 February 1856, p. 8, col. 2

Baines, Thomas, 1859. *Liverpool in 1859: The Port & Town of Liverpool, and the harbour, docks, and commerce of the Mersey in 1859* (London), 53.

Bateman, D. 2013. "The time ball at Greenwich and the evolving methods of control - Part I," *Antiquarian Horology*, 34, 2, 198–218.

Burney, William. 1830. *A new and universal dictionary of the marine* (London, 1830).

Dick, S. J. 1980. "How the U.S. Naval Observatory began, 1830–1865", *Sky and telescope,* (1980), 466–71.

Evans, David S. et al (eds), *1969. Herschel at the Cape: Diaries and correspondence of Sir John Herschel, 1834–38.* Austin and London, 16–17.

Gill, David. 1913. *A history and description of the Royal Observatory Cape of Good Hope* (London, 1913), cxliii.

Gilliss, J. M. 1845. "A report of the plan and construction of the depot of charts and instruments, with a description of the instruments, &c", 28th Congress, 2nd session, Senate Report 114, 18 February 1845, 77pp.

Gilliss, J. M. 1862. *Astronomical and meteorological observations made at the United States Naval Observatory during the year 1861* (Washington, 1862), p. x.

Gosse, Philip. 1938. *St. Helena, 1502–1938* (London), 331–4.

Gould, R. T. 1923.*The marine chronometer: Its history and development* (London, 1923), 253.

Howse, D. 1975. *Greenwich Observatory: Origins and early history, 1635–1835* (London, 1975), 150.

Howse, D. *1980. Greenwich time and the discovery of the longitude* (Oxford, 1980).

Laurie, P. S. 1958. "The Greenwich time ball", *The observatory,* lxxvii, 113–15.

Liddell, James. 1835. "Communicating longitude at sea", *Nautical magazine,* iv, 401–3.

Liddell, James. 1837. "St. Helena time-ball," *Nautical magazine*, vol. vi, 1837, 364–65.

Maskelyne, Nevil. 1767. *An account of the going of Mr. John Harrison's watch, at the Royal Observatory, from May 6, 1766, to March 4, 1767* (London), l–li.

Maury, M. F. 1846. *Washington astronomical and meteorological observations for 1845* (Washington, 1846), title page.

Nautical Magazine. 1833. "Chronometer rates", ii, 488 and 680.

Nautical Magazine. 1835. "The time-ball of St. Helena," vol. iv, 658–60

Nautical Magazine. 1836, 463 and *passim.*

Pritchett, H. S. 1881. "The Kansas City electric time ball", *Kansas City review of science and industry,* iv, 720–23

Raper, H. 1866. *The practice of navigation and nautical astronomy.* 9th edn, London, 275.

Shadwell, C.F. 1861. *Notes on the management of chronometers and the measurement of meridian distances* (new edn, London, 1861).

Stephens, W. Barclay. 1961. "Time balls", *Bulletin of the National Association of Watch and Clock Collectors,* x (Philadelphia), 3–10.

Thompson, Malcolm. 1978. *The beginning of the long dash:* A *history of timekeeping in Canada.* Toronto, 4.

Tuck, O. 1924. "The old telegraph", *The fighting forces,* i, 465–81.

Upton, Winslow. 1881. "Information relative to the construction and maintenance of time-balls", United States of America, War Department, Professional Papers of the Signal Service, no. 5 (October 1881), 12.

U. S. National Archives. 1829–1830. Record Group 45, Naval Records Collection of the Office of Naval Records and Library; Entry 222, Records of the Board of Navy Commissioners, Letters Received from the Secretary of the Navy, v (1829–30), 207, 211–19.

U. S. National Archives. 1838–1884. Record Group 78, Records of the Naval Observatory, Letters Received, 1838–84, Box 2: letter of John Y. Mason to M. Maury; letter of J. Coffin to Maury.

U. S. National Archives. 1842–1862. Record Group 78, Records of the Naval Observatory, Letters Sent, July 1842–November 1862, vol. i, p. 317.

U. S. Naval Observatory. 1885. *Report of the Superintendent of the United States Naval Observatory for the year ending June 30, 1885* (Washington).

Warner, B. 1979. *Astronomers at the Royal Observatory Cape of Good Hope.* Cape Town.

Wauchope, R. 1830a. "Plan for ascertaining the rates of chronometers by signal," Edinburgh new philosophical journal, viii, 160–162.

Wauchope, R. 1830b. "Description of the apparatus or signal-post for regulating chronometers," Edinburgh New Philosophical Journal, viii, 289–291.

Wauchope, R. 1834. "Establishment of Captain R. Wauchope's signal for ascertaining the rates of chronometers at the Royal Observatory, Greenwich," *Edinburgh New Philosophical Journal*, vol. xvi, 157–60.

Wauchope, R. 1836. "Time signals for chronometers," vol. v, 460–64.

Chapter 27
The U. S. Naval Astronomical Expedition of James Melville Gilliss in the Southern Hemisphere, 1849–1852

Abstract In early 1849, the U. S. Navy was making feverish preparations for a bold expedition to Chile. A Congressional appropriation had been secured, and hopes were high that the expedition would help solve one of the main problems in astronomy—determining the distance from the Earth to the Sun, today known as the astronomical unit. This chapter covers the preparations and results of the expedition, and places it in the context of the history of science, with glimpses of how the expedition sheds light on the early relations between the United States and Chile. The expedition is a particularly robust example of how science and culture interact, not only in carrying out international expeditions, but also as embodied in the six volumes Gilliss published detailing the geography, climate, meteorology, mineral and agricultural resources, and culture of Chile, one of the earliest descriptions of the newly formed country.

27.1 James Melville Gilliss and the U. S. Naval Observatory

The U. S. Naval Observatory was founded as a Depot of Charts and Instruments in 1830, transformed into the Naval Observatory in 1842, and eventually many of its functions were split off to form other agencies in the U. S. government (Dick 2003). The United States was still a young country, and the Depot and Observatory were formed in the era before specialization in science—and where the United States was trying to demonstrate to the world that it could contribute to the scientific enterprise. The Depot was a place where navigational instruments were kept, and especially where chronometers were rated (kept running accurately), because they were needed for longitude determinations. In order to rate the chronometers, to tell how fast or slow they were running, a standard of time was needed, and that standard was determined by observations of the stars based on the Earth's rotation. This is how the Navy became involved in astronomy.

James Melville Gilliss (Fig. 25.2), who would later become the leader of the Chile expedition, joined the U. S. Navy at the age of 15 and was only 25 years old when he was put in charge of the Navy's Depot of Charts and Instruments. It was Gilliss who

Address given in March 2016 at the Pontificia Universidad Católica Chile, published here for the first time.

S. J. Dick, *Space, Time, and Aliens*, https://doi.org/10.1007/978-3-030-41614-0_27

in 1842 secured from Congress a $25,000 appropriation for an observatory building, books and instruments in the area of Washington known as "Foggy Bottom," not a very auspicious name for a place where astronomical observations would be made. It was on the banks of the Potomac river, and fog often surrounded the observatory.

In any case when it came to appointing the first Superintendent of the U. S. Naval Observatory when the building was completed in 1844, Gilliss was the obvious choice, but the Secretary of the Navy chose instead another naval officer, Matthew Fontaine Maury, later known as the father of oceanography. This choice was mostly due to politics, since Maury was from Virginia, also the home state of the Secretary of the Navy. In any case this freed up Gilliss for other endeavors, including the expedition to Chile only 5 years later. The Naval Observatory would assist the Chilean expedition in other ways. Maury himself, for example, wrote a circular "To the Friends of Astronomical Science" announcing the expedition to the astronomical world, asking astronomers to make observations, and detailing what kinds of observations would be necessary.

Both the Naval Observatory and the Chilean national observatory that would be formed as a result of the expedition were part of a national observatory movement, part of a pattern of observatories established by national governments for practical reasons (see Chap. 24). The national observatory movement is characterized by three eras: the first era, in which the prototype observatories were founded in Paris, Greenwich, Berlin and St. Petersburg; the second era characterized by offshoots from previous national observatories, and by new observatories of younger nations, such as U. S. Naval Observatory and the Chilean national observatory; and the third era, post-World War II, characterized by national or international consortia, large budgets, and increasingly large and sophisticated telescopes. Chile also came to play a very important role in this third era with the Cerro-Tololo Inter-American Observatory about 500 km north of Santiago, and of a very strong pattern of astronomy that continues today.

27.2 The Southern Hemisphere Expedition to Chile

Why a southern hemisphere expedition, why Chile, and why Santiago in particular? The immediate reason for the expedition was to make observations to determine the distance between the Earth and the Sun, known as the astronomical unit. Surprisingly, at this time only the relative distances of the planets were known—Jupiter was known to be more distant than Mars, for example, but the absolute distances of the planets were not well known. The distance to the Sun was the key to knowing the all distances in the Solar System, in other words to determine the scale of the Solar System. This was a kind of Holy Grail for astronomy in the nineteenth century, and there were several methods for doing it. One was to observe transits of Venus as it crossed the face of the Sun. This, however, was a very rare occurrence, occurring in pairs only twice in about 120 years. Many nations mounted expeditions to observe this rare phenomenon, but the next transit of Venus pair would not occur until 1874 and 1882 (see Chap. 28). The 1882 transit was observed by both Americans and Belgians at stations in Santiago.

During Gilliss's time these transits of Venus were several decades in the future, and scientists were not willing to wait another generation to find the scale of the Solar System. So when in 1847 the German astronomer Christian Gerling wrote to Gilliss proposing another method of determining the Sun's distance, Gilliss was receptive, and saw it as another way to showcase young American science, and the U. S. Navy. The new method involved observing the parallax of Venus and Mars, in other words, how much those two planets shifted against the background of fixed stars. Here is the key: the observations had to be made by two widely separated observatories at approximately the same longitude. The method had been used already in the seventeenth century with Paris and Cayenne, the latter in the northern part of South America. The results were not very good but with more modern equipment, hopes were high. The Naval Observatory would be one observing station (at 77° west), and Santiago the other (at 70.7° west). They were separated by 5000 miles. Gilliss was well connected politically, and immediately secured the backing of the U. S. Coast Survey and the American Philosophical Society, and most importantly, secured $5000 in funding from the US Congress. The Chilean government welcomed the proposal and offered any unoccupied public land to erect the observatory. Wanting to establish a permanent facility, Gilliss chose Santiago, the seat of the government and home to the University of Chile founded only a few years earlier.

After months of frantic preparation gathering instruments and staff, in July 1849 Gilliss's staff for the expedition departed the port of Baltimore with the instruments bound for Chile via Cape Horn. Gilliss himself, still a youthful 38 years old, followed in August via Panama, arriving in Valparaiso in late October, 1849. Travelling to Santiago, he soon chose Santa Lucía (Fig. 27.1), a rocky hill 629 m in height in the central part of the city, as the site for the Observatory. Chilean independence leader Bernardo O'Higgins had previously contemplated this site for an Observatory. The instruments included two equatorial telescopes, a meridian instrument, a clock, and three chronometers. The equatorial telescopes were a Fitz 6.5-in. refractor of 8.5-foot length, and a 5-foot Fraunhofer. The meridian circle, which could only observe objects as they crossed the north-south meridian, was of German construction, a Pistor and Martins with 4-in. aperture. It is the same kind of instrument still used 150 years later to determine very precise star positions, but superseded in the 1990s by satellite observations with Hipparcos and later Gaia. Gilliss used it for exactly that, to determine star positions as part of his secondary goals.

In December and January Gilliss and his staff began a series of observations of Mars, then in opposition, using the equatorial telescope. Gilliss writes that it was common in the evening for residents of Santiago to climb the summit of Santa Lucia and look through the telescope. From October 1850 to February 1851 he and his colleagues undertook observations of Venus, then in inferior conjunction. Then a second series of Mars observations was undertaken from December to March 1852. And finally another series on Venus was undertaken from late May to September 1852, when the observatory was turned over to the government of Chile. The government of Chile also appointed three Chileans—one professor of mathematics and two of his best students, to learn astronomy and the use of the instruments.

Fig. 27.1 Gilliss's Observatory on Santa Lucia, central Santiago, Chile

The main results of the expedition were published in six volumes, today rare brittle volumes housed only in specialized libraries such as the Naval Observatory (Fig. 27.2). The contents are quite amazing and much broader than astronomical observations. In 1855 the first two volumes were issued and did not cover astronomical observations, which took a long time to analyze. Rather, the first volume was entitled simply "Chile" and examined "the geography, climate, earthquakes, governments, social conditions, mineral and agricultural resources, commerce, etc." (Gilliss 1855). In the usual fashion these reports were transmitted from Gilliss to the Secretary of the Navy, then to the President, and from the President to Congress, which had funded the expedition. Gilliss's Prefatory comments are notable for explaining American interest in Chile: Gilliss wrote that "At that period the young republic of Chile was daily becoming of greater consequence to our people." American travelers on their way West saw it as a stopping point for supplies of bread and fruit; merchants saw it as a rapidly growing market for their shiploads of goods; and statesmen "greeted cordially the nation every day giving stronger proofs of is ability for self-government. Within the preceding quarter of a century Chile has advanced for more rapidly than any other nation of Spanish America in intelligence, good order, agricultural and mineral wealth, and commercial importance," Gilliss wrote (Gilliss 1855, Preface). Yet, there was a large gap in the knowledge of Chile, and Gilliss aimed to fill this gap with these two volumes, with information gathered when he was not observing. Volume one includes 14 stunning illustrations,

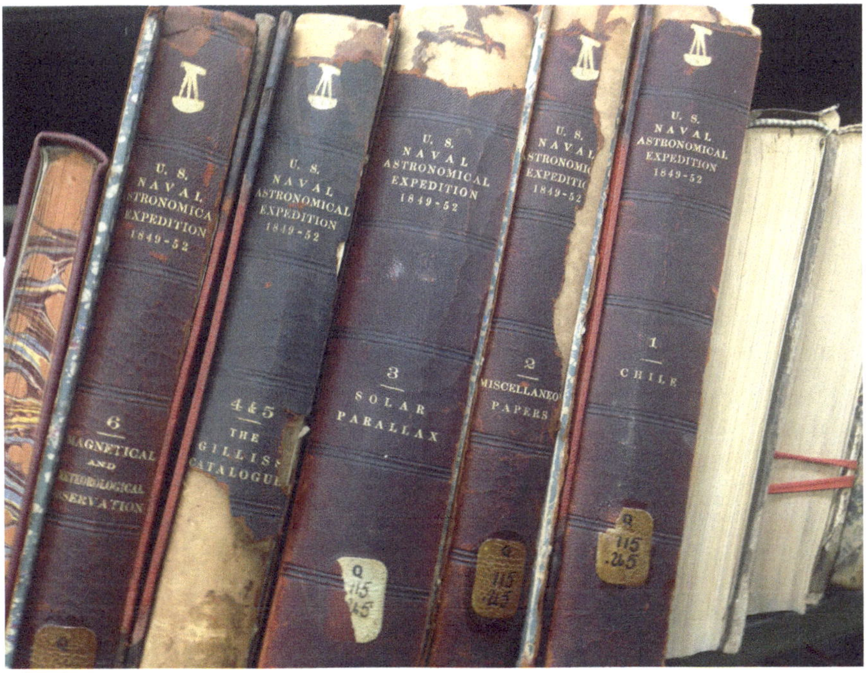

Fig. 27.2 The six Gilliss volumes on Chile as they now appear in the library of the U. S. Naval Observatory. (Photo by Steven J. Dick)

including a map of the Republic of Chile, a panoramic view from the summit of Santa Lucía, and a portrait of an Araucanian Chief, a member of a group of South American Indians that still live in the fertile valleys and basins of South-Central Chile. Of interest to historians of Santiago, and urban planning, are a map of Santiago, the Palace, and the Mint. Volume two includes "miscellaneous papers" written by a variety of authors. They cover everything from the Andes to minerals, Indian artifacts, and animal life. This volume is also richly illustrated with color prints ranging from Indian artifacts to birds and native Indians (Figs. 27.3 and 27.4).

The remaining volumes of the Gilliss expedition cover the astronomical results. Volume three, "Solar Parallax," refers to the attempts to determine the parallax of Venus and Mars, which would then give the solar parallax, and thus the distance between the Earth and the Sun (Gilliss 1856). Although this was the largest volume of the six, it was also the least successful for several reasons. The most common interpretation has been that not enough observations were made in the northern Hemisphere, usually blaming Matthew Maury. But more recent analysis has claimed "the impracticality of the expedition's observational program," especially the difficulties of the method (Huffman 1991, 208). In any case, the value for the solar parallax (8.5 arcseconds, as later determined by astronomer Benjamin Gould), was off the charts compared to later values from the transits of Venus. The determination of solar parallax was too low, meaning the estimate of the distance to the Sun was

Fig. 27.3 Indian artifacts in Chile, a sampling of many color illustrations in Gilliss's Chile volumes depicting cultural and natural history

Fig. 27.4 Araucanian Indian chief depicted in Gilliss's Chile volumes

too high by about three million miles. Today the Earth-Sun distance of about 93 million miles is known at any given time from radar measurements to within a few meters.

Volumes four and five consisted of star catalogs and were not published until 1870 and 1895. These volumes contain hundreds of pages of star positions, and are not of much interest today, since they have been superseded by more accurate methods. Volume six consisted entirely of magnetic and meteorological observations, surely among the earliest systematic observations of their kind for Chile.

27.3 Importance of the Expedition

Upon Gilliss's departure in the summer of 1852 the Observatory was turned over to Chile. The Chilean government had earlier expressed interest in acquiring the building and instruments as a national observatory for Chile, and the U. S. Navy was willing. Despite some difficulties in the transition from President Andres Bello (1841–1851) to President Manuel Montt (1851–1861) in 1851, the Chilean government bought the instruments and buildings for about 7800 pesos. El Observatorio Nacional de Santiago officially opened the day the title was transferred, August 17, 1852, with a new director (Carlos Moesta) and a staff of several Chileans who had helped Gilliss with observations. The Santa Lucía observatory remained at this site only 10 years; in 1859 the Observatorio Nacional was removed from Santa Lucia to a site outside Santiago—the grounds of the Quinta Normal de Agricultura—the lowest and foggiest part of the city, reminiscent of Foggy Bottom in Washington. The observatory eventually moved to higher ground at Lo Espejo south of the city, and between 1956 and 1963 to the present site on Cerro Calan, run by the Department of Astronomy at the University of Chile. On its website the Observatorio Nacionale still gives Aug 17, 1852 as its founding date.

As for Gilliss, he would return to South America only once, to observe an eclipse in Peru in 1858. He would finally become Superintendent of the U. S. Naval Observatory during the American Civil War years, 1861–1865. This is because his rival Maury, who was from the Southern state of Virginia, defected to the Confederacy during the War. During that time Gilliss—even as he met the unusual naval demands of the Civil War that threatened to tear the country apart—would transform the Naval Observatory by upgrading its instruments and programs, and by adding young staff who would usher in a golden era for the institution for the remainder of the century (Dick 2003). He would also undertake observations in cooperation with the Chilean National Observatory, in particular observations for the parallax of Mars again at its close approach in 1862. This time the results of the Sun-Earth distance were considerably more accurate than Gilliss's previous attempt. He died from a stroke in February 1865 just as the Civil War was nearing its end, at the relatively young age of 53, the day after his son returned from a Confederate prison.

The results of the U. S. Naval Astronomical expedition are quite remarkable. It is true the scientific results were a partial failure—the distance to the Sun was not improved, but a catalogue of southern stars was observed. Most importantly Gilliss left behind the seeds for the Chilean National Observatory which still thrives today. He left the earliest description of the culture, geography, meteorology and magnetic observations of Chile, and his volumes on Chile are recently the subject of renewed interest in Chile (Hermosilla 2017). And his expedition represents a very early example of cooperative relations between two young American nations, the United States and Chile. Today, because of its clear skies and dry conditions in places like the Atacama Desert, Chile has become the astronomical capital of the world, with almost half of the world's astronomical instruments. It all began with the events we have described here, the work of James Melville Gilliss, and his successors at the Observatorio Nacional in Santiago.

27.4 Commentary 2020

This paper was given in March 2016 at the Pontificia Universidad Católica Chile, at the invitation of Professor Germán Hidalgo Hermosilla in the School of Architecture. It had an unusual origin: rather than astronomical in nature, Professor Hermosilla and his colleagues were working on a project focusing on the history of Santiago around 1850, funded by the Chilean equivalent to the National Science Foundation in the United States. They were particularly interested in the urban views of Chile that appeared in Gilliss's volumes (Fig. 27.5). Since I had written the history of the U. S. Naval Observatory (Dick 2003), I served as the international collaborator for this project.

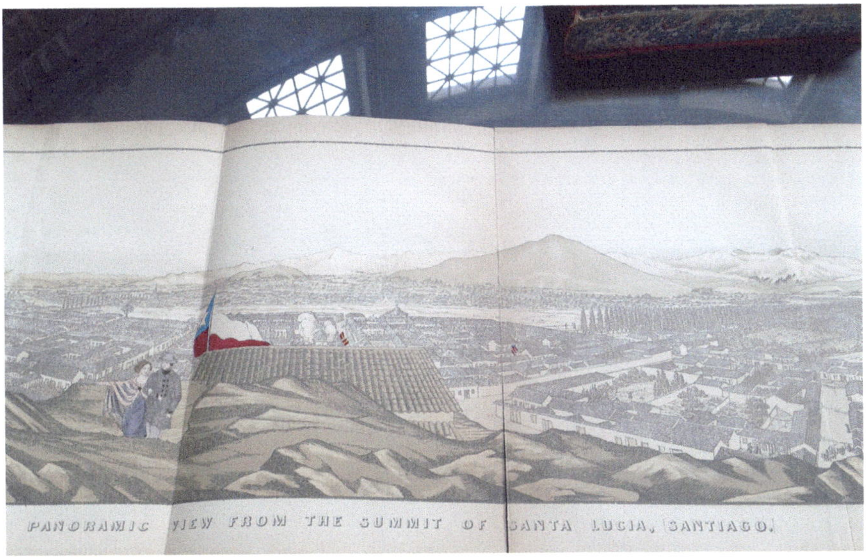

Fig. 27.5 One of several views of Santiago from Santa Lucía, as published in Gilliss's volume. Taken in the Library of the U. S. Naval Observatory. The library windows are in the upper part of the photo, taken by Steven Dick

During the trip I was able to climb Santa Lucía, take in the views of the now greatly expanded central city, and learn about Santa Lucia's history. Of particular interest is the fact that Charles Darwin climbed Santa Lucía on August 27th, 1834—15 years before the arrival of Gilliss—during a layover of his famous Voyage of the Beagle. A plaque commemorating this event, written in Spanish, quotes Darwin as writing in his journal "An inexhaustible source of pleasure it is to climb Santa Lúcia, a little rocky hill that rises in the center of the city. From there the view is truly impressive and unique." Indeed we find this passage in his diary for August 28th (Keynes 1988, 256). For more on Gilliss's expedition to Chile see also Schrimpf (2014). The six Gilliss volumes from the expedition to Chile deserve further study, along with the better- known volumes from another naval expedition—the United States Exploring Expedition of 1838–1842, commanded by Lt. Charles Wilkes.

References

Dick, S. J., 2003. *Sky and Ocean Joined. The U.S. Naval Observatory 1830–2000*. New York, Cambridge University Press.

Gilliss, J. M., 1855. *U.S. Naval Astronomical Expedition to the Southern Hemisphere during the Years 1849-´50-´51-´52*. Vol. 1, Washington, A.O.P. Nicholson Printer.

Gilliss, J. M., 1856. *U.S. Naval Astronomical Expedition to the Southern Hemisphere during the Years 1849-´50-´51-´52*. Vol. 3, Washington, A.O.P. Nicholson Printer.

Hermosilla, Germán Hidalgo. 2017. Revisiting J. M. Gilliss' Astronomical Expedition to Chile during in 1849–1852, *Journal of Astronomical History and Heritage*, 20, 161–176

Huffman, W. 1991. "The United States Naval Astronomical Expedition (1849–1852) for the solar parallax," *Journal for the History of Astronomy*, 22, 69, 208–220.

Keynes, R. D., ed. 1988. *Charles Darwin's Beagle Diary*. Cambridge University Press, Cambridge.

Schrimpf, Andreas. 2014. "An international campaign of the 19th century to determine the solar parallax: The US Naval expedition to the southern hemisphere 1849–1852," European Physics Journal H., 1–20, https://www.academia.edu/11482246/An_international_campaign_of_the_19th_century_to_determine_the_solar_parallax

Chapter 28
Measuring the Astronomical Unit: The American Transit of Venus Expeditions of 1874 and 1882

Abstract When in 1874 and 1882 Venus passed in front of the face of the Sun, most countries with a scientific reputation to keep or to gain made plans to observe the great event. The United States was no exception. With a 177,000 dollar Congressional appropriation for the 1874 event, and 78,000 dollars for 1882, the Americans sent out eight well-equipped expeditions for each transit. Under the U. S. Transit of Venus Commission, the responsibility fell to the U. S. Naval Observatory (Dick, Cambridge University Press, Cambridge: 238–273, 2003). Relying heavily on photographic methods, the Americans returned 350 plates in 1874, and 1380 measurable plates in 1882. Simon Newcomb grew skeptical of the results, but in 1894 William Harkness produced a final value, after adjustments with other constants, of 8.809 arcseconds, with a probable error of .0059 arcseconds, yielding an Earth-Sun distance of 92,797,00 miles, with a probable error of 59,700 miles. How important were the transit of Venus observations? In the end it was Newcomb who had the final say, for it was his system of astronomical constants that was adopted internationally at a Paris conference in 1896. In determining a final value for the solar parallax from all methods, Newcomb gave all photographic observations of the 1874 and 1882 transit a weight of 2, compared to a weight of 40 for Pulkovo Observatory's determination of solar parallax from the constant of aberration.

28.1 Introduction

As the nineteenth-century transits of Venus approached, the phenomenon had been observed only three times in recorded history—1639, 1761, and 1769. By the nineteenth century the main interest in observing the transits of Venus—aside from their extreme rarity—was in determining the solar parallax, and thereby the scale of the Solar System. At stake was reducing an uncertainty in the Earth-Sun distance by several million miles. Every difference of one hundredth of an arcsecond in the solar parallax, in other words, from 8.79″ to 8.80″, translated into approximately 100,000 miles. The methods had been worked out in the eighteenth century by

First published in D.W. Kurtz, ed., *Transits of Venus: New Views of the Solar System and the Galaxy*, IAU Colloquium 196 (Cambridge University Press, 2005), pp. 100–110.

S. J. Dick, *Space, Time, and Aliens*, https://doi.org/10.1007/978-3-030-41614-0_28

Edmond Halley, Joseph-Nicolas D'Isle, and others, and required observations from widely varying geographic locations. This inspired expeditions to remote locations around the world, including eight American expeditions in 1874 and eight more in 1882.

Aside from their disputed scientific results, these transit of Venus expeditions are of historical interest for the international disagreements over techniques and instruments, as an early example of international cooperation and rivalry in astronomy, and for their place in two broader historical trends: the determination of the fundamental astronomical constants, and the great scientific voyages of the nineteenth century. A great deal was at stake, not only for science but also for national interests. This is why in 1874 alone the British would have 12 expeditions, the Russians 26, France and Germany six each, Italy three and Holland one. As the nineteenth-century historian Agnes Clerke put it, when in 1874 Venus passed in front of the face of the Sun, "every country which had a reputation to keep or to gain for scientific zeal was forward to cooperate in the great cosmopolitan enterprise of the transit" (Clerke 1902). The United States was no exception; it had a growing reputation in science and was anxious to accelerate that growth. The nineteenth-century transit of Venus pair offered a unique opportunity for the country to showcase its rising scientific talent—one that would not come again until the distant twenty-first century in the year 2004.

28.2 Organizing in the USA

In the United States, Naval Observatory astronomer Simon Newcomb began the discussion of the transits of Venus with a paper in 1870, followed by a resolution before the National Academy of Sciences in April of that year (Newcomb 1872). At the recommendation of the National Academy, in 1871 Congress approved $2000 for "preparing instruments." The same bill also provided for Transit of Venus Commission, to be composed of the Superintendent and two professors of mathematics of the Navy attached to the Naval Observatory, the President of the National Academy of Sciences, and the Superintendent of the Coast Survey. In creating a government commission, the United States followed the lead of Germany (1869) and France (1870), while in Britain the responsibilities were shared by the Royal Observatory at Greenwich and the Royal Astronomical Society. The U. S. Transit of Venus Commission as originally constituted included Benjamin F. Sands, Benjamin Peirce, Joseph Henry (President of the National Academy at the time), Newcomb, and William Harkness. Only the Newcomb and Harkness would survive the full term of the Commission, which held its last meeting and effectively disbanded in 1891, after more than two decades. And of those two, only Harkness believed to the end that the method produced a valuable result.

28.3 Instruments and Methods

Among the first and most crucial decisions to be made by the American Commission, and by other countries, were the method of observing and the instruments required. The measurement of the relative position of the center of the planet and the center of the Sun was the method of choice, but there was nothing to lose in attempting to measure the exact moment when the planet came into contact with the limb of the Sun. Either method might be attempted visually or photographically, and with photography still in its infancy, especially as applied to astronomy, that decision between visual and photographic methods was by no means a foregone conclusion. But at the urging of Newcomb, from early on the U. S. Commission was drawn to the photographic method, although it hedged its bet by also making visual observations with small refractors.

The photographic method proposed by Newcomb was unique. Unlike most of the European participants, who also opted for the photographic method but devoted their attention to securing the best photographs, the American method proposed by Newcomb concentrated also on the problem of measuring the photograph. Because the measurements on the photograph were made in inches and fractions, Newcomb reasoned, and because the quantity to be determined was in minutes and seconds of arc, a precise knowledge of the scale factor was necessary in order to convert from linear to angular measurement. This conversion, according to Newcomb, was "the greatest difficulty which the photographic method offered." To meet it, Newcomb proposed an instrument conceived by Joseph Winlock and already in operation at Harvard College Observatory. This was a fixed horizontal telescope of nearly 40-foot focal length, through which sunlight was directed by a heliostat, a slowly turning mirror that kept the Sun's image stationary with respect to the telescope (Fig. 28.1). The lens and heliostat mirror—a piece of finely polished but unsilvered glass that reflected about 1/20th of the sunlight into the lens—were mounted on a 4 foot-high iron pier embedded in concrete. The lens, designed to give the best photographic image, formed that image 4 in. in diameter about 38.5 feet away on the photographic plate. The plate itself was held vertically on another iron stand next to a grid that was overlaid on each photograph for purposes of measurement. A special device consisting of 5-foot lengths of pipe was used to measure the distance from the lens to the plate within a hundredth of an inch—the crucial measurement on which the scale factor depended. To complete the setup, a transit instrument was to be used to align the system North-South, as well as for other purposes. This photographic method was to be used both for photographs of contact and of Venus as it moved across the face of Sun, with the latter believed to hold the most promise (Newcomb 1872).

At the request of Naval Observatory Superintendent Benjamin Sands, Congress appropriated \$50,000 in June 1872 for the purchase and preparation of instruments, and another \$100,000 in 1873 for the actual expeditions. The following year another \$25,000 was appropriated to complete work and return parties home, for a grand total of \$177,000 for the 1874 event alone, not including salaries and the use of

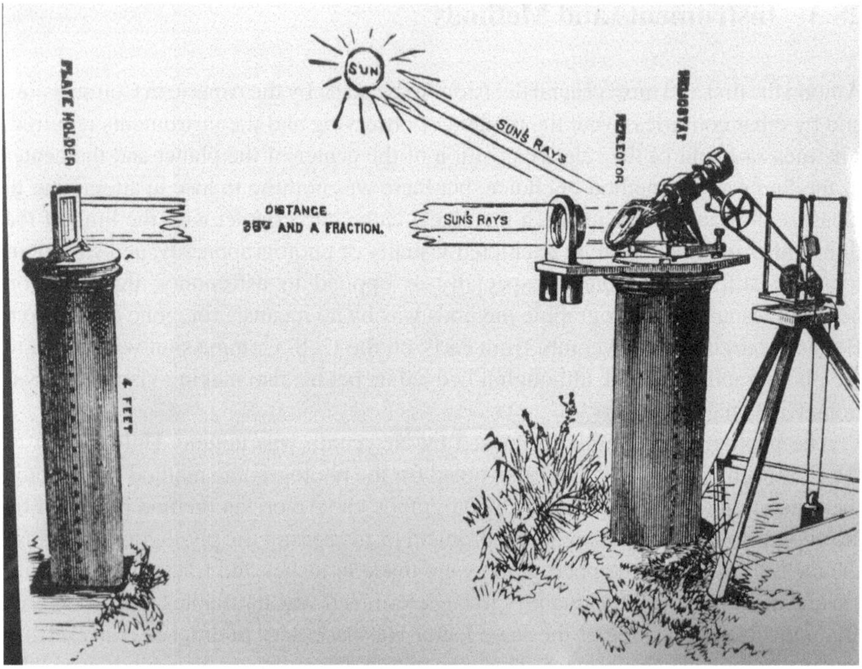

Fig. 28.1 The fixed horizontal telescope known as a photoheliograph, in which a weight-driven heliostat directs the Sun's rays through a lens, which focuses the image on a photographic plate 38.5 feet away. The long focal length was necessary because the scale factor needed to be known very accurately in order to obtain maximum accuracy for the parallax. The method was used by American and French observers. (From Simon Newcomb, *Popular Astronomy* (New York, 1878), 186)

Navy facilities and ships. This was a munificent sum indeed, one from which the country might well expect a decent scientific return.

With the first $50,000 in hand, the Commission held its first meeting on July 22, 1872, and went to work in earnest on the matters of instruments, choice of stations and organization of the expeditions and their personnel. The Naval Observatory, with the majority of the members of the Transit of Venus Commission including its President (the Superintendent) and Secretary (Newcomb), was authorized to take charge of the details of the expeditions. Harkness drew up the specifications for most of the instruments, eight sets of which had to be manufactured, since the Commission had decided that the appropriation was enough to equip eight American parties. The instruments were constructed by a variety of makers. But for the most crucial of these instruments the Commission turned to Alvan Clark & Sons, then constructing the Observatory's Great 26-in. Refractor, used by Asaph Hall to discover the two moons of Mars in 1877. The firm not only made the 5-in. refractors for the visual observations (Fig. 28.2), but also the 5-in. 40-foot photoheliograph lenses and the heliostat mirrors crucial for the photographic method, as well as the chronographs for precise registration of time. The polishing of the 7-in. mirrors,

Fig. 28.2 Five-inch Alvan Clark refractor, one of eight manufactured for the American transit of Venus expeditions. The portable telescope was adjustable to any latitude from the North Pole to the South Pole. Several remain on display at the U. S. Naval Observatory today. (Pictured here is astronomer Alfred Mikesell with the instrument in 1965. U. S. Naval Observatory)

Newcomb recalled, was the most difficult part of the whole apparatus. The accuracy needed to be such that "if a straight edge laid upon the glass should touch at the edges, but be the hundred-thousandth of an inch above it at the centre, the reflector would be useless."

While the visual 5-in. refractors and the 40-foot photoheliographs were the most important equipment, a broken-tube transit instrument and sidereal clock were also crucial, as well as sidereal and mean time chronometers, and a chronograph for registering time. The transit instrument, designed by Harkness and built by Stackpole, was used not only for determining latitude, longitude and time but also to insure that the central vertical line of the photographic plate holder could be very near the meridian. The clock used with the transit instrument, built by the Howard Clock Company of Boston, was designed for the rugged fieldwork and therefore not particularly elegant. One of the best photographs of the layout of all this instrumentation was taken at the U. S. Centennial Exhibition in Philadelphia, where a full-scale display of the transit of Venus equipment was mounted, a showcase for the progress of astronomy in the United States.

28.4 Stations and Personnel

Once the methods were decided it was necessary to choose the stations. Although the entire transit would last about 4 h, a very long time compared to the few minutes of totality for a solar eclipse, optimal weather was a prime consideration. To observe the parallax effect, both Northern and Southern Hemisphere stations were required, and after studying weather records it was decided to have three northern and five southern stations. In order to choose the stations, Newcomb began heavy correspondence with U. S. consulates and astronomers around the world. At the suggestion of Struve, Vladivostok was chosen as one northern station, with Nagasaki and Peking the other two. The southern stations chosen were Crozet Islands, Kerguelen Island, Hobart Town (Tasmania), Bluff Harbor (New Zealand), and the Chatham islands off New Zealand. In the end the Crozet Islands site was abandoned when the ship could not land due to severe weather, and two American stations would be located in Tasmania. Practical adjustments were not unusual; in the end, C.H.F. Peters led his party from Bluff Harbor to Queenstown, New Zealand, some 200 km away.

The personnel of each station consisted of one chief of party, one astronomer, one chief photographer and two assistant photographers, with a few parties having an additional astronomer and in one case, an instrument maker. The members of the parties were chosen with the greatest care, especially the chiefs and the all-important chief photographer. Two of the chiefs-of-party were from the Naval Observatory, two from the Coast Survey, one from the Army, one from the Navy, and two from outside the government. The chief photographers were all professionals in photography, but the assistants were for the most part "young gentlemen of education, recent graduates of different colleges, who had been practiced in chemical and photographic manipulation."

28.5 Practice

For practicing visual observations, beginning in May 1873 an artificial Sun and Venus apparatus was mounted on a building near the War Department, about two thirds of a mile from the Observatory. Using the 9.6-in. refractor, a 5-in. telescope and a 4-in. comet seeker from the dome atop the Observatory, Newcomb, Harkness and Hall repeatedly observed the small black dot representing Venus impinging on the artificial Sun, which was a white circular disk. In the spring of 1874, many of the participants gathered on the grounds of the Observatory to practice, with goal of improving the accuracy with which contacts could be observed (Fig. 28.3). At the same time, the photographic apparatus was set up and the photographic process rehearsed, with Henry Draper offering his services to the Commission for several weeks.

Fig. 28.3 Spring 1874 practice for transit of Venus on the grounds of the Naval Observatory. Standing at left nearest the photoheliograph is Admiral C.H. Davis, founder of the American Nautical Almanac Office and President of the Transit of Venus Commission at this time. Standing in front of him are Henry Draper and C.H.F. Peters (with hat); seated is Simon Newcomb. Asaph Hall is the tall man in front of the ladder, with hat; at far right (with stovepipe hat) is the Observatory's instrument maker, William F. Gardner. (From a set of stereo views of Washington published by J.F. Jarvis, Washington D.C., courtesy U. S. Naval Observatory)

28.6 Results

A great deal could be said about the colorful details of the expeditions, the travel problems, the weather problems, the triumphs and heartbreak. But in terms of science, the most important detail is the result. Now began the saga of analyzing the observations. A total of some 350 plates were returned from the 1874 American expeditions. In October, 1875 Newcomb still expressed optimism based on what he knew at that point about the observations. The optical observations of contacts made by the observers of all nations would, he believed, "by their combination give a value of the solar parallax of which the probable error will lie between 0.02″ and 0.03″ (Recall that means 200,000–300,000 miles). The American photographs alone, he further felt, "will give a result at least as accurate as this, and probably more so." However, Newcomb cautioned, "it is not to be disguised that there is a possibility of unforeseen perturbing causes being brought to light by a comparison of all the observations which will upset all our *a priori* estimates of probable error" (U. S. Naval Observatory 1875).

Newcomb's cautionary statement proved prophetic. Reporting on the results of the 1874 expeditions 8 years later, Harkness recalled that after the parties returned, attention was first turned to the visual contact observations as the easiest to analyze, but "it was soon found that they were little better than those of the eighteenth century." Around the world the result was the same: The problem was that.

> ...the black drop, and the atmospheres of Venus and the Earth, had again produced a series of complicated phenomena, extending over many seconds of time, from among which it was extremely difficult to pick out the true contact. It was uncertain whether or not different observers had really recorded the same phase, and in every case that question had to be decided before the observations could be used. Thus it came about that within certain rather wide limits the resulting parallax was unavoidably dependent upon the judgment of the computer, and to that extent was mere guesswork. (Harkness 1883)

The photographic observations were thus all the more important, but here again disappointment was widespread. Harkness recalled that "it soon began to be whispered about that those taken by European astronomers were a failure." The official British report declared that "after laborious measures and calculations it was thought best to abstain from publishing the results of the photographic measures as comparable with those deduced from telescopic view." The problem was the Sun itself: "however well the sun‘s limb on the photograph appeared to the naked eye to be defined, yet on applying to it a microscope it became indistinct and untraceable, and when the sharp wire of the micrometer was placed on it, it entirely disappeared."

All hope focused on the American expeditions, which had returned with about 200 measurable plates in 1874 taken with the long-focus photoheliographs. In June 1875 the Commission charged Harkness with measuring these plates, and he devised a machine especially for this purpose. Harkness reported that 221 photographs yielded "excellent results" for the period between second and third contact when the planet was on the face of the Sun. However, those taken between first and second, and again between third and fourth contacts "proved of no value" because of the infamous black drop problem. In other words, while even long-focus photographic contact observations were no better than visual ones, there was reason for hope in obtaining results from the photographs of Venus fully upon the face of the Sun.

In the end, however, no result of the 1874 American transit of Venus expeditions was ever officially published, although in 1881 David P. Todd (then an assistant in the Nautical Almanac Office) did publish a brief three-page article in which he determined a provisional value of $8.883 \pm 0.034''$ (Todd 1881). By the eve of the 1882 transit the official American results remained uncertain, and it was a heated question whether parties should even be dispatched for the 1882 transit. They were—eight more expeditions were sent around the world.

To make a long story short, relying heavily on photographic methods, the Americans returned 350 plates in 1874, and 1380 measurable plates in 1882. And contrary to the common opinion today, a result was produced. Simon Newcomb grew skeptical of the results, but William Harkness (Fig. 28.4) produced a final value, after adjustments with other constants, of 8.809 arcseconds, with a probable error of .0059 arcseconds, yielding an Earth-Sun distance 0f 92,797,00 miles, with a probable error of 59,700 miles (Harkness 1891, 1894).

Fig. 28.4 William Harkness, who led the American efforts for the 1882 transit of Venus, and almost single-handedly achieved the final American result. (U. S. Naval Observatory)

The publication of the official reports of the American Transit of Venus Commission, however, was another matter. Recalling the impressive tomes published by other nations, Harkness argued that the full publication of the American results was essential. His hope in this respect, however, was not realized. Aside from the *Papers Relating to the Transit of Venus,* and the *Instructions for Observing* it in 1874 and 1882, only Part I of a projected four parts of *Observations* was published. Part II, two volumes of reports of the eight 1874 parties and consisting of some description and much data, but no results, reached the page proof stage, and today exist as only a single copy of 564 pages in the Naval Observatory Library. Parts III and IV, which were supposed to be the results, were never published.

The reasons for this failure to publish again were bureaucratic. With the passing of the 1882 expeditions, the Transit of Venus Commission decided it would be best to combine the results of both transits into one report. By 1891, however, despite a recognition of the need "that the United States keep pace with other governments in publishing the results of its observations of these important transits," no further action was taken. The failure to publish an official American report, however, must be distinguished from the American result, which was not only obtained but published and discussed in the context of the other astronomical constants by Harkness in the *Washington Observations* for 1885 (Harkness 1891).

28.7 Significance

How important were the transit of Venus observations? In answering this question we need to recall that the transits of Venus were only one method for determining the solar parallax. Ironically, just as the transit of Venus observations were producing an improved result, other methods became practical that gave even better results. This is true of asteroid parallaxes, but especially of the method involving the aberration of light. From the measurement of the aberration of light one can produce the light time; combined with newly accurate measurements of the speed of light, in which both A.A. Michelson and Newcomb were involved, an accurate distance can be determined. In the end it was Newcomb who had the final say in which methods were applied to the final solution for the astronomical unit, for it was his system of astronomical constants that was adopted internationally at a Paris conference in 1896. In determining a final value for the solar parallax from all methods, Newcomb gave all photographic observations of the 1874 and 1882 transit a weight of 2, compared to a weight of 40 for Pulkovo Observatory's determination of solar parallax from the constant of aberration. Considering the probable errors, Newcomb's system and Harkness's system actually overlapped in their values for solar parallax, and Newcomb came closest to overlapping the modern value of 8.794146 (Fig. 28.5).

By act of Congress dated July 26, 1886, the instruments and records of the Transit of Venus Commission were turned over to the Secretary of the Navy. In that year the instruments, valued at some $30,000, were made the property of the Naval Observatory, and put into the hands of the Observatory's instrument maker. On April 25, 1891 the Commission held its last meeting, where Harkness reported on the status of the reductions. Along with the effective demise of the Commission, an interesting episode in astronomy passed into history. Today some of the instruments and records remain at the Naval Observatory, and memorial plaques still mark several of the sites of the observations (Fig. 28.6).

The nineteenth-century transits of Venus thus took their place in the long history of attempts to determine the astronomical unit, one of the fundamental constants of astronomy. Without vastly superior methods, Harkness cautioned in his 1894 paper, the value of the solar parallax was not likely to be improved from where it stood after the nineteenth-century transits of Venus. Harkness did live to see the close passage of the minor planet Eros to Earth in 1900–1901, a method that gave a considerable improvement then, and again in 1930–1931.

Just before the 1882 observations, Harkness wrote the following poetic lines:

> We are now on the eve of the second transit of a pair, after which there will be no other till the twenty-first century of our era has dawned upon the earth, and the June flowers are blooming in 2004. When the last transit season occurred the intellectual world was awakening from the slumber of ages, and that wondrous scientific activity which has led to our present advanced knowledge was just beginning. What will be the state of science when the next transit season arrives God only knows. Not even our children's children will live to take part in the astronomy of that day. As for ourselves, we have to do with the present...

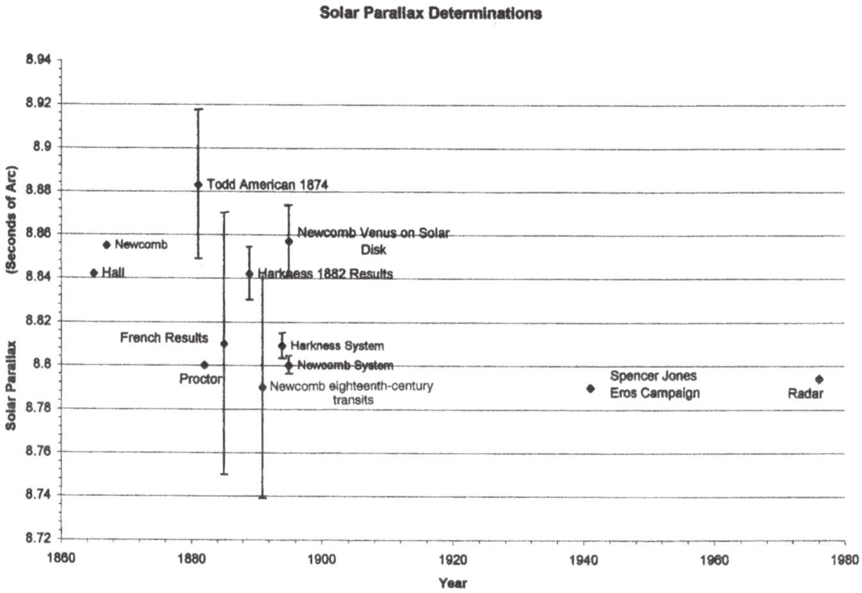

Fig. 28.5 Selected solar parallax determinations, 1860–1976. Aside from the results of Hall and Newcomb in 1865 and 1867 from observations of the meridian parallax of Mars, all nineteenth-century observations shown were derived from the transits of Venus. Two twentieth-century results are shown for comparison. (From Dick (2003))

Fig. 28.6 Transit of Venus plaques marking the sites of worldwide observations exist around the world. This one commemorates the U. S. Naval Observatory observations at Queenstown, New Zealand in 1874. (Photo by Steven Dick)

Harkness would surely marvel that radar methods have now determined the solar parallax to six significant figures, and that the mean distance to the Sun is known within a few meters. And he would surely marvel that when the flowers were blooming in June 2004, we would be observing transits of planets around other stars, with plans afoot to launch spacecraft just for that purpose. What the state of science will be with the next transit pair in 2117 and 2125, we cannot predict, any more than Harkness could have predicted orbiting spacecraft, nanotechnology, or electronic computers.

28.8 Commentary 2020

This chapter was given at International Astronomical Union Colloquium 196, held at Preston, United Kingdom from 7–11 June, 2004, and published in the Proceedings of the meeting (Kurtz 2005). The discerning reader will note that this was exactly the time that Harkness foretold in the last paragraph of this chapter "when the flowers are blooming in June 2004." The location of the meeting was also crucial. Any given transit of Venus is only visible from certain parts of the world (thus the need for expeditions), and on June 8, 2004, I was able to view this extremely rare phenomenon in its six-hour entirety near the site in central Lancashire, UK, near where Jeremiah Horrocks first observed such a transit in 1639. The site was actually about a 30-min drive from Much Hoole, where the building still stands where Horrocks is believed to have made the first observation of a transit of Venus. This was a big gamble, since the weather in the United Kingdom is usually cloudy this time of year. I also published an article on the subject in May 2004 issue of *Scientific American* (Dick 2004), was in charge of the Transit of Venus working Group of the IAU, and gave several presentations on the subject at IAU General Assemblies.

Transits of Venus always occur in pairs separated by 8 years, and June 2012 found me in Washington, DC for the last event of the century. This time it was totally cloudy, but I was at NASA Goddard Space Flight Center, where we were able to view the transit in real time on a large screen with images from the SOHO spacecraft. The next pair of transits of Venus will not occur until 2117 and 2125. Transits of Mercury occur on a more regular schedule, but for technical reasons cannot be used for determining the astronomical unit.

As revealed by the title of the proceedings, the subject of the meeting was much broader than the transits of Venus, and that event was used as a springboard to discuss progress in measuring the universe in a variety of ways.

References

Clerke, A. 1902 *History of Astronomy during the Nineteenth Century*, chapter 6, A. and C. Black.

Dick, S.J. 2003. *Sky and Ocean Joined: The U. S. Naval Observatory, 1830–2000*, 238–273, Cambridge University Press, Cambridge.

Dick, S. J. 2004. "The Transit of Venus," *Scientific American*, 290, no. 5 (May, 2004), 99–105.

Harkness, W. 1883 Address by William Harkness, *Proceedings of the AAAS 31st Meeting, August 1882*, 77.

Harkness, W. 1891. The Solar Parallax and its Relation Constants in *Washington Observations for 1885*, Appendix III.

Harkness, W. 1894. On the Magnitude of the Solar System, *Astronomy and Astro-Physics*, 13, 605–626.

Kurtz, D. W. 2005. *Transits of Venus: New Views of the Solar System and Galaxy*. Cambridge University Press, Cambridge.

Newcomb, S. 1870 On the mode of Observing the coming transits of Venus, *American Journal of Science*. 50, 74–83.

Newcomb, S. 1872 On the application of photography to the observation of the transits of Venus In *Papers Relating to the Transit of Venus in 1874*, Part I, pp. 14–25. U. S. Transit of Venus Commission.

Todd, D. 1881. The Solar Parallax as derived from the American photographs of the transit of Venus, 1874, *American Journal of Science*, 21, 491–493.

U. S. Naval Observatory 1875 *Annual Report*, 80–81

Chapter 29
Geodesy, Time, and the Markowitz Moon Camera Program: An Interwoven International Geophysical Year Story

Abstract The Markowitz Moon camera program, originated at the U. S. Naval Observatory in Washington, D.C. in 1952 for the determination of "ephemeris" time, was transformed a few years later into one of the many programs of the International Geophysical Year (IGY). The Moon camera's stated IGY goal was to improve geodesy—the study of the gravity field, shape, and size of the Earth. Geodetic positions are related to time through longitude; 360° of longitude is 24 h of time, so 1° at the equator is equal to 4 min of time, 69 statute miles or 60 nautical miles. Although in the end the Moon camera program failed in its ambitious geodetic goals, the history of this particular program emphasizes that not all IGY programs were successful. Moreover, the Moon camera highlights an important aspect of making science global that historians usually ignore—the determination and dissemination of time worldwide. Without accurate timing and synchronization of observations on a global scale, much of the data from the IGY would have been compromised or useless. The Moon camera program was undertaken in the midst of rapid changes in timekeeping and time dissemination, events in which the eponymous William Markowitz played an important role.

Given the lack of historical treatment of this program compared to the Smithsonian's Moonwatch/Baker-Nunn program with its similar aims, in this paper I will answer several questions: Who was William Markowitz and what was the context in which the Markowitz Moon camera program emerged? What were the results of the Markowitz Moon camera program? Where does the program fit in the context and history of the IGY? The answers to these questions will help us assess the broader social and intellectual significance of the program. And along the way we will come to appreciate the importance of time measurement and synchronization for IGY global observations.

First published in *Globalizing Polar Science: Reconsidering the Social and Intellectual Implications of the International Polar and Geophysical Years*, edited by Roger D. Launius, James R. Fleming, and David H. DeVorkin (Palgrave MacMillan: New York, 2010)

S. J. Dick, *Space, Time, and Aliens*, https://doi.org/10.1007/978-3-030-41614-0_29

29.1 Context

Geodesy is characterized for most of its history by triangulation networks such as pioneered by Karl Friedrich Gauss beginning in 1821 (Dunnington 1955).[1] These networks were gradually improved and expanded, and provided increasingly accurate geographic positions over increasingly wide areas; as geodetic maps over time show, it was truly a literal example of a science that had to be made global by definition. Chronometer trips during the nineteenth century and the advent of powerful radio transmitters had even enabled some intercontinental longitude determinations to be made, notably during the World Longitude Operations of 1926 and 1933, but not enough for geodetic accuracy (Dick 2003: 472–473).[2] By the 1950s, driven by the needs of intercontinental ballistic missile systems, satellite tracking, and global navigation, the continents needed to be geodetically tied together by a World Geodetic System. Triangulation of satellites against the starry background was the revolutionary new way toward that old problem of geodesy, and it proved very successful. This paper examines an innovative ground-based method for geodesy that was carried out as part of the IGY just before the Space Age, and that illustrates how difficult the problem of geodesy would have been in the absence of the Earth-orbiting satellites.

NOAA (National Oceanic and Atmospheric Administration) historian John Cloud has delineated three eras of geodetic positional accuracy: From 1957 to 1970 ground-based systems observed satellites visually or by Doppler shifts, providing 10-m accuracy and the revelation of a "pear-shaped" Earth in 1958. During the years 1970–1990, laser-ranging satellites such as LAGEOS, and Very Long Baseline Interferometry (VLBI), as well as satellite radar altimetry from GEOS-3 (1975), Seasat (1978), and TOPEX-Poseidon (1992) provided accuracies to 1 meter, allowing instantaneous determination of dynamic sea level heights. The current era began around 1990, when an array of ground and space-based systems yielded positional accuracies of a centimeter or less, allowing observations of plate tectonics. VLBI, satellites such as CHAMP, and pairs of satellites such as GRACE (Gravity Recovery and Climate Experiment), were used to determine the structure of the Earth's gravity field with more resolution (Cloud 2010).[3]

The first World Geodetic System (WGS), known as the Department of Defense World Geodetic System, was achieved in 1960; Deborah Warner has detailed how it was the result of an intense rivalry between the Army and the Air Force. Many WGS's have been developed since then, culminating in WGS 84, which has kept that name despite numerous updates since it was developed in 1984 (Warner 2000a, b; Cloud 2002).[4]

Within Cloud's tripartite scheme, the IGY is situated firmly at the very beginning of what he defines as the first era. The International Council of Scientific Unions (ICSU) proposed the IGY in 1952, with the Comité Spécial de l'Année Géophysique Internationale (CSAGI) as the governing body for all IGY activities. "Longitude and Latitude" was one of eleven subject areas, and the location of "World Data Center A" for this area was the U. S. Naval Observatory in Washington, D.C. The

Longitude and Latitude program for the IGY was twofold, using different methods: to obtain *astronomical* longitude and latitude (a process affected by the deflection of the vertical defined by a plumb line), and to obtain *geodetic* (or geocentric) longitude and latitude.[5] It is the latter we concentrate on here in the context of the Markowitz Moon camera program. In addition to its geodetic ambitions, the stated goals of the Latitude and Longitude program were to determine accurate coordinates of participating observatories, to improve terrestrial time determination, knowledge of the irregularities of the Earth's rotation, and the improvement of star catalogues.

Two programs for obtaining geocentric coordinates on the Earth for determining the shape of the Earth (the geoid), were planned for the IGY (Markowitz 1958a, b). One, the Smithsonian Astrophysical Observatory's (SAO) optical Moonwatch program organized by Fred Whipple, has received considerable attention from historians (McCray 2006). It and the SAO's complementary Baker-Nunn program involved optical and photographic observations of artificial satellites. While the primary goal of the SAO program was satellite tracking, such tracking was the means to several scientific ends, including geodetic data. The other program, the Markowitz Moon camera (Fig. 29.1), originated and administered from the U. S. Naval Observatory in the early 1950s as part of its mission to determine time, is less well known but just as interesting from an international point of view, even

Fig. 29.1 William Markowitz and his Moon camera. U. S. Naval Observatory

though it had better success in the determination of time than for any direct geodetic results. These timing achievements of the Markowitz Moon camera, were, however, in themselves important for geodesy, not to mention broader aspects of science. Conceived and implemented five years before the launch of the first artificial satellite, the Markowitz program for time determination was based on observations of the real Moon, not an artificial one. It was made possible in the 1950s by improved star catalogs against which the Moon's position could be measured, an improved lunar ephemeris, and, above all, a new photographic technique invented by William Markowitz, whose career we describe in the next section. Only in 1954 was geodesy made one of its goals, when Markowitz sent a proposal to the U. S. National Committee for the IGY for "a program to measure the geodetic distance between continents, the size and shape of the Earth, and the variations in its speed of rotation" (Markowitz 1954a).

The expectations for the Markowitz Moon camera were high. Almost all accounts of the IGY, both before and after it took place, devote a few paragraphs to this program. As early as 1954 Joseph Kaplan, first Chair of the U. S. National Committee for the IGY, wrote:

Highly precise longitude and latitude measurements, important in surveying, mapping, and navigation, can now be made by direct photography of the Moon. The new technique, developed by William Markowitz, makes it possible also to measure changes in the speed of rotation of the earth to ten times the existing precision, and the observational material obtained will shed new light on the inner constitution of the earth. It is planned to photograph the Moon and stars nearby every possible night, using the newly designed cameras attached to existing telescopes at astronomical observatories. Analysis of the data obtained will yield longitude and latitude at all stations, and, using existing geodetic nets, the exact distance in miles between all stations can be determined with a precision of 90 feet. During the International Geophysical Year it will be possible for the first time to triangulate the whole earth, because observations will be taken simultaneously all over the earth. Observations will be made at more than fifteen stations distributed over the Earth. England and France, for example, contemplate studies at Greenwich, the Cape of Good Hope, and Paris. The United States expects to make observations at Washington, D.C.; San Diego, California; and Hawaii. (Kaplan 1954).

The year following the IGY, Sydney Chapman, then Chair of the US National Committee for the IGY, described geodetic methods and then the Moon camera. Chapman wrote:

Seen from different observatories at the same instant, the Moon occupies different positions relative to the stars. Each observatory makes two or more such observations during the night. Calculations from such observations will determine the geometry of the observatory distribution over the earth, independent of gravity. Intercontinental distances will be measured with errors of the order of no more than 100–200 feet. At present the uncertainty is greater. The positions of some ocean islands are uncertain by a mile. This program will also give an improved orbit of the Moon, and a more accurate standard of time measurement. (Chapman 1959).

After the IGY was completed, the Moon camera program is still mentioned, but with a notable lack of discussion of results, at least relating to geodesy. Writing in the *Annual Report of the Smithsonian Institution* for 1959, before pointing out the value of artificial satellite observations (such as Vanguard) for geodesy, Elliott B. Roberts of the Coast and Geodetic Survey wrote,

The framework of international cooperation established for the IGY was seen at the outset to favor establishing a new and better measure of the longitude differences between continents and major isolated island groups, such as Hawaii. This was realized through use of new instruments and techniques, including the American dual-rate [Markowitz] Moon camera, which provided new precision in the relation of terrestrial positions to the celestial firmament. Better absolute knowledge of geographic locations of the earth's landmasses was obtained, with advantages in mapping, operation of earth satellites, scientific studies of the earth, and the mechanics of its rotation, including problems of timekeeping (Roberts 1959).

In his well-known popular account of the IGY, *Assault on the Unknown* (1961), New York Times science writer Walter Sullivan juxtaposed the Markowitz Moon camera program with the Smithsonian's Baker-Nunn program, and remarked "In addition to the prototype at the Naval Observatory, cameras of this design were delivered to the observatories at Greenwich, Paris, and seventeen other points around the world" (Sullivan 1961). But again, no results were reported. We are thus confronted with the mystery of the significance of the Markowitz Moon camera given its high expectations.

29.2 William Markowitz and the Origins of the Moon Camera Program

William Markowitz (1907–1998) obtained his PhD in astronomy from the University of Chicago in 1931, working under W. D. MacMillan and writing his dissertation of the statistics of binary stars. During this period he also worked at Yerkes Observatory under the noted instrument maker Frank Ross. He joined the U. S. Naval Observatory in Washington, D.C. in 1936 and remained there until his retirement in 1966, working under Paul Sollenberger and with Gerald Clemence, spending almost his entire career in the Time Service Department. One of Markowitz's early duties was operating the Photographic Zenith Tube (PZT), designed by Ross and used at the Naval Observatory beginning in 1915 to observe the variation of latitude, and since 1934 to determine Universal Time (Dick and McCarthy 1999; Dick et al. 2000).[6] One of Markowitz's early articles was on latitude and longitude determinations (Markowitz 1945).

The variation of latitude was one of the chief interests of Markowitz throughout his career; it was the analysis of this data that led to his contributions to the study of polar motion (Chap. 30). Markowitz directed the Time Service Department from 1953 to 1966, a period during which methods for more uniform and accurate time were being devised. In addition to time determination, Markowitz was fundamentally concerned with improvements to the dissemination of time. At the International Astronomical Union meeting in Dublin in 1955 he proposed the system of UT0, UT1, and UT2, which corrected observed universal time successively for polar motion and seasonal variations in the Earth's rotation (see below). The system went into effect within months, with UT2 being disseminated internationally. Markowitz later participated in experiments in synchronizing time around the world using artificial satellites and portable atomic clocks transported by airplanes; relativistic corrections were not yet determined until a famous experiment conducted by his Naval Observatory colleagues and published in 1972. He contributed to the Navy system of time and frequency transmissions, and the control of precise time via Loran-C and the Transit navigation satellite.

Markowitz was heavily engaged in international cooperation in time determination and dissemination. He served as President of the commission on Time of the IAU from 1955 to 1961, and was also active in the International Union of Geodesy and Geophysics (IUGG), the American Geophysical Union (AGU) and the International Consultative Committee for the Definition of the Second. In short, he was well connected with the American and international scientific communities, connections that helped him implement the Moon camera program.

In examining the reasons for the origins of the Moon camera program, one must keep in mind the varieties of time evolving in the 1950s. It was a time of extraordinary advances in time determination, timekeeping and time dissemination, and in setting standards that are largely still with us today. Universal time was based on the rotation of the Earth, and came in the three flavors (UT0, UT1, and UT2) that had been defined by Markowitz and the IAU in 1955. As the accuracy of quartz crystal clocks, and then atomic clocks, improved, the rotation of the Earth was shown to be irregular at the level of several tens of milliseconds, thus the need for a more uniform time (Dick 2003: 482 ff).[7] That more uniform time came to be known as *Ephemeris time*, because it was based on the orbit of the Earth as reflected in the ephemerides of the Sun and Moon. The difference between Ephemeris Time and Universal Time (specifically UT1) gave *delta T*. Why was delta T important? Because it gave the most uniform time possible, and for historians and scientists alike, an accurate and uniform timescale was highly desirable. If an historian or astronomer wanted to study ancient eclipses, for example, or any phenomenon that required accurate timing over long periods, uniform time was indispensable for accurate results (Stephenson 1997, 2007).[8] Aside from ET and UT, atomic clocks were being developed in the 1950s; the relation between Ephemeris time and atomic clocks proves essential to our story.

Given Markowitz's background and his central role in the determination and dissemination of time, the origin of the Markowitz Moon camera is not surprising. Because since the seventeenth century official time had been based on the rotation of the Earth (for UT and its later refinements) or the orbit of the Earth (since 1960 for ET), astronomers had controlled it for centuries. The Moon camera now became part of a larger struggle in the 1950s to determine whether physicists or astronomers would be the arbiters of time. In two oral history interviews, Markowitz recalled clearly the origins of the Moon camera well before the IGY came on the scene. At the June 1951 meeting of the American Astronomical Society (AAS) at the National Bureau of Standards (NBS) in Washington, D.C., papers were presented related to the subject of atomic time. As Markowitz recalled, "this was shortly after Harold Lyons and coworkers at the NBS had constructed an atomic clock, which was operated by the principle of absorption of ammonium microwaves. At the meeting of the AAS there were some glowing accounts of the accuracy that the atomic clocks would achieve, and that it would eliminate the necessity for astronomical time." "Shortly thereafter," Markowitz further recalled,

> [Gerald] Clemence, who was then the Director of the Nautical Almanac Office, called me in to discuss the question of time. Ephemeris time was obtained from lunar occultations, but with a long delay of probably two or three years or more from observing occultations and getting the resulting ephemeris time, say a value of delta T. Clemence thought the claims were a little overblown—there was still a necessity for astronomical time of the nature of what is called Ephemeris time. What he was afraid of was that the physicists on their own might adopt some value for the frequency of an atomic clock that did not correspond to the Ephemeris time. That is to say, the two kinds of seconds would differ in value, and this would cause considerable confusion and other problems. What he wanted was a method of obtaining Ephemeris Time rapidly. If this could be done then we could give physicists the value needed for obtaining the frequency of cesium [which had taken over from ammonium as the favored atomic standard], so that an atomic second and an ephemeris second would be the same. The result was that he asked me if there were some way of getting Ephemeris time rapidly (Markowitz 1986, 1987; Dick 2003: 480).[9]

It did not take Markowitz long to take up the challenge: "I returned the next day to Clemence and said that, yes, I had a way that I thought would work." He then laid out the problem and a relatively simple solution:

> The problem in getting accurate positions of the Moon by photography was that if one took a short exposure, then the seeing or the scintillation would cause displacements of the stars and Moon. Whereas, if one took a long exposure, say of about 20 or 30 s, then there would be a blurred image of either the stars or the Moon. So I told Clemence that I had a method in mind, and I proposed to track both the Moon and the stars for about 20 s, each at their own rate. He was astounded at this proposal, and asked how I intended to accomplish it. The explanation was very simple. One would use a glass filter to cut down the light of the Moon, and one would tilt it during the exposure. The tilting of the glass would move the image of the object being photographed and it was a simple matter to calculate the rate so as to keep it stationary. He was enthusiastic, and proposed immediately that we set about with the construction of a camera."

Thus, because the Sun itself could not be accurately observed, and because the planets moved too slowly in their orbits for accurate time determination, Markowitz proposed to use the Moon, and it was for this purpose the dual-rate Moon camera was developed. Markowitz designed the instrument himself, and it was built in the USNO machine shop under the direction of Mr. George Steinacker. The dual-rate camera was attached to the 12-inch refractor at the Naval Observatory, and a regular observing program began June 2, 1952. The 12-inch refractor was soon dismounted and the Moon camera was attached to an 8-inch telescope (Fig. 29.2) in the same dome atop the Observatory's main building (Markowitz 1954a, b, 1960).[10]

Fig. 29.2 The Markowitz Moon camera attached to the 8-inch telescope at the U. S. Naval Observatory. (Credit: U. S. Naval Observatory)

As with the navigational method of lunar distances, the Moon was to serve as the hands on the face of a clock. However, these much more accurate observations were to be made from the Observatory, not from shipboard. The tilting dark-glass filter effectively held the Moon fixed relative to the background stars during a simultaneous exposure of 20 s. The result was an image of the Moon that could be accurately measured with respect to the background stars (Fig. 29.3). The Moon camera did indeed succeed in determining ET, delta T, and the calibration of the atomic second. The latter, 9,192,631,770 vibrations of radiation in cesium (resulting from the hyperfine transition due to the electron-spin flip), is the number still used for the official definition of the second (Markowitz et al. 1958). Those contributions are documented and undisputed, and would themselves have justified the camera. But our question in this paper is about the Moon camera's contributions to geodesy, linked to time through longitude.

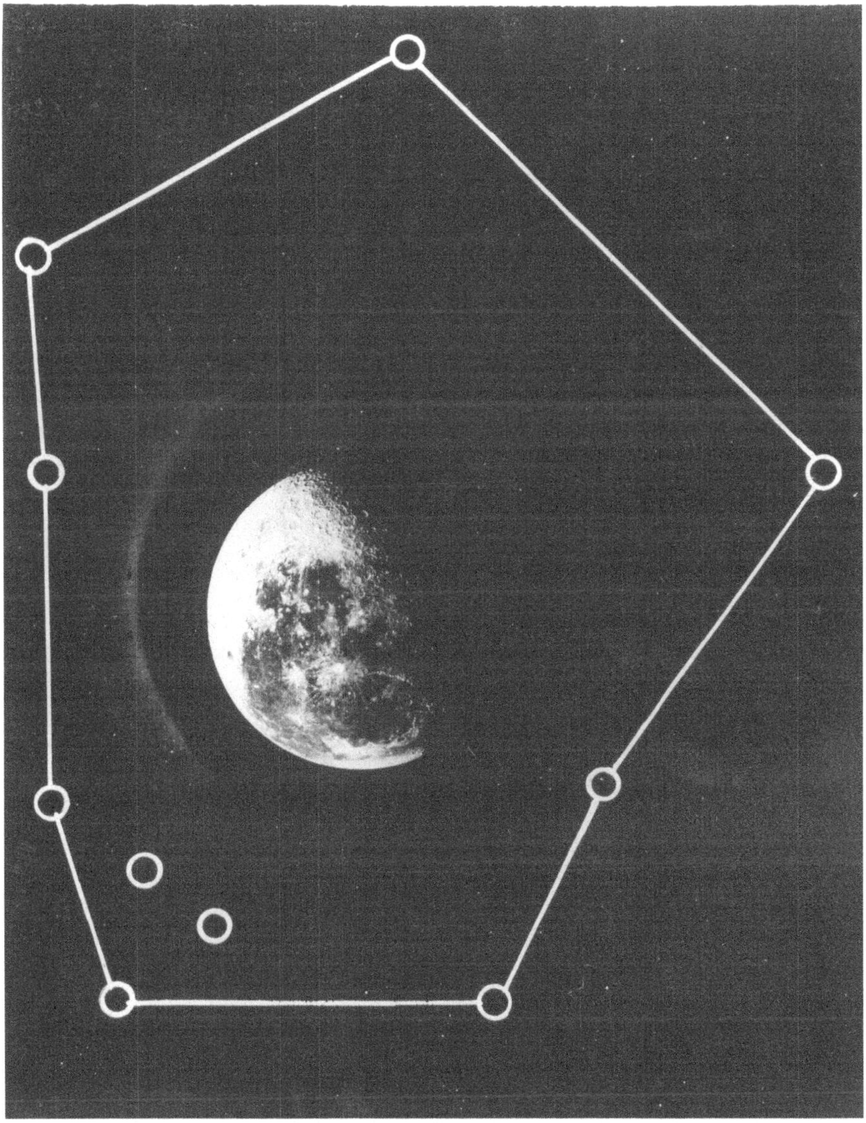

Fig. 29.3 The position of the Moon, held fixed by a tilting dark-glass filter of the Moon camera, could be accurately measured with respect to reference stars. (Credit: U. S. Naval Observatory)

29.3 The Moon Camera, the IGY and Geodesy

In 1952 the International Council of Scientific Unions (ICSU) appointed a committee for the International Polar Year (IPY), renamed the effort the International Geophysical Year (IGY) to cover the whole Earth, and fixed the IGY to cover the period from July 1, 1957 through December 31, 1958. Eventually it would embrace 67 nations. Detailed national proposals were due by May, 1954, and so it was that in late 1953 or early 1954 the Naval Observatory sent a proposal to the US National Committee for the IGY outlining a program to measure the geodetic distance between continents, the size and shape of the Earth, and the variations in its speed of rotation. The proposal, undoubtedly written by Markowitz, pointed out the difficulties of extending triangulation nets across the oceans. Although in recent years advances in precision of observation and in knowledge of the motion of the Moon had made possible the use of the Moon as a triangulation point, those methods used total solar eclipses and occultations of stars, rare phenomena that did not allow much progress and had their own accuracy problems (Markowitz 1954a).[11]

The proposal pointed out that during the past 20 years the Moon had also been used as a clock to measure variations in speed of the rotation of the Earth. Now Markowitz outlined a new soon-to-be published technique to photograph the Moon, a technique that he claimed, with a well-distributed system of observatories, could accomplish in a single year what might take a century by older methods. The probable error of a single observation would be 0.15 s of arc, about 900 feet on the surface of the Earth, but many observations could reduce it to 90 feet, with undetermined systematic errors.[12]

The proposal noted that the equipment required was a refractor telescope with minimum aperture of 10 inches and focal length 10–20 feet, standard equipment at many observatories, plus a special camera that would cost about $700 at each site. Also required was a more precise ephemeris of the Moon, which the proposal promised was to come with support of the Office of Naval Research, the Army Map Service, the U. S. Naval Observatory, and the British Nautical Almanac Office. Ten to 20 observatories, preferably at the extremities of existing continental triangulation nets, would be required. Each observatory would measure and reduce its own observations, but the Naval Observatory would render assistance. The total cost for the Moon cameras was estimated at $12,000 for 20 copies, $40,000 for four measuring machines, plus personnel costs. Beyond the IGY, the Naval Observatory also foresaw a permanent Moon position program for determining delta T (Markowitz 1954a).[13]

By way of preparation, the first meeting of the CSAGI in Brussels in July, 1953 had already adopted a resolution recommending "that for longitude determinations observations shall be made by the method of Dr. Markowitz at a number of fundamental stations, and that the International Association of Geodesy shall, through its appropriate section, study the corrections required in order to refer the determinations to a common system of reference" (Annals 1959, 22).[14] No details were given, and Markowitz's first paper to describe the technique and its application to geodesy appeared in the *Astronomical Journal* for March, 1954 (Markowitz 1954b).

1954 was a crucial year for latitude and longitude program preparations, as well as for the other programs of the IGY. As with those programs, the progress of the Moon camera program can be followed in the volumes of the *Annals* of the IGY, at the USNO archives, and in correspondence now in the IGY archives at the U. S. National Academy of Sciences in Washington, D.C. It was at the second meeting of the CSAGI in Rome (30 September–4 October, 1954), that detailed programs were laid out, including determination of longitudes and latitudes. As mentioned above, it is important to remember that there were two parts to the IGY program, the first to determine astronomical latitude and longitude with the Danjon astrolabe and the PZT, the second to determine geodetic latitude and longitude with the Moon camera (Dick 2003: 258–61).[15] The first (for which the central agency was the Bureau Internationale de l'Heure (BIH) in Paris) was to be undertaken with standard programs planned at as many as 39 stations, while the Moon camera was "an auxiliary lunar observation campaign … envisaged (by the Markowitz method), which has the double aim: (a) To improve certain tabular data on lunar movement as well as the definition of uniform time; (b) To determine, for the observatories participating in these observations, the variation between the vertical and the normal to a conventional ellipsoidical surface." In two separate annexes to the report, the plans for the Markowitz Moon camera were laid out, specifically mentioning the desirability of extending triangulation nets across oceans. The U. S. National Committee commented on this need "There are not enough islands to use as stepping stones, nor has it been possible to connect the separate networks of longitudes and latitudes with each other more accurately than to 200 or 300 ft. The location of some islands is uncertain by as much as a mile" (Annals 1959: 139–45).

The Moon camera, the report stated, could overcome this deficiency. The plan was as follows:

> The United States proposes to employ the new technique for photographing the Moon, and also to determine astronomical longitudes and latitudes using the new Danjon astrolabe, at three stations: Washington, San Diego and Hawaii. It will supply dual-rate cameras suitable for attachment to existing long-focus telescopes at about seventeen astronomical observatories strategically located in other countries. It will also undertake to measure approximately one-fourth of the photographic plates with the twenty cameras, and will supply three measuring engines of special design to observatories in other countries willing to undertake the remaining measures.

Lists were given of participating observatories, and the four with measuring engines were specified as Washington, the Cape (S. Africa), Paris and Greenwich. The values of delta time (ET – UT) were to be published by the measuring centers; the reduction of the plates was to be completed by 1959, and the general solution by 1960.

Annex B to the same report from the Rome CSAGI meeting, undoubtedly largely written by Markowitz, gave the most detailed description of the theory and objectives of the Moon camera program:

The aims of the International Geophysical Year program, which are of astronomical, geophysical and geodetic interest, include the determination of the following: (a) changes in the speed of rotation of the Earth; (b) the size and shape of the earth, independent of gravity; (c) corrections to the orbital elements of the Moon, the lunar parallax, and the distance to the Moon; (d) the three rectangular geocentric co-ordinates of each observatory, to about 40 m probable error; (e) deflections of the vertical to about 1″ [one arcsecond] probable error.

Markowitz left no doubt about the ultimate goal: "The above data will allow geodesists to form a World Geodetic System. The size and shape of the earth determined geometrically from this program may be combined with the determinations given by the usual methods to obtain the most probable size and shape" (Annals 1959: 143–144).[16]

On April 29, 1955 the Technical Panel on Longitude and Latitude met at the USNO, with Gerald Clemence, Markowitz, and Donald Rice (U. S. Coast and Geodetic Survey) present. Markowitz was elected Chairman of the Panel. After discussing the two main elements of the program, they affirmed that the general solution from the measurement of the Moon camera plates would yield values of delta T, rotation of the Earth, and geodetic results; there was no discussion of attainable accuracies. The overall budget from the IGY was only $16,300 for the operations of the Coast and Geodetic Survey operations in Hawaii; all other U. S. operations were funded by the Office of Naval Research.[17]

On November 30, 1956 the USNO received official notice from the U. S. National Committee of the National Research Council that the USNO would be designated the U.S. Data Center for latitude and longitude.[18] Markowitz was eventually appointed chair of the Latitude and Longitude section of IGY.

On May 1, 1957 Markowitz issued the first Circular on the Moon position program. He reported that the 20 cameras were being finished in the instrument shop of the USNO, and four measuring engines had been completed and tested. Markowitz's deputy, astronomer Glenn Hall, was to carry out the computer reductions with the IBM 650, and for this purpose the Yale Zone star catalogues were placed on punch cards. The cameras were to be distributed by naval attaches nearest the participating observatory, beginning May 3 with most of the others to follow by June, 1957. Plate specifications were given, with an estimate of 60 dozen fast plates and 60 dozen slow plates for the entire program, with addresses for two suppliers (Kodak and Ilford).[19] Detailed "Instructions for Operation of the Moon-Position Camera" were published in 1957, along with instructions for the other methods by astrolabe and PZT (Annals 1957: 195–196).

In a progress report dated January, 4, 1958, Markowitz noted the program would commence about January 1958 and continue for 18 months, in accordance with decisions reached in Barcelona. He reported 20 cameras had been distributed around the world, and that both Moon cameras and Danjon astrolabes had been installed at the three U. S. stations: USNO, San Diego State College, and U. S. Coast Survey station in Hawaii, nine miles east of Honolulu.[20] The data indeed came rolling in, as evidenced by numerous monthly reports and files of data still in the USNO archives. The question is, what were the results?

29.4 Moon Camera Results

In discussing Moon camera results, it should be noted that the USNO agreement with the IGY was to provide, for each plate, apparent right ascension and declination of the Moon and the universal time of observation in the *Astronomical Journal* or in the USNO circular series.[21] This goal was fulfilled. Beyond that, one searches in vain for any geodetic results from the Moon camera program in the published reports or the archives of the IGY at the National Academy of Sciences in Washington, D.C, the archives of the Naval Observatory, or professional publications such as the *Astronomical Journal* where one would expect to find them. My conclusion is that none were published. As late as 1958, Markowitz stated in an article that the USNO hoped to complete a general solution that would yield not only corrections to the orbit of the Moon and Ephemeris time but also the size and shape of the geoid (Markowitz 1958b: 47). In 1959 a summary of Moon camera observations was indeed published in the IGY General Report series, giving numbers of plates received, measured and reduced, but no results, nor were results given in updates in January and July 1961. IGY data collection officially terminated December 1, 1961, and the final Catalogue of Data in the World Data Centers, covering July, 1, 1957 through December, 31, 1959, appeared in 1963.[22] Final results were published in the IGY Annals for 1964, but they were not results that would yield geodetic accuracies. In fact, they were no more accurate than the Danjon astrolabe determinations (Annals 1964: 294). There was a statement that more would be published in 1965. But Markowitz retired in 1966, and, although the program for Ephemeris time continued successfully until 1974, no geodetic results ever appeared.

The interesting historical question is what happened to a program so highly touted in the IGY literature by its leaders, including Sydney Chapman and Joseph Kaplan, by its participants, including Elliott B. Roberts, and by its popularizers, represented by Walter Sullivan? First of all, scientists' queries as to exactly how geodetic results were to flow from the Moon camera observations demonstrate there was some skepticism about the method; this precipitated Markowitz to write an article in 1958 with more details, supplemented by articles from others he had solicited (Markowitz 1958a, b).[23] Secondly, USNO astronomer David K. Scott, who had a background and considerable work in the field of geodesy and worked with the Moon camera reductions at USNO, stated that he believed geodetic results from the Moon camera were not possible in principle because of uncertainties in the motion of the Moon that could not be disentangled from the uncertainties in the observers' positions on Earth: "…the corrections to the Moon's position, which was essentially Delta T, were 100% correlated with the corrections to the observer's position, so the scheme would not work."[24] It was later found that artificial satellites were close enough that Doppler techniques could be used to solve the problem where the Moon camera could not. Markowitz himself had remarked on the difficulties of using the Moon compared with an artificial satellite. And lest we jump to the conclusion that Markowitz was pushing his Moon camera invention beyond its capabilities, Scott also stated that "the Air Force thought that we could get Earth-centered geodetic positions from Markowitz's Moon camera. Well, Markowitz was not particularly interested. He wanted to get Delta T from it."

Finally, and most compellingly, a single handwritten letter by Fred Whipple in the USNO archives titled "For Markowitz by Whipple" and dated July 25, 1960, indicates that Markowitz's method was being superseded by Space Age results.

Whipple (head of the SAO Moonwatch/Baker-Nunn program) wrote that "The SAO has been preparing for the precise determination of geoid shape and world-wide ties among geodetic systems by observations of artificial satellites. A major effort is under way and should begin to produce results during the year 1960."[25] Whipple went on to describe the 12 stations around the world where photographic observations were being made of artificial satellites since mid-1958, and that some 3000 "precision observations" had been reduced by mid-1960, with an expected accuracy of 10–12 m. The Whipple archives at the Smithsonian Institution contain ten items of correspondence between him and Markowitz from 1956 to 1961. While there is no direct discussion of the Moon camera, Markowitz was keenly aware of Whipple's potentially competing program, and asked him in February 1956 to participate in a meeting to describe his proposed work on artificial satellites. Later correspondence between the two was limited to more perfunctory matters, such as Markowitz asking for timing accuracies required for Moonwatch, a natural question from the Director of Time Service at USNO, the nation's timekeeper. (Whipple replied in 1961 that he needed .001 s (1 ms) immediately, and half that (.0005 s) within the next year.)[26] The bottom line is that while Moonwatch and Baker-Nunn results became part of the National Satellite Geodetic Program and were used to create the SAO Standard Earth Model,[27] the Markowitz Moon camera data was not accurate enough to be incorporated into any geodetic system.

Neither correspondence at USNO or at the National Academy of Sciences IGY archives contains any smoking gun document stating that geodetic results from the Moon camera were a failure. Pembroke Hart at the National Academy wrote to Markowitz in August 1959 simply that his obligations would be discharged by updating his summary of Moon camera observations in a fourth six-monthly catalog.[28] This was perhaps a de facto recognition that not every program proposed for the IGY would or could be successful.

As for post-IGY observations with the Moon camera, in a letter from Markowitz to Odishaw dated December 4, 1958 Markowitz wrote that a recent Moscow meeting had decided that astronomical observatories would in general continue their latitude and longitude observations throughout the world.[29] In March, 1960 he reported that longitude and latitude observations with the PZT at USNO up to October 1959 had been published in Time Service Bulletins A 130 to A 170, and that astrolabe results would be sent to Paris as part of the broader latitude and longitude program. As to the Moon program, Markowitz wrote "The loan of Moon cameras will be continued to those observatories which wish to keep on making observations. The measurement and reduction of the plates are proceeding at a moderate but satisfactory rate. It is expected that corrections to the limb of the Moon will be utilized during 1960 in order to obtain the apparent positions of the Moon as observed."[30] The program for Ephemeris time continued until 1974, employing Time Service personnel at the U. S. Naval Observatory and graduate students at participating observatories around the world.

The Moon camera program, as well as the broader IGY latitude and longitude program, also left another legacy. Because timing was essential for latitude and longitude observations, already at the first CSAGI meeting in Brussels in 1953, it was recommended that all observatories associated with the broader latitude and longitude program make concurrent observations of time and latitude; that "two-way intercommunication of time signals shall be arranged during the IGY, for the precise determination of the transmission time of the signals and of the variations of this transmission time;" and that the measurement of the transmission times between stations be made with utmost precision, including ionospheric measurements along the propagation path to determine conditions affecting propagation time. Furthermore, the CSAGI recommended that URSI send the following communication to all concerned organizations:

CSAGI, considering (a) the necessity to ensure the utmost precision of international longitude measurements during the IGY; (b) the advantages afforded for the interchange of time between astronomical observatories, by the emission at frequent intervals of time signals capable of being received over the entire surface of the earth, including the southern hemisphere, draws the attention of the responsible organizations, and especially of CCIR [Comité Consultatif International des Radio Communications, International Radio Consultative Committee] and URSI [Union Scientifique Radio Internationale, International Union of Radio Science], to the importance of organizing a world system of emission of time signals during the IGY, and asks that this be established and in working order before the IGY begins (Annals 1959: 22).

The Latitude and Longitude program, of which the Markowitz Moon camera was a part, thus emphasizes that behind the scenes of the IGY was an alphabet soup of international organizations—including the French Bureau Internationale de l'Heure (BIH), the International Latitude Service (ILS), the CCIR, URSI, and the CSAGI itself—involved in ensuring that time was properly measured and disseminated, not only for latitude and longitude, but for all IGY observations.[31]

29.5 Summary and Conclusions

The conclusion, then, is that the Markowitz Moon camera was successful for time determination and calibration of the atomic second, but not for geodesy. So why study a program that failed? As Peter Galison has emphasized in his pioneering book *How Experiments End*, history is full of failures, and in the realm of science and technology historians are interested not only in successful programs, but also in programs that failed because they were too ambitious, or conceptually flawed, or overtaken by events.[32] In this case all three played their role. The Moon camera was too ambitious because its intended accuracies could not be reached, in part because only 12 of 20 observatories reported results due to technical difficulties, personnel shortages or funding problems. It was conceptually flawed because in the end it appears the uncertainties in the Moon's complicated motion could not be separated from uncertainties in the observer's position. And it was overtaken by events when the Space Age made possible satellite geodesy. None of this could have been totally foreseen before the effort was made. Sometimes an experiment has to be tried before success or failure becomes clear. Failed experiments are an important part of the history of science and technology, and it is important to realize that not all the ambitious plans for the IGY succeeded.[33]

The scope of the Markowitz Moon camera program was also considerable, at least in terms of the Navy's involvement in time and money, an indication of its importance to military needs. Some 20 Moon cameras were constructed in the Naval Observatory instrument shop, as well as four measuring engines. Some of these Moon cameras are still at the Naval Observatory and elsewhere, as are thousands of the photographic plates, and voluminous data records and correspondence—all testimony to a once-vibrant program.

The international scope of the program was also impressive. Securing the cooperation of 20 observatories around the world was no small task, and the overall latitude and longitude program of the IGY was even larger, involving some 39 observatories using astrolabes, transit circles, or photographic zenith tubes. The continuing international nature of the program is evident from the results reported to the National Academy of Sciences in 1970. Results poured in from 12 observatories, together with the number of plates measured at the Cape, Greenwich and Paris. More generally, the BIH, ILS, CCIR, and URSI were involved with the IGY latitude and longitude program.[34]

The technology and technique for the Moon camera program were also innovative. Markowitz developed a new technology that had not been built before, though he cited earlier attempts by H. N. Russell 50 years earlier using a similar technology at a single station (Markowitz 1954b). In addition, the program involved long-standing astronomical problems analogous to the method of lunar distances three centuries earlier: accurate star catalogues, accurate theory, and observation of the motion of the Moon, and robust computing techniques, though now at an order of magnitude or more better accuracy (Dick 2003: 16–18). Though the accuracies were still not enough to address geodetic demands, the program nonetheless shows science marching toward ever greater accuracies required for its increasingly refined problems.

Finally, the Markowitz Moon camera was in many ways a success, certainly in terms of the determination of Ephemeris time, delta T and the calibration of the atomic second to astronomical reality. For six years from 1960 to 1967 Ephemeris Time was basis for the official international definition of the second. The Moon camera continued to determine Delta T until 1974. And even when Atomic Time superseded Ephemeris Time in 1967, its definition of the second as 9,192,631,770 vibrations of the cesium atom had been calibrated to the real world of astronomy through the Moon camera. The Moon camera was not, however, accurate enough to produce the hoped-for geodetic results such as the determination of the geoid and detection of continental drift; for the latter no one at the time knew the magnitude of the movements, and it was three decades before the actual centimeter level shifts of the continents could be measured. Even the early satellite results, while revealing the slight pear-shape of the Earth, could not detect continental drift (Warner 2000b: 404–405).

Finally, it turns out the Markowitz Moon camera results were important for geodesy in other ways, again related to time. In his book on *Satellite Geodesy* published in 1964, Ivan Mueller wrote that "two kinds of time are required for satellite tracking. The geocentric motion of any satellite, the data in the satellite ephemerides, is expressed as a function of Ephemeris Time. For this in practice Atomic Time is substituted" (Mueller 1964: 290–293). The Moon camera was important for both of these, determining Ephemeris time and calibrating atomic time, the latter a community response to the challenge posed by the atomic clocks of the physicists. The other kind of time important for satellite tracking was Universal time in the form of UT1, which takes into account polar motion. Again, Markowitz's work was important, and indeed it was he who had set up the system of UT0, UT1 and UT2 at the IAU meeting in Dublin in 1955.[35]

In the end, the Markowitz Moon camera was an interesting ground-based attempt at geodesy implemented as part of the IGY. Its successes and failures highlight the importance of the new methods that superseded it, methods made possible only by the Space Age. The geodetic failure of the program highlights NASA's important early role in geodesy with the launch of geodetic satellites like Anna 1B (1962), LAGEOS (1976, 1992) and others. The extent of the subsequent multi-institutional satellite effort in geodesy is evident in the massive and detailed volumes of the *National Geodetic Satellite Program,* which in themselves could form the basis for a history of space-age geodesy.[36] NASA (and the U. S. Naval Observatory) remained very much involved in geodesy with programs in Satellite Laser Ranging (SLR) and Very Long Baseline Interferometry (VLBI), as well as its use of data from the DoD's Global Positioning System (GPS). In light of these programs, the once ambitious hopes for geodetic results using the ground-based Markowitz Moon camera faded into history.

29.6 Commentary 2020

Versions of this paper were given at the General Assembly of the International Union of Geodesy and Geophysics in Perugia, Italy in July, 2007 and at a meeting at the Smithsonian Institution in the Fall of 2007 on the occasion of the 50th anniversary of the International Geophysical Year and the opening of the fourth International Polar Year (Launius et al. 2010).

This chapter demonstrates how our concepts of time have grown increasingly sophisticated since the 1950s. That subject is too extensive to discuss here even as an overview, but see Dick (2003) for the context of how one institution, the U. S. Naval Observatory representative of many others, was involved in the changing concepts of time and time dissemination during the twentieth century. The discovery of polar motion, and its role in time and other astronomical and geophysical programs, is discussed in Chap. 32.

Acknowledgments I wish to thank Dennis McCarthy (U. S. Naval Observatory), Brent Archinal (Astrogeology Research Program of the U. S. Geological Survey), Patrick McCray (University of California Santa Barbara) and David Devorkin (National Air and Space Museum) for their very useful comments, as well as Sally Bosken and Gregory Shelton at the U. S. Naval Observatory Library, Janice Goldblum at the National Academy of Sciences archives, and Mary Markey at Smithsonian Institution archives in Washington, D.C.

Notes

1. Geodesy is a huge area in need of a critical history. Some of the general history is given along with the scientific concepts in Smith (1997).
2. The first Washington-Paris radio determinations of longitude were in 1913.
3. John Cloud, one of the few historians of geodesy, describes the development and work of perhaps the most significant early geodetic sciences group during the first period, that at Ohio State University, in (Cloud, 2000).
4. For a list of "Geodetic Reference Spheroids" preceding the WGS 84, and the current context in which WGS 84 is used, see U. S. Naval Observatory (2004).
5. These concepts are defined, with diagrams, in Smith (1997), especially pp. 85–87. For a succinct discussion of the relationship between geocentric and geodetic coordinates, see any edition of the U. S. Naval Observatory's *Astronomical Almanac*, for example, the 2006 edition, "Reduction of Terrestrial Coordinates," pp. K11–K13.
6. On Markowitz's work at the USNO see Dick (2003), pp. 480–85.
7. As Paul Forman (1985, 1182) has written, "The seasonal variations in the rate of the earth's rotation, which the pendulum was barely able to detect, were made fully apparent by quartz oscillators in the years before the Second World War. Still, the evidence tended rather to be overlooked by both astronomers and horologists unready for the conceptual revolution which it implied." It was only in the 1950s that they took action.
8. Values of Delta T are tabulated in the *Astronomical Almanac*, for example on pp. K8–K9 of the 2006 edition.
9. Paul Forman (1985, 1182) states: "In the mid 1950s, astronomers remained disinclined to cede primacy in time determination to physicists and engineers. The steady improvement of quartz oscillators and the advent of the first atomic frequency standards pushed astronomers to try to reestablish uniformity in astronomical time by shifting from the earth's daily rotation to its annual revolution to define the second; 'Ephemeris Time,' as it was called was formally adopted in 1960 by the General Conference of Weights and Measures."
10. Markowitz (1954b) cites (p. 69) a similar program by A. S. King and H. N. Russell from 1911–17, but here a snapshot of the Moon was taken in the middle of a 10-minute exposure. On the history of the 12-inch see Dick (2003), pp. 374 ff., and Rafferty (1981)
11. The document is undated, but internal evidence indicates late 1953 or early 1954; it preceded publication of the technique in the March 1954 *Astronomical Journal* (Markowitz 1954b). On the use of solar eclipses and occultations for geodesy, see (Mueller, 1954). As mentioned above, the Moon had been used for three centuries to determine geographical longitude through the famous "lunar distances" method. But the results were of low precision.
12. This stated accuracy, or at least the conversion from arcseconds to feet also quoted in the IGY Annals, vol. IIA, p. 142 (Annals 1959), does not make sense. 0.15 arcseconds is about 15 feet on the surface of the Earth. 0.15 arcminutes (9 arcseconds) is 900 feet, but accuracies should have been better than that. Other descriptions (e.g. Chapman 1959, pp. 18–19) state distances on the Earth's surface will be measured "with errors of the order of no more than 100 to 200 feet.

13. The IGY Moon camera proposal was one part of the plans for USNO participation in IGY. The first program was the Moon camera, which would produce geocentric latitude and longitude, and changes in the rate of rotation of the Earth. A second program was the determination of astronomical latitudes and longitudes using the prismatic astrolabe invented by Danjon. The third program was observation of solar activity, already long underway at USNO. Total cost was estimated at $165,000, of which the Office of Naval Research had already funded $49,200 for FY 1954 and 1955.
14. Markowitz (1954b), p. 73, references this Brussels meeting.
15. The best contemporary overview of the PZT is Markowitz (1960).
16. This volume was not published until 1959, but shortly after the Rome meeting Markowitz wrote Odishaw of these results, Markowitz to Odishaw, October, 29, 1954, National Academy of Sciences (NAS) IGY archives, Washington, DC, series 5.8. Odishaw was the Executive Secretary of the U. S. National Committee of the IGY; this was the first letter in a series of correspondence on this subject between them. Appendix B describing the Moon camera program was originally written by Markowitz and dated 10 September, 1954, with copies both at the National Academy and USNO archives. Andre Danjon was the president of the Latitude-Longitude group in 1954, and circulated the adopted resolutions and Markowitz's appendix B shortly after the meeting.
17. Minutes of the First Meeting, NAS IGY archives, series 4. Rice was the Chief of the Gravity and Astronomy branch of the Coast and Geodetic Survey. The Survey had participated in the two previous world longitude campaigns, operating the station in Hawaii in both 1926 and 1933. It would now operate the Hawaii station (known as NIU longitude, nine miles east of Honolulu) for both IGY programs, astronomical and geodetic latitude and longitude. The Survey also operated two of the five permanent observatories for determining variation of latitude under the International Latitude Service. See NAS IGY archives, series 8.1, folder on latitude and longitude, project 7.1, 1956–57.
18. Hugh Odishaw to Captain Charles Freeman, Superintendent USNO, Nov. 20, 1956, USNO on-site archives, Moon camera files, and NAS archives, IGY records, series 8.1, World Data Centers, USNO, 1956–1959.
19. Markowitz, Letter Circular No. 1, Moon Position Program of the IGY, 1 May, 1957, NAS IGY archives.
20. Markowitz to Odishaw, 4 January, 1958, NAS IGY archives, series 8.1.
21. C. L. Freeman, Superintendent USNO, to U.S. National Committee for IGY, Feb 13, 1957, NAS archives, series 8.1. E. O. Hulbert, senior scientist for USNC-IGY to Chairman and Executive Committee, USNC-IGY, Feb 14, 1957. See also IGY *Annals*, vol. VII, Parts I, II, and III (London, 1959), p. 283, where the USNO also hoped to discuss results in its Publications. A full list of latitude and longitude observatories is given on pp. 286–288, broken down by astronomical vs. geodetic observations.
22. *Interim Catalogue of Data in IGY World Data Center A*, IGY General Report Series, Number 7, Nov. 30, 1959, 78. Results are reported for 1 July, 1957–30 September, 1959. Also *Seventh 6-monthly Catalogue of Data in IGY World Data Center A*, IGY General Report, Number 13, January, 1961, pl. 98. And eighth catalogue, July 1961. The final *Catalogue of Data in the World Data Centers* was published (Pergamon Press, 1963).
23. Markowitz (1958a, b). Markowitz wrote that "Although much has been published which describes the various techniques it is not clear to geodesists just how the Moon is to be used to determine the size and shape of the earth. Prof. P. Tardi requested that at Toronto specific details be given to show just how this was to be done. Accordingly, I invited several of my colleagues to prepare reports covering eclipses, occultations, and the earth satellite, while I undertook to describe the lunar photographic method" (Markowitz 1958a). The reports Markowitz refers to were Goldstein et al. (1958), Markowitz (1958b), O'Keefe (1958), and Whipple and Hynek (1958). Markowitz further noted that these reports would concentrate on their respective programs because "It would be assumed in each case that observations could be made to the degree of precision anticipated" (Markowitz 1958a, 34).

24. Steven J. Dick OHI with David K. Scott, 26 Feb, 1988, p. 19, 27–33, USNO library.
25. "For Markowitz by Whipple," July 25, 1960, USNO on-site archives, Moon camera files. The manuscript letter is erroneously dated "July 25, 190," but internal evidence makes it clear that the year is 1960.
26. Fred L. Whipple to Wm. Markowitz, January 19, 1961, Unit 7431, Fred Lawrence Whipple papers, series 1 (General Correspondence), Box 3, Accession 98–132, Smithsonian Institution Archives. There are 10 pages of Whipple/Markowitz correspondence from 1956–1961 in box 3.
27. SAO results are given in *National Geodetic Satellite Program* (NASA: Washington, DC, 1977), Pt. II, p. 885. NASA SP-365.
28. Pembroke Hart to Markowitz, August 7, 1959, NAS IGY archives, series 8.1.
29. Markowitz to Odishaw, Dec 4, 1958, NAS IGY archives, series 5, latitude and longitude, 1954–60.
30. Markowitz to Hart, 25 March, 1960, "Data Center A Longitude and Latitude," NAS IGY archives, series 8.1, folder WDC project 45.10, 1960."
31. On the history of the ILS and BIH see Dick, McCarthy and Luzum (2000), especially Part II.
32. "Historians of science, principally those interested in the development of ideas, have often insisted that any historical reconstruction that ignores what seems in retrospect to be erroneous will be an inadequate account" (Galison 1987, 5). Galison cites Alexandre Koyré and Thomas Kuhn in this regard, noting that when they refer to historically held beliefs as "mistaken," is not to judge past theories anachronistically, but rather to appreciate forgotten frames of mind that may reveal precepts were held to be of central importance in their time. With regard to experiments and observations being superseded, I was involved in one such situation at the U. S. Naval Observatory when a 10-year transit circle program to produce fundamental positions for Southern Hemisphere stars at the level of a few hundredths of an arcsecond was superseded even before publication by the Hipparcos satellite, which produced positions at the milliarcsecond level, ten to a hundred times better accuracy. The satellite almost failed to achieve orbit, however, and if it had not worked the transit circle results, using new electronic "image dissector" technology, would have been considered cutting edge science. See Dick (2003, 239–41).
33. While the Moon program was only a small part of the IGY, it would be interesting to know how many other IGY programs did not fully attain their goals.
34. Report of the Moon Program, 1 January, 1957 to 1 January, 1970, R. Glenn Hall to Richard Y. Dow, 6 February, 1970, USNO on-site archives, Moon camera files.
35. For a real-life story of the effect of the failure to take into account polar motion in spacecraft navigation see Muller (2000) in Dick, McCarthy (2000)
36. Fred L. Whipple to Wm. Markowitz, January 19, 1961, Unit 7431, Fred Lawrence Whipple papers, series 1 (General Correspondence), Box 3, Accession 98–132, Smithsonian Institution Archives.

References

Annals of the International Geophysical Year. 1957. vol. IV, parts I, II and III. London

Annals of the International Geophysical Year, A H. Shapley et al., eds. 1959, vol. IIA. London: Pergamon Press

Annals of the International Geophysical Year. 1964. vol. XXXVI. Oxford. Pergamon Press

Chapman, Sydney. 1959. *IGY: Year of Discovery. The Story of the International Geophysical Year* (Ann Arbor: University of Michigan Press [1959], 926–931: 929.

Cloud, John. 2000. "Crossing the Olentangy River: The Figure of the Earth and the Military-Industrial-Academic Complex, 1947-1972," *Studies in the History and Philosophy of Modern Physics*, vol. 31, no. 3, pp. 371–404.

Cloud, John. 2002. "American Cartographic Transformations during the Cold War," *Cartography and Geographic Information Science*, vol. 29, no. 3, pp. 261–282.

Cloud, John. 2010. "World Geodetic System," in Stephen B. Johnson, General Editor, *Space Exploration and Humanit*y: A Historical Encyclopedia. Santa Barbara, ABC-Clio, vol. 1.
Dick, Steven J. 2003. *Sky and Ocean Joined: The U.S. Naval Observatory, 1830-2000* Cambridge: Cambridge University Press, pp. 472–73.
Dick, Steven J. and Dennis D. McCarthy. 1999. "William Markowitz, 1907-1998," *Bulletin of the American Astronomical Society*; vol. 31, no. 4 (1999), p. 1605
Dick, Steven J., Dennis McCarthy and Brian Luzum, eds., 2000. *Polar Motion: Historical and Scientific Problems*. San Francisco: Astronomical Society of the Pacific
Dunnington, G. Waldo. 1955. *Gauss: Titan of Science*. New York, 1955, reprinted 2004, especially 113–138.
Forman, Paul. 1985. "Atomichron: The Atomic Clock from Concept to Commercial Product," Proc. IEEE, 73, 1181–1204. http://ieeemilestones.ethw.org/images/8/8d/Forman_Proc_IEEE_1985.pdf
Galison, Peter. 1987. *How Experiments End*. University of Chicago Press: Chicago and London.
Goldstein, A., O.F. Mattingly and F.J. Heyden, S.J. 1958. "On the Geodetic Application of a Solar Eclipse," *Bulletin Geodesique, 49*
Kaplan, Joseph. 1954. "The Scientific Program of the International Geophysical Year," in *Symposium on Scientific Aspects of the International Geophysical Year, 1957-1958*, PNAS, vol. 40 (1954), pp. 926–931: 929.
Launius, Roger D. et al. eds. 2010. *Globalizing Polar Science: Reconsidering the Social and Intellectual Implications of the International Polar and Geophysical Years*. Palgrave MacMillan: New York
McCray, W. Patrick. 2006. "Amateur Scientists, the International Geophysical Year, and the Ambitions of Fred Whipple," *Isis*, 97, 4, 634–658.
Markowitz, William. 1945. "Redeterminations of Latitude and Longitude," *Transactions of the AGU*, 26, no. 2 (October, 1945), 197.
Markowitz, William. 1954a. "Proposal for the consideration of the USA National Committee on the Third Geophysical Year, 1957-58." USNO archives on-site, Moon Camera files.
Markowitz, William. 1954b. "Photographic Determination of the Moon's Position, and Applications to the Measure of Time, Rotation of the Earth, and Geodesy," *Astronomical Journal*, 59 (1954), 69–73: 70.
Markowitz, William. 1958a. "Use in Geodesy of the Results of Lunar Observations and Eventual Observations of Artificial Satellites," *Bulletin Geodesique*, 49, 33–40.
Markowitz, William. 1958b. "Geocentric Co-Ordinates from Lunar and Satellite Observations," *Bulletin Geodesique*, 49, 41–49.
Markowitz, William. 1960. "The Photographic Zenith Tube and the Dual-Rate Moon-Position Camera," in Gerard P. Kuiper and Barbara M. Middlehurst, eds., *Telescopes*. University of Chicago Press: Chicago, pp. 88–114, vol. 1 of *Stars and Stellar Systems*, G. Kuiper, General Editor.
Markowitz, William. 1986. "Reminiscences of the U.S. Naval Observatory, 21 October, 1986," oral history interview by Dennis McCarthy. U. S. Naval Observatory Library.
Markowitz, William. 1987. Oral history interview by Steven J. Dick, 18 August, 1987. U. S. Naval Observatory library.
Markowitz, William, R. Glenn Hall, L. Essen and J. V. L. Parry. 1958. "Frequency of Cesium in Terms of Ephemeris Time," *Physical Review Letters*, 1 (August 1, 1958), 204L, 1-2 and 105–106.
Mueller, Ivan. 1964. *Introduction to Satellite Geodesy*. Frederick Ungar: New York. Part I.
Muller, Paul M. 2000. "Time and Polar Motion in Early NASA Spacecraft Navigation," in Dick, McCarthy and Luzum (2000), pp. 215–19.
O'Keefe, J. A.. 1958. "The Occultation Method of Line Measurements," *Bulletin Geodesique, 49*
Rafferty, Theodore J. 1981. "Refurbishing the US Naval Observatory's 1892 Saegmuller 12-inch Refractor," *Telescope Making*, 15, pp. 24–29.

Roberts, Elliott B. 1959. "The IGY in Retrospect," *Annual Report of the Smithsonian Institution for 1959*, p. 282.

Smith, James R. 1997. *Introduction to Geodesy: The History and Concepts of Modern Geodesy*. New York, John Wiley and Sons, Inc.

Stephenson, F. R. 1997. *Historical Eclipses and Earth's Rotation*. Cambridge: Cambridge University Press

Stephenson, F. R. 2007. "Variations in the Earth's Clock Error Delta T Between AD 300 and 800 as Deduced from Observations of Solar and Lunar Eclipses, *Journal of Astronomical History and Heritage*, vol. 10, 211–220

Sullivan, Walter. 1961. *Assault on the Unknown: The International Geophysical Year*. McGraw Hill, New York, 395.

U. S. Naval Observatory. 2004. *The Astronomical Almanac for the Year 2006*. U.S. Government Printing Office, "Reduction of Terrestrial Coordinates," p. K13.

Warner, Deborah Jean. 2000a. "Political Geodesy: The Army, the Air Force and the World Geodetic System of 1960," *Annals of Science*, 59, 363-89

Warner, Deborah Jean. 2000b. "From Tallahassee to Timbuktu: Cold War Efforts to Measure Intercontinental Distances, *Historical Studies in the Physical Sciences*, vol. 30: 2, 393–415.

Whipple, F. L. and J.A. Hynek. 1958. "The IGY Optical Satellite Tracking Program as a Source of Geodetic Information," *Bulletin Geodesique, 49*

Part V
Discovering, Classifying, and Understanding the Cosmos

Part V Frontispiece Pluto, as seen by the New Horizons spacecraft on July 13, 2015, the day before closest approach. The large white area at bottom right is a nitrogen ice plain known as Sputnik Planitia, and the dark brown areas at bottom left are rugged highlands with numerous impact craters. Almost nothing was known about Pluto's surface features before New Horizons. The discovery and classification of celestial objects plays an important role in ordering and understanding the universe around us. (NASA/SwRI)

The real art of discovery consists not in finding new lands, but in seeing with new eyes.

Marcel Proust, The Prisoner, *Remembrance of Things Past*, 1923

Part V encompasses a broad array of related subjects ranging from discovery and classification of astronomical objects to our changing cosmological worldviews. Again, at first glance this may seem to have little to do with cosmos and culture, but Chap. 30 quickly disabuses us of that idea. It begins with the great Pluto debate, which began in 2006 when astronomers at the General Assembly of the International Astronomical Union voted to reclassify (some said "demote") Pluto from a planet to a dwarf planet, which they declared was not a planet at all! Much to the surprise of the astronomers, not only some scientists but also much of the general public viewed this move with disdain. How could there be only eight planets in the Solar System when they had been taught in school that there were nine? The ensuing discussion, which continues only slightly abated today, is a lesson in the cultural impact of astronomy and how attuned the public can be to their astronomical ambience. Chapter 30 then carries the discussion beyond Pluto to more general issues of discovery and classification in astronomy. The Pluto debate raised the question of how an astronomer knows when he or she has discovered a new class of astronomical object. Who decides if a dwarf planet, or a quasar or pulsar is a new class of object? The takeaway from my personal experience with the Pluto debate (I was one of those voting astronomers) was that discovery and classification are concepts that need to be studied, dissected, analyzed and put back together again. This is what I do briefly in Chap. 30, and in a much more detailed book on the subject, *Discovery and Classification in Astronomy*, published in 2013. My conclusion is that discovery is an extended process consisting of detection, interpretation, and understanding, with pre-discovery and post-discovery phases, and technological, conceptual, and social roles at each stage, while classification is a deep philosophical problem.

This sets the stage for Chap. 31, where I set forth the first comprehensive classification system for all classes of astronomical objects according to a set of consistent principles. This begs the question of "what is a class," and how we can put together a classification system for astronomy, just as has been done in biology (the Linnaean or the Three Domain system), chemistry (the Periodic Table), physics (the Standard Model), and other sciences. The system described here consists of the Three Kingdoms of planets, stars, and galaxies, 18 Families, and 82 Classes of objects. Gravitation is the defining organizing principle for the Families and Classes, and the physical nature of the objects is the defining characteristic of the Classes. Whether this system becomes useful for scientific or pedagogical purposes remains to be seen, but it starkly illustrates the problems and promise of classification.

In the next three chapters we turn to three distinct discoveries. Chapter 32 describes the discovery of the polar motion of the Earth, a phenomenon that affects not only astronomy and geophysics, but also practical problems like spacecraft navigation and positioning. Chapter 33 analyzes the discovery of the Leonid meteors as an astronomical phenomenon. Although meteors had been seen from ancient times, only in the nineteenth century were they proven to be astronomical in origin. Astronomical events such as the Leonids have played an important role in cultural history. One can imagine a spectrum of effects through history, depending upon whether the event was a meteor shower, a meteor storm, the fragmentation of a giant comet, or an event of the Tunguska type. Aside from the public spectacle, the ability to predict these phenomena is now of more than passing interest, as meteor storms pose a potential threat to spacecraft. Chapter 34 maintains the theme of discovery in the nineteenth century, although with a quite different subject: the two moons of Mars. The discovery of Phobos and Deimos was made in 1877 by Asaph Hall with the giant 26-in. refractor at the U. S. Naval Observatory in Washington, D.C., just blocks from the White House. It was made despite frequent previous attempts and declarations that there were no moons of Mars. How this occurred is a story of persistence, luck, and skepticism of previous opinion. A renaissance in moons of Mars studies came with robotic space missions beginning in the 1970s, and these peculiar potato-shaped tiny moons continue to challenge our understanding of planetary satellites. It has often been suggested they may well play a role as the target of human exploration in themselves or as outposts as we explore Mars.

The final two chapters of this section tackle two very different worldviews and illustrate how quickly our worldviews can change. The naturalist Alfred Russel Wallace, cofounder with Darwin of the theory of natural selection, drew upon the best astronomical opinion of his time to produce a cosmological worldview very different from ours today. Wallace set the dimensions of the entire universe at 3600 light years, placed the Sun at the outer margins of what he called the solar cluster, and put it in orbit around the center of gravity of the cluster. In other words, the Solar System was practically the center of the universe, and through a process of elimination Wallace came to the conclusion that Earthlings were the only inhabitants in this universe. His book *Man's Place in the Universe: A Study of the Results of Scientific Research in Relation to the Unity or Plurality of Worlds* (1903) is a case study of how a scientist can take advantage of all the data available to the time, but come to a terribly wrong conclusion due to metaphysical predispositions, in Wallace's case a dominant anthropocentric worldview endowed with purpose.

Chapter 36, however, shows how quickly cosmological worldviews can change. In the 1920s the astronomer Edwin Hubble first observationally proved that many of the nebulous objects seen with the telescope were gigantic star systems beyond our own Milky Way Galaxy, in effect galaxies of their own. While Wallace's worldview proved to be very "wrong," Hubble's proved to be very "right" in the sense that his work is the foundation of our modern cosmological worldview—the one we all live with in the back of our brains as we go about our busy lives. Although Hubble was skeptical about whether or not the universe was actually expanding (which we now know to be the case, even accelerating in its expansion), we know that Hubble's universe consists of hundreds of billions of galaxies, graphically revealed by various iterations of the Hubble Deep Field. Their classification and nature is a major preoccupation of astronomers, and the eponymous Hubble Space Telescope continues to reveal their true nature in ever-greater detail.

Chapter 30
Pluto, Discovery, and Classification in Astronomy

Abstract The debate over the demotion of Pluto to a "dwarf planet" in Prague at the IAU General Assembly in 2006 showed that classification is far from a boring subject, and in fact opens a window on a little-known aspect of the history of astronomy over the last 400 years. How does an astronomer know when he or she has discovered a new class of astronomical object? Who decides if a dwarf planet, or a quasar or a pulsar, is a new class of object? To put it more broadly, how does an astronomer know if he or she has discovered something new, especially to the extent that it is declared a totally new class of object? This chapter begins with the Pluto debate, but quickly broadens out to the discovery of other new classes of objects over the last 400 years of telescopic astronomy. Classification is an open-ended process. Even as the vote was taken in Prague, NASA's New Horizons spacecraft was speeding toward the edge of the Solar System, some three billion miles away, to explore the nature of the object so much in dispute. The discoveries made there revitalized the debate over its classification status, as did the arrival of NASA's Dawn spacecraft at another newly designated dwarf planet, the asteroid Ceres.

30.1 The Pluto Affair

The standard story is that Pluto was discovered in 1930 when Clyde Tombaugh noticed that a single object on his photographic plates had moved over a series of days. But what was it? From its slow motion it appeared to be on the outer edge of the Solar System, and from its albedo and color it was assumed to be about the mass of the Earth. So the discovery of this object was not widely believed at the time to constitute the discovery of a new class of objects, uncertainties in its mass notwithstanding. Rather, it seemed to fit into the class of planets; the discovery was, after all, made after a deliberate search for such a planet, focusing on a specific prediction, even if that prediction is now known to be spurious.

The uncertainty in Pluto's mass was not resolved until 50 years later, when James Christy discovered a moon around Pluto (Dick 2003: 423–429). With the simple application of Kepler's laws to two bodies in orbit, the resulting mass was only

Condensed from portions of Steven J. Dick, *Discovery and Classification in Astronomy: Controversy and Consensus* (Cambridge: Cambridge University Press, 2013)

S. J. Dick, *Space, Time, and Aliens*, https://doi.org/10.1007/978-3-030-41614-0_30

1/400th of the Earth and a diameter of less than 1500 miles, considerably smaller than our Moon. Plotting determinations of the mass of Pluto over time, two witty scientists jokingly predicted the planet would disappear by 1984! In a way, it did, but 12 years later by the dictate of the IAU. By allowing an accurate mass for Pluto to be calculated, Charon began Pluto's decline from planetary status. In the end it was the discovery of a new class of Trans-Neptunian objects beginning in 1992 that precipitated the change in Pluto's class status, especially when in 2003 one of these, dubbed 2003 UB 313, was determined to be larger than Pluto. It turns out that there is a whole class of objects beyond Pluto in what is known as the Kuiper Belt.

It was this situation that resulted in its "demotion" in 2006 by vote of the International Astronomical Union. And, in a linguistic contradiction truly to be regretted, a dwarf planet was ruled not to be a planet. My book *Discovery and Classification in Astronomy* (Dick 2013) details what happened at that meeting; suffice it to say that much more than science was involved; politics, personalities, and the predilections of different scientific communities were all in the mix. Even in the scientific component, much more was involved than pointing a telescope and "detecting" a new class of object; in this case it was many years before the new class was inferred and then declared.

The vote unleashed a reaction the likes of which the IAU had never seen on any issue. On the plane returning to Washington, I was astonished that the onboard TV news clips highlighted the IAU vote. Back home, all forms of media gave Pluto the attention normally reserved for politics, sports and Hollywood stars. Walking the streets of Berkeley the casual stroller down Telegraph Street would have spotted the Mars Vintage Thrift Shop with the curiously melancholy marquee pleading "I Want Pluto To Be A Planet Again." Nor was the reaction short-lived. A few weeks later the California Assembly, noting the millions of textbooks that would have to be revised and that Pluto shared its name with California's famous Disney dog, passed a resolution that "condemns the International Astronomical Union's decision to strip Pluto of its planetary status for its tremendous impact on the people of California and the state's long term fiscal health." Seven months later the legislature of New Mexico, home to Pluto's discoverer Clyde Tombaugh, passed a more sober bill declaring that "as Pluto passes overhead through New Mexico's excellent night skies, it be declared a planet and that March 13, 2007 be declared 'Pluto Planet Day' at the legislature." And the debate among scientists continued long after the vote, with some threatening to overturn the decision. That, however, has not happened, and now that the textbooks have been changed, I don't think it will.

It is important to note that historically, the Pluto situation has occurred before in the Solar System. Two centuries ago, in 1806, William Herschel's discovery of the seventh planet, Uranus, was exactly 25 years in the past. But astronomers were rejoicing in the discovery of three new planets in the last 3 years, Ceres in 1801, Pallas in 1802, and Juno in 1804. And Vesta was about to be discovered in 1807. So in 1806, astronomers thought there were 11 planets. Astronomer James Hilton has shown how for almost 50 years the Nautical Almanacs listed 12 planets, including Vesta (Hilton 2017). Then, 39 years after those four new planets had been discovered, came a problem: in 1847 three new one were found, and by the end of 1851 there were 15. Only by the mid-nineteenth century, "once their numbers grew too large to fit the existing scheme of classification," were minor planets or asteroids accepted as a class of their own. We now face a similar problem: if Pluto is a planet, then immediately so are a dozen or more objects, and by the end of the decade the Solar System will have 50 or more planets. We will be adding several planets per year to the Solar System. What will the school kids say about that? Or, we can reclassify Pluto as the IAU has done. Textbooks had to change in the eighteenth and nineteenth centuries with the discoveries of Uranus and Neptune. Surely we should not be less flexible now, though for sentimental reasons, subtracting a planet seems to be more traumatic than adding.

My conclusion from the Pluto debate was that discovery is a concept that needs to be studied, dissected, analyzed and put back together again. That is what I did in *Discovery and Classification in Astronomy*, the highlights of which are given in the rest of this chapter.

30.2 The Meaning of Discovery: The Ideas of Extended and Collective Discovery

If we look at episodes of discovery over the last 400 years, one of my primary conclusions is that discovery is an extended process (Fig. 30.1). As we saw with Pluto, discovery is not an event at a discrete moment of time. It has a macrostructure that consists of detection, interpretation, and multiple stages of understanding (Dick 2013). It also has a microstructure, consisting of technological, conceptual, and social roles. Discovery is often preceded by a pre-discovery phase (as in the case of both Uranus and Neptune), and is always followed by a post-discovery phase, both

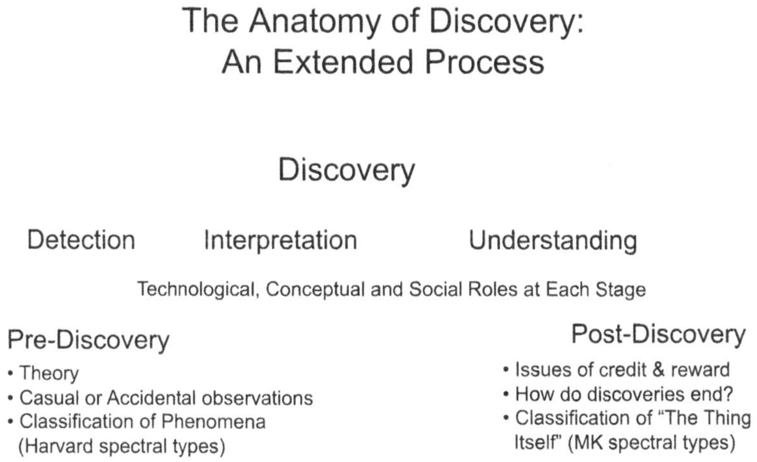

Fig. 30.1 The extended structure of discovery, showing its three stages of detection, interpretation and basic understanding, as well as the pre-discovery and post-discovery stages. Discovery also has a microstructure consisting of conceptual, technical, and social roles. From Dick (2013)

of which delimit the structure of discovery itself in both time and space. Pre-discovery can include theory, observation, n and classification of phenomena, while post-discovery includes issues of credit and reward, how discoveries end, and classification of "the thing itself."

A classic case of the extended nature of discovery is the discovery of Saturn's rings, which has been discussed in detail by the historian of astronomy Albert van Helden (Van Helden 1974a, b). Galileo certainly detected them in 1610, and his confusion is evident in his first description. But it was only Christiaan Huygens who correctly interpreted them in 1655, and an understanding of their physical and dynamical nature took much longer, beginning with Maxwell's demonstration that they could not be a single solid body. In the twentieth century we discovered rings around the other outer gas giant planets, but this form of discovery—of new members of an already known class—was quite different, because we already knew what to look for.

But when *new* classes of astronomical objects are discovered the extended nature of discovery is evident again and again. A classic case is William Herschel's discovery of nebulae. Separating these nebulae into classes is one of the great stories in the history of astronomy (Dick 2013 and Chap. 36 of this volume). To take only one example from the many objects represented in Herschel's drawings, in 1610 Peiresc detected the diffuse cloud known now known as the Orion Nebula; with more evidence Herschel adduced it was indeed a nebulosity rather than an unresolved group of stars. Sir William Huggins proved this spectroscopically in 1864, and a number of astronomers in the 1920s including Henry Norris Russell finally identified these kinds of objects as composed mostly of ionized hydrogen, what we now call HII regions (ionized hydrogen). And so on down the line, in the pattern that one sees throughout discovery in astronomy.

I also want to emphasize the idea of collective discovery, which historians of science have recently investigated (Caneva 2005). To put the question in more concrete terms in our case, did Galileo, Huygens, Maxwell, or the Voyager spacecraft "discover" the rings of Saturn? Did Peiresc, Herschel, Huggins, or Russell discover gaseous diffuse nebulae? And did Messier, Kant, Rosse, or Hubble discover galaxies? One philosopher concluded that the concept of "discovery" is so problematic that he seriously suggested the word should be banned from the language. This is not going to happen, but I *would* argue that the problem of discovery can be advanced by using more precise language. If discovery begins with detection, continues with data-gathering and its interpretation, and ends with a basic understanding of the object discovered, then the role of the discoverer in each of these phases must be made clear. In advocating an extended structure of discovery, the best solution would seem to be an embrace of the idea of "collective discovery." Such a view constitutes a more sophisticated picture of what discovery actually entails as an extended process, and it avoids what are usually counterproductive arguments about the true discoverer. The concept of collective discovery does not help Nobel Prize committees, but it *would* make them more honest. This is true not only because of the extended nature of discovery over time, but also because of the size of modern research teams.

The concept of collective discovery, however, is not an easy sell to astronomers, the issue of fairness notwithstanding. At a special session on "Discovery and Classification in Astronomy" at the International Astronomical Union's General Assembly in Beijing, China in 2012, after an explanation of the idea of collective discovery astronomers proceeded to argue about who "really" discovered quasars and other new classes of objects. Whether by temperament, training, or human nature, the desire to pinpoint "*the* discoverer" seems to be strongly ingrained among scientists.

Before I leave this discussion about the nature of discovery, let me ask two questions: Has dark matter been discovered? And has the Higgs boson been discovered? Dark matter is an example of what I call an "indirect discovery," in other words it cannot be directly detected. Yet by its indirect effects, it is today not only believed to be as real as planets, stars, and galaxies, but also to make up 23% of the matter-energy in the universe, with all the objects we have discussed heretofore, including black holes, comprising less than 5%. The mysterious dark energy comprises the remaining 72%. But have dark matter and dark energy really been discovered, any more than when nineteenth-century physicists pronounced the luminiferous ether to be "real" because of its perceived necessity based on its presumed effects? The anomalous rotation rate of galaxies indicates dark matter should exist in order to explain gravitational effects. The accelerating universe indicates dark energy "must exist" to account for the acceleration. But neither has been detected, interpreted or understood, only inferred, and arguably inferred in a much weaker sense than, say, the standard classes of stars. Indirect discovery likely has a big future, but it needs to be approached with caution.

The Higgs boson has been called "the most sought-after particle in physics and the key to physicists' explanation of how all particles get their mass." On July 4, 2012, physicists at CERN's Large Hadron Collider made the announcement of the Higgs boson, but still laced with caution: "To the layman I now say, I think we have it," said Rolf-Dieter Heuer, director general of CERN, "…we have a discovery. We have discovered a new particle consistent with the Higgs boson. It's a historic milestone today" (Vastag and Achenbach 2012). Two separate experiments at the Large Hadron Collider had indeed detected a signal "just shy of particle physicists' 5-sigma standard for declaring discovery," the journal *Science* reported (where *sigma* represents a multiple of uncertainty) (Cho 2012). While the announcement was met with universal applause, the result was at the lower end of the accepted statistical standard. Others pointed out that it was still unknown whether the new particle was the theorized Higgs of the Standard model, or merely "Higgs-like;" months later *Physics Today* headlined "The Higgs particle, or something much like it, has been spotted." Properties, such as spin and its interaction with other particles in the manner expected of the theorized Higgs, remained to be determined. Similarly CERN theorist John Ellis was quoted as saying "There is no doubt that something very much like the Higgs boson has been discovered (Miller 2012)." Clearly, the process of detection itself may be extended, mingling with the pre-discovery phase.

30.3 The Problem of Class

Now let me turn to the problem of class and classification. My work focuses on the discovery of new classes of astronomical objects, which begs the question of "what is a new class" of astronomical object, when did astronomers begin to think in these terms, and how does an astronomer know when a new class of object has been discovered? (Dick 2013, Chap. 8). This is a very interesting historical problem far beyond the Great Pluto Debate, and it requires that we separate the use of "class" as a general wishy-washy colloquial concept from the use of "class" as a specific taxon in a classification system. Galileo realized he had discovered something new in a general sense, but I would argue that he didn't think of Jupiter's satellites, or what turned out to be Saturn's rings, or star clusters, in terms of a new "class" of objects. This clearly had changed by the late eighteenth century with Herschel's classes of nebulae, and by the early nineteenth century with the discovery of asteroids. By 1802, after the discovery of Pallas, William Herschel tried to convince Piazzi that Piazzi had discovered a new class of objects that Herschel dubbed "asteroids" rather than just another planet, and that such a new class was a much more important discovery than another member of an already known class.

Many people consider classification systems boring. But consider the words of Linnaeus, the founder of Linnaean classification in biology. So classification is important, and the proof is that we are all familiar with classification systems in science, ranging from the Periodic Table of chemistry to the Standard Model of Particle Physics. In biology especially there are huge battles over classification systems, namely the Five Kingdoms versus the Three Domains of Carl Woese. The initial problem for all these systems is the same as the Pluto problem: namely, what characteristics do you use to classify?

As Stephen Jay Gould has said in regard to biology, "taxonomies are reflections of human thought. They express our most fundamental concepts about the objects of our universe. Each taxonomy is a theory of the creatures it classifies" (Gould 1988). This is no less true in astronomy. I would say that although the idea of "class" was thus used in astronomy by the early nineteenth century, it was still not a taxon in the strict Linnaean sense—in other words, not part of a classification system. I would argue that only when with a classification system can one define class as a taxon to be distinguished from another level such as Family.

Unfortunately—and rather amazingly—astronomy has no comprehensive classification system, only bits and pieces, as in the case of stars and galaxies. What would a comprehensive system look like for astronomy? As an exercise in classification I've constructed one such system, dubbed the "Three Kingdom" system (Chap. 31). As in biology, the system has definite taxa, is hierarchical, and proceeds from Kingdoms to Families to Classes. There are 3 Kingdoms, 18 Families, and 82 Classes. The Kingdoms are those enshrined in virtually all astronomy textbooks over the last century. Gravity is the basis for defining the Families. And the 82 classes are based primarily on physical characteristics. Because of this—and because gravity is the basis for defining the Families—I would claim the 3K system is a "Natural Classification System," in other words, one that gets to the heart of "the thing itself."

Only now, with such a defined system, can we see the problems in a more focused way of deciding "what is a class?" Let me give a few examples:

1. When did comets become a new class of astronomical object? I would argue not until 1577 when Tycho Brahe proved, by determining their parallax and thus distance, that they were not atmospheric phenomenal.
2. When was the class of gas giants distinguished from terrestrial planets? There is no simple answer to that question. The giant planets were indeed separated from the terrestrial planets already in the mid-seventeenth century with the micrometer measurements of Christiaan Huygens and others. But the title "gas giant" implies a much more sophisticated understanding than the simple title "giant planet" because size is much easier to determine than structure and composition. To be sure, Jupiter's density was known with considerable accuracy even in the eighteenth century. But that was only an average density. The differential rotation of Jupiter's atmospheric features could be measured, a remarkably rapid 10 h. But these might be only atmospheric features covering a solid surface, masking what, by analogy with the terrestrial planets, would be its underlying rocky composition. How deep was the atmosphere? There is a long history here; suffice it to say that only by applying the new arsenal of spectroscopy, thermodynamics and kinetics to the mystery of Jupiter's structure and composition over many decades, employing many types of evidential reasoning, was the nature of the outer planets revealed. The term "gas giant" appears to have first been used by science fiction writer James Blish in his story "Solar Plexus," as anthologized in 1952. It was not present in the original 1941 version of the story. Thus, in Blish's mind, undoubtedly a reflection of the new scientific realities, the outer planets as gas giants became a common concept only at mid-twentieth century, even if their exact composition was still unknown. With the large numbers of exoplanets being discovered, this problem will only be multiplied. If you base your classification of planets on physical characteristics rather than dynamical ones, it's not clear that a new class of planets has yet been discovered among the exoplanets. Hot Jupiters are still gas giants, as far as we know.
3. What about exoplanets? Here again there is along and tortured history, largely centered on the second half of the twentieth century, illustrating the problem of evidential reasoning that we've been discussing at our meeting this week. Van de Kamp's claimed astrometric detection of a planet around Barnard's star in 1963 proved premature. It would be more than 30 years later that Mayor and Queloz announced, using the radial velocity method, what is widely accepted as the discovery of the first planet around a solar-type star. Even in this case there is controversy over whether David Latham and colleagues discovered an earlier exoplanet. And you will have heard about the Kepler spacecraft, which has found several thousand candidate planets, even around binary stars, and some Earth-sized.

4. How about the realm of the stars? There is a long story here of how giant stars were separated from dwarf stars and so on, based on the Hertzsprung-Russell diagram which plots temperatures (Harvard spectral classes) vs absolute magnitudes. By determining distances, and thus luminosities of the stars, stellar classes were separated into the classes evident on the H-R Diagram.
5. Each of the 82 classes has its own story—pulsars, planetary nebulae, open clusters, globular clusters, spiral galaxies, ellipticals, lenticulars, and so on. The problems of separating stars and galaxies into physically based classes illuminates the history of astronomy in a new way. In short, every class in the 3K system has its own story, and those stories let us see astronomy from a new angle.

30.4 Discovery Over the Last 450 Years in Astronomy: The End of Discovery?

I want to end by asking whether our discussion of new classes of astronomical objects sheds any light on a subject question of burning interest to astronomers: Are we at the end of discovery or only the beginning, or somewhere in between? A long history of speculation on this general question has been less than enlightening. One is reminded of the end of the nineteenth-century declaration that all discoveries in physics had been made, a declaration followed in short order by the revolutionary discoveries of radioactivity, relativity, and quantum physics.

Looking back over the last 450 years at the discovery of new classes of astronomical objects, we can easily see a number of patterns if we plot the date of first detection/inference/declaration of our 82 new classes (Fig. 30.2). The most obvious feature of the graph is the mountain of discoveries in the twentieth century, three times greater than the sum of the previous 350 years. To put it another way, three-quarters of all discoveries of new classes occurred in the twentieth century. Only a few other peaks stand out as oases, and they are only decadal peaks: the 1610s following Galileo's application of the telescope to the heavens, the 1780s following the construction of William Herschel's large telescopes, and the 1860s, when both meteors and meteoroid streams were discovered as astronomical phenomena. All else preceding the twentieth century is a virtual desert during which much astronomy was accomplished in the form of theoretical work, data gathering, and the discovery of new types of objects at a level below the class taxonomic level, and therefore by definition of less importance. Notably, to the far left in the histogram are the only pre-telescopic discoveries of new classes: Tycho's determination of the celestial nature of comets and novae.

What the graph does not show is the discovery of new phenomena such as Hubble's velocity-distance relation, the expanding and accelerating universe, cosmic evolution, and so on, all of which are considerably more complicated than the discovery of new classes. But the strength of the graph lies in the systematic definition of classes and the resulting relative homogeneity of the plotted data. The

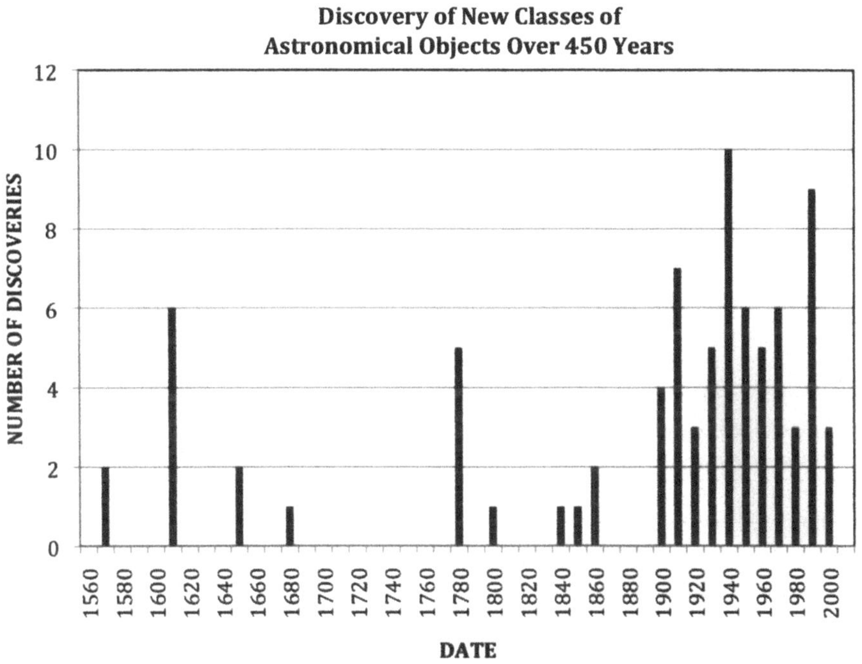

Fig. 30.2 Distribution of discoveries of new classes of astronomical objects over the past 450 years. The Galilean and Herschelian peaks are evident around 1610 and the 1780s amid the general desert of discoveries, but the most obvious feature is the mountain of discoveries during the twentieth century. Since discovery is an extended process, what is actually plotted here is the year of first detection, inference or declaration of each of the classes listed in Fig. 31.1. From Dick (2013)

question about the end of discovery is sharpened when put in terms of new classes of astronomical objects, especially as systematically defined in the Three Kingdom System. Are we near the end of discovery of new classes of astronomical objects? After all, surely a finite number of such classes must exist. We see in Fig. 30.2 that there has been no decline of discoveries in the decade of the 1990s, which indeed stands out for its discovery of nine new classes, including such well-known ones as exoplanetary systems, pulsar planets, protoplanetary systems, and the first Kuiper belt objects. On the other hand, the first decade of the twenty-first century can boast only two new classes (Oort cloud objects and protogalactic clouds) having been discovered, and one (dwarf planets) declared.

It is true that some of these classes were inferred or declared rather than detected, as was the case of dwarf planets, ice giant planets, and bright giant stars. It is also true that some classes might be questioned, for example, three classes of cosmic rays, one for each Kingdom, even though perhaps justified by their different origin, energy, abundance, and composition—and by aesthetic considerations of symmetry in the Three Kingdom system. And it is true that some classes may turn out to collapse into a single class, as in the case of the active galaxies, which may yet turn out to be the same kind of object viewed from different angles (yet for all that may remain classes if the distinctions prove useful to astronomers). But such declarations or changes in class status will not change the basic form of the histogram. They occur because more information becomes available, never a bad thing. Neither astronomers nor the public should feel cheated when such changes occur; this is a normal part of science.

We hesitate to repeat the mistake of those in the nineteenth century who proclaimed the end of physics, in particular with the mysteries (we prefer not to say "discoveries") of dark matter, dark energy, and talk of a possible multiverse. It is too soon to come to such an important conclusion based on only on the data of the last decade, especially considering the graphical mountain of discoveries in the twentieth century—and the fact that the low number of discoveries in the 1920s, for example, presaged the largest number of discoveries in astronomical history. The argument that there can only be a finite number of classes is compelling from a common-sense point of view, but whether we are entering a temporary valley, or whether the next decades will feature new peaks, only the future will tell. It is difficult to see how a new Kingdom could be discovered, unless it is the universe itself as part of the conjectured multiverse. On the other hand it is easy to see how many more types of objects could be discovered at taxonomic levels below the class level, the domain of the many types of exoplanets, variable stars, binary stars, and variations on galaxies.

But as to whether there are more classes to add to Astronomy's Three Kingdoms, we'll have to wait and see. It is always possible that—as in the case of the electromagnetic spectrum—new views of the universe will be opened up that we cannot yet guess.

30.5 Summary and Conclusions

1. Discovery is a central concept in astronomy. This may seem obvious, but the point needs to be made because discovery is not often emphasized as a problematic concept in the history of science. Thomas Kuhn's work in the 1960s emphasized revolutions more than discovery, but discovery is much more common in science and deserves center stage. Martin Harwit was the pioneer in bringing this subject to the fore 30 years ago for astronomy with his pioneering book *Cosmic Discovery*. But, aside from an IAU Symposium "Accelerating the Rate of Astronomical Discovery," little work has been done on the subject since Harwit's work, especially in astronomy.[1]
2. Discovery is an extended process. A discovery is not an event at a discrete moment of time. It has a macrostructure that consists of detection, interpretation, and multiple stages of understanding. It also has a microstructure, consisting of technological, conceptual, and social roles. Finally, discovery is sometimes preceded by a pre-discovery phase (as in the case of both Uranus and Neptune), and is always followed by a post-discovery phase, both of which delimit the structure

of discovery itself in both time and space. Pre-discovery can include theory, observation and classification of phenomena, while post-discovery includes issues of credit and reward, how discoveries end, and classification of "the thing itself." Kuhn explicitly made this point about the extended nature of discovery based on only a few examples from science, and it is evident in Al Van Helden's work on the discovery of Saturn's rings. This pattern is repeated again and again in astronomy when new classes of astronomical objects are being discovered, not least in William Herschel's discovery of nebulae.

This idea of discovery as an extended process has been largely accepted by those few who have studied the nature of discovery over the last 50 years. The details of the structure, however, remain open to discussion, and the structure presented here would seem to follow more closely what most astronomers actually do, and see themselves as doing, without the loaded Kuhnian language of normal vs. revolutionary science. In fact in his 1981 study *Cosmic Discovery*, Martin Harwit already asserted that "in astronomy the major discoveries seem to differ qualitatively from the revolutions that Kuhn describes." I agree with that assessment.

3. I want to emphasize that my work focuses on the discovery of new *classes* of astronomical objects, which begs the question of "what is a new class" of astronomical object, and how does an astronomer know when one has been discovered? This is a very interesting historical problem, and it requires that we separate the use of "class" as a general wishy-washy colloquial concept from the use of "class" as a specific taxon in a classification system. Galileo realized he had discovered something new in a general sense, but didn't think in terms of a new "class" of objects. This clearly had changed by the early nineteenth century. By 1802, after the discovery of Pallas, William Herschel tried to convince Piazzi that Piazzi had discovered a new class of objects that Herschel dubbed "asteroids" rather than just another planet, and that such a new class was a much more important discovery than another member of an already known class. Though the idea of "class" was thus used in astronomy by the early nineteenth century, it was still not a taxon in the strict Linnaean sense—in other words, not part of a classification system. A century later when Hertzsprung and Russell distinguished giant from dwarf stars, they certainly had the concept of class in mind, though again not as a strict taxon, despite the delineation of types in spectral classification. And although Harwit tried to define class by defining it as an object different in some property by more than 1000 times from another object, I would argue that only when you have a classification system can you define class as a taxon to be distinguished from another level such as Family. At the same time, as Stephen Jay Gould has said, "taxonomies are reflections of human thought."
4. Classification plays its role in discovery, sometimes problematically if attempted before understanding has reached a mature level, as notoriously in the case of Pluto, or even in the case of the Harvard spectral types, which had to be rearranged to represent an evolutionary sequence. On the other hand, and counterintuitively, classification has historically been both a pre-discovery and a post-discovery activity, corresponding to phenomenology in the first case and

"the thing itself" in the second. Thus the Harvard spectral types represented a pre-discovery activity with respect to the physical classes of dwarfs, giants, supergiants and so on, while the MK system represents a post-discovery classification activity representing the thing itself, that is, the physical stellar luminosity classes. The same can be said for galaxies.

5. The idea of collective discovery is essential. Did Galileo, Huygens, Maxwell, or the Voyager spacecraft "discover" the rings of Saturn? Did Peiresc, Herschel, Huggins or Russell discover gaseous diffuse nebulae? And did Messier, Kant, Rosse, or Hubble discover galaxies? One philosopher concluded that the concept of "discovery" is so problematic that he seriously suggested the word should be banned from the language. But the problem of discovery can be advanced by using more precise language. If discovery begins with detection, continues with data-gathering and its interpretation, and ends with a basic understanding of the object discovered, then the role of the discoverer in each of these phases must be made clear. In advocating an extended structure of discovery, the best solution would seem to be an embrace of the idea of "collective discovery." As historian Ken Caneva put it, this is the concept that "scientific discovery … must be understood as an extended process by which a diversity of individuals develop and come collectively to accept as true a statement about the physical world." Such a view not only provides a more sophisticated picture of what discovery actually entails, it also avoids what are usually counterproductive arguments about the true discoverer.
6. Finally, our analysis of new classes of astronomical objects informs a question of burning interest to astronomers, if not to historians: Are we at the end of discovery or only the beginning, or somewhere in between? A long history of speculation on this general question has been less than enlightening. The question is, however, sharpened when put in terms of new classes of astronomical objects. After all, surely a finite number of such classes must exist. Harwit argued that during the twentieth century astronomical discovery accelerated exponentially, largely due to new technology, and that by 1980 some 43 cosmic phenomena were known. His definition of discovery included both new classes of objects, defined as those differing in some property by more than 1000 times, and new astronomical phenomena such as the expanding universe. Using a statistical technique, he predicted that the total number of astronomical phenomena is about 130, so that two-thirds remained to be discovered. He envisioned a bell-shaped curve falling off rapidly after the year 2000.

New Zealand astronomer John Hearnshaw, on the other hand, concluded in a much briefer and more cursory study that important astronomical discoveries of the twentieth century, if plotted by decade, showed peaks in the 1920s and 1960s, have declined since 1975, and exhibit only a "dismal number" in the 1990s. His list also included both classes of objects such as quasars and pulsars, and new phenomena, but inexplicably excluded the Solar System. His overall conclusion was that "astronomers are running out of new things to discover," at least things in the same league as the top 50 discoveries of the twentieth century.

We must emphasize several things about these studies, tentative as they are. First, the answer in each case is highly sensitive to the definition of "important discovery," and to the definition of "class." Secondly, the studies do not take into account the extended nature of discovery. These shortcomings are addressed in some measure by using the classes systematically defined in the Three Kingdom System.

In the end "Accelerating the Rate of Astronomical Discovery," the title of the IAU Symposium in 2009 mentioned above, will remain a goal devoutly to be desired. Whether or not it is reached depends not only on public policy, individual and collective innovation, and networked science, but also on the nature of a universe we do not yet fully understand. There are many more aspects to discovery, including the extent to which telescopes are engines of discovery, the role of theory, and the social aspects of discovery.

30.6 Commentary 2020

The reviews of *Discovery and Classification in Astronomy* on which this paper is based stimulated considerable interesting commentary. One of the most substantive (Nasim 2014) pointed out that the concept of discovery itself has evolved over the centuries, a point well taken and that has since been elaborated in David Wooten's book *The Invention of Science*, which includes a long chapter on "Inventing Discovery" (Wooten 2015: 57–109). This realization indeed adds another dimension to the idea of discovery, with nuances that need to be taken into account in any analysis of the concept. In my view, however, this still leaves my major conclusions intact. Other commentators (Gingerich 2014) failed to realize that my focus was on the discovery of new *classes* of astronomical objects, not on discoveries in astronomy in general, though I did contrast the discovery of new classes of objects with the discovery of new astronomical phenomena such as the expansion of the universe or gravitational waves or the cosmic microwave background. Still other commentators complained that the Three Kingdom system of classification delineated in the next chapter did not contain their favorite objects, such as variable stars. But variable stars are Types of various classes of stars, and thus fall a taxon below the class level. The system could indeed be elaborated to include various Types within the classes. Reviews from philosophers of science (Bhakthavatsalam 2015) were generally positive but pointed to a great deal of work still to be done on the subject. It is my hope this will indeed happen.

As for Pluto, the New Horizons spacecraft was fantastically successful, making its closest flyby of Pluto on July 14, 2015. As a result our knowledge of Pluto has increased tremendously, including a treasure trove of views of its surface. The Frontispiece for Part V reveals that Pluto was much more diverse than anyone expected. As Alan Stern and David Grinspoon summarized these phenomena in their riveting description of the New Horizons mission, Pluto showed "Ground fogs, high-altitude hazes, possible clouds, canyons, towering mountains, faults, polar caps, apparent dune fields, suspected ice volcanoes, glaciers, evidence of flowing (and even standing) liquids in the past, and more" (Stern and Grinspoon 2018). Surprisingly, Pluto was found to be geologically active, with both ancient cratered and fresh-looking landscapes, and a 1000-km wide churning nitrogen glacier dubbed Sputnik Planitia. Even more surprising, there are indications of an interior liquid water ocean, analogous to those thought to exist inside Europa and Enceladus. "Could there be Plutonian life forms swimming deep underneath the planet's icy surface?" Stern and Grinspoon provocatively ask. Only further exploration will tell.

Pluto is now known to have five moons, with Charon by far the largest at 751 miles (1212 km) in diameter. Hydra and Nix are about 25 miles (40 km) along their longest axis, while Kerberos and Styx are only about seven miles (12 km) along the longest axis, comparable to the two moons of Mars discussed in Chap. 34. New Horizons imaged all of these moons, but discovered no new ones.

NASA's Dawn mission to the dwarf planet Ceres also proved very successful, and entered orbit around Ceres on March 6, 2015. It revealed a combined water ice and rocky surface, a differentiated rocky core and icy mantle, and perhaps even an internal ocean of liquid water. Several mysterious bright spots were seen inside a crater, believed to be highly reflective ice or salts, or a cyrovolcanic feature.

Surprisingly, the Great Pluto Debate continues—New Horizon's Principal Investigator Alan Stern, for one, has vociferously argued that Pluto should be classified as a planet. Perhaps a compromise is in order. After all, to say that a "dwarf planet" is not a planet is a linguistic barbarity. Why could a dwarf planet not be a class of planet, like the terrestrial, gas giant, and ice giant planets? Suffice it to say there are some celestial politics involved, another manifestation of cosmos and culture. Time will tell if the IAU reconsiders.

Note

1. The IAU Symposium is online at https://pos.sissa.it/cgi-bin/reader/conf.cgi?confid=99#session-810.

References

Bhakthavatsalam, Sindhuja. 2015. *Philosophy of Science*, 82, no. 3, 520–524.

Caneva, Ken. 2005. "Discovery as a Site for the Collective Construction of Scientific Knowledge," *Historical Studies in the Physical Sciences*, 35, 175–291.

Cho, Adrian. 2012. "Higgs Boson Makes Its Debut After Decades-Long Search," *Science*, 143, 141–143

Dick, Steven J. 2003. *Sky and Ocean Joined: The U. S. Naval Observatory, 1830-2000*. Cambridge University Press.

Dick, Steven J. 2013. *Discovery and Classification in Astronom*y: Controversy and Consensus. Cambridge, Cambridge University Press.

Gingerich, Owen. 2014. *Isis*, vol. 105, 864–865.

Gould, Stephen Jay. 1988. Foreword to Lynn Margulis and Karlene V. Schwartz, *Five Kingdoms: An Illustrated Guide to the Phyla of Life on Earth.* New York: W. H. Freeman, 2nd edition, p. x

Hilton, James. 2017. "When did the Asteroids become Minor Planets?" https://aa.usno.navy.mil/faq/docs/minorplanets.php

Miller, Johanna. 2012. "The Higgs particle, or something much like it, has been spotted," *Physics Today*, 65, 12–15.

Nasim, Omar W. 2014. "Discovery in Science," *Metascience*, published online 18 June 2014.

Stern, Alan and David Grinspoon. 2018. *Chasing New Horizons: Inside the Epic First Mission to Pluto*. New York, Picador.

Van Helden, Albert. 1974a. "Saturn and his Anses," *Journal for the History of Astronomy*, 5 (1974), 105–121.

Van Helden, Albert. 1974b. "Annulo Cingitur: the Solution of the Problem of Saturn," *Journal for the History of Astronomy*, *5*, 155–174.

Vastag, Brian and Joel Achenbach. 2012. "Scientists Search for Higgs Boson Yields New Subatomic Particle," *Washington Post,* July 4, 2012.

Wooten, David. 2015. *The Invention of Science: A New History of the Scientific Revolution* (UK: Penguin Random House)

Chapter 31
Astronomy's Three Kingdoms: A Comprehensive Classification System of Celestial Objects

Abstract Although classification has been an important aspect of astronomy since stellar spectroscopy in the late nineteenth century, to date no comprehensive classification system has existed for all classes of objects in the universe. Here we present such a system, and lay out its foundational definitions and principles. The system consists of the Three Kingdoms of planets, stars and galaxies, 18 Families, and 82 Classes of objects. Gravitation is the defining organizing principle for the Families and Classes, and the physical nature of the objects is the defining characteristic of the Classes. The system should prove useful for both scientific and pedagogical purposes.

31.1 Introduction to the Three Kingdom System

This article introduces a classification system of celestial objects developed by the author. In contrast to biology, physics and chemistry, and despite a long and distinguished history of classifying specific objects such as stars and galaxies, astronomy lacks a comprehensive classification system for what has become a veritable celestial zoo. What would such a system look like, and based on what principles? Here we present a system devised for pedagogic use over the last several decades (Fig. 31.1), but that will also be useful for scientific purposes. This so-called "Three Kingdom" system begins with the three Kingdoms of planets, stars and galaxies, stipulates 6 Families for each Kingdom, and distinguishes 82 distinct classes of astronomical objects. Like biology, it is hierarchical, extending from Kingdom to Family to Class, with the possible extension to further categories lower in the hierarchy such as Type and Subtype. As in biological classification it occasionally adds an intermediate Subfamily level wherever useful. With the benefit of hindsight, and with utility in mind, the system incorporates some classes as they have historically been defined, and adds others as they might be defined in a more coherent and consistent system.

In constructing such a system one immediately runs into the problem of how to define the categories of Kingdom, Family and Class. The three Kingdoms adopted here (planets, stars, galaxies) are the three canonical divisions adopted in astronomy

First published as "Astronomy's Three Kingdom System" (International Society for Knowledge Organization, *Encyclopedia of Knowledge Organization*, 2019)

S. J. Dick, *Space, Time, and Aliens*, https://doi.org/10.1007/978-3-030-41614-0_31

Astronomy's 82 Classes

Kingdom of the Planets	Kingdom of the Stars	Kingdom of the Galaxies
Family: Protoplanetary	Family: Protostellar	Family: Protogalactic
Class P 1: Protoplanetary Disk	Class S 1: Protostar	Class G 1: Protogalaxy
Family: Planet	Family: Star	
Class P 2: Terrestrial (rocky)	Subfamily: Pre-Main Sequence	Family: Galaxy
Class P 3: Gas Giant	Class S 2: T Tauri	Subfamily: Normal
Class P 4: Ice Giant	Class S 3: Herbig Ae/Be	Class G 2 Elliptical
Class P 5: Pulsar Planet	Subfamily: Main Sequence	Class G 3 Lenticular
Family: Circumplanetary	(H burning - Luminosity Class V)	Class G 4 Spiral
Class P 6: Satellite	Class S 4: Dwarf	Class G 5 Irregular
Class P 7: Ring	Class S 5: Subdwarf	Subfamily: Active
Class P 8: Radiation Belt	Subfamily: Post-Main Sequence	Class G 6 Seyfert
Family: Subplanetary	(He burning and higher elements)	Class G 7 Radio Galaxy
Class P 9: Dwarf Planet	Class S 6: Subgiant (Luminosity Class IV)	Class G 8 Quasar
Class P 10: Meteoroid	Class S 7: Giant (Luminosity III)	Class G 9 Blazar
Subfamily: Small Bodies of Solar System	Class S 8: Bright Giant Class II)	Family: Circumgalactic
Class P 11: Minor Planet/	Class S 9 Supergiant (Lumin. Class I)	Class G 10 Satellites and
Asteroid	Class S 10 Hypergiant (Lumin. Class 0)	Stellar Streams
Class P 12: Comet	Subfamily: Evolutionary Endpoints	Class G 11 Galactic Jet
Class P 13: Trans-Neptunian	Class S 11 Supernova	Class G 12 Galactic Halo
Objects	Class S 12 White Dwarf	Family: Subgalactic
Family: Interplanetary Medium	Class S 13 Neutron Star/Pulsar	Class G 13 Subgalactic Object
Class P 14: Gas	Class S 14 Black Hole	Family: Intergalactic Medium
Class P 15: Dust	Family: Circumstellar	Subfamily: Gas
Subfamily: Energetic Particles	Class S 15: Debris disk	Class G 14 Warm Hot IGM
Class P 16: Solar Wind	Class S 16: Shell (dying stars)	Class G 15 Lyman alpha blobs
Class P 17: Anomalous Cosmic	Class S 17: Planetary Nebula	Subfamily: Dust
Ray	Class S 18: Nova Remnant	Class G 16 Dust
	Class S 19: Core Collapse Supernova	
	Remnant	
Family: Systems	Class S 20: Stellar Jet	Subfamily: Energetic Particles
Class P 18: Planetary Systems/	Class S 21: Herbig-Haro Object	Class G 17 Galactic Wind
Exoplanets	[See also Protoplanetary Disk (P 1);	Class G 18 Extragalactic
Class P 19: Asteroid Groups	Planetary System, (P 18)	Cosmic Rays
Class P 20: Meteoroid streams	Kuiper Belt (P 21)	Family: Systems
Subfamily: Trans-Neptunian Systems	Oort Cloud (P 22)]	Class G 19 Binary
Class P 21:Kuiper Belt	Family: Substellar	Class G 20 Interacting
Class P 22: Oort Cloud	Class S 22: Brown dwarf	Class G 21 Group
	Family: Interstellar Medium	Class G 22 Cluster
	Subfamily: Gas (99%]	Class G 23 Supercluster
	Class S 23: Cool Atomic Cloud (H I)	Class G 24 Filaments & Voids
	Class S 24: Hot Ionized Cloud (H II)	
	Class S 25: Molecular Cloud (H2)	
	Class S 26: White Dwarf Supernova	
	Remnant	
	Subfamily: Dust (1%)	
	Class S 27: Dark Nebulae	
	Class S 28: Reflection Nebulae	
	Subfamily: Energetic Particles	
	Class S 29: Stellar Wind	
	Class S 30: Galactic Cosmic Rays	
	Family: Systems	
	Class S 31: Binary Star	
	Class S 32: Multiple Star	
	Class S 33: Association (OB)	
	Class S 34: Open Cluster	
	Class S 35: Globular Cluster	
	Class S 36: Population	

Fig. 31.1 The Three Kingdom System delineating 82 Classes of objects

textbooks for almost a century, since it became clear that galaxies were indeed a separate realm from our Milky Way Galaxy, as determined by the American astronomer Edwin Hubble in the early 1920s. For each Kingdom six astronomical Families are delineated, based on the object's origin (Proto-), location (Circum- and Inter-), subsidiary status (Sub-), and tendency to form systems (Systems), in addition to the "Central" Family (planet, star, or galaxy) with respect to which the other Families are defined. These considerations give rise to astronomy's 18 Families, and the symmetry of the 6 Families of each Kingdom reflects their physical basis in gravity's action in all 3 Kingdoms.

For a more general introduction to astronomical classification and its issues see Buta et al. (2007), DeVorkin (1981), Dick (2013, 2018), Feigelson (2012), Gray and Corbally (2009), Morgan (1937, 1988), Morgan and Keenan (1973), and Sandage (2005).

31.2 Defining Astronomy's 82 Classes

But this begs the question: How does one define a class of astronomical objects? More specifically, how does one recognize a new class of objects? We have tackled these questions in previous books, including *Discovery and Classification in Astronomy: Controversy and Consensus* (Dick 2013), and *Classifying the Cosmos: How We Can Make Sense of the Celestial Landscape* (Dick 2018), in which the Three Kingdom system is laid out in full and the history and science of each class is described.

One way of approaching the question of the definition of Class is by looking at history, where (exceptions like stars and galaxies notwithstanding) classification has often been ad hoc, haphazard, and historically contingent on circumstance. If astronomical history demonstrates anything, it is that the classification of astronomical objects has been based on many characteristics, depending on the state of knowledge and the needs of a particular community at the time. For example planets could be divided according to their physical nature (terrestrial, gas giant, and ice giants) or as the recent discovery of planetary systems has taught us, by orbital characteristics (highly elliptical or circular), proximity to their parent star ("hot Jupiters") and so on. Historically, binary stars have often been classified by the method of observation as visual, spectroscopic, eclipsing, and astrometric, or (after more information became known) by the configuration or contents of the system, such as a white dwarf binary, or by the dominant wavelength of its electromagnetic radiation, as in an X-ray binary. While these overlapping systems have served astronomers well and illustrate how the same object may be classified in many ways, such designations are the source of much confusion among students, not to mention indecipherable to the public.

History also demonstrates that at the time of discovery, by the very nature of the problem, it is sometimes difficult to decide if a new class of object has been discovered. Perhaps by analogy with the Earth's Moon Galileo decided relatively quickly that the four objects he first saw circling Jupiter in 1610 were satellites, proof that the Moon was not unique, but a member of a class of circumplanetary objects (even if he did not speak in terms of "class"). But the object he first saw surrounding Saturn was not at all obviously a ring, and awaited the interpretation of Christiaan Huygens more than 40 years later. Even in the late twentieth century it was not immediately evident that pulsars were neutron stars, or that quasars were active galactic nuclei, both qualifying in the end for new class status.

Inconsistency notwithstanding, the criterion that astronomers have most often used in the astronomical literature for determining class status—and the one we adopt for the Three Kingdom system—is the physical nature of the object. In the planetary Kingdom, for example, rather than orbital characteristics the definition of planetary classes in our own Solar System has been based on their physical characteristics as rocky, gaseous, or icy in composition; pulsar planets have also been distinguished by being inferred as physically very different again due to the extreme nature of their environment and probable different origin. As we have noted, new classes of planets will undoubtedly be uncovered as observations of extrasolar planets progress, but thus far not enough is known about their physical nature to do so. Many of the extrasolar planets discovered so far are believed to be gas giants; many are close to their stars and thus called "hot Jupiters." The first terrestrial extrasolar planets have also been claimed, in the form of "Super-Earths" and the first rocky transiting system, known as CoRoT-7b.

This history indicates that a comprehensive classification system for astronomy can perhaps do no better than to use the typological definition of "class" largely discarded by biologists:

> Membership in a class is determined strictly on the basis of similarity, that is, on the possession of certain characteristics shared by all and only members of that class. In order to be included in a given class, items must share certain features which are the criteria of membership or, as they are usually called, the "defining properties." Members of a class can have more in common than the defining properties, but they need not. These other properties may be variable – an important point in connection with the problem of whether or not classes may have a history. (Mayr 1988, 337)

But what is the unit of classification for astronomy? For physics it is elementary particles. For chemistry it is the elements, defined by atomic number in the Periodic Table. For biology it is species at the macro level, giving rise to biology's "Five Kingdoms," still favored by some macrobiologists, and genetic sequences of 16S ribosomal RNA at the molecular level, giving rise to Carl Woese's "Three Domains" of Archaea, Bacteria and Eucarya—favored by most molecular biologists.[1] For astronomy, the unit of classification adopted here is the astronomical object itself, and with some theoretical justification. For as strong and weak forces are dominant in particle physics, and as the electromagnetic force is dominant in chemistry (except for nuclear chemistry), so in astronomy is it the weakest but most far-reaching force of gravity that predominantly acts on and shapes these astronomical objects. Though other considerations such as hydrostatics and gas and radiation pressure come into play, gravity is the determining factor for the structure and organization of planets, stars, and galaxies, their Families and Classes of objects. To put it another way, the strong interaction holds protons and neutrons together and allows atoms to exist; the electromagnetic interaction holds atoms and molecules together and allows the Earth to exist; and the gravitational interaction holds astronomical bodies together and allows the Solar System, stellar systems and galactic systems to exist.[2] Gravity is thus a prime candidate—the one adopted here—to serve as the chief organizing principle for a comprehensive classification system for all astronomical objects.

Where does such a definition of class lead in the construction of a classification system? In the Kingdom of the Stars stellar spectra were first classified on what turned out to be a temperature sequence, a system devised at Harvard in the late nineteenth century with its familiar O, B, A, F, G, K, and M stars and so on. Spectra were later classified on a luminosity scale, devised at Yerkes Observatory in the 1940s, the so-called MKK (Morgan-Keenan-Kellman) system with its dwarfs, giants, and supergiants.[3] Which to choose to delineate "classes" for stars in a more comprehensive system for astronomical objects? We have adopted the Yerkes/MKK system (now known as the MK system) as a more evolved two-dimensional system based on spectral lines sensitive not only to temperature, but also to surface gravity (g) and luminosity. As astronomers Richard Gray and Christopher Corbally recently put it in their magisterial volume *Stellar Spectral Classification* in connection with the luminosity classes, "Stars readily wanted to be grouped according to gravity as well as according to temperature, and this grouping could be done by criteria in their spectra" (Gray and Corbally 2009, 10). The resulting luminosity classes (main sequence, subgiant, giant, bright giant, and supergiant labeled from Roman numeral V to I respectively), together with the stellar endpoint classes (supernova, white dwarf, neutron star, and black hole) not only have significance in the evolutionary sequence, but also have a real history of discovery that can be uncovered. W. W. Morgan delineated these luminosity classes to begin with because he realized each grouping of stars formed a sequence of near constant log g (surface gravity) (Gray and Corbally 2009, 9–10 and Morgan 1937, 380 ff.). Thus gravity as a sculpting force for stars was recognized already by the founders of the MKK system as the dominating force for the luminosity classes.

The choice of luminosity for stellar classes does not subordinate the Harvard system of spectral types. To the contrary, Harvard spectral types are still an integral part of the system. As the originators of the Yerkes/MKK system argued, it is simply the case that their system contains more information and better represents the physical nature of stars, as astronomers gradually separated them (over the 30 years from 1910 to 1940) into supergiants, bright giants, giants, and subgiants. In other words, since 1943 with the Yerkes/MKK system, modern astronomy has a formal two-dimensional temperature-luminosity system with distinct classes, building on the Hertzsprung-Russell diagram, which was literally a two-dimensional plot of temperatures versus luminosities when it was first constructed around 1914. Both the Harvard and the Yerkes systems are represented in the full designation of a star, as in Sirius (A1V) as a main sequence star with Harvard spectral type A1.

Thus, choices for class status become more clear-cut once there is a guiding principle such as physical meaning, which goes to the heart of Morgan's quest for "the thing itself." Again in the stellar Kingdom, for the interstellar medium instead of "diffuse nebulae" (a morphological classification), classes in the Three Kingdom system are distinguished according to physical constitution of the nebulae: gas (cool atomic neutral hydrogen, hot ionized hydrogen, and molecular), and dust (reflection nebulae). These categories are used in astronomy and subsume classifications based on morphology that are historically contingent. In the galactic kingdom galaxy morphologies (elliptical, lenticular, spiral, barred spiral, and irregular) laid out by Edwin Hubble in the 1920s also reflect compositional differences (as Morgan's galaxy classification system showed), so the principle of physical meaningfulness still holds.

31.3 Classification Principles in the Three Kingdom System

As we have stipulated, by definition Kingdoms are delineated by the three central prototypes of objects in the universe—planets, stars, and galaxies, as enshrined in canonical textbooks since the 1950s. Families are delineated by the various manifestations of the gravitational force acting on astronomical objects, e.g. protoplanetary, planetary, circumplanetary, subplanetary, interplanetary, and systems. As in any classification system, there will be ambiguities of placement in lower taxon levels. These can be mitigated by a system of classification principles. For the Three Kingdom system these include the following when it comes to the determination of classes and the placement of objects in classes:

1. Classes are delineated based on the physical nature of the object, defined as physical composition wherever possible.
2. An object should always be placed in its most specific Class.
3. To the extent possible, Classes already in use are retained, as in the luminosity Classes of the MK system and the Hubble classes for galaxies, supplemented by new knowledge.
4. The recommendations of the International Astronomical Union are followed; e.g. a dwarf planet is not a Class of planet.
5. Potential, but unverified, classes are not included.

Figure 31.1 is the result of applying these principles to astronomical objects. For those who do not recognize their favorite objects, it is likely because they exist at a taxonomic level below that of "Class." The plethora of variable stars, for example, are not classes of objects in this system, on the same level as giant and dwarf stars, and so on. Rather, they are Types of these stars that could be elaborated in a more complete system.

It is important to emphasize that classification in astronomy has similarities and differences with classification in biology, chemistry, and physics. The most obvious difference between the classes (species) in biology and the classes in astronomy, at least as depicted in our Three Kingdom system, is the sheer number of species. E. O. Wilson, the Harvard naturalist who is one of the chroniclers of the diversity of life, has estimated that by 2009, 150 years after Darwin's *Origin of Species*, some 1.8 million species had been discovered and described, out of perhaps tens of millions that now exist. And this does not include what Wilson (in a rare astronomical analogy employed in the domain of biology) calls the dark matter of the microscopic universe, which could be tens or hundreds of millions of species of sub-visible organisms.[4]

The number of "species" or classes in astronomy is obviously put to shame by the effusive and creative diversity of biology, no matter how one defines Class or what classification system one uses. In terms of number, astronomy's classes, at least as defined in the Three Kingdom system, are more comparable to elements in chemistry (93 natural and 15 artificial), or to the phyla (32) and classes (90) in just one of Lynn Margulis's Five Kingdoms (*Animalia*) of biology, which contains almost a million species by itself. Any such comparison depends not only on how one defines a class of astronomical objects, but also whether the classes as defined here in the Three Kingdom system are really analogous to species in the biological hierarchy of classification, or to elements in the linear classification. That is also a matter of definition, and in part a subjective matter based on relation to higher and lower categories in the system. One can argue whether a giant star of Luminosity Class III in the MK system should be called a Class or a Type, but one cannot argue that a particular member of the Class, a Type of giant star such as an RR Lyrae, for example, should be placed at a higher level in the system than the class of which it is a member.

This classification exercise also illustrates a problem that astronomical taxonomy has in common with biological taxonomy: classification characteristics do not necessarily conform to evolutionary relationships. The class of giants as defined by the MK system definition was not precisely the same as the class of giants that Henry Norris Russell declared about 1910, nor is it entirely coextensive with the evolutionary states of the giant stars as known today. Russell's definition (and the Mt. Wilson system) was based on size and luminosity, as determined by their distances and apparent magnitudes, which could be converted to luminosity. The MKK definition was based on spectroscopy, in particular "line ratios" defined by standard stars. If an unclassified star matched the standard in a spectroscopic sense, it became a member of that class, such as a giant, without regard to its internal structure or evolutionary status. While luminosities and MK definitions are still used, today astrophysicists often think of giant stars and other stellar classes in terms of their evolutionary state, which for a giant is normally undergoing core helium fusion, but varies depending on the star's mass and where it stands in the spectral temperature sequence. Moreover a particular class may be adjusted based on new data; in the early 1990s the Hipparcos satellite determined distances ten times more accurate than ground-based parallaxes, and correspondingly more accurate luminosities. The data showed that many of the luminosities were in error, and in the post-Hipparcos, and now the Gaia spacecraft era, the modern concept of a giant star (core helium fusion with shell hydrogen burning via the CNO cycle) is by no means co-extensive with MK class III defined by spectral line ratios. Nevertheless, the general classes of stars remain, but with a broader definition than determined by the MK system.

In short, astronomical classes have evolved in a way analogous to biology, where "the way it looks" (the phenotype) was primary in the Five Kingdom classification embraced by zoologists, as opposed to the deeper structure based on genetic makeup (the genotype). But whereas in biology Woese's Three Domain system caused an uproar in biology with its finding of a completely new Domain of life and different relationships for parts of the classification system, the classification of stars by how they physically operate rather than by how they appear has thus far led to broader thinking with only minor adjustments.[5]

31.4 Uses of the System and Future Development

A good classification system must not only be useful, but should also lead to deeper understanding and advance its subject. The uses of the Three Kingdom system are at least threefold, all of which may potentially lead to deeper understanding for different audiences.

First, for scientific purposes, as a comprehensive system for all astronomical objects based on consistent physical principles, the Three Kingdom system brings a consistent set of classification principles to discussions such as the status of Pluto as a planet. It suggests that the definition of a planet should not be based primarily on hydrostatic equilibrium, or roundness, or dynamical considerations, but on physical constitution—just as stellar classification was based on consistent physical principles as determined by spectroscopy. Other criteria may indeed enter any classification decision, but they should be secondary. The Three Kingdom system thus brings consistency to astronomical classification, and more clarity in making classification decisions. In the process it might also, over the longer term, bring consistency to astronomical nomenclature as far as taxa such as Class and Type are concerned.

Secondly, again for scientific purposes, the symmetric structure of the Three Kingdom system facilitates comparisons at three different scales. In the comparison of Families across Kingdoms, one can ask, for example, how the interplanetary, interstellar and intergalactic media compare, and analyze what this tells us about the nature of the cosmos. Similarly for protoplanetary, protostellar, and protogalactic processes, and so on. Such comparisons are sometimes already made, but the Three Kingdom system cries out for such comparison in a systematic way. Comparisons of Classes across Kingdoms may also prove enlightening. Planetary rings, stellar rings, and galactic rings in the form of stellar streams have much in common as broken up remains, but at vastly different scales and energies. Similarly for planetary, stellar, and galactic jets, or subgalactic, substellar, and subplanetary objects. However, since the bedrock definition of a Class is that at least one representative object must have been observed, we have not included a Class of planetary jets, even though the discovery of brown dwarf jets in 2007 led to speculation that planetary jets might exist during the accretion phase of gas giants. Based on symmetry among Families in the Three Kingdoms, we might also predict the existence of such jets, as well as other objects. While some might argue that volcanic eruptions or water spouts from Europa or Enceladus might qualify as jets, this does not seem to me quite analogous to stellar and galactic jets formed by energetic processes. But one could argue.

Thirdly, there is an educational advantage for the teaching of astronomy. The Three Kingdom system allows students to perceive immediately where an object fits in the scheme of astronomical objects. In assessing a new discovery, for example, whether the object is a Type, Class, Family, or Kingdom should help a student to see its relative importance in the astronomical zoo. Thus definitive proof of a new Kingdom in astronomy would be vastly more important than, say, a new Type of subgiant star. Moreover, the decision as to whether a particular Class should be placed in a particular Family can lead to fruitful discussion among students, and maybe even scientists. For example, the question of whether a globular cluster is circumgalactic or not will lead students to realize that these objects are not found just surrounding the Galaxy, but also within the Galaxy, and so on.

Finally, as new discoveries are made in astronomy the Three Kingdom system may well be elaborated. For the most part the additions and revisions will be made at the Class and Type level, for example, as new classes of planets are discovered, or new classes of baryonic dark matter objects are revealed, or newly detected objects are analyzed such as the mysterious "G objects" at the center of our Galaxy that look like gas clouds but behave like stars (W. M. Keck Observatory 2018). It is not out of the question that a new Family could be added, though this seems unlikely given our definition of Family. At the Kingdom level, surprisingly, one can already glimpse a possible new entry: the universe itself may be one of a class of objects in what has been called the multiverse. Because this is a Kingdom that so far we have not seen, but only inferred from concepts like the anthropic principle, it has not been included in the Three Kingdom system at present. Only time will tell. More fundamentally we must always remember we are classifying baryonic objects composed of protons, electrons and neutrons, and that baryonic matter constitutes only 4.6% of the matter and energy content of the universe. Non-baryonic dark matter is 23%, and dark energy (believed to be responsible for the accelerating universe) is 72%. But we have no idea what that dark matter and dark energy may be. Classification of the objects that we know notwithstanding, plenty of work remains for future astronomers based on what we do not yet know.

Finally, it is essential to emphasize that because all classes and classification systems are socially constructed, the Three Kingdom system for astronomy is not the only system that could be proposed. But in the end, like the other classification systems, its raison d'être and its staying power are dependent on its accuracy, simplicity and utility, both in scientific and pedagogical terms. Such features are an asset for astronomical classes and classification systems in general.

31.5 Commentary 2020

The idea underlying this chapter took root in 1995 after my work on *The Biological Universe* was complete. It is nothing less than an attempt at a classification scheme for all classes of astronomical objects. Surprisingly, no such comprehensive system existed in astronomy, only bits and pieces for objects like stars and galaxies. How would one construct a comprehensive system, I asked myself? I did this mainly for my own curiosity and to try to order the vast and growing menagerie of astronomical objects in a way that made sense to me. After many false starts, I eventually settled on what I called the *Three Kingdom* (3K) system described in this chapter. I first broached the idea publicly in a very crude form as part of the "Dibner Lecture" at the Smithsonian in 2000 (available at https://library.si.edu/digital-library/book/extraterrestria00dick). I delivered it in much more substantial form as my lecture for the LeRoy E. Doggett Prize of the American Astronomical Society in Washington, D.C. in 2006. The system and the underlying idea of classification in astronomy snowballed, and this snowball effect resulted in two books, Dick (2013) and Dick (2018).

Classification in astronomy is only a small part of the classification of knowledge in general. As evidence of this, the present chapter was written at the request of the International Society for Knowledge Organization (ISKO, https://www.isko.org/index.php) for its ISKO Encyclopedia of Knowledge Organization. At this site the reader can see the enormity of the issues associated with classification and knowledge organization.

Notes

1. On the Three Domain versus Five Kingdom controversy in biology see especially Sapp (2009). On classification in physics and chemistry see Gordin (2004), Pickering (1984), and Gell-Mann (1994).
2. Davies (2007), especially Chap. 4. Isaac Asimov has made the same point in his popular books; for example, Asimov (1992, 263).
3. For more on these classification systems for stars see Dick (2013), Chap. 4. A recent popular account of the development of the Harvard system is Sobel (2016).
4. Wilson (2010, xi). In 2011 a group of biologists using a novel analysis estimated 8.7 million eukaryotic species exist, give or take a million. Eukaryotic species contain a nucleus, in contrast to prokaryotes (Strain 2011).
5. Taxonomy has also evolved, see Mayr (1982, 145), for stages in classification, and microtaxonomy vs. macrotaxonomy.

References

Asimov, Isaac. 1992. *Atom: Journey Across the Subatomic Cosmos*. New York: Penguin.

Buta, Ronald J., Harold G. Corwin, Jr., and Stephen C. Odewahn. 2007. *The de Vaucouleurs Atlas of Galaxies*. Cambridge: Cambridge University Press.

Davies, Paul. 2007. *Cosmic Jackpot: Why Our Universe is Just Right for Life*. Boston and New York: Houghton-Mifflin.

DeVorkin, David H. 1981. "Community and Spectral Classification in Astrophysics: The Acceptance of E. C. Pickering's System in 1910," *Isis*, 72, 29–49.

Dick, Steven J. 2013. *Discovery and Classification in Astronomy: Controversy and Consensus*. Cambridge: Cambridge University Press.

Dick, Steven J. 2018. *Classifying the Cosmos: How We Can Make Sense of the Celestial Landscape*. New York: Springer.

Feigelson, Eric. 2012. "Classification in Astronomy: Past and Present." In *Advances in Machine Learning and Data Mining for Astronomy*. London: CRC Press, Taylor & Francis Group, eds.: Michael J. Way, Jeffrey D. Scargle et al., 3–10.

Gell-Mann, Murray. 1994. *The Quark and the Jaguar: Adventures in the Simple and the Complex*. New York: W. H. Freeman.

Gordin, Michael D. 2004. *A Well-Ordered Thing: Dmitrii Mendeleev and the Shadow of the Periodic Table*. New York: Basic Books.

Gray, Richard O. and Christopher J. Corbally. 2009. *Stellar Spectral Classification*. Princeton and Oxford: Princeton University Press.

W. M. Keck Observatory. 2018, "More Mystery Objects Detected Near Milky Way's Supermassive Black Hole," *News Release*, June 6–7, 2018, https://phys.org/news/2018-06-mystery-milky-supermassive-black-hole.html

Mayr, Ernst. 1982. *The Growth of Biological Thought: Diversity, Evolution and Inheritance*. Cambridge, MA: Harvard University Press.

Mayr, Ernst. 1988. *Toward a New Philosophy of Biology*. Cambridge, MA: Harvard University Press.

Morgan, William Wilson 1937. "On the Spectral Classification of the Stars of Types A to K". *Astrophysical Journal* 85, no. 5: 380–97.

Morgan, William Wilson. 1988. "A Morphological Life," *Annual Reviews of Astronomy and Astrophysics*, 26, 1–9.

Morgan, William Wilson and P. C. Keenan. 1973. "Spectral Classification," *Annual Reviews of Astronomy and Astrophysics*, 11, 29–50.

Pickering, Andrew. 1984. *Constructing Quarks: A Sociological History of Particle Physics*. Edinburgh: Edinburgh University Press.

Sandage, Alan. 2005. "The Classification of Galaxies: Early History and Ongoing Developments," *Annual Reviews of Astronomy and Astrophysics*, 43, 581–624.

Sapp, Jan. 2009. *The New Foundations of Evolution*. Oxford: Oxford University Press.

Sobel, Dava. 2016. *The Glass Universe: How the Ladies of the Harvard Observatory Took the Measure of the Stars*. New York: Viking.

Strain, Daniel. 2011. "8.7 Million: A New Estimate for All the Complex Species on Earth." *Science* 333, no. 6046: 1083.

Wilson, Edward Osborne. 2010. "Foreword". In *Kingdoms and Domains: An Illustrated Guide to the Phyla on Earth* by Lynn Margulis and Michael Chapman. Amsterdam: Elsevier, lxi–lxii.

Chapter 32
The Discovery of Polar Motion and Its Importance

Abstract The study of polar motion, termed "variation of latitude" at the time of its discovery in the late nineteenth century by Karl Friedrich Küstner and Seth Chandler, is of historical interest for many reasons. From a scientific viewpoint, its discovery must be seen in the context of positional astronomy, geodesy and Earth rotation studies. From an institutional viewpoint, the founding of the International Latitude Service (ILS) in 1899 represents an early case of international cooperation in astronomy, preceding the International Astronomical Union by two decades. In addition to discussing these themes in this review, we analyze in some detail the Gaithersburg (USA) station of the ILS as a case study of the early problems and promise of polar motion studies. With milliarcsecond accuracies, polar motion studies are important today not only for astronomy and geophysics, but also for practical problems like spacecraft navigation and positioning.

32.1 Introduction: The Context of Polar Motion Studies

The discovery and subsequent refined measurement of polar motion over the last century must be seen in several contexts: astrometrically, in terms of the evolution of instrumental accuracies; conceptually, in terms of its place in astronomy, geodesy and geophysics, and Earth rotation studies; and internationally in terms of the evolution of international cooperation in science. Seen in these terms, the past century of polar motion research is a rich subject for historical study.

A broad view of astrometric accuracies (Fig. 23.2) shows where matters stood at the end of the nineteenth century when polar motion was unambiguously discovered—halfway between the landmark star catalogues of Friedrich Argelander in 1842 and Benjamin Boss's General Catalogue (GC) in 1937. Table 23.1 reveals three eras: the naked eye, telescopic, and interferometric/space eras, corresponding roughly to the minute of arc, second of arc, and milliarcsecond accuracy regimes. Polar motion studies encompass the last two eras, with accuracies ranging from a few tenths of an arcsecond at the beginning of the century to one milliarcsecond at

First published as "Polar Motion: A Historical Overview on the Occasion of the Centennial of the International Latitude Service," in *Polar Motion: Historical and Scientific Problems*, S. J. Dick, D. McCarthy and B. Luzum, eds. (San Francisco, 2000), 3–23

S. J. Dick, *Space, Time, and Aliens*, https://doi.org/10.1007/978-3-030-41614-0_32

its end; the field now has the promise of making further discoveries by exploiting the microarcsecond regime. Table 23.1 also shows that both visual and photographic methods were capable of reaching these accuracies of a few tenths of an arcsecond, and indeed one of the continuing debates through much of twentieth-century polar motion studies is the comparative value of visual versus photographic methods.

A glance at Table 23.1 also shows that stellar parallax, an effect of a few tenths of an arcsecond, was detected in 1838. An immediate historical question is why polar motion was not detected and announced earlier. Indeed, Verdun and Beutler (2000) discuss possible earlier determinations beginning in the 1840s. Of course, the detection of such small effects is related to methodology. Markowitz (1976), in perhaps the best succinct history of polar motion studies, noted possible reasons for the failure of an unambiguous detection and announcement of a variation in latitude prior to the late nineteenth century. He emphasized especially Küstner's use of the Talcott method with a zenith telescope, a method that required only micrometer readings, no circle readings with all their attendant problems. But this does not solve the problem, since Andrew Talcott had designed and used his zenith telescope as early as 1834.

Moreover, the astronomer and optics expert Frank Ross, whom we will see later in connection with the Gaithersburg latitude station, noted that the Airy reflex zenith tube "had been accumulating observations continuously for a period of more than 30 years before the discovery of the latitude variation, which remained a mine of hidden wealth until opened up by Dr. S. C. Chandler." Chandler, he noted, analyzed these observations, traced the variation of latitude effect, and "removed the stigma of inefficiency from an instrument which had been tried, patiently and tenaciously, and apparently found wanting." Ross also noted that although the zenith telescope, using the Horrebow-Talcott method was the most accurate way to determine variation of latitude, this variation had also been detected by meridian circles, visual and photographic almucantars, prime vertical transit instruments, as well as the Airy reflex zenith tube (Ross 1915, p. 9). In fact, in his seminal paper on variation of latitude, Chandler (1891b) discusses how perplexed astronomers at the Naval Observatory worried about anomalies in observations made with the prime vertical instrument from 1862 to 1867 to determine the aberration constant (Hall 1888). So I pose the first of several historical problems:

Historical Problem #1: Why did it take so long to discover polar motion, and how did astronomers explain discrepancies prior to its announcement?

Part of the goal of this meeting is to shed light on these and other historical problems.

A second context is the scientific effects on astronomy, geophysics and Earth rotation. In a general sense polar motion affects nearly everything having to do with absolute measurement in astronomy. A knowledge of polar motion is an essential part of today's International Terrestrial Reference Frame (ITRF), so the history of polar motion is essential to the study of the evolution of more precise reference frames. With regard to Earth rotation, polar motion of course affects the measurement of Universal Time. Although polar motion amounts to several hundredths of a second of time, it was not until the second half of the century that this quantity was incorporated into the daily dissemination of time. Polar motion could only be determined months after the fact, and for many years the Bureau International de l'Heure (BIH) applied polar motion corrections only in its annual analysis of time signals. The Royal Greenwich Observatory first applied polar motion corrections as a daily practice in 1947.

With the IAU's establishment of a Rapid Latitude Service in 1955, however, and its provision for the BIH to extrapolate Earth rotation corrections a year in advance, the timing community was poised to incorporate polar motion and seasonal variations in Earth rotation into the disseminated time in a coordinated and systematic way. At the meeting of the IAU in Dublin in 1955, William Markowitz, the incoming President of Commission 31 on Time and the Director of Time Service at the Naval Observatory, introduced four resolutions, including one instructing the BIH to compute for those observatories cooperating in international time service the longitude corrections due to polar motion (Oosterhoff 1957). The result was that, beginning January 1, 1956 Universal Time (UT), as observed by the visual and photographic instruments of the time, was designated UT0, while UTl was corrected for polar motion and UT2 was further corrected for seasonal variations in the Earth's rotation. In effect each step from UT0 to UT2 produced a more uniform time scale. Guinot (2000) provides more detail.

Finally, we should recall that it was geodesists, not astronomers, who began the International Latitude Service (ILS), and that today the measurement of both polar motion and the length-of-day aspect of Earth rotation provide important data for geophysics. Sir Harold Jeffreys (1952) provided an early overview, and Munk and MacDonald (1960) pioneered in synthesizing the geophysical aspects, both in terms of excitation causes and consequences. Their work has been extended by many researchers, synthesized in several excellent treatments, including Lambeck (1980), and will be the subject of more papers at this meeting (Dick et al. 2000). So polar motion is important in the history of geodesy and geophysics.

A third context is international cooperation, particularly as it related to the International Association of Geodesy (IAG) and its predecessors. Stimulated by the need for cooperation in the measurement of the central European arc, the inaugural conference of the Central European Arc Measurement (Miteleuropäische Gradmessung), the forerunner of the IAG, met in Berlin in 1862, and by the end of that year 15 European states were participants (Höpfner 1999; Levallois 1980; Mueller 1990). It became the "Internationale Erdmessung" in 1886, known in its French and English translations as the Association Internationale de Geodesie and the International Geodetic Association, and since 1932 officially the International Association of Geodesy. Thus when variation of latitude was verified, the international scientific community was in a good position to act, and it did with the IAG's founding of the ILS (Torge 1993).

At the founding of the International Astronomical Union in 1919, Standing Committee 19 on Latitude Variation was appointed with Hisashi Kimura as President (Blaauw 1994). Following the IAU's first General Assembly in 1922 the ILS work was put in charge of a joint Commission of the IAU and the International Geodetic and Geophysical Union (Lambert et al. 1931, p. 272). The IAU's Commission 19 on variation of latitude maintains its identity now as the Commission on Earth Rotation.

Table 32.1 Landmarks in polar motion studies

1765	Euler's prediction of polar motion with period of 305 days due to rotation of a rigid body about its center of gravity
1842–1844	Bessel (Königsberg) and Peters (Pulkovo) observations show possible variation of latitude
1862	Inaugural Conference of Central European Arc Measurement (Mitteleuropäische Gradmessung), forerunner of International Association of Geodesy, in Berlin
1873	Nyren determines statistically significant values for amplitude of polar motion
1884–1885	Chandler's Harvard observations Küstner's Berlin observations Küstner announces variation of latitude of .2 arcseconds/year
1891–1892	Chandler's discovery that polar motion results from two periodic components, a 428-day free circulation, and an annual forced motion
1891	Crucial test: simultaneous measurement at Waikiki and Berlin show curves in opposite phase
1899	International Latitude Service (ILS) founded with six stations
1901	Chandler announces discovery of a 436-day component of polar motion, whose amplitude is much smaller than the 428-day component
1917	B. Wanach finds evidence for secular polar motion of 0.003 arcseconds, based on ILS data
1919	International Astronomical Union founded, including Standing Committee 19 on Latitude Variation, with Kimura as first President Bureau. International de l'Heure (BIH)
1955	BIH begins independent observations for polar motion
1962	International Polar Motion Service (IPMS) started
1983	MERIT begins main campaign
1980s	SLR, LLR, VLBI observations for polar motion begin; order of magnitude improvement in accuracy
1988, Jan.	International Earth Rotation Service (IERS) replaces IPMS and Earth-rotation section of BIH
1994	International GPS Service for Geodynamics created by IUGG

Following recommendations at IAU Symposium 13 on "The Future of the ILS," held in Helsinki in 1960, in 1962 the International Polar Motion Service (IPMS) began using both the ILS and independent latitude stations (Melchior 1961). Both the IPMS and the Earth rotation section of the BIH have now been replaced by the International Earth Rotation Service (IERS), to which the International GPS Service (IGS) makes invaluable contributions. It is therefore appropriate that all of these organizations are sponsoring this colloquium.

32.2 Landmarks in Polar Motion Studies

So as not to lose sight of the big picture, I want to give an overview of the landmarks of polar motion studies (Table 32.1). Many of these landmarks will be elaborated in the course of the meeting. As is well known, Leonhard Euler (Euler 1765) was the first to discuss theoretically the possible variation of latitude due to the rotation of a rigid body about its center of gravity; he predicted the period of this variation should

Fig. 32.1 The almucantar, the instrument Chandler used to measure differences in the latitude of Cambridge, Massachusetts in 1885. (From Chandler 1887)

be about 10 months. Verdun and Beutler (2000) have shown how Bessel suspected an effect in 1843 based on observations of C. A. F. Peters, and furthermore that Magnus Nyren in 1873 determined statistically significant values for the amplitude of polar motion, and even a secular variation in polar motion, based on observations made by Peters, Johann Gylden and Nyren. In terms of a correct interpretation of the data, however, all were misled by the "Eulerian paradigm" prediction of a 10-month period. The Königsberg-Dorpat-Pulkovo tradition that these astronomers represent places in new perspective the work of the German astronomer Friedrich Küstner in Berlin and the American astronomer Seth Chandler.

There remain differences of opinion as to the relative role of Küstner, Chandler and the International Association of Geodesy in the discovery of variation of latitude; the papers in Dick et al. (2000) help to clarify some of these differences, while still demonstrating that the concept of discovery is not always a simple one. It was while trying to determine the constant of aberration via a series of observations beginning in 1884 that Küstner noticed his data could not be explained unless a variation of latitude was included, which he announced in 1888 in the amount of a few tenths of an arcsecond per year (Küstner 1888, 1890). Chandler in 1885 had noted differences in the latitude of Cambridge, Massachusetts using his almucantar (Fig. 32.1; Chandler 1885, 1887), but remarked in 1891 that an actual variation of

Table 32.2 International latitude service stations

Station	Nominal longitude	Nominal latitude	Period of observation	Instrument
Mizusawa	141°07′51″	(39°8′) 3.602″	1899–1984	VZT[a]
Carloforte	8°18′44″	8.941″	1899–1942	VZT
			1946–1984	
Tschardjui	63°29′	10.662″	1899–1909.5	VZT
			1909.5–1920	
Ukiah	123°12′35″	12.096″	1899–1984	VZT
Gaithersburg	−77°11′57″	13.202″	1899–1914	VZT
			1932–1984	
Cincinnati	−84°25′	19.364″	1899–1915	VZT
Kitab	66°52′51″	1.850″	1922–1984	VZT

Adapted from Henriksen (1977) (″ = arcseconds)
[a]*VZT* visual zenith tube

latitude "seemed at the time too bold an inference to place upon record" (Chandler 1891a). By November 1891, however, Chandler not only accepted variation of latitude based on new data, he also announced a period of 427 days (Chandler 1891b). This was 40% more than Euler's classic period, an anomaly explained by Newcomb (1891) as due to the elasticity of the Earth. Chandler followed up on his work with a whole series of papers in the *Astronomical Journal* in the late nineteenth century (Carter and Carter, 1995, 2000). While others had determined the amplitude of the variation of latitude, to Chandler goes the undisputed achievement of first determining the true period.

In 1899 the ILS was set up, with observations beginning in September, just 100 years ago. The 12th General Conference of the Internationale Erdmessung (already referred to by some as the International Geodetic Association) had been held in Stuttgart in 1898 and decided to finance the organization of four stations, all located at latitude 39°, 8 min: Carloforte (Italy), Gaithersburg, Maryland (USA), Mizusawa (Japan), and Ukiah (California, USA). Cincinnati was on the same parallel and offered to cooperate, as did the Russians with a station at Tschardjui. Some of these stations changed over the years. Gaithersburg was closed in 1914 for reasons of economy, but reopened in 1932 for 50 more years of observations. Cincinnati dropped out permanently in 1915. The Russian station was replaced in 1930 with a station some 3° to the west at Kitab, near Samarkand (Ehgamberdiev 2000; Lambert et al. 1931, 255). Carloforte did not observe from 1943 to 1946 due to World War II. Although Mizusawa and Ukiah observed continuously, the number of ILS stations varied from three during the period 1927–1935, to six from 1901 to 1906 (Table 32.2).

The primary source for the history of polar motion is found in the published volumes of the ILS data. These volumes delineate several eras in terms of ILS data: the German era under Albrecht et al. (1900–1922); the Japanese era under Kimura (1922–1935); the Italian era under Camera and Ceccini (1935–1962); and a second Japanese era from 1962 to 1987 as the Central Bureau of the International Polar

Table 32.3 Publication of polar motion results

Vol	Author	Title	Publication	Data for
	GERMAN			
1	Albrecht	Resultate des Internationalen Breitenclienstes	1903	1899–1902
2	Albrecht	Same	1906	1902–1904
3	Albrecht	Same	1909	1899–1905
4	Albrecht	Same	1911	1906–1908
5	Albrecht	Same	1916	1908–1912
6	Wanach /Mahnkopf	Ergebnisse des Internationalen Breitenclienstes	1932	1912–1922
	JAPANESE			
7	Kimura	Results of the International Latitude Service	1935	1922–1931
8	Kimura	Same	1940	1922–1935
	ITALIAN			
9	Carnera	Resultati del Servizio Internationale della Latituclini	1957	1935–1941
10	Nicolini	Same	1970	1941–1948
11	Fichera Ceccini	Results of the International Latitude Service	1973	1949–1962
	JAPANESE			
12	Yumi Yokoyama	Same	1978	1962–1967
	Central Bureau Of the IPMS	Annual Report of the IPMS	Annual	1962–1979
	Mizusawa			
	Central Bureau Of the IPMS Mizusawa	Monthly Notices of the IPMS	Monthly	1977–1987
	FRANCE and U. S. IERS Central Bureau, Paris Observatory Sub-Bureau, NEOS	Bulletins A, B, C, D	1988–present	

Motion Service (IPMS) (Table 32.3). A landmark was reached in 1955, when the Bureau International de l'Heure (BIH) in Paris began an independent set of polar motion observations. By 1962 this was joined by the International Polar Motion Service (IPMS), with some 50 stations. The Main Campaign of MERIT (Monitor Earth Rotation and Intercompare the Techniques of observation and analysis) began in 1983, and MERIT continued until January 1, 1988, when the International Earth Rotation Service (IERS) replaced IPMS and the Earth-rotation section of the BIH. The Central Bureau of the IERS is at Paris Observatory, and the sub-bureau in the United States (U. S. Naval Observatory, National Oceanic and Atmospheric Administration, and the National Aeronautics and Space Administration) comprise the National Earth Orientation Service.

This simple recitation shows the tremendous international cooperation required for polar motion studies, and masks many interesting stories and much science

Table 32.4 Polar motions

Frequency	Amplitude (arcseconds)	Detected	Discovery technique	Cause
Decadal (Markowitz)	Few hundreths	1960	VZT[a]	?
427 days (Chandler)	Few tenths	1890s	VZT transit	Free oscillation
365.25 days (annual)	Few tenths	1890s	VZT	Meteorological
Monthly	Few thousandths	1980s	Space Geodetic VZT/PZT	Meteorological/ Oceanss
<10 days	One thousandth	1990	Space Geodetic	Atmospheric
1 day/0.5 days	Few thousandths	1990s	Space Geodetic	Tidal forcing of oceans
Hourly	Few millionths	?	?	?

[a]*VZT* visual zenith tube. Discovery based on data using a variety of instruments

politics. The end result was more and more accurate measurements of polar motion, and an increase in the timeliness of dissemination and the accuracy of predictions.

One of the main goals of polar motion studies has been a determination of its periodicities. Table 32.4 is an overview of the periodic polar motions now known. A striking feature of the table is the remarkable increase in accuracies attained in the 1980s and 1990s by the new techniques of Very Long Baseline Interferometry (VLBI), Satellite Laser Ranging (SLR), and Lunar Laser Ranging (LLR). The question of excitation of the periodic motions remains open in many cases (McCarthy 2000).

What is not shown in the Table is the problem of secular polar motion, first suspected by Chandler already in 1892, reported by B. Wanach in 1917 in the amount of 0.003 arcseconds/year toward 55° West longitude based on ILS data, studied by Lambert, Kimura and others in the 1920s, and still a research problem today. Markowitz (1960a) provided both a history and an analysis of the data at the beginning of the Space Age, concluding that the mean pole had moved about .2 arcseconds from 1900 to 1959, and that it was likely a real motion. Lambeck (1980, p. 91) put the secular drift at the same magnitude as Wanach, but in a direction 65–75° West longitude, with an uncertainly of 20% in rate and 10° in direction, and questioned whether the motion was apparent or real. McCarthy and Luzum (1996) find a secular motion of 0.333 arcseconds per century in the direction of 75° West longitude.

32.3 The Gaithersburg Station as a Case Study

I want now to turn in more detail to the institutional response, particularly as seen in the Gaithersburg case of the ILS. The Gaithersburg, Maryland station of the ILS, about 20 miles north of Washington, D.C., became one of the first 6 latitude

Fig. 32.2 The Wanschaff Visual Zenith telescope (VZT) employed for observation of variation of latitude as stations of the International Latitude Service. (From Albrecht 1902)

observatories of the International Latitude Service by an agreement with the U.S. Coast and Geodetic Survey dated January 23, 1899. (The Naval Observatory, a logical place to locate such a station, has a latitude of 38° 55.3 min, and so was slightly out of the latitude band of 39° 8 min for the ILS stations.) Edwin Smith and Frank Schlesinger, in charge of the stations at Gaithersburg and Ukiah respectively, described the site, buildings and instrument in the U. S. Geodetic Survey report for 1900 (Smith and Schlesinger 1901). A map of the location and layout of Gaithersburg and the other ILS stations is also found in the first volume of ILS observations (Albrecht 1903).

The instrument employed for observations of variation of latitude at Gaithersburg and other ILS stations was the Wanschaff [Visual] Zenith Telescope (Fig. 32.2). This instrument, however, was known to have small errors due to personal equation, and at thc instigation of Frank Ross, at the IGA meeting in Cambridge, England in September, 1909 the Superintendent of the Survey submitted plans for the construction and operation of a photographic reflex zenith tube that would remove this personal equation. The idea was to compare the visual and photographic observations with each other and with the other stations, to discover any short-period or diurnal terms in latitude variation, and as a byproduct, to determine a value for the constant of aberration. The IGA (headquartered in Potsdam, Germany) voted 10,000 marks for the construction of the instrument, and an additional 2000 marks annually for plate measurement and reduction.

The Photographic Zenith Tube (PZT) was built by William Gaertner and Company in Chicago, and from June, 1911 to October 1914 some 6944 stars down to magnitude 8.5 were photographed on 450 nights, an average of some 15 per night (Ross 1915, p. 7). After a detailed analysis of these observations Ross concluded that "the accuracy of the individual latitudes obtained with the photographic instrument appears to be considerably greater than that obtained with the visual instrument. For the study of the minute quantities here dealt with this is a matter of considerable importance. That there is also greater freedom from systematic errors has been proved for some of the classes [of stars], and is probably true for all" (1915, p. 127).

Despite the apparent success of the PZT, it was not adopted at Gaithersburg, in fact, the Gaithersburg station was closed in 1915 for reasons of economy. Nor did the ILS adopt the photographic method at its latitude stations, possible because of the expense, and the desire to keep the observations on a uniform system with a single set of instruments. The PZT, however, did not go to waste. In 1915 the U.S. Naval Observatory purchased the Ross PZT when Ross became an employee of the Observatory, and began observations for latitude with it in October, 1915 (Markowitz 1960b). This instrument, known as PZT 1, had an interesting history; not only was it used continuously for latitude determination from 1915 until 1955, F. B. Littell and J. E. Willis (Fig. 32.3) also modified it for the determination of time beginning in 1934 (Littell and Willis 1929; Markowitz 1960b; Dick 2003). This was the first time a PZT was used for time determination, a task typically reserved for the transit instrument.

Surprisingly, the use of the PZT for time and latitude determination spread only gradually; in 1955 small transit circles were still the instrument of choice for time determination, and visual zenith telescopes still the instrument of choice for latitude observations. By 1960, nine or ten PZTs were in use around the world (Spencer Jones 1955, 1958). Another instrument for latitude determination was the Danjon astrolabe (Danjon 1958, 1960). An idea of the distribution of instruments is given by the fact that during the International Geophysical Year (1957–1958), 22 visual zenith telescopes, 10 PZTs, and 16 Danjon astrolabes were in operation at over 30 latitude stations (Munk and MacDonald 1960, p. 61). This gives rise to.

Historical Problem # 2: What were the factors over the last century influencing the choice of instrumentation for time and latitude determination?

One would think that accuracy would be the chief factor, but especially given instruments of approximately equal accuracy, it is no surprise that funding, nationalism, and personal preference were important factors. It is interesting, however, that when the accuracy could be increased by an order of magnitude or more in the 1980s, the transition to the new techniques was rapid and almost universal. The PZT at USNO was superseded by VLBI for time determination in 1984 (Archinal 1993).

As reported in the official volumes of the ILS, the Gaithersburg station was reopened in April, 1932, still under the auspices of the U.S. Coast and Geodetic

Fig. 32.3 J. E. Willis with the Photographic Zenith Tube (PZT) originally used at Gaithersburg, and shown here about 1932 at the U. S. Naval Observatory in Washington with modifications for determining time in addition to latitude. USNO archives

Survey. Visual latitude observations continued at Gaithersburg until 1982, and in 1990 the Gaithersburg Latitude Observatory was named a National Historic Landmark (Butowsky 1989). The story of some of the other ILS stations is told in Dick et al. (2000). Collectively they represent the longest-running series of focused observations in the history of astronomy.

32.4 Summary: The Importance of Polar Motion

To non-scientists the rigorous study of polar motion over a circular area with a diameter of only about 10 m, equivalent to the length of two tennis courts, may seem much ado about little. Such was the advancement of astronomy, however, that already at the end of the nineteenth century its importance was undisputed. As Chandler (1893) pointed out, all determinations of celestial coordinates and of astronomical constants had been made on the assumption of an invariable zenith. Not only would variation of latitude need to be taken into account in the future, but also, he noted,

> ...nearly everything that concerns absolute astronomical measurement must have been to greater or less extent vitiated – equator-point, equinoctial point, obliquity of the ecliptic, aberration coefficient, absolute measurements of parallax, absolute systems of right ascensions and declinations, and even, possibly by sensible amounts, the refraction, nutation and precession constants.

We have already mentioned how polar motion was eventually incorporated into Earth rotation studies, and into daily time dissemination. Of course, polar motion results also had to be incorporated into meridian circle catalogues, and Rafferty (1982) and others have shown how not only polar motion, but different sources of polar motion data, could have a significant effect on individual observations as well as the overall catalogues of star positions. Today, the IERS is responsible for the maintenance of the International Terrestrial and Celestial Reference Frames, and in the era of milliarcsecond accuracies polar motion is important not only to astronomy and geophysics but also for practical problems like spacecraft navigation, unforeseen a century ago.

Finally, polar motion is important to geophysics in a variety of ways that will be elaborated (Dick et al. 2000). The simultaneous needs of astronomers, the Earth rotation community, and geophysicists gives rise to a final historical problem I wish to pose:

> *Historical Problem # 3: What have been the factors driving the increase in polar motion accuracy? More specifically, what has been the interface between polar motion data providers, and polar motion data users?*

As we celebrate the centennial of the International Latitude Service and the first internationally coordinated observations of polar motion, I cannot help but think how amazed Euler, Küstner, Chandler, and the other pioneers in the field would be at the modern work on polar motion, and how it has become integrated into the broader programs of astronomy and geophysics. No matter how much progress has been made a century from now, surely the 1980s and 1990s will stand as a turning point in polar motion studies. I hope this IAU Colloquium and its Proceedings (Dick et al. 2000) will serve to document the progress in polar motion, and stand as a pivot point between the past and the future.

32.5 Commentary 2020

This paper was given as the opening keynote lecture at a meeting sponsored by the International Astronomical Union in Cagliari, Sardinia in September, 1999, for which I served as Chair of the Scientific Organizing Committee (Fig. 32.4). The occasion was the centennial of the International Latitude Service. Cagliari was particularly appropriate as the venue because it was one of the original ILS stations, as described above. Much more information on polar motion can be found in the Proceedings of the meeting, published in Dick et al. (2000). Further context, including the role of the Photographic Zenith Tube in time determination, photographs and schematics of PZT 1 and 2, and the context for time concepts such as UT0, UT1, UT2 and today's Universal Time Coordinated (UTC) is found in Chap. 11 of Dick (2003).

Although the concept of latitude has not achieved the fame of its sister concept longitude as exemplified in Dava Sobel's bestseller *Longitude* (Sobel 1995), this

Fig. 32.4 Poster for the meeting on polar motion, held in Cagliari, Sardinia on the occasion of the centennial of the International Latitude Service

chapter shows that latitude has its own interesting story. That story has been told for a more popular audience in Bill and Merri Sue Carter's excellent book *Latitude* (Carter and Carter 2002).

This chapter could just as easily have been included in Part IV on "Measuring the Universe," since it shows that in order to measure the universe accurately we need to know in minute detail how the Earth itself is moving. But it fits equally well in this section on "Discovery" because, as emphasized in Sect. 30.2, it demonstrates how discovery is not a simple concept, either in terms of the discoverers or in trying to pinpoint a date of discovery. It demonstrates that discovery is an extended process not only for the discovery of discrete classes of objects, as emphasized in Chaps. 30 and 31, but also for the discovery of astronomical phenomena such as polar motion.

The Gaithersburg station today is a National Historic Landmark (Butowsky 1989); the nomination form and much historic information can be found at https://npgallery.nps.gov/NRHP/GetAsset/NHLS/85001578_text, and visitors are welcome to the original site. Some of the other sites are also preserved.

Acknowledgments The author wishes to thank Dennis McCarthy for discussions and comments on this paper.

References

Albrecht, T. 1902. *Anleitung zum Gebrauche des Zenitteleskops auf den Internationalen Breitendienstes* (Berlin)

Albrecht, T. 1903. *Resultate des Internationalen Breitendienstes* (Berlin).

Archinal, B. 1993. "Summary of Photographic Zenith Tube Observations of the U.S. Naval Observatory," manuscript, U.S. Naval Observatory Library

Blaauw, A. 1994. *History of the IAU* (Kluwer, Dordrecht), pp. 7–14.

Butowsky, H.1989. "Gaithersburg Latitude Observatory," *Astronomy and Astrophysics National Historic Landmark Theme Study* (National Park Service, Washington, D.C.), 187–208.

Carter, M. S. and W. E. Carter. 2000. "Seth Carlo Chandler, Jr.: The Discovery of Variation of Latitude," in Dick, McCarthy, and Luzum (2000), pp. 109–119.

Carter, W. E. and M. S. Carter. 1995. "Seth Carlo Chandler, Jr.," *Biographical Memoirs of the National Academy of Sciences,* 66 (National Academy Press, Washington, D.C.)

Carter, W. E. and M. S. Carter. 2002. *Latitude: How American Astronomers Solved the Mystery of Variation*. Naval Institute Press.

Chandler, S. C. 1885. "On the Latitude of Harvard College Observatory," *Astronomische Nachrichten,* 112, no. 2672, 113–120.

Chandler, S. C. 1887. "The Almucantar: An Investigation Made at the Observatory in 1884-1885," *Annals of the Astronomical Observatory of Harvard College*, 17, 1–222.

Chandler, S. C. 1891a. "On the Variation of Latitude," *Astronomical Journal* [hereafter AJ], 11, 59–60.

Chandler, S. C. 1891b. "On the Variation of Latitude," *AJ*, 11, 65–70.

Chandler, S. C., 1893. "On the Influence of Latitude-Variations Upon Astronomical Constants and Measurements," *AJ*, 12, 153–155.

Danjon, A. 1958. "The Contribution of the Impersonal Astrolabe to Fundamental Astronomy," *Monthly Notices of the RAS*, 118, 411–431.

Danjon, A. 1960. "The Impersonal Astrolabe," in *Telescopes,* ed. Gerard P. Kuiper and Barbara M. Middlehurst (University of Chicago Press, Chicago), 115–137, vol. 1 of *Stars and Stellar Systems*, G. Kuiper, General Editor.

Dick, S. J. 2003. *Sky and Ocean Joined: The U. S. Naval Observatory, 1830-2000.* Cambridge University Press, Cambridge

Dick, S. J, D. McCarthy and B. Luzum. 2000. *Polar Motion: Historical and Scientific Problems.* Astronomical Society of the Pacific, San Francisco.

Ehgamberdiev, S., 2000. "Kitab as One of the Five Stations of the ILS: History and Present," in Dick, McCarthy and Luzum, 2000, pp. 163–168.

Euler, L. 1765. *Theoria motus corporum solidorum seu rigidorum* (Greifswald).

Guinot, B. 2000. "History of the Bureau International de L' Heure," in Dick, McCarthy, and Luzum 2000, pp. 175–184.

Hall, A. 1888. "The Constant of Aberration," *AJ,* 8, 1–5 and 9–13.

Henriksen, S. W. 1977. *Systems for the Determination of Polar Motion*, NOAA Technical Report NOS 72 NGS 7 (U.S. Dept. of Commerce, Washington, D.C.)

Höpfner, J. 1999, "On the Contribution of the Geodetic Institute Potsdam to the International Latitude Service," Geoforschungzentrum Potsdam, Scientific Technical Report STR99/08; abridged version in Dick, McCarthy and Luzum, 2000, pp. 139–146.

Jeffreys, H. 1952. *The Earth* (Cambridge University Press, Cambridge).

Küstner, F. 1888. *Neue Methode zur Bestimmung der Aberrations-Constante nebst Untersuchungen uber die Veranderlichkeit der Polhhe* (Berlin).

Küstner, F. 1890. "Ueber Polhhen-Aenderungen beobachtet 1884 bis 1885 zu Berlin und Pulkowa," *Astronomische Nachrichten*, 125, no. 2993, 272–278.

Lambeck, K. 1980. *The Earth's Variable Rotation: Geophysical Causes and Consequences* (Cambridge University Press, Cambridge).

Lambert, W. D., F. Schlesinger and E. W. Brown. 1931. "The Variation of Latitude," *Bulletin of the National Research Council*, 78 (Washington, D. C.), 245–277.
Levallois, J. J. 1980. "The History of the International Association of Geodesy," *Bulletin Geodesique*, 54, 249–313.
Littell, F. B. and J. E. Willis. 1929. "A New Method of Determining Time," *AJ*, 40, 7–9.
Markowitz, W. 1960a. "Latitude and Longitude, and the Secular Motion of the Pole," in *Methods and Techniques in Geophysics*, ed. S. K. Runcorn (Interscience, London), 325–361.
Markowitz, W. 1960b. "The Photographic Zenith Tube and the Dual-Rate Moon-Position Camera," in *Telescopes*, ed. Gerard P. Kuiper and Barbara M. Middlehurst (University of Chicago Press, Chicago), 88–114, vol. 1 of *Stars and Stellar Systems*, G. Kuiper, General Editor.
Markowitz, W. 1976. "Polar Motion: History and Recent Results," *Sky and Telescope*, 52 (August), 99–108.
McCarthy, D. D. 2000. "Polar Motion – An Overview," in Dick, McCarthy and Luzum, 2000, pp. 223–236.
McCarthy, D. D. and B. Luzum. 1996. "Path of the Mean Rotational Pole from 1899 to 1994," *Geophysical Journal Int.*, 125, 623–629.
Melchior, P., 1961. *The Future of the International Latitude Service Bulletin Geodesique* no. 59.
Mueller, I., 1990. "125 Years of International Cooperation in Geodesy", IAG Symposium 102 (Springer Verlag, New York), 421–432.
Munk, W. H. and G. J. F. MacDonald. 1960. *The Rotation of the Earth: A Geophysical Discussion* (Cambridge: Cambridge University Press).
Newcomb, S. 1891. "On the Periodic Variation of Latitude, and the Observations with the Washington Prime-Vertical Transit," *AJ*, 11, 81–82.
Oosterhoff, T. ed. 1957. *Transactions of the IAU*, vol. 9 (Cambridge University Press, Cambridge), 452 and 456–459.
Rafferty, T. 1982. "Effect of different Sources of Variation of Latitude Data on Meridian Circle Catalogues," *Astronomy and Astrophysics Supplement Series*, 50, 27–47.
Ross, F. 1915. *Latitude Observations with Photographic Zenith Tube at Gaithersburg, MD* (U. S. Coast and Geodetic Survey Special Publication No. 27, Washington).
Smith, E. and F. Schlesinger. 1901. *Report of the Superintendent of the U. S. Coast and Geodetic Survey showing the progress of the Work from July 1, 1899 to June 30, 1900* (Washington, 1900).
Sobel, Dava. 1995. *Longitude: The True Story of a Lone Genius Who Solved the Greatest Scientific Problem of His Time.* Walker and Company, New York.
Spencer Jones, H. 1955, *Transactions of the IAU* 1955 (Dublin), 446.
Spencer Jones, H. 1958, *Transactions of the IA U* 1958 (Moscow), 488.
Torge, W. 1993. "The International Association of Geodesy," https://iag.dgfi.tum.de/fileadmin/IAG-docs/Torge_1996_IAG_more_than_130_years.pdf
Verdun, A. and G. Beutler. 2000. "Early Observational Evidence of Polar Motion," in Dick, McCarthy and Luzum, 2000, pp. 67–81.

Chapter 33
Observation and Interpretation of the Leonid Meteors Over the Last Millennium

Abstract With a possible "storm" of Leonid meteors due in November 1998 or 1999, interest in the Leonids is once again at a peak. The history of the Leonids is of particular importance, not only because they are closely associated with the origins of meteor science, but also because historical observations extending back a millennium are a substantial aid in increasing our knowledge of the Leonid meteor stream. Leonid history is thus a prime example of applied historical astronomy. In this review paper, we recount the origins of meteor science with the Leonids, the discovery of the historical observations and their scientific and cultural interpretations, and the application of this information to characterize the meteor stream and to predict the strength of the 1998–1999 event. These predictions are now of more than passing interest, as meteor storms pose a potential threat to spacecraft.

33.1 Introduction

Of the classes of Solar System objects known today, meteors were among the last to be recognized as astronomical in origin. The wandering planets had been known since antiquity; comets were recognized as astronomical rather than meteorological after Tycho Brahe and others placed the great comet of 1577 beyond the Moon; the abundance of circumplanetary objects became known with Galileo's discovery of the Jovian satellites in 1610; and Giuseppe Piazzi discovered the first asteroid on the first day of the nineteenth century, January 1, 1801. If one considers the recently detected Kuiper belt objects and the supposed Oort cloud objects to constitute new classes of Solar System bodies, like the meteors discussed here they have a likely cometary connection since they are believed to be the sources of short-period and long-period comets rather than cometary debris. Oort cloud and Kuiper belt objects, however, remain at a safe distance until they are perturbed into the inner Solar System, giving us not only cometary phenomena, but also eventually the meteor phenomena described here.

The phenomenon of "shooting stars" has been widely observed throughout history; catalogues of meteors record observations dating back at least to the seventh

First published in *Journal of Astronomical History and Heritage*, 1 (1998), 1–20.

S. J. Dick, *Space, Time, and Aliens*, https://doi.org/10.1007/978-3-030-41614-0_33

century BC, and they were, of course, observed long before that. The Roman poet Virgil, in Book I of the *Georgics,* 11. 365–367 (30 BC), wrote

> Oft you shall see the stars, when wind is near
> Shoot headlong from the sky and through the night
> Leave in their wake long whitening seas of flame

Some have even claimed that meteor storms and larger cometary debris have had a strong impact on historical events (Bailey 1996; Bailey et al. 1989; Clube 1996; Clube and Napier 1990). Although the suspicion that meteors were of cosmic origin dates at least to Edmond Halley in the eighteenth century (Hughes 1982), it was not until 1863 that they were definitely proven to be astronomical. This proof was based largely on observations of the Leonid meteors, so-called because their "radiant point" was in the constellation Leo. The celestial origin of the Leonids, the determination of their periodic nature, the recognition that they resulted from an orbiting stream of objects, and the identification of this stream with a parent comet, are all landmark events that take on added significance because they represent the origin of the relatively recent science of meteor studies. Although the "August meteors" (now known as the Perseids) also played a concurrent role (Littman 1996), they were not so important as the "November meteors" (later known as the Leonids), which periodically tended to storm, and thus demanded an immediate explanation.

Since the recognition of their celestial origin in the nineteenth century, records of Leonid observations have been discovered over the last millennium, dating back at least to AD 902. Historical records of meteors in general are of more than passing interest; indeed, they have proved essential to meteor astronomy by making possible conclusions about the orbits of the meteor streams and their parent bodies. They were of critical importance to the birth of meteor science in the nineteenth century, and they remain no less important today.

33.2 The Leonids in the Last Two Centuries

It was the Leonid storm of November 11–12, 1799, observed by the German naturalist Alexander von Humboldt among others, that first established the simultaneous geographical extent of the meteor phenomenon. From his location while on travel in Cumana, east of Caracas, van Humboldt wrote that towards the morning of November 12, after half past two, "... thousands of halides and falling stars succeeded each other during four hours. Their direction was very regularly from north to south ... All these meteors left luminous traces from five to ten degrees in length" (von Humboldt and Bonpland 1818:331). Curious as to how widespread the phenomenon was, van Humboldt gathered reports from South America and elsewhere. The result was that meteors were reported across 90° of longitude from South America to Germany, and across 60° of latitude from South America to Greenland. Von Humboldt also heard reports of a similar event in South America in 1766. Significant meteor showers are possible in years preceding and following meteor storms; Steel (1998) has argued that a noteworthy Leonid shower in England on the night of November 12–13, 1797, may have inspired Samuel Taylor Coleridge's *The Rime of the Ancient Mariner,* one of the greatest poems of the English language.

The geographic extent of the 1799 storm was a good clue that the phenomenon might be celestial, but the origins of meteor studies awaited the great storm of 1833.

33.2.1 The Leonids and the Origins of Meteor Studies

Although a storm of meteors occurred in 1832 in Eastern Europe and the Middle East, only the 1833 storm inspired astronomers to action. The latter meteor storm peaked in the Eastern part of North America (Fig. 33.1), and it is no coincidence that this was the birthplace of modem meteor studies, though there were antecedents (Hughes 1982). Denison Olmsted, Professor of Mathematics and Natural Philosophy at Yale University in New Haven, Connecticut, was the crucial figure in this birth. His connection to meteor studies has been detailed in Hoffleit (1992:24–32) in the context of his work at Yale. Olmsted (1834:363) captured the uniqueness of the 1833 November 12/13 event when he wrote.

> Probably no celestial phenomenon has ever occurred in this country, since its first settlement, which was viewed with so much admiration and delight by one class of spectators, or with so much astonishment and fear by another class. For some time after the occurrence, the "Meteoric Phenomenon" was the principal topic of conversation in every circle, and the descriptions that were published by different observers were rapidly circulated by the newspapers, through all parts of the United States.

His interest ignited by the phenomenon, Olmsted collected and published in *The American Journal of Science and Arts* twelve descriptions of the event as seen from Massachusetts to Georgia. From these and other sources he collated data relating to weather, time and duration, number, variety, sound, and apparent origin. After considering accounts of other meteor showers, Olmsted concluded that the meteors of 1833 originated beyond the Earth's atmosphere, became luminous upon entering the atmosphere at high velocities up to four miles per second, and "appeared to proceed from a fixed point in the heavens ... Those who marked its position among the fixed stars, observed it to be in the constellation Leo, in which it appeared stationary, accompanying that constellation in its diurnal progress" (Olmsted 1834:394). With that statement, confirmed by the independent accounts of other observers during the storm, Leo was established as the "radiant" of these meteors. From this time, increasingly the November meteors were referred to as the Leonid meteors. Olmsted (1834:172) also conjectured that "the meteors of Nov. 13th consisted of portions of the extreme parts of a nebulous body, which revolves around the sun in an orbit interior to that of the earth, but little inclined to the plane of the ecliptic, having its aphelion near to the earth's path, and having a periodic time of 182 days, nearly."

Though Olmsted even identified the nebulous body as possibly a comet, he was wrong in believing that its orbit was between the Earth and the Sun. Thus, Olmsted

Fig. 33.1 This view of the 1833 Leonid storm, probably the most widely seen depiction of a meteor event, first appeared in April, 1888. It was produced by the Swiss artist Karl Jauslin (1842–1904) and engraved by Adolf Vollmy (1864–1914). (See Hughes 1995 for the history of the illustration.)

is generally credited with establishing the Leonid radiant, and beginning the study of meteors as a science. Alexander C. Twining, a civil engineer at West Point, New York who published his ideas in the same journal in 1834, is often also given credit for determining the radiant point, and the same claim is even made for von Humboldt (Hughes 1982). One may conclude that there is nothing like an impressive natural phenomenon to stimulate scientific thinking, at least if cultural conditions are conducive to such thinking. And in the United States in the 1830s, science was budding into a much more substantial enterprise, as exemplified by the *American Journal of Science and Arts* itself and a variety of other activities (Dupree 1957; Struik 1962).

The subsequent history of meteor showers and their eventual connection with comets has been recounted in Yeomans (1991:188–201), but we may here summarize the highlights based upon the original sources. Despite Olmsted's conclusion of a cosmic origin for meteors, even that much was not as yet certain in the first half of the nineteenth century. So great an authority as Brussels Observatory Director, Lambert-AdolpheJacques Quetelet, equivocated in the chapter on meteors in his influential *Sur la Physique du Globe* (Quetelet 1861), causing Hubert A. Newton (1863) to insist that Quetelet's own chapter gave a very strong argument that star-showers, and probably sporadic meteors as well, "... are caused by the entrance into our atmosphere of bodies revolving about the sun."

The clinching argument for the cosmic origin of meteors came when Newton determined that the cycle repeated in intervals of sidereal years, not tropical years. Hoffleit (1992:47–56) has also placed Newton's work in the context of nineteenth-century astronomy at Yale University, where he was a Professor of Mathematics. If meteors were due to terrestrial phenomena such as magnetism, heat, or electricity, he reasoned, meteor events should repeat in intervals of the tropical year. But Newton (1863) cited historical dates of known meteor showers to "... show quite clearly that the true period is not widely different from the sidereal year." Moreover, based upon his historical data, Newton calculated the interval between Leonid showers to be 33.25 years, and predicted that the meteor shower would return in 1866. Finally, Newton speculated that the meteor shower was caused by small bodies in elliptical orbit around the Sun. He determined five possible periods for the orbit of these bodies, including 33.25 years, but did not calculate a definitive orbit.

Newton's prediction proved true, but the peak of the 1866 storm occurred in Europe rather than in the United States. And sure enough, this time astronomers in four European countries were inspired to solve the riddle of the orbit of the meteor stream and its parent body. The largely independent and almost simultaneous work in 1866–1867 of John Couch Adams in England, Giovanni Schiaparelli in Italy, U. J. J. Le Verrier in France, and Theodor von Oppolzer and C F W. Peters in Germany, are often cited together in this respect. But in his paper "On the Orbit of the November Meteors," Adams (1867), famous for his earlier work leading to the discovery of Neptune, provides a contemporary account of the chronology and a hint of the international rivalry that must have taken place during a few crucial months in 1866–1867.

Adams tells us that Schiaparelli, the Director of the Milan Observatory, showed in four letters to Angelo Secchi (Schiaparelli 1866–1867) that the orbits of meteor streams around the Sun are very elongated, as are those of the comets, and that "... both these classes of bodies originally come into our system from very distant regions of space." More specifically, in Schiaparelli's last letter, dated December 31, 1866, he remarked on the very close agreement in the orbital elements of the August meteors (now known as the Perseids) and Comet II 1862. To the Italian astronomer thus goes the credit of showing that the comet now known as 109P/Swift-Tuttle is the source of the Perseids. Schiaparelli also attempted to find a comet with elements similar to those of the November meteors, but failed because the inaccurate radiant point he used yielded erroneous orbital elements for the meteor stream.

According to Adams, on January 21, 1867, Le Verrier (1867) communicated to the French Academy of Sciences a theory similar to Schiaparelli's, but with more accurate elements, including a period of 33.25 years. It was left to Peters (1867) of Altona, Germany, to notice 1 week later that Le Verrier's elements agreed closely with a certain Comet I 1866, discovered by Tempel on December 19, 1865, and which passed perihelion in early January 1866. The elements of what is today known as Comet 55P/Tempel-Tuttle had been determined by Theodor von Oppolzer (1867), who calculated a period of 33.18 years for the comet. A few days after Peters's announcement, Schiaparelli (1867) independently noticed the same agreement between Le Verrier's elements of the meteor stream and the comet. Thus, in the space of only 1 month, the connection that we accept today between comets and meteors was established beyond doubt.

Adams's own paper appeared in the April 12, 1867 issue of *Monthly Notices of the Royal Astronomical Society.* Referring to the papers of 1863 and 1864, where Newton identified 13 displays of the Leonids, Adams recalled Newton's estimate of 33.25 years for the recurrence of the displays. He also recalled Newton's conclusion that the November meteors "... belong to a system of small bodies describing an elliptic orbit about the Sun, and extending in the form of a stream along an arc of that orbit which is of such a length that the whole stream occupies about one-tenth or one-fifteenth of the periodic time in passing any particular point" (Adams 1867:248). In 1 year, Newton had concluded, this group must revolve about the Sun in 2 ± 1/33.25 revolutions, or 1 ± 33.25 revolutions, or 1/33.25 revolutions. In other words, the period corresponded to either 180 days, 185.4 days, 354.6 days, 376.6 days, or 33.25 years. Adams's own contribution was to show that the actual period was the last of these, 33.25 years. This he did by first assuming that period, and showing that during this time the longitude of the node would be increased by 20 arcseconds by the perturbation of Jupiter, seven more arc seconds by Saturn, and one arcsecond by Uranus, for a total of 28 arcseconds compared to the 29 arcseconds actually observed. This remarkable accordance between the results of theory and observation allowed Adams to then determine independently elements very similar to those of Le Verrier.

We thus see during the nineteenth century the progression in understanding of meteors from establishing the radiant and a probable celestial origin in 1834, to a definite celestial origin and an accurate interval between showers with Newton (1863), a definite period for the meteor stream (Adams 1867), and finally identification of meteors with comets as parent bodies, first by Schiaparelli (1866–1867), who identified the August Perseids with the comet now known as 109P/Swift-Tuttle, then by Peters (1867), who identified the November Leonids with the comet now known as 55P/Tempel-Tuttle. Only a few months later, the Irish astronomer, G. Johnstone Stoney (1867), wrote a paper on the connection between meteors and comets, the beginning of a long series of such studies that has continued to the present day. Further significant showers in North America in 1867 and 1868, which Mason (1995) classifies as storms, inspired E. L. Trouvelot to create a memorable artistic rendering.

But there was still more to learn about the Leonids, in particular about the distribution of the material in the meteor stream. Just when astronomers thought they had a good understanding of the November meteors, the predicted storm of 1899 failed to appear. In 1925 the astronomer Charles Olivier called this "… the worst blow ever suffered by astronomy in the eyes of the public…" (Olivier 1925:38). This is a public relations problem which astronomers can still sympathize with today.

33.2.2 The Twentieth Century

Leonid studies in the twentieth century were marked by good showers in 1930–1932, the strongest storm witnessed in modern times in 1966, and steady progress in understanding meteors and meteor streams with the help of new techniques of observation.

The 1930–1932 showers were best seen from North America, and exhibited rates of about 240 per hour (Kronk 1988; Mason 1995). Following the disappointments of 1899 and 1930–1932, Lovell (1954:338) concluded that "It now seems certain that the main part of the Leonid orbit has been removed from the Earth's orbit by successive perturbations, and the recurrence of the tremendous meteoric storms of the Leonids in the future seems unlikely."

Many were surprised, then, when a storm was visible in Europe in 1965, and even more so when an estimated 100,000 meteors per hour were observed at the peak of the 1966 Leonid storm in the southwestern United States (Kronk 1988; Mason 1995; Milon 1967). The descriptions were less hysterical than in the previous centuries, but the event was clearly awe-inspiring nonetheless. "The sky began to rain shooting stars," one observer wrote. "By 11:30 there were several hundred a minute. A quarter-hour later, the meteors were so intense that we were guessing how many could be seen in a one-second sweep of the observer's head. The fantastic rate of 40 per second was reached at 11:54, difficult to gauge but the consensus of our observing group" (Great Leonid meteor shower…, 1967:5).

Visual, photographic and radar techniques helped make the 1965–1966 storms the most studied on record. The techniques available by that time are systematically summarized in Millman and McKinley (1963). The development of radar techniques, which detect the ionization trail of meteors, was intimately connected with the development of radar in World War II (Lovell 1954; Butrica 1996). Pioneered by the group of radio astronomers led by Bernard Lovell at Jodrell Bank, the *radio-echo technique* not only gave a scientific record of observations, but for the first time allowed observations in daylight. The technique also refined Newton's thesis that meteors were of cosmic origin. By showing there was no significant hyperbolic

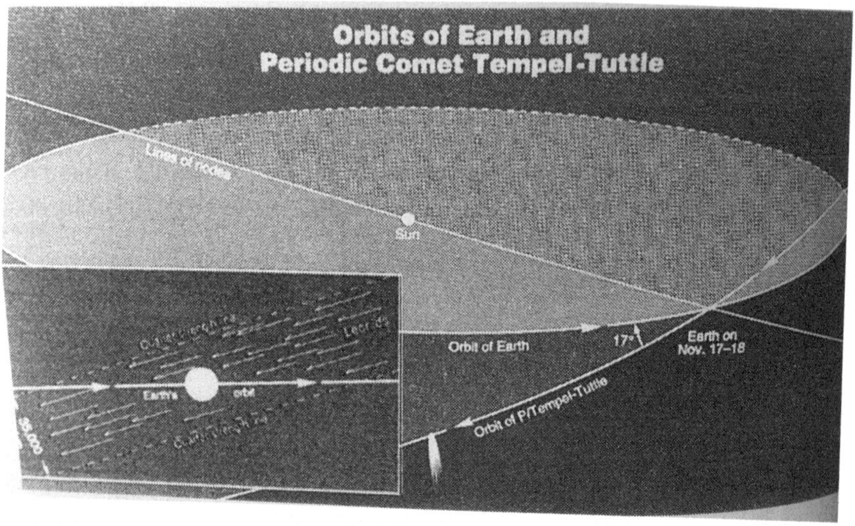

Fig. 33.2 Orbits of Earth and comet 55P/Tempel-Tuttle. Because the orbit of the comet is retrograde compared to Earth's orbit, Earth moves head-on into the stream. The particles are scattered all around the orbit of the comet, but are particularly dense within a few astronomical units of the comet. The width of this dense swarm is only about 35,000 km. (Sky Publishing Corp)

velocity component, meteors were determined to be orbiting the Sun rather than of interstellar origin.

Photographic techniques were systematically used by the Harvard University projects led by Fred Whipple during the 1930s and 1940s (Lovell 1954), and became widely successful with the development of the Baker Super Schmidt cameras. The primary purpose of these cameras, developed by James G Baker in the 1950s, was to give excellent image-definition over a wide field up to 55 degrees. The context of Whipple's work is described in Doel (1996).

Our modern view of the Leonids, shaped over the last two centuries, thus envisions them as a stream of particles in the orbit of Comet 55P Tempel-Tuttle (Fig. 33.2), a retrograde orbit (compared to Earth's) that has its aphelion just beyond the orbit of Uranus. The comet has spewed out these particles along its orbit in a complex distribution pattern, the outlines of which were recognized already at the end of the nineteenth century by Stoney and Downing (1899). Some, now widely scattered all along the orbit of the comet, are called the clino-Leonids and produce the weak annual Leonid showers. But a dense swarm of particles, known as the ortho-Leonids, remains within a few astronomical units of the comet. The apparent weakness of the clino-Leonids and the density of the ortho-Leonids are anomalous for such an old meteor stream, and Williams (1997) has proposed that this may be due to perturbations by Uranus. In any case, it is the length of the dense swarm that gives us the chance for greatly increased activity in November, over several sequential years. The width of the narrow swarm, however, is perhaps only 35,000 km, so that the activity will last only a few hours—and will be visible only for that part of the Earth that happens to be turned in the right direction, speeding head-on into the meteor stream. We observe the meteors as they ignite in the atmosphere at an average height of 70–110 km.

A great deal of work has subsequently gone into the study of meteors, and the volumes of Kresak and Millman (1968) and Stohl and Williams (1993) are representative of the research that has refined our knowledge of the nature of meteor streams and their relationship to comets.

33.2.3 Assessment: Why the Nineteenth Century?

We may well ask why meteor science began only in the nineteenth century, and why meteors took so long to be recognized as astronomical. For most of the period prior to the seventeenth century, the answer is that acceptance of the celestial-terrestrial dichotomy of Aristotelian cosmology prevented the identification of meteors as cosmic in origin. Aristotle believed the heavens were unchangeable, and that any observed change had to be Earthly or meteorological. Thus, comets were long held to be atmospheric phenomena, and there was a crisis when Tycho Brahe proved the comet of 1577 to be located beyond Earth.

On the assumption that a major storm was necessary to trigger a new explanation of meteors rather than the few sporadic meteors occasionally observed, one may wonder about the other Leonid storms during and after the seventeenth century. We note from Table 33.1 that the seventeenth century Leonid storms of 1601–1602 and 1666 were visible only in China, where the Scientific Revolution under way in Europe had yet to take place (for Chinese natural philosophers still labored under Aristotelian assumptions). The 1698 Leonid storm, by contrast, was visible in Europe and Japan. One might have thought this would trigger a search for a celestial origin, at least in Europe, but of this there is no record. And in the eighteenth century, when Leonid storms were reported in 1766 in South America, and in 1799 in eastern North and South America, astronomy had not yet advanced far enough in those locales to trigger a crisis that demanded explanation in scientific terms.

The storm of 1832, visible in Eastern Europe and the Middle East, might have triggered renewed attempts at explanation. In his pioneering account of historical Leonids, Newton (1863) noted that descriptions of the 1832 display were given in many newspapers and scientific journals throughout Europe, and that Le Verrier himself saw them and noted "… it would have taken several hours to count those visible at one instant, supposing them fixed." Why there was not more of an attempt to provide a scientific explanation of this impressive display in sophisticated Europe remains a mystery. For whatever reasons, that role was left to the storm of 1833, when fledgling American science, represented by Olmsted and Twining, precipitated the events we have described earlier.

Table 33.1 Probable Leonid meteor storms excerpted from Mason (1995)

	Date	Where observed
902	Oct 12–13	Southern Europe and North Africa
934	Oct 13–14	Europe, North Africa, China
	Oct 14–15	
1002	Oct 14–15	China and Japan
1202	Oct 18–19	Middle East and China
1237	Oct 18–19	Japan
1238	Oct 18–19	Japan
1366	Oct21–22	Europe and China
1532	Oct 24–25	China and Korea
	Oct 25–26	
1533	Oct 24–25	Europe, China, Korea, Japan
	Oct 25–26	
1566	Oct 25–26	Korea and China
	Oct 26–27	
1601	Nov 5–6	China
1602	Nov 6–7	China
1666	Nov 6–7	China
1698	Nov 8–9	Europe and Japan
1766	Nov 11–12	South America
1799	Nov 11–12	Eastern parts of North and South America
1832	Nov 12–13	Eastern Europe and Middle East
1833	Nov 12–13	Eastern parts of North America
1834	Nov 12–13	North America
1866	Nov 13–14	Europe
1867	Nov 13–14	Eastern parts of North America
1868	Nov 13–14	North America
1965	Nov 16–17	Eastern Europe
1966	Nov 16–17	Central/Southwest North America

33.3 Historical Observations of the Leonids Prior to 1799

33.3.1 Early Meteor Catalogues and Cultural Effects of Meteors

The observations over the last two centuries might have been the only Leonid observations known, had not scholars begun systematically examining the historical literature. Catalogues of falling stars were compiled before they were studied as groups of particular showers or storms. Among the earliest of these was Adolphe Quetelet's 1839 *Catalogue des Principales Apparitions d'Etoiles Filantes*, which was widely used for subsequent meteor studies. It was followed by the catalogues of Yale librarian and amateur astronomer, Edward Herrick (1841); the Frenchman,

Edouard C. Biot (1841); the French astronomer and popularizer, Francois Arago (1860); and by Quetelet's own update, in 1861. Biot's catalogue covered 24 centuries of Chinese observations, from 687 BC to AD 1644. Arago's catalogue, in his *Astronomie Populaire,* was entirely a compilation from other catalogues, listed by month, from which the prominence of the August Perseids and the November Leonids was evident, but still unexplained.

But it was only in the 1860s that H. A. Newton set out to identify Leonid events more systematically, using previous catalogues and the historical record. He found six possible Leonid events (Newton 1863), including the 1799 and 1833 apparitions described by Humboldt and Olmsted. In 1864 he extended these to 13 Leonid events, dating between AD 902 and 1833 (Newton 1864). Most of Newton's accounts were cited in previous catalogues, but he took the precaution of going back to the original sources wherever possible.The identification of these early accounts of the Leonids naturally raises the question of their effect on the cultures of the times, for given the impact of the Leonids in 1833, one can imagine the effect that such storms had on the population in pre-scientific cultures. The interpretation of these early events was not physical, but cultural, political, and religious. Even more than for comets, which appeared relatively calm and distant by comparison, such apparitions were often associated with great, and usually catastrophic, events. This was the case for the earliest of these accounts, in AD 902, which was visible in Southern Europe and North Africa, where Islamic civilization was in full bloom. For this event, Newton (1864:378) translated from the original Spanish of Conde's *Historia de la Dominacion de las Arabes en Espana,* where the dates were based on the Islamic calendar reckoned from the start of the Hegira (July 16 or 17, AD 622): "In the month Dhu-1-Ka'dah of this same year (289 A.H.) died king Ibrahim bin Ahmad, and that night there were seen, as it were lances, an infinite number of stars, which scattered themselves like rain to right and left, and that year was called the year of the stars." Dhu-1-Ka'dah is the tenth month of the Islamic calendar. And regarding the same event, from a history of the Saracens:

> In the year 286 [Newton believes it was AH 289, and thus the same year as in the above account] there happened in Egypt an earthquake, on the Fourth Day [of the week], on the 7th of Dhu-1-Ka'dah, lasting from the middle of the night until morning; and so-called flaming stars struck one against another violently, while being borne eastward and westward, northward and southward; and no one could bear to look toward the heavens, on account of this phenomenon. (Newton 1864:380)

Not only the death of a king and an earthquake, but also the razing of the Italian city of Castellum Lucullanum from fear of the Saracens, the removal of its population to Naples, and various Christian events are associated with this astronomical event, which Newton (1864:381) dates as October 13, 902.

The 934 event, classed as a Leonid storm by Mason (1995), and visible in Africa, China, and a Europe still deep in the so-called Dark Ages, was also associated with an earthquake: "And there was an earthquake, in Egypt, on the third day of Dhu-1-Ka'dah of the year [AH 323]; and flaming stars struck against one another violently" (Newton 1864:382).

The 1002 event was seen in China and Japan, and refers with some precision to positions of the meteors in the sky:

> Period Khien-ping, fifth year, ninth month, 35th day of the cycle (October 14th) there were seen moreover thousands of small stars, which appeared in the group alpha, gamma, delta Cancri, and went as far as the group lambda, mu, Ursae Majoris. Generally a large star was seen followed by a half score of small stars. Among them were seen two stars as large as a quart measure; these went, one to the star Sirius, the other to the group phi, rho, tau Sagittarii, and vanished. (Newton 1864)

Had this event been viewed in the Western world, it would undoubtedly have been imbued with millennial fear; even though no catastrophic connection is specified in this account, we may well imagine that such connections were made.

An Arab account of the AD 1202 storm, visible in the Middle East and China, refers to a religious reaction:

> And in the year 599 [AH], on the night of Saturday, on the last day of Muharram, stars shot hither and thither in the heavens, eastward and westward, and flew against one another, like a scattering swarm of locusts, to t he right and left; this phenomenon lasted until day-break; people were thrown into consternation, and cried to God the Most High with confused clamor; the like of it never happened except in the year of the mission of the Prophet, and in the year 241. (Newton 1864:383)

The 1366 event, also believed to have been a Leonid storm, was visible in Europe and China. The original source was Duarte Nunez do Liao, *Chronicas dos Reis de Portugal Reformadas* (Lisbon, 1600), and was quoted by von Humboldt in his *Kosmos* (1850), whence Newton (1864:384) took it:

> In the year 1366, and xxii days of the month of October being past, three months before the death of the King Don Pedro (of Portugal), there was in the heavens. a movement of stars, such as men never before saw or heard of. From midnight onward, all the stars moved from the east to the west; and after being together they began to move, some in one direction, and others in another. And afterward they fell from the sky in such numbers, and so thickly together, that as they descended low in the air, they seemed large and fiery, and the sky and the air seemed to be in flames, and even the earth appeared as if ready to take fire. That portion of the sky where there were no stars seemed to be divided into many parts, and this lasted for a long time. Those who saw it were filled with such great fear and dismay, that they were astounded, imagining they were all dead men, and that the end of the world had come.

The 1533 Leonid storm is of interest because it was visible not only in China, Korea, and Japan, but also in Europe, where Copernicus was about to shock the world with his *De Revolutionibus.* Newton (1864) gives only one account, and it is from Asia:

> Period Kia'tsing, twelfth year, ninth month, the 13th day of the cycle (October 24th) … from the fourth to the fifth watch (from 2 to 4 am), in the four parts of the heavens, there were innumerable shooting stars, great and small, moving together in straight and oblique lines. This continued until daylight.

One may wonder whether some of the early figures of the scientific revolution in the western world viewed this event.

The 1602 storm, which might have been of interest for the reaction of a scientifically enlightened individual in Western Europe like Kepler, was visible only in

China, as was the 1666 event. Even the 1833 storm had cultural implications, for renditions of the falling stars from this storm were used to illustrate the Day of Judgment, an allusion that undoubtedly came to mind for more than a few individuals who observed the event.

The cultural effects of the Leonid meteor storms and other meteor events may be put in context by reference to the claims of Bailey et al. (1989), and Clube and Napier (1990), who have suggested that high points in the meteor flux caused by the debris of giant comets may have affected historical events more than we might think. As Bailey (1996:659) most recently puts it, "Episodes of bombardment … may provide an explanation for periods of global cooling as registered in the historical record, even for the strong interest displayed by most early civilizations in celestial phenomena, providing a possible common origin for myths and legends from around the world." More than that, Clube (1996) speculates that transmutations in human society, documented by historians, such as Spengler and Toynbee, are due to cometary fragmentation events.

One need not fully embrace this stimulating but unproven hypothesis in order to assert that astronomical events, such as the Leonids, have played an important role in cultural history. The question is how great the effect has been, and one can imagine a spectrum of effects through history, depending upon whether the event was a meteor shower, a meteor storm, the fragmentation of a giant comet, or an event of the Tunguska type. The logical extension of this punctuated equilibrium backward through geological time is manifested in the record of major extinctions of terrestrial life, including the one believed to have caused the demise of the dinosaurs.

33.3.2 Modern Catalogues and the Problem of Interpretation

The pioneering modern catalogue of meteor showers is that of Imoto and Hasegawa (1958), first published in Japanese in 1956. It contained 118 meteor showers recorded in the Orient in the last 25 centuries, including 18 Leonid events. In 1993, Hasegawa updated his earlier catalogue, incorporating Chinese records (Beijing Observatory 1988; Tian-Shan 1977) and European records (Dall'olmo 1978). His new catalogue (Hasegawa 1993) includes 331 meteor showers, 48 of which are Leonid events.In 1992, Rada and Stephenson examined medieval Arab chronicles, and Kidger (1993) used their data to identify seven possible Leonid events, including one in AD 855, 14 years before a Leonid storm would be due. Kidger remarks that this could have been an anomalously strong annual Leonid shower, which, if true, would make it the earliest recorded Leonid event yet found. Hasegawa (1996) makes further refinements in the Rada and Stephenson data, based upon the solar longitudes at the time of the meteor shower maximum. In his comprehensive review of all Leonid data, Mason (1995) found 58 Leonid events, including 23 probable storms.

Table 33.2 summarizes the growth in our knowledge of the historical Leonids. While in 1863 only six historical Leonid events were known, 58 are now on record,

Table 33.2 Growth in knowledge of historical leonids

Author	Year	# of Leonid events identified
Newton	1863	6
Newton	1864	13
Imoto and Hasegawa	1956	18
Hasegawa	1993	48
Mason	1995	58

thanks to the detective work of many astronomers and historians. Most of that increase has come since 1956, and we can be sure that many more events remain to be discovered. The discovery of meteor events in the historical literature remains a promising field, one that will shed more light on the Leonids as we approach future potential Leonid storms.

As the early accounts given above clearly illustrate, historical accounts of meteor observations are subject to various problems. Not the least of these is dating, including conversions among the calendars of different cultures, an essential determination if an event is to be classified as a Leonid. In classifying an event as a shower or a storm, subjective accounts are not always reliable. Moreover, astronomical considerations must also be taken into account, for the difference between the sidereal year and the tropical year causes a 1.4 day per century delay in the maxima of Leonid events, while the nodal advancement due to planetary perturbations causes another 1.4 day delay per century, giving a total of 2.8 days per century. Thus, as we see in Table 33.1, the 902 event took place on October 12–13, compared to November 16–17 in 1966, when the last Leonid storm occurred. The data contained in Table 33.1 are thus a triumph not only of observation, but also of interpretation over the last 130 years.

33.4 Applications of Historical Data

As we have seen, historical observations of the Leonids were used by Newton and others to determine the basic elements of the meteor stream orbit in the nineteenth century. Such observations continue to be used for modern scientific purposes. The volumes of Kresak and Millman (1968) and Stohl and Williams (1993), for example, illustrate how these observations have contributed to our knowledge of the dynamical and physical nature of meteor streams.

It was Yeomans (1981), however, who used the full range of Leonid meteor shower data from the period 902–1969 to map the distribution of dust surrounding the parent comet Tempel-Tuttle, to predict the strength of the Leonid event in 1998–1999, and to re-determine the orbit of Tempel-Tuttle. Using these historical data, Yeomans graphically presented the dust distribution (Fig. 33.3), and demonstrated that most of the dust resides outside the orbit of the comet, and behind or just slightly ahead of it. This led him to conclude from the position of the dust that

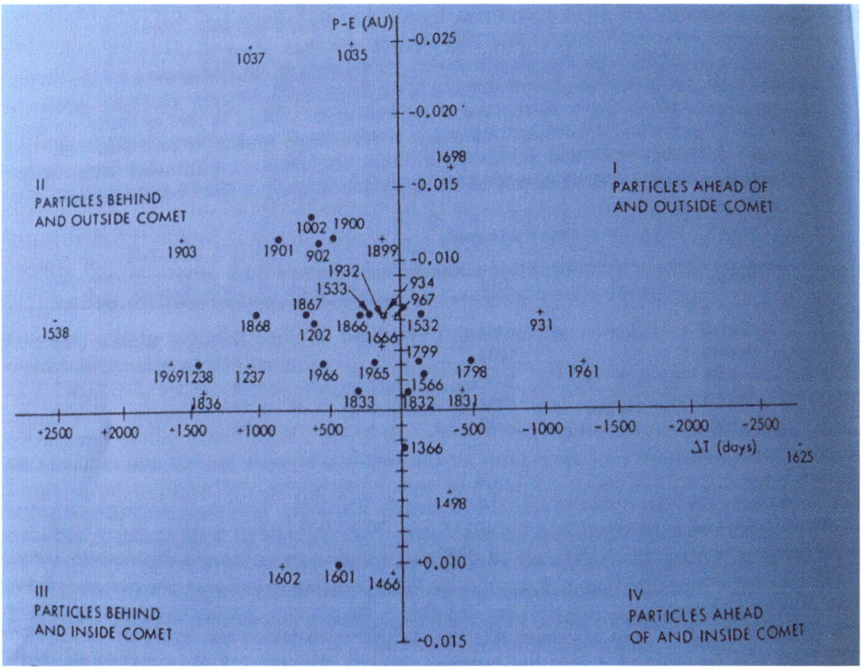

Fig. 33.3 Distribution of dust around Comet Tempel-Tuttle, from the study of Yeomans (1981). The vertical line represents the distance in astronomical units that the Leonid particles were inside or outside the orbit of the parent comet, and the horizontal line gives the times in days that these particles lag or lead the parent comet. A + sign represents a meteor shower, while a filled circle represents a meteor storm. (From *Icarus* **47** (1981), 492–499, by permission of Academic Press, Inc. and D. K. Yeomans)

radiation pressure and planetary perturbations, rather than ejection processes, control the dynamic evolution of the Leonid stream. Plotting past Leonid events on this graph, Yeomans concluded that "… the likelihood of an unusual Leonid shower event in 1998 and 1999 is very good but by no means certain" (Yeomans 1981:498–499). Certainty cannot be obtained because of the uneven distributions within the larger body of dust.

Mason undertook a similar study of the historical evidence, which resulted in more precise predictions. His study found that "Of the 58 Leonid displays researched, 27 of the 35 outstanding showers, and all 23 meteor storms, have occurred between 750 days before and 1750 days after the parent comet's passage through the descending node…" (Mason 1995:219). Of those, "A total of 19 meteor storms took place between 250 days before and 750 days after the comet's nodal passage" (Mason 1995:232). The study also showed that most of the dust was concentrated outside the comet's orbit, where 41 of the 58 documented Leonid events, including 18 of the 23 storms, occurred. Based upon these findings, Mason believes the circumstances for 1997–2000 are comparable to those of 1865–1868. He concludes (Mason 1995:234) that there is " … an excellent chance of enhanced Leonid activity … "

between 1996 and 2002. More specifically, "A Leonid meteor storm is most likely, but by no means certain, in November 1998 or 1999 or both, with probable noteworthy showers in 1997 and 2000" (Mason 1995:219). For November 17, 1998, the Leonid events are likely to peak in eastern and central Asia, and in the early morning hours of November 18, 1999, they will peak in eastern and central Europe and in North Africa.

These expectations are being carefully monitored by both amateur and professional astronomers (Rao 1995). Since 1991, the International Meteor Organization has sponsored a coordinated International Leonid Watch (Jenniskens and Butow 1998; MacRobert 1995, 1996). Actual observations of the Leonids in the last few years have shown modestly increased activity; both the predictions and observations are updated on the World Wide Web (Jenniskens and Butow 1998). A large audience attended a Joint Discussion on "The Leonid Meteor Storms: Historical Significance and Upcoming Opportunities" at the IAU General Assembly in Kyoto in 1997 August, where a variety of observing techniques was also discussed and at least one attendee vividly recalled his observations of the 1966 storm. The concerns, however, are more than academic; in light of the potential Leonid event, a Conference on "Leonid Meteoroids Storm and Satellite Threat" was held in California in 1998.

Those who view the Leonids as a threat to satellites may hope for little activity, but far more numerous are those who wish to see a rare astronomical event. If it does not materialize, humanity may have to wait a long time for another display equal to the great storms of the past. Mason (1995) concludes that significant Leonid activity during the period 2029–2033 is unlikely. Yeomans et al. (1996) go even further, stating that because of planetary perturbations on the Leonid stream, significant Leonid events are not likely for another century after 1999 (Rao 1996). This is graphically apparent in the plot of minimum distances between Comet Tempel-Tuttle and Earth at the time of the comet's passage through its descending node (Yeomans et al. (1996).

And what of the comet itself, the cause of all this agony and ecstasy? Observations of Comet 55P/Tempel-Tuttle have been much rarer than observations of its debris. Yeomans et al. (1996) have recently recomputed its orbit, using the only known observations of the comet as it appeared in 1366, 1699, 1865–1866 and 1965. As Yeomans et al. show in a revealing plot, on most of Tempel-Tuttle's returns it has been too faint to reach naked-eye visibility. In 1997 March, it was recovered at 22.5 magnitude, on its way to perihelion passage on February 28, 1998. At its minimum geocentric distance in 1998, it will be under tenth magnitude, and so easily within reach of moderate telescopes, but not the naked eye.

Whether or not the Leonid predictions for 1998 and 1999 are borne out will soon be known. In any event, meteors are an important case study of the growth in understanding of one class of Solar System objects, a story that is unique and equally interesting for each class of astronomical bodies. Perhaps for few other classes of objects, however, have historic observations played such an important role in both discovery and elaboration, although we know from the work of Clark and Stephenson (1977), Stephenson (1978), and others that they do play a significant role in understanding such diverse phenomena as comets, eclipses, and supernovae. The role of historical data in meteor studies is thus only one example of how the history of astronomy demonstrates its usefulness to science. And the very existence of these data reminds us of how much astronomical phenomena have affected the civilizations of the past (Schaefer 1992, 1994, 1997a, b), perhaps more than we yet realize.

33.5 Commentary 2020

This paper was given at the Joint Discussion on "The Leonid Meteor Storms: Historical Significance and Upcoming Opportunities" at the IAU General Assembly in Kyoto in August, 1997. Excitement was high at this time just preceding the predicted meteor storms of 1998–1999. The predictions of Mason and Yeomans proved accurate for some sites, though the latter hedged his bets when he said "the likelihood of an unusual Leonid shower event in 1998 and 1999 is very good but by no means certain" (Yeomans 1981:498–499). In fact the peak of the 1999 Leonid storm was predicted to within 5 min and produced spectacular meteor rates of 3500 per hour (one per second) from Europe and the Middle East. However, outside of these regions, the rates were much less. So in a sense Yeomans was right.

Occasionally meteoroids interact with humans in a more direct way. On January 18, 2010 a half-pound meteorite, travelling 200 miles per hour, crashed into a doctor's office in Lorton, Maryland near Washington, D.C. The "Lorton meteorite" splintered into several pieces, and became the object of a dispute over ownership rights. It was estimated to be worth $50,000 in the meteorite market, and although the doctors wanted to donate it, the landlords weren't so sure. Pending a decision on these rights, it was placed into the custody of the Smithsonian's National Museum of Natural History, which holds about 15,000 meteorites, about half of the world's collections. A year after the fall, the landlords dropped their claim, the Smithsonian paid the doctors $10,000. The doctors then donated it to Doctors Without Borders, and the Lorton meteorite took its permanent place in the museum.

The International Meteor Organization maintains an extensive and regularly updated website on meteoroids, meteors, fireballs, and related phenomena at http://www.imo.net/, including photos and videos, observing methods, fireball reports, and upcoming meteor showers. The American Meteor Society also hosts similar information at https://www.amsmeteors.org/home.html. On meteor showers and storms see entry P20 (Meteoroid streams).

The phenomenon of meteors and meteor storms are an example of the much broader field of applied historical astronomy, which includes the use of past astronomical observations from many cultures in drawing conclusions relevant to the present, as well as the role in society of phenomena such as comets, lunar and solar eclipses, and supernovae (Steele 2004; Stephenson 1978). Aside from these regularly and sporadic occurring phenomena, other unexpected phenomena occasionally pop up in the heavens. An unusual example is comet Shoemaker-Levy 9, a comet that in summer 1994 smashed into Jupiter with spectacular effect. In Washington, the event drew media coverage from all over the region, especially because the Naval Observatory hosted both Gene Shoemaker and David Levy looking through

Fig. 33.4 Artist rendition of comet Shoemaker-Levy 9 impacting Jupiter. The poster has been signed by Gene and Carolyn Shoemaker and David Levy. Even though not visible to the naked eye, such events impact the general public

one of its telescopes as fragments of the comet visibly changed the face of Jupiter as they impacted (Fig. 33.4). The 1990s proved to be spectacular decade for comets. In March 1996 Comet Hyakutake appeared very bright in the sky. In 1997 Comet Hale Bopp was visible in the sky for almost 18 months and was a public hit, but not nearly of the magnitude of Shoemaker-Levy. Because it was visible so long (twice as long as the previous record holder, the Great Comet of 1811), it was dubbed "the Great Comet of 1997."

For succinct histories of the classes of objects mentioned in this article including planets, asteroids, comets, the Kuiper belt and the Oort cloud, see the classification system in Dick (2019), succinctly summarized in Chap. 31.

Acknowledgements This paper is based on one given in Joint Discussion 23, "The Leonid Meteor Storms: Historical Significance and Upcoming Opportunities," at the 1997 General Assembly of the International Astronomical Union in Kyoto, Japan. I am particularly indebted to Yeomans (1991) and Mason (1995) cited below, as well as to I. P. Williams for organizing the session and inviting my participation. Also my thanks to Don Yeomans, David Hughes, and Wayne Orchiston for their comments on the paper.

References

Adams, J.C., 1867. On the orbit of the November meteors. *Monthly Notices of the Royal Astronomical Society*, 27:247–252.

Arago, D.F.J., 1860. *Astronomie Populaire*. Gide, Paris, Volume 4, pp. 292–314.

Bailey, M. E., 1996. Recent results in cometary astronomy: implications for the ancient sky. *Vistas in Astronomy*, 39:647–671.

Bailey, M.E., Clube, S.V.M. and Napier, W.M. 1989. *The Origin a/Comets*. Pergamon, Oxford.

Biot, E.C., 1841. *Catalogue General des Etoiles Filantes et des Autre Meteores Observes en Chine Pendant Vingt-quatre Siecles*. Imprimerie Royale, Paris.

Butrica, A J., 1996. *To See the Unseen: A History of Planetary Radar Astronomy*. NASA SP-4218, Washington, D. C., Chapter 1.

Clark, D.H. and Stephenson, F.R., 1977. *The Historical Supernovae*. Pergamon, Oxford.

Clube, S.V.M. 1996. The nature of punctuational crises and the Spenglerian model of civilization, *Vistas in Astronomy*, 39:673–698.

Clube, S.V.M. and Napier, W.M. 1990. *Cosmic Winter*. Basil Blackwood, Oxford.

Dall'olmo, U. 1978. Meteors, meteor showers and meteorites in the Middle Ages: from European medieval sources. *Journal for the History of Astronomy*, 9:123–134.

Dick, Steven J. 2019. *Classifying the Cosmos: How We Can Make Sense of the Celestial Landscape*. Springer: Switzerland.

Doel, R. 1996. *Solar System Astronomy in America: Communities. Patronage, and Interdisciplinary Science, 1920-60*. Cambridge University Press, Cambridge.

Dupree, A.H., 1957. *Science in the Federal Government*. Harvard University Press, Cambridge (Mass.), Chapter 3.

Great Leonid meteor shower of 1966. *Sky and Telescope*, 33:4–10 (1967).

Hasegawa, I. 1993. Historical records of meteor showers. In J. Stohl and LP. Williams (eds.), *Meteoroids and Their Parent Bodies*. Astronomical Institute, Slovak Academy of Sciences, Bratislava, pp. 209–233.

Hasegawa, I. 1996. Further comments on the identification of meteor showers recorded by the Arabs. *Quarterly Journal of the Royal Astronomical Society*, 37:75–78.

Herrick, E.C. 1841. Contributions towards a history of the star-showers of former times. *American Journal of Science and Arts* (first series), 40:349–365.

Hoffieit, D. 1992. *Astronomy at Yale, 1701-1968*. The Connecticut Academy of Art and Sciences, New Haven.

Hughes, D.W. 1982. The history of meteors and meteor showers. *Vistas in Astronomy*, 26:325–345.

Hughes, D.W. 1995. The world's most famous meteor shower picture, *Earth, Moon and Planets,* 68:311–322.

Imoto, S. and Hasegawa, I. 1958. Historical records of meteor showers in China, Korea, and Japan. *Smithsonian Contributions to Astrophysics*, 2(6):131–144 (originally published in Japanese, in 1956).

Jenniskens, P. and Butow, S. 1998. International Leonid Watch. World Wide Web: http://wwwspace.arc.nasa.gov/~leonid/watch.html

Kidger, M.R. 1993. Some comments on the identification of medieval meteor showers recorded by the Arabs. *Quarterly Journal of the Royal Astronomical Society,* 34:331–334.

Kresak, L. and Millman, P.M. 1968. *Physics and Dynamics of Meteors.* Springer- Verlag, New York.

Kronk, G.W., 1988. *Meteor Showers: A Descriptive Catalog.* Enslow, Hillside.

Le Verrier, U.J.J., 1867. Sur les etoiles filantes du 13 Novembre et du 10 Aout. *Comptes Rendus,*64:94–99.

Littman, M., 1996. The discovery of the Perseid meteors. *Sky and Telescope,* 92 (2):68–71.

Lovell, A.C.B. 1954. *Meteor Astronomy.* Oxford, Clarendon Press.

MacRobert, A. 1995. Planning your Leonid watch. Sky and Telescope, 90(5):32– 34.

MacRobert, A. 1996. The Leonids: waiting and watching. *Sky and Telescope,* 92(5):72–73.

Mason, J.W. 1995.The Leonid meteors and comet 55P/Tempel-Tuttle. *Journal of the British Astronomical Association.* 105:219–235.

Millman, P.M. and McKinley, D.W.R. 1963. Meteors. In B. Middlehurst and G. Kuiper (eds.), *The Solar System Volume IV. The Moon, Meteorites and Comets.* University of Chicago Press, Chicago, pp. 674–773.

Milon, D. 1967. Observing the 1966 Leonids. *Journal of the British Astronomical Association,* 77:89–93.

Newton, H.A.1863. Evidence of the cosmical origin of shooting stars derived from the dates of early star-showers. *American Journal of Science and Arts* (2nd series) 36:145–149.

Newton, H.A.1864. The original accounts of the displays in former times of the November star-shower; together with a determination of the length of its cycle, its annual period, and the probable orbit of the group of bodies around the sun. *American Journal of Science* (2nd series), 37:377–389, and 38:53–61.

Beijing Observatory. 1988. *General Compilation of Chinese Ancient Astronomical Records.* Jiangsu, Beijing.

Olivier, C.P. 1925. *Meteors.* Williams and Wilkins, Baltimore.

Olmsted, D. 1834. Observations on the meteors of 13 Nov. 1833. *American Journal of Science and Arts,* 25:363–411 and 26:132–174.

Peters, C.F.W. 1867. Letter dated January 29. *Astronomische Nachrichten,* 68: 287–288.

Quetelet, A. 1839. *Catalogue des Principales Apparitions d'Etoile Filantes.* M Hayez, Brussels.

Quetelet, A. 1861. *Sur la Physique du Globe.* M. Hayez, Brussels, Chapter 4.

Rada, W.S. and Stephenson, F.R., 1992. A catalogue of meteor showers in medieval Arab chronicles.

Rao, J., 1996. The Leonids' last hurrah? *Sky and Telescope,* 92(5):74.

Schaefer, B.E. 1992. Lunar eclipses that changed the world. *Sky and Telescope,* 84:639–642.

Schaefer, B.E. 1994. Solar eclipses that changed the world. *Sky and Telescope,* 87:36–39.

Schaefer, B.E. 1997a. Sunspots that changed the world. *Sky and Telescope,* 93(4):34–38.

Schaefer, B.E. 1997b. Comets that changed the world. *Sky and Telescope,* 93(5):46–51.

Schiaparelli, G.V. 1866–1867. Intorno al corso ed all'origine probabile delle telle meteoriche. *Bullettino meteorologico dell' osservatorio del Collegio Romano,* 5(8,10, 11,12) and 6(2).

Schiaparelli, G.V. 1867. Sur la relation qui existe entre Jes cometes et Jes etoiles filantes. *Astronomische Nachrichte,* 68:331–334.

Steel, D. 1998. The Leonid meteor showers and the genesis of the *Ancient Mariner. Astronomy and Geophysics,* 39 (1): 1.20–1.23.

Steele, J. M. 2004. "Applied Historical Astronomy: An Historical Perspective. *Journal for the History of Astronomy*, vol. 35, 337–355.

Stephenson, F.R. 1978. *Applications of Early Astronomical Records.* Oxford University Press, New York.

Stohl, J. and Williams, I.P. (eds.). 1993. *Meteoroids and Their Parent Bodies.* Astronomical Institute, Slovak Academy of Sciences, Bratislava.

Stoney, G.J. 1867. On the connexion between comets and meteors. *Monthly Notices of the Royal Astronomical Society,* 27:291–295.

Stoney, G.J. and Downing, A.M.W. 1899. Perturbations of the Leonids. *Proceedings of the Royal Society,* 64:403–409.

Struik, D.J. 1962. *Yankee Science in the Making.* MacMillan, New York.

Tian-Shan, Z. 1977. Ancient Chinese records of meteor showers. *Chinese Astronomy,* **1**:197–220. Trouvelot, E.L. 1882. *The Trouvelot Astronomical Drawing Manual.* Charles Scribner's Sons, New York, pp. 115–127, 166.

Twining, A.C. 1834. Investigations respecting the meteors of Nov. 13th, 1833. Remarks upon Prof. Olmsted's theory respecting the cause. *American Journal of Science and Arts,* 26:320–350.

von Humboldt, A.1850. *Kosmos.* Stuttgart and Tubingen, Volume 3, p. 621.

von Humboldt, A. and Bonpland, A. 1818. *Personal Narrative of Travels to the Equinoctial Regions of the New Continent.* H. M. Williams, trans. Longmans, London, Volume 2, pp. 331–346.

Oppolzer, T. von. 1867. Bahnbestimmung des Cometen I 1866. *Astronomische Nachrichten,* 68:241–250.

Williams, I. P. 1997. The Leonid meteor shower: why are there storms but no regular annual activity? *Monthly Notices of the Royal Astronomical Society,* 292:L37–L40.

Yeomans, D.K. 1981. Comet Tempel-Tuttle and the Leonid meteors. *Icarus,* 47:492–499.

Yeomans, D.K. 1991. *Comets: A Chronological History of Observation, Science, Myth, and Folklore.* John Wiley, New York.

Yeomans, D.K., Yau, K.K. and Weisman, P.R. 1996. The impending appearance of comet Tempel Tuttle and the Leonid Meteors. *Icarus,* 124:407–413.

Chapter 34
The Discovery and Exploration of the Moons of Mars

Abstract This chapter describes the discovery of the two moons of Mars, Phobos and Deimos, made in 1877 by Asaph Hall using the giant 26-in. refractor at the U. S. Naval Observatory. Subsequent research over the next century is briefly highlighted, most notably the secular acceleration of Phobos, leading to claims of its artificial nature. A renaissance in moons of Mars studies came with robotic space missions beginning in the 1970s. The description of the discovery of Phobos and Deimos presented here is based on a manuscript discovered by the author in the U. S. Naval Observatory archives, as well as on Hall's published work. Hall read the manuscript before Washington Philosophical Society on February 16, 1878, 6 months after the discovery of the two moons.

34.1 Discovering the Moons of Mars, 1877

As the last quarter of the nineteenth century began, the conventional wisdom was that Mars had no moons. There had been speculation about Martian moons since the time of Galileo and Kepler, and in *Gulliver's Travels* (1726) Jonathan Swift had spoken of the scientists of Laputa discovering "two lesser stars, or satellites, which revolve about Mars" (Gingerich 1970; Gould 1945). There had been many searches, but they all failed. What led Asaph Hall in the summer of 1877 to search for the moons of Mars, and to find them where others had failed, is an intriguing scientific detective story.

34.1.1 Prelude to Discovery

Hall (Fig. 34.1) had arrived at the Observatory in 1862 and was elevated to the same level as his colleague Simon Newcomb, Professor of Mathematics in the U. S. Navy, the following year. Hall was the eldest son of a clockmaker from Goshen,

Keynote address at the First International Conference on the Exploration of Phobos and Deimos, held November 5–7, 2007 at NASA Ames Research Center in Mountain View, California, published here for the first time.

S. J. Dick, *Space, Time, and Aliens*, https://doi.org/10.1007/978-3-030-41614-0_34

Fig. 34.1 Asaph Hall about 1875, 2 years before his discovery of the moons of Mars. (U. S. Naval Observatory)

Connecticut, and learned carpentry before entering Central College in McGrawville, New York.[1] This school was not renowned for its teaching, and when Hall left in 1855 after little more than a year, perhaps the most valuable asset he took with him was Chloe Angeline Stickney (Fig. 34.2), who as a student 2 years ahead of Hall, tutored him in mathematics. In 1856 they were married in Wisconsin, where they had gone to look for work near Miss Stickney's relatives. Not finding any they moved on to Ann Arbor, Michigan. Here, having apparently decided to become an astronomer, Hall studied under Franz Brünnow for several months. Lacking resources to continue, he then headed east with his wife, and (having made some money applying his carpentry skills) in 1857 entered the Lawrence Scientific School at Harvard. Here he attended Benjamin Pierce's lectures briefly before deciding they were too theoretical for his interests. He worked for William C. Bond making astronomical observations, coming to the Naval Observatory in 1862, in the midst of the Civil War (Hill 1908; Gingerich 1972; Hall 1945).[2]

When Hall arrived at the Observatory, the largest instrument was a 9.6-in. refractor. That changed only in 1873, when the 26-in. refractor was mounted as the largest telescope in the world (Fig. 34.3).

One might have expected that the general background of speculation, the approaching favorable opposition of Mars in 1877, a new 26-in. refractor at the U. S. Naval Observatory, and the need for an improved mass of Mars would be all

Fig. 34.2 Chloe Angeline Stickney Hall, wife of Asaph Hall. (From her obituary in the *Journal of the National Science Club*, March, 1899)

the incentive needed. But the more proximate cause of the decision to search was an immediate result of Hall's work on Saturn, commenced 2 years earlier on the 26-in. refractor—now only 4 years old and one of the largest in the world. Consciously retracing his steps in discovering the two moons of Mars, Hall wrote within a few months of that discovery that Saturn was the principal factor:

> In December, 1876, while observing the satellites of Saturn I noticed a white spot on the Ball of the planet, and the observations of this spot gave me the means of determining the time of the rotation of Saturn, or the length of Saturn's day, with considerable accuracy. This was a simple matter, but the resulting time of rotation was nearly a quarter of an hour different from what is generally given in our text books of astronomy: and this discordance, since the error was multiplied by the number of rotations and the ephemeris soon became utterly wrong, set before me in a clearer light than ever before the careless manner in which books are made, showed the necessity of consulting original papers, and made me ready to doubt the assertion one reads so often in the books, "Mars has no moon." (Hall 1878a, b)[3]

The crucial seed of doubt planted and the opposition of Mars fast approaching, Hall began in the spring of 1877 seriously to ponder the possibility that Mars might have a moon. So slight did the chances seem at the outset, Hall later wrote, "that I might have abandoned the search had it not been for the encouragement of my wife" (Hall 1878b, 5). Looking at past searches, Hall was surprised that almost no one had searched for Martian moons since William Herschel in 1783, with the exception of Heinrich d'Arrest, Director of the Observatory at Copenhagen, who had made a search during the opposition of 1862. That failure was discouraging, but "remembering the power and excellence of our glass there seemed to be a little hope left," though this hope was further diminished by the southern declination of Mars at this

Fig. 34.3 The 26-in. Great Equatorial Telescope, soon after its mounting in 1873. Simon Newcomb is at the eyepiece and Admiral Benjamin F. Sands, Superintendent of the Naval Observatory, is seated in the background. (Plate IV from *Instruments and Publications of the U. S. Naval Observatory*).

time, a circumstance that favored the giant new 4-foot reflecting telescope at Melbourne (Hall 1878b, 5).[4]

34.1.2 The Discovery of Phobos and Deimos

With this faint hope, Hall began his search in early August. Hall clearly realized from theoretical considerations that it would be useless to search beyond 70 arc-minutes from the planet. He doesn't say where he started, only that his attention was first directed to faint objects "at some distance from the planet," all of which turned out to be fixed stars. On August 10, Hall began to examine the region within the glare of light of the planet. "This was done by keeping the planet just outside the field of view, and turning the eyepiece so as to pass completely around the planet." On this night the image of the planet was "blazing," the sky "too thick" and Hall saw nothing. Hall tried again early the next night and found nothing, "but trying again some hours later I found a faint object on the following side and a little north of the planet. I had barely time to secure an observation of its position when fog from the River stopped the work. This was at half past two o'clock on the night of the 11th." In his logbook Hall wrote "Seeing good for Mars. The edge of the White spot [the Martian South Pole] has two notches near the center of its outline. (A faint star near Mars)." Although Hall later wrote in ink "this proves to be satellite 1," with only one set of measures he had no way of knowing at the time whether the object was moving with Mars or was just a faint background star (Hall 1878b, 6).[5]

One can imagine Hall's state of mind as cloudy weather intervened for the next several days. On the 15th Hall slept at the Observatory, and when the sky cleared at 11 o'clock he returned to the telescope. After observing Saturn's ring, and the Saturnian moons Titan and Iapetus, he returned to Mars. But the seeing was very bad, and (as Hall later discovered) the moon was so close to the planet as to be invisible, so again Hall went home disappointed. Here he received continuing encouragement. Son Angelo recalled that "Mrs. Hall was full of enthusiasm. Each night she sent her husband to the observatory supplied with a nourishing lunch, and each night she awaited developments with eager interest. I can well remember the excitement at home. There was a great secret in the house, and all the members of the family were drawn more closely together by mutual confidence" (Hall 1908, 112).

Finally, on Thursday the 16th the weather cooperated and Hall was able to try again. Mars rose about 8:25 in the evening, preceded 20 min before by Saturn. At 11:42 pm he noted a 12th magnitude "star near Mars" and measured its distance from Mars with the micrometer. By 1:30 am, after more micrometer measures, he had concluded that it was moving with the planet, which was itself moving at the rate of 15 arcseconds per hour and therefore would quickly have left behind any fixed stars. Although still referred to in the logbook as the "Mars Star," Hall could barely contain his excitement:

> Until this time I had said nothing to anyone at the Observatory of my search for a satellite of Mars, but on leaving the Observatory after these observations of the 16th, at about three o'clock in the morning, I told my assistant, George Anderson, to whom I had shown the object, that I thought I had discovered a satellite of Mars. I told him also to keep quiet as I did not wish anything said until the matter was beyond doubt. (Hall 1878a, 4–5)

Anderson did faithfully keep quiet, but not Hall: "the thing was too good to keep and I let it out myself. On August 17th between one and two o'clock, while I was reducing my observations, Professor Newcomb came into my room to eat his lunch and I showed him my measures of the faint object near Mars which proved that it was moving with the planet." Hall himself spilled the beans, and to none other than his arch-rival, Simon Newcomb! Examining Hall's measures, Newcomb estimated its period of revolution around Mars, and pointed out that if it were a satellite it would not be visible most of the following night, but would reappear toward morning.

The night of Friday the 17th was therefore filled with uncertainty and surprise. Early in the night Hall measured a "Mars Star" which was "a fixed star and not the object observed last night;" as Newcomb had predicted, the Martian satellite was not visible. At four o'clock in the morning, however, not only did the satellite emerge from behind the planet as predicted, but while watching for the outer moon, Hall discovered the inner one. Within a few minutes Hall had measured the separations and position angles of both, still referred to as "Mars Star," and with daylight quickly approaching noted in his logbook (Fig. 34.4) "Both the above objects faint but distinctly seen by G. Anderson and myself."[6]

On Saturday the 18th notice of the discoveries was telegraphed to Alvan Clark & Sons at Cambridgeport, who that night verified Hall's discovery, as did Pickering and his assistants with the Harvard 15-in.. That same night Newcomb, astronomer David P. Todd, and William Harkness joined Hall and Anderson in the dome. Already at 9:40 Hall saw the satellite immediately. First Newcomb and then Harkness measured the satellite positions, with Hall writing down the micrometer readings for separation and position angle. With Harkness's observations, "Mars Star" becomes for the first time "Mars satellite." The measures of this and the preceding evening had "put beyond doubt the character of these objects." Not one to waste a good night, after midnight Hall turned to observations of Saturn and its moons, returning to Mars at 2:30 to check on his newly discovered satellite.[7]

August 19, normally a quiet Sunday, was the day to tell the world. The discovery was announced to the scientific world and the general public by Admiral Rodgers, via a communication to the Smithsonian Institution, which in turn released the following dispatch: "Two satellites of Mars discovered by Hall, at Washington. First, elongation west, August 18, 11 h, Washington time. Distance eighty seconds. Period, 30 h. Distance of second, fifty seconds." Still, Hall remained puzzled by the inner moon; it would appear on different sides of the planet on the same night, leading Hall to believe there might be two or three inner moons. On the nights of the 20th and 21st he watched the inner moon for long intervals, "and saw that there was in fact but one inner moon which made its revolution around the primary in less than one third the time of the primary's rotation—a case unique in our solar system." On

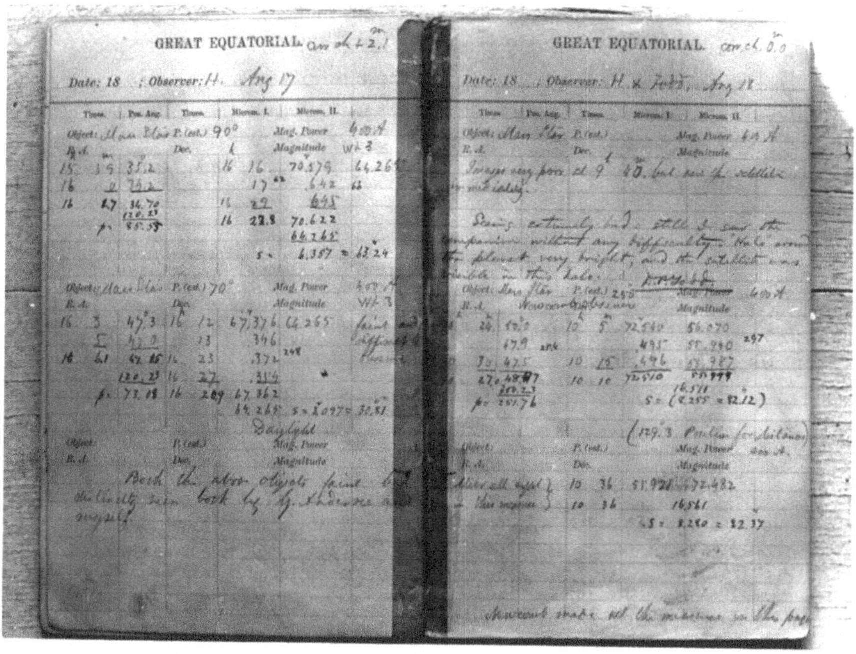

Fig. 34.4 Hall's log book of the Martian observations with the discovery of the two moons of Mars, August 17, 1877. At the bottom left, the log reads "Both the above objects faint but distinctly seen both by G. Anderson and myself." On the right, the next night David P. Todd and Newcomb have both seen the newly discovered moon. (U. S. Naval Observatory)

August 20 Hall wrote in his notebook, "The observations of Mars Sat. 2 of tonight show that this satellite has a period of about 6 h 20 m. I had a good opportunity to compare the brightness of the two satellites. The inner one is the brighter" (Hall 1878b, 6).[8]

The news took both the public and the astronomical world by storm. The *Washington Evening Star* for August 20, under the heading of "Political News from the Executive Mansion and Departments" of the government, reported "Glorious news from the skies—Brilliant Discovery with the Great Telescope—Two satellites of Mars found by Prof. Hall." The British journal *Nature* briefly announced the "extraordinary discovery"in its August 23 issue (having heard from the Smithsonian via Leverrier in Paris), and its September 6 issue published Newcomb's calculations of the first 10 days' measures. *Nature* characterized the discovery as "in the highest degree an honor to American science," the instrument, of American construction, being "a masterpiece of mechanical skill." Similarly, *Scientific American* called the discovery "a triumph both for Professor Hall and for Mr. Alvan Clark," and looked forward to the even more colossal telescope under consideration for Lick Observatory. Leverrier, in the last few weeks of his life, called the find "one of the most important discoveries of modern astronomy." Visitors great and small streamed to the telescope even more than before, including the Vice President of the United States, who obligingly came on November 3 on a night, Hall duly reported, "good for naught but visitors" (Nature 1878a, b, c; Scientific American 1877).

Remarkably, aside from the Great Reflector in Melbourne, no one else had seriously attempted to search for Martian satellites, and there were no rival claims to the discovery. To Hall would go the credit for discovering the two—and only two—satellites of Mars.

Hall continued his observations of the new satellites with the 26-in. until October 31st; these observations alone gave enough data for a good determination of their orbits. By this time observations had been made at many other observatories, including those of E.C. Pickering and his assistants with the 15-in. at Harvard and of H.S. Pritchett with the 12-in. at the Morrison Observatory in Missouri. Hall and his Naval Observatory colleagues J. R. Eastman and H.M. Paul even saw the outer satellite with the Observatory's 9.6-in. refractor. Six months after the discovery, several observatories in Europe had seen the outer satellite, but, so far as Hall knew, not the inner one. By this time photometric observations at Harvard had already shown that the diameter of the outer moon was about six miles, and the inner moon seven miles. Hall calculated the period of the outer satellite at 30 h 18 min at a distance from the surface of Mars of 12,500 miles, and the period of the inner satellite as 7 h 39.5 min at a distance of 3760 miles. When discovered they were only 85 and 34 arc seconds from Mars, with the inner and outer satellites at magnitude 10 and 12, respectively. Hall delighted in describing the peculiar appearance that the two moons would present to a hypothetical Martian. The inner moon would rise in the west and set in the east, and pass through all its phases in 11 h. Such a moon could be of very practical use, Hall the practical astronomer notes, but Martian astronomers might be hindered by a dense atmosphere and large refraction effects.[9]

Hall chose to name the inner satellite Phobos (Flight) and the outer satellite Deimos (Fear), based on the suggestion of a Mr. Madan of Eton, England. The names were taken from the 15th book of Homer's Iliad. Hall noted that as personified by Homer, Phobos and Deimos are the attendants, or sons, of Mars (Hall 1878b, 6).[10]

The question of why the moons of Mars had not been discovered long before was one that naturally interested Hall. He calculated that for 42 days from September 4 to October 16 during the favorable opposition of 1862, the moons were brighter than they were in 1877, when viewed with a small 9.6 in. refractor like the one at the Naval Observatory. Moreover, the planet was 15° further north in 1862 than in 1877, and so higher in the sky in the Northern Hemisphere. During this time, Hall pointed out, the moons were within easy reach of the Harvard 15-in. refractor. "The reason therefore why they were not found before is that astronomers did not search for them in the right place and in the right manner. Probably every one who searched looked too far away from the planet, and the fact appears to be that in recent times very few searched at all." A good telescope was certainly helpful in the discovery of the moons of Mars. But even more essential were curiosity, persistence and the need to doubt conventional wisdom.

34.2 Discovery of the Secular Acceleration of the Moons of Mars. Are the Satellites Artificial?

Because of the small size of the Moons of Mars, prior to the space age there was very little more that could be learned about them, except for observing their motions and refining their orbits. The furor over Percival Lowell and the canals of Mars notwithstanding, Pascu (1978) has noted that the first century of observation of the moons of Mars can be divided into two distinct periods. The first is what he calls the "classical period" from 1877 to 1928, characterized by visual observations using a micrometer. These observations were motivated by an interest in the dynamical properties of Mars. Most of these 1500 observations for each satellite were made at USNO, Lick and Pulkovo observatories.

The second period is from 1928 to the present, characterized by photographic techniques (Pascu 1978). Its impetus was claims of the secular acceleration in the orbital longitude of the moons of Mars. Not surprisingly, since the Naval Observatory had a proprietary interest in the moons, that institution also made this discovery. This discovery actually involved both visual and photographic techniques, pitting the old 26-in. refractor against the new 40-in. reflector. Naval Observatory astronomer Henry E. Burton had noted already in 1929 that the orbital longitude of Phobos was not well-behaved (Burton 1929), and at the next close opposition of Mars in 1939 Burton and his colleague Bevan Sharpless undertook an observational program to confirm this. Burton made visual observations with the 26-in. refractor, while Sharpless undertook photographic observations with the 40-in. reflector. Although the visual observations were superior to the photographic ones, Sharpless nevertheless had enough data to publish his secular acceleration results in 1945 (Sharpless 1945). The results remained controversial, and George Wilkins' claim in 1964 that he could not confirm this motion set off a new program for astrometric observations of the moons at the USNO.

Although this might seem an esoteric subject, curiously the study resulted in one of the more frequently cited papers from the Naval Observatory, especially after the well-known Russian astrophysicist I. S. Shklovskii used it in 1966 to claim that the satellites of Mars might be artificial (Shklovskii and Sagan 1966, 362–376).[11] The results for Phobos have been confirmed (at about one-third the value of Sharpless), but are now considered to be due to gravitational interactions with Mars, known as *solid body tides*, rather than to hollow artificial satellites (Sinclair 1978; Burns 1978). Still, the whole episode gave renewed interest to planetary satellite work.[12] One fairly certain prediction is that Phobos will crash into Mars in less than 100 million years!

34.3 The Moons of Mars and the Space Age

A new era in the moons of Mars began with spacecraft observations, which allowed physical studies of the moons. Mariner 9 in 1971 and 1972 first revealed the surface features of the moons. Mariner 9 obtained 32 high-resolution pictures of Phobos with 200-m resolution, and nine of Deimos. Because the spacecraft orbit was inside the orbit of Deimos, coverage was about 80% complete for Phobos, but for the outer satellite Deimos coverage was restricted to the Mars-facing side (Veverka et al. 1974). Phobos was found to be 27 × 21 × 19 km across, and Deimos about half that size at 15 × 12 × 11 km (Ververka in Seidelmann, 163). Phobos was found to be dominated by three craters, Roche (5-km diameter), Hall (6 km) and Stickney (10 km). And as expected, both satellites were found to have spin rotational periods synchronous with their orbital periods (7.6 for Phobos and 30.3 h for Deimos).

Much greater resolution was achieved when the two Viking Orbiters flew within a few hundred kilometers of the surfaces of the moons in the summer and fall of 1976, then as close as 100–300 km in February and May 1977, detecting features as small as 10–200 m. In October 1977 Viking Orbiter II flew less than 30 km from Deimos. As Tom Duxbury pointed out at the time, the low resolution photos were useful in determining "the sizes, shapes, photometric functions, rotational properties, libration amplitudes, libration periods and ephemerides of the Martian satellites as well as the gravity field, spin axis direction and internal structure of Mars itself." Viking's higher resolution photos could detect "small scale features such as surface fractures, crater chains, wall slumping of craters, base-surge deposits, pool melting, and boulders" (Seidelmann 1978, 149). It was during the 880-km flyby of Phobos in late 1976 that numerous grooves were discovered on the surface (Fig. 34.5). More study showed they were associated with the crater Stickney, and may come from deep fractures under the surface. Deimos has no such grooves (Fig. 34.6).

A major objective of Vikings observations of the moons was to determine their composition through color measurements and mass determination. In 1977 Veverka and colleagues noted that the surface of Phobos was similar to Type I or II carbonaceous chondrites, and Ceres. From this researchers, including Veverka (Veverka, in Seidelmann 1978, 189), have inferred that Phobos and Deimos may have originated in the asteroid belt and were captured by Mars. Almost exactly one century after Hall had discovered the moons, enough information had been gleaned to speculate on their origin.

The first spacecraft launched toward Mars after Viking were Russian, not American. Their specific goal was to orbit Mars, explore Phobos photographically, and drop a "hopper" rover and small lander. After a hiatus of 13 years from Viking, Phobos 1 was launched on July 7, 1988, the most favorable opposition since 1971 and Mariner 9. Owing to a command error, it was lost on September 2, 1988, just 2 months into its 200-day journey to Mars. Phobos 2 was launched on July 12, 1988, arrived safely at Mars and was inserted into Mars orbit on January 30, 1989. It failed in late March, before its planned close rendezvous with Phobos. This spacecraft carried a plaque with the discovery page from Asaph Hall's notebook dated August 17, 1877. Unfortunately, the mission was not a success, returning only a few photos.

Further reconnaissance of the moons of Mars were undertaken by the Mars Global Surveyor in 1998 and 2003, and by the European's Mars Express beginning in 2004. The latter made sporadic close flybys of Phobos with high-resolution imaging. The Mars Reconnaissance Orbiter is also studying the spectral properties of the

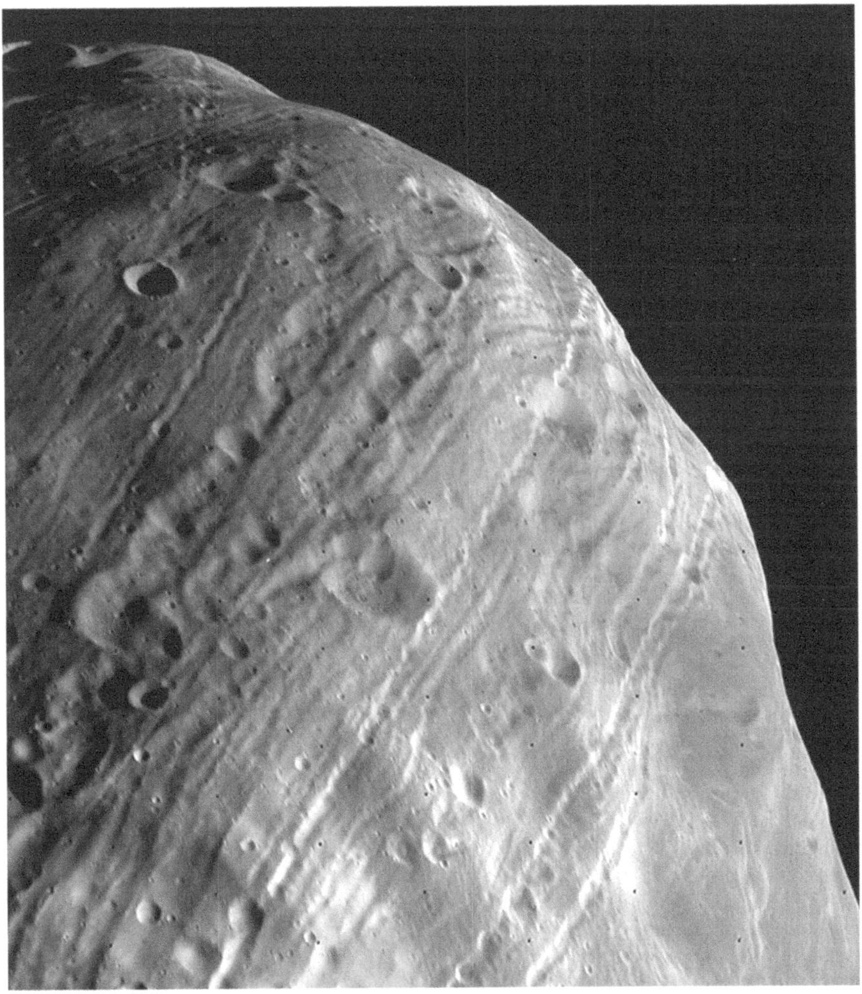

Fig. 34.5 Viking 1 Orbiter image of Mars' satellite Phobos. The linear features and crater chains from lower left to upper right are believed to be results of the impact which formed Stickney crater, which is just off the image to the bottom left. The image is 9 km across and north is at 10:30. The small crater at the center of the image is located at about 20 N, 15 W. (Viking Orbiter 343A15, NASA)

Martian moons as part of its larger Mars mission. Thanks to the Mars Exploration Rovers Spirit and Opportunity, we even know how the moons appear from the surface—as tiny dots due to their small size (Fig. 34.7). Unlike the Earth, the Martian surface will not be moonlit, despite having two moons instead of one.

Despite these advances, a great deal remains unknown about the moons of Mars, including even their basic composition and their origin. In 2007 the attendees at the First International Conference on the Exploration on Phobos and Deimos discussed what was known, what was unknown, and how to find out more. Most continue to

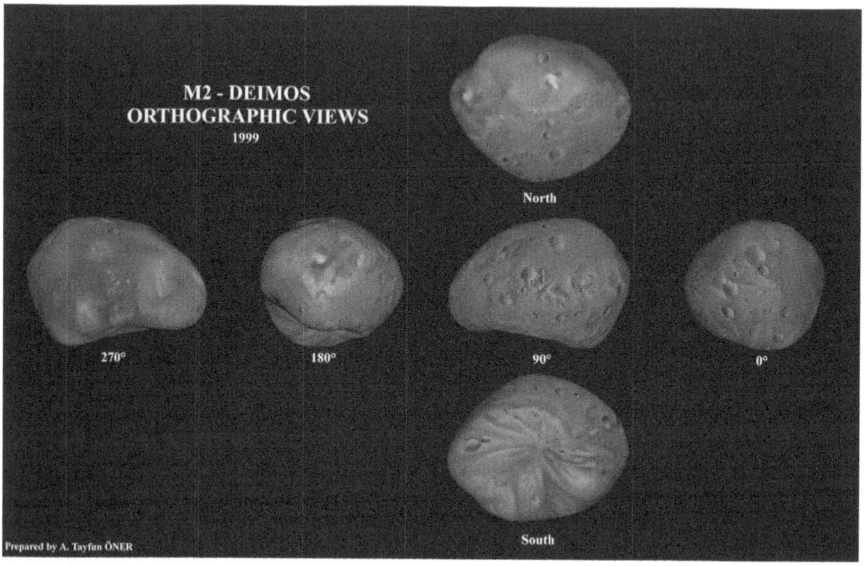

Fig. 34.6 Six different orientations of Deimos from Viking images. The images were rendered by using Peter C. Thomas' model and a combination of the USGS airbrush map and a photomosaic of Deimos. (NASA)

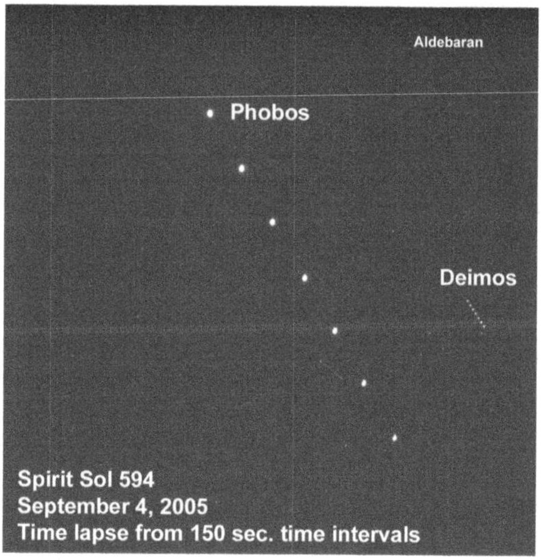

Fig. 34.7 Taking advantage of extra solar energy collected during the day, NASA's Mars Exploration Rover Spirit photographed the two moons of Mars as they crossed the night sky. Spirit took a succession of images at 150-s intervals from a perch atop "Husband Hill" in Gusev Crater on Martian day, or sol, 594 (Sept. 4, 2005), as the faster-moving Martian moon Phobos was passing Deimos in the night sky. (NASA/JPL/Cornell/ Texas A&M)

believe the moons are captured asteroids, but the capture mechanism is unknown and the scenarios unlikely. Some believed rings of Mars might exist—just as they do around Jupiter, Saturn, Uranus and Neptune. Surprisingly, the fact they have not yet been seen is no proof of their absence—gossamer dust rings might be so thin that even spacecraft instruments have not yet seen them. In any case, it seems the two moons may hold the secret to the evolution of small bodies in the Solar System. Conferees discussed the possibility of both human and robotic missions to the moons, and their possible use as outposts for the study of Mars close-up. As various studies have emphasized since 1981, landing humans on the moons of Mars would save a considerable amount of money because the liftoff for return to Earth is much easier from such small bodies with less gravity.

34.4 Commentary 2020

This paper was delivered as the keynote talk at the First International Conference on the Exploration of Phobos and Deimos, held November 5–7, 2007 at NASA Ames Research Center in Mountain View, California. Since that time two more International Conferences on the subject have taken place at NASA Ames, in 2011 and 2016, with the latest including two science sessions, a robotic reconnaissance session, and a human exploration session. Observations of Phobos and Deimos continue to be analyzed from the Mars Express, MAVEN, and other spacecraft in orbit or on the surface of Mars. Future exploration plans, including a Japanese Martian moons exploration mission, and a Phobos sample return mission by the Russia-European Space Agency mission to the moons of Mars program, continue to be studied. The agenda of the latest meeting can be found at https://sservi.nasa.gov/articles/live-now-3rd-international-conference-on-the-exploration-of-phobos-and-deimos/

Phobos and Deimos have been considered as possible stepping stones in the human exploration of Mars. NASA's *Humans to Mars* monograph details several of these prior to the year 2000 (Portree 2001). As early as 1963 Lockheed studied a manned interplanetary mission showing how a manned Mars orbiter might study Phobos and Deimos. At a "Case for Mars" conference in 1981, Fred Singer proposed a ten billion dollar expedition that would use Deimos as a base of operations. It proposed no Mars landing, thus reducing the spacecraft weight. Rather, astronauts would remote control 10–20 surface rovers. Two astronauts would also visit Phobos. The mission would have launched in 1990–1991, with a 2-year duration. In 1988 NASA's Office of Exploration examined a human mission to Phobos, which was used as the basis for case studies (Exploration Studies Technical Report, 1988). In 1989 President George W. Bush's plan to return to the Moon and go to Mars proposed three astronauts land on Phobos to make spacecraft propellants. In 2005 Pascal Lee and his colleagues proposed "Mars Indirect."

The moons of Mars have also often been featured in popular culture. Aside from Jonathan Swift's Voyage to Laputa in *Gulliver's Travels* (1726), both Phobos and Deimos are featured in several of the novels of Edgar Rice Burroughs at the beginning of the twentieth century. In 1955 science fiction writer Donald Wollheim published his bracing adventure novel *The Secret of the Martian Moons*, which stimulated my interest in all things Martian as an early teenager, along with the books of Patrick Moore. In Kim Stanley Robinson's Mars trilogy, the first colonists on Mars build a city within the crater Stickney on Phobos, and a manned landing

Fig. 34.8 Plaque at Asaph Hall's residence in Georgetown, where he resided during the discovery of the moons of Mars. (Photo by Steven Dick)

takes place on Deimos. More recently Ty Drago wrote his debut novel *Phobos,* and there have even been video games featuring Phobos.

Asaph Hall's discoveries continue to capture the imagination more than 140 years later, ranging from a plaque erected at the site of his Georgetown home (Fig. 34.8) to Thomas Mallon's novelization of the discovery in a fanciful way in his book *Two Moons* (2000). And today, 130 years after Hall's discovery, Phobos and Deimos still hold their secrets. As missions to Mars continue in the twenty-first century, they are likely to receive increased attention from both science and the general public. And these are only two of the 189 known moons of the Solar System—each with its own history and mystery, each awaiting exploration.

Notes

1. The future astronomer's father, sometimes designated Asaph Hall II, died in 1842. His grandfather, who served in the Revolutionary War, was Asaph Hall I. The astronomer was therefore Asaph Hall III, and his son (also to become an astronomer) was Asaph Hall IV, although neither used the nomenclature. Asaph Hall IV was appointed an Assistant Astronomer at the USNO on July 27, 1889. He would become a Professor of Mathematics in February, 1908.
2. Among the numerous obituaries see especially (Pritchett 1908). For a view of Asaph and his wife, see the biography of Mrs. Hall by their son Angelo Hall (Hall 1908, as well as Percival Hall (1945), and *Journal of the National Science Club* (March, 1899), 1–6. The Asaph Hall papers, comprising six boxes of material and approximately 1000 items, are located at the Library of Congress.
3. The Hall manuscript (Hall 1878a) was first reported in (Dick 1988). The published reports in Hall 1878b are the account-of-record, but the manuscript gives a fuller and more informal account. Modern accounts of the discovery of the moons of Mars are given by Gingerich (1970) and Pascu (1978).
4. The report of the Melbourne Observatory for this period shows that its "Great Reflector" was used in a search for satellites of Mars, but the observers did not see them, *MNRAS*, 38 (1877–78), 188–89.
5. See also (Hall 1878a, 4) and the 26-in. log book, Aug. 10–11, 1877, Naval Observatory Library.
6. 26-inch logbook, August 17, 1877, USNO Library.
7. 26-inch logbook, August 18, 1877, USNO Library; Hall (1878b, 6).
8. Hall to Rodgers, Aug. 19, 1877, USNO Archives. Next to the date, Hall wrote "1600 hours," which, since the astronomical day was then reckoned from noon, refers to 4 am.
9. Clearly the discovery caught the imagination of the normally staid Hall. Noting that the angle subtended from Earth by the outer satellite of Mars was about.031 s of arc, Hall calculated that at the distance of our Moon this angle would correspond to 187 ft. on the Moon's surface. "Hence, if we assume that the diameter of six miles is nearly correct it appears that the proposition of a German astronomer to establish on the plains of Siberia a system of fire signals for communicating with the inhabitants of the Moon is by no means a chimerical project." In the year that Schiaparelli would begin the great Mars controversy with his observations of *canali*, it is interesting to see Hall sizing up the prospects of this scheme, attributed to K. F. Gauss, for communication with lunar inhabitants!
10. While one might think (as most astronomers do) that Phobos is "fear" and Deimos "flight," Blunck (1982) has the definitions the other way around. In fact classicist sources such as *Crowell's Handbook of Classical Mythology* agree that Phobos means "panic," "flight," or "rout," while Deimos means "fear" or "terror." Thus there is some confusion between astronomers and classicists on this matter.
11. Shklovskii first proposed the idea in 1959.
12. Inexplicably, Sharpless does not cite Nautical Almanac Office Director Edgar W. Woolard's article of the previous year (Woolard 1944), nor did Woolard cite the work of his Naval Observatory colleagues Burton and Sharpless. See Pascu (1978). In contrast to Phobos, Sinclair finds the results for an acceleration of Deimos "barely significant."

References

Blunck, Jurgen. 1982. *Mars and its Satellites: A Detailed Commentary on the Nomenclature*. Exposition Press, Smithtown, N.Y., p, 151

Burns, J. A. 1978. "The Dynamical Evolution and Origin of the Martian Moons," in Symposium on the Satellites of Mars, *Vistas in Astronomy*, 22, 193–210.

Burton, H. A. 1929. "Elements of the Orbits of the Satellites of Mars," AJ, 155–164.

Dick, Steven J. 1988. "Discovering the Moons of Mars," *Sky and Telescope* (September, 1988), 242–243.

Gingerich, Owen. 1970. "The Satellites of Mars: Prediction and Discovery," JHA, 1 (Aug., 1970), 109–115

Gingerich, Owen. 1972. "Asaph Hall," *Dictionary of Scientific Biography*, Charles Gillespie, ed., vol. 6, 48–50

Gould, S. H. 1945. "Gulliver and the Moons of Mars," *Journal of the History of Ideas*, vi (1945), 91–101.

Hall, Angelo. 1908. *An Astronomer's Wife: The Biography of Angeline Hall*. Baltimore

Hall, Asaph. 1878a. "Discovery of the Satellite of Mars," manuscript in U. S. Naval Observatory Archives, read before Washington Philosophical Society February 16, 1878, p. 1.

Hall, Asaph. 1878b. *Observations and Orbits of the Satellites of Mars, with Data for Ephemerides in 1879*. Washington, U. S. Naval Observatory.

Hall, Asaph. 1878c. MNRAS, volume 38 (1877-8), pp. 205–209.

Hall, Percival 1945. *Asaph Hall: Astronomer, A Biographical Sketch of Asaph Hall, 3d* [1945], privately printed and distributed by its author.

Hill, George W. 1908. "Biographical Memoir of Asaph Hall, 1829-1907," BMNAS, 6, Washington, 243–309.

NASA. 1988. *Exploration Studies Technical Report*, Washington, DC: NASA https://ntrs.nasa.gov/archive/nasa/casi.ntrs.nasa.gov/19890007336.pdf

Nature, Aug 23, 1877a, 341

Nature. Sept 6, 1877b, 397–98

Nature, Sept. 13, 1877c, 427–28

Pascu, Dan. 1978. "A History of the Discovery and Positional Observation of the Martian Satellites, 1877-1977, *Vistas in Astronomy,* 22, 141–148.

Pritchett, Henry S. 1908. *Popular Astronomy* (Feb, 1908), 66–70

Scientific American, 1877. Sept. 8, 1877, p. 1.

Seidelmann, P. K., ed. 1978. *The Satellites of Mars, special issue of Vistas in Astronomy, vol. 22.*

Sharpless, Bevan P. 1945. "Secular Accelerations in the Longitudes of the Satellites of Mars," AJ, 51, 185–186.

Shklovskii, I. S. and Carl Sagan. 1966. "Are the Moons of Mars Artificial Satellites?", in *Intelligent Life in the Universe*. Holden-Day: San Francisco, pp. 362–376.

Sinclair, A. T. 1978. "The Orbits of the Satellites of Mars," *Vistas in Astronomy*, 22, 133–140

Veverka, J., M. Noland, C. Sagan, J. Pollack et al. 1974. *A Mariner 9 Atlas of the Moons of Mars*, *Icarus*, vol. 23, Oct. 1974, p. 206–289

Woolard, Edgar W. 1944. "The secular Perturbations of the Satellites of Mars," AJ, 51 (August, 1944), 33–36

Chapter 35
The Universe and Alfred Russel Wallace

Abstract Alfred Russel Wallace is best known for his work as a naturalist and evolutionist, and for his general interest in life on Earth, taking "life" in the broadest sense of the word to include both biology and culture. So it comes as a surprise to many to learn that he had an early interest in astronomy, remained "deeply interested" in astronomical discoveries throughout his life, and late in life wrote two books on one of the most sensational and important aspects of the subject—life on other worlds. This raises interesting questions for the historian of science: Were Wallace's astronomical views incidental or fundamental to his life and thought? If the former, why did he write those books? And if the latter, just how did his astronomical ideas fit into his own worldview? And did the influence in developing his worldview go from astronomy to biology or from biology to astronomy?

In part the answer to these questions is to be found in Wallace's ideas on purpose in the universe, ideas that relate to what we would today call anthropic reasoning and intelligent design. These ideas would culminate with his book *Man's Place in the Universe: A Study of the Results of Scientific Research in Relation to the Unity or Plurality of Worlds*. Published in 1903, that book sheds unique light on Wallace's anthropocentric worldview, and illuminates his entire career, including his belief that the human mind must be set off from animals in the evolutionary process. In accordance with the worldview expressed there, in 1907 Wallace also wrote a more narrowly focused book on the habitability of Mars. Taken together, these works tell us about Wallace's views of the universe and its relation to humans, placing his better-known work in a much broader context. And the considerable reaction to those works reveal how others viewed Wallace and his worldview, both in it terrestrial and extraterrestrial aspects.

35.1 Wallace and Astronomy

In Chap. 8 of his autobiography (Wallace 1908, 101–102), Wallace spoke of an early interest in practical astronomy. He had learned the use of the sextant for surveying, aided by a book on Nautical Astronomy borrowed from his brother. When

First published in *Natural Selection & Beyond: The Intellectual Legacy of Alfred Russel Wallace*, Charles H. Smith and George Beccaloni, eds. (Oxford University Press, 2008), pp. 320–340.

S. J. Dick, *Space, Time, and Aliens*, https://doi.org/10.1007/978-3-030-41614-0_35

Fig. 35.1 Alfred Russel Wallace in 1895, age 72. First published in *Borderland Magazine*, April 1896

he was about 18, he had constructed a small telescope, with which he observed the Moon, Jupiter's satellites and star clusters. These activities, he noted, "served to increase my interest in astronomy, and to induce me to study with some care the various methods of construction of the more important astronomical instruments; and it also led me throughout my life to be deeply interested in the grand onward march of astronomical discovery." But, Wallace continued in the next sentence, "what occupied me chiefly and became more and more the solace and delight of my lonely rambles among the moors and mountains, was my first introduction to the variety, the beauty, and the mystery of Nature as manifested in the vegetable kingdom."

Thus did Wallace concentrate on his great life's work in biology and natural history, work that at times was physically rigorous and required the vigor and carefree attitude of youth. Contemplating the bigger picture of the universe as an octogenarian did not require fieldwork, simply a view of the heavens overhead, a voracious appetite for reading about the latest developments in astronomy, and correspondence with some of the leading astronomers of the day (Fig. 35.1).

Wallace's numerous recent biographers barely mention his interest in astronomy, yet hint at the importance it might have held for his overall worldview. In a single paragraph, Raby (2001, 271–272) notes:

> All Wallace's subjects, however disparate they might seem, or driven by circumstances, were closely connected, first teasing out the relationship between man and the rest of animate life, and then moving on to consider the physical conditions that gave rise to life in time and space; or seeking systems to improve the ways individuals and races shared the earth's resources more fairly, so that moral and intellectual progress, and happiness, could follow.

It was logical, Raby continues, for Wallace to address the bigger issues of astronomy that he had broached in *The Wonderful Century: Its Successes and its Failures* (Wallace 1898), and this he did in *Man's Place in the Universe* (Wallace 1903c), published 10 years before his death in 1913. The thesis of the latter book, Raby tells us, is "in essence, a philosophical or theological, assertion of intelligent cause and design over chance … Wallace would not accept that life was accidental, because he refused any explanation whose corollary was that man would die out by the continued operation of the same laws that had allowed man to evolve in the first place." As a kind of afterthought, or specific instance of this belief, Raby notes, Wallace's much slimmer volume *Is Mars Habitable?* argued against Percival Lowell's claim that intelligence exists on Mars.

Slotten (2004, 457–461) is only a bit more expansive in his biography of Wallace, devoting a half dozen of its 500 pages to Wallace and astronomy. After reviewing Wallace's arguments in *Man's Place in the Universe,* Slotten summarizes its thesis as follows:

> …that human beings, the culmination of conscious organic life, had arisen on our planet alone in the whole vast material universe. If human beings were the unique and supreme product of this vast universe, then some controlling Mind or Intelligence had conceived us for this very purpose. The immensity of the stellar universe, the long and slow and complex progress of nature, and the vast aeons of time that had passed before our development served as the raw materials and the spacious workshop for a Mind that 'produced' the planet that eventually resulted in humankind.

Despite the rejection of Wallace's arguments by some astronomers, Slotten suggests that the state of astronomy was unsettled enough at the time to make Wallace's position plausible, thereby giving solace to Wallace's refusal to believe that "humanity, with its faculties, aspirations, and powers for good and evil, was a simple by-product of random forces – that human beings were merely animals of no importance to the universe and requiring no great preparation for their advent." Like Raby, Slotten sees Wallace's Mars treatise as an expansion of these arguments. Underlying Wallace's argument against life on Mars, he says,

> …was an idiosyncratic view of the origin of the universe – one might call it teleological evolution, the belief that a controlling Mind or Intelligence manipulated natural laws for distinct ends. It was this belief, more than anything else, that had made so many of his scientific contemporaries uncomfortable. Essential to his theory was his anthropocentrism – a belief in humanity's unique position in the universe. Nothing, barring the discovery of intelligent life elsewhere in the universe, could change his mind on that issue. (Slotten 2004, 474–475)

Yet another recent biography (Shermer 2002) is similarly brief on Wallace and astronomy, but its author emphasizes how the basis for *Man's Place in the Universe* stretched back to Wallace's early days, and stretched forward to one of his last books, *The World of Life: A Manifestation of Creative Power, Directive Mind and Ultimate Purpose* (Wallace 1910). Wallace's raison d'etre, in Shermer's view, was

> …a belief in a purposeful cosmos that under the direction of a higher intelligence inexorably led to the appearance of humans who were capable of perfectibility and would, in time, achieve immortality of spirit. It was a consilient worldview that tied together his many and diverse interests and commitments, ideologies and philosophies, and was ultimately grounded in a unique form of Wallacean scientism. (Shermer 2002 230–231)

Like Raby and Slotten, Shermer connects *Is Mars Habitable?* with his earlier work: "Because Wallace had already committed himself several years earlier to the position that humans are unique in the cosmos, he could not let such apparent contradictory evidence be presented without a challenge" (Shermer, 2002, 294).

Considering the fundamental nature of the questions these biographers raise in connection with Wallace's astronomical writings, it is surprising those works have not been subjected to more detailed analysis. Though these writings came late, we shall argue that the ideas expressed in his two astronomy books are integral to his thought, and that the influence was from biology to astronomy rather than the reverse. Despite the fact that *Man's Place in the Universe* was written shortly after Wallace had moved houses, was desperate for money and verging on bankruptcy (Raby 2001; Slotten 2004)—and that a book on the popular subject of life on other worlds would help him financially—much more than money was involved. As Slotten noted, though the fragile state of Wallace's finances served as the proximate inspiration for his writings on astronomy, this was true of much of his work. Wallace himself noted in his autobiography "I feel that without the spur of necessity, I should not have done much of the work I have done" (Slotten 2004, 457–458). For Wallace the heavens and the Earth were united by an anthropocentric worldview, teleological evolution, and a special role for humans in the great chain of being.

35.2 Man's Place in the Universe[1]

In tackling the astronomical problem of life on other worlds, Wallace was tapping into a tradition well known in the world of the naturalist. "The question of questions for mankind – the problem which underlies all others, and is more deeply interesting than any other," T. H. Huxley wrote with regard to Darwin's theory of evolution, "is the ascertainment of the place which Man occupies in nature and of his relations to the universe of things" (Huxley 1863). Wallace certainly knew of this passage in Huxley's famous work *Man's Place in Nature*. And he certainly knew also of the larger plurality of worlds tradition that placed this question in the context of the larger universe. This controversy (Crowe 1986; Dick 1982) has a long tradition, and had raged especially in Britain since the publication of William Whewell's book on the subject at mid-century (Whewell 1853). That work went through five editions by 1859, with another printing in 1867. Concerned about the implications for Christian doctrine—especially Redemption and Incarnation—Whewell had argued that the Earth was the only world with life. Quickly disposing of one of the chief philosophical arguments of the pluralists, the teleological argument that the vast spaces of the universe must have some purpose, Whewell argued that confining intelligence to the "atom of space" that was the Earth was no worse than confining humanity to the "atom of time" that geology revealed intelligence had existed on Earth. On the more empirical side he argued that no proof existed of other solar systems, that the stars might not be exactly similar to our Sun, and that in any case many of them were binary stars whose putative planets would therefore not have conditions conducive for life. In our own Solar System, he argued, only Mars approached the conditions of the Earth, and it was just as likely as not that Mars was still in a condition of "preintelligence." Finally, Whewell cautioned against the unbridled use of the analogy argument in science. Crowe (1986) has discussed Whewell's pluralism in great detail, and Heffernan (1978) has compared the arguments of Whewell and Wallace. Although the nineteenth-century philosopher and the nineteenth-century naturalist differed in some of their arguments, particularly those having to do with purpose in the universe, they agreed in their conclusion that humans were likely the only intelligent life in the universe.

Both Whewell's and Wallace's antipluralist positions were very much in the minority in the context of nineteenth century science and natural theology. Whewell's treatise generated a tremendous amount of debate, but did little to weaken support for a plurality of worlds among scientists or the religious. Crowe (1986) documents twenty books and some 54 articles and reviews written in response to Whewell; of these about two-thirds still favored pluralism despite Whewell's arguments. Treatises such as Sir David Brewster's *More Worlds Than One: The Creed of the Philosopher and the Hope of the Christian* (1854) continued to be driven by an attachment to teleology and natural theology. It went through ten printings by 1871, and a third English edition in 1874 was reprinted in 1876 and 1895. By that time at least seven editions of Richard Proctor's *Other Worlds than Ours* (1870) had appeared. Although reconciliation with the doctrines of incarnation and redemption was never achieved, for most the benefits to natural theology overwhelmed Whewell's objections.

All of this, and its connections to the problem of anthropocentrism, did not escape Wallace's notice. As an evolutionist it was only natural that Wallace, like Huxley, should be interested in the issue of humanity's place in nature, and the plurality of worlds tradition allowed him to explore the subject on the broader canvas. Wallace had become aware of recent astronomical advances while writing four chapters for *The Wonderful Century* (Wallace 1898), and it is in this work that we find the source of his anthropocentric view of the universe. Here Wallace related his amazement at discovering the view of John Herschel, Simon Newcomb, and Sir Norman Lockyer that our Sun was situated near the center of the Milky Way system. Other research had shown that this system was finite, implying that our Sun and its accompanying planets were situated in the center of the entire universe. The startling fact of this privileged position, together with the indication based on planetary environmental conditions that the Earth is the only inhabited planet in the Solar

System, led Wallace to wonder if the Earth is the only inhabited planet in the whole universe. In addition, Wallace added,

> For many years I had paid special attention to the problem of the measurement of geological time, and also that of the mild climates and generally uniform conditions that had prevailed throughout all geological epochs; and on considering the number of concurrent causes and the delicate balance of conditions required to maintain such uniformity, I became still more convinced that the evidence was exceedingly strong against the probability of any other planet being inhabited. (Wallace 1903c, v–vi; see also Dawson 1903)

Wallace was particularly influenced by the British astronomer and historian Agnes Clerke's books *The System of the Stars* (1890) and *History of Astronomy* (1885), the American astronomer Simon Newcomb's *The Stars* (1902), Sir John Herschel's *Outlines of Astronomy* (1869 edition), and by the work of Lord Kelvin. As early as March 1901 Wallace had corresponded with Clerke about the possibility of a book on the subject of other worlds. But first would come an article published simultaneously in Britain and the United States in February and March 1903 (Wallace 1903a, b), 6 months before the book appeared in the autumn of that year. Beginning with the Copernican system and continuing with the vast Newtonian universe and the revelations of larger telescopes, Wallace argued that the importance of the Earth and its inhabitants had diminished. He argued that our Sun is located in the center of a cluster of suns, itself at the center of a finite stellar universe, and that only this central position in the stellar universe is suitable for life. His striking conclusion was

> ...that our position in the material universe is special and probably unique, and that it is such as to lend support to the view, held by many great thinkers and writers today, that the supreme end and purpose of this vast universe was the production and development of the living soul in the perishable body of man. (Wallace 1903b, 474)

By the end of August 1903 Wallace's book was ready for the press. The manuscript of the book (Wallace 1903d), preserved in the British Library in London, tellingly shows that the original title *Universe for Man* was replaced with *Man's Place in the Universe*. The book's "connected argument" for the Earth as the unique home of life in the universe began only after five chapters on the background of contemporary astronomy, chapters that Clerke later praised as a brilliant summary of the latest results in astronomy (Clerke 1904). From that point on, Wallace's main argument for the uniqueness of the Earth is based on three indispensable arguments: (1) life can exist only around our Sun or the cluster of suns surrounding it; (2) no life exists on planets around other suns in the solar cluster; and (3) no life exists in our Solar System beyond the Earth.

It is no exaggeration to say that the first argument was the dominant and catalyzing argument for the whole book. In supporting it, Wallace claimed that he was simply espousing the view of the most eminent astronomers of his day, a claim exaggerated but not completely wide of the mark. When Wallace wrote in 1903, all stars, and indeed all observable phenomena in the universe, were widely believed to be part of a single system, physically associated through the gravitational force, perhaps several thousand light years in diameter (compared to the 100,000 light years now estimated), with the Sun in a nearly central position. The *island universe theory*, which postulated many such systems, had been in gradual decline since the 1860s and had completely fallen from favor by the late 1880s (Smith 1982; Berenzden et al. 1976).

It is therefore not surprising that Wallace viewed the universe as a single system of stars with our Solar System at the approximate center. From the equality of star counts on both sides of the plane formed by the Milky Way—"the fundamental phenomenon upon which the argument set forth in this volume primarily rests"—Wallace argued that we reside in the central part of this plane and that the stellar system is spherical. Embedded in this general spherical structure of stars was a "solar cluster" consisting of a group of several hundred to several thousand stars surrounding our sun, and which seemed to form a condensed group of stars separate from the rest of the Milky Way. Wallace set the dimensions of the entire system at 3600 light years, placed the Sun at the outer margins of the solar cluster, and put it in orbit around the center of gravity of the cluster (Fig. 35.2).

The centrality of the Sun was "the very heart of the subject" of Wallace's inquiry for more than philosophical reasons, for on that position rested much of his argument against other worlds. The importance of centrality for Wallace was that only in such a position could a star such as the Sun generate a uniform heat supply over the long period needed for the original development of life on Earth, which he put at several million years. The origin of the Sun's power was a longstanding problem whose solution lay two decades in the future with an understanding of atomic processes in the Sun. Wallace's own theory must be judged to be as plausible as any other at the time. Drawing on Lord Kelvin's meteoric hypothesis, he argued that the solar cluster would, through gravitation, draw inflowing matter toward it from the outer region of the star system. This inrushing matter would furnish the energy for the formation and maintenance of luminous suns. Because this process was gravitationally most efficient near the center of the system, stars in other regions would presumably have much shorter lives, not sufficient for the development of life. This view of the universe thus reduced to perhaps several hundred the number of candidate stars that might harbor life-bearing planets.

In arguing against planetary systems around stars of the solar cluster, Wallace played a game of successive elimination. Some stars would be too small or too large for the appropriate long-term heat required for the development of life. Others were just in the process of forming. Others, recent research had shown, were gravitationally bound double stars inimical to the development of life. Even if planets existed around a few stars, they might not be the proper distance from the Sun, or have the proper mass. Looking at all these required conditions for life, Wallace concluded, "I submit that the probability is now all the other way" (Wallace 1903c, 278–285).

This brought the problem down to our own Solar System, and here Wallace launched into a detailed discussion of the complexity of life on Earth, which he saw as a prerequisite for understanding the possibility of life beyond Earth. Here Wallace brought to bear some of the problems on which he had spent a lifetime of contemplation, thus offering a unique view. As Clerke commented, "No great biologist has ever before seriously considered the possibilities of cosmic life, and they can only be fully discussed in the light of expert biological science" (Clerke 1904). This was

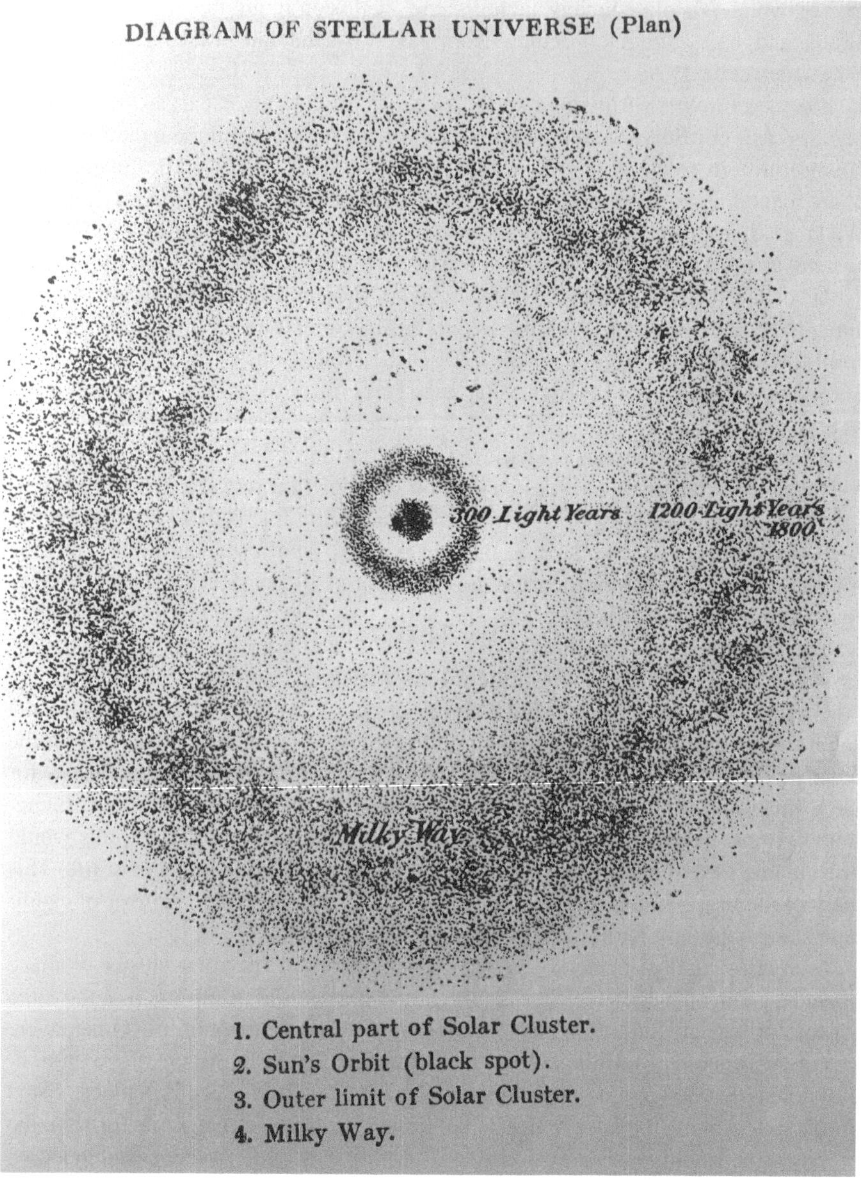

Fig. 35.2 Diagram of the stellar universe from Wallace's book *Man's Place in the Universe* (1904)

a difficult task; in many ways biology did not even become a science until the nineteenth century, in part through events surrounding Darwin and Wallace themselves, and the study of physiology was still at the organic rather than the microscopic level (Allen 1975).

Nevertheless, Wallace plunged ahead with one of the greatest questions of all: what is life? Rejecting philosophical definitions given by Aristotle, Spencer and others as vague and abstract, and dissatisfied as well with definitions based on general properties, Wallace quickly focused on protoplasm, which Huxley had called "the physical basis of life." He marveled that it is composed basically of only the four elements hydrogen, carbon, nitrogen and oxygen, and that the variety of life is produced by the amazing ability of carbon to form compounds. The chemical reactions in the protoplasm, Wallace argued, determined the physical conditions necessary for the development and maintenance of life on Earth: a regular heat supply, resulting in a limited range of temperatures; a sufficient amount of solar light and heat; water in great abundance and universally distributed; an atmosphere of sufficient density; and alterations of day and night to keep the temperature in balance.

Life, Wallace believed, must have a temperature range from 32 to 104 °F, while solar light is necessary for its essential role in decomposing carbon dioxide into carbon and oxygen in plants. Water not only constitutes a large proportion of the living body of both planets and animals, but it is also essential in producing a limited range of temperatures. And the atmosphere must be of sufficient density to store heat and to supply oxygen, carbon dioxide, and water vapor for life; he estimates that a reduction of the density by one-fourth would probably render the Earth uninhabitable. Thus, any change in this great variety of conditions would be inimical to life. In order to achieve these conditions on a planetary scale, the essential factors were the distance from the Sun, the obliquity of the ecliptic (causing the seasons), the persistence of a mild climate through geological time, and the distribution of water.

The question, then, is whether this complex combination of conditions is found beyond the Earth. Wallace leaves no doubt of his opinion: considering all the causes, "it seems in the highest degree improbable that they can *all* again be found combined either in the solar system or even in the stellar universe"(Wallace 1903c, 310). Pointing out that there was no reason to believe life could thrive except under conditions similar to those on Earth, Wallace concluded, "We may therefore feel it to be an almost certain conclusion that – the elements being the same, the laws which act upon and combine and modify those elements being the same – organized living beings wherever they may exist in this universe must be fundamentally, and in essential nature, the same also." And as a corollary, "Within the universe we know, there is not the slightest reason to suppose organic life to be possible, except under the same general conditions and laws which prevail here" (Wallace 1903c, 182–189).

Turning to the planets of our Solar System, Wallace next argued that planetary habitability depends primarily on the mass of the planet, for this determines whether or not it can retain the molecules that compose and atmosphere. Thus Mercury and Mars are too small to retain water vapor. Because of their low density the larger planets "can have very little solid matter" on which life might develop. This left only Venus, but Venus was believed to always keep the same face toward the Sun, so one side is too cold, the other too hot. (This is now known not to be true; Venus is uniformly hot at a temperature of about 900 °F due to the greenhouse effect of its atmosphere—thus ruling out life in any case.) Moreover, the lifetime of the Sun renders it impossible that these planets now unsuitable for life might have been suitable in the past or could be in the future:

> We are, therefore, again brought to the conclusion that there has been, and is, no time to spare; that the whole of the available past life-period of the sun has been utilized for life-development on the earth, and that the future will be not much more than may be needed for the completion of the grand drama of human history, and the development of the full possibilities of the mental and moral nature of man. (Wallace 1903c, 258–277)

Although this ended Wallace's "connected argument," he also put forth very briefly a philosophical argument from purpose. If the Earth is the only inhabited planet in the universe, Wallace asserted, this may be seen either as a coincidence or as a very important conclusion indicating that the universe was brought into existence for the ultimate purpose of the development of man on Earth. While Wallace believed the majority of scientific men would call it coincidence, he left no doubt that he was of the opposite opinion, and that this opinion was contrary neither to science nor to religion. Life on every planet

> …would introduce monotony into a universe whose grand character and teaching is endless diversity. It would imply that to produce the living soul in the marvelous and glorious body of man … was an easy matter which could be brought about anywhere, in any world. It would imply that man is an animal and nothing more, is of no importance I the universe, needed no great preparations for his advent, only perhaps, a second-rate demon, or a third or fourth-rate earth. (Wallace 1903c, 317–318)

This conclusion was diametrically opposed to natural theology, which had consistently argued for two centuries that other worlds would declare the glory of the Creator. By contrast, Wallace believed the immensity of space and time "seem only the appropriate and harmonious surroundings, the necessary supply of material, the sufficiently spacious workshop for the production of that planet which was to produce, first, the organic world, and then, Man" (Wallace 1903c, 317–318). Although isolated and briefly stated, this argument from purpose perhaps played a greater role as a driving force than Wallace would admit. As Heffernan (1978, 96) concluded, for both Whewell and Wallace "their private interest in plurality sprang from extra-scientific convictions as to the purpose of the universe." The same is true not only of their interest, but also of their conclusions.

A final argument, "An Additional Argument Dependent on the Theory of Evolution," was added to the 1904 fourth edition of Wallace's book. Especially interesting because Wallace was so closely involved with the evolution arguments of his day, it is independent of the three connected scientific arguments and may be seen as another aspect leading to the same conclusion. Wallace argued that since humanity is the result of a long chain of modifications in organic life, since these modifications occur only under certain circumstances, and since the chances of the same conditions and modifications occurring elsewhere in the universe were very small, the chances of beings in human form existing on other planets was very small. Moreover, since no other animal on Earth, despite the great variety and diversity of forms, approaches the intelligent or moral nature of humanity, Wallace concluded that intelligence in any other form was also highly improbable:

> If the physical or cosmical improbabilities as set forth in the body of this volume are somewhere about a million to one, then evolutionary improbabilities now urged cannot be considered to be less than perhaps a hundred millions to one; and the total chances against the evolution of man, or an equivalent moral and intellectual being, in any other planet, through the known laws of evolution, will be represented by a hundred million of millions to one (Wallace 1904, Appendix, 334–335).

By its deletion of the idea of purpose from his central argument and by its inclusion of biological aspects, Wallace's book marked a signal advance in the debate about other worlds. It asked some of the same fundamental questions about life that are still asked today in the field of astrobiology (Dick and Strick 2004). Its approach to eliminating other life sites based on conditions believed necessary for life, and concluding that Earth was unique or rare, was one adopted almost a century later by reputable scientists (Ward and Brownlee (2000), and his view that life might exist, but not intelligence, was similar to their conclusion.

Yet, the book's thesis about the location of humanity at the center of the universe was recognized as a mistake even before the book was published. Reviewing the article on which the book was based, the British astronomer E. Walter Maunder concluded that "Every one … of Dr. Wallace's demonstrations falls to the ground" (Maunder 1903). H. H. Turner, Savilian Professor of Astronomy at Oxford, charged the Wallace "seems to me to have unconsciously got his facts distorted, and to indicate practically nothing wherewith to link them to his conclusions" (Turner 1903). In the United States Harvard astronomer and Lowellian W. H. Pickering rejected Wallace's argument, characterizing it as "not science, and …not very satisfactory" (Pickering 1903). And in France, pluralist par excellence Camille Flammarion surprised no one with his conclusion that "An examination of Mr. Wallace's plea in favor of his geocentric and anthropocentric theory has not convinced me; on the contrary, it seems to me to give a more solid basis than ever to the opposite opinion" (Flammarion 1903). Wallace's book received more than 40 reviews, most of them, with the exception of a few such as that of Agnes Clerke, highly skeptical.

Though *Man's Place in the Universe* went through seven editions by 1908 and another in 1914, and was translated into German in 1903 and French in 1907, it had little influence beyond the second decade of the twentieth century. The reason is not far to seek. Within 15 years of Wallace's death in 1913, most of his central assumptions had been rendered obsolete by an emerging new cosmology. In 1918 the American astronomer Harlow Shapley reported, based on his study of the distribution of globular clusters of stars, that our Solar System was located in a very eccentric position in the galaxy, at its periphery rather than its center. By 1924 Edwin P. Hubble had demonstrated to the satisfaction of most astronomers that many other galaxies exist outside our own, galaxies that he showed a few years later are fleeing from one another in an "expanding universe"(Smith 1982). And beginning in the 1920s Arthur S. Eddington and others devised energy-producing mechanisms for stars that swept away any need for an infalling matter theory such as Wallace proposed. We now know that we live in a universe billions of light years in extent, characterized by an interrelation among parts and the whole that astronomers characterize by the term *cosmic evolution*. Though Wallace recognized the evolution of the stars based on the contemporary work of astronomers (Wallace 1903c, 128–134), neither he nor they could have known the extent of full-blown cosmic evolution, ranging from the Big Bang to the present and covering some 13.7 billion years of time (Chaisson 2001).

In the end, the failure of *Man's Place in the Universe* is marked by the dominance of the anthropocentric worldview over all other arguments. Despite his best intentions and protestations to the contrary, in the end Wallace's theory was simply too ad hoc, less driven by probabilities than by deep-seated anthropocentrism and teleology. Convinced by 1898 of the nearly central position of the Sun, Wallace first sought and found the significance of this fact in the uniqueness of life, and then adduced arguments in favor of the view that advanced life was found beyond the Earth neither in our Solar System nor in others. Although his theory was based on what he believed to be the fact of the Sun's centrality, the pioneer of evolution nevertheless fell victim to the effect of a worldview with insufficient proof. In his concept of the plurality of worlds, no less than in his ideas about the evolution of humanity, he found it necessary to set humanity apart. Although adopting a scientific approach, Wallace's book serves as a lesson on the limits of science when worldviews dominate empirical evidence. It is a lesson we still need to remember in the twenty-first century.

35.3 Life on Mars

As Wallace's biographers hint, his treatise on Mars was a more specific instance of his broader claims for an anthropocentric universe, which meant a universe without intelligent life beyond Earth. Indeed, Wallace himself makes the connection in the opening words of *Is Mars Inhabitable?*: "This small volume was commenced as a review article on Professor Percival Lowell's book, *Mars and its Canals*, with the object of showing that the large amount of new and interesting facts contained in this work did not invalidate the conclusion I had reached in 1902, and stated in my book on *Man's Place in the Universe*, that Mars was not habitable" (Wallace 1907). These words, and the book's subtitle, "A Critical Examination of Professor Lowell's Book 'Mars and Its Canals,' With an Alternative Explanation," place the volume squarely in the plurality of worlds tradition and more specifically in the context of the claims of its most spectacular adherent, Percival Lowell. The treatise is of interest not only for these reasons, but also because it illustrates Wallace's insistence on physical evidence for a theory, and for its criticism of Lowell's predisposition toward an idea and his use of the concept of purpose, both of which led him astray. Ironically, despite Wallace's laudable insistence on physical evidence, it was the latter two ideas, in the form of anthropocentrism and the purpose of the universe, that (we now know) also led him astray in *Man's Place in the Universe.*

Percival Lowell the man and his ideas have by now been well researched. His most recent (and best) biographer, David Strauss, has emphasized how Lowell was one of the principal American disciples of Herbert Spencer, Wallace's British contemporary, principally known for his evolutionary theories applied to culture (Strauss 2001, 6, 97–113). Wallace was also a Spencerian in many ways, and even named his first son Herbert Spencer Wallace (Shermer 2002, 240–242). Strauss shows in detail how Lowell's work was "shaped by his lifelong commitment to the realization of Herbert Spencer's cosmic evolutionary project. Lowell embraced Spencer's concept of a cosmos governed by the law of evolution to explain the development of both the natural and the social world. He applied Spencer's evolutionary scheme during his travels in East Asia and again as he launched his search for extraterrestrial life (Strauss 2001, 97)."

That search for life beyond Earth centered on Mars and its supposed canals. The controversy has its roots in the observations of the Italian astronomer Giovanni Schiaparelli, made during the favorable opposition of Mars in 1877 (the same year its two moons were discovered), that the planet was crisscrossed by a network of straight lines. The Schiaparellian Mars that Lowell inherited was a planet with two polar caps composed of snow and ice, seas and continents arranged very differently from those on Earth, and an atmosphere rich in water vapor. The vaporous atmosphere he believed was supported not only by changes in the polar caps requiring a transportation mechanism for water vapor, but also by spectroscopic observations. It was also a planet of change, for the melting polar caps seemed to produce a temporary sea around the northern cap. Schiaparelli believed this water was distributed over great distances by "a network of canals, perhaps constituting the principle mechanism (if not the only one) by which water (and with it organic life) may be diffused over the arid surface of the planet." How such a system of lines could originate had led some to see them as the work of Martians. Schiaparelli was inclined to consider them as produced by the evolution of the planet, similar to features such as the English Channel on Earth. But he left the door open to the artificial hypothesis: "I am very careful not to combat this supposition, which includes nothing impossible" (Dick 1996, 69–70).

Lowell walked through this open door without hesitation. In January 1894 Lowell made the decision to finance an expedition to Arizona, where he founded the Lowell Observatory. Its sole purpose was to study the Mars problem, and in particular the origin of the canals with the possibility of demonstrating their artificial nature. August of that year was another particularly favorable opposition of Mars, and Lowell made the best of it. Already in 1895 Lowell published his first book on the subject, entitled simply *Mars*, where he argued that Mars was not only habitable but inhabited, by intelligent Martians who had built the canals. This was followed by *Mars and its Canals* (1906)—which precipitated Wallace's book, and then by *Mars as the Abode of Life* (1908) and numerous other publications in which Lowell attempted to prove the artificial nature of the canals. The details of that controversy have been given elsewhere (Hoyt 1976; Crowe 1986; Dick 1996), but let us now return to Wallace and his place in the debate.

Wallace's reaction to Lowell's work is characterized by its insistence on good physical evidence and sound arguments based on that evidence. We need not go through those arguments in detail, but Wallace's summary at the end of his volume will suffice to show the nature of his arguments:

> (1) All physicists are agreed that, owing to the distance of Mars from the sun, it would have a mean temperature of about −35 °F (= 456 °F abs.) even if it had an atmosphere as dense as ours.
>
> (2) But the very low temperatures on the earth under the equator, at a height where the barometer stands at about three times as high as on Mars, proves, that from scantiness of atmosphere alone Mars cannot possibly have a temperature as high as the freezing point of water; and this proof is supported by Langley's determination of the low *maximum* temperature of the full moon.
>
> The combination of these two results must bring down the temperature of Mars to a degree wholly incompatible with the existence of animal life.
>
> (3) The quite independent proof that water-vapor cannot exist on Mars, and that therefore, the first essential of organic life – water – is non-existent.
>
> The conclusion from these three independent proofs, which enforce each other in the multiple ratio of their respective weights, is therefore irresistible – that animal life, especially in its higher forms, cannot exist on the planet.
>
> Mars, therefore, is not only uninhabited by intelligent beings such as Mr. Lowell postulates, but is absolutely UNINHABITABLE.

Wallace's insistence that evidence be primary is ironic because Lowell prided himself on his empirical observations of Mars. As detailed by both Dick (1996) and Crowe (1986), the problem in the canals of Mars controversy is not with the lack of observations but with their difficulty and interpretation. As Lowell's recent biographer has pointed out, Lowell's observations were theory driven, and thus

> …reflected in important ways deep-seated intellectual preferences on his part. To be sure, Lowell undertook painstaking empirical investigations of the canals of Mars over twenty-two years … And he was certainly following accepted scientific practice in using his data to confirm a theory. What distinguished Lowell from some of his counterparts, however, was his plan to use the data to support Spencer's all-embracing evolutionary scheme. In attempting to link data and theory, Lowell frequently engaged in thinking of an imaginative sort. Thus, he confirmed Spencer's theory to his own satisfaction, but he acquired a reputation among his peers for selectivity in collecting and interpreting data. (Strauss 2001, 97).

The canals-of-Mars debate embodies one of the great problems in history and philosophy of science: the relation between theory and observation. Wallace was no stranger to this relationship, nor was Darwin. They also had to adduce a mass of facts in constructing their theory of evolution by natural selection. Their theory, however, was at a different level than Lowell's, and therefore of a different nature, and the differences highlight the ambiguities of the word "theory." Lowell's theory of artificial canals on Mars has been definitively disproven by better observations—spacecraft to Mars have shown no features resembling canals, much less artificial constructions (the "face on Mars" notwithstanding). The theory of evolution by natural selection, while based on a mass of empirical facts, is much more subtle in not representing a single planetary object at a given time, but numerous biological outcomes developed over a great length of time.

There is another problem that Wallace detects in Lowell's argument, the problem of teleology. Speaking of Lowell's supposed canals Wallace notes

Again, he urges the "purpose" displayed in these "canals." Their being *all* so straight, *all* describing great circles of the "sphere," all being so evidently arranged (as he thinks) either to carry water to some "oasis" 2000 miles away, or to reach some arid region far over the equator in the opposite hemisphere! But he never considers the difficulties this implies. Everywhere these canals run for thousands of miles across waterless deserts, forming a system and indicating a purpose, the wonderful perfection of which he is never tired of dwelling upon (but which I myself can nowhere perceive) (Wallace 1907, 103)

In this criticism Wallace was correct. But it is indeed ironic in light of Wallace's own tendencies toward a teleological worldview on a much grander scale, which we have seen displayed in *Man's Place in the Universe*, and to which we now return.

35.4 Wallace and Purpose in the Universe

Although Wallace's anthropocentric geography (or uranography) of the universe was quickly outdated, his more general idea of purpose in the universe—represented in his statement that "the supreme end and purpose of this vast universe was the production and development of the living soul in the perishable body of man"—rather surprisingly made a comeback in a different guise in scientific circles during the last quarter of the twentieth century. Moreover, in the last chapter of *Man's Place in the Universe* Wallace wrote as follows:

> Lastly, I submit that the whole of the evidence I have here brought together leads to the conclusion that our earth is almost certainly the only inhabited planet in our solar system; and, further, that there is no inconceivability – no improbability even – in the conception that, in order to produce a world that should be precisely adapted in every detail for the orderly development of organic life culminating in man, such a vast and complex universe as that which we know exists around us, may have been absolutely required. (Wallace 1903c, 306).

These ideas relate to what today is called anthropic reasoning and its counterpart, intelligent design. In order to see just how they relate we need to understand the modern idea of the anthropic principle. The basis for the principle is the increasing realization, founded on modern observational data, that the universe appears to be fine-tuned for life. In other words, if the values of some of the physical constants, the strengths of the fundamental forces, and the masses and charges of subatomic particles were even slightly different, stars and planets could not form, and could not give rise to the conditions necessary for life. This leaves two possibilities: either an intelligent entity (which may or may not be the "God" of terrestrial history) designed the universe to be this way (intelligent design), or there is some natural explanation. The latter is often framed in terms of what is called the multiverse (Davies 2007; Rees 1997, 1999, 2001), namely that our universe is just one of an ensemble of universes, some of which are suitable for life, and others of which are not. We happen to be in one that is so finely tuned; in the words of Davies (2007) we have hit the "cosmic jackpot," and so we should not be surprised. Each of these two scenarios has its weaknesses: the first is a supernatural explanation (if one invokes a supernatural entity), while the second is natural, but thus far just as empirically unverified as intelligent design.

These themes are controversial for both scientific and religious reasons. As Wilson (1993, 11) wrote:

> What makes the anthropic principle approach provocative is that it hints at the existence of deep connections between the universe and humanity, makes striking claims about mankind's place in the universe, and is reminiscent of classical arguments for a divine cosmic designer. The principle involves additional philosophical and religious themes as well: the significance and value of human life, the goal-directedness of nature, and the degree to which the universe is anthropocentric.

The concepts of an anthropic principle, anthropocentrism, and design in nature each have their own history, but at the same time are related to each other. One can be anthropocentric (human-centered) without invoking design or the anthropic principle. It is difficult, however, to invoke the anthropic principle and not be anthropocentric, unless one plausibly redefines it as a biocentric principle (see below). And the anthropic principle implies design to some people, especially the theologically inclined (Harris 1991, 1992; Ross 1993), who use the fine-tuned universe to argue for design by a creator God. Others, however, look for a natural, rather than a supernatural explanation for such a universe.

The design argument is very old indeed, as Barrow and Tipler (1986) detail in their massive volume on the anthropic cosmological principle. It was often used in the natural theology arguments of the seventeenth through nineteenth centuries, and is commonly invoked even today to explain those things not fully understood in astronomy or biology, as in the idea of *irreducible complexity* (Behe 1996). The more specific idea that the world is fine-tuned for life is not so old. It was anticipated by Wallace's American contemporary, the Harvard biochemist Lawrence J. Henderson, who marveled at the fitness of the terrestrial environment for life, especially because of the properties of water (Fry 1996). In the final paragraph to his *The Fitness of the Environment*, Henderson (1913, 312), even extended his argument to the universe at large:

> The properties of matter and the course of cosmic evolution are now seen to be intimately related to the structure of the living being and to its activities; they become, therefore far more important in biology than previously suspected. For the whole evolutionary process, both cosmic and organic, is one, and the biologist may now rightly regard the universe in its very essence as biocentric.

The *Biocentric Principle* would have been a better term for the fitness of the universe for life, since it does not imply that terrestrial lifeforms, including humans, could be the only form of life. The universe was fine-tuned for life, including beings of higher consciousness, not only for humans, even if the only intelligent life in the universe we yet know is human. But when these questions were once again pondered six decades later, the term *anthropic principle* was coined, almost by accident. This occurred in a path-breaking article on the subject by astronomer Brandon

Carter (1974), who first formulated what he called the weak anthropic principle: "what we can expect to observe must be restricted by the conditions necessary for our presence as observers." Thus the gravitational constant is constrained by the fact that we exist, otherwise the conditions for life could never have arisen. Carter went on to expound a "strong anthropic principle," namely, "that the universe (and hence the fundamental parameters on which it depends) must be such as to admit the creation of observers within it at some stage," a much more problematic claim. Although such reasoning turned the deductive method on its head, Carter argued that it might be considered a kind of explanation for why our universe is the way it is.

These two versions of the anthropic principle embody two interpretations of the old idea of "final causes:" is the universe just biofriendly, or is life written into the laws of the universe so that the existence of life and intelligence is predetermined? Moreover, they also embody the age-old question of chance and necessity in the workings of Nature. The question is still a very active one in astrobiology, cosmology, origins of life studies, and science in general, especially since the French biologist and Nobelist Jacques Monod stated in 1971 that "The universe was not pregnant with life, nor the biosphere with man. Our number came up in a Monte Carlo game" (Monod 1971). To which another Nobelist replied almost 25 years later:

> The Earth is part, together with trillions of other Earth-like bodies, of a cosmic cloud of "vital dust" that exists because the universe is what it is. Avoiding any mention of design, we may, in a purely factual sense, state that the universe is constructed in such a way that this multitude of life-bearing planets was bound to arise··· The universe is not the inert cosmos of the physicists, with a little life added for good measure. The universe *is* life, with the necessary infrastructure around; it consists foremost of trillions of biospheres generated and sustained by the rest of the universe. (De Duve 1995)

There is no doubt that Wallace denied chance in connection with the origin and fate of humans. And with all his discussion about the conditions necessary for life, Wallace would have embraced the weak anthropic principle, though he had no idea of the quantitative underpinnings in the physical constants. But he never came close to stating the strong version, that the universe could not exist without life. In fact, because Wallace believed he had proved the Earth to be the only abode for life, the term "anthropic" is much more aptly applied to his beliefs than to today's formulation, which is really a biocentric principle. A universe fit for life does not necessarily imply that life is the purpose of the universe, as Wallace stated. But Barrow and Tipler (1986) in a later version of the anthropic principle made exactly this teleological argument: that habitability is the goal of the universe. In this version, they argued, life could never die out, because then the goal of the universe would cease to exist.

Because Wallace opted for a final causes explanation for the purpose he saw in the universe, it might seem he would also have accepted the idea now known as intelligent design (Numbers 2006, 373 ff), the modern version of the design argument. That such is not the case seems clear from his "Notes Added to the Second Edition of *Contributions to the Theory of Natural Selection*"(Wallace 1871). Here

Wallace explicitly distinguishes between a supernatural God or Deity and a superior or controlling intelligence, especially when it comes to the origins of humans:

> Now, in referring to the origin of man, and its possible determining causes, I have used the words "some other power" – "some intelligent power" – "a superior intelligence" – "a controlling intelligence," and only in reference to the origin of universal forces and laws have I spoken of the will or power of "one Supreme Intelligence." These are the only expressions I have used in alluding to the power which I believe has acted in the case of man, and they were purposely chosen to show, that I reject the hypothesis of "first causes" for any and every *special* effect in the universe, except in the same sense that the action of man or of any other intelligent being is a first cause. In using such terms I wished to show plainly, that I contemplated the possibility that the development of the essentially human portions of man's structure and intellect may have been determined by the directing influence of some higher intelligent beings, acting through natural and universal laws.

In this opinion, Wallace anticipated the view of one of the giants of twentieth century astronomy, Fred Hoyle, who made a similar argument in his book *The Intelligent Universe: A New View of Creation and Evolution* (Hoyle 1983). Here Hoyle, a self-proclaimed atheist, concluded that the simplest explanation for the biofriendly universe was that a natural (not supernatural) superintellect had engineered the universe to make it fit for carbon-based life and intelligence. This view has recently been elaborated in two books by complexity theorist James Gardner, whose subtitles reveal their thesis: *Biocosm – The New Scientific Theory of Evolution: Intelligent Life is the Architect of the Universe* (Gardner 2003), and *The Intelligent Universe: AI, ET, and the Emerging Mind of the Cosmos* (Gardner 2007). It is thus clear that the concept of God as a supernatural entity and the concept of superior intelligence can be separated. Alternately, they could also be united by the unconventional idea of a *natural God* (Dick 2000), though Wallace himself never took this step. Although written more than three decades earlier, Wallace's 1871 "Note" needs to be kept in mind in the context of his later astronomical volumes. In an interview 1 month after the publication of *Man's Place in the Universe*, Wallace was asked if he agreed with the view of Sir Oliver Lodge that "the attempt to explain the universe by chance has absolutely failed. It must have had a designer," which the interviewer described as an Infinite Being. "Certainly," Wallace replied. "My whole argument tends in that direction, though my object in writing 'Man's Place in the Universe' was purely scientific, not religious" (Dawson 1903). Again, Wallace's appeal to science rather than religion indicates this "designer" may be seen as a non-supernatural superior intelligence

Nevertheless, in the same interview, as elsewhere in his books on natural selection, Wallace clearly set himself apart from Darwin, who believed evolution applied to all aspects of humans, including the physical, mental, and moral. Wallace even had other exceptions to Darwinian thinking: "I do not think it is possible to form any idea beyond this, that when man's body was prepared to receive it, there occurred an inbreathing of spirit—call it what you will. I believe this influx took place at three stages in evolution – the change (1) from the inorganic to the organic, (2) from the plant to the animal, (3) from the animal to the soul of man. Evolution seems to me to fail to account for these tremendous transitions" (Dawson 1903). These ideas were part and parcel of his anthropocentric teleological worldview. This is clear even from the title of Wallace's 1910 volume *The World of Life: A Manifestation of Creative Power, Directive Mind and Ultimate Purpose* (Wallace 1910).

Intelligent design must also be starkly distinguished from the idea of late-twentieth century creationism (today called "scientific creationism"), which implies that the world was created in less than 10,000 years (Numbers 2006, 8–9). With his vast knowledge of biology and geology—and evolution—Wallace knew better than that, putting the time scale for the history of life at about 100 million years. Nonetheless, it seems today ironic, in a world where evolutionists are often set against the intelligent design movement, that one of the founders of evolution by natural selection should embrace final causes, though neither intelligent design in its modern meaning, nor creationism.

35.5 Conclusion: Wallace and the Connections Between Biology and Cosmology

Lest there be any doubt of the connections between Wallace's own work on biology and astronomy, we may compare two passages of his writings. The first comes from his book *Darwinism* (Wallace 1889), which Shermer calls Wallace's "definitive statement" on the subject, written when he was 66. Wallace rails against those who think humans "are but the products of blind eternal forces of the universe," who came into being by chance and would once again disappear when the Sun lost its heat energy:

> As contrasted with this hopeless and soul-deadening belief, we, who accept the existence of a spiritual world, can look upon the universe as a grand consistent whole adapted in all its parts to the development of spiritual beings capable of indefinite life and perfectibility. To us, the whole purpose, the only raison d'etre of the world – with all its complexities of physical structure, with its grand geological progress, the slow evolution of the vegetable and animal kingdoms, and the ultimate appearance of man – was the development of the human spirit in association with the human body. (Wallace 1889, 476–477; see also Shermer 2002, 230–231, and see also p. 33).

The second passage we have already cited from the article "Man's Place in the Universe", written 14 years later and providing the foundation for the book by the same name. There Wallace argued that modern astronomy now showed

> That our position in the material universe is special and probably unique, and that it is such as to lend support to the view, held by many great thinkers and writers today, that the supreme end and purpose of this vast universe was the production and development of the living soul in the perishable body of man. (Wallace 1903b, 474)

These passages clearly show that by the end of his life, Wallace's biology and astronomy were part and parcel of the same anthropocentric and teleological world-view—one that incorporated final causes, but not first causes in the sense of a super-natural deity. Despite a lifelong interest in astronomy, because his astronomical writings came so late in life, it seems likely that his biological worldview affected his cosmology rather than the reverse. But in either case the publication of Wallace's two books on astronomy reveals a consistent and consilient worldview, one that must have been a great satisfaction to him in his final years. Within a few years after his death developments in astronomy would shatter his anthropocentric worldview of man's place in the universe. But his teleological worldview remains very much alive in current discussions of the anthropic principle and intelligent design, both—undoubtedly to the delight of the now disembodied Wallace—still considered out of the mainstream of modern science.

In the end the significance of Wallace's astronomical writings is not in their failure to grasp the geography of the universe. In that he was a man of his times, grappling with the limited observations available at the turn of the twentieth century. The significance is rather in his joint treatment of astronomy and biology, not only in anticipating some of the arguments and content of the current discipline of astrobiology, but also the profound questions of the cosmological connections between the two, bearing on whether the universe is by its very nature biocentric. The concept of a biological universe (Dick 1996) encompasses both the question of the existence of life in the universe, and why the universe should be so biofriendly. Today, even as astrobiologists search for life, others ponder the mysteries of why life can exist at all. British Astronomer Sir Martin Rees (1997, 1999, 2001), physicist and astrobiologist Paul Davies (1982, 2007) and many others are now exploring the ramifications of a possibly deep and profound relationship between biology and cosmology.

In this connection twenty years ago the physicist Freeman Dyson called for the need to build bridges between biology and cosmology. "Only a few heretics … dare to express the view that the structure of the universe may not be unambiguously reducible to a problem of physics," he wrote in his iconoclastic book *Infinite in All Directions*. "Only a few romantics like me continue to hope that one day the links between biology and cosmology may be restored." He suggested further that

> The prospects are bright for a future-oriented science, joining together in a disciplined fashion the resources of biology and cosmology. When this new science has grown mature enough to differentiate itself clearly from the surrounding farrago of myth and fiction, it might call itself "cosmic ecology," the science of life in interaction with the cosmos as a whole. Cosmic ecology would look to the future rather than to the past for its subject matter, and would admit life and intelligence on an equal footing with general relativity as factors influencing the evolution of the universe. (Dyson 1988)

The heretical Wallace might have agreed with this sentiment had it been limited to humans, which he believed to be the sole intelligence in the universe. As astronomers search the universe for life, we cannot be so confident of such an anthropocentric and teleological world view today. We do not know if the universe is full of intelligence, or if humans are its ultimate creation—and therefore whether the connections between biology and cosmology, if they exist, will reveal an anthropic or a biocentric universe. But if and when such a science of cosmic ecology does develop, Wallace—along with his many other roles—may be seen as a significant early precursor, even if his belief in final causes may not lead to final answers.

35.6 Commentary 2020

This chapter appeared in Charles Smith and George Beccaloni's collection of essays on Wallace (Smith and Beccaloni 2008). That volume covers Wallace's remarkable career in many areas, not just as a naturalist exploring the idea of natural selection, but also as a world traveler, social justice warrior, and many other aspects of his full career. Much of interest about Wallace, including news, images, resources, bibliographies, books, and links, may be found at two websites founded by Beccaloni and Smith: http://wallacefund.info/users/george-beccaloni and http://people.wku.edu/charles.smith/index1.htm.

Note

1. Parts of this section are reprinted with permission of Cambridge University Press from Chap. 2 of Dick (1996).

References

Allen, G. 1975. *Life Science in the 20th Century.* New York: Cambridge University Press.

Barrow, J. D, and F. J. Tipler. 1986. *The Anthropic Cosmological Principle.* New York: Oxford University Press

Behe, M. J. 1996. *Darwin's Black Box: The Biochemical Challenge to Evolution.* New York: Free Press.

Berenzden, R., R. Hart and D. Seely. 1976. *Man Discovers the Galaxies.*

Carter, B. 1974. "Large Number Coincidences and the Anthropic Principle in cosmology," in M. S. Longair, ed. *Confrontation of Cosmological Theories with Observational Data* (Dordrecht, 1974), 291–298.

Chaisson, E. 2001. *Cosmic Evolution: The Rise of Complexity in Nature.* Cambridge, Mass.: Harvard University Press.

Clerke, A. 1904. "Life in the Universe," *Edinburgh Review*, 200, p. 60.

Crowe, M. J. 1986. *The Extraterrestrial Life Debate 1750-1900: The Idea of a Plurality of Worlds from Kant to Lowell.* Cambridge: Cambridge University Press.

Davies, P. 1982. *The Accidental Universe.* Cambridge: Cambridge University Press.

Davies, P. 2007. *The Cosmic Jackpot. Why our Universe is Just Right for Life.* Boston: Houghton-Mifflin

Dawson, A. 1903. "A visit to Dr. Alfred Russel Wallace," An interview by Albert Dawson printed in *The Christian Commonwealth* of 10 December 1903. online at http://www.wku.edu/~smithch/wallace/S741.htm

De Duve, Christian.1995. *Vital Dust: Life as a Cosmic Imperative.* New York: Basic Books.

Dick, S. J. 1982. *Plurality of Worlds: The Origins of the Extraterrestrial Life Debate from Democritus to Kant.* Cambridge: Cambridge University Press.

Dick, S. J.1996. *The Biological Universe: The Twentieth Century Extraterrestrial Life Debate and the Limits of Science.* Cambridge: Cambridge University Press.

Dick, S. J. 2000. "Cosmotheology: Theological Implications of the New Universe," in Steven J. Dick, ed, *Many Worlds. The New Universe, Extraterrestrial Life and the Theological Implications.* Philadelphia and London: Templeton Foundation Press, 191–210

Dick, Steven J. and J. E. Strick, 2004. *The Living Universe: NASA and the Development of Astrobiology.* New Brunswick: Rutgers University Press.

Dyson, F. 1988. *Infinite in all Directions.* New York: Harper and Row, pp. 50–51.

Flammarion, C. 1903. "The Earth and Man in the Universe," *The Independent*, 55, 958–968.

Fry, I. (1996). "On the Biological Significance of the Properties of Matter: L. J. Henderson's Theory of the Fitness of the Environment," *Journal of the History of Biology*, 29, 155–196.

Gardner, J. N. 2003. *Biocosm: The New Scientific Theory of Evolution: Intelligent Life is the Architect of the Universe*. Maui, HI: Inner Ocean Publishing.

Gardner, J. N. 2007. *The Intelligent Universe: AI, ET, and the Emerging Mind of the Cosmos.* Franklin Lakes, N.J., New Page Books.

Harris. E. E. 1991. *Cosmos and Anthropos: A Philosophical Interpretation of the Anthropic Cosmological Principle.* New Jersey and London: Humanities Press International

Harris. E. E. 1992. *Cosmos and Theos: Ethical and Theological Implications of the Anthropic Cosmological Principle.* New Jersey and London: Humanities Press International

Heffernan, William C. 1978. "The Singularity of our Inhabited World" William Whewell and A. R. Wallace in Dissent," *Journal of the History of Ideas*, 42, pp. 81–100.

Henderson, L. J. 1913. *The Fitness of the Environment: An Inquiry into the Biological Significance of the Properties of Matter*. Cambridge, Mass.: Macmillan, reprinted in 1970 with an introduction by George Wald, Gloucester, Mass: Peter Smith.

Hoyle, F. 1983. *The Intelligent Universe: A New View of Creation and Evolution.* New York: Holt, Rinehart and Winston.

Hoyt, W. G. *Lowell and Mars.* Tucson, Arizona, 1976.

Huxley, T. H. 1863. *Man's Place in Nature*. 1971 edition. Ann Arbor, Michigan, p. 71.

Maunder, W. 1903. "The Earth's Place in the Universe," *The Journal of the British Astronomical Association*, 13, 227–234.

Monod, Jacques. 1971. *Chance and Necessity*. New York: Collins.

Numbers, R. 2006. *The Creationists: From Scientific Creationism to Intelligent Design.* Cambridge, Mass.: Harvard University Press.

Pickering, W. H. 1903. "Man's Place in the Universe," *The Independent*, 55, 597–600

Raby, P. 2001. *Alfred Russel Wallace: A Life*. Princeton and London: Princeton University Press.

Rees, M. 1997. *Before the Beginning: Our Universe and Others*. Reading, Mass.: Addison-Wesley

Rees, M. 1999. *Just Six Numbers: The Deep Forces That Shape the Universe*. New York: Basic Books

Rees, M. 2001. *Our Cosmic Habitat.* Princeton: Princeton University Press

Ross, Hugh. 1993. *The Creator and the Cosmos: How the Greatest Scientific Discoveries of the Century Reveal God.* Colorado Springs: NavPress.

Shermer, M. 2002. *In Darwin's Shadow: The Life and Science of Alfred Russel Wallace*. Oxford, Oxford University Press.

Slotten, H. 2004. *The Heretic in Darwin's Court: The Life of Alfred Russel Wallace.* New York: Columbia University Press.

Smith, Charles and George Beccaloni. 2008. *Natural Selection and Beyond: The Intellectual Legacy of Alfred Russel Wallace*. Oxford, Oxford University Press.

Smith, R. 1982. *The Expanding Universe: Astronomy's Great Debate, 1900-1931.* Cambridge: Cambridge University Press.

Strauss. D. 2001. *Percival Lowell: The Culture and Science of a Boston Brahmin* Cambridge, MA and London: Harvard University Press.

Turner, H. H. 1903. "Man's Place in the Universe: A Reply to Dr. Wallace," *Fortnightly Review*, 73, 598–605.

Wallace, A. R. 1871. *Contributions to the Theory of Natural Selection. A series of Essays.* London and New York: MacMillan.

Wallace, A. R. 1889. *Darwinism: An Exposition of the Theory of Natural Selection with some of its Applications.* New York and London: MacMillan. Pp. 476–77, cited in Shermer, 230–231.

Wallace, A. R.1898. *The Wonderful Century: Its Successes and its Failures.* London: Swan Sonnenschein.

Wallace, A. R.1903a. "Man's Place in the Universe, as Indicated by the New Astronomy," *The Fortnightly Review*, March 1, 1903, 396. Online at http://www.wku.edu/~smithch/wallace/S602.htm

Wallace, A. R.1903b. "Man's Place in the Universe," *The Independent*, February 26, 1903.

Wallace, A. R.1903c. *Man's Place in the Universe: A Study of the Results of Scientific Research in Relation to the Unity or Plurality of Worlds*. New York, McClure, Phillips and Co. Online at http://www.wku.edu/~smithch/wallace/S728-1.htm

Wallace, A. R.1903d. "Man's Place in the Universe", manuscript of the 1903 book by the same title, British Library, London. ADD MSS 46420, 263 folios.

Wallace, A. R.1904. *Man's Place in the Universe: A Study of the Results of Scientific Research in Relation to the Unity or Plurality of Worlds*. London.

Wallace, A. R. 1907. *Is Mars Habitable? A Critical Examination of Professor Lowell's Book "Mars and Its Canals," With an Alternative Explanation.* London, MacMillan. Online at http://www.wku.edu/~smithch/wallace/S730.htm

Wallace, A. R.1908. *My Life: A Record of Events and Opinions*. London: Chapman and Hall.

Wallace, A. R.1910. *The World of Life: A Manifestation of Creative Power, Directive Mind and Ultimate Purpose*. London: Chapman and Hall.

Ward, P. and D. Brownlee. 2000. *Rare Earth: Why Complex Life is Uncommon in the Universe*. New York: Copernicus.

Whewell, W. 1853. *Of the Plurality of Worlds: An Essay*. London: J. W. Parker. Facsimile reprint, Chicago and London: University of Chicago Press, 2001.

Wilson, P. 1993. "Anthropic Principle," in *Encyclopedia of Cosmology: Historical, Philosophical, and Scientific Foundations of Modern Cosmology*, Norriss Hetherington, ed. New York and London: Garland, pp. 11–17: 11.

Chapter 36
Discovering a New Realm of the Universe: Hubble, Galaxies, and Classification

Abstract In the course of the eighteenth century, Charles Messier and William Herschel catalogued nebulae that comprise what we today recognize as six distinct classes of objects in the realm of the stars, as defined by their physical composition: gaseous nebulae including planetary nebulae and hot hydrogen clouds; dusty clouds delineated as dark nebulae and reflection nebulae; and open and globular star clusters actually composed of stars. That a seventh class turned out to be an entirely new realm of objects—galaxies—as far removed from our own galaxy as the sidereal realm was from the planets, was beyond Herschel's considerable abilities to prove even with the power of his new telescopes. It was only in the twentieth century that Edwin Hubble solved this problem. His research reveals much about the nature of discovery and classification.

36.1 Introduction

As early as the mid-eighteenth century, astronomers had speculated that some of the nebulae might lie beyond our own stellar system. But Thomas Wright of Durham's ideas in 1750 were primarily motivated by theology, and the budding philosopher Immanuel Kant's views 5 years later, though more grounded in observation, were still largely speculation. Both were ignored, though Kant exhibited the spirit, and even the terminology, later used by Herschel: "Here a wide field is open for discovery, for which observation must give the key. The nebulous stars, properly so called, and those about which there is still dispute as to whether they should be so designated, must be examined and tested under the guidance of this theory." Kant's theory was that different types of nebulae require different explanations—not much of a theory, but a prescient statement if ever there was one. Though ignored at the time, the two eighteenth-century thinkers grasped the basic idea that later became known as the *island universe theory* (Wright 1750; Kant 1755; Berenzden et al. 1976, 6–10; Whitney 1971, 77–86).[1]

First published in Steven J. Dick, *Discovery and Classification in Astronomy: Controversy and Consensus* (Cambridge: Cambridge University Press, 2013).

S. J. Dick, *Space, Time, and Aliens*, https://doi.org/10.1007/978-3-030-41614-0_36

Over the next century-and-a-half, opinion oscillated between the twin poles of near or distant, single system or multiple island universes. The resolution of some nebulae into stars by Lord Rosse in the 1840s tended to support the island universe theory, particularly with the (erroneous) claim of the resolution of the Orion Nebula into stars. On the other hand William Huggins' spectroscopic demonstration in 1864 that some nebulae were composed of luminous gas tended to sway opinion in the other direction. This inside-our-system theory received strong support from the detection in 1885 of the extremely bright "nova" in Andromeda, since if the Andromeda nebula were remote it meant this star would have to be shining with the equivalent of 50 million suns, inconceivable at the time. By the late 1880s the island universe theory had fallen completely from favor, and in 1890 the historian Agnes Clerke wrote that "No competent thinker, with the whole of the available evidence before him, can now, it is safe to say, maintain any single nebula to be a star system of coordinate rank with the Milky Way," an assertion she repeated in 1905 (Clerke 1890; Smith 1982, 2008, 2009).

Meanwhile, the numbers of observed nebulae had increased dramatically as the natural history of the heavens continued. William Herschel's sweeps of the sky with telescopes of his own construction had resulted in some 2500 nebulae being catalogued by the early nineteenth century. By 1864 his son John had doubled the number to 5079 nebulae in his *General Catalogue of Nebulae and Clusters of Stars*, fondly known as the GC. The endeavor could no longer be confined to the Herschel family, and when in the 1880s J. L. E. Dreyer combined the GC with other observations to produce the *New General Catalogue* (NGC), the number of objects rose to 7840. Supplemented by two more catalogues, the NGC by 1908 contained more than 13,000 objects. But of these only a few had been resolved into stars.[2]

The biggest advances in understanding the true nature of nebulae came with the application of photography to astronomy. As Allan Sandage has remarked, the faint structural features of galaxies could be detected only when photographic surveys came into general use about 1890, and "these features proved to be decisive in the classification problem because the presence or absence of spiral arms is what divides galaxies into two major groups," the ellipticals and spirals (Sandage 1975, 2). Even this bipartite division was far from obvious at the turn of the twentieth century, as witness the German astronomer Max Wolf's eclectic forms of nebular objects based on photographs taken at Heidelberg (Wolf 1909). Wolf's (1909) drawings based on these photographs are reminiscent of Herschel's but considerably more regular in form. More importantly, Wolf's scheme was the first to display nebular forms in a sequence ranging from amorphous blobs to well-developed spiral arms. But he could not and did not tackle the distance problem based solely on natural history methodologies.

36.2 Enter Edwin Hubble

As Wolf published his drawings in 1909 in a far-off Germany heading into war, Edwin Hubble (Fig. 36.1) was a young American lad of 19 in the midst of studying mathematics and astronomy at the University of Chicago. There he was influenced

Fig. 36.1 Edwin Hubble, classifier of "nebulae." In 1924 he demonstrated that most nebulae outside the plane of our Milky Way Galaxy are extragalactic systems of stars. In doing so, he added a new "Kingdom" to astronomy (along with stars and planets), which he divided into four classes. Classes of active galaxies would be discovered and delineated later. (Hale Observatories, courtesy AIP Emilio Segrè Visual Archives)

most during his undergraduate years by Forest Ray Moulton, who with Chicago geologist T. C. Chamberlin had published their *planetesimal hypothesis* in 1904. Chamberlin and Moulton had drawn attention to the spirals as possible solar systems in formation, based on recent photographs taken by Lick Observatory astronomer James Keeler, who believed they were not only the most abundant form of nebulae, but also likely proto-solar systems illustrating Laplace's nebular hypothesis (Dick 1996, 166–170). It was these very spiral nebulae that would prove crucial to Hubble's career, stepping stones to proof that, far from being solar systems inside our stellar system, some nebulae were actually beyond our own galaxy. After graduation from Chicago in 1910, Hubble went to Oxford as a Rhodes Scholar, from 1910 to 1913 studying Spanish and law and receiving his BA in jurisprudence (Christianson 1995).

After a brief stint teaching in Indiana, in 1914 Hubble returned to the University of Chicago and its Yerkes Observatory, where he began his life's work. Laboring under the minimal supervision of the nearly blind astronomer Edwin Frost, Hubble produced his thesis "Photographic Investigations of Faint Nebulae" in 1917. A scant 17 pages including numerous Tables and two Plates, the document is nevertheless revealing as a marker of where things stood and what was to come. In the first paragraph of his thesis Hubble noted that about 17,000 nebulae had been catalogued, and an estimated 150,000 were within reach of existing instruments, namely the

Mt. Wilson 60-in. and 100-in. telescopes. In the next two paragraphs he summarized his understanding of the state of knowledge at the time, which was clearly very confusing and therefore worth quoting in full (Hubble 1920, 1)[3]:

> Extremely little is known of the nature of nebulae, and no significant classification has yet been suggested; not even a precise definition has been formulated. The essential features are that they are situated outside our solar system, that they present sensible surfaces, and that they should be unresolved into separate stars. Even then an exception must be granted for possible gaseous nebulae which appear stellar in the telescope, but whose true nature is revealed by the spectroscope. It may well be that they differ in kind and do not form a unidirectional sequence of evolution. Some at least of the great diffuse nebulosities, connected as they are with even naked-eye stars, lie within our stellar system; while others, the great spirals, with their enormous radial velocities and insensible proper motions, apparently lie outside our system. The planetaries, gaseous but well defined, are probably within our sidereal system, but at vast distances from the earth. In addition to these classes are the numberless small, faint nebulae, vague markings on the photographic plate, whose very forms are indistinct. They may give gaseous spectra, or continuous; they maybe planetaries or spirals, or they may belong to a different class entirely. They may even be clusters and not nebulae at all. These questions await their answers for instruments more powerful than those we now possess.

This statement is notable, among other reasons, for its clear statement that spirals "apparently lie outside our system." For his dissertation Hubble photographed seven fields with the Yerkes 24-in. reflector during the years 1914–1917. Altogether they showed some 588 nebulae, and Hubble attempted to sort them according to Wolf's classification system, which he said "while admittedly formal, offers an excellent scheme for temporary filing until a significant system shall be constructed." Having done this, Hubble found that

> ...the most striking feature is the great predominance of the classes e and f. These two classes form a continuous sequence from the brightest in the list to the very limits of the plates, where they are but mere faint markings on the films. Eleven are clearly spirals, and the spindles are unexpectedly common. These results are typical
>
> ... As far as telescopes of moderate focal length are concerned, the predominant form of nebulae as we know them at present is not the spiral, but is this same "e, f" class, described as round or nearly so, brightening more or less gradually toward the center, and devoid of detail. (Hubble 1920, 3–4).

Indeed, 429 of the 588 nebulae shown on Hubble's plates belonged to Wolf's e and f classes. As was apparent from Wolf's drawings that Hubble appended to his dissertation, the e and f classes were what Hubble later classed as *ellipticals*, the terminology still used today. Hubble believed his dissertation was not particularly original, a fair assessment. But it served the important purpose of letting him explore for himself the realm of the nebulae, seeing firsthand what problems needed solving.

After a stint in the United States infantry during WW I, in September 1919 Hubble went to Mt. Wilson in the San Gabriel Mountains near Los Angeles. There he had access not only to the 60-inch telescope, with which he did work on reflection nebulae as described in Chap. 3, but also access to the world's largest telescope, the Hooker 100-in., which had begun regular observations only in 1919 just as Hubble arrived. It was with the latter telescope that Hubble could examine the nebulae outside the plane of the Milky Way Galaxy with the best instrument in the world, seeing fainter and farther than anyone before him.[4]

Given what was to follow, it is somewhat ironic that Hubble's first major paper, published in 1922, was "A General Study of Diffuse Galactic Nebulae," in other words, a study of those nebulae known to be within our galaxy, not at all the subject of his dissertation or the work for which he was to become most famous. But it was in this paper that Hubble first ventured, by way of introduction in the paper's inaugural section, his own attempt at a classification system of nebulae, much simpler than Wolf's. "There appears to be a fundamental distinction between galactic and non-galactic nebulae," Hubble noted. "This does not mean that the latter class must be considered as 'outside' our galaxy, but that its members tend to avoid the galactic plane and to concentrate in high galactic latitudes." What is notable here is that the term "non-galactic" first denoted a location outside of the plane of the Milky Way Galaxy, not outside of the galaxy itself. Hubble also noted a physical difference: galactic nebulae were associated with stars, while non-galactic nebulae were not (Hubble 1922, 162–199: 166).

There was therefore no factor of distance involved when Hubble ventured a classification system based on nebular morphology, playing the biologist's role of natural historian. The great bipartite division was between galactic nebulae, consisting of planetary and diffuse nebulae, and non-galactic nebulae, divided into spiral, elongated (spindle and ovate), globular, and irregular morphologies. The following year, in unpublished materials prepared for the International Astronomical Union, Hubble simplified his scheme and defined only three categories of non-galactic nebulae: ellipticals, spirals, and irregulars. This became the terminology employed in his classic paper of 1926, the foundation of the galaxy classification system still used today, the so-called *Hubble sequence* (Fig. 36.2). Although the scheme closely followed James Jeans' sequence based on nebular evolution theory, Hubble held that "a conscious attempt was made to ignore the theory [of Jeans] and arrange the data purely from an observational point of view" (Hart and Berenzden 1971).[5]

Hubble's classification system for non-galactic nebulae therefore was in place well before the distances to these objects had been determined, in fact even before it was known whether they were galactic or extragalactic. But more than natural history was needed to solve the problem of non-galactic nebulae; for that, distance techniques were essential. As Hubble himself noted in his retrospective book *The Realm of the Nebulae* (1936), three landmarks stand out in the road that eventually revealed the nature of the non-galactic nebulae: the first radial velocity measurements of a galaxy in 1912, the discovery of novae in 1917, and his own discovery of Cepheid variable stars in the Andromeda nebula and elsewhere.

In demonstrating the existence of a class of nebulae beyond our own star system (ironically not yet known to be a spiral itself), spiral nebulae were to prove crucial in all three of these landmarks.[6] William Parsons, the third Earl of Rosse in Ireland, had first detected the spiral form of nebulae in 1845 when he turned his 6-ft-diameter speculum metal reflector, twice the size of William Herschel's, to the 51st object in Messier's catalogue. His observations and sketches of M51 (Fig. 36.3), now known as the Whirlpool Galaxy, were presented to the Royal Society of London in 1850,

CLASSIFICATION OF NEBULAE

	Symbol	Example
I. Galactic nebulae:		
A. Planetaries	P	N.G.C. 7662
B. Diffuse	D	....
1. Predominantly luminous	DL	N.G.C. 6618
2. Predominantly obscure	DO	Barnard 92
3. Conspicuously mixed	DLO	N.G.C. 7023
II. Extra-galactic nebulae:		
A. Regular:		
1. Elliptical (n = 1, 2, ..., 7 indicates the ellipticity of the image without the decimal point)	En	N.G.C. 3379 E0; 221 E2; 4621 E5; 2117 E7
2. Spirals:	Symbol	Example
a) Normal spirals	S	....
(1) Early	Sa	N.G.C. 4594
(2) Intermediate	Sb	2841
(3) Late	Sc	5457
b) Barred spirals	SB	....
(1) Early	SBa	N.G.C. 2859
(2) Intermediate	SBb	3351
(3) Late	SBc	7479
B. Irregular	Irr	N.G.C. 4449

Extra-galactic nebulae too faint to be classified are designated by the symbol "Q."

Fig. 36.2 Hubble's 1926 classification of nebulae divided them into galactic and extragalactic. The latter term at first signified outside the plane of our Milky Way Galaxy, and, after 1924, outside of the galaxy itself. Only after Hubble's death in 1953 did the term "galaxy" come into widespread use. (From Hubble 1926a, reproduced by permission of the American Astronomical Society)

with the following unequivocal words: "It will be at once remarked, that the spiral arrangement so strongly developed in 51 Messier, is traceable, more or less distinctly, in several of the sketches" (Rosse 1850).[7] It is notable that Rosse used the word "detected" rather than "discovered" with regard to the spiral form, and in the following spring he "detected" another spiral, M99, and suspected several more from John Herschel's southern catalogue. By the time of the publication of his results in 1850, Rosse and his observing colleagues had detected 14 spirals, though he cautioned they were "comparatively difficult to be seen, and the full power of the instrument is required to bring out the details … 51 Messier is the most conspicuous object of that class." Moreover, he cautioned "we are in the habit of calling all objects spirals in which we have detected a curvilinear arrangement not consisting of regular re-entering curves; it is convenient to class them under a common name, though we have not the means of proving that they are similar systems." Despite

Fig. 36.3 Lord Rosse's drawing of the spiral nebulae M51, today known as the Whirlpool Galaxy, 1850. Rosse called M51 "the most conspicuous object of that class," among 14 he had observed beginning in 1845. The class of spiral nebulae would become spiral galaxies after 1925, when Hubble proved they were independent systems of stars outside our galaxy, though Hubble himself persisted in calling them spiral nebulae. Compare to the Hubble Space Telescope image in Fig. 20.2

these laudable cautions Rosse was nevertheless the first to detect and declare the spiral form as a new class of nebulae. Today we know that 27 out of 110 objects in Messier's extended catalogue are spiral galaxies.

Rosse himself, however, did not yet know the true nature of his spiral structures as conglomerations of stars. Even in 1887, when Isaac Roberts made the first long-exposure photographs of the Andromeda Nebula and revealed its spiral structure, its distance was so uncertain that Roberts and other astronomers believed it to be a solar system in formation, in line with the nebular hypothesis. And when in 1899 the German spectroscopist Julius Scheiner identified familiar stellar features in the spectrum of the same object, and declared that "the previous suspicion that the spiral nebulae are star clusters is now raised to a certainty," the idea of the Andromeda Nebula being a system of stars was still too radical for most (Scheiner 1899).[8]

The three landmarks that led to the extragalactic distance determinations all employed these spiral nebulae whose form Rosse first detected. In 1912 V. M. Slipher at the Lowell Observatory began his spectroscopic radial velocity observations of spirals, and within 2 years showed that that spectral lines in 14 spiral nebulae were redshifted, an observation he interpreted as meaning they are moving away from us. Ejnar Hertzsprung and other astronomers took these velocities, ranging up to 1100 km/s, as evidence that these objects could not be gravitationally bound to our own system of stars: "It seems to me, that with this discovery the great question, if the spirals belong to the system of the Milky Way or not, is answered with great certainty to the end, that they do not" (Smith 1982, 22).

The second event was the discovery of faint novae in spirals in 1917. Bright novae, known today to be supernovae, had been observed in spirals before, in Andromeda in 1885 and Centaurus in 1895. But it was George W. Ritchey's photographic detection of a nova in the spiral galaxy NGC 6946, as well as Lick Observatory astronomer Heber D. Curtis's discovery of three more in the same year, followed by even more on old photographic plates, that allowed crude distances to be determined to these spirals. By comparing the brightness of novae in spirals to their brightness in our galaxy, Curtis calculated an average distance to spiral nebulae of 20 million light years, clearly far outside our galaxy. By 1917, when Hubble completed his dissertation on the nebulae, the island universe theory was the leading theory of spirals, the one toward which Hubble himself tended.

The island universe theory was not, however, accepted by all. The so-called *Great Debate* on the scale of the universe, argued between Mt. Wilson astronomer Harlow Shapley and Curtis and held in 1920 in Washington, D.C. under the auspices of the National Academy of Sciences, clearly showed the differences that still existed. Using his nova results and a variety of other methods, Curtis argued that nebulae were island universes. Shapley, on the other hand, was fresh from his discovery that globular clusters are arrayed at considerable distances around our own stellar system, in which we hold an eccentric position. He argued that our galaxy was larger, and the spirals nearer, than believed, estimating the galaxy's diameter at 300,000 light years, ten times the size accepted by most astronomers. If true, it meant the Magellanic Clouds were inside our galaxy. Moreover, Shapley's colleague at Mt. Wilson, Adriaan van Maanen, had measured internal motions in several spirals including M101, motions which, if true, were much too large to be seen if the spirals were distant island universes. While the island universe theory was still ascendant in 1920, doubts remained (Hoskin 1976; Smith 1982, 77–99; Trimble 1995).

The third event was the one in which Hubble himself proved central. His method used Cepheid variables instead of novae, and was much more accurate and robust than the nova method. This was because in 1908 Henrietta Leavitt at the Harvard College Observatory had discovered and published a period-luminosity relation for what became known as Cepheid variable stars that she had observed in the Magellanic Clouds. Their periods ranged from 1.25 to 127 days, and the longer the periods the greater the luminosities. Therefore if the period could be determined, so could the luminosity, and if one had the luminosity the star's apparent magnitude would yield its true distance. The 1908 paper went largely unnoticed, but Leavitt's 1912 paper gave the results in graphical form and was noticed to great effect, despite its three-page brevity. Shapley used this method in 1918 for some of the nearer globular clusters in his famous work showing our eccentric position in the galaxy, and Hertzsprung had used it before him. Thus, when in late 1923 Hubble found a Cepheid in the Andromeda Nebula, it afforded an accurate method (at least accurate relative to the time) for measuring the distance to Andromeda. Hubble calculated that distance at about one million light years, well beyond our own star system. In 1924 Hubble found several more Cepheids in Andromeda and its companion M 32 (Leavitt 1908; Leavitt and Pickering 1912; Smith 1982, 59–77).[9]

The first public announcement of Hubble's Cepheid work came in the *New York Times* on November 23, 1924. But, still worried about van Maanen's contrary measurements of internal motions of spirals, Hubble's results were not *formally* announced until New Year's Day, 1925 at a joint meeting of the AAS and AAAS in Washington, D.C. Even then Hubble himself was not present, but sent his paper to be read (Berenzden 1967).[10] From a total of 12 Cepheids in Andromeda and 22 in M 33, Hubble announced a distance to the spirals of roughly 285,000 parsecs, placing them even outside Shapley's large version of our galaxy. Thus, the first day of 1925 marks the official declaration of final proof that spiral nebulae, in particular the Andromeda Nebula M31 and its companion M33, were beyond our galaxy, and consisted of stars, not gas. Step by step, Hubble could proceed deeper and deeper into space using the Cepheid method; by the end of 1925 he reported 11 Cepheids in the object known as NGC 6822, and showed it also to be beyond our Milky Way Galaxy.[11] Despite a few holdouts due mainly to van Maanen's claim of internal rotation of galaxies (later shown to be erroneous), opposition to the island universe theory collapsed. In addition to the realm of the planets and the realm of the galactic stars, astronomy now had the realm of the galaxies.

In his 1926 paper significantly titled "Extra-galactic Nebulae" Hubble (1926a) put the pieces together. It was here that Hubble first published the classification system for galaxies as we now know it. He pointed out that it had first been presented in the form of a memorandum distributed to all members of the Commission on Nebulae of the IAU in 1923. It was discussed at a meeting in Cambridge in 1925 and was published in an account of the meeting in *L'Astronomie*. The question of the adoption of the system was left to a subcommittee of the Commission, "with a resolution that the adopted system should be as purely descriptive as possible, and free from any terms suggesting order of physical development." The Commission also indicated a preference for the term "extragalactic" in place of the necessarily non-committal "non-galactic." The title of Hubble's 1926 paper marked the all-important shift: non-galactic had become extragalactic, a true landmark in the history of astronomy.

Still, one of the curiosities of the history of astronomy testifying to social factors, at least in nomenclature matters, is that Hubble persisted in calling the spirals extragalactic nebulae, and his popular book on the subject even in 1936 remained *The Realm of the Nebulae*. This terminology was likely due to Harlow Shapley, who had urged Hubble in October 1924 to call his objects galaxies rather than the unwieldy "non-galactic nebulae." Shapley had also urged Slipher, the head of the Commission on Nebulae to adopt the term "galaxy." Neither did so. As Gale Christianson, Hubble's biographer, put it, "Hubble was in no mood to be dictated to by the man whose universe he had demolished." "Extra-galactic nebulae" was the phrase used by Hubble in his landmark 1926 paper, and by Mt. Wilson astronomers until Hubble's death in 1953, when the term "galaxies" was universally adopted. Alan Sandage, Hubble's successor at Mt. Wilson, confirms that "within a week of

Hubble's death in September, 1953, 'galaxy' became the standard name, even at Santa Barbara Street," headquarters for Mt. Wilson staff (Hubble 1926a, 323; Sandage 2004, 487).[12]

As we shall see in Chap. 8, it is notable that Knut Lundmark also put forth a similar classification system for galaxies in 1926, and Harlow Shapley a very different one in 1927—not to mention that the British industrialist John H. Reynolds devised a similar classification system in 1920, 6 years before Hubble's scheme was published. But it was Hubble's that won the day, both for reasons of simplicity and social factors. Thus, just as Herschel and then Huggins were central in proving definitively that nebulous matter exists in the universe, so Edwin Hubble became the central figure in finally proving that galaxies exist beyond our own. As more and more galaxy distances were determined, later in the decade he would propose a velocity-distance relationship, the more distant the object as shown by Cepheids, the faster was it moving away from us as shown by red-shifted spectra, a relation that came to be known as *Hubble's law*. For most (though not Hubble himself) this was evidence that the universe was expanding, another pillar of modern cosmology.

36.3 The Role of Classification

With the discovery of the true nature of the spiral nebulae, extragalactic astronomy—and the study of the separate classes of galaxies—could begin in earnest. The extragalactic nature of the spirals had been shown, but how about the other morphological forms in Hubble's classification system, the ellipticals and irregulars? In his landmark 1926 paper Hubble noted that definite distances were known for only six systems including the two Magellanic Clouds, which were then considered irregulars, not spirals. However, he continued, "the similar nature of the countless fainter nebulae has been inferred from the general principle of the uniformity of nature" (Hubble 1926a, 322).[13] In other words, Hubble was inferring that all the objects in his non-galactic nebulae category were actually outside the galaxy. With this leap "Non-galactic" became "extra-galactic," not only for the spirals but also for ellipticals and irregulars (Hubble 1926a, 1936, 39–48, Lang and Gingerich 1979).

As we have seen, elliptical galaxies had long been observed as fuzzy objects inseparable from many other types of nebulae. We now know that Charles Messier's 1781 catalogue of nebulae and star clusters, for example, contained six of them, and four more were later added to the extended catalog of 110 objects. But they could not be directly proven to be extragalactic because they could not be resolved as stars as was the case for many spirals. It was recognized early on that, in stark contrast to spiral galaxies like our own, ellipticals are almost featureless aggregations distinguished only by their ellipsoidal or spherical morphology. As Hubble wrote even in 1936, "Elliptical nebulae are highly concentrated and show no indications of resolution into stars. The luminosity falls rapidly away from bright, semistellar nuclei to undefined boundaries … Small patches of obscuring material are occasionally silhouetted against the luminous background, but otherwise these nebulae present no

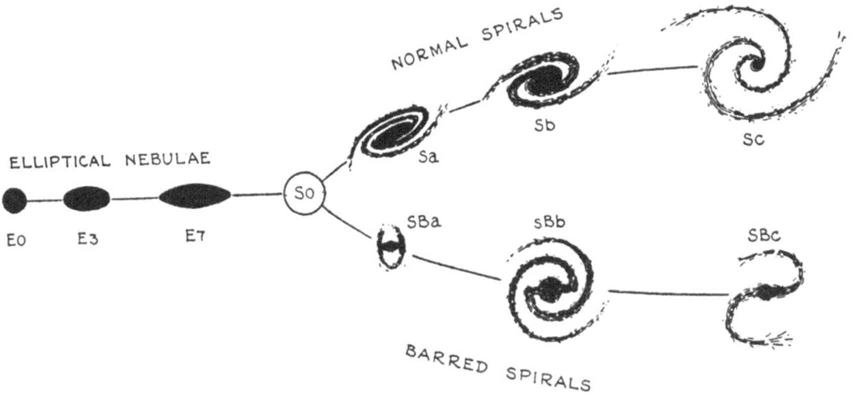

Fig. 36.4 Hubble's "tuning fork" diagram of galaxy classification, from *The Realm of the Nebulae*, 1936, p. 45. Hubble referred to the diagram as "the sequence of nebular types ... a schematic representation of the sequences of classification." For Hubble the diagram represented an evolutionary sequence, moving from left to right. The diagram illustrates the difficulties of delineating classes; Hubble considered E7 lenticulars a "transition stage ... more or less hypothetical," between ellipticals and spirals. The existence of a lenticular class of galaxies remains controversial to this day, when they are more often identified with Hubble's S0 galaxies

structural details" (Hubble 1936, 39). Ellipticals were clearly extragalactic in Hubble's original sense of outside the plane of our galaxy, but only by the uniformity of nature could they be declared extragalactic in the sense of being placed outside our galaxy.

Therefore, assuming ellipticals to be outside our galaxy, further inferences as to their nature could only be made based on their morphology. In his 1926 classification system Hubble had divided extragalactic nebulae into regular and irregular galaxies, classifying ellipticals and spirals as regulars, those defined as having "rotational symmetry about dominating non-stellar nuclei." These two classes, he estimated, comprised 97% of all galaxies. But Hubble went further than the delineation of classes to specify types; depending on their ellipticity he arranged elliptical galaxies from E0 (spherical like M87) to E7 (elongated), shading into what he called "a limiting lenticular figure" at the junction of the spirals. Spiral galaxies were typed as early, intermediate, and late "normal" spirals, as well as early, intermediate and late "barred" spirals, the latter having a central bar structure not evident in the normal galaxies. The significance of the "early" and "late" terms is graphically evident in the famous "tuning fork" diagram of galaxy types first published 10 years later in *The Realm of the Nebulae* (Fig. 36.4), where Hubble placed ellipticals on the handle of the fork because he thought, influenced by Jeans theory, they were early objects before acquiring their more mature morphologies as spirals. In the 1926 paper, however, he continued to insist on his independence from Jeans, while remarking on the suggestive parallels:

Although deliberate effort was made to find a descriptive classification which should be entirely independent of theoretical considerations, the results are almost identical with the path of development derived by Jeans from purely theoretical investigations. The agreement is very suggestive in view of the wide field covered by the data, and Jeans's theory might have been used both to interpret the observations and to guide research. It should be borne in mind, however, that the basis of the classification is descriptive and entirely independent of any theory.

Clearly, in Hubble's mind the tuning fork also represented an evolutionary sequence, although he realized the different forms of the ellipticals could be due to projection effects due to viewing angle.

Irregular galaxies originally comprised all those galaxies that did not fit into other classification categories that Hubble defined in 1926, amounting to only about 3%. In his book *Realm of the Nebulae* (1936), Hubble, still using his favored term nebulae rather than galaxies, commented that whereas the "regular nebulae"—spirals and ellipticals—"are characterized by rotational symmetry around dominating central nuclei … about one nebula in forty is irregular in the sense that one or both characteristics are absent" (Hubble 1936, 40). He recognized the Magellanic Clouds as conspicuous examples, already proven extragalactic using the Cepheid method.

We now know that elliptical galaxies are dominated by older, low mass, and metal poor population II stars moving in elongated randomly oriented orbits, and that ongoing star formation is rare, ellipticals having mostly depleted their interstellar matter. Although they are typically surrounded by numerous globular clusters, because they are structureless they seem simpler than other classes of galaxies. And because of their older stars, ellipticals are still sometimes referred to as early-type galaxies, as opposed to late-type spirals, harking back to Hubble's original designation but for different reasons. Round elliptical galaxies are believed to be the oldest ellipticals, forming from galaxy mergers 2 to 3 billion years ago and exhibiting no star formation. While small ellipticals are the most common galaxies in the universe, particularly in clusters of galaxies, ellipticals in general represent only ten to 15% of galaxies in the local universe. And we know that in contrast to spirals, they were less common in the early universe than now.

Today irregulars are defined as galaxies that either have little structure or lack any structure whatsoever, and, having a much larger sample than in Hubble's day, about one-third of all galaxies are classified as irregular. They are more common at higher redshift, as one moves back to earlier epochs in the universe. Irregular galaxies vary in mass from 100,000 to 10 million solar masses, and in size from a thousand to tens of thousands of light years. They are often rich in gas and dust, and may represent ellipticals or spirals that have been distorted over time by interactions with other galaxies. Irregular galaxies are well represented among the nearest galaxies in our Local Group. The Magellanic Clouds were for a long time considered the nearest irregular galaxies to our own. Based on Harlow Shapley's work on Cepheids, Hubble considered the Large Magellanic Cloud (LMC) to be at about 85,000 light years, and the Small Magellanic Cloud (SMC) at about 95,000 light years from the Milky Way. We now know them to be at about 160,000 and 200,000 light years, respectively. And only recently they have been shown to have some spiral characteristics, a paean to the difficulties of classification.

Thus the delineation of the basic classes of galaxies, and their placement in a separate realm beyond the stars, was essentially complete already in 1926. Hubble's classification system did not change much in the next 10 years. In his 1936 book *Realm of the Nebulae*, he still identified lenticulars with the most flattened form of ellipticals, known as E7, and hypothesized what he called an S0 class between ellipticals and spirals. It is this S0 class we today we refer to as lenticulars—an interesting case of adjusting a classification system as more data became available.

That is not to say that disputes about classes have all been resolved. Even today some dispute whether lenticulars should be considered a separate class of galaxies. At a different level Hubble's prototype of an E7 elliptical, NGC 3115 (also known as the Spindle Galaxy), is now recognized as an early S0 system because new observations clearly reveal a thin disk. In fact 3 of 12 of Hubble's prototypes have been reclassified since 1936 based on better evidence.[14] Moreover, it is notable that when Hubble proposed an intermediate S0 class of galaxies, none had been observed. But eventually they were observed, and the types, if not the classes themselves, adjusted accordingly, as in Gerard de Vaucouleurs' 1959 extension of the Hubble sequence of galaxies, where lenticulars are further typed as barred (SB0) or unbarred (SA0). De Vaucouleurs later considered all S0 galaxies to be lenticulars. Nevertheless, even today it is uncertain whether galaxies such as M84 and M86 should be classified as highly elongated E7 ellipticals or S0 lenticulars. Such are the problems faced by any classification system, whether biological or astronomical. In the case of astronomers, however, their objects are located millions of light years distant.

Similarly, while Hubble originally classed the Magellanic Clouds as "typical irregular nebulae—highly resolved, with no nuclei and no conspicuous evidence of rotational symmetry," DeVaucouleurs recognized rotational symmetry and some spiral structure. He suggested they were in fact part of the spiral sequence, and his classification system added the Sm and Im stages to account for these characteristics in the Magellanic Clouds as well as in other similar galaxies. In his system they are thus typed as Irr/SB (s)m, the "B" indicating characteristics of a barred spiral, and the "s" indicating an s-shaped spiral emerging directly from a central bulge or the ends of a bar.[15]

The DeVaucouleurs system and others greatly refined Hubble's classification work, leaving the original classes intact, the ambiguous lenticulars notwithstanding. Quite aside from these developments, great surprises were in store in the form of the discovery of galaxies so different from the "normal" varieties that they are considered to be an entirely new family of classes, known as active galaxies. But that is another story (Dick 2013).

Notes

1. The phrase "the nebulous stars, properly so-called" was later used by William Herschel. Hoskin discusses Wright in the Introduction to the facsimile edition, and in Hoskin (1970). Relevant parts of Wright and Kant are reprinted in Bartusiak (2004, 168–188).

2. On the history of the GC and NGC see Steinicke (2020).
3. The dissertation has dual pagination, running from pages 1 to 17 for Hubble's dissertation, corresponding to pages 69–85 of the Yerkes Publications, plus two plates. See also Christianson (1995, 102–103).
4. On Hubble's arrival at Mt. Wilson, see Christianson (1995, p. 20 ff).
5. The Jeans work is Jeans (1919), updated as *Astronomy and Cosmogony* in 1929. Hubble's 1926 paper is (Hubble 1926a).
6. Hubble (1936, 84) himself recalled that planetary and diffuse nebulae were considered galactic, while the spirals had been the subject of the island universe controversy, Although we now know we live in a spiral galaxy, this is far from obvious. The Milky Way as seen in the night sky—now known to be the "backbone" of our galaxy—was one of the first phenomena Galileo observed with his telescope in 1610, resolving it into stars. But the first to speculate that this phenomenon was due to a large flat disk of stars were Wright and Kant in the eighteenth century. A few decades later William Herschel was the first to base this conclusion on actual star counts. Even then, the spiral nature of our home galaxy would not be uncovered until the mid-twentieth century, and today it is believed to be a barred spiral.
7. For context see Hoskin (1990) and Dewhirst and Hoskin (1991).
8. On Scheiner's observations see Scheiner (1899)
9. In 1908 Leavitt had only determined periods for 16 of the 1777 variables, but it was enough to see the pattern. The 1912 paper reported 25 periods. On Shapley's use of this method see (Gingerich 1992, 238–245).
10. Even by the time of the public announcement in the November 23 *New York Times* (p. 6), Hubble's Cepheid work was well known to other astronomers.
11. The M 31 and M 33 results were reported in Hubble (1925) and Hubble (1926b), Also Hubble, "NGC 6822, A Remote Stellar System," ApJ, 62 (1925, 409–433). For further context see Smith (1982, 111–136), and Smith (2009, 71–107).
12. It should also be pointed out that some supposed galaxies actually were nebulosities, and that centers of spirals like M 31 had not yet been resolved.
13. The Magellanic Clouds are today classified as spirals in the De Vaucouleurs system, the LMC as SBm (irregular in appearance with no bulge) and the SMC as Im (highly irregular). Some still classify Im's as irregular (Irr). See Buta et al. (2007), especially p. 16.
14. Hubble (1936, 40) and Buta et al. (2007, 8–9).
15. Hubble (1936, 131–137) and Buta et al. (2007, 15–16).

References

Bartusiak, Marcia. ed.. 2004. *Archives of the Universe: A Treasury of Astronomy's Historic Works of Discovery*. New York: Pantheon Books

Berenzden, Richard. 1967. "Hubble's Announcement of Cepheids in Spiral Nebulae," PASP, 10, pp. 425–440.

Berenzden, Richard, Richard Hart and Daniel Seeley. 1976. *Man Discovers the Galaxies* New York: Science History Publications

Buta, Ronald J. Harold G. Corwin, Jr., and Stephen C. Odewahn. 2007. *The De Vaucouleurs Atlas of Galaxies*. Cambridge: Cambridge University Press

Christianson, Gale E. 1995. *Edwin Hubble: Mariner of the Nebulae*. New York: Farrar, Straus and Giroux

Clerke, Agnes. 1890. *The System of the Stars*. London: Longmans, Green and Co. p. 368

Dewhirst, D. W. and M. Hoskin. 1991. "The Rosse Spirals," JHA, 22, 257–265.

Dick, Steven J. 1996. *The Biological Universe: The Twentieth Century Extraterrestrial Life Debate and the Limits of Science*. Cambridge: Cambridge University Press

Dick, Steven J. 2013. *Discovery and Classification in Astronomy: Controversy and Consensus.* Cambridge University Press, Cambridge

Gingerich, Owen. 1992. "Harlow Shapley and the Cepheids," in *The Great Copernicus Chase*. Cambridge: Cambridge University Press, pp.238–245.

Hart R. and R. Berenzden. 1971. "Hubble's Classification of Non-Galactic Nebulae," JHA, 2, 109–119: 113.

Hoskin, M. A. 1970. "The Cosmology of Thomas Wright of Durham," in *JHA*, 1, 44–52.

Hoskin, Michael. 1976. "The 'Great Debate': What Really Happened," JHA, 7, 169–182;

Hoskin, Michael. 1990. "Rosse, Robinson and the Resolution of the Nebulae," JHA 21, 331–44.

Hubble, Edwin 1920. "Photographic Investigations of the Faint Nebulae," *Publications of the Yerkes Observatory*, 4, pt. 2.

Hubble, Edwin. 1922. "A General Study of Diffuse Galactic Nebulae," ApJ, 56, 162–199

Hubble, Edwin. 1925. "Cepheids in Spiral Nebulae," *Popular Astronomy*, 33 (1925), 252 ff., reprinted in part in Bartusiak, *Archives*, pp. 407–414

Hubble, Edwin. 1926a. "Extra-Galactic Nebulae," ApJ, 64, 321–372, reprinted in part in Lang and Gingerich (1979), pp. 716–724;

Hubble, Edwin. 1926b. "A Spiral Nebula as a Stellar System, Messier 31," ApJ, 63 (1926), 236–274.

Hubble, Edwin. 1936. *The Realm of the Nebulae*. New Haven: Yale University Press

Jeans, James. 1919. *Problems of Cosmogony and Stellar Dynamics*. Cambridge: Cambridge University Press

Kant, Immanuel. 1755. *Allgemeine Naturgeschichte und Theory des Himmels*, English edition, *Universal Natural History and Theory of the Heavens*, Stanley L. Jaki, translator. Norwich: Scottish Academic Press, 1981.

Lang, Kenneth and Owen Gingerich, eds., 1979. *A Source Book in Astronomy and Astrophysics, 1900-1975*. Cambridge, MA.: Harvard University Press

Leavitt, Henrietta. 1908. "1777 Variables in the Magellanic Clouds," *HCO Annals*, 60 87–108

Leavitt, Henrietta and E. C. Pickering. 1912. "Periods of 25 Variable Stars in the Small Magellanic Cloud," *HCO Circular* 173

Rosse, Earl of. 1850. "Observations on the Nebulae," PTRSL, 140, 499-514, reprinted in part in Bartusiak, 191–195.

Sandage, Allan. 1975. "Classification and Stellar Content of Galaxies," in *Galaxies and the Universe*, Allan Sandage, Mary Sandage and Jerome Kristian, eds. Chicago and London: University of Chicago Press. Online at http://ned.ipac.caltech.edu/level5/Sandage/paper.pdf

Sandage, Allan. 2004. *The Mount Wilson Observatory: Breaking the Code of Cosmic Evolution*, vol. 1 of the Centennial History of the Carnegie Institution of Washington Cambridge: Cambridge University Press, 2004).

Scheiner, J. 1899. "Über das Spectrum des Andromedanebels," AN, 148 (1899), 325, English translation in Scheiner, "On the spectrum of the great nebula in Andromeda" ApJ, 9, (1899), 149–150.

Smith, Robert. 1982. *The Expanding Universe: Astronomy's 'Great Debate,' 1900-1931* Cambridge: Cambridge University Press

Smith, Robert. 2008. "Beyond the Galaxy: The Development of Extragalactic Astronomy 1885–1965, Part 1," *Journal for the History of Astronomy*, 39, 91–119.

Smith, Robert. 2009. "Beyond the Galaxy: The Development of Extragalactic Astronomy 1885–1965," Part 2, *Journal for the History of Astronomy*, 40, 71–107.

Steinicke, Wolfgang. 2020. *Observing and Cataloguing Nebulae and Star Clusters: From Herschel to Dreyer's New General Catalogue*. Cambridge: Cambridge University Press.

Trimble, Virginia. 1995. "The 1920 Shapley-Curtis Discussion: Background, Issues and Aftermath," PASP, 107, 1133–1144.

Whitney, Charles A.. 1971. *The Discovery of Our Galaxy*. New York: Alfred A. Knopf

Wolf, Max. 1909. "Die Klassifizierung der kleinen Nebelflecken," *Publikationen des Astrophysikalischen Instituts Königstuhl-Heidelberg* 3, No. 5, pp. 109–112.

Wright, Thomas.1750. *An Original Theory or New Hypothesis of the Universe*, facsimile reprint, introduction and transcription by Michael Hoskin. The Netherlands: Elsevier, 1971

Part VI
The Philosophy of Astronomy, Cosmology, and Astrobiology

Part VI Frontispiece The Juggler of Universes, symbolizing the many issues that arise in the philosophy of astronomy and cosmology, ranging from the epistemological and metaphysical to the societal and sociological. (Illustration by J. J. Grandville from *Un Autre Monde,* Paris, 1844)

Come, my friends,
'T is not too late to seek a newer world.
Push off, and sitting well in order smite
The sounding furrows; for my purpose holds
To sail beyond the sunset, and the baths
Of all the western stars, until I die.
It may be that the gulfs will wash us down:
It may be we shall touch the Happy Isles,
And see the great Achilles, whom we knew.
Tho' much is taken, much abides; and tho'
We are not now that strength which in old days
Moved earth and heaven, that which we are, we are;
One equal temper of heroic hearts,
Made weak by time and fate, but strong in will
To strive, to seek, to find, and not to yield.

Tennyson, Ulysses

Many chapters in this volume have already embodied what we term here "the philosophy of astronomy"—the meta-study of astronomy in the sense of its underlying foundations, assumptions, and interpretations.

In Part VI we address the subject head on, beginning with Chap. 37, which lays out for the first time in a systematic manner the general issues in the philosophy of astronomy and cosmology, examines the differences and similarities between the two, and provides a case study using astrobiology, which can be considered a forefront subfield of both astronomy and cosmology. While robust fields of philosophy of physics, chemistry, and biology already exist, not to mention the philosophy of the social sciences and religion, the philosophy of astronomy has been largely neglected as a coherent field, but is ripe for study.

Chapters 38 and 39 then focus on the philosophical issues in astrobiology. The first lays out epistemological, metaphysical, societal, and sociological issues, all framed by the concept of cosmic evolution. It makes the case that science and philosophy are inseparable, especially in the form of epistemology and metaphysics. It therefore behooves us to have a more robust understanding of issues that fall under the realm of philosophy of astronomy. Chapter 38 also argues that as in the philosophy of science in general, and in the philosophies of particular sciences, critical issues in the philosophy of astrobiology are best illuminated by history. We therefore lay out an historically informed philosophy of astrobiology, drawing on the philosophy of biology and other sciences and informed by real-world issues.

Chapter 39 looks at possible lessons learned from the history of the twentieth-century extraterrestrial life debate. We address three overlapping areas: the problematic nature of evidence and inference, and its relation to scientific preconceptions; the role of theory in raising expectations, interpreting observations, and generating conclusions; and an evaluation of the success or failure of some of the debate's most general arguments, including the principles of plenitude and mediocrity and "Goldilocks-type" arguments that life occurs under such tight constraints that it is rare in the universe. The underlying assumption of the chapter is that lessons can be learned from history, even if one of the biggest lessons of history is that such lessons are seldom learned.

Chapter 40 focuses in on one critical issue: how biology and cosmology may be intertwined in unexpected ways. Even as astrobiologists search for life in the universe, scientists and philosophers alike are pondering the possibility of a deep and profound relationship between cosmology and biology. The universe appears in its very essence to be biocentric, in that the laws of nature and the physical constants are finely tuned for life, giving rise to what has been termed the *anthropic principle*. This in turn has given rise to the idea of an ensemble of universes, or multiverse, as an explanation for why we happen to be in a universe particularly suited for life. In his book *Infinite in All Directions*, physicist Freeman Dyson speculated that the prospects are bright for a future-oriented science, joining together in a disciplined fashion the resources of biology and cosmology. Fine-tuning and the multiverse are two concepts that may prove central to this task. In such a *cosmic ecology*, life and intelligence would play a central role in the evolution of the universe, no less than its physical laws.

The astrobiological examples presented in this section provide only a glimpse of the kinds of studies that could be undertaken in the much broader fields of astronomy and cosmology. To some extent these studies have been undertaken sporadically and independently, but not as part of a coherent discipline in the same way as has happened in other disciplines in the natural and social sciences. A more coherent endeavor would utilize and synergize approaches and results from philosophy of science and of related sciences such as physics and biology, including discussions on the nature of space and time and the nature of life and intelligence. Part VI may be seen as a call for philosophers, historians, scientists, and scholars in science studies and the social sciences to join this new endeavor.

Chapter 37
The Philosophy of Astronomy, Cosmology, and Astrobiology: A Preliminary Reconnaissance

Abstract Although the philosophy of science and of specific sciences such as physics, chemistry, and biology are well-developed fields with their own books and journals, the philosophy of astronomy and cosmology have received little systematic attention. It is time for coherent action in creating this new discipline. At least six categories of issues may be identified in the astronomical context: (1) the nature of reasoning, including the roles of observation, theory, simulation, and analogy, as well as the limits of reasoning, starkly evident in the anthropic principle, fine-tuning, and multiverse controversies; (2) the problematic nature of evidence and inference, especially since the objects of astronomical interest are for the most part beyond experiment and experience; (3) the influence of metaphysical preconceptions and non-scientific worldviews on astronomy, evidenced, for example in the work of Arthur S. Eddington and many other astronomers; (4) the epistemological status of astronomy and its central concepts, including the process of discovery, the problems of classification, and the pitfalls of definition (as in planets); (5) the role of technology in shaping the discipline of astronomy and our view of the universe; and (6) the mutual interactions of astronomy and cosmology with society over time. In addition to astronomy and cosmology in general, astrobiology is highlighted as a specific case study of the issues at hand.

Discussion of these issues should draw heavily on the history of astronomy as well as current scientific research, and may reveal an evolution in approaches, techniques, and goals, perhaps with policy relevance. This endeavor should also utilize and synergize approaches and results from philosophy of science and of related sciences such as physics, including discussions on the nature of space and time. A coherent discipline of philosophy of astronomy and cosmology is desirable for the same reason that the philosophy of the biological and other physical sciences not only exist but flourish: to better understand the foundations, assumptions, and goals of the science. This chapter is therefore a call for philosophers, historians, scientists, and science studies scholars to join in this new endeavor.

Published here for the first time.

S. J. Dick, *Space, Time, and Aliens*, https://doi.org/10.1007/978-3-030-41614-0_37

37.1 Introduction

Much of the well-developed field of philosophy of science deals with technical aspects like confirmation theory, the nature and purpose of science, and the claims of science to be a special kind of knowledge. There are also well-developed fields of philosophy of physics, chemistry, and biology, each with its own books and journals. But there has been no systematic discussion of philosophy of astronomy and cosmology. There have been sporadic attempts, especially at specific venues. The biennial History of Astronomy workshops at Notre Dame University have featured several sessions on the subject at its 2009 and 2015 meetings, a testimony to the interest of historians of astronomy in philosophical aspects. A collaboration between Oxford, Cambridge, and U. S. universities pioneered an initiative called "Establishing a Philosophy of Cosmology," which led to a conference on the subject in Tenerife in the Fall of 2014 and a volume on the *Philosophy of Cosmology* (Chamcham et al. 2017). An anthology compiled 25 years ago by the Canadian philosopher John Leslie, *Modern Cosmology and Philosophy* (Leslie 1998) provides a good starting point for the subject. And a few scholars such as cosmologist George Ellis (Ellis 2006; Ellis et al. 2004) and philosopher George Gale (Gale 1993) have written on the philosophy of cosmology.

This chapter constitutes a reconnaissance of issues in the philosophy of astronomy and cosmology, and one of its fields, astrobiology, which spans astronomy and biology. What might a field of philosophy of astronomy and cosmology look like? We can gain some insight from the philosophies of other sciences. Philosophy of science looks at how science works, the relation between theory and observation, and the status of its claims to objective knowledge (Curd et al. 2012). Philosophy of physics examines such concepts as the nature of space, time and causality, the reality of electric, magnetic and gravitational fields in relation to action-at-a-distance, and the ontological status of mass (Lange 2012). Philosophy of chemistry examines concepts like atom, element, the chemical bond, and the underlying foundations of the periodic table (Scerri 2008; Scerri and McIntyre 2015; Scerri and Fisher 2016). Philosophy of biology examines the concepts of species and speciation, classes and classification, and much else (Mayr 1988). And philosophy of technology addresses issues such as technology and culture, technocracy, the role of scientific instruments in observation, and technological determinism. All these studies are not part of science itself, but are important for understanding the basis for science and any claims it may have to privileged knowledge as opposed to other areas of human knowledge.

37.2 The Philosophy of Astronomy

We can identify at least seven categories of critical issues for a philosophy of astronomy, issues that also apply in many ways to cosmology and astrobiology. We shall examine some of the differences among these three fields later in this chapter.

1. What are the metaphysical foundations of astronomy and how have they evolved? The influence of metaphysics on science has been recognized at least since the

philosopher E. A. Burtt wrote his *Metaphysical Foundations of Modern Physical Science* (1924) almost a century ago, a volume that still resonates and that has been reinforced and refined in the modern context (Daston 1991; Fry 2012). Burtt argued that the Scientific Revolution represented a shift in our conception of the fundamental entities that exist in the universe, for example, space and time, mass, force, action-at-a-distance, and causality, as well as the mathematization of Nature. In other words the Scientific Revolution characterized by Galileo, Kepler, Newton and all the rest we know so well, represents a change in our conception of the essential nature of reality. Since then, astronomy has been revolutionized again by other changes in our conception of reality, according to general relativity (where gravity is not a force but a warp in the fabric of spacetime) and quantum mechanics. Today, reality is again being revolutionized with the introduction of the concepts of dark matter and dark energy into our cosmological ontology. What are the foundations of these changes, how justifiable are they, and might they someday be altered or replaced? Metaphysics also encompasses foundational questions such as the one that puzzled Einstein, and that Eugene Wigner expressed in his article "The Unreasonable Effectiveness of Mathematics in the Physical Sciences" (Wigner 1960). Given our evolutionary history optimized for survival, why the human mind can understand the cosmos at all is perhaps the mystery of mysteries.

2. What are the roles of metaphysical preconceptions and non-scientific worldviews on astronomy? In his book *Practical Mystic: Religion, Science and A. S. Eddington*, Matthew Stanley has shown how Eddington's "innovative methodology was based on values that he carried with him from his Quaker faith," in particular his belief that "scientists should not obsess over the absolute certainty of their physics [as was the method of the day in Britain], but instead work with a spirit of exploration that relied on physical intuition and observation" (Stanley 2007). It is safe to say that all astronomers—indeed all scientists—have in one way or another metaphysical preconceptions before they begin their empirical and theoretical work. As the cosmologist George Ellis has said "Some cosmologists tend to ignore the philosophical choices underlying their theories; but simplistic or unexamined philosophical standpoints are still philosophical standpoints!" Ellis (2006, 33). Perhaps the ultimate metaphysical question is "Why does the universe exist?" (Holt 2012).
3. What is the nature of reasoning in astronomy and cosmology, and what are the limits of reasoning? By the nature of reasoning I mean the relative roles of observation and theory, as well as the role of arguments such as simulation, modeling, and analogy. This is essentially a question of epistemology, a question of what are the legitimate sources of knowledge in astronomy. The relative roles of observation and theory first became obvious to me looking at the history of the search for other planetary systems in the twentieth century. For the first half of the century observation played very little role, and the *Jeans-Jeffreys tidal hypothesis* ruled, whereby very rare collisions of stars were necessary for solar system formation. But at the first hint of observational evidence of extrasolar planets in the 1940s, the theories began to change, and the *nebular hypothesis*

came into favor (Dick 1996 and Chap. 39). So a good specific philosophy of astronomy research question is "how did observation and theory interact in the search for planetary systems?" It does not take much imagination to see that the interplay of theory and observation is a question that frequently arises in much astronomical research, ranging from the discovery of planets to the existence and nature of black holes. The relative roles of theory and observation are evident even in the most fundamental theories of the origin of the universe, the decades-long debate over the steady state versus the Big Bang cosmology (Kragh 1996).

Side-by-side with the nature of reasoning is the question of the limits of reasoning. This is starkly evident in subjects such as the anthropic principle, fine-tuning, and multiverse controversies (Barrow and Tipler 1986; Carr 2007, and Chap. 40). Some question whether those concepts are even science, since they may not be verifiable or falsifiable in principle. The limits of reasoning are also evident in cosmological ideas about dark matter and dark energy.

4. How problematic is the nature of observation in astronomy and cosmology? Empiricism is usually regarded as the primary epistemology of science. But for at least the last half-century philosophy of science has emphasized the problematic role of observation, and how it is often theory-laden. More than 50 years ago, the philosopher Norwood Russell Hanson in his book *Patterns of Discovery* demonstrated how preconceptions are embedded in observations (Hanson 1958). In astronomy John Herschel was acutely aware of the problematic nature of early parallax observations, writing that the determination of stellar parallax was a goal that seemed within reach of the astronomer, "Only to elude his seizure when apparently just within his grasp, continually hovering just beyond the limits of his distinct apprehension, and so leading him on in hopeless, endless, and exhausting pursuit" (Herschel 1841). Peter van de Kamp's problematic observations of planets around Barnard's star are legendary (see Chap. 39). Numerous other examples are given in Norriss Hetherington's book *Science and Objectivity*, including the canals of Mars, spurious rings around Uranus and Neptune, and the purported measurement of the rotation of spiral nebulae (Hetherington 1988). William Sheehan makes the same point in *Planets and Perception* (Sheehan 1988), analyzing the perceptual and cognitive factors that enter into astronomical observation and interpretation. The problematic nature of observations is also evident even when the object is in hand, as in the controversy over the famous ALH84001 Mars rock (see Chap. 39). There are even more recent examples—the BICEP-2 experiment to observe gravitational radiation supporting the inflation theory, for example, announced positive results in March 2014 but the observations did not survive 6 months of scrutiny by the scientific community, in the end believed to be produced by cosmic dust (Overbye 2014). So the epistemological question of the reliability and interpretation observations would be a major area of research in a philosophy of astronomy.

5. How can we clarify astronomy's central concepts? Here I have in mind not only entities like black holes, but also astronomers' daily activities like discovery, classification, and definition (of planets, for example), and their use of evidence, laws, models, and so on. I have tackled some of these problems in my recent

book *Discovery and Classification in Astronomy*, where I show that discovery is an extended process consisting of detection, interpretation, and understanding (Dick 2013). Certainly in the area of classification astronomers have much to learn from biologists, where people like Harvard biologist Ernst Mayr have elaborated the problems of classification (Mayr 1988). The entire debate over Pluto as a planet might have been more systematic and less contentious had astronomers been more aware of the philosophical problems of classification in biology, e.g. the case of archaeopteryx. The explicit examination and clarification of central concepts is a vital part of philosophy of astronomy.

6. What role does methodology play? For example, what is the role of technology, both as a driver of discovery, and in shaping the discipline of astronomy and our view of the universe? My conclusion in *Discovery and Classification* was that new technology did not play a decisive role, at least in the case of discoveries of new classes of objects. But Martin Harwit's recent book *In Search of the True Universe* (Harwit 2013) asks a more fundamental question: what role does our technology play in determining our view of the universe? He shows convincingly that World War II technologies adapted for astronomy have produced our conception of the universe today. How, he asks, might our conception of the universe today differ if postwar efforts over the last seven decades had focused instead on dark matter and dark energy rather than on exploring information brought to us by the full spectrum of electromagnetic radiation. That is a question in philosophy of astronomy with a difference, since it makes us realize we really only know about the 4% of the current universe composed of baryonic matter and radiation.
7. What are the mutual interactions of astronomy and cosmology with society over time? One might well ask whether this is really philosophy of astronomy, but I would argue that it is, since astronomy throughout history has provided worldviews in a way that other sciences like biology and chemistry have not. In the western world the geocentric universe provided the worldview for the Christian Middle Ages (as in Dante's *Inferno*); the heliocentric worldview overturned everything from religion to literature (Milton's *Paradise Lost*); and what I have called the biological universe already pervades Western society in the form of astrobiology and SETI programs, UFOs, science fiction literature, and the view of our place in the universe. Chinese cosmology, Mayan cosmology, and other Native American cosmologies were all central to their cultural worldviews at the time (Aveni 1989; Pankenier 2013). This can all be seen as part of the astronomy and society component of the philosophy of astronomy and cosmology.

There is a long tradition of astronomers constructing or informing worldviews, ranging from the cosmic perspective in Harlow Shapley's *Of Stars and Men* (Shapley 1958; Palmeri 2009; Dick and Lupisella 2009) to current scientists who write on cosmos and culture in the context of cosmic evolution (Zakariya 2017). Philosophers are also becoming more aware of the cosmological perspective; most recently the young Belgian philosopher Clement Vidal's book *The Beginning and the End: the Meaning of Life in a Cosmological Perspective* (Vidal 2014) provides a deep examination of the subject. This is, however, a subject of considerable controversy, with some emphasizing that scientists should not impose their worldviews on parts of society, as in the case of religious naturalism. The renowned historian of science John Greene, for example, cautioned in no uncertain terms that naturalism is just another world view, no more privileged than supernaturalism or anything else to serve as the source of value and meaning in human life. He pointed out that scientists such as Ernst Haeckel, Julian Huxley, Ralph Burhoe, and E. O. Wilson "all typify the scientist-ideologue bent on saving society by promulgating a new ethics and a new religion claiming the sanction of science." The pattern is "depressingly familiar," in Greene's view:

> When will scientists and others learn that naturalism is a philosophical point of view with no more claim to the status of science than any other philosophical viewpoint, whether Marxian, Freudian, Russellian, Whiteheadian, or whatever. Scientists have as good a right to expound their philosophical, ethical and religious views as anyone else, but they have no right to palm these off as the findings of science. (Moore 1989, 404)

In my view this attitude is both instructive and dead wrong. Instructive in the sense that a science-oriented theology is far from the sole source of meaning and value in life. And dead wrong in the sense that science has *at least as much* claim to infuse theological thought as any other worldview, and perhaps more, since being grounded in the natural world gives it a pillar of support that supernatural religions do not have, numerous attempts at natural theology notwithstanding (see Chap. 13).

In summary, the issues discussed here are metaphysical/ontological, epistemological, methodological, conceptual, and societal in nature. While a considerable amount of scattered work has been done on this subject, a field of philosophy of astronomy and cosmology has not yet congealed. My purpose here is to call attention to this fact, and to emphasize that recognizing all these issues as part of a connected whole can have a unifying and synergizing effect, providing a new dimension to studies of our cosmos just as philosophical issues have in other fields of science.

37.3 The Philosophy of Cosmology

In many ways cosmology—the study of the origin, evolution, and structure of the universe—may be seen a branch of astronomy in the sense that both encompass the study of the cosmos. Both endeavors, it need hardly be said but is often forgotten, are undertaken by people with preconceptions, biases, emotions, and all the faults of other humans, even if they are in the end seeking "the truth" (Hetherington 1993; Lightman and Brawer 1990; Overbye 1991). But at the same time, because of its expansive scope, techniques, and theories, the issues in any philosophy of cosmology seem quite different from, say, those involved in measuring the diameters of planets or the motions of stars or observing discrete astronomical objects from our Solar System to the farthest galaxies. Because some work has been done in philosophy of cosmology, one way to probe the differences and similarities with the philosophy of astronomy is to ask what issues the philosophers and scientists in the past have seen as the main issues in the field. Though one hesitates to call philosophy of cosmology a coherent discipline at this time, let us focus first on a pioneering individual study, and then on two volumes compiled several decades apart, the first by the Canadian philosopher John Leslie (Leslie 1998), first published in 1990, and the second a landmark volume titled *The Philosophy of Cosmology* (Chamcham et al. 2017).

The pioneering individual is George F. R. Ellis, a cosmologist at the Department of Mathematics and Applied Mathematics at the University of Cape Town, South Africa. Ellis is a collaborator with Stephen Hawking, a prolific author, and the winner of the 2004 Templeton Prize for fostering dialogue between science and religion. In his pioneering, comprehensive, and systematic article "Issues in the Philosophy of Cosmology" (Ellis 2006), Ellis distinguishes eight issues that any analysis must address: the uniqueness of the universe, the large scale of the universe in space and time, the unbound energies in the early universe resulting in the cosmos we see today, the origins of the universe, the universe as the background for existence in the sense of determining the initial conditions for all of science, the philosophical basis underlying cosmological theory, the questions of fine-tuning and the anthropic principle, the possible existence of multiverses, and the nature of existence. Though highly mathematical in places, Ellis's 2006 article and others that preceded and followed it (Ellis 1998a, b, 2017) are good entrées into the issues in philosophy of cosmology.

The first of the edited volumes on the subject, John Leslie's *Modern Cosmology and Philosophy* (1990, 2nd edition, 1998) provides some indication of what issues he considered important at the time, well before Ellis's more comprehensive survey. Tellingly, the volume consists of papers by 21 cosmologists and only five philosophers. Among the philosophers, the Notre Dame University Catholic priest Ernan McMullin addressed the stark question "is philosophy relevant to cosmology?" It is notable that McMullin's article first appeared in the *American Philosophical Quarterly* for 1981 (McMullin 1981), and that by his count out of 10,000 philosophical articles published in west European languages over the previous 3 years, only a dozen were related to cosmology in any way. After pointing out that some scientists such as Steven Weinberg and Paul Davies have given a resounding answer of "no!" to his self-posed question of philosophical relevance, McMullin counters by answering "yes." He helpfully distinguishes three areas of relevance: ontological, methodological, and conceptual, the latter including questions of time and space, as well as the anthropic principle (the latter just coming into prominence, as seen in Chap. 40). He lays out in a systematic way some of the problems associated with these areas, altogether a pioneering, if brief foray into this subject. Three of the other philosophers, Richard Swinburne, George Gale, and Leslie himself, focused on theories of multiple universes (the ensemble of which is now known as the multiverse) and related arguments of the anthropic principle and fine-tuning of the constants that make our universe possible. And the fifth philosopher, Adolf Grunbaum, addressed what he called "the pseudo problem of creation in physical cosmology," concluding that neither the steady-state cosmogony nor the Big Bang cosmogony validates the traditional cosmological arguments for divine creation. All of these themes were subsumed in Ellis's later survey in one form or another.

The 21 cosmologists of Leslie's volume are all distinguished scientists, some of them long gone, who have raised philosophical questions in the course of their scientific work. They range from George Gamow, Herman Bondi and Martin Rees to Ellis, Brandon Carter, and John Wheeler. In general these cosmologists do not address philosophical issues head on; rather their work raises such issues that are ripe for analysis.

A quarter century later, nine of the 25 authors represented in *The Philosophy of Cosmology* were philosophers, clear evidence of increasing interest among philosophers in issues related to cosmology. Several other points are clear. First, the book, although based on workshops and conferences sponsored by the Templeton Foundation, nevertheless largely steers clear of religious issues. Secondly most of the papers, even by the philosophers, are very technical and therefore their conclusions have not penetrated far beyond a small circle of specialists. And thirdly, as in Leslie's volume, many of the papers consist of cosmologists presenting the results of their own research, with little explicit reference to philosophical aspects. The implicit philosophical aspects are presumably to be left to others. At the outset the editors briefly ponder the fundamental question of the relevance of philosophy to cosmology. After noting that no less a cosmologist than Stephen Hawking declared "philosophy is dead" in the sense that cosmologists no longer need it given recent advances, they argue that cosmology has long been a testing ground for philosophy, and that the central problem of cosmology, "the problem of understanding the world, including ourselves and our knowledge," is essentially philosophical no matter how much science advances. Nevertheless they correctly declare that "philosophy of cosmology as a body of philosophical literature engaged with contemporary cosmology does not yet exist" (Chamcham et al. 2017, xi). They view their volume as marking a beginning.

And an ambitious beginning it is. The editors divide their book into five parts: (1) issues in the philosophy of cosmology, (2) structures in the universe and the structure of modern cosmology, (3) foundations of cosmology in the form of gravity and quantum theory, (4) quantum foundations and quantum gravity, and (5) methodological and philosophical issues. Each of these parts opens a window into what the editors and authors considered an integral to the philosophy of astronomy. Significantly, Part 1, and therefore the entire volume, begins with a paper by George Ellis, arguably recognized by his peers as the cosmologist with the most experience in dealing with philosophical aspects of cosmology, dating back at least to the 1990s and leading to his 2006 comprehensive article described above. Ellis's paper focuses on the testing of cosmological theories. He notes the uniqueness of the universe is one of the foundational issues in any philosophy of cosmology; since we have only one example of a universe, the unobservable multiverse notwithstanding, we cannot compare similar objects to find regularities as is usually the case in science.

Moreover, we have to extrapolate physical theories beyond the bounds of where they can be tested in the laboratory, even beyond the most powerful particle accelerators. Theory is therefore pushed beyond testable limits, raising questions about the scope and goals of theory, the nature of causality and explanation, and the limits of data when coming to any conclusions. Because of these difficulties, the question is posed whether a "post-empirical phase" is allowed in cosmology. Ellis's paper is supplemented by cosmologist Bernard Carr on three problems at the limits of science (Black holes, cosmology and the passage of time), by mathematician Roderich Tumulka on the question of why there exists something rather than nothing, and by philosopher Claus Beisbart exploring the boundaries between philosophy and cosmology.

Part 2 on structures includes the formation of galaxies and higher clustering, and raises such questions as whether stars and galaxies are inevitable given the initial conditions of our universe, whether the laws of physics we currently know applied in the early universe, and the methodological question of to what extent computer simulations can approximate reality. Parts 3 and 4, with their emphasis on quantum theory in a cosmic context, demonstrate that philosophy of cosmology has much in common with philosophy of physics (Lange 2002). And yet, the editors note, "where philosophers of physics have made inroads on conceptual questions in physics, they have tended to focus on cleanly defined theories treated in isolation." By contrast synthetic theories, in the complex applications of cosmology, "are messy and ill suited to rigorous analysis, axiomatisation, or regimentation by other means—the standard tools of philosophy." Part 5 on methodological and philosophical issues includes Einstein's philosophy of cosmology, testing the multiverse, and how much we can infer in cosmology beyond the data.

Like philosophy of physics, philosophy of cosmology clearly is engaged with concepts such as space and time, fields, energy, and mass. One way to analyze the universality of such concepts is through a thought experiment: imagining the perceptions of space and time, for example, by conscious/cognitive/intelligent beings other than ourselves, and realizing (based on terrestrial examples of complex life) that they may be very different from ours. Surely the fundamental basis of life's interaction with the external world in any environment will depend on how any given life form sees that world in terms of geometry and mathematics (if it sees it at all in those terms), and how it employ its perhaps very different sense of time. This is difficult for humans to visualize, but a valiant and delightful attempt at the spatial dimension was made already 130 years ago when the British schoolmaster Edwin A. Abbott penned *Flatland: A Romance in Many Dimensions* (1884). There, a creature named "A Square" visited pointland, lineland, flatland, and spaceland, representing zero, one, two, and three dimensions respectively. The creatures in each dimension thought it ridiculous that any other dimensions could exist.

Similarly, fundamental questions about time arise when contemplated from an alien point of view. We are not talking about "mechanical time," which in the case of Earth involves timekeeping based on planetary rotation being divided into 24 h, 1440 min, or 86,400 s. That is an interesting history about the representation of Earth's rotation through the painstaking construction of mechanical clocks, ranging from pendulum clocks to chronometers for longitude and navigation to the intricacies of atomic clocks, with all the associated problems of leap seconds to keep those timekeepers synchronized with astronomical reality (Dick 2003). If aliens keep time also based on planetary rotation (though surely not with our hours, minutes and seconds and perhaps not even with their equivalent in some other number system), that would be an interesting story too. But here we are talking here about the perception of time, sometimes called *subjective time* or *psychological time*, related to how any given nervous system processes sensory information. We know that on Earth animals have very different perceptions of time, and that this perception is related to body mass: the smaller the animal, the faster its metabolic rate and the slower time passes. A mouse has a very different perception of time than an elephant (Reas 2014; Davies 1995; Eagleman 2009). The same may be true with aliens, or their body clocks may operate according to some other variable.

This difference in time perception in the terrestrial context is the subject of much research, and has occasionally been the subject of science fiction. Most notably for one extreme case is the physicist Robert Forward's imaginative novel *Dragon's Egg* (1980), where the sesame-seed-sized intelligent cheela live out their lives on a neutron star with 67 billion times the surface gravity of Earth. Their biology utilizes nuclear rather than chemical reactions, so they think and develop a million times faster; 1 min for humans is many years for a cheela. Their civilization therefore develops very rapidly as the humans watch with astonishment. In another example of a different variety the aliens in Kurt Vonnegut's *Sirens of Titan* do not distinguish among past, present, and future, just as the heptapods in the movie *Arrival* do not (see below). Other examples could be multiplied.

What is clear from an analysis of these volumes is that cosmology is fertile ground for philosophers, that philosophers must have some grounding in the actual science of cosmology and vice versa in order to make robust contributions to the field, and that because of the difficulties of collaboration between these two fields, philosophy of cosmology is still in an embryonic stage. But new collaborations are cause for hope, and no field better illustrates the possibilities of interdisciplinary collaboration than astrobiology, which over the past two decades has become a robust science of its own, even now drawing in not only natural scientists, but also scholars in the humanities, social sciences, and yes—philosophy (see Chap. 16). As a new discipline necessitating collaboration between astronomers, biologists, and many other fields, astrobiology can serve as a case study of how these disciplines have interacted over the last few decades to produce tangible results—though not yet life itself beyond Earth.

37.4 The Philosophy of Astrobiology

Having looked at some general issues in philosophy of astronomy and cosmology, we now turn to a specific case study, that of astrobiology, a field that studies the origin, distribution, and future of life in the universe. The general philosophical issues with astrobiology have been brilliantly laid out in Milan Ćirković's stimulating book *The Astrobiological Landscape: Philosophical Foundations of the Study of Cosmic Life* (Ćirković 2012), as well as in Dunér et al. 2012, 2013. Ćirković highlights some of the same issues we have emphasized above: methodological, epistemological, and metaphysical problems, while also tackling the relationship between biological complexity and cosmological realities such as the fine-tuning of constants that make life possible, and the scientific status of the idea that we are part of a multiverse. Moreover, he addresses the so-called *continuity thesis*: whether physical, chemical, biological, and cultural evolution (evident, for example, in the Drake Equation) are parts of the same evolutionary continuum. This raises all sorts of deep philosophical questions. Here, however, we focus on only one small but important part of astrobiology: the origins of life.

37.4.1 What is Life?

One of the earliest themes of the U. S. space program was what was then known as *exobiology*—the search for life in the universe—thanks to pioneers like Joshua Lederberg, Jerry Soffen, and Carl Sagan (Dick 1996). One could easily argue that the search for life has been an important driver of the space program, certainly when it comes to the exploration of Mars, and the oceans on some of the Jovian and Saturnian satellites (Dick and Strick 2004). From the beginning, before expanding in the 1990s into the much broader field of astrobiology, exobiology focused largely on the origins of life and raised important philosophical questions. What is life? Are the underlying foundations for life universal? What is the role of chance and necessity? Here we examine some of these issues with an eye to philosophical issues.

Let me begin with a conceptual issue that has always been at the heart of exobiology, and the broader discipline known as astrobiology. In order to search for life on Mars, the early exobiology pioneers had to have some idea of what to search for, and so the first of the philosophical issues was raised almost immediately: "What is Life?" That is the question in the background behind all searches for life, whether on Earth, in the laboratory, or in space. Someone recently quipped that life is what is left over after you eliminate everything that's not alive. That, of course, does not solve the problem. The question has been contemplated by biologists, biochemists, and even physicists (Schrodinger 1944) over several centuries and has been the subject of a special issue of the journal *Astrobiology* (Deamer 2010). That history reads like a mirror of each period when the question was asked, as biologists tried to determine the physical basis of life, with attention turning successively from protoplasm to the protein enzyme, the virus, the gene and the cell as the unit of life. By 1937 the British biologist Norman W. Pirie argued that the terms "life" and "living" were meaningless, and that the transition from nonliving to living was comparable to the gradual transition from green to yellow in the spectrum or from acid to alkaline in chemistry, except that life could not be defined in terms of only one variable (Pirie 1937, 1975).

Today the debate about the origin of life is at the level of the molecular assembly of life. We often see the NASA definition of life as a "self-sustaining chemical system capable of Darwinian evolution" (Joyce 1994, Deamer and Fleischaker 1994). But there is consensus that no single definition of life is totally satisfactory. Some philosophers and scientists such as Carol Cleland and Chris Chyba, beginning in a series of papers in 2002, have repeatedly argued that all attempts to define life are fundamentally misguided in the absence of a general theory of living systems (Cleland and Chyba 2007; Sullivan and Baross 2007). Other philosophers such as Kelly Smith call this stance "definitional pessimism," and argue that one need not have a universal definition of life to have a working definition for any given circumstance (Smith 2016). In other words, there must be a balance between definitional pessimism and pragmatism.

I tend to come down on Smith's side for several reasons: (1) we cannot help ourselves from defining life, just as we cannot help ourselves from defining other things, like "planet," to take a recent example from astronomy; (2) institutions like NASA need an operational definition of life if they are going to search for it; the latest Astrobiology Strategy Document (NASA 2015, 155) notes that definitions of "life," "function," and "evidence" shape our research, and therefore we need them, not to say we should obsess over them; (3) definitions are invariably problematic in many areas, but that does not prevent us from using them as a benchmark for discussion.

In astronomy we think most recently of the controversy surrounding the question "Is Pluto a planet?" This depends on definition, and the International Astronomical Union (IAU) decided in 2006 that the answer was no, Pluto was not a planet, based on a rather arbitrary set of conditions, a compromise among astronomical communities that defined Pluto as a dwarf planet, and then declared that a dwarf planet was really not a planet at all (Chaps. 30 and 31). As a member of the IAU, at its 2006 General Assembly in Prague I voted for dwarf planet status, but went on to vote that a dwarf planet should simply be a class of planet, like terrestrial planets, gas giants, and ice giants. But the point is that the concept of a planet *was* defined, in a way that reflected the different astronomical communities, even though it disappointed some scientists like Alan Stern (later of New Horizons fame), not to mention school kids who had been taught that there were nine planets in the Solar System.

These arguments over definition did not keep New Horizons from having a spectacular mission; as Stern likes to point out, Pluto doesn't care if it's a planet or a dwarf planet. The same is true of the definitional problems with species, a common one being that any organisms that can interbreed belong to the same species. But we now know that *homo sapiens* and Neandertals interbred, and most of us are from 1 to 3% Neandertal. Instead of throwing out the concept of species, we adjust its definition with new information. And the same is true of the idea of intelligence, which SETI practitioners skirt by talking about the search for technology instead. I once asked Berkeley philosopher John Searle about the concept of intelligence. "Forget intelligence," he said. "Consciousness is the important thing!" But, famously, there is no consensus definition of consciousness either. So we settle for operational definitions that apply to the specific task at hand.

Of course, operational definitions can get you into trouble, especially if you choose the wrong definition for any given circumstance, as some think was the case for the Viking biology experiments, all of which relied on a definition of life based on metabolism. On the other hand, one of Viking's principal investigators, Gil Levin, still believes he found life on Mars, a claim that is no longer completely dismissed given what we now know about the surface chemistry on Mars. For many years that result was trumped by the results from the GCMS (Gas Chromatograph Mass Spectrometer), which showed no organic molecules on the surface at the two sites of the Viking landers. No organic molecules, no life, so the argument went. In effect a biochemical definition of life trumped all of the other metabolic experiments—until we found out more about the role of perchlorates on the surface of Mars. The lesson of Viking seems to be that experiments should be designed with multiple definitions of life in mind (see Chap. 39).

The best way to define life would, of course, be to find some—on Mars, Europa, or elsewhere and compare it to Earth. In that case, if it was truly a second genesis, we might have the beginnings of a universal biology, similar to a universal physics. And that is the second issue I want to address—aspirations for universality—by which I mean two things: Is there a universal biology as there is a universal physics? And is life universal, in other words, is it spread throughout the universe?

37.4.2 Aspirations for Universality

There is no doubt that the space program catalyzed aspirations for a universality in both ways—the possibility of a universal biology, and the search for life in the universe. Implicitly or explicitly, those at the intersection of exobiology and origins of life expressed interest in the desire for a universal biology, one that might be more general than terrestrial life. In the Soviet Union, Alexander Oparin called exobiology and the origins of life "the inseparable connection," and Stanley Miller said almost exactly the same thing (Oparin 1964; Miller and Urey 1959). Origins of life studies became the central focus of exobiology, bringing into the field biologists such as Melvin Calvin and his student Cyril Ponnamperuma. According to Freeman Quimby, the first head of NASA's exobiology office, Calvin did more than anyone else to establish the posture of the nation in bracing itself for the search for extraterrestrial life. "There was no question in my mind about the Darwinian behavior of molecules," Calvin said. "And if that were true, then that meant it must be true in the entire universe" (Calvin 1981). Ponnamperuma agreed: "If we can understand how life began on Earth, we can argue that the sequence of events which lead to the appearance of terrestrial life may be repeated in the staggering number of planetary systems in our universe" (Ponnamperuma 1972). These lines were written at a time when no other solar systems were known beyond ours. We now know they are everywhere. In short, the space program catalyzed the integration of origin of life and extraterrestrial life studies, as theories and experiments previously viewed in the context of life's beginning on Earth were increasingly seen in an extraterrestrial context, integrating the two fields of study.

Still, in contrast to the universality of the physical sciences, biology remains an open question in this respect, and one of the main questions astrobiology hopes to address. Unlike physics and chemistry, with their remote observations using spectroscopy, life is much more difficult to detect remotely. Indeed, the earliest searches for life were undertaken by sending spacecraft such as Viking to make observations in situ on the surface of Mars, or from orbit around Mars, Jupiter and Saturn. Only recently has the method of biosignatures held out hope of remote detection of life in planetary atmospheres or surfaces. Even then, the detection of life in this manner is likely to be ambiguous, and the proof of universal biological principles far from immediate.

There is, however, another approach, based not on observation but on principles—the universal principles of evolution by natural selection. In the absence of other evidence, it seems eminently reasonable to believe natural selection operates throughout the universe, whether carbon and DNA based, or silicon based with some other genetic system. Indeed, I believe most scientists would argue this is the case, and many have actively done so. Almost 40 years ago, in his path-breaking book *The Selfish Gene*, evolutionary biologist Richard Dawkins argued that any life we find beyond Earth would surely be strange and unearthly. But, he asked, "Is there anything that must be true of all life, wherever it is found, and whatever the basis of its chemistry? If forms of life exist whose chemistry is based on silicon rather than carbon, or ammonia rather than water… if a form of life is found that is not based on chemistry at all but on electronic reverberating circuits, will there still be any general principle that is true of all life?" His unequivocal answer was spoken like a true Darwinian: "If I had to bet, I would put my money on one fundamental principle. This is the law that all life evolves by the differential survival of replicating entities." (Dawkins 1989, 191–192). In other words, evolution by natural selection. Dawkins went on to argue for the "meme" as the analog to the gene for cultural evolution, and more generally for the idea of universal Darwinism in a number of fields. But the important point here is that evolution by natural selection provides a principle for the elaboration of universal biology, in the absence of observational evidence beyond the Earth.

Universal biology as a robust research program is just beginning to evolve. Perhaps its most important impetus comes from astrobiology, and it has been an explicit goal of NASA's astrobiology program. Its 2008 roadmap stated that "we must move beyond the circumstances of our own particular origins in order to develop a broader discipline, 'Universal Biology'." In doing so the authors of the roadmap proposed to use the universal laws of physics and chemistry to understand biological processes such as polymer formation, self-organization, energy utilization, information transfer, and Darwinian evolution. Accordingly, the University of Illinois at Urbana-Champaign, one of the research teams of the NASA Astrobiology Institute, harbors the Institute for Universal Biology at its Institute for Genomic Biology, as part of its research on the origin and evolution of life in the universe. It uses the fields of microbiology, geobiology, computational chemistry, and genomics to find general principles for living organisms. "It is important to develop the field of universal biology, because we may never find traces of life on other planets. But if we understand that life is generic, maybe even an expected outcome of the laws of physics, then we'll know for sure that we are not alone," wrote the Institute Director, Nigel Goldenfeld.[1]

37.4.3 Chance and Necessity

These conclusions harbor a deep metaphysical preconception: the question of chance and necessity in the origin of life. It is a tribute to many of the scientists involved that they recognized their preconceptions. In fact the philosophical question that dominated the problem of the origin of life beyond Earth was exactly that: the role of chance and necessity in the formation of life. On this question hinged the whole enterprise of exobiology, because if life was a chance occurrence with very low odds, it might never happen again no matter how big the universe and how expansive he time available. The two prototypes for the metaphysical preconceptions of chance and necessity are the French biologist Jacques Monod on the one hand, and the Belgian biologist Christian De Duve on the other. Monod, awarded the Nobel Prize in 1965 for explaining the genetic regulation of protein synthesis in cells, drew on the latest results in molecular biology to illustrate how chance predominated in all aspects of life's development. In his classic volume *Chance and Necessity* Monod famously wrote "The universe was not pregnant with life, nor the biosphere with man. Our number came up in a Monte Carlo game" (Monod 1970). Monod's view was based on what he called the "revelation and disappointment" that the sequences composing the insulin protein showed no general law of assembly, only the blind workings of chance. Many scientists at that time agreed with this view about the role of chance, including some astronomers like Fred Hoyle, who argued that an origin of life by chance was less likely that a 747 aircraft being spontaneously generated from a junkyard. He thought that it was vastly more probable that life had been assembled by a "cosmic intelligence," though before the theologically inclined could get too excited, he emphasized that in his opinion it was not a supernatural intelligence. In his book *The Astonishing Hypothesis* advocating panspermia, Francis Crick tended to agree that life could not have arisen by chance (Crick 1995).

A quarter century after Monod's pronouncement, in his book *Vital Dust: Life as a Cosmic Imperative*, Christian DeDuve argued just the opposite, that life was a cosmic imperative. DeDuve, who won the 1974 Nobel Prize for discoveries related to the structural and functional organization of the cell, was most impressed with the organic molecules then being discovered in space. He wrote,

> In this organic cloud, which pervades the universe, life is almost bound to arise in a molecular form not very different from its form on Earth, where physical conditions are similar to those that prevailed on our planet some four billion years ago. This conclusion seems to me

inescapable. Those who claim that life is a highly improbable event, possibly unique, have not looked closely enough at the chemical realities underlying the origins of life. (De Duve 1995)

In coming to his conclusion de Duve was looking at the 20% of interstellar dust—"vital dust"—that was composed of organic compounds—the "organic cloud" he is talking about here. De Duve was joined in his conclusion by others for separate reasons, including Richard Dawkins, who declared that "climbing Mt. Improbable" was not impossible (Dawkins 1996).

It is one of the hallmarks of exobiology, and now astrobiology, that its practitioners place their faith in the necessity, or at least the high probability, of the origin of life under proper conditions. It is their most basic metaphysical presupposition, otherwise they would not be searching for life. It seems to me a plausible presupposition, partly because life arose so quickly on Earth—just recently new evidence has surfaced of life from 3.7 billion years ago, fossilized stromatolites from the Isua rocks on the southwest coast of Greenland, dating from less than a billion years after Earth's formation. That seems also to argue for the possibility of panspermia, the spreading of life throughout the universe, or at least giving life a head start by spreading complex organic molecules, which we know are ubiquitous in interstellar molecular clouds up to the level of amino acids.

Beyond life's rapid origin on Earth, the idea of necessity seems also to be reinforced by the fact that the physical laws of nature are universal, and that what happened here ought to happen everywhere under proper conditions. That presupposition certainly worked for planets: In early 1995 no planets were known to exist beyond our Solar System, but the idea of universal laws, that what happens here happens there, argued they should exist. We now know they are everywhere, even around a red dwarf star Proxima Centauri, the closest star to us. That doesn't mean they all have life. In fact most of the exoplanetary systems are not similar to ours. But it does mean that where the conditions are right in the habitable zone, life will arise. That is the metaphysical presupposition. And with the discovery of extremophiles over the last half century, we know that "proper conditions" are much broader than we thought in the past. We also know that the idea of a *habitable zone* has been expanded to the extent that some of the moons of Jupiter and Saturn are now considered to be in a habitable zone, thanks to tidal heating mechanisms that allow liquid oceans to exist. That is not to say that life will be like ours—far from it. There will be diversity, just as there is a tremendous diversity in exoplanets. So this idea of universal laws producing similar outcomes, with diversity, has been proven time and again. It is of course a presupposition, but a good one. Some presuppositions are better than others.

This gives rise to even deeper metaphysical questions: Why is the universe biofriendly to start with? Do we just happen to live in a universe where the constants of nature are such that galaxies, planets, stars, and life can exist? Is life in some way one of the fundamental entities of the universe, due to what some call the anthropic principle, or as I prefer to call it the biocentric principle? Is there a deep connection between life and the very structure of the universe? Such questions (see Chap. 40) lend all the more importance to the current disciplines of astrobiology and SETI, which will determine just how abundant life is in the universe. Some argue that the biological universe may be not the result but the cause of the physical universe, part of its very fabric, and more surprisingly fundamental than we have ever realized. As Freeman Dyson—another physicist who has dabbled in origins of life studies—has said, "It almost seems as if the universe knew we were coming … The prospects are bright," he argued, "for a future-oriented science, joining together in a disciplined fashion the resources of biology and cosmology … 'Cosmic ecology' would look to the future rather than to the past for its subject matter, and would admit life and intelligence on an equal footing with general relativity as factors influencing the evolution of the universe" (Dyson 1988).

In the end we still do not know whether or not life is abundant in the universe. It remains perhaps the greatest question in science. But if there is life out there, it is prudent that we prepare for that discovery, which in the long term could change our worldview in the same way the Copernican and Darwinian worldviews have changed us. And this may be true even if the universe is "only" full of microbes. This challenge of the societal impact of astrobiology is a very large topic, one that is now beginning to be addressed (Dick 2015, 2018).

37.5 Putting It All Together: An Emerging Discipline?

The ideas in all three of the previous sections are of course related, and there is no profit in trying to make detailed artificial distinctions. Nevertheless, it may be useful to highlight several broad distinctions in the philosophical issues among the three related areas. Certainly, some of the issues listed in Sect. 37.2 cross all three disciplines, including metaphysical, epistemological, and methodological regimes. Regarding differences between philosophical issues in astronomy and cosmology, we note first that cosmology as a mathematized science is much younger than astronomy, most historians and cosmologists dating the beginning of cosmology only to Einstein's general theory of relativity in 1915, and the application of that theory of gravity, space, and time to the universe as a whole (North 1965; Kragh 1996). Secondly, Ellis's point about having only one example of a universe to study sets fundamental limits to reasoning that differ greatly from astronomy, which typically has many examples to compare. And thirdly, while a post-empirical phase of cosmology due to the difficulties of observation and other usual modes of reasoning may be entertained, that idea would certainly be foreign to most of the concerns of astronomy.

Although Ellis does not specifically address astrobiology, he does make a distinction that is useful, separating the usual physical concerns of cosmology from what he calls "cosmologia," dealing with "the major themes of origin of life and the nature of existence that are raised because the physical universe characterized by cosmology is the context for our physical being … Cosmologia necessarily considers major themes in philosophy and metaphysics, perhaps relating them to issues of meaning and purpose in our lives, and is of great public interest and concern" (Ellis 2017, 4). Astrobiology is of course anchored to the nature of our physical universe, in ways small and large (fine-tuning issues), but the search for life beyond Earth crosses over into the well-known issues in the established field of philosophy of biology. Issues of chance and necessity, for example, are endemic to physics no less than to biology.

The question remains of how to crystallize philosophy of astronomy and cosmology as a discipline, just as philosophy of science and many of its splinter disciplines have done. First, the importance of core research centers is essential. Many philosophy of astronomy papers have been presented at history meetings, in my view appropriately so because the philosophy of astronomy is closely related to the history of astronomy. In fact academics of many persuasions have realized conversely that the history of science needs to be seen in the context not only of philosophy, but also of sociology and broader science studies, and vice versa. Papers have also been presented at scientific meetings such as the American Astronomical Society and the International Astronomical Union. Institutions harboring these academic disciplines are therefore the prime candidates for core research centers to foster philosophy of astronomy and cosmology. But who will take up the challenge of cohering the new discipline, whether historians, philosophers, scientists, or some combination thereof, remains to be seen.

Secondly, just as happened in astrobiology and other new disciplines in recent history, a *Journal for the Philosophy of Astronomy and Cosmology* (JPAC) could help focus attention on these studies as a unified discipline. There is always a question of the "balkanization" of knowledge, which in many ways these journals represent. But there is no denying that *Astrobiology* and the *International Journal of Astrobiology* have helped to cohere the field. The same has happened with journals such as *Philosophy of Science* (the official journal for the Philosophy of Science Association), *The British Journal for the Philosophy of Science*, and *Studies in the History and Philosophy of Science*. And this cohering effect has also been evident in more specialized fields such as the journal *Biology and Philosophy* and *HYLE: International Journal for Philosophy of Chemistry*. Balkanization notwithstanding, the role of journals in cohering new disciplines and communities should not be underestimated.

37.6 Commentary 2020

This chapter represents an amalgam of papers presented at four venues that have been particularly important for the embryonic field of history and philosophy of astronomy, cosmology, and astrobiology. Early versions of the chapter were presented at special sessions organized at the biennial History of Astronomy Workshops at Notre Dame University in 2009 and 2015 (Notre Dame History of Astronomy Workshops 2009, 2015); at the Historical Astronomy Division of the American Astronomical Society meeting in Kissimmee, Florida, Jan 5, 2016; and at the

History of Astronomy Commission at the General Assembly of the International Astronomical Union in Beijing, China in 2012. Section 37.3 was presented at a Library of Congress Symposium organized by Nathaniel Comfort, the Baruch S. Blumberg NASA/Library of Congress Chair in Astrobiology on September 15, 2016.

The 2009 Notre Dame workshop featured in its philosophy of astronomy session two philosophers (George Gale and Don Howard) and three historians (Owen Gingerich, Steven Dick, and Matthew Stanley). In a 2015 session of that workshop on "Epistemology and Authority in Cosmology," speakers addressed philosophical aspects of the multiverse, the failure of the BICEP-2 observations, and the role and status of cosmological simulations.

Many aspects of this paper could be elaborated, for example the metaphysical problem that Eugene Wigner enunciated, mentioned in Sect. 37.2. The formalist position of mathematics, which states that numbers are the creation of the human mind, leaves us with the problem that Wigner framed as "the unreasonable effectiveness of mathematics in the natural sciences." Galileo 400 years ago marveled that the laws of nature are written in the language of mathematics. But why? More to the point, why should our minds, shaped by natural selection, be able to fathom laws of nature in mathematical terms that we invented? Wigner pointed out that Newton's laws of planetary motion (the universal law of gravitation), which Newton verified at an accuracy of 4%, has now proved accurate beyond all reasonable expectations, at less than a ten-thousandth of a percent in 1960. It is verified with even more accuracy now. He went on to point out that the same is true in elementary quantum mechanics, the mathematics of complex spectra, and in quantum electrodynamics. So the question is, why does this invented mathematics work so well in our analysis of the universe? Wigner dubbed the uncanny accuracy of our mathematical abilities to match nature's reality "the empirical law of epistemology," but for him it remained a mystery, "a wonderful gift which we neither understand nor deserve." More than once, he referred to it as a "miracle" that we have come to take for granted because it works so well. The statement resonates with Albert Einstein's more general remark that the most incomprehensible thing about the world is that it is comprehensible.

Responses to Wigner came from many fields, including the physicist Max Tegmark and the mathematician Ivor Grattan-Guinness. Tegmark argued that mathematics is so effective because the world not only is described by mathematics, but *is* mathematics, and it is therefore not surprising it is so effective since we are gradually uncovering its structure—Platonism with a vengeance (Tegmark 2014). Grattan-Guinness argued that since science motivates much mathematics and both use theories and analogies, the close correspondence is no surprise at all (Grattan-Guinness 2008). Such explanations have not satisfied everyone, and the problem is still a burning question in the philosophy of mathematics, which underlies all of astronomy and cosmology. The embodied mind theories espoused by George Lakoff and others from cognitive science largely adopt the formalist view, and argue that mathematics works so well because the human brain evolved in response to the physical world around it. The concept of numbers arises because we count concrete objects, and mathematics works because the brain evolved to understand its surroundings. Mathematics is a construction of the brain, not something to be discovered (Lakoff 2003). There are also positions in between the Platonist and the formalist. Mathematician Carl DeVito thinks natural numbers do exist independently of us, while the rest of mathematics is invented. Thus, for Plato and DeVito, 2 + 2 does equal 4 for aliens, but for the formalist this may not be true, since the alien may not have invented the concept of number, or if it has, may not have invented the same numbers (DeVito 2013). Wigner's famous paper continues to generate discussion and controversy about one of the most fundamental questions of science: how is it that the human mind can comprehend the universe at all, especially through mathematics? (Nicholson 2012; Fillion 2012; Aguirre et al. 2016).

Note

1. See http://astrobiology.illinois.edu/ for the Institute for Universal Biology.

References

Aguirre, Anthony, Brendan Foster and Zeeya Merali, eds. 2016. *Trick or Truth? The Mysterious Connection Between Physics and Mathematics*. Switzerland: Springer International.

Aveni, Anthony. 1989. *Empires of Time: Calendars, Clocks, and Cultures*. New York: Basic Books.

Barrow, John D. and Frank J. Tipler. 1986. *The Anthropic Cosmological Principle*, Oxford University Press, Oxford

Burtt, Edwin A. 1924. *Metaphysical Foundations of Modern Physical Science*

Calvin, Melvin. 1981. Interview with Calvin in David Swift, SETI Pioneers. Tucson, 1990, 120–122.

Carr, Bernard. 2007. *Universe or Multiverse?* Cambridge: Cambridge University Press

Chamcham, Khalil, Joseph Silk, John D. Barrow, and Simon Saunders eds. 2017. *The Philosophy of Cosmology*. Cambridge University Press: Cambridge

Ćirković, Milan. 2012. *The Astrobiological Landscape: Philosophical Foundations of the Study of Cosmic Life,* Cambridge: Cambridge University Press.

Cleland, Carol E. and C. Chyba. 2007. "Does 'Life' Have a Definition? in Sullivan and Baross (2007) pp. 119–131.

Crick, F. 1995. *The Astonishing Hypothesis: The Scientific Search of the Soul*, Scribner, New York.

Curd, Martin, J. A. Cover and Christopher Pincock. 2012. *Philosophy of Science: The Central Issues*. W. W. Norton and Company: New York

Daston, Lorraine. 1991. "History of Science in an Elegiac Mode: E. A. Burtt's *Metaphysical Foundations of Modern Physical Science* Revisited," *Isis*, 82, 522–531.

Davies P, *About Time: Einstein's Unfinished Revolution* (New York: Simon and Schuster, 1995), pp. 272ff.

Dawkins, Richard. 1989. *The Selfish Gene*. Oxford: Oxford University Press, 1st ed., 1976.

Dawkins, Richard. 1996. *Climbing Mount Improbable*. New York: Norton.

De Duve, Christian. 1995. *Vital Dust: Life as a Cosmic Imperative.* Basic Books: New York.

Deamer, David W. 2010. "Special Collection of Essays: What is Life?" *Astrobiology*, 10, 1001–1002 and following.

Deamer, David W. and Fleischaker, G. 1994. *Origins of Life: The Central Concepts*. Jones and Bartlett, Boston.
DeVito, Carl, 2013. *Science, SETI and Mathematics*. Berghahn Books.
Dick, Steven J. 1996. *The Biological Universe: The Twentieth Century Extraterrestrial Life Debate and the Limits of Science*. Cambridge: Cambridge University Press
Dick, Steven J. 2003. *Sky and Ocean Joined: The U. S. Naval Observatory, 1830-2000*. Cambridge University Press, Cambridge
Dick, Steven J. 2013. *Discovery and Classification in Astronomy*. Cambridge University Press: Cambridge.
Dick, Steven J. 2015. *The Impact of Discovering Life Beyond Earth*. Cambridge University Press: Cambridge.
Dick, Steven J. 2018. *Astrobiology, Discovery, and Societal Impact*: Cambridge: Cambridge University Press.
Dick, Steven J. and Lupisella, M. L., editors. 2009. *Cosmos & Culture: Cultural Evolution in a Cosmic Context*. Washington, DC: NASA, online at http://history.nasa.gov/SP-4802.pdf
Dick, Steven J. and Strick, J. E. 2004. *The Living Universe: NASA and the Development of Astrobiology*. New Brunswick: NJ Rutgers University Press
Dunér, David, J. Parthemore, E. Persson, and G. Holmberg, eds. 2013. *The History and Philosophy of Astrobiology: Perspectives on Extraterrestrial Life and the Human Mind*. Newcastle-upon-Tyne: Cambridge Scholars Publishing.
Dunér, David, E. Persson, and G. Holmberg, eds. 2012. *The History and Philosophy of Astrobiology*, special issue of *Astrobiology*, 12, pp. 901–1016.
Dyson, Freeman. 1988. *Infinite in all Directions*. New York: Harper and Row.
Eagleman, David. 2009. "Brain Time," at https://www.edge.org/conversation/brain-time
Ellis, G. F. R. 1998a. "Cosmology and Verifiability," in Leslie (1998), 119–126
Ellis, G. F. R. 1998b. "Emerging Questions and Uncertainties," in Leslie (1998), 273–288.
Ellis, G. F. R. 2006. Issues in the philosophy of cosmology. In *Philosophy of Physics*, edited by J. Butterfield and J. Earman, pp. 1183–1285. North Holland, online at http://arxiv.org/pdf/astro-ph/0602280v2
Ellis, G. F. R. 2017. "The Domain of Cosmology and the Testing of Cosmological Theories," in Chamcham et al., 2017, 3–39.
Ellis, G. F. R., U. Kirchner and W. R. Stoeger. 2004. "Multiverses and physical cosmology," MNRAS, 346, 921–936.
Fillion, Nicolas. 2012. "The Reasonable Effectiveness of Mathematics in the Natural Sciences," University of Ontario PhD Thesis. https://ir.lib.uwo.ca/cgi/viewcontent.cgi?article=2265&context=etd
Fry, Iris. 2012. "Is Science Metaphysically Neutral?" *Studies in the History and Philosophy of Biological and Biomedical Sciences*, 43, 665–673.
Gale, George. 1993. "Philosophical Aspects of the Origin of Modern Cosmology," in *Encyclopedia of Cosmology: Historical, Philosophical and Scientific Foundations of Modern Cosmology* Garland: New York and London, pp. 481–495.
Grattan-Guinness, I. 2008. "Solving Wigner's mystery: The reasonable (though perhaps limited) effectiveness of mathematics in the natural sciences". *The Mathematical Intelligencer*, 30: 7–17, online at http://www.sfu.ca/~rpyke/cafe/reasonable.pdf.
Hanson, Norwood Russell. 1958. *Patterns of Discovery: An Inquiry into the Conceptual Foundations of Science*. Cambridge: Cambridge University Press.
Harwit, Martin. 2013. *In Search of the True Universe: The Tools, Shaping, and Cost of Cosmological Thought*. Cambridge: Cambridge University Press.
Herschel, John. 1841. Presidential Address on the occasion of awarding the Gold Medal of the Society to Bessel, *Monthly Notices of the Royal Astronomical Society of London*, vol. 5 no. 12, February 12, 1841, 90.
Hetherington, Norriss. 1988. *Science and Objectivity*. Iowa State University Press.
Hetherington, Norriss. 1993. *Cosmology: Historical, Literary, Religious, and Scientific Perspectives*. New York and London, Garland.

Holt, James. 2012. *Why Does the World Exist? An Existential Detective Story*. Liverright Publishing, New York.

Joyce, G. G. 1994. Foreword to Deamer and Fleischaker (1994).

Kragh, Helge. 1996. *Cosmology and Controversy: The Historical Development of Two Theories of the Universe*. Princeton, Princeton University Press.

Lakoff, George. 2003. "How to Live With an Embodied Mind: When Causation, Mathematics, Morality, the Soul, and God are Essentially Metaphorical Ideas," in Anthony Sanford, ed., *The Nature and Limits of Human Understanding*. London and New York, pp. 75–108: 90–96.

Lange, Marc. 2012. *Introduction to the Philosophy of Physics: Locality, Fields, Energy and Mass*. Wiley-Blackwell: New York

Leslie, J. ed. (1998. *Modern Cosmology & Philosophy*. Prometheus Books, New York.

Lightman, Alan and Roberta Brawer. 1990. *The Lives and Worlds of Modern Cosmologists*. Cambridge, MA: Harvard University Press

Mayr, Ernst. 1988. *Toward a New Philosophy of Biology*. Cambridge, MA: Harvard University Press

McMullin, Ernan. 1981. "Is Philosophy Relevant to Cosmology?" *American Philosophical Quarterly*, 18, no. 3, 177–189.

Miller, Stanley and Harold Urey. "Organic Compound Synthesis on the Primitive Earth," *Science*, 130 (July 31, 1959, 245–251: 251.

Monod, Jacques. 1970. *Chance and Necessity*. Knopf, New York, pp. 144–146.

Moore, James R. 1989. *History, Humanity and Evolution: Essays for John C. Greene*. Cambridge: Cambridge University Press.

NASA. 2015. *Astrobiology Strategy* online at https://nai.nasa.gov/media/medialibrary/2016/04/NASA_Astrobiology_Strategy_2015_FINAL_041216.pdf

Nicholson, Jason Scott. 2012. "A Perspective on Wigner's 'Unreasonable Effectiveness of Mathematics," *Notices of the AMS*, 59, 1, 38–42. https://www.ams.org/notices/201201/rtx120100038p.pdf

North, John. 1965. *The Measure of the Universe: A History of Modern Cosmology*. Oxford: Oxford University Press

Notre Dame History of Astronomy Workshop. 2009. https://www3.nd.edu/~histast/workshops/2009ndix/program.shtml

Notre Dame History of Astronomy Workshop. 2015. https://www3.nd.edu/~histast/workshops/2015ndxii/program.shtml

Overbye, Dennis. 1991. *Lonely Hearts of the Cosmos: The Story of the Scientific Quest for the Secret of the Universe*. New York, Harper Collins.

Overbye, Dennis. 2014. "Astronomers Hedge on Big Bang Detection Claim," *New York Times*, 19 June, 2014. https://www.nytimes.com/2014/06/20/science/space/scientists-debate-gravity-wave-detection-claim.html

Palmeri, JoAnn. 2009. "Bringing Cosmos to Culture: Harlow Shapley and the Uses of Cosmic Evolution," in Dick and Lupisella, *Cosmos and Culture*, pp. 489–521.

Pankenier, David W. 2013. *Astrology and Cosmology in Early China: Conforming Earth to Heaven*. Cambridge, Cambridge University Press.

Pirie, Norman W. 1937. "The Meaninglessness of the terms 'life' and 'living,' in *Perspectives in Biochemistry*, edited by J. Needham and D. R. Green, Cambridge: Cambridge University Press, pp. 11–22.

Pirie, Norman W. 1975. *A discussion on the recognition of alien life*. London: Royal Society.

Ponnamperuma, Cyril. 1972. *Exobiology*. Amsterdam.

Reas, Emilie. 2014. "Small Animals Live in a Slow-Motion World," *Scientific American Mind*, July 1, 2014, https://www.scientificamerican.com/article/small-animals-live-in-a-slow-motion-world/. See

Scerri, Eric. 2008. *Collected Papers on Philosophy of Chemistry*. Imperial College Press, London

Scerri, Eric, and Grant Fisher. 2016. *Essays in the Philosophy of Chemistry*. Oxford University Press, New York

Scerri, Eric, and Lee McIntyre. 2015. *Philosophy of Chemistry: Growth of a New Discipline*. Springer, Dordrecht, Berlin

Schrodinger, E. 1944. *What is Life? The Physical Aspect of the Living Cell*, Cambridge University Press, Cambridge.
Shapley, Harlow. 1958. *Of Stars and Men,* Beacon Press, Boston,104–114.
Sheehan, William. 1988. *Planets and Perception: Telescopic Views and Interpretations.* Tucson, University of Arizona Press.
Smith, Kelly. 2016. "Life is Hard: countering definitional pessimism concerning the definition of life," *International Journal of Astrobiology* 15 (4): 277–289
Stanley, Matthew. 2007. *Practical Mystic: Religion, Science and A. S. Eddington.* University of Chicago Press: Chicago
Sullivan, Woodruff T. III and John A. Baross, eds. 2007. *Planets and Life: The Emerging Science of Astrobiology.* Cambridge: Cambridge University Press.
Tegmark, Max. 2014. *Our Mathematical Universe: My Quest for the Ultimate Nature of Reality.* New York: Knopf
Vidal, Clement. 2014. *The Beginning and the End: The Meaning of Life in a Cosmological Perspective.* New York: Springer.
Wigner, Eugene. 1960. "The Unreasonable Effectiveness of Mathematics in the Natural Sciences," in *Communications in Pure and Applied Mathematics*, vol. 13, No. 1. New York: John Wiley & Sons
Zakariya, Nasser. 2017. *A Final Story: Science, Myth and Beginnings.* Chicago: University of Chicago Press.

Chapter 38
Critical Issues in the History, Philosophy, and Sociology of Astrobiology

Abstract 50 years after serious scientific research began in the field of exobiology, and 40 years after serious historical research began on the subject of extraterrestrial life, this paper identifies and examines some of the most important issues in the history, philosophy, and sociology of what is today known as astrobiology. As in the philosophy of science in general, and in the philosophies of particular sciences, critical issues in the philosophy and sociology of astrobiology are both stimulated and illuminated by history. Among those issues are (1) epistemological issues such as the status of astrobiology as a science, the problematic nature of evidence and inference, and the limits of science; (2) metaphysical/scientific issues including the question of defining the fundamental concepts of life, mind, intelligence, and culture in a universal context; the role of contingency and necessity in the origin of these fundamental phenomena; and whether or not the universe is in some sense fine-tuned for life and perhaps biocentric; (3) societal issues such as the theological, ethical, and worldview impacts of the discovery of microbial or intelligent life; and the question of whether the search for extraterrestrial life should be pursued at all, and with what precautions; and (4) issues related to the sociology of scientific knowledge, including the diverse attitudes and assumptions of different scientific communities and different cultures to the problem of life beyond Earth, the public "will to believe," and the formation of the discipline of astrobiology. All these overlapping issues are framed by the concept of cosmic evolution—the 13.7-billion-year Master Narrative of the Universe—which may result in a physical, biological, or postbiological universe and determine the long-term destiny of humanity.

38.1 Introduction

My purpose in this paper is to give a broad overview of what the history, philosophy, and sociology of astrobiology can mean as an intellectual field of study. 40 years ago extraterrestrial life was considered taboo as an historical or philosophical subject, virtually nothing substantial having been written on either aspect. That changed on the historical side in the 1980s with the publication of three serious histories of the extraterrestrial life debate, together covering the period from the ancient Greeks

First published in *Astrobiology, 12,* (2012), 906–927

S. J. Dick, *Space, Time, and Aliens*, https://doi.org/10.1007/978-3-030-41614-0_38

to Percival Lowell (Dick 1982; Guthke 1983; Crowe 1986), followed in the next decade by histories extending the story through the twentieth century (Dick 1996, 1998), and then by a more detailed history of the development of exobiology and astrobiology within NASA (Dick and Strick 2004), as well as an historically nuanced reader on the subject (Crowe 2008). Although a few studies come across as polemic rather than history (Basalla 2006), the first duty of the historian must be to place the subject in proper historical context. Although many peculiar ideas exist in the history of the debate over other habitable worlds, it is not the business of the historian to ridicule them from the viewpoint of the present. That is not to say that the reasoning was always good by contemporary or present standards; to the contrary the arguments used in the debate, placed in proper context, are fair game for the critical study of both the history and philosophy of astrobiology (Sheridan 2009).

The cumulative effect of these historical studies, which in many ways constitute only a preliminary reconnaissance, has been to show that the extraterrestrial life debate has a robust history very much connected to science, or "natural philosophy" as it was known prior to the mid-nineteenth century when William Whewell coined the term "scientist." The belief in or rejection of other inhabited worlds was part and parcel of the cosmologies of the ancient Greek atomists, Aristotle, Copernicus, Descartes, and Newton, and a substantial tradition developed based on the ideas of these early thinkers (Dick 1982). In what historians of science today label the early modern era, Copernicus made the Earth a planet, and the planets potential Earths; to what degree they are Earth-like is a theme still playing out today within and beyond our Solar System. Descartes with his vortex cosmology believed every star was a Sun surrounded by planets, a Brunonian idea popularized by the French litterateur Bernard le Bovier de Fontenelle in one of the bestselling works of the seventeenth century (Fontenelle 1686). Cartesian physics thus graphically carried the subject beyond the Earth, but was soon replaced by Newtonian physics, which made no immediate pronouncements on the subject from a scientific point of view, but provided the foundation for future progress on the development of planetary systems—again a theme that reached fruition only in the last few decades. Newton himself believed the universe was populated, providing an argument from natural theology that kept God in a universe that no longer needed divine assistance in perpetuating planetary motion. A similar pattern of connections to scientific ideas is found in the nineteenth century, with clear and substantive connections to science, philosophy, and intellectual thought in general, too numerous to mention here but discussed at length in Crowe (1986). The same is true of the twentieth century (Dick 1996, 1998), which finally injected a robust empiricism into the debate, including in situ observations of the planets with the beginning of the Space Age. In a controversy stretching back to the historian Arthur O. Lovejoy (1936), a major historiographic point of contention has been the relative role of philosophical versus more properly scientific components in the debate. As we shall see, however, science and philosophy are inseparable, especially in the form of epistemology and metaphysics.

This progress in historical studies of the debate has not been paralleled by deep philosophical studies. In 1971, about the same time as serious historical studies were beginning, the president of the Eastern Division of the American Philosophical

Association drew attention to the need for philosophers to consider the subject (Beck 1985). Admitting that "our Association is not hospitable to cosmological speculations," Lewis White Beck remarked that had he not been president, his subject would likely have been rejected by the Program Committee. "There are new sciences like exobiology whose foundations are in need of philosophical scrutiny," he wrote, but "when the National Academy of Sciences explicitly calls attention to the philosophical dimensions and ramifications of the problem, it seems to me we philosophers should relax our ban on cosmological speculation and think about possible worlds that may actually exist" (National Academy of Sciences 1962; Beck 1985). The intervening 40 years, however, has drawn only sporadic interest from a few philosophers (Regis 1985; Rescher 1985; Ruse 1985; Leslie 1998; Cleland 2001, 2002; Cleland and Chyba 2007; Bedau 2010). This neglect is surprising considering the well-developed field of philosophy of science (Curd and Cover 1998; Rosenberg 2000), and doubly so because the philosophy of biology is a well-established field (e.g. Hull 1974; Mayr 1988; Ruse 1989; Hull and Ruse 2007; Ayala and Arp 2009), boasting at least three journals and a well-organized international society. It would not seem too large a leap from biology to astrobiology, but even philosophers of biology have yet to make that leap, thus leaving wide open an entire field of study, the efforts of a few scientists notwithstanding (Jakosky 2000; Grinspoon 2003; Ćirković 2012).

In what follows we use the term *astrobiology* in the sense first developed in the late 1990s in the Astrobiology Roadmap and subsequent versions (Des Marais et al. 2008), to include the past, present, and future of life, embracing elements of planetary science, planetary systems science, origins of life, and SETI, the Search for Extraterrestrial Intelligence (see Figs. 2.3 and 16.2). (Although SETI is not a programmatic part of astrobiology, the NASA program having been terminated by Congress in 1993, it is certainly an intellectual part, and so is discussed in this paper.) Earlier uses of the term "astrobiology" were sporadic (Lafleur 1941; Struve 1955; Wilson 1958; Strughold 1959), and when the study of life beyond Earth began as a serious endeavor with the dawn of the space age, the term *exobiology* (Lederberg 1960), still in use today both as a NASA program and a field of study, denoted a much narrower field substantially limited largely to origins of life studies, especially in the laboratory. This was mostly a matter of operational capabilities at NASA at the time, and the development of the broader field of astrobiology more than three decades later was also a matter of organizational and intellectual currents within NASA (Dick and Strick 2004). Thanks to the NASA Astrobiology Institute (Blumberg 2003) and independent efforts in other countries (National Research Council 2003), astrobiology is now a flourishing endeavor, even if one may still argue whether it is, or should be, a true discipline (Dick 1996; Strick 2004). The term *bioastronomy*, while little used at funding agencies like NASA, is the name given to one of the Commissions of the International Astronomical Union, and is favored by many astronomers, who see their subject as a branch of astronomy rather than biology. This is an interesting example of different perspectives by different scientific communities, not surprising since the Commission is dominated by astronomers interested in planetary systems rather than biologists interested in origins of life. Even so, meetings organized by astrobiologists and bioastronomers alike draw substantial numbers of both biologists and astronomers as well as the other natural sciences and a few from the social sciences, all of whom patiently struggle to understand each other in their common interest of shedding light on one of the most profound issues in the history of science.

As in similar efforts in other areas (Claggett 1957; Dick and Launius 2006), critical issues as defined here represent a limited number of problematic issues in the field of astrobiology, and differ in character from the much larger domain of historical themes that have been, and could be, elaborated in historical studies of the debate. Here we combine history, philosophy, and sociology to engage in a deeper study than any of them alone could achieve. As in the philosophy of science in general, and in the philosophies of particular sciences, critical issues in the philosophy of astrobiology are best illuminated by history. Here we lay out a historically informed philosophy of astrobiology, drawing on the philosophy of biology and other sciences and informed by real-world issues, rather than the esoteric approach philosophers are sometimes accused of employing. We proceed under four main headings: epistemology, metaphysics, astrobiology and society, and sociology of scientific knowledge.

38.2 Epistemological Issues

Epistemology is the theory of knowledge, including its nature, extent and justification; its practitioners undertake to determine "the standards to which genuine knowledge should conform" (Harré 1971). Empiricism, the idea that experience and observation justify knowledge, emerged substantially with the Scientific Revolution of the seventeenth century; was analyzed by philosophers such as John Locke, George Berkeley and David Hume in the eighteenth century; and served as the undisputed epistemology of science through much of the twentieth century, supplemented by advances in logic, probability theory and statistical inference (Rosenberg 2000). Thomas Kuhn's work on scientific revolutions (Kuhn 1962, 1977) introduced into epistemology aspects of psychology, sociology and history, and led some to question the objectivity of science, with overzealous Kuhnians carrying his ideas to postmodern relativist extremes that Kuhn himself decried. But Kuhn did raise fundamental questions about the status of scientific knowledge, questions strikingly relevant to astrobiology in many ways. If it is the job of the epistemologist "to show how knowledge can be distinguished from true belief, and certainty from probability" (Harré 1971, p. 5), the extraterrestrial life debate is fertile ground for epistemologists, especially where extraterrestrial intelligence is concerned. I will discuss two issues under epistemology: the status of astrobiology as a science, and the problematic nature of evidence and inference, reserving for the next section the question of objective knowledge in an extraterrestrial context.

Just as the status of the extraterrestrial life debate as a valid subject for historical research was questioned 40 years ago, its status as a science—and before that as a part of natural philosophy—was also questioned, most notably by scientists themselves. This skepticism extends to the period 50 years ago when exobiology was born. In 1964 the evolutionary biologist George Gaylord Simpson, pointing to the long history of the debate, wrote that "There is even increasing recognition of a new science of extraterrestrial life, sometimes called exobiology—a curious development in view of the fact that this 'science' has yet to demonstrate that its subject matter exists!" Simpson noted that this supposed new science was very expensive, and called exobiology "a gamble at the most adverse odds in history," resembling "more a wild spree more than a sober scientific program" (Simpson 1964). Simpson concluded with a plea "that we invest just a bit more of our money and manpower, say one-tenth of that now being gambled on the expanding space program," on studying the systematic and evolution of earthly organisms—that is to say, his own field! An interesting case of the rhetoric of science (Gross 2006), clearly Simpson had an ulterior motive in declaring that exobiology was not a science. But with Isaac Asimov's article in the *New York Times Magazine* the following year entitled "A Science in Search of a Subject" (Asimov 1965), the phrase was too good to ignore as a kind of mindless meme deployed innumerable times the following decades, despite the article's positive assessment of exobiology (Strick 2004).

Even a minimal consideration of this idea suffices to show it is a misrepresentation of science, even if admittedly a catchy phrase. One could say the search for gravitational waves, or the Higgs boson, or planetary systems, are, or were, "sciences without a subject." But this hardly seems a productive way of approaching the problem. Every science is looking for a subject until it finds it (planetary systems), thinks it may have found it (the Higgs boson), or does not find it (gravitational waves, at least so far). From an epistemological point of view, the methods of astrobiology are as empirical as in any historical science such as astronomy or geology (Jakosky 2000; Cleland 2001, 2002), though it is true that astrobiological observations and experiments are often especially difficult, and the inferences more tenuous. Therefore, it is in my view long past time to retire the "science without a subject" meme, and long overdue to study the issue in a more nuanced way. We frame this as

Critical Issue # 1: What is the status of astrobiology as a science? How have its problems and methods compared to other sciences in historical context?

Although Simpson criticized the pioneer in the field, Joshua Lederberg, by claiming that exobiology was not strictly biology because its techniques differed (Wolfe 2002), certainly astrobiologists today would be surprised to learn they are not doing science; from their point of view their endeavors constitute not only science, but cutting-edge science. Related to this issue, but not coextensive with it, is the formation of astrobiology as a discipline (Wells et al. 2007), for it is difficult to admit a scientific discipline as long as it is not considered a science. While more than one practitioner early on heralded astrobiology or its equivalent as a new scientific discipline (Shklovskii 1965; Billingham 1981), these claims may have been premature (Dick 1996, pp. 475–478). Moreover, being labeled a discipline may be good or bad in terms of "Balkanization" and isolation from broader parent fields, such as was contemplated, but did not happen, in the case of radio astronomy in relation to astronomy as a whole (Sullivan 2009, pp. 435–438). A historical comparison of discipline formation in other fields such as biochemistry (Kohler 1982), molecular biology (Abir-Am 1992), and geophysics (Good 2000) would help illuminate the problem for astrobiology.

This brings us to the second, related, epistemological issue: the problematic nature of evidence and inference. This is true whether dealing with the nearest planets of the Solar System, the search for more distant planetary systems, or theories and experiments about the origin and evolution of life. And it is true whether using ground-based telescopes or spacecraft that actually land and undertake in situ observations. It applies equally to the visual, photographic, electronic, photometric, and spectroscopic techniques of astronomy, as well as to the variety of techniques employed in studies of the origin and evolution of life. There is no lack of exemplars. In planetary science the most notorious historical case is the canals of Mars, which cannot simply be chalked up to the imagination of Percival Lowell himself, since many other astronomers claimed to have seen them (Crowe 1986; Dick 1996; Strauss 2001). In the modern era the problematic nature of evidence and inference was on stark display when the Viking spacecraft landed on Mars in 1976 and yielded different interpretations of the experimental results, with one of the Principal Investigators insisting life had been found (Levin and Straat 1976; DiGregorio et al. 1997). Though a consensus seemed to be reached for several decades that life (indeed even organics) had not been found, the issue was reopened especially after the Phoenix lander discovered perchlorates on Mars in 2008, and it remains open among prominent researchers today (Navarro-González et al. 2010), with Levin more than ever convinced he discovered life on Mars (Levin 2011). Problems are not confined to remote sensing, but are present even when the evidence is in hand, and all the analytical techniques available to terrestrial laboratories can be applied to the problem, as in the case of the claims for microfossils in the Mars rock ALH84001 (McKay et al. 1996), known with certainty to be from Mars but now believed unlikely to harbor such fossils (Dick and Strick 2004; Sawyer 2006). That conclusion also remains very much open for some researchers, including the original investigators. The more recent debate over the possible discovery of arsenic life on Earth, in the sense of arsenate replacing phosphate in the nucleic acids and proteins of the GFAJ-1 Halomonadaceae bacterium, continues the pattern (Wolfe-Simon et al. 2010, 2011).

Problems of evidence and inference are redoubled in the elusive search for planetary systems beyond our own; there is a long history in the twentieth century of observational claims for such extrasolar planets, most notoriously in the case of Peter Van de Kamp and Barnard's star, announced in 1963 (Van de Kamp 1963; Dick 1996) (Fig. 38.1). It would be more than three decades before more definitive claims were made for planets around Sun-like stars, not with van de Kamp's astrometric technique measuring mere microns of stellar perturbations on photographic plates, but with the spectacular application of radial velocity techniques measuring a few meters per second variation in stellar line-of-sight motion as a planet tugged on its parent star (Mayor and Queloz 1995; Marcy and Butler 1996). This method, too, has its pitfalls, as evidenced by its own occasionally spurious claims as

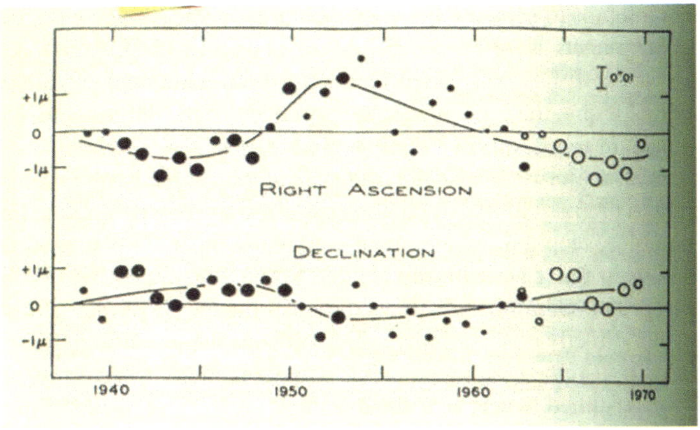

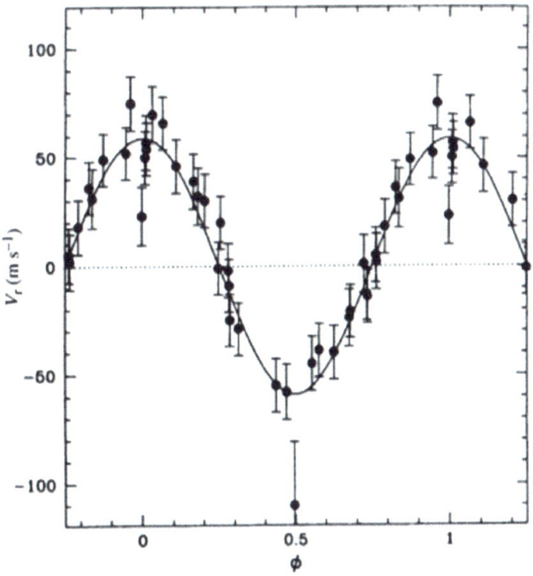

Fig. 38.1 Problematic nature of evidence and inference. (Above) Peter van de Kamp's astrometric evidence for a planet around Barnard's star, 1963. The data plotted represent measurements of the minute gravitational perturbations of the star by the supposed planet, amounting to a few hundredths of an arcsecond over a period of decades. The scale is ±1 μm as measured on photographic plates taken with a long-focus refractor. This data was proven spurious, but only after several decades. (From Van de Kamp (1963), reproduced by permission of the American Astronomical Society). (Below) Radial velocity variations of the star 51 Peg, interpreted to be caused by the planet 51 Peg b, amounting to ±59 m/s. The period is 4.2 days, indicating the planet orbits very close to its parent star. This data was real, and marked the beginning of the era of extrasolar planet discoveries around Sun-like stars. (From Mayor and Queloz (1995). A previous announcement by Latham et al. (1989) remains controversial)

astronomers attempt to separate the signal from the noise. In any case, more than 400 extrasolar planets were known before the launch of the Kepler spacecraft in 2009; that spacecraft has now added more than 1200 additional candidates, using yet another technique, the photometric measurement of the dimming of starlight as a planet transits in front of it.

Thus the vast majority of extrasolar planets are not directly observed, but only inferred from their gravitational or photometric effect on their parent star. Fomalhaut b is one notable exception among about a dozen companion planets detected by direct imaging, in this case with the Hubble Space Telescope, and in other cases with the VLT, Keck-Gemini, and a few other instruments (Perryman 2011, p. 168). But even when directly observed there often remains the question of the true planetary mass, as is also the case in the radial velocity method, where only "msin i" is empirically determined, where *m* is the mass and *i* is the inclination of the planet's orbit to our line of sight. When the planetary orbit is edge on the full radial velocity signal is measured, but when it is face on, no radial signal is measured. What is measured is therefore the minimum mass, and because of this issue the first extrasolar planet discovered around a Sun-like star may not have been 51 Peg b in 1995, but HD 114762 b, discovered by Latham and his colleagues in 1989. In that case the minimum mass is 11 times the mass of Jupiter, and a higher masses would make it a brown dwarf sub-stellar object or a low mass red dwarf star. The authors were equivocal in their conclusions, reflecting the ambiguities of the data, which they stated "leads to the suggestion that the companion is probably a brown dwarf, and may even be a giant planet. However, because the inclination of the orbit to the line of sight is unknown, the mass of the companion may be considerably larger than this lower limit" (Latham et al. 1989). The question of how to separate brown dwarfs from planets at the high mass end of planets is no less interesting than the problem of how to separate planets from dwarf planets at the low mass end. The latter gave rise to the infamous Pluto issue (Tyson 2009; Brown 2010), a revealing classification problem in the history and philosophy of astronomy, analogous to classification issues prominent in the history of biology (Mayr 1988). All these issues of observation and inference may also be discussed in relation to the philosophical problem of epistemological realism and anti-realism (Kosso 2006).

The epistemological questions extend also to the nature of the arguments for and against extraterrestrial life. In the case of the Mars rock, do four independent but (critics say) weak arguments—from the morphology of the nanostructures, carbonate globules, the presence of magnetite and polycyclic aromatic hydrocarbons—add up to a strong argument for biogenesis? The authors of the discovery paper thought so, ending their paper with the argument "Although there are alternative explanations for each of these phenomena taken individually, when they are considered collectively, particularly in view of their spatial association, we conclude that they are evidence for primitive life on early Mars" (McKay et al. 1996). Critics, including paleobiologist J. William Schopf, thought not, arguing that "spatial association" held no persuasive value at all, and citing Carl Sagan's dictum "extraordinary claims require extraordinary evidence" (Dick and Strick 2004). In another arena we may question the role of statistical arguments, for example, the epistemological status of

Drake Equation, SETI's central icon, in which probabilities are piled on probabilities. Drake himself has repeatedly stated the equation was, and is, merely a heuristic device, created as an agenda at the pioneering Green Bank meeting on interstellar communication in 1961. But others invested it with more meaning, updating the parameters as more data became available, for example, on extrasolar planet frequency, and making it the central icon of cosmic evolution (Dick 1996). On the reverse side of the coin, just how strong is SETI's supposed nemesis, the Fermi paradox argument (Hart 1975; Dick 1996; Webb 2002), which states that if extraterrestrials exist, given the vast scale of time they should be here, but they are not, therefore they do not exist, therefore why waste money looking? These are not purely academic questions; the U. S. Congress was impressed enough with the Fermi Paradox argument to terminate NASA's SETI program in 1993. As a final example, in astrobiology's classical arena, just how relevant are experiments on the origins of life to the problem of extraterrestrial life? The famous Miller-Urey experiment was conducted under supposed primitive Earth conditions; but those conditions may not in fact have prevailed on the early Earth, calling into question their relevance not only for other planets but even for Earth. While such experiments may well continue to be relevant (Wills and Bada 2000), the broader philosophical question is to what extent experiments on Earth can be transferred to conclusions about other planets.

The profusion of questions is no cause for despair. To the contrary such problems of evidence, inference and objectivity are not limited to astrobiology, but are well known in the history of science (Daston and Galison 2010). Nevertheless, it is true that astrobiology has remained at the very limits of science in the sense that the required techniques have been barely sufficient to the task, leading to problematic evidence and stretching the normal principles of scientific inference, propelled by public interest in one of the greatest remaining questions in science. All these ruminations may be summarized as

> *Critical Issue # 2: How have observation, evidence, and inference been deployed in arguments in astrobiology, and how does this deployment compare to their use in other sciences?*

This remains a rich area for further research, and can be further pursued through many avenues, including historical analogs, the philosopher's penchant for logical analysis, and the sociologist's attention to human factors.

38.3 Metaphysical/Scientific Issues

While epistemology deals with the theory of knowledge, metaphysics traditionally sought to describe basic entities, including the nature of matter, the existence of god and his attributes, and the existence of the soul. Philosophers were convinced that human reason is capable of getting answers to ontological questions, what exists independently of us. Empiricists, notably Hume, called into question the possibility of knowing what is not based in principle on our experience. Logical positivism at the end of the nineteenth century and the beginning of the twentieth century went further in repudiating metaphysics. But here, metaphysics and science are logically placed under the same heading because almost a century ago, the American philosopher Edwin Arthur Burtt showed how science and metaphysics were inseparable in the transition from the medieval world to the Scientific Revolution of the seventeenth century, when the prevailing conceptions of reality and causality, and the relation of the human mind to nature, were forever transformed (Burtt 1924; Daston 1991). Indeed, Burtt insisted such metaphysical changes in these conceptions were the core of the scientific revolution. Today, this close connection between metaphysics and science is widely accepted by philosophers, if not always made explicit in scientific work. Philosophers investigate the most basic concepts in science, including space, time, and cause in physics (Lange 2006); number, proof, and logic in mathematics (Brown 1999); and atom, element, and bond in chemistry (Baird et al. 2006). Just as the cosmologist G. F. R. Ellis has recently insisted that philosophical choices necessarily underlie cosmological theory, and that unexamined philosophical standpoints are still philosophical standpoints (Ellis 2006), so it is in astrobiology.

If we apply such questions to the life sciences, it means examining the changes in our core conceptions of life, mind, and intelligence, among other categories such as "species," and methods such as those used in systematics. For astrobiology it means putting these venerable questions in a cosmic context. The question "What is Life?" is an old one, contemplated by biologists, biochemists and even physicists (Schrodinger 1944) over several centuries (Fry 2000; Tirard et al. 2010). In the most general sense, the overarching theme of vitalism versus mechanism echoes through biological history, and was still very much alive at the beginning of the twentieth century, influencing not only the origins of life issue, but also many other aspects of biology. For mechanists such as T. H. Huxley and E. B. Wilson, who believed life could be reduced to physics and chemistry, the problem was to determine the physical basis of life (Dick 1996, pp. 332–337). The concept of protoplasm as the unit of life held sway until almost the turn of the twentieth century, when the rise of colloidal biochemistry and the discovery of viruses turned the focus to smaller entities. Attention turned successively to the protein enzyme, the virus, the gene and the cell as the unit of life. By 1939 the British biologist Norman W. Pirie, in a widely cited article, reviewed the definitions of life and argued that the terms "life" and "living" were meaningless, and that the transition from non-living to living was like the transition from green to yellow in the spectrum, or from acid to alkaline in chemistry (Pirie 1939).

Today the debate about the origin of life (Shapiro 1987; Deamer and Fleischaker 1994; Davies 1998; Fry 2000; Bedau and Cleland 2010) is at the level of the molecular assembly of life. Experiments on Darwinian evolution in the laboratory at the level of nucleic acids were first carried out in the 1960s by Sol Spiegelman and his colleagues at the University of Illinois, using Qβ virus RNA carried through repeated cycles of replication, mutation and selection (Spiegelman 1967). Such abiotic "evolution in a test tube" has proven to be an illuminating technique, and by 1994 a committee assembled by NASA defined life as a "self-sustaining chemical system capable of Darwinian evolution," where Darwinian evolution at the molecular level

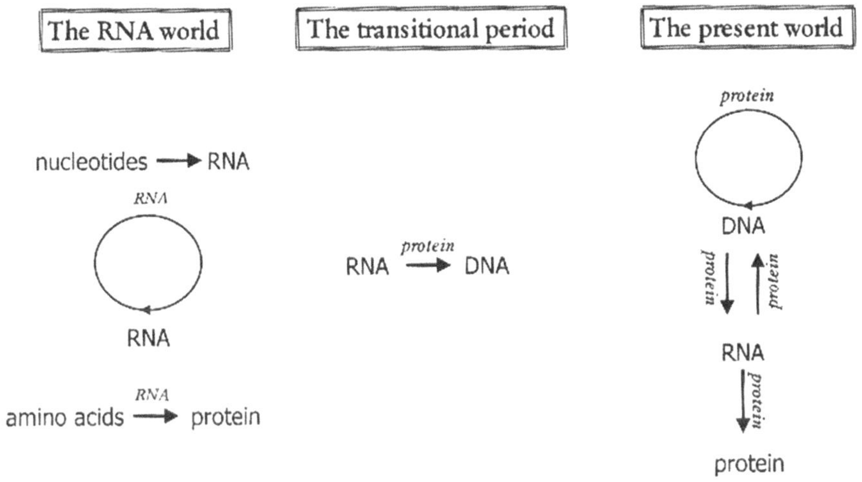

Fig. 38.2 The RNA World and transition to the current DNA-RNA-protein world. Left: In this scenario RNA synthesized on the primordial Earth functioned as both a carrier of information and an enzyme that catalyzed its own replication and the synthesis of proteins. Center: Proteins previously synthesized with the help of RNA enzymes catalyzed the transition from RNA to DNA. Right: In life as we know it, protein enzymes catalyze the replication of DNA. They also catalyze the transcription of DNA into RNA and the transcription of messenger RNA into proteins. (Reprinted by permission from Fry 2000, p. 139)

refers to descent with modification by natural selection in a replicating genetic system such as DNA on Earth (Joyce 1994). Under this definition a major task has been to identify the origin of the first replicable molecules, with some favoring the "RNA world" (Fig. 38.2) in which RNA is able to both store information and catalyze reactions, acting as an enzyme (Gilbert 1986). The "self-sustaining" part of the NASA definition is one of many difficulties remaining, as well as detailed knowledge of the nature of primitive Earth conditions.

Other definitions of life are possible, none totally satisfactory, and among the few philosophers who have entered the fray, attention has turned again to the question of whether life can be defined at all. Some philosophers have concluded the answer is no, that all attempts to define life are fundamentally misguided in the absence of a general theory of living systems (Cleland and Chyba 2007). Scientists are hardly satisfied with this answer, since an operational definition of life is a very practical problem, as in the search for life on Mars; the Viking landers in 1976 embodied implicit ideas about the nature of life as a metabolic process, in order to build and undertake the biology experiments. Attempts to define life will therefore continue, whether in terms of Darwinian processes, metabolism, energy, and thermodynamics, complexity theory, cybernetics, or some new insight (Benner 2009, 2010; Bedau 2010; Deamer 2010; Tsokolov 2009, 2010). We thus arrive at a cluster of questions that constitute

Critical Issue # 3: "What is life when considered in the universal context of astrobiology? How does astrobiology change our core conceptions of life? Is there a general theory of living systems, a universal biology as there is a universal physics? What is the origin of replicating systems, under what conditions will life arise, and what is the role of chance and necessity in the origin of life in the universe? What are the metaphysical assumptions that underlie our concepts of life?

One helpful approach to highlighting how our conceptions of life have been shaped by history and metaphysics is to study not only how they have changed through various periods (Farley 1977; Kamminga 1980; Fry 2000; Wills and Bada 2000; Strick 2002, 2010), but also how they have been affected by political and scientific worldviews, such as Oparin's pioneering origins of life work in the Soviet Union, which germinated in the midst of the philosophy of dialectical materialism (Graham 1989). Our changing concepts of the evolution of life can be illuminated further by considering how new perspectives and discoveries such as extremophiles have resulted in a massive rethinking of the categories of classification, as exemplified in the debate over the Five Kingdoms of macrobiology versus the Three Domains of molecular biology (Sapp 2009; Margulis and Chapman 2010). The search for "shadow life" on Earth gives an added dimension to the problem (Davies 2011). But these aspects are only the beginning, for the conditions in the Solar System, much less the rest of the universe, give vast scope for natural selection to work its will on the universal scale, broadening even further the scope of what "life" may mean, both inside and outside the carbon-based paradigm (Ward 2005; Ward and Benner 2007). It is for this reason that the search for life on Mars has proven so stimulating to the previously earthbound problem of defining life, and why the search for life on Jovian and Saturnian moons and exoplanet environments will continue to enrich research in this area in the future.

The role of chance and necessity in the origin of life is at the core of the extraterrestrial life debate. Does life arise wherever conditions are favorable? Or is life a fluke, a chance occurrence or happy accident in which Earth was lucky? (Davies 2007). The two points of view are classically represented by the French biologist and Nobelist Jacques Monod on the one hand, and the Belgian-American biochemist and Nobelist Christian DeDuve on the other. In his classic work *Chance and Necessity* (Monod 1971) argued "the universe was not pregnant with life, nor the biosphere with man. Our number came up in the Monte Carlo game." Nor was Monod the only one to favor chance; the astronomer Fred Hoyle agreed that the chance of a random shuffling of amino acids producing a workable set of enzymes was miniscule, and went one step further in asserting that life must have been assembled by a "cosmic intelligence," though not necessarily the supernatural intelligence of Christianity (Hoyle 1983). DeDuve, on the other hand, argued just the opposite, declaring Monod wrong and viewing life as a "cosmic imperative," while Richard Dawkins argued that "climbing Mt. Improbable" was not impossible (De Duve 1995; Dawkins 1997).

This debate, of course, opens an entire range of theological and philosophical issues, with intense public input fueled by vested religious interests. As Fry (Fry 2000, 2010) has emphasized, an historical perspective is essential to any discussion of these metaphysical issues. She insists the question involves not only empirical knowledge but also philosophical considerations, but denies the origin of life was a happy accident, a miracle, or a supernatural production as favored by creationists. Rather, she argues:

> In agreement with most researchers in the origin-of-life field today, it is my contention that within the realistic confines of space and time of our universe, the emergence of life could not have been the result of chance, nor was it a miracle fostered by a supernatural power. Rather it involved the working of physical and chemical mechanisms responsible for the self-organization of matter into living systems. Such mechanisms, given the appropriate environmental conditions, could have produced similar results elsewhere in the universe.

In other words, she opts for a naturalistic, evolutionary world view "that denies the separation of living systems and inanimate matter into two unbridgeable categories," an evolutionary worldview that "provides an implicit guideline for origin-of-life research." Though many unanswered questions remain, she argues, this is no reason "to forsake the naturalistic worldview for the creationist one." Indeed even most theologians have given up on the idea of a "God of the gaps," a supernatural *Deus ex machina* that supplies the answer to anything that remains unexplained, thus explaining nothing. As Fry points out, the historical perspective is required to see the different epistemological status of empirical statements, scientific theory and philosophical commitment (Fry 2000, p. 212). It follows that history and philosophy should play a leading role in any attempt to define life.

All these questions in the origin of life arena are multiplied when it comes to the nature of consciousness, mind and intelligence, subjects on which Darwin and A. R. Wallace, the co-discoverers of natural selection, parted company (Shermer 2002a). Our understanding of the relationship among these three concepts, and their relation to the brain, has been problematic throughout history. The mind-body problem, for example, is one of the long-standing issues in philosophy, dating back at least to Cartesian dualism, which most of the general public in the Western world still accept, as well as a few scientists—notably Nobelist John Eccles (Eccles 1989). The concept of consciousness is clouded by definitional problems, with some ascribing lower consciousness to animals, and the evolution of a higher-order consciousness, perhaps with higher primates, still a subject of great debate (Dennett 1991). One philosopher quipped that "a newborn is barely conscious, if that; a philosopher is fully conscious – too much so, sometimes" (Ruse 2006). What is certain is that the concepts of mind and consciousness are increasingly amenable to research by neuroscientists, paleobiologists, anthropologists, and psychologists studying both animals and humans. Today almost without exception, professional experts in philosophy, psychology, cognitive science, neurobiology and artificial intelligence accept some version of Darwinian materialism, the "astonishing hypothesis" whereby consciousness and intelligence are seen to arise from the workings of the brain, itself a product of natural selection, even if the mechanism is still under dispute (Donald 1991; Crick 1995; Parker and McKinney 1999; Edelmann 2004; Searle 2004). It is here that even Wallace could not fathom the brain as a product of natural selection, as Darwin insisted, producing perhaps his greatest scientific heresy among many others (Smith and Becalloni 2008). Still, in many ways defining "intelligence" remains more problematic than defining "life," with many different possible approaches undertaken in a very large literature (Sternberg 2000, 2002). To frame it another way, there is no "general theory of intelligence" or even of human brain function, much less a general theory of intelligence in a cosmic context. Moreover, all the problems of chance and necessity that apply to the origin of life debate also apply to the debate over the origin and evolution of intelligence.

Most of the discussion of intelligence has understandably been confined to the origin and evolution of intelligence on Earth. Carl Sagan's Pulitzer Prize-winning book *The Dragons of Eden* (Sagan 1977), updated by Skoyles and Sagan (2002), was one widely read attempt to synthesize what was known about the evolution of intelligence on Earth, featuring the triune brain consisting of the primal reptilian R-complex, the mammalian limbic system and the primate neocortex. Mithen (1996) offers a more influential scenario grounded in paleoanthropology, in which a variety of specialized intelligences, which he terms social, technical and natural history intelligences, were integrated over millions of years, giving rise to a "cognitive fluidity" that allowed for language and a more generalized intelligence leading to tools, religion, art and the modern mind less than 100,000 years ago. Other scenarios exist from the point of evolutionary psychology (Barkow et al. 1992; Pinker 2002), mapping the mind (Hampden-Turner 1981), climate and brain-cooling (Falk 2004), the computational metaphor (Turing 1950), and the memory and pattern analysis employed in the neocortex of the human brain as understood by neurobiology (Hawkins and Blakeslee 2004), among others. None of these have as yet proven definitive, but all have in common the Darwinian concept that intelligence is a product of cognitive evolution brought about by natural selection acting on the environment. It may be that some combination of these approaches will eventually result in a general theory of intelligence.

How does this research apply in the cosmic context, in other words, to the concerns of astrobiology? As might be predicted, the problems become more complicated, but not intractable. One of Sagan's motivations for writing *Dragons of Eden* was to gain "hints or insights" into extraterrestrial intelligence; he concluded that "once life has started in a relatively benign environment and billions of years of evolutionary time are available, the expectation of many of us is that intelligent beings would develop. The evolutionary path would, of course, be different from that taken on Earth … But there should be many functionally equivalent pathways to a similar end result. The entire evolutionary record on our planet, particularly the record contained in fossil endocasts, illustrates a progressive tendency toward intelligence" (Sagan 1977, p. 230). That conclusion embodies many assumptions questioned by others. Evolutionists such as Simpson (1964) and Dobzhansky (1972), for example, had already argued just the opposite, and Harvard evolutionist Ernst Mayr also differed strongly with Sagan, arguing that intelligence (by his definition) had emerged only once on Earth (Mayr 1985, 1988). Outspoken Harvard evolutionist Stephen Jay Gould agreed with the nonprevalence of *humanoid* intelligence, arguing in an entire book on the Burgess Shale fossils of the Cambrian explosion that if we "Wind back the tape of life to the early days of the Burgess Shale; let it play again from an identical starting point, and the chance becomes vanishingly small that anything like human intelligence would grace the replay" (Gould 1989). By contrast evolutionary paleobiologist Simon Conway Morris (Conway Morris 1998, 2003) has argued from the same evidence, and others, that evolutionary convergence applies not only to morphology, but also to intelligence, if only the conditions are present. He is, however, skeptical that the proper conditions often obtain, summarizing his position in the subtitle of his 2003 book *Life's Solution: Inevitable Humans in a Lonely Universe*. In this he reached the same conclusion as had Ward and Brownlee (2000), who argued, Wallace-like, that complex life and thus intelligence in the universe will be rare, not from a lack of convergence but because so many factors must come together in order for it to exist.

Given the number of extrasolar planets now known, admittedly mostly gas giants sprinkled with super-Earths, but with Earths likely still to come, what is obviously needed is more knowledge of conditions on those planets, conditions on which natural selection might act in producing not only life but also intelligence. Work on biosignatures, either in planetary materials or in the atmospheres of extrasolar planets, is a step in this direction (Pilcher 2003; Des Marais et al. 2008; Ohmoto et al. 2008; Kaltenegger et al. 2010). Lacking this knowledge at present, Bogonovich (2011) has emphasized that one way forward in studying the problem of intelligence is to understand the evolution of intelligence on Earth in the most general sense. Indeed, while most research has focused on the evolution of intelligence in hominids on Earth over the last 5–7 million years, for astrobiology the evolution of intelligence in a more general sense is of primary importance. Such research has been undertaken especially on dolphins and other mammals (e.g. Marino 2002; Reiss 2011), and some of the results have been sporadically summarized and applied in an extraterrestrial context at astrobiology meetings, particularly in terms of dolphin communication (Reiss 1988); brain encephalization, a normalized measure of brain size in a variety of animals evolved over the last 500 million years (Marino 1995; Jerison 2000); convergent brain evolution on Earth (Falk 1995); and progressive trends in the history of life (Russell 1983, 1995). Bogonovich outlines ten hypotheses related to both human intelligence and a broader definition of intelligence in animals, and calls for a research program to determine what might be relevant to the evolution of a more generalized intelligence in the astrobiological context.

One thing seems likely: evolutionary convergence notwithstanding, the specific sequence of events seen on Earth will not reappear on another planet, and natural selection will accordingly affect both morphology and forms of intelligence. Loren Eiseley's famously poetic dictum "In the nature of life and in the principles of evolution we have had our answer. Of men elsewhere, and beyond, there will be none forever" (Eiseley 1957) may be true in a specific sense. But the jury is still out in the most general sense of whether intelligence is a trait selected for competitive advantage that would lead to its widespread existence in the cosmos, and whether such intelligence once formed would be similar to ours or vastly different in its physical foundation, internal workings, sensory output, and ability to communicate. This cluster of issues may be summarized as

Critical Issue # 4: What is intelligence and its relationship to consciousness and mind when considered in the universal context of astrobiology? How does astrobiology change our core conceptions of intelligence? Under what conditions does intelligence develop, and what is the role of chance and necessity in the origin and evolution of intelligence? What are the metaphysical assumptions that underlie our concepts of intelligence? Is it possible that other minds and other intelligences are so different from ours so as to be incommensurable, and thus to preclude communication?

The relationship of consciousness and intelligence to communication again opens a wide field for philosophical and scientific study. The role of human language, for example, was problematized long ago by the philosopher Ludwig Wittgenstein, and is widely recognized today as a critical problem in philosophy (Rosenberg 2000; Coleman and Welty 2010). Attempts have been made at studying these relationships using animals, with some concluding that communication among animals is a window into their cognition and consciousness (Griffin 2001). Reiss (1988) suggests "there may be a convergence or continuity in the communication and cognitive abilities in animals from different evolutionary paths," and Conway Morris (2003, p. 252) cites this idea favorably as fitting in with his work on convergence. But a great deal more research needs to be done before such far-reaching conclusions can be accepted.

In the extraterrestrial realm, the questions run even deeper, for problems of language and communication depend on the possible domains of mental structure. As Dunér has pointed out, a theory of situated cognition holds that the environment has an active role in driving cognitive processes, producing what has been called "the extended mind" (Clark and Chalmers 1998): "Aliens, developed in and adapted to an entirely different physical and cultural environment would probably have very different modes of thinking. Thinking in the universe could be very different from what we are used to in our westernised, anthropocentric and earthbound human culture" (Dunér 2011, p. 123). Following this line of thinking, in the coarsest sense we may envision three cases when comparing terrestrial and extraterrestrial minds (Fig. 38.3). In case 1 our mental structures and modes of perceiving and thinking may overlap entirely, in which case dialogue may be relatively "easy." In case 2 they may overlap only partially, yielding some common basis for dialogue. In case 3 there is no overlap at all, thus no dialogue at all. But all hope is not lost, for in case 4, a "dialogue chain" of partially overlapping mental structures may eventually enable dialogue (Billingham 2000a, b; Dick 2000a). These Venn diagrams of "mental structures" at the physiological level, or "modes of thinking" at the output level, embody many problems and require more research in many areas before any conclusions can be reached about their effect on communication. Meanwhile, from the point of view of artificial intelligence, Marvin Minsky has concluded that communication will be possible because we will think in similar ways (our case 1 or 2), because "all intelligent problem-solvers are subject to the same ultimate constraints—limitations on space, time and materials" (Minsky 1985). His arguments, however, apply "only to those stages of mental evolution in which beings are still concerned with surviving, communicating, and expanding their control of the physical world. Beyond that, we may be unable to sympathize with what they come to regard as important," perhaps leading to our cases 3 and 4. In any event, contact with

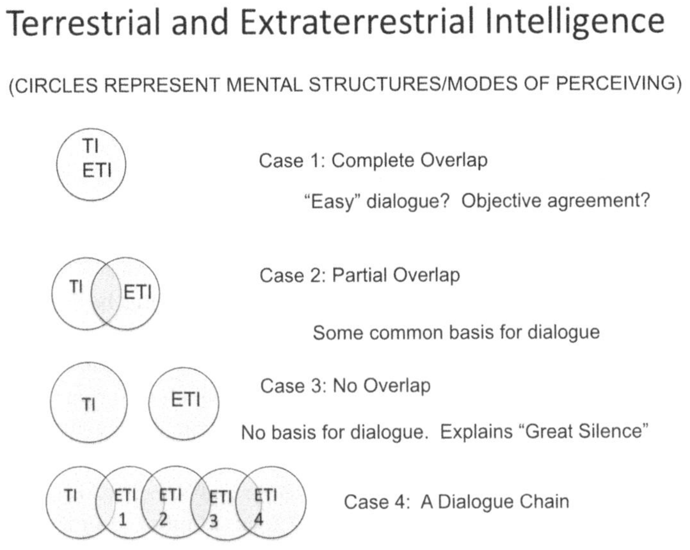

Fig. 38.3 Possible scenarios in the relationship between extraterrestrial intelligence (ETI) and terrestrial intelligence (TI). The Venn diagrams may be taken as mental structure or modes of perceiving and thinking. In case 1 these overlap entirely, in which case dialogue may be relatively "easy." In case 2 they may overlap only partially, yielding some common basis for dialogue. In case 3 there is no overlap at all, in which case there is no overlap at all and thus no dialogue at all. In case 4 a "dialogue chain" of partially overlapping mental structures may eventually enable dialogue. (Drawing by Steven Dick)

extraterrestrials holds open the possibility of solving one of the longest standing epistemological problems of philosophy: the problem of objective knowledge (Popper 1979). For by comparing knowledge derived from many independently evolved mental structures, what remains would surely constitute objective knowledge on a higher level than fathomable by terrestrial standards alone.

From an operational point of view psychologist DouglasVakoch has pioneered the field of possible methods of encoding communications with extraterrestrial intelligence. Reaching beyond the usual scientific and mathematical approaches, he suggests ways of communicating our altruism, our music, even our aspirations, arguing that if the principles of science are universal, extraterrestrials will already know about science and mathematics, and probably much more if they are more advanced (Vakoch 1998, 1999). Extraterrestrial historians and anthropologists, however, will be interested in our level of science and mathematics, and Vakoch also recommends communicating our understanding of cosmic evolution in interstellar messages (Vakoch 2009). His work has inspired others to enter the field of extraterrestrial communications and semiotics (Vakoch and Harrison 2011).

No critical history of ideas of the origin and evolution of intelligence exists analogous to Fry's detailed and nuanced analysis of issues in the origins of life debate (Fry 2000); this constitutes a fundamental lacuna in the field, all the more so because SETI practitioners typically fail to consider the nature of intelligence in their work.

In fact, they skirt the idea by talking about searching for extraterrestrial technology rather than intelligence. As SETI pioneer Jill Tarter puts it, "SETI is a misnomer because there is no known way to detect intelligence directly across interstellar distances. Even on Earth we argue about exactly what constitutes intelligence, and we have no reliable way of measuring it at a distance (either spatial or temporal). In the case of extraterrestrial intelligence, the best we can do is to search for some manifestation of another technology" (Tarter 2007). Such technologies could take many forms (Dyson 1966; Ćirković 2006), but it is not surprising that for operational purposes radio astronomers search for radio signals, observing the effects of technology, not the intelligence behind the technology. But this begs another question: does intelligence necessarily lead to technology?

The issues surrounding the relationship between intelligence and technology are in reality problems of cultural evolution, a realm in which historians, anthropologists, and other social scientists dominate rather than paleobiologists and neurobiologists (Basalla 1988). These issues were broached already in an astrobiological context when NASA was forming its SETI program and Nobelist Joshua Lederberg led a 2-day "Workshop on Cultural Evolution," focused more specifically on "evolution of intelligent species and technological civilizations" (Morrison et al. 1977). They are occasionally found in subsequent SETI literature (Harrison et al. 2000; Basalla 2006; Dick 2006; Denning 2011), and most recently were highlighted in an entire volume, *Cosmos and Culture: Cultural Evolution in a Cosmic Context* (Dick and Lupisella 2009). Such studies, however, have been hampered by our understanding of cultural evolution on Earth, which can only be described as rudimentary by comparison with our knowledge of astronomical and biological evolution. As with life and intelligence, the idea of culture is difficult to define; nevertheless anthropologists and others have spent a great deal of time attempting to do so. 50 years ago, two anthropologists collapsed 164 distinct definitions of culture into one: "[C]ulture is a product; is historical; includes ideas, pattern, and values; is selective; is learned; is based upon symbols; and is an abstraction from behavior and the products of behavior" (Kroeber and Kluckhohn 1952). Anthropologist Clifford Geertz defined culture more intuitively as "an historically transmitted pattern of meanings embedded in symbolic forms by means of which men [people] communicate, perpetuate and develop their knowledge about and attitudes toward life" (Geertz 1973; Kuper 1999). If, as Harvard biologist E. O. Wilson logically asserts (Wilson 1998), each society creates culture and is created by it, the idea of culture is a moving target, evolving with time and in space. Astrobiology vastly expands that possible cultural space. In the cosmic context, our terrestrial ideas of culture will surely be broadened if we discover cosmic civilizations, in which case the natural history of culture and its theoretical underpinnings will be taken to a new level.

The same holds true for cultural evolution. Social scientists have in general judged as too simplistic Spencerian models of cultural evolution, based on the nineteenth-century work of Herbert Spencer, which views society as evolving through well-defined stages from simple to complex. Darwinian models have proliferated in recent decades but have been highly controversial. Attempts to Darwinize culture include sociobiology (Wilson 1975), gene-culture co-evolution (Boyd and

Richerson 1985; Richerson and Boyd 2006), universal Darwinism (Dennett 1996) and memetics (Blackmore 1999; Aunger 2000). All these models have considerable problems, and some social scientists still resist evolutionary models of culture altogether (Lalande and Brown 2002). A more robust knowledge of the idea of culture, and the mechanisms of cultural evolution, is nevertheless central to the SETI endeavor (Denning 2009), arguably more central than SETI practitioners realize.

One example will suffice to indicate the power and relevance of cultural evolution to astrobiology. It has recently been argued that the laws of physics may prevent the human brain from increasing its intelligence even over long time scales (Fox 2011). Long-lived technical civilizations, for whom the improvement of intelligence may be a driving force of cultural evolution (Dick 2003), may therefore seek to increase intelligence through other means. One scenario is that given the long time scales in the universe, on the order of billions of years since life has been possible after the creation of heavy elements and rocky planets, cultural evolution may have given rise to artificial intelligence. This is an event that some see as happening in a few generations on Earth (Moravec 1988, 1999; Kurzweil 1999, 2005); if so it might have already happened long ago elsewhere in the universe, perhaps yielding a postbiological universe (Dick 2003) rather than a biological universe. While this may seem like science fiction, the concept arises and is given force by taking cultural evolution no less seriously than physical and biological evolution in the course of cosmic evolution. Aside from the broader implications of this scenario, a postbiological universe would likely affect the nature of the signal and the methods of communication, important considerations for SETI. On the one hand, as an extrapolation of a contemporary terrestrial theme, the greatest weakness of the postbiological scenario is that it is not bold enough, and perhaps a failure of imagination. Other means may exist to advance intelligence rather than through artificial intelligence. On the other hand, all of this presupposes that, when all is said and done, science and technology progress over long time scales, an Enlightenment idea that is itself problematic, even in the terrestrial context (Laudan 1977; Nisbet 1980; Billings 2007; Dark 2007).

This complex of technology, culture, and cultural evolution form another critical issue, for in order to have technology, culture must arise (though the converse—that every culture must have technology, especially radio communicative technology, is not necessarily true, unless one defines culture by its tool-making capacity). This cluster of issues may be summarized as.

> *Critical Issue # 5: What is the relationship among intelligence, technology and culture? What are the mechanisms in the evolution of culture, and what are the possible outcomes, including postbiological? Are there universal aspects to the evolution of culture? What are the metaphysical assumptions underlying our concepts of culture? And, how does the evolution of culture affect methods of communication with extraterrestrial intelligence, and ultimately our chances of successful communication?*

As with the concepts of life and intelligence, this critical issue is likely to constitute a long and complex research program, with results that are both useful and enriching if cultures are found beyond the Earth. Along with the literature on culture and cultural evolution, and the literature on the history of technology, culture in a cosmic context stands to learn much from work in the philosophy of technology, which among other aspects probes the foundational concepts of technology and its relation to technology to culture (Dusek 2006).

Finally, under the purview of metaphysics we may consider the universe itself—"all that is or ever was or ever will be" (Sagan 1980)—as a mental and scientific category. One is immediately led in this direction when asking the clearly astrobiological question whether or not the universe is in some sense fine-tuned for life and perhaps even biocentric, implying a link between biology and cosmology. Over the last four decades some scientists have come to question why the laws of nature and the physical constants appear to be *biofriendly*, giving rise to what has been termed the *anthropic principle*. The principle has many variants, all having to do with the apparent fine-tuning of the physical constants for life (Carter 1974; Barrow and Tipler 1986; Carr 2007). The phrase is a spectacular misnomer, and the term *biocentric principle* is much preferred, since in the context of astrobiology the universe appears to be friendly to life, and the very question to be answered is whether humans are the only intelligent life (Davies 2007). The prospect of a fine-tuned universe has given rise to the idea of an ensemble of universes, a *multiverse*, as an explanation for why we happen to be in a universe particularly suited for life (Carr 2007). In this case, *pace* Sagan, the universe would not constitute "all that is or ever was or ever will be."

Whether or not we invoke the multiverse, the physicist Freeman Dyson has suggested that the prospects are bright for a future-oriented science, joining together in a disciplined fashion the resources of biology and cosmology (Dyson 1988). In such a *cosmic ecology*, life and intelligence would play a central role in the evolution of the universe, no less than its physical laws. In two provocative books, complexity theorist James Gardner (2003, 2007) evokes a cosmological worldview that places life and intelligence at the center of cosmology, specifically arguing that "the ongoing process of biological and technological emergence, governed by still largely unknown laws of complexity, could function as a von Neumann controller and that a cosmologically extended biosphere could served as a von Neumann duplicating machine in a conjectured process of cosmological replication." Gardner speculated that the replication process could occur through the fabrication of baby universes by a highly evolved intelligence, in which case physical laws and constants would serve as a kind of cosmological DNA, providing a recipe for the birth and evolution of intelligent life (Gardner 2009). In such a *selfish biocosm* scenario, it is logical that the physical laws and constants would be rigged in favor of the emergence of life. Gardner offers three falsifiable tests of his hypothesis: SETI research, artificial life evolution, and the emergence of computer intelligence. His conjectures render relatively tame physicist Lee Smolin's idea of collapsing black holes birthing new universes, which then undergo natural selection based on their ability to reproduce, for unlike Gardner (and Hoyle), Smolin invokes no celestial intelligence (Harrison 1995; Smolin 1997, 2007). These speculations lead to

Critical Issue # 6: Is there a link between biology and cosmology? Is the universe indeed finely tuned for life? Is the universe in some sense biocentric, and if so, why?

We should hasten to add that some consider the anthropic principle and the idea of the multiverse not to be science (Pagels 1985; Smolin 2007), while others consider the arguments to be completely misguided (Stenger 2011). If the sole epistemological method for science is empiricism, there is some justification for this skepticism, for the anthropic principle is clearly not empirical, at least at this stage and perhaps in principle, and the multiverse by definition can most likely never be observed. In Popperian terms, it is thus unfalsifiable. One might counter that empiricism is no longer the sole criterion for science, which accepts the atom, elementary particles, dark matter and dark energy as real based on their effects but without actually seeing them. And there are theoretical reasons, from the inflationary theory of Big Bang cosmology to black hole generation, string theory, and the many worlds interpretation of quantum mechanics, to believe that many universes are possible. Still, the multiverse idea is problematic as science, since, unlike atomic theory and dark matter theory, the multiverse has no observable "effect" and may not be confirmed even in principle. As Ellis (2007) puts it,

> There can be no direct evidence for the existence of other universes in a true multiverse, as there is no possibility of even an indirect causal connection. The universes are completely disjoint and nothing that happens in one can affect what happens in another. Since there can be no direct or indirect evidence for such systems, what weight does the claim for their existence carry?

Ellis concludes that the concept of the multiverse has explanatory power, but is a metaphysical concept. From this point of view the multiverse may exist, but we will never know, and thus we will never know whether our universe is "all that is or ever was or ever will be." Others are not so sure that it is completely unfalsifiable (Aguirre 2007; Page 2007; Stoeger 2007), and even Smolin offers his theory of cosmological natural selection of universes as falsifiable in the case the ensemble of universes is not randomly sprinkled in parameter space and has certain other properties (Smolin 2007). The topic has been discussed now for more than three decades (Dick 1996, 2008), is being robustly debated by both seasoned scientists and young scholars entering the field (Carr 2007; Vidal 2010; Lanza and Berman 2009; Ćirković, forthcoming, 2012), and shows no signs of disappearing.

It should be clear by now that one of the advantages of the extraterrestrial life debate is that it forces us to think in broader categories than usual about these fundamental issues of life, mind, intelligence, and culture, in other words, to contemplate the metaphysical and epistemological foundations of knowledge. Even if extraterrestrial life is not found, the debate will have been worth it for this fundamental reevaluation alone.

38.4 Astrobiology and Society: Ethical and Impact Issues

In the first article of the new journal *Astrobiology*, published at the opening of the new millennium in 2001, astronomer David Morrison wrote "Astrobiology research has implications that are felt beyond the confines of the laboratory. As our understanding of living systems and the physical universe increases, we will confront the implications of this knowledge in more than just the scientific and technical realms … Together we will explore the ethical and philosophical questions related to the existence of life elsewhere, the potential for cross-contamination between ecosystems on different worlds, and the implications of future long-term planetary habitation and engineering" (Morrison 2001). The theme of astrobiology and society covers an enormous area and has recently developed its own roadmap (Race et al. 2012). While too broad to explore here in its entirety, at least two issues may be categorized as explicitly philosophical: the societal impact of the discovery of extraterrestrial life, and the ethical/public policy questions surrounding the search for microbial life and communication with extraterrestrial intelligence. The first of these can be formulated simply as

Critical Issue # 7: What are the theological, ethical, philosophical and worldview impacts of the discovery of microbial or intelligent life? What are the most productive approaches to addressing these impacts?

The issue of cultural impacts of astrobiology has received increasing attention since John Billingham, the head of NASA's SETI program at the time, convened a series of workshops on "The Cultural Aspects of SETI" (CASETI) on the eve of the inauguration of the NASA SETI program in 1992 (Dick 1995; Billingham et al. 1999). During the formulation and initiation of the first Astrobiology Roadmap in 1998, calls were made for the study of cultural impacts of astrobiology (Dick 2000b), and in 1999 NASA Ames Research Center organized a workshop on societal implications (Harrison and Connell 2001). Other organizations, including the Templeton Foundation and the Foundation for the Future, organized meetings on the subject at about the same time (Dick 2000c; Harrison and Dick 2000; Tough 2000). Interest has increased in the last decade, notably with the American Association for the Advancement of Science (AAAS) series of workshops sponsored by its program on Dialogue on Science, Ethics and Religion (Bertka 2010), and several meetings at the Royal Society of London (Dominik and Zarnecki 2011). While most of the attention has focused on the impact of the discovery of extraterrestrial intelligence, the AAAS volume also addresses the quite different scenario of the impact of discovery of microbial life. Most recently Race (2012) has taken the lead in marshalling the astrobiology, social sciences and humanities communities to address these issues in the context of the latest Astrobiology Roadmap, with the support of the NASA Astrobiology Institute. There have also been individual studies concentrating on different aspects of the problem (White 1990; Almar 1995; Davies 1995; Randolph et al. 1997; Achenbach 1999; Vakoch 2000; Harrison et al. 2000; Michaud 2007, Arnould 2008), including a comparison to the impact of other scientific endeavors such as biotechnology (Race 2007). Issues of theological impact have received particular attention, with some concluding that Abrahamic religions with a godhead might be more affected by contact with extraterrestrials than Eastern religions (Dick 1996, 2000c; Crowe 1997; Vakoch 2000; Bertka 2010).

The results of these studies have been to demonstrate the serious impact that the discovery of extraterrestrial life could have on society, especially in the case of extraterrestrial intelligence. These studies have also demonstrated the number of approaches that might be taken in studying possible impacts of extraterrestrial life. These include the study of historical analogs to the impact of new worldviews, to historical culture contact on Earth, and to the transmission of new ideas among cultures through time. One notable finding is that physical cultural contacts, typically disastrous in the history of the Earth, are not good analogs, since contact with extraterrestrial intelligence is unlikely to be physical, science fiction notwithstanding. Rather, those studying the impact of astrobiology on society would do better to concentrate on analogs of the transmission of new ideas, such as the transmission of Greek knowledge to the Latin West by way of Islamic civilization in the eleventh and twelfth centuries (Dick 1995; Billingham et al. 1999). The impact of new worldviews such as the Copernican and Darwinian could also yield insights based on history. This is true particularly because the biological universe has achieved the status of a worldview, a *biophysical cosmology* that asserts the importance of both the physical and biological components of the universe (Dick 1989, 1996, 1997). The construction of worldviews and their influence on our thinking is a deep philosophical problem (Vidal 2007, 2011), and this particular worldview of the biophysical cosmology is increasingly testable, indeed, it is the long-term goal of astrobiology.

Related to this issue of impacts is the question of how philosophy itself would change if we confirm the existence of extraterrestrial intelligence. An historical approach to this question might ask how philosophers viewed the possibility of extraterrestrial life in the context of their philosophy, or conversely how a belief in extraterrestrial life has historically affected philosophy. The first major work of Immanuel Kant, for example, was not his critical philosophy, but his *Allgemeine Naturgeschicte and Theorie des Himmels* [Universal Natural History and Theory of the Heavens, 1755], in which extraterrestrials played a central role (Kant 1981; Shea 1986). Could this "extraterrestrial perspective" have affected Kant's unusually generalized yet deeply critical philosophy? (Crowe 1986, pp. 54–55). Similar historical questions may be asked of other philosophers who have also tackled the question, especially in the nineteenth century (Crowe 1986). A more future-oriented study would ask how venerable philosophical questions such as objective knowledge and the mind-body problem would be expanded by an extraterrestrial perspective.

A second quite different critical issue in the domain of Astrobiology and Society may be formulated as

> *Critical Issue # 8: What are the ethical issues in pursuing the search for extraterrestrial life and communication with extraterrestrial intelligence? How do we balance planetary protection and stewardship with the human exploration imperative, of which astrobiology is a significant part?*

Here we come face-to-face with planetary protection issues, which gave birth to exobiology 50 years ago (Dick and Strick 2004). In part because NASA has for decades had a Planetary Protection Officer, these issues have been at the forefront of both human and space science missions, and the subject of numerous studies and technical conferences (Meltzer 2011). This is no surprise, for when it came to the search for extraterrestrial life, there was the real possibility that contamination would destroy the very object of the search, or result in ambiguity at the very least. Still, because the details of planetary protection are mission-specific, the problems are far from resolved and will be the subject of more studies in the future. Though historically a source of friction between scientists, engineers and cost-cutters in the past, the gravity of these issues can hardly be overemphasized, not only because of the real prospect of forward contamination but also because a single "Andromeda Strain" scenario would be enough to jeopardize the terrestrial biosphere through back contamination.

In quite a different arena we come to a real-world issue that has percolated at a low-level throughout the extraterrestrial intelligence debate, and has recently heated up in the SETI community and the media with Stephen Hawking's pronouncement that messaging to extraterrestrials could be dangerous (BBC News 2010). In venues such as The Royal Society of London and the Permanent SETI Committee of the International Academy of Astronautics, this subject has also played a contentious role, with arguments from both sides well-represented (Brin 2011; Zaitsev 2011). Opponents of messaging argue that history demonstrates "shouting in the jungle" can be dangerous, and protestations that it is too late for action because our military, radio and TV signals are inexorably spreading among the stars, are met by arguments that these signal are unlikely to be decipherable. Others counter that the carrier signal of such information-laden transmissions would be quite enough to reveal our presence, and point out that in any case directed radar beams mapping the asteroids could be intercepted at interstellar distances, and are much stronger in signal strength than any messages sent or contemplated. Questions such as who speaks for Earth, with what message, and toward what end, are likely to be overwhelmed by practicalities such as lack of international enforcement power.

More broadly the question arises "should mankind hide?" from the cosmos, in fear of what may lie in wait in reaction to any message. This question was raised by the *New York Times* in the wake of Nobelist Martin Ryle's negative reaction to the first deliberate interstellar radio message, sent in 1974 with the Arecibo telescope to the 300,000 stars of the globular cluster M13 by Frank Drake and his colleagues (*New York Times* 1976; Dick 1996). It is here that one must weigh the balance between planetary protection writ large and human exploration. Admittedly, human exploration on Earth has not always been a pretty picture, especially with respect to culture contacts. Certainly we would not wish to be treated as we have treated ourselves. But as Nobelist and former Astrobiology Institute Director Baruch Blumberg liked to emphasize in comparing astrobiology to the Lewis and Clark expedition, our exploration imperative is strong and likely unquenchable. Whether extraterrestrials would have learned altruism we cannot know; Hawking and others are skeptical.

Such questions as posed in this section have no easy answers, but are properly placed in the ethical domain of philosophy, where they deserve close scrutiny considering the potential consequences for humanity.

38.5 Sociology of Scientific Knowledge

A final set of issues, distinguishable from ethical issues or the impact of astrobiology on society, stems from the sociology of scientific knowledge, the study of science as a social activity. Historians and social scientists have called attention to the fact that most astronomers seem to be much more optimistic about the chances of extraterrestrial life, while biologists, especially evolutionary biologists such as George Gaylord Simpson, Ernst Mayr and Theodosius Dobzhansky, are much more critical (Barrow and Tipler 1986, pp. 132–133; Dick 1996; Harrison and Connell 2001). Is this difference to be found in their training, in their different approaches to science, or in the cognitive content of their respective sciences? Others have called attention to the fact that the extraterrestrial life debate has been largely the purview of the Western world, stretching from Leucippus, Democritus, Epicurus, and Aristotle in Greece to Lucretius in the Roman world, entering the Latin West by way of commentaries on Aristotle, and then embarking on its remarkably robust career described in the now standard histories of the debate. Other cultures have had only sporadic interest, or no interest at all in the debate through most of history (Schneider 2010). This leads to

Critical Issue # 9: What are the sources of the diverse attitudes and assumptions of different scientific communities to the problem of life beyond Earth? And why is the debate more prominent in some cultures than others?

There is, of course, a great deal of variation of belief among individuals on the subject, although it must be said that, in the western world at least, the vast majority of people (at least 60%) believe that extraterrestrial life could exist (Bainbridge 2011). That is a rather astonishing fact, and very different from belief in Darwinian evolution in the sense that many people reject the latter despite a large body of evidence, while many people believe in the existence of extraterrestrial life even without strong evidence. Why? Are extraterrestrials any less threatening to core religious beliefs than Darwinian evolution is perceived to be? Is it simply the power of suggestion because of the popularization of extraterrestrials in science fiction and film? Or are alien science fiction and film popular for some deeper reason? Perhaps, as some have suggested, humans are prone to belief in imaginary beings (Planck 1968). Or, as others have suggested, are extraterrestrials surrogate "deities for atheists?" (Shermer 2002b, 2006; Basalla 2006). Or is our search for life the most exciting part of the exploration of space, and in the end a search for ourselves? (Jakosky 2000). Such musings give rise to

Critical Issue # 10: What is the source of the public "will to believe" in extraterrestrial life, at least in the Western world, and how does it compare to other cultures?

These critical issues in the sociology of scientific knowledge are related to a larger question central to the "culture wars" over the last few decades: to what extent are our beliefs socially constructed? Despite the metaphysical foundations of modern science, the epistemology of modern science seems very different from the epistemology of theology, and the epistemic status of the results is therefore very different. But the concept of social construction takes on new meaning in an extraterrestrial context, where mental structures may be very different. One thing is certain: if there are extraterrestrials and science is socially constructed, it will be revealing to see what their science is like. At least one philosopher concludes it will be nothing like ours (Rescher 1985). And if their science is different, does that mean their technology is different (Basalla 2006), and what does that imply for communication (see issues 4 and 5)?

38.6 Cosmic Evolution: The Master Narrative of the Universe

Finally, this reconnaissance would be incomplete without emphasizing that many of the critical issues of astrobiology—the evolution of life, intelligence and culture—are embedded in the larger process of cosmic evolution, the 13.7-billion-year Master Narrative of the Universe (see Fig. 2.4). The concept has its roots in the eighteenth and nineteenth centuries, but only became widely accepted and a major driver for research programs in the last half of the twentieth century (Dick 2009; Zakariya 2010). I have argued elsewhere (Dick 2004) that the outcome of cosmic evolution may result in a physical, biological, or postbiological universe, in other words, a physical universe composed of planets, stars and galaxies in which life is a fluke; a biological universe full of carbon-based life; or a postbiological universe in which cultural evolution has resulted in a universe full of artificial intelligence. These outcomes determine the long-term destiny of humanity, and because the scope of astrobiology as set down in the Astrobiology Roadmap applies not only to the past and present, but also the future, the destiny of humanity falls within the purview of the philosophy of astrobiology.

This expansive view of the universe as an evolving process also has its issues. For example, is physical, biological, and cultural evolution really a continuum, despite very different evolutionary mechanisms? The philosopher Karl Popper held that the problem of the origin of life was "an impenetrable barrier to science and a residue to all attempts to reduce biology to chemistry and physics" (Popper 1974; Ruse 2006). That a substantive continuum does in fact exist is what Iris Fry calls the "continuity thesis" in the context of the origin of life and the transition from the physical to the biological (Fry 1995). But it applies even more to intelligence and culture, where our ideas are even less robust than in the origins of life debate. I frame this issue as the *Cosmic Evolution Continuity thesis*:

Critical Issue # 11: Despite the very different mechanisms of physical, biological and cultural evolution, can and should cosmic evolution be seen as a single, coherent, scientific and productive concept?

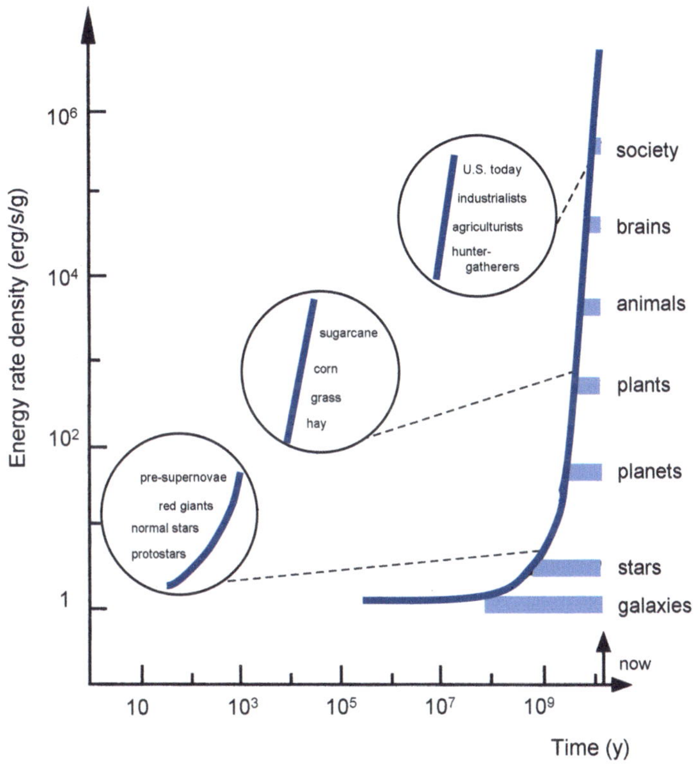

Fig. 38.4 The evolution of complexity and the unification of cosmic evolution. According to the cosmic continuity thesis, all forms of cosmic evolution—physical, biological and cultural—are united, in this formulation by the concept of energy flow, more specifically quantified as the "energy rate density," the energy rate per unit time per unit mass calculated for a variety of open, organized, non-equilibrium systems. Here the increase in energy rate density is plotted as histograms starting at those times when various open structures dominantly emerged in nature. The increase has been especially rapid in the last billion years. (By permission from Chaisson 2009)

Astronomer Eric Chaisson has argued in the affirmative, postulating that all these forms of evolution are united by the concept of energy flow, more specifically quantified as the "energy rate density," the energy rate per unit time per unit mass calculated for a variety of open, organized, non-equilibrium systems (Fig. 38.4). He finds this concept of an energy rate density to be "a single, measurable, and unambiguous quantity uniformly characterizing Nature's many varied complex systems, potentially dictating their natural selection on vast spatial and temporal scales" (Chaisson 2001, 2010, 2011).

The questions do not end there. An evolutionary universe need not have a purpose but inquiring minds want to know: Is the universe "evolving in the direction of greater complexity, consciousness, and culture?" (Lupisella 2009; Vakoch and Harrison 2011). If so, is there a teleological goal, as the biocentric principle implies, as Gardner implies when he proposes that the cosmic utility function (the outcome being maximized) is "the propagation of baby universes exhibiting the same life-friendly physical qualities as their parent universes" (Gardner 2003, 2007), and as Dyson implies when he muses that "As we look out into the universe and identify the many accidents of physics and astronomy that have worked together to our benefit, it almost seems as if the universe must in some sense have known that we were coming" (Dyson 1979, p. 250)?

This leads to a final, science-fiction-like set of questions:

> *Critical Issue # 12: Are cosmic evolution and cosmic natural selection goal-oriented, giving a teleological aspect to the universe? If so, what is the goal? Is cosmic evolution really linear as usually depicted—"merely" the unfolding of cosmic time from 13.7 billion years ago to the present? Or is there some more complex structure, involving the very nature of space-time, the creation of baby universes via black holes, perhaps by way of some cosmic natural selection of universes and their constants, if not by a cosmic natural intelligence?*

In the twentieth century, philosophers such as Henri Bergson, Alfred North Whitehead, and Teilhard de Chardin proposed philosophical systems infused with teleology (Barrow and Tipler 1986, pp. 123–218). And though teleology has long been anathema to biology and science in general for centuries, philosophers of biology have emphasized that it is far from totally excised (Hull 1974). Harvard evolutionary biologist Ernst Mayr concluded that "the use of so-called teleological language by biologists is legitimate; it neither implies a rejection of physicochemical explanation nor does it imply noncausal explanation" (Mayr 1988). But Mayr found that for biology "it is illegitimate to describe evolutionary process or trends as goal-directed," a conclusion with which most biologists would agree. The question remains whether cosmic evolution can be teleological, as some have suggested (see issue 6). Although scientists are naturally reluctant to pose teleological questions, which may seem to border on the precepts of the Intelligent Design Movement (Gonzalez and Richards 2004), there is a critical difference between the supernatural intelligence of religion, and natural laws that guide events toward an ultimate goal, or even the actions of a natural intelligence that may be the product of natural selection over very long periods of time. Only natural law and a natural intelligence are within the province of science. Here we come full circle to critical issues 1 and 2, for the biocentric principle, the multiverse, and cosmic teleology represent the ultimate in the problematic nature of evidence and inference. If universe, life and mind are in some deep cosmological sense connected, the subject of critical issue 6, human minds naturally want to know the goal and our role in that connection. It seems the stuff of science fiction, but work is being done on the subject.

At a time when it is clear there is a great deal we do not know about the universe at a deep level, it would seem prudent to keep an open mind on the great questions of cosmology and life. As Richard Feynmann, Carl Sagan and others have emphasized, however, our minds should not be so open that our brains fall out. Or, as Sagan has more eloquently said, "what is called for is an exquisite balance between two conflicting needs: the most skeptical scrutiny of all hypotheses that are served up to us and at the same time a great openness to new ideas" (Sagan 1987).

The 12 critical issues in astrobiology enunciated here comprise a long-term research program leading in many directions. And—because it encompasses physical, biological and cultural evolution—it is a research program that must be robustly informed by history, philosophy and the social sciences, as well as the natural sciences. It should also be informed by the well-developed field of the philosophy of biology, and by the philosophy of astronomy, a field that still exists only in embryonic form except for a few promising forays into philosophy of cosmology (Leslie 1998; Ellis 2006). In short, as Nobelist Baruch Blumberg never tired of saying, astrobiology counters the usual academic trends by drawing in numerous disciplines rather than compartmentalizing knowledge and increasing specialization. Wide as our scope has been in this overview, we have yet undoubtedly failed to enunciate some critical issues, perhaps even entire categories, and have certainly not exhausted the possibilities even in the issues enumerated here. But perhaps we have provided a framework, enough to glimpse how rich field the history, philosophy, and sociology of astrobiology can be.

38.7 Commentary 2020

This chapter is based on a keynote paper given at a conference in an unusual but memorably unique location: the Swedish island of Hven, where the Danish noble Tycho Brahe made his pioneering observations in the late sixteenth century at his own observatory, prior to the telescopic era of astronomy (Fig. 38.5). The meeting, held in 2011 and titled "The History and Philosophy of Astrobiology," was

Fig. 38.5 Astronomers Chris Impey and Tycho Brahe look heavenward on Tycho's island of Hven, Sweden. (Photo by Steven Dick)

testimony to the fact that interest in astrobiology is international; it was organized by an interdisciplinary group at Lund University in Sweden. The Proceedings of the meeting, including this paper, were published in the premier journal of the field, *Astrobiology* (Dunér et al. 2012), yet another indication of scientists' interest in the philosophical aspects of their subject.

Acknowledgments This article is dedicated to the memory of Baruch Blumberg, founding Director of the NASA Astrobiology Institute and an inspiration for the breadth and depth of his thinking. I wish to thank Michael Crowe, Gregory Dick, Iris Fry, and Clément Vidal for comments, as well as David Dunér and the organizers of the conference "The History and Philosophy of Astrobiology," held on Tycho Brahe's island of Hven in September, 2011.

References

Abir-Am, P. G. 1992. The politics of macromolecules: molecular biologists, biochemists and rhetoric. *Osiris* 7: 164–191

Achenbach, J. 1999. *Captured by Aliens: The Search for Life and Truth in a Very Large Universe*, Simon and Schuster, New York.

Aguirre, A. 2007. Making predictions in a multiverse: conundrums, dangers, coincidences, in Carr (2007), pp. 367–386.

Almar, I. 1995. The consequences of a discovery: different scenarios, in Shostak (1995), pp. 499–505.

Arnould, J. 2008. Does extraterrestrial intelligent life threaten religion and philosophy? *Theology & Science* 6: 439–50.

Asimov, I. 1965. A science in search of a subject. *New York Times Magazine*, 23 May, pp. 52–58.

Aunger, R. ed. 2000. *Darwinizing Culture: The Status of Memetics as a Science*, Oxford University Press, Oxford.

Ayala, F. J. and Arp, R. 2009. *Contemporary Debates in Philosophy of Biology*, Wiley-Blackwell, London.

Bainbridge, W. S. 2011. Cultural beliefs about extraterrestrials: A questionnaire study. In *Civilizations Beyond Earth: Extraterrestrial Life and Society*, Douglas A. Vakoch and Albert A. Harrison (2011., pp. 118–140.

Baird, D., Scerri, E. R. and McIntyre, L, editors. 2006. *Philosophy of Chemistry: Synthesis of a New Discipline*, Springer, Dordrecht.

Barkow, J., Cosmides, L. & Tooby, J., editors. 1992. *The Adapted Mind: Evolutionary Psychology and the Generation of Culture*. Oxford University Press, New York.

Barrow, J. D. and Tipler, F. J. 1986. *The Anthropic Cosmological Principle*, Oxford University Press, Oxford.

Basalla, G. 1988. *The Evolution of Technology*, Cambridge University Press, Cambridge.

Basalla, G. 2006. *Civilized Life in the Universe: Scientists on Intelligent Extraterrestrials*, Oxford University Press, Oxford.

BBC News. 2010. Stephen Hawking warns over making contact with aliens, 25 April, online at http://news.bbc.co.uk/2/hi/uk_news/8642558.stm

Beck, L. W. 1985. Extraterrestrial intelligent life, in Regis (1985), pp. 3–18; Presidential address delivered before the Sixty-eighth Annual Eastern Meeting of the American Philosophical Association in New York City, December 28, 1971.

Bedau, M. A. 2010. An Aristotelian account of minimal chemical life, *Astrobiology* 10: 1011–1020.

Bedau, M. A. and Cleland, C. E., editors. 2010. *The Nature of Life: Classical and Contemporary Perspectives from Philosophy and Science*, Cambridge University Press, Cambridge.

Benner, S. 2009. *Life, the Universe and the Scientific Method* (Ffame Press, 2009..
Benner, S. 2010. Defining life, *Astrobiology* 10: 1021–1030.
Bertka, C., editor. 2010. *Exploring the Origin, Extent, and Future of Life: Philosophical, Ethical and Theological Perspectives,* Cambridge University Press, Cambridge
Billingham, J. 1981. *Life in the Universe,* MIT Press, Cambridge, MA.
Billingham, J. 2000a. Summary of results of the seminar on the cultural impact of extraterrestrial contact, in Lemarchand and Meech, editors. (2000), pp. 667–675.
Billingham, J. 2000b. Who said what: A summary and eleven conclusions, in Tough (2000), pp. 33–39.
Billingham, J. Heyns, R., Milne, D., Doyle, S., Klein, M., Heilbron, J., Ashkenazi, M., Michaud, M., Lutz, J., and Shostak, S. 1999. *Social Implications of the Detection of an Extraterrestrial Civilization,* SETI Press, Mountain View, Ca.
Billings, L. 2007. Ideology, advocacy, and spaceflight: Evolution of a cultural narrative, in (Dick and Launius, 2007), pp. 483–500.
Blackmore, S. 1999. *The Gene Machine*, Oxford University Press, Oxford.
Blumberg, B. S. 2003. The NASA Astrobiology Institute: Early history and organization, *Astrobiology*, 3, 463–470.
Bogonovich, M. 2011. Intelligence's likelihood and evolutionary time frame, *International Journal of Astrobiology*, 10: 113–122.
Boyd, R. and Richerson, P. J. 1985. *Culture and the Evolutionary Process*, University of Chicago Press, Chicago.
Brin, D. 2011. A contrarian perspective on altruism, in Shuch (2011), pp. 429–450.
Brown, J. R. 1999. *Philosophy of Mathematics*, Routledge, London and New York.
Brown, M. 2010. *How I Killed Pluto and Why it Had it Coming*, Spiegel and Grau, New York.
Burtt, E. A. 1924. *Metaphysical Foundations of Modern Physical Science*, rev. ed., 1932; first US edition, 1954.
Carr, B., editor. 2007. *Universe or Multiverse?* Cambridge University Press, Cambridge.
Carter, B. 1974. Large number coincidences and the anthropic principle in cosmology, in M. S. Longair ed. *Confrontation of Cosmological Theories with Observational Data*, Dordrecht, 291–298.
Chaisson, E. 2001. *Cosmic Evolution: The Rise of Complexity in Nature*. Harvard University Press, Cambridge, Mass.
Chaisson, E. 2009. Cosmic Evolution: State of the Science, in Dick and Lupisella, eds. *Cosmos and Culture*.
Chaisson, E. 2010. Energy rate density as a complexity metric and evolutionary driver. *Complexity*, 16, issue 3, pp. 27–40
Chaisson, E. 2011. Energy rate density. II. Probing further a new complexity metric, *Complexity*, 17, issue 1, pp. 44–63.
Ćirković, M. 2006. Macroengineering in the Galactic Context: A New Agenda for Astrobiology, in *Macro-Engineering: A Challenge for the Future*, edited by V. Badescu, R. B. Cathcart and R. D. Schuiling, Springer Verlag, New York, pp. 281–300
Ćirković, M. 2012. *The Astrobiological Landscape: Philosophical Foundations of the Stud of Cosmic Life,* Cambridge University Press, Cambridge.
Claggett, M., ed. 1957. *Critical Problems in the History of Science*, reprinted by University of Wisconsin Press, Madison, 2005.
Clark, A. and Chalmers, D. 1998. The extended mind. *Analysis* 58: 7–19.
Cleland, C. E. 2001. Historical science, experimental science, and the scientific method, *Geology* 29, 978–990.
Cleland, C. E. 2002. Methodological and epistemic differences between historical science and experimental science, *Philosophy of Science*, 69: 474–496.
Cleland, C. E. and Chyba, C. 2007. Does 'Life' Have a Definition? in Sullivan and Baross (2007), 119–131.
Coleman, A. and Welty, I. 2010. Ordinary language philosophy: A reappraisal, *Essays in Philosophy,* online at http://commons.pacificu.edu/eip/vol11/iss2/

Conway Morris, S. 1998. *The Crucible of Creation*, Oxford University Press, Oxford.
Conway Morris, S. 2003. *Life's Solution: Inevitable Humans in a Lonely Universe*, Cambridge University Press, Cambridge.
Crick, F. 1995. *The Astonishing Hypothesis: The Scientific Search of the Soul*, Scribner, New York.
Crowe, M. J. 1986. *The Extraterrestrial Life Debate, 1750-1900: The idea of a Plurality of Worlds from Kant to Lowell*, Cambridge University Press, Cambridge.
Crowe, M. J. 1997. A history of the extraterrestrial life debate, *Zygon* 32: 147–62.
Crowe, M. J, ed. 2008. *The Extraterrestrial Life Debate, Antiquity to 1915:* A Source Book, University of Notre Dame, Notre Dame, IN.
Curd, M. and Cover, J. A. 1998. *Philosophy of Science: The Central Issues*, W. W. Norton and Co., New York.
Dark, T., III 2007. Reclaiming the future: Space advocacy and the idea of progress, in Dick and Launius (2007), pp. 555–572.
Daston, L. 1991. History of science in an elegaic mode: E. A. Burtt's *Metaphysical Foundations of Modern Physical Science* revisited, *Isis*, 82: 522–531.
Daston, L. and Galison, P. 2010. *Objectivity*, Zone Books, New York.
Davies, P. C. W. 1995. *Are We Alone? Philosophical Implications of the Discovery of Extraterrestrial Life,* Basic Books, New York.
Davies, P. C. W. 1998. *The Fifth Miracle: The Search for the Origin of Life*. London: Penguin Press.
Davies, P. C. W. 2007. *Cosmic Jackpot: Why Our Universe Is Just Right for Life*, Houghton Mifflin, Boston.
Davies, P. C. W. 2011. Searching for a shadow biosphere on Earth as a test of the cosmic imperative, in Dominik and Zarnecki (2011), pp. 624–632
Dawkins, R. 1997. *Climbing Mount Improbable,* W. W. Norton, New York.
De Duve, C. 1995. *Vital Dust: Life as a Cosmic Imperative*. Basic Books, New York.
Deamer, D. W. 2010. Special Collection of Essays: What is Life? *Astrobiology* 10: 1001–1002.
Deamer, D. W. and Fleischaker, G. 1994. *Origins of Life: The Central Concepts*. Jones and Bartlett, Boston.
Dennett, D. 1991. *Consciousness Explained*, Little, Brown and Company, Boston.
Dennett, D. 1996. *Darwin's Dangerous Idea*. Simon and Schuster, New York.
Denning, K. 2009. *Social evolution: State of the field*, in Dick and Lupisella (2009), pp. 63–124.
Denning, K. 2011. Being technological, in Shuch (2011) pp. 477–496.
Des Marais, D., Nuth, J. A. III, Allamandola, L. J., Boss, A. P., Farmer, J. P., Hoehler, T. M., Jakosky, B. M., Meadows, V. S., Pohorille, A., Runnegar, B., and Spormann, A.M. 2008. The NASA Astrobiology Roadmap, *Astrobiology*, 8: 715–730.
Dick, S. J. 1982. *Plurality of Worlds: The Origins of the Extraterrestrial Life Debate from Democritus to Kant*. Cambridge University Press: Cambridge.
Dick, S. J. 1989. The concept of extraterrestrial intelligence – An emerging cosmology, *Planetary Report*, 9: 13–17
Dick, S. J. 1995. Consequences of success in SETI: lessons from the history of science, in Shostak (1995), pp. 521–532.
Dick, S. J. 1996. *The Biological Universe: The Twentieth Century Extraterrestrial Life Debate and the Limits of Science*. Cambridge University Press, Cambridge.
Dick, S. J. 1997. *The Biophysical cosmology: The place of bioastronomy in the history of science*, in Astronomical and Biochemical Origins and the Search for Life in the Universe, edited by C. B. Cosmovici, S. Bowyer and D. Werthimer, Editrice Compositori, Bologna, Italy, pp. 785–788.
Dick, S. J. 1998. *Life on Other Worlds*. Cambridge University Press, Cambridge.
Dick, S. J. 2000a. Extraterrestrials and objective knowledge, in Tough (2000), pp. 47–48.
Dick, S. J. 2000b. Cultural aspects of astrobiology: A preliminary reconnaissance at the turn of the millennium, in Lemarchand and Meech (2000), Astronomical Society of the Pacific, San Francisco, pp. 649–659.
Dick, S. J. 2000c. *Many Worlds: The New Universe, Extraterrestrial Life and the Theological Implications*. Templeton Press, Philadelphia

Dick, S. J. 2003. Cultural evolution, the postbiological universe and SETI, *International Journal of Astrobiology,* 2: 65–74; reprinted as Bringing culture to cosmos: The postbiological universe, in Dick and Lupisella (2009), pp. 463–488.

Dick, S. J. 2004. The new universe, destiny of life, and the cultural implications, in *Life in the Universe: From the Miller Experiment to the Search for Life on other Worlds*, edited by J. Seckbach, Julian Chela-Flores, Tobias Owen and Francois Raulin, Kluwer Academic Publishers, Dordrecht, pp. 319–326.

Dick, S. J. 2006. Anthropology and the search for extraterrestrial intelligence, *Anthropology Today* 22 (2), 3–7.

Dick, S. J. 2008. Cosmology and biology, *Proceedings of the 2008 Conference of the Society of Amateur Radio Astronomers*, National Radio Astronomy Observatory, Green Bank, West Virginia, Amateur Radio Relay League, pp. 1–16

Dick, S. J. 2009. Cosmic evolution: History, culture and human destiny, in Dick and Lupisella (2009), pp. 25–59.

Dick, S. J. and Launius, R. D., editors. 2006. *Critical Issues in the History of Spaceflight*, NASA, Washington, D.C., available online at http://history.nasa.gov/series95.html

Dick, S. J. and Lupisella, M. L., editors. 2009. *Cosmos & Culture: Cultural Evolution in a Cosmic Context*, NASA, Washington, DC, online at http://history.nasa.gov/SP-4802.pdf

Dick, S. J. and Strick, J. E. 2004. *The Living Universe: NASA and the Development of Astrobiology*, Rutgers University Press, New Brunswick, NJ.

DiGregorio, B. E., Levin, G. V and Stratt, P. A. 1997. *Mars: The Living Planet*, Frog Books, Berkeley, CA.

Dobzhansky, T. 1972. Darwinian evolution and the problem of extraterrestrial life, *Perspectives in Biology and Medicine*, 15: 157–175

Dominik, M. and Zarnecki, J. C. 2011. The detection of extra-terrestrial life and the consequences for science and society, *Philosophical Transactions of the Royal Society A*, 369, 1936, Table of Contents and abstracts online at http://rsta.royalsocietypublishing.org/content/369/1936.toc.

Donald, M. 1991. *Origins of the Modern Mind: Three Stages in the Evolution of Culture and Cognition*, Harvard University Press, Cambridge, MA.

Dunér, D. 2011. Astrocognition, in *Humans in Outer Space – Interdisciplinary Perspectives*, edited by U. Landfester, N. L. Remuss, K. Schrogl & J. C. Worms, Springer Verlag, Wien, pp. 117–140.

Dunér, David, E. Persson, and G. Holmberg, eds. 2012. *The History and Philosophy of Astrobiology*, special issue of *Astrobiology*, 12, pp. 901–1016.

Dusek, V. 2006. *Philosophy of Technology: An Introduction*, Malden, MA and Oxford.

Dyson, F. 1966. The search for extraterrestrial technology. In *Perspectives in Modern Physics*, edited by R. E. Marshak, Interscience Publishers, New York, pp. 641–655.

Dyson, F. 1979. *Disturbing the Universe*, Harper and Row, New York, p. 250.

Dyson, F. 1988. *Infinite in All Directions*, Harper and Row, New York.

Eccles, J. C. 1989. *Evolution of the Brain, Creation of the Self*, Routledge, London and New York.

Edelman, G. M. 2004. *Wider than the Sky: The Phenomenal Gift of Consciousness*, Yale University Press, New Haven.

Eiseley, L. 1957. *The Immense Journey*, Random House, New York, pp. 162–163.

Ellis, G. F. R. 2006. Issues in the philosophy of cosmology. In *Philosophy of Physics*, edited by J. Butterfield and J. Earman, pp. 1183–1285. North Holland, online at http://arxiv.org/pdf/astro-ph/0602280v2

Ellis, G. F. R. 2007. Multiverses: Description, uniqueness and testing, in Carr (2007), pp. 387–409.

Falk, D. 1995. Brain evolution in dolphins, humans and other mammals: Implications for ETI, in Shostak (1995), 41–52.

Falk, D. 2004. *Braindance: New Discoveries about Human Origins and Brain Evolution*, University Press of Florida, Gainesville.

Farley, J. 1977. *The Spontaneous Generation Controversy from Descartes to Oparin*, Baltimore and London.

Fontenelle, B. 1686. *Conversations on the Plurality of Worlds*, trans. H. A. Hargreaves, University of California Press, Berkeley, CA

Fox, D. 2011. The limits of intelligence, *Scientific American* (July), 36–43.

Fry, I. 1995. Are the different hypotheses on the origin of life as different as they seem? *Biology and Philosophy*, 10 (4), 389–417, reprinted in Bedeau and Cleland, eds. (2010), pp. 137ff.

Fry, I. 2000. *The Emergence of Life on Earth: A Historical and Scientific Overview.* Rutgers University Press, New Brunswick, NJ.

Fry, I. 2010. Philosophical aspects of the origin-of-life problem: the emergence of life and the nature of science, in Bertka (2010), pp. 61–79.

Gardner, J. 2003. *Biocosm – The New Scientific Theory of Evolution: Intelligent Life is the Architect of the Universe*, Inner Ocean Publishing, Makawao, Maui, Hawaii.

Gardner, J. 2007. The *Intelligent Universe: AI, ET and the Emerging Mind of the Cosmos*, New Page, Franklin Lakes, NJ

Gardner, J. 2009. The intelligent universe, in Dick and Lupisella (2009), pp. 361–382: 375–376.

Geertz, C. 1973. *The Interpretation of Cultures*, Basic Books, New York

Gilbert, W. 1986. The RNA world, *Nature*, 319, 618.

Gonzalez, G. and Richards, J. 2004. *The Privileged Planet: How our Place in the Cosmos is Designed for Discovery*, Regnery Publishing, Washington, D.C.

Good, G. A. 2000. The assembly of geophysics: scientific disciplines as frameworks of consensus, *Studies in the History and Philosophy of Modern Physics*, 31, 259–292.

Gould, S. J. 1989. *Wonderful Life: The Burgess Shale and the nature of history*, W. W. Norton, New York.

Graham, L. R. 1989. *Science, Philosophy and Human Behavior in the Soviet Union*, Columbia University Press.

Griffin, D. R. 2001. *Animal Minds: Beyond Cognition to Consciousness*, University of Chicago Press, Chicago.

Grinspoon, D. 2003. *Lonely Planets: The Natural Philosophy of Alien Life*, Harper Collins, New York.

Gross, A. G. 2006. *Starring The Text: The Place of Rhetoric in Science Studies*, Southern Illinois UP, Carbondale.

Guthke, K. S. 1983. *Der Mythos der Neuzeit. Das Thema der Mehrheit der Welten in der Literatur- und Geistesgechichte von der kopernikanischen wende bis zur Science Fiction*, Francke, Bern; English translation *The Last Frontier: Imagining other Worlds from the Copernican Revolution to Modern Science Fiction*, Cornell University Press, Ithaca, N.Y. 1990.

Hampden-Turner, C. 1981. *Maps of the Mind: Charts and Concepts of the Mind and its labyrinths*. Macmillan Publishing Company, New York.

Harré, R. 1971. *The Philosophies of Science*, Oxford University Press, Oxford.

Harrison, A. and Dick, S. 2000. Contact: Long-term implications for humanity, in Tough (2000), pp. 7–31.

Harrison, A. A., Billingham, J., Dick, S., Finney, B., Michaud, M., Tarter, D. E., Tough, A., and Vakoch, D. A. 2000. The role of social science in SETI, in Tough (2000), pp. 71–85.

Harrison, A. A. and Connell, K. 2001. *Workshop on the Societal Implications of Astrobiology*, held at NASA Ames Research Center, November 16–17, 1999. Moffett Field: NASA Ames Research Center. Online at http://www.astrosociology.org/Library/PDF/NASA-Workshop-Report-Societal-Implications-of-Astrobiology.pdf

Harrison, E. R. 1995. The natural selection of universes containing intelligent life, *QJRAS*, 36, 193–203.

Hart, M. 1975. An Explanation for the absence of extraterrestrials on Earth, QJRAS, 16, 128–135, reprinted in Zuckermann and Hart (1996).

Hawkins, J. and Blakeslee, S. 2004. *On Intelligence*, Henry Holt and Co., New York.

Hoyle, F. 1983. *The Intelligent Universe: A New View of Creation and Evolution*, Holt, Rinehart and Winston, New York.

Hull, D. L. 1974. *Philosophy of Biological Science*, Prentice–Hall, Englewood Cliffs, N.J.

Hull, D. L., and Ruse, M. 2007. *The Cambridge Companion to the Philosophy of Biology*, Cambridge University Press, Cambridge.

Jakosky, B. M. 2000. Philosophical aspects of astrobiology, in Lemarchand and Meech (2000).

Jerison, H. J. 2000. The evolution of intelligence, in Sternberg (2000), pp. 216–244.

Joyce, G. G. 1994. Foreword, in Deamer and Fleischaker (1994).

Kaltenegger, L., Selsis, F., Fridlund, M., et al. 2010. Deciphering spectral fingerprints of habitable exoplanets. *Astrobiology* 10(1): 89–102.

Kamminga, Harmke. 1980. *Studies in the History of Ideas on the Origin of Life from 1860*, PhD thesis, University of London.

Kant, I. 1981. *Universal Natural History and Theory of the Heavens*, translated by Stanley L. Jaki, Scottish Academic Press

Kohler, R. E. 1982. *From Medical Chemistry to Biochemistry*, Cambridge University Press, Cambridge.

Kosso, P. 2006. Detecting Extrasolar Planets, *Studies in the History and Philosophy of Science*, *37*, 224–236.

Kroeber, A. L and Kluckhohn, C. K. M. 1952. *Culture: A Critical Review of Concepts and Definitions*, The Peabody Museum, Cambridge, MA, p. 656.

Kuhn, T. 1962. *The Structure of Scientific Revolutions*, University of Chicago Press, Chicago.

Kuhn, T. 1977. *The Essential Tension*, University of Chicago Press, Chicago.

Kuper, A. 1999. *Culture: The Anthropologists' Account*, Harvard University Press, Cambridge, MA

Kurzweil, R. 1999. *The Age of Spiritual Machines: When Computers Exceed Human Intelligence*, Viking, New York

Kurzweil, R. 2005. *The Singularity is Near: When Humans Transcend Biology*, Viking, New York.

Lafleur, L. J. 1941. Astrobiology, *Astronomical Society of the Pacific Leaflets, vol. 3, No. 143* (January), 333–340

Lalande, K. N. and Brown, G.R. 2002. *Sense and Nonsense: Evolutionary Perspectives on Human Behaviour*, Oxford University Press, Oxford.

Lange, M. 2006. *An Introduction to the Philosophy of Physics*, Blackwell, Oxford.

Lanza, R. and Berman, B. 2009. *Biocentrism: How Life and Consciousness are the Keys to Understanding the True Nature of the Universe*, BenBella Books, Dallas.

Latham, D., Stefanik, R. P., Mazeh, T, Mayor, M. and Burki, G. 1989. The unseen companion of HD114762: A probable brown dwarf, *Nature*, 339, 38–40.

Laudan, L. 1977. *Progress and Its Problems: Towards a Theory of Scientific Growth*, University of California Press, Berkeley.

Lederberg, J. 1960. Exobiology: Approaches to life beyond the Earth, *Science*, 132, 393–400; reprinted in L. V. Berkner and H. Odishaw, *Science in Space*, New York, 1961, 407–425.

Lemarchand, G. A. and Meech, K. J. editors. 2000. *Bioastronomy '99: A New Era in Bioastronomy*, Astronomical Society of the Pacific, San Francisco, CA.

Leslie, J. editor. 1998. *Modern Cosmology & Philosophy*, Prometheus Books, New York.

Levin, G. 2011. private communication.

Levin, G. and Straat, P. 1976. Viking labeled release biology experiment: Interim results. *Science*, 194, December, 1323.

Lovejoy, A. O. 1936. *The Great Chain of Being*, Harvard University Press, Cambridge, MA

Lupisella, M. 2009. Cosmocultural Evolution: The Coevolution of Culture and Cosmos and the Creation of Cosmic Value, in Dick and (Lupisella 2009), pp. 321–356.

Marcy, G. and Butler, P. 1996. A Planetary Companion to 70 Virginis, ApJ, 464: 147.

Margulis, L. and Chapman, *M. J.* 2010. *Kingdoms & Domains: An Illustrated Guide to the Phyla of Life on Earth*, Elsevier, Amsterdam.

Marino, L. 1995. SETI begins at home: Searching for terrestrial intelligence, in Shostak (1995), pp. 73–82.

Marino, L. 2002. Convergence of complex cognitive abilities in cetaceans and primates, *Brain, Behavior and Evolution* 59, 21–32.

Mayor, M. and Queloz, D. 1995. A Jupiter-Mass Companion to a Solar-Type Star. *Nature*, 378, 355.

Mayr, E. 1985. The probability of extraterrestrial intelligent life, in Regis (1985), pp. 24–30; reprinted with updated notes in Mayr (1988), pp. 67–74.
Mayr, E. 1988. *Toward a New Philosophy of Biology: Observations of an Evolutionist*, Harvard University Press, Cambridge, MA.
McKay, D. S., Gibson, E. K., Jr., Thomas-Keprta, K. L, Vali, H., Romanek, C., Clemett, S., Chillier, X. D. Maechling, C. R., and Zare, R. N. 1996. Search for past life on Mars: Possible relic biogenic activity in Martian meteorite ALH84001, *Science*, 273: 924–930.
Meltzer, M. 2011. *When Biospheres Collide: A History of NASA's Planetary Protection Programs*, NASA, Washington, D.C., online at http://history.nasa.gov/SP-4234.pdf
Michaud, M. A. G. 2007. *Contact with Alien Civilizations: Our Hopes and Fears about Encountering Extraterrestrials*, Copernicus, New York.
Minsky, M. 1985. Why intelligent aliens will be intelligible, In Regis (1985), 117–128.
Mithen, S. 1996. *The Prehistory of the Mind: The Cognitive Origins of Art, Religion and Science*, Thames and Hudson, London.
Monod, J. 1971. *Chance and Necessity*, Knopf, New York, pp. 144–146.
Moravec, H. 1988. *Mind Children: The Future of Robot and human Intelligence*, Harvard University Press, Cambridge, MA.
Moravec, H. 1999. *Robot: Mere Machine to Transcendent Mind*, Oxford.
Morrison, D. 2001. The NASA astrobiology program, *Astrobiology*, 1, 3–13.
Morrison, P. M., Billingham, J. and Wolfe, J., eds. 1977. *The Search for Extraterrestrial Intelligence*, NASA SP 419: Washington, D.C.
National Academy of Sciences, Space Science Board. 1962. *A Review of Space Research*, 9–2, 9–3.
National Research Council. 2003. *Life in the Universe: An Assessment of the U. S. and International Programs in Astrobiology*, National Academies Press, Washington, D.C.
Navarro-González, R., Vargas, E., de la Rosa, J., Raga, A.C., and McKay, C.P. 2010. Reanalysis of the Viking results suggests perchlorate and organics at mid–latitudes on Mars, *Journal of Geophysical Research*, 115: E12010.
New York Times. 1976. Should mankind hide? November 22, p. A24.
Nisbet, R. 1980. *History of the Idea of Progress*, Basic Books, New York.
Ohmoto, H., Runnegar, B., Kump, L., Fogel, M., Kamber, B., Anbar, A., Knauth, P., Lowe, D., Sumner, D., and Watanabe, Y. 2008. Biosignatures in Ancient Rocks: A Summary of Discussions at a Field Workshop on Biosignatures in Ancient Rocks, *Astrobiology*, 8, 883–907.
Page, D. N. 2007. Predictions and tests of multiverse theories, in Carr (2007), pp. 411–429.
Pagels, H. 1985. *Perfect Symmetry: the Search for the Beginning of Time*, Simon and Schuster, New York.
Parker, S. T and McKinney, M. L. 1999. *Origins of Intelligence: The Evolution of Cognitive Development in Monkeys, Apes, and Humans*, Johns Hopkins University Press, Baltimore.
Perryman, M. 2011. *The Exoplanet Handbook*, Cambridge University Press, Cambridge.
Pilcher, C. B. 2003 Biosignatures of early earths, *Astrobiology*, 3, 471–476.
Pinker, S. 2002. *The Blank Slate: The Modern Denial of Human Nature*, Viking Penguin, New York.
Pirie, N. 1939. The Meaninglessness of the Terms 'Life' and 'Living. In *Perspectives in Biochemistry*, edited by J. Needham and D. R. Green, Cambridge University Press, Cambridge, pp. 11–22.
Planck, R. 1968. *The Emotional Significance of Imaginary Beings: A Study of the Interaction Between Psychopathology, Literature, and Reality in the Modern World*, Charles C. Thomas, Springfield, Ill.
Popper, K. 1974. Darwinism as a metaphysical research program. In *The Philosophy of Karl Popper*, edited by P. A. Schilpp, vol. 1, Open Court: LaSalle, Ill., 133–143.
Popper, K. R. 1979. *Objective Knowledge: An Evolutionary Approach*, Clarendon Press, Oxford.
Race, M. S. 2007. Societal and Ethical Concerns, in Sullivan and Baross (2007), pp. 483–497.
Race, M. S., Denning, K., Bertka, C., Dick, S., Harrison, A., Impey, C. Mancinelli, R. and Workshop Participants. 2012. Astrobiology and society: Building an interdisciplinary research community, *Astrobiology*, *Astrobiology*, 12, number 10, 958–965.

Randolph, R., Race, M., and McKay, C. 1997. Reconsidering the theological and ethical implications of extraterrestrial life, *Center for Theology and Natural Sciences Bulletin,* 17, no. 3, 1–8.

Regis, E. Jr. editor.1985. *Extraterrestrials: Science and Alien Intelligence*, Cambridge University Press, Cambridge.

Reiss D. 1988. Can we communicate with other species on this planet? In Bioastronomy – The next steps, edited by G. Mark, Kluwer, Dordrecht, pp. 253–264: 253

Reiss, D. 2011. *The Dolphin in the Mirror: Exploring Dolphin Minds and Saving Dolphin Lives*, Houghton, Mifflin, Harcourt, New York

Rescher, N. 1985. Extraterrestrial science, in Regis 1985), 83–116.

Richerson, P. J. and Boyd, R. 2006 *Not by Genes Alone: How Culture Transformed Human Evolution*, University of Chicago Press, Chicago

Rosenberg, A. 2000. *Philosophy of Science: A Contemporary Introduction*, Routledge, London and New York.

Ruse, M. 1985. Is Rape Wrong on Andromeda? An Introduction to Extraterrestrial Evolution, *Science* and Morality. In Regis (1985), 43–78.

Ruse, M. 1989. *The Philosophy of Biology Today,* State University of New York Press, Albany, N.Y.

Ruse, M. 2006. *Darwinism and its Discontents*, Cambridge University Press, Cambridge, pp. 52–71.

Russell, D. 1983. Exponential evolution: implications for intelligent extraterrestrial life, *Advances in Space Research*, 3, 9, 95–103

Russell, D. 1995. Biodiversity and time scales for the evolution of extraterrestrial intelligence, in Shostak (1995), pp. 143–151.

Sagan, C. 1977. *The Dragons of Eden: Speculations on the Evolution of Human Intelligence*, House, New York.

Sagan, C. 1980. *Cosmos*, Random House, New York.

Sagan, C. 1987. The burden of skepticism, Pasadena Lecture.

Sapp, J. 2009. *The New Foundations of Evolution on the Tree of Life*, Oxford University Press, Oxford.

Sawyer, K. 2006. *The Rock from Mars: A Dectective Story on Two Planets*, Random House, New York.

Schneider, J. 2010. The extraterrestrial life debate in different cultures, in *Towards Habitable Planets,* Vincent Coudé du Foresto, Dawn M. Gelino, and Ignasi Ribas, eds., Astronomical Society of the Pacific, San Francisco, pp. 324–330; preprint online at http://arxiv.org/pdf/1003.0277v1

Schrodinger, E. 1944. *What is Life? The Physical Aspect of the Living Cell*, Cambridge University Press, Cambridge.

Searle, J. R. 2004. *Mind: A Brief Introduction*, Oxford University Press, Oxford.

Shapiro, R. 1987. *Origins: A Skeptic's Guide to the Creation of Life on Earth*, Summit Books, Toronto and New York.

Shea, W. 1986. Filled with wonder: Kant's cosmological essay, the Universal Natural History and Theory of the Heavens. In *Kant's Philosophy of Physical Science*, edited by Robert Butts, Kluwer, Boston.

Sheridan, M. 2009. SETI: A Critical History, Ph.D dissertation, Drew University

Shermer, M. 2002a. *In Darwin's Shadow: The Life and Science of Alfred Russel Wallace*, Oxford University Press, Oxford, pp, 158 ff.

Shermer, M. 2002b. Shermer's Last Law: Any sufficiently advanced extraterrestrial intelligence is indistinguishable from God, *Scientific American*, January.

Shermer, M. 2006. Deities for Atheists, *Science*, 311: 1244; online at http://www.skeptic.com/eskeptic/06-03-09/

Shklovskii, I. S. 1965. Multiplicity of inhabited worlds and the problem of interstellar communications, in *Extraterrestrial Civilizations*, Erevan; English translation, Jerusalem, 1967.

Shostak, G. S. 1995. *Progress in the Search for Extraterrestrial Life*, Astronomical Society of the Pacific, San Francisco.

Shuch, P. 2011. *Searching for Extraterrestrial Intelligence: SETI Past, Present and Future*, Springer, New York.

Simpson, G. G. 1964. The Non-Prevalence of Humanoids, *Science*, 143, February 21, 769–775, reprinted in *This View of Life: The World of an Evolutionist*, Harcourt, Brace & World, New York, pp. 253–271.

Skoyles, J. and Sagan, D. 2002. *Up from Dragons: The Evolution of Human Intelligence*, McGraw-Hill, New York.

Smith, C. H and Becalloni, G. 2008. *Natural Selection & Beyond: The Intellectual Legacy of Alfred Russel Wallace*, Oxford University Press, Oxford.

Smolin, L. 1997. *The Life of the Cosmos*, Oxford University Press, Oxford.

Smolin, L. 2007. Scientific alternatives to the anthropic principle, in Carr (2007), pp. 323–366.

Spiegelman, S. 1967. An *in vitro* analysis of a replicating molecule, *American Scientist*, 55, 221–264.

Stenger, V. J. 2011. *The Fallacy of Fine-Tuning: Why the Universe is Not Fine-Tuned for Us*, Prometheus Books, New York

Sternberg, R. J. editor. 2000. *Handbook of Intelligence*, Cambridge University Press, Cambridge.

Sternberg, R. J. 2002. The search for criteria: why study the evolution of intelligence. In *The Evolution of Intelligence*, edited by R. J. Sternberg and J. C. Kaufman, Lawrence Erlbaum Associates, Mahwah, N. J, pp. 1–8

Stoeger, W. R. 2007. Are anthropic arguments, involving multiverses and beyond, legitimate? In Carr (2007), pp. 445–458.

Strauss, D. 2001. *Percival Lowell: the Culture and Science of a Boston Brahmin.*, Harvard University Press, Cambridge, MA.

Strick, J. E. 2002. *Sparks of Life: Darwinism and the Victorian Debates over Spontaneous Generation*, Harvard University Press, Cambridge, MA.

Strick, J. E. 2004. Creating a Cosmic Discipline: The Crystallization and Consolidation of Exobiology, 1957–1973,' *Journal of the History of Biology*, 37: 131–80.

Strick, J. E. 2010. From Aristotle to Darwin, to Freeman Dyson: changing definitions of life viewed in a historical context, in Bertka (2010), pp. 47–60

Strughold, H. 1959. Advances in astrobiology, *Proceedings of the Lunar and Planetary Exploration Colloquium*, 1, no. 6 (April), 1–7.

Struve, O. 1955. Life on other worlds, *Sky and Telescope*, 14: 137–146.

Sullivan, W. T. III, and Baross, J. A., editors. 2007. *Planets and Life: The Emerging Science of Astrobiology*, Cambridge University Press, Cambridge.

Sullivan, W. 2009. *Cosmic Noise: A History of Early Radio Astronomy*, Cambridge University Press, Cambridge.

Tarter, J. C. 2007. Searching for extraterrestrial intelligence, in Sullivan and Baross (2007), pp. 513–536: 513.

Tirard, S., Morange, M., and Lazcano, A. 2010. The definition of life: A brief history of an elusive scientific endeavor, *Astrobiology* 10: 1003–1009.

Tough, A., ed. 2000. *When SETI Succeeds: The Impact of High-Information Contact*. Foundation for the Future, Bellevue, Washington

Tsokolov, S. A. 2009. Why is the definition of life so elusive? Epistemological considerations, *Astrobiology*, 9:401–412.

Tsokolov, S. A. 2010. A Theory of Circular Organization and Negative Feedback: Defining Life in a Cybernetic Context, *Astrobiology*, 10: 1031–1042.

Turing, A. M. 1950. Computing machinery and intelligence, *Mind*, 59, 433–460.

Tyson, Neil DeGrasse 2009. *The Pluto Files: The Rise and Fall of America's Favorite Planet*, W. W. Norton, New York.

Vakoch and Harrison, eds. 2011. *Civilizations Beyond Earth: Extraterrestrial Life and Society*, Berghahn Books, New York and Oxford.

Vakoch. D. 1998. Constructing Messages to Extraterrestrials: An Exosemiotic Perspective. *Acta Astronautica*, 42: 697–704.

Vakoch, D. 1999. The view from a distant star: Challenges of interstellar message-making, *Mercury*, 28, 26–32.
Vakoch, D. 2000. Roman Catholic views of extraterrestrial intelligence: Anticipating the future by examining the past, in Tough (2000), 165–174.
Vakoch, D. 2009. Encoding our origins: Communicating the evolutionary epic in interstellar messages, in Dick and Lupisella (2009), pp. 415–439.
Van de Kamp, P. 1963. Astrometric study of Barnard's Star from plates taken with the 24-Inch Sproul refractor, *Astronomical Journal*, 68, 515–521.
Vidal, C. 2007. An enduring philosophical agenda: Worldview construction as a philosophical method, online at http://cogprints.org/6048
Vidal, C. 2010. "Computational and biological analogies for understanding fine-tuned parameters in physics," in *Foundations of Science* 15 (4): 375–393, online at http://arxiv.org/abs/1002.3905 .
Vidal C. 2011. Metaphilosophical criteria for worldview comparison, *Metaphilosophy*.
Ward, P. 2005. *Life as We Do Not Know It*. Viking Penguin, New York.
Ward, P. and Benner, S. 2007. Alien Biochemistries, in Sullivan and Baross (2007), pp. 537–544.
Ward, P. and Brownlee, D. 2000. *Rare Earth: Why Complex Life is Uncommon in the Universe*, Springer-Verlag, New York.
Webb S. 2002. *Where is Everybody? Fifty Solutions to the Fermi Paradox and the Problem of Extraterrestrial Life*, Praxis, New York.
Wells, L., Armstrong, J. and Huber, J. A. 2007. Disciplinary aspirations and educational opportunities, in Sullivan and Baross 2007., pp. 547–557.
White, F. 1990. *The SETI Factor*, Walker and Co., New York.
Wills, C. and Bada, J. 2000. *The Spark of Life: Darwin and the Primeval Soup*, Perseus Publishing, Cambridge, MA.
Wilson, A. G. 1958. Problems common to the fields of astronomy and biology, *Publications of the Astronomical Society of the Pacific*, 70, 41–78.
Wilson, E. O. 1975. *Sociobiology: The New Synthesis*, Harvard University Press, Cambridge, MA.
Wilson, E. O. 1998. *Consilience: The Unity of Knowledge*, Knopf, New York.
Wolfe, A. 2002. Germs in space: Joshua Lederberg, exobiology, and the public imagination, 1958-1964, *Isis*, 93, 183–205.
Wolfe-Simon, F., Jodi Switzer Blum, Thomas R. Kulp, Gwyneth W. Gordon, Shelley E. Hoeft, Jennifer Pett-Ridge, John F. Stolz, Samuel M. Webb, Peter K. Weber, P. C. W. Davies, Ariel D. Anbar and Ronald S. Oremland. 2010. A bacterium that can grow by using arsenic instead of phosphorus, 332 (6034): 1163–6.
Wolfe-Simon, F.; Blum; Kulp, Thomas R.; Gordon, Gwyneth W.; Hoeft, Shelley E.; Pett-Ridge, Jennifer; Stolz, John F.; Webb, Samuel M. et al. 2011. Response to comments on A bacterium that can grow using arsenic instead of phosphorus, 27 May, 2011, online at http://www.sciencemag.org/content/early/2011/05/26/science.1202098.full.pdf
Zaitsev, A. 2011. METI: Messaging to extraterrestrial intelligence, in Shuch (2011), p. 399–428.
Zakariya, N. 2010. *Towards a Final Story: Time, Myth and the Origins of the Universe*, PhD dissertation, Harvard University.
Zuckerman, B. and Hart, M., eds. 1996. *Extraterrestrials: Where are They?* Cambridge University Press, Cambridge.

Chapter 39
Lessons Learned from the Twentieth-Century Extraterrestrial Life Debate

Abstract Now that historians have completed surveys of the extraterrestrial life debate such as Dick's *Plurality of* Worlds and *The Biological Universe*, and Crowe's, *The extraterrestrial life debate, 1750–1900*, we can begin to study the possible lessons learned from that history. In this chapter we make that attempt in three overlapping areas: (1) the problematic nature of evidence and inference, and its relation to scientific preconceptions; (2) the role of theory in raising expectations, interpreting observations, and generating conclusions; and (3) an evaluation of the success or failure of some of the debate's most general arguments, including the principles of plenitude and mediocrity and "Goldilocks-type" arguments that life occurs under such tight constraints that it is rare in the universe.

39.1 Introduction

Whether or not there are lessons to be learned from history is a subject of some contention among historians. It is, of course, a dangerous game, with some (politicians in particular) reading into history whatever lessons they want to learn based on their own ideology. My attitude is more optimistic: lessons may be ambiguous, but they are there and can be debated and deployed. After all, not without reason does there exist a National Archives in the United States with the words "What is Past is Prologue" scrolled along the top of its impressive facade, a building whose function is duplicated in all civilized countries of the world. Not without reason did the Columbia Accident Investigation Board devote an entire chapter to history in its official report on the Space Shuttle's demise, and conclude that "history is not just a backdrop or a scene-setter, history is cause" (Columbia Accident Investigation Board 2003, p. 195). And not without reason does every high school, college and university teach history, ever hopeful that at the very least it will provide context, if not lessons, for students as they enter a complex world.

My optimism in this regard holds despite the fact that many thinkers—from Samuel Taylor Coleridge to Georg Wilhelm Friedrich Hegel, from Aldous Huxley

First published as part of "The Twentieth Century Extraterrestrial Life Debate: Major Themes and Lessons Learned," in *Astrobiology, History and Society: Life Beyond Earth and the Impact of Discovery* Douglas A. Vakoch, ed. (Springer: Heidelberg, 2013), pp. 133–175.

S. J. Dick, *Space, Time, and Aliens*, https://doi.org/10.1007/978-3-030-41614-0_39

to scholars today—have concluded that the main lesson of history is that the lessons of history are either misused or never learned. Thus Coleridge: "If men could learn from history what lessons it might teach us! But passion and party blind our eyes, and the light which experience gives is a lantern on the stern, which shines only on the waves behind us!" (Coleridge 1831). Hegel: "What experience and history teach is this – that people and governments never have learned anything from history, or acted on principles deduced from it" (Hegel 1832). Aldous Huxley: "That men do not learn very much from the lessons of history is the most important of all the lessons of history" (Huxley 1959). Or, as a recent author put while contemplating Herodotus's ancient message about intercultural understanding: "it goes unheeded, as it always has and it always will, because history teaches us that we do not learn from history, that we fight the same wars against the same enemies for the same reasons in different eras, as though time really stood still and history itself as moving narrative was nothing but artful illusion" (Marozzi 2008). With such cautions in mind, we nevertheless proceed to examine possible lessons to be learned from the history of the extraterrestrial life debate, in the (perhaps misguided) hope that scientists are more receptive to lessons learned than politicians.

39.2 Evidence, Inference and Preconceptions

Evidence, inference, and interpretation are problems in all areas of science, not to mention broader areas such as law, where 5-4 votes are not uncommon on the Supreme Court, based on interpretation of the best available evidence. Ask any three people to describe in detail *any* event they have just witnessed, much less something unexpected and emotional like a UFO event, and the likely outcome is three different answers. Scientists are trained in gathering evidence and making conclusions from that evidence, so one would hope their record would be better. Sometimes it is, but often not, especially when pushed to the limits of science, as is certainly the case in astrobiology. The episode surrounding the canals of Mars centered around the beginning of the twentieth century is the most infamous example (Crowe 1986; Dick 1996). But we need not set our gaze back that far. Here we examine four episodes in the second half of the twentieth century that we have already mentioned: William Sinton's claim of spectroscopic evidence for supposed Martian vegetation; Peter van de Kamp's claim of planets around Barnard's star; the ambiguities of the Viking experimental results; and the controversy surrounding the Mars rock ALH84001. While many more exemplars could be used, these will suffice to illustrate the problems of evidence, inference, and preconceptions.

39.2.1 *Vegetation on Mars?*

As the favorable close approaches of Mars neared in 1954 and 1956, interest in the red planet was increasing, driven by the Martian vegetation hypothesis based on seasonal changes. As the 1956 opposition approached, Harvard astronomer William Sinton planned a direct search for vegetation by spectroscopic methods. Keenly aware that previous tests for infrared reflectivity characteristic of plants had been negative, Sinton's own search had a new element: it depended on the fact that organic molecules have absorption bands at about 3.4 μm in the infrared part of the spectrum—beyond Gerard P. Kuiper's work that had been done in the 1–2.5 μm region leading to the discovery of carbon dioxide on Mars (Kuiper 1949). Sinton still used a lead sulfide photoconductive cell, as had Kuiper, but now cooled to 96 K with liquid nitrogen to increase its sensitivity to 3.6 μm. The difficulties of the observations can be appreciated from the fact that the sensitive area of this cell was only .16 mm square, and the diameter of Mars was less than 1 mm. Nevertheless, after four nights of observations Sinton believed he had enough evidence for his conclusion that the probability was "very high that an organic spectrum is required to account for the data" (Sinton 1957).

Sinton was very much aware of previous visual evidence for vegetation in the form of seasonal changes in the size and shape of the Martian dark areas. In fact he saw the dip at 3.4 microns as "additional evidence for vegetation," and concluded that "this evidence, together with the strong evidence given by the seasonal changes, makes it seem extremely likely that plant life exists on Mars"(Sinton 1957). Thus, his claim of infrared absorption did not constitute direct visual confirmation, but depended on the interpretation of spectrograms—an interpretation undoubtedly affected by preconceived ideas. The result caused considerable excitement, especially when Sinton confirmed it with equipment ten times more sensitive on the 200-in. Palomar telescope during the 1958 opposition. Although the image of Mars was only 2 mm, this time Sinton separated the dark areas from the bright areas on Mars and confirmed his previous conclusion of absorption bands near 3.5 μm (Fig. 39.1). Again he concluded that "the observed spectrum fits very closely … that of organic compounds and particularly that of plants" (Sinton 1959). In addition a 3.67- μm absorption band was confirmed, which Sinton attributed to carbohydrate molecules in plants, analogous to tests on plants on Earth.

Although Sinton's results were widely hailed and cited in the literature, they were open to interpretation: not only were other biological interpretations possible, by 1963 researchers had done extensive work on infrared reflection spectra of terrestrial compounds and were critical of Sinton's interpretation (Rea et al. 1963). And by 1965 Sinton himself suggested that two of the Sinton bands were due to heavy deuterated water (HDO) in the Earth's atmosphere, with the remaining band still possibly organic (Rea et al. 1965). In the end, Sinton's refined methods had been mitigated by refined problems. Just as V. M. Slipher a half-century before believed he had found oxygen and water vapor on Mars only to find that his results were contaminated by the Earth's own atmosphere, Sinton's results too were contaminated, this time by heavy water, despite his attempt to separate analysis of Martian dark areas from its bright areas.

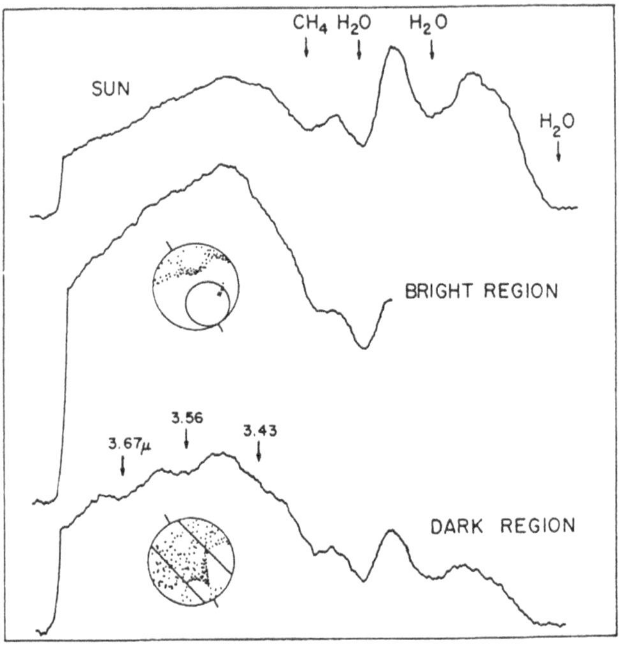

Fig. 39.1 Sinton's infrared spectroscopic evidence for vegetation on Mars, obtained on the Palomar 200-in. telescope. The top curve shows a solar spectrum, with superimposed absorptions by methane and water in the Earth's atmosphere. The middle curve shows a spectrum of a bright desert area of Mars, where no vegetation was expected. The bottom spectrum, obtained when the spectrograph slit was placed over one of the dark areas of Mars, shows three apparent absorption features (indicated by arrows) that were interpreted as due to vegetation. The evidence turned out to be spurious; the absorptions were actually due to deuterated water (HDO) in the Earth's atmosphere, as Sinton himself published 6 years later. (With permission, from Sinton (1959: 1234). Copyright 1959 AAAS)

As with the canals of Mars, the search for Martian vegetation demonstrates again differences in approach and world view among scientists, with one extreme much more likely to go out on a limb and to extrapolate than the other. Some astronomers probing the physical conditions on Mars presented their data and left it at that. Others used their data—indeed, were probably first inspired to gather their data—in the service of the question of extraterrestrial life. Still others rendered no opinion at all. In his book *Physics of the Planet Mars*, astronomer Gerard de Vaucouleurs only rarely mentioned the problem of life because "It is our belief that such a problem is still, to a large extent, beyond the limits of our positive knowledge and can only be the subject – either way – of vague speculations in which general 'principles' of a metaphysical nature have always to be taken as a guide" (de Vaucouleurs 1954).

To many, such a cautious attitude was not satisfying. They undoubtedly realized that the stakes in the debate extended far beyond Mars: as Kuiper wrote, "If life truly exists on the only two planets of the Solar System that are at all suitable to sustain it, it is tempting to conclude that, after enough time has elapsed, it will develop spontaneously wherever conditions permit. Since planetary systems are presumed to be very numerous, life would then be no exception in the universe"(Kuiper 1955).

At the dawn of the Space Age, then, the canal controversy had receded, and much was known about the physical conditions of the planet Mars. Vegetation of some sort was still a very real possibility, dependent to some extent on what one saw as the limits to the adaptability of life. Vegetation did not have the popular appeal of intelligence, but to the scientist it was still a holy grail that held the promise of revealing the secrets of life. That promise was to play no small role in making Mars and important target for interplanetary probes of the Space Age.

39.2.2 Organics on Mars? The Viking Experiments

The culmination of the twentieth century search for life in the Solar System was the landing of two Viking spacecraft on the surface of Mars in 1976, surely one of the great adventures in the history of science and technology (Ezell and Ezell 1984). The Viking project, initiated in 1968 after the demise of the Mars Voyager project and now managed by NASA's Langley Research Center, was an example of "big science" at its best in terms of budget, staff, goals, and results. The cost of the Viking spacecraft, including the orbiters, landers, and support (but not launch vehicles) was $930 million. Although the usual funding hurdles had to be overcome and many critics answered, in the end two Viking orbiters arrived at the planet on June 19 and August 7, 1976. After suitable reconnaissance, as the United States celebrated its Bicentennial back on planet Earth, two Viking landers set down on Mars in July and September. Under the guidance of project scientist Gerald A. Soffen, thirteen teams with a total of 78 scientists undertook 13 separate investigations, including three mapping experiments from the orbiter, one atmospheric experiment, one radio and radar experiment, and eight surface experiments. The total costs for development and execution of these experiments was another $227 million. The results increased knowledge of Mars far beyond all previous investigations combined, finally providing definitive answers to age-old questions, including the issues of temperature, atmospheric composition, and pressure so crucial to life.

From beginning to end, though the various science teams grappled with the myriad problems of meteorology, seismology, chemistry, imaging, and physical properties of the planet Mars, the Viking biology experiments were the driving force behind the project, as evidenced by both budget and public, Congressional and even scientific interest. $59 million was spent on the Viking biology package and another $41 million on the molecular analysis experiment that was relevant to the question of life because of its ability to detect organic molecules. Harold P. Klein of NASA's Ames Research Center headed the Viking biology science team; Klaus Biemann of MIT headed the separate molecular analysis team. While the results of several of the teams were relevant to the question of Martian biology, these 2 out of the 13 teams were most directly relevant.

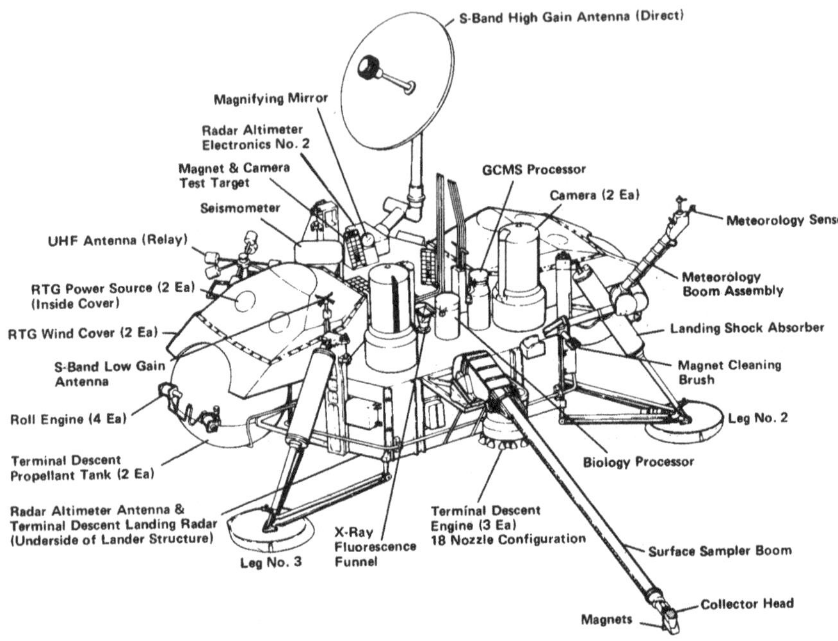

Fig. 39.2 The Viking lander, a complex machine incorporating eight experiments, landed on the surface of Mars in July, 1976, followed by another lander in September. The biology processor (labeled to the lower right) was within a small canister of volume 0.03 m^3. Nearby is the Gas Chromatograph Mass Spectrometer (GCMS), which detected no organic molecules at the two landing sites, down to parts per billion. Its conclusions have recently been called into question, and the biology experiment results are being reevaluated in the light of new evidence of the nature of the Martian surface. (For scale: diameter of the lander body is ~3 m.) (NASA)

The Viking biology package (Fig. 39.2) embodied in one piece of technology the most sophisticated thinking of the twentieth century on the subject of extraterrestrial life in the Solar System. The assumptions behind its experiments, the results obtained, and the ensuing controversies over the interpretation of these results are therefore of considerable importance. The diverse ideas about the nature of Martian life led to three different biology experiments aboard Viking, each representing a different approach to the problem of life. Indeed, biology team leader Klein later stated that had it not been for the constraints of 15 kg weight and about 1 cubic foot volume for the biology package, even more of the approaches conceived during the previous two decades would have been included on the spacecraft. The idea was that the three experiments, singled out and recommended by the Space Science Board of the National Academy of Sciences in 1968, would test for life using different philosophies, environmental conditions, and detectors.

One approach, which came to be known as the "labeled release" experiment, was developed by Gilbert Levin, who had spent much of the 1950s trying to improve methods for the detection of bacterial contaminants in city water supplies, and believed his method could be applied to the search for life on Mars. He was awarded a NASA contract for his "Gulliver" concept in 1961, and was reporting on his experimental apparatus already in the early 1960s. Levin's approach assumed that any Martian microorganisms, like those on Earth, would assimilate (eat) simple organic compounds, decompose them, and produce gases such as carbon dioxide, methane or hydrogen as end products. For this reason a dilute aqueous solution of seven such organic compounds, radioactively labeled for detection purposes, was added to the incubation chamber containing the Mars soil sample. The experiment tested for the expected "labeled release" of the gas produced as any organisms ate the organics and breathed out the decomposition products. The output was in the form of radioactive disintegrations, measured by a carbon-14 detector, in counts per minute.

The second biology test, the "gas exchange" experiment, was developed by Vance Oyama of NASA Ames Research Center, a veteran of life detection experiments on Apollo lunar samples. The gas exchange experiment tested for life under two different conditions. In the first mode, it was assumed that any organism in the dry Martian environment would be stimulated to metabolic activity by the addition of slight water moisture, and would give off a gas that could be detected by chromatography in the area immediately above the sample. In the second "wet nutrient" (or chicken soup) mode a rich nutrient of 19 organic compounds was added as an additional stimulus to metabolic activity, the products to be detected in the same manner. In both cases, the liquid added did not come into contact with the soil, but was added underneath the cell in which the soil "incubated." Water vapor gradually seeped up through the porous bottom of the incubation chamber, creating gradations of moisture through the soil. Experiments were also undertaken without the addition of any moisture.

The "pyrolytic release" experiment (also called the carbon assimilation experiment) was headed by Norman Horowitz of Caltech. Horowitz, a member of the WESTEX group in 1959, had cooperated with Levin's project in the early 1960s, but after Mariner IV showed that liquid water could not exist on the planet, he split with Levin and became convinced that it was best to test for Martian organisms under conditions known to exist on Mars when the experiment was designed. Thus to the small sample of Martian soil Horowitz proposed in his experiment to add only carbon dioxide and carbon monoxide, gases known to exist in the Martian atmosphere and now radioactively "tagged" for detection purposes. It was assumed that any organism on Mars would have developed the ability to assimilate these gases and convert them to organic matter. After 120 h of incubation, the soil chamber was to be heated to 635 °C to pyrolyze the organic matter and release the volatile organic products, thus the name "pyrolytic release." A radiation counter yielded disintegrations per minute.

All three experiments sought to detect metabolic activities. Of the experiments, Oyama's "wet nutrient" mode was the most Earth-like approach, in that it added rich terrestrial organics to stimulate any Martian organisms. Horowitz's was the most Mars-like, making few assumptions about Martian life except that it would be carbon based. Levin's, with his weak organic nutrient, fell in between. Levin and Oyama's experiments attempted to detect life by the decomposition of organics into gas during metabolism (a universal property of terrestrial organisms), while

Horowitz sought to synthesize organic matter, which he would then pyrolyze in order to be able to detect. For detection purposes both Levin and Horowitz made use of standard techniques of radioactive carbon-14 as a "tracer," a method that did not change the chemistry, but provided a means of distinguishing atmospheric carbon from metabolized carbon. Oyama made use of the well-known method of gas chromatography for detection, as did Biemann (in conjunction with a mass spectrometer) for the organics experiment, which had nothing to do with metabolism. Ignorant of the nature of Martian life, the fondest hope of all the experimenters was that at least one of the experiments—hopefully their own—would turn up something.

Summer 1976 finally brought the day that Lowell, Kuiper and a host of scientific ghosts would have savored: the landing of two spacecraft on the surface of Mars to test for life in situ. They would not have been disappointed: Viking 1 landed successfully on the Chryse plain on July 20, and the first results of the biology experiments returned from Viking were exciting, to say the least. Although no visible life forms walked across the field of view of the camera, once the soil samples were collected on July 28, the biology experiments quickly began to return major surprises. Levin's experiment evolved gas into the chamber after the nutrient was added, then the reaction tapered off. Horowitz's pyrolytic release test was also positive, and Oyama's gas exchange experiment evolved not only CO2 but also oxygen, the latter a reaction never before seen in tests on terrestrial or lunar soils. Because of the speed and course of the latter reaction, Oyama's experiment was not believed to be biological in nature. In short, two of the three biology experiments gave "presumptive positive results" for biology, and the third gave evidence of an oxidizing material in the surface at the Viking site. There was only one problem: in another unexpected finding, Biemann's organic analysis showed no organic molecules present to the level of a few parts per billion, a result later called "probably the most surprising single discovery of the mission." As Klein has subsequently recounted, these first results caused the carefully laid out experimental strategy to be abandoned, as the scientists attempted to discover whether chemical or biochemical reactions were taking place (Klein 1977; Dick and Strick 2004).

By eight and a half months after the first Viking had landed, 26 biological experiments had been carried out, and the first relatively complete results were reported, along with other Viking experiments, in *The Journal of Geophysical Research*. By then, shortly before the biological experiments were terminated in May, 1977, Klein's considered judgment was that the positive result from Horowitz's pyrolytic release experiment was probably non-biological in origin, while Levin's labeled release experiment remained ambiguous (Klein 1977). Ironically, the gas exchange experiment of Oyama—the Viking scientist most optimistic about Martian life—showed no evidence at all for biological activity. Oyama and most of his colleagues concluded that the spontaneous evolution of oxygen was due to a chemical reaction involving "superoxides" such as hydrogen peroxide, perhaps by the effect of solar radiation on the small amount of water vapor in the upper atmosphere of Mars. "It's like the three bears," Klein later said. "Not too much water, not too little water, just the right amount of water in its atmosphere to produce something like this. This is one of the big mysteries, and any future missions to Mars have to find out what this stuff is" (Klein 1977; Dick1996, p. 155).

In the end, there was not complete consensus among the experimenters themselves. Writing for *Scientific American*, Horowitz concluded that although "it is not easy to point to a nonbiological explanation for the positive results" of his pyrolytic release experiment, "it appears that the findings of the pyrolytic-release experiment must also be interpreted nonbiologically," mainly because the reaction was less sensitive to heat that one expected from a biological process (Horowitz 1977). Levin, however, did not agree; for decades he continued to argue forcefully that a biological interpretation of his data was still possible (Levin and Straat 1976; DiGregorio et al. 1997).

Clearly sensitive to their own assumptions, the Viking biologists continued to ponder the strategy of their experiments. What if their assumptions about Martian life, on which the biology experiments were based, were not correct? With this in mind Klein concluded his summary of Viking biology results with the astonishing remark that "we must not overlook the fact, in assessing the probabilities of life on Mars, that all of our experiments were conducted under conditions that deviated to varying extents from ambient Martian conditions, and while we have accumulated data, these and their underlying mechanisms may all be coincidental and not directly relevant to the issue of life on that planet"(Klein 1977; Dick 1996, p. 157).

Ten years later, contemplating the experiments conducted for some 10 months on the surface of Mars, Horowitz remained convinced that they not only proved the absence of life on Mars, but by extension "Since Mars offered by far the most promising habitat for extraterrestrial life in the solar system, it is now virtually certain that the earth is the only life-bearing planet in our region of the galaxy" (Horowitz 1986). Although most scientists were not ready to make that quantum leap, it is also fair to say that they were much less optimistic about life on Mars in the aftermath of Viking. The Viking results were impressive enough that most scientists shifted the focus of their biological Martian interests to either past Martian history, or to different Martian environments such as rocks, polar caps, subsurface soil, volcanic regions, and the ancient river valleys.

That, however, is not quite the end of the Viking story. Though a consensus seemed to have been reached for several decades that life (indeed even organics) had not been found on the Martian surface at the two Viking landing sites, the issue was reopened especially after NASA's Phoenix lander discovered perchlorates on Mars in 2008. Some well-known and indisputably reputable scientists argued that such perchlorates would have destroyed any organics present in the Martian soil when it was heated during the Viking experiments (Navarro-González et al. 2010). The issue remains open among prominent researchers today, with Levin more than ever convinced he discovered life on Mars (Levin 2011; Bianciardi et al. 2012). Surely, the extended discussion of the Viking results provides a cautionary note on the need for sensitivity to the preconceptions that enter into the design of experiments, and the difficulties of interpretation of the resulting evidence.

39.2.3 The Mars Rock

As the twentieth century approached its end, it appeared that the Viking landers had written the last chapter in the search for life on Mars. But almost exactly 20 years after the Viking landings, the world was startled with the announcement that organic molecules, possibly biogenic minerals, and even microfossils may have been found in a meteorite that originated on Mars. The result was controversial, though one might have thought that the inconclusive evidence would be balanced to some extent by the fact that the Martian meteorite could now be examined, not with the limited resources of a spacecraft on the surface of Mars, but with the full power of analytical techniques in many laboratories on Earth. A new era in Martian life studies had begun.

Meteorites had long been associated with the question of extraterrestrial life, but those meteorites were a special variety known as carbonaceous chondrites, and their parent body had not been identified. Only in the post-Viking era was a new category of extremely rare meteorites identified, and a case slowly built that they had originated on Mars. Known as SNC meteorites after the locations of their three types (shergottites, nakhlites and chassignites), they were also stony meteorites, but "achondrites," because they exhibited none of the millimeter-size embedded mineral spheres characteristic of chondrites. They were known to have come from Mars not only because of their chemical composition, but also because the gases trapped in them were precisely the same composition and proportions as those of the Martian atmosphere, as determined by the Viking landers. Thus, although Viking did not unambiguously find life itself, ironically it enabled the identification the SNC meteorites as Martian in origin.

The surprising announcement in the summer of 1996 centered on the Martian meteorite known as Allan Hills 84001, believed to have fallen on the ice fields of the Antarctic 13,000 years ago. The first meteorite found in the Antarctic during an NSF-sponsored search season in 1984 (thus the name ALH84001), it was not identified as Martian in origin until 1994. One of only 12 such meteorites identified at the time, the 4.5 pound (1.9 kg) softball-sized rock was by far the oldest of the 12, estimated to have formed about 4.5 billion years ago, from a period when Mars was warmer and had water and an atmosphere. It was hypothesized that a meteorite impact on Mars fractured the rock about 3.6 billion years ago, and that another impact about 16 million years ago launched the rock into space, where it eventually intercepted the Earth.

The evidence, announced by a NASA team led by David McKay of NASA Johnson Space Center in Houston (Fig. 39.3), consisted of four parts (McKay et al. 1996). None of these parts, the participants pointed out, were conclusive in themselves, but taken together they could be interpreted as biogenic. First, the multidisciplinary science team reported, the fractured surfaces of the rock contained large complex organic compounds in the form of polycyclic aromatic hydrocarbons (PAHs). This was already a step beyond what the Viking landers had found, but even though the NASA team undertook analysis that showed to their satisfaction that the

Fig. 39.3 Press conference August 7, 1996 at NASA Headquarters announcing the possibility of fossilized life in the Mars rock ALH84001. The photo, taken by NASA photographer Bill Ingalls, stunningly captures the excitement and media frenzy likely to occur with any announcement of life beyond Earth. (NASA/Bill Ingalls)

PAHs were not contamination from Earth, this was not proof of life, since organic molecules could have originated by non-biogenic processes on Mars. But then the plot thickened: in the fractures the team also discovered carbonates and magnetite, minerals that are produced (among other ways) by certain "magnetotactic" bacteria on Earth. Finally, using a high resolution scanning electron microscope the team suggested the existence of microfossils in the carbonates and other mineral grains (Fig. 39.4); at only 20–100 nm they were 100 times smaller than the smallest known bacteria on Earth.

Less than 2 months later, a British team of scientists led by Colin Pillinger of the Open University announced independent evidence of possible traces of life, both in ALH 84001 and in a much younger Martian meteorite known as Elephant Moraine 79001 (EETA 79001, again named after the location of its discovery in the Antarctic). The latter meteorite was only 175 million years old, and was blasted from Mars only 600,000 years ago. This was so recent, geologically speaking, that it held open the possibility that life might still exist on Mars.

As in past controversies over Martian life, the stakes were high and the skeptics numerous. One of the chief objections came from Ralph Harvey of Case Western and Harry Y. McSween of the University of Tennessee, who had reported in *Nature* shortly before the NASA announcement that their analysis of the same meteorite showed that the carbonates formed not as a result of microbial life, but during the asteroid impact when carbon dioxide combined with the rock at temperatures of

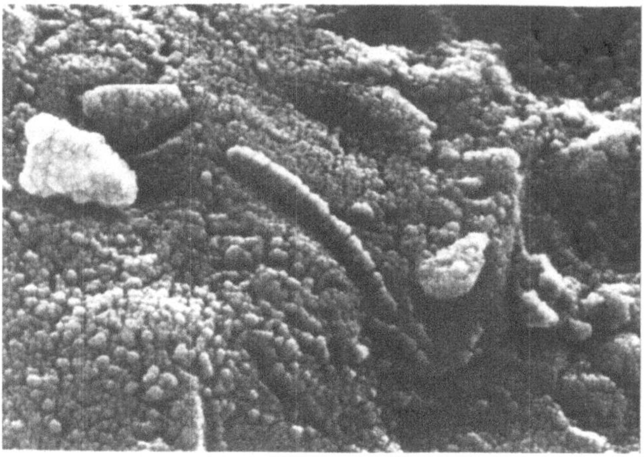

Fig. 39.4 High-resolution scanning electron microscopic image showing an unusual tube-like structure less than 1/100th the width of a human hair, found in Martian meteorite ALH 84001 and interpreted by some to be evidence of fossil life on Mars. Such morphological evidence was challenged by paleobiologist J. William Schopf, among other critics. (NASA)

1200 °F. Such temperatures are inimical to life; if this method of carbonate formation was confirmed it would cast severe doubt on the claims of past Martian life. Others, however, argued for low-temperature formation of the carbonates, one that did not rule out life.

More general questions of inference from evidence were also asked. For example, do four independent but (critics said) weak arguments—from the morphology of the nanostructures, carbonate globules, the presence of magnetite and polycyclic aromatic hydrocarbons—add up to a strong argument for biogenesis? The authors of the discovery paper thought so, ending their paper with the argument "Although there are alternative explanations for each of these phenomena taken individually, when they are considered collectively, particularly in view of their spatial association, we conclude that they are evidence for primitive life on early Mars" (McKay et al. 1996). Critics, including paleobiologist J. William Schopf, thought not, arguing that "spatial association" held no persuasive value at all, and citing Carl Sagan's dictum "extraordinary claims require extraordinary evidence" (Dick and Strick 2004).

Definitive proof of past life on Mars would come only by sectioning thin sections of the microfossils to search for cell walls, DNA or other structures unambiguously linked to life. In the years following the announcement many teams did precisely that, but with still ambiguous results. Though there is now consensus that Martian nanofossils have likely not been found, some (including most scientists who made the original claims) have not given up. Thus, the pattern is similar to that of the claims of Gilbert Levin about extant life on Mars based on the Viking experiments.

39.2.4 *Planetary Systems?*

If the evidence from the relatively nearby Solar System proved problematic, the evidence for other much more distant solar systems would be even more difficult, if of an entirely different type. It is some measure of the difficulty of the search for planetary systems that the Space Age did not bring immediate advances in the problem. Unlike Solar System studies, where planetary spacecraft brought immediate and revolutionary progress in our knowledge of the planets, no such prospect was in store for planetary systems. It is true that increased knowledge of our own planetary system provided voluminous data for the refinement of theories of the origin of the Solar System, which by the usual gross analogies could be applied to other solar systems. But, although substantial, these refinements changed little the fortunes of planetary systems. Perhaps the largest impact of the Space Age on planetary systems science was the infusion of funds from space agencies such as NASA, which displayed an interest in both observational and theoretical aspects of the subject almost from the beginning, but with delayed results.

We should therefore not be surprised that, while most astronomers in the second half of the twentieth century were optimistic about other planetary systems, observational proof of their existence through the 1970s remained entirely dependent on the old astrometric technique. That technique, the results of which remained elusive in many cases, created a public and scientific sensation with the announcement in the 1960s of the detection of several planetary systems. The promise and limitations of this technique, and the difficulties of tackling a problem at the limits of science, may best be seen in the famous case of Barnard's star. The central figure in the case is astronomer Peter van de Kamp, who had begun his search for low mass companions at the Sproul Observatory of Swarthmore College in Pennsylvania in 1937. Such is the long-term nature of the problem of determining perturbations in stellar motions that only 25 years later was Van de Kamp beginning to announce results with planetary companions.

Barnard's star was a star of 9.5 magnitude, so called after Barnard's discovery in 1916 of its enormous proper motion of about 10.3 arcseconds per year. This meant that it was a close star (the closest known after the Alpha Centauri system), and it was immediately placed on observational programs, including the parallax program at Sproul in 1916–1919. In 1938 van de Kamp had placed it back on the Sproul parallax program with his arrival as Director in 1937, and by 1944 he announced a low-mass companion stellar in nature. Over the next 20 years, as Kuiper and others were predicting an abundance of planetary systems based on their own work, and as theory once again made plausible abundant planetary systems, van de Kamp patiently collected data on Barnard's star and other nearby stars.

There is no doubt that van de Kamp was sensitive to the question of whether low-mass companions were "stars or planets," at least since his 1944 article on the subject, an article undoubtedly stimulated by the observational claims of 1943. In a progress report on "Planetary Companions of Stars" in 1956, van de Kamp pointed out that while numerous unseen objects had been detected over the last two decades with masses .05 of the Sun's or greater, it was "extremely likely" that these objects were stars. "There are tentative indications of unseen companion objects with about .01 solar masses or more, and these may be planetary companions. However, definitive interpretation can hardly be reached at present, partly due to limitations of accuracy," he wrote in that year (Van de Kamp 1956). In particular, van de Kamp pointed out that the 1943 claims for a planetary companion of 70 Ophiuchi had not been confirmed as of 1952; and he held out hope only for the claim in 1943 for a companion of 61 Cygni of .016 solar masses, well into the planetary range, as confirmed by two other observers. As for his own program, now almost two decades old, van de Kamp claimed only that the companion to Lalande 21185 was probably a star of low luminosity, and that no satisfactory explanation existed for some perturbations seen in the motion of Barnard's star. Almost two decades after the start of his observational program at Sproul, no one could accuse van de Kamp of rushing to judgment on planetary companions.

All of this was to change in the 1960s. First, in 1960 Sarah Lippincott, van de Kamp's colleague at Sproul, announced that the companion of Lalande 21185 had a mass of only .01 that of the Sun. Although this was at within planetary range (recall that the 1943 claims for planetary companions to 61 Cygni and 70 Ophiuchi gave them about this same mass), Lippincott's technical article in the *Astronomical Journal* made no mention of the word "planet," and perhaps for this reason her announcement did not raise much of a stir. But van de Kamp's 1963 article with the mundane title "Astrometric Study of Barnard's Star from Plates Taken with the 24-inch Sproul Refractor" created a sensation (Van de Kamp 1963). In it he announced the discovery of a companion to Barnard's star with a mass of only .0015 the mass of the Sun, only 1.6 times the size of Jupiter, which he specifically characterized as a planet. He further found the distance of the planet from Barnard's star to be similar to that of Jupiter from the Sun, and its surface temperature about 60 K compared to 120 K for Jupiter.

Van de Kamp's claim was based on 25 years of photographic observations, using three types of photographic emulsions, and fifty observers, yielding 2413 plates. To the extent that the public was aware of such details, they might have been persuaded by this alone that such an intensive scientific effort must have yielded a definitive result. But they would have not been aware of the subtleties of the technique, which included taking into account a variety of insidious errors that could affect the results. Having taken into account all known sources of error as best as he could, van de Kamp found a perturbation in the motion of Barnard's star with a period of about 24 years (Fig. 39.5). In order to come up with an actual mass for the companion, he further had to carry out a "dynamical interpretation" calculation, using the mass of Barnard's star. By Kepler's law, once this mass was known, and the period of the orbiting body, the mass of the latter could be calculated. It was here that van de Kamp finally came to the figure of .0015 times the mass of the sun for his new planet: "The orbital analysis leads, therefore, to a perturbing mass of only 1.6 time the mass of Jupiter. We shall interpret this result as a companion of Barnard's star, which therefore appears to be a planet, i.e., an object of such a low mass that it would not create energy by the conventional nuclear conversion of hydrogen into helium" (Van de Kamp 1963).

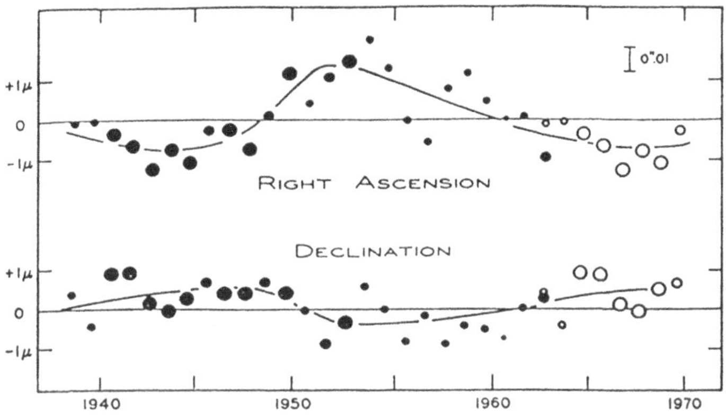

Fig. 39.5 Peter Van de Kamp's evidence for a planet around Barnard's star made use of the classical astrometric method for planet detection. Van de Kamp reported that the star underwent minute, periodic gravitational perturbations of a few hundredths of an arcsecond over three decades. The two plots show the star's measured east-west and north-south relative positions (in microns) on photographic plates taken with a 24-inch refractor. The data was proven spurious, but only after several decades. (From van de Kamp 1963, reproduced by permission of the American Astronomical Society)

Like the announcements 20 years before, the reaction to van de Kamp's result was swift. From *Time* magazine to popular science magazines and more sober scientific journals, countless reports of van de Kamp's results hailed the discovery of another planetary system. Independent verification of the result, on the other hand, was more difficult, since the observations were very specialized and required decades to reach a result—van de Kamp had been at it for a quarter century. It is not surprising, therefore, that van de Kamp himself was the first to reinforce his own result. In 1969, with 5 more years of photographic measures of Barnard's star, he reiterated his claim that a planetary companion existed around that star, with a slightly revised mass of 1.7 times that of Jupiter. In the same year, Van de Kamp proposed an alternate analysis of his data that held out the possibility that two planets orbited Barnard's star, with periods of 26 and 12 years, and masses of 1.1 and .8 times Jupiter.

But trouble was around the corner, and the 1970s saw serious questions raised about van de Kamp's momentous result. In 1973, John Hershey, one of van de Kamp's own student's, found that changes made to the Sproul telescope, in particular a change of lens cell in 1949, caused jumps in the data at that point and may have affected the results for Barnard's star. In the same year George Gatewood of the Allegheny Observatory in Pittsburgh and Heinrich Eichhorn of the University of South Florida concluded "with disappointment", based on an independent analysis of 241 photographic plates, that no perturbations existed in the motion of Barnard's star. Attempting to explain their result, they pointed to the disadvantages of van de Kamp's analysis technique, to changes in the optical system of van de Kamp's telescope over the extended period of time of his study; and to the fact that his claimed perturbation was just "on the verge of significance," a status similar to claimed measurements of parallax before Bessel (Gatewood and Eichhorn 1973). A similar analysis by Gatewood published the following year gave the same null result for Lippincott's 1960 claim of a planetary companion around Lalande 21185. Two other studies of van de Kamp's data in 1973 were more favorable to his claim of one or more planetary companions, but a decade after the first announcement by the van de Kamp group, planetary systems were once again in trouble.

Van de Kamp understandably did not take lightly this negation of the main result of his work of 25 years. In order to take the objections into account, especially the finding that changes in the instrument might have affected positional measurements, van de Kamp re-measured his plates on a new machine, and included only material from 1950 onwards. He confirmed the existence of the shorter period planet, with a mass now .4 Jupiter, while the second planet was "less well determined." In 1977, 60 years after the discovery of Barnard's star, van de Kamp took the occasion to reassert his belief in the reality of its planetary companions. In addition to the now familiar scientific defense, the article concluded with a Rembrandt etching on the appearance of Christ to Thomas, with a caption "blessed are they that have not seen, and yet have believed," suggesting a religious invocation of faith undoubtedly seen by some as not readily transferable to the scientific realm (Van de Kamp 1977). Van de Kamp's last paper based on new data, published in 1982, again supported the conclusion of two planets around Barnard's star, a conclusion he never relinquished.

The Barnard's star episode was only the most notorious of several claims made for planetary companions by the mid-1970s, all subject to the same limitations of technique and inference. Although the assault on Barnard's star continued to receive the greatest attention, in the field of astrometric perturbations it was not unique, and thus could not be written off as a fluke. In an extensive 1975 review of the subject of unseen astrometric companions van de Kamp could list 17 "well-established perturbations" of stars by unseen companions, including Barnard's star and three others with possible planetary companions. Another 14 stars, including the famous 61 Cygni, were listed with "perturbations of provisional, suspected, or uncertain nature" (Van de Kamp 1975).

Not everyone was convinced that even those stars showing well-established perturbations necessarily harbored planets, for this depended on theoretical ideas about the cutoff point for stable hydrogen burning in stars. While CalTech geochemist Harrison Brown supported the idea of numerous planets by an extension of the "luminosity function" (the distribution of the stars with their visual magnitudes) to low masses, S. S. Kumar, for example, argued that all of the objects claimed as planets were probably very low mass "degenerate" objects that he termed "black dwarfs" (Kumar 1967). The dividing line between stars and planets remained the subject of vigorous discussion, and the accompanying search for what came to be known as "brown dwarfs" (objects not massive enough to sustain nuclear fusion) remained almost as elusive as the search for planets themselves.

Numerous other instances could be examined, including the controversy surrounding the claims of extrasolar planets in the 1980s and 1990s (subsequently demonstrated to be true), the claims of protoplanetary systems found by the Infrared Astronomical Satellite (IRAS) and the Hubble Space Telescope, and the recent reaction to the claims of arsenic life in the Halomonadaceae bacterium. In fact, it is safe to say, controversies about the interpretation of evidence are the central core of science, the rule rather than the exception.

All these examples also bear on the role of preconceptions, manifesting themselves especially in the form of prior assumptions feeding into the design of experiments, the interpretation of observations and the definitions of life and intelligence. For example, the Viking experiments were designed with metabolic conceptions of life, as opposed to other possibilities such as Darwinian processes, metabolism, energy and thermodynamics, complexity theory, cybernetics, or some new insight. This was in part due to the constraints of spacecraft investigation; nevertheless with the perspective of four decades we can look back on these experiments and ask how they might have been done differently. Certainly had scientists known about perchlorates on the surface of Mars, the experiments would have had a different design.

But what are the lessons learned from these representative case studies? Scientists need hardly be reminded to be careful with the interpretation of data. But the general public can never be reminded too often that this is the nature of science. The lesson is not that science should be mistrusted or abandoned because of its imperfections, but that despite the difficulties history shows that something approaching "the truth" eventually emerges as nature is continuously interrogated at increasingly more subtle levels, even if the final outcome may take decades. The lesson is both pedestrian and profound: with all its personal and cultural biases, science is the best way we have of interrogating nature, and even in extremely difficult areas such as astrobiology progress can be made, even if we have to resort to analogies, which much of microbial astrobiology does. As scientists know—but as much of the public doesn't seem to understand—science, including astrobiology, is a series of trial and error observations, subject to constant interpretation and re-interpretation, and therefore to constantly changing ideas, asymptotically approaching the truth—until the truth may suddenly change with a new way of looking at things. In this it in no way resembles religion, as is sometimes charged. The difficult and expensive nature of science, especially space exploration, makes experimental and observational iterations very extended. But in the end, once again, most scientists would say the objective truth is out there (though this is a deep philosophical problem), and remarkably the human mind can eventually ferret it out.

39.3 The Role of Theory

The role of theory in science is a huge and complex subject. We confine ourselves here to one example that demonstrates that theory can serve both as catalyst and as hindrance: the case of the origin of solar systems. As we have seen, during the first half of the twentieth century, the primary and most widely accepted theory of the origin of solar systems was the Chamberlin-Moulton and the Jeans-Jeffreys tidal theory, whereby solar systems arise by material tidally pulled out during the close encounters of stars. This theory lowered expectations for observing other solar systems, since encounters between stars were believed to be extremely rare. Only in the 1940s, when the tidal theory was replaced by a revived nebular hypothesis, which postulated solar systems as a common byproduct of stellar evolution, did the expectations radically change, providing the backdrop first to Van de Kamp's work, and to the real discoveries of exoplanets 30 years later. So theory can affect both the undertaking and the interpretation of observations (Dick 1996).

In the absence of decisive observational evidence for planetary systems, one might expect that theories of solar system formation would play an especially important role, at least in determining the plausibility of such systems. This had indeed been the case for the nebular hypothesis, which favored abundant planetary systems because the formation of planets from a rotating gaseous disk was assumed to be a universal process. But as we have briefly seen in the first part of this chapter, that hypothesis was under serious attack in 1900, and for the first two decades of the twentieth century the new theory, to the limited extent that it addressed the issue at all, gave conflicting indications about the possibility of other planetary systems. Developed by T. C. Chamberlin, chairman of the Geology Department at the University of Chicago, and F. R. Moulton, an astronomy graduate student at the same university, the Chamberlin-Moulton hypothesis sought to surmount the technical weaknesses of the nebular hypothesis by proposing instead that solar systems were formed by the close encounter or actual collision of stars in space. According to this hypothesis, the close encounter caused material to be ejected from the Sun. The passing intruder then caused the ejected material to form spiral arms. These arms contained knots of denser material that condensed into nuclei, which in turn grew into planets and satellites by the capture of planetesimals, cold particles in the nebula. The spiral nebulae recently observed in the heavens, they believed, might be evidence of such collisions and of solar systems in formation. Curiously, rarity or abundance of planetary systems does not seem to have been an issue for Chamberlin or Moulton. To the extent that their rarity or abundance was an issue at all, it oscillated between the twin pillars of the *planetesimal hypothesis*: the spiral nebulae, which implied abundance, and stellar encounters, which implied rarity. With the gradual realization that spiral nebulae were too large to represent planetary systems in formation, the stellar encounter aspect of the theory was free to gain the upper hand—and with it the implication of the rarity of planetary systems.

This, in fact, is precisely what occurred, not in America but in Britain, where in the tradition of William Whewell and A. R. Wallace, the scientific community seemed more skeptically inclined toward planets and life. It was at the hand of the British mathematical physicist and astronomer James Jeans that the question of other solar systems would become closely linked with the rarity of planets and life in the universe. Jeans, a 1903 graduate of Trinity College, Cambridge, had done important work on atomic theory and statistical mechanics prior to 1914. After 1914 he turned from the microscopic to the macroscopic, from atoms to astronomy, and specifically to cosmogony. Jeans' attention was at first devoted to the stability of rotating bodies, on which subject he published two lengthy papers in 1915 and 1916. This work he applied to cosmogony in 1916 with reference to tidally distorted masses, in other words, determining how a rotating astronomical body would be affected by tidal forces raised by another passing astronomical object, as would happen in the case of a close stellar encounter. In a paper read before the Royal Astronomical Society in 1916 and published in the Society's Memoirs in the following year, Jeans dealt not only with the origin of solar systems, but also with binary star formation and spiral nebulae. In contrast to the binaries and spirals, Jeans concluded that the Solar System might well have been formed from a tidally distorted mass, in particular by another star approaching our sun. Unlike the Chamberlin-Moulton hypothesis, however, Jeans' analysis showed that neither spiral nebulae nor planetesimals played a role in planet formation, and he thus emphasized that for solar systems "the origin which seems most probable is not that of the planetesimal hypothesis"(Jeans 1917). Instead, his analysis showed that rather than the streams of gas torn from the Sun condensing into numerous small cold planetesimals that in turn accreted to form the planets, a single cigar-shaped filament of hot gas would be ejected and condense directly into the planets. As the theory was later elaborated, he pointed out that the largest planets would form near the center where the filament was thickest, and the smaller ones at each end, giving the distribution of planets observed in our Solar System.

The central question in determining whether this mathematical conclusion could really occur in nature was the frequency of close stellar encounters. It is clear at the outset of the paper that Jeans was already thinking in these more general terms, not only with regard to the origin of our Solar System, but in connection with the frequency of planetary systems. In his earliest statement on what would become a lifelong contentious issue, he wrote: "We have absolutely no knowledge as to whether systems similar to our Solar System are common in space or not. It is quite possible, for aught we know to the contrary, that our system may have been produced by events of such an exceptional nature that there are only a very few systems similar to ours in existence. It may even be that our system is something quite unique in the whole of space" (Jeans 1917, p. 46).

Jeans' analysis showed that the issue of abundance was very sensitive to the assumptions made about a variety of parameters, including the density of stars in the universe, the velocity of the stars in space, the age of the stars and of the universe, and the size and mass of the stars at time of encounter. All of these parameters were subject to change in the discussion that ensued over the next three decades. For now, using the best estimates known in 1916 and assuming stellar masses and velocities similar to the Sun, Jeans found that at most one star in 4000 might have experienced a "non-transitory" encounter at the distance of Jupiter in a lifetime of 10 billion years, the upper limit that he placed on the age of the universe. If the encounter distance were a hundred times greater and the other parameters adjusted accordingly, one star in three might have experienced such an encounter, and "we may, without postulating anything very improbable, suppose our system to have experienced an encounter as close as this…" (Jeans 1917). However, Jeans clearly did not think all these conditions would ensue at one time, and in the end he labeled these occurrences as "somewhat improbable" and systems similar to our own "somewhat rare," but in general the entire process not "impossible or very improbable." Given the number of parameters and their uncertainty, Jeans' waffling is not surprising.

But he emphasized that no reasonable choice of parameters was likely to alter the result that only very few stars can have experienced non-transitory encounters. And most importantly, Jeans stressed, the theory violated no quantitative criterion.

In his classic work *Problems of Cosmogony and Stellar Dynamics* (1919), Jeans discussed the problem in more detail, and ended with results even more pessimistic: only one encounter in 30 billion years, a situation so improbable in the present universe as to cast doubt on the validity of the close encounter hypothesis. Pointing out that the parameters were not well known, Jeans concluded that while tidal breakup by a passing star was hardly a likely event, its improbability was not grounds for rejecting the tidal theory. In whatever case one adopted, the Solar System seemed to be very exceptional, "and for aught we know may be unique" (Jeans 1919). In his 1923 lecture "The Nebular Hypothesis and Modern Cosmogony," Jeans carried his train of thought one step further, arguing that it was just possible, though not probable, that only the earth could support life in the universe. "Astronomy does not know whether or not life is important in the scheme of nature, but she begins to whisper that life must necessarily be somewhat rare" (Jeans 1923).

In the hands of Jeans, this whisper soon grew to a crescendo. In both his technical and popular publications by the late 1920s, Jeans spread his view far and wide. The numbers varied somewhat, but always present was the basic scenario that the stars are sparsely scattered in space, close encounters exceedingly rare, and the conditions for life very exacting. "All this suggests," Jeans inevitably concluded, "that only an infinitesimally small corner of the universe can be in the least suited to form an abode of life" (Jeans 1930). In his popular works this view of the disruptive approach of stars was vividly drawn, and the rarity of such approaches and their ensuing solar systems was an integral part of this picture—clear even to the public.

For two decades the Jeans tidal theory—with contributions by Sir Harold Jeffreys—was widely accepted, and when the beginning of the end came in 1935 it was once again because of problems with physical principles. This time it was the Americans' turn again, in the form of Henry Norris Russell, who criticized the tidal hypothesis because it could not account for the present orbits of the planets. Russell could not see how a close stellar encounter would remove the planets so far from the Sun and give them most of the angular momentum of the system rather than the Sun, which was a thousand times more massive (Russell 1935). He also could not see how the planets could condense out of the high-temperature matter ejected from the Sun, an objection given definitive form by Russell's student Lyman Spitzer 4 years later. In their discussion the possibility of other planetary systems played no role, but their fatal objections left science without a workable theory of the origin of the Solar System, and by association placed in limbo the idea that such systems were rare.

The nineteenth-century view of abundant planetary systems was thwarted for decades by the tidal theory of Jeans and Jeffreys. Far from the teleological view of R. A. Proctor or J. E. Gore, Jeans' colleague Sir Arthur Eddington asked "How many acorns are scattered for one that grows to an oak? And need she be more careful of her stars than of her acorns? If indeed she has no grander aim than to provide a home for her greatest experiment, Man, it would be just like her to scatter a million stars whereof one might haply [sic] achieve her purpose" (Eddington 1929). To have provided a theoretical underpinning for this startlingly different worldview was no small part of the legacy of James Jeans.

But alas, this world view had no more claim to objective truth than the nineteenth-century belief in abundant planetary life, for if the early observational claims for planetary systems at the turn of the century had yielded no definitive result, by 1940 neither had theory solved the problem—nor could it—especially with the departure of spiral nebulae as confirming evidence. The discredited nebular hypothesis had been superseded by the planetesimal hypothesis of Chamberlin and Moulton and then the tidal theory of Jeans and Jeffreys, only to have Russell and Spitzer overturn the latter, leaving only the void. Reviewing the collisional and nebular hypotheses in 1938, Lick Observatory Director Emeritus Robert G. Aitken still saw the development of planetary systems as an "exceptional event." "Exceptional" did not mean unique to Aitken, who pointed out that even if only one star in a million had planets, there would still be 30,000 solar systems in the Milky Way Galaxy—and two million galaxies were within the range of current telescopes (Aitken 1938).

The 15 years between 1943 and 1958 saw a remarkable turning point in the fortunes of planetary systems. It had begun with Russell's criticism of the Jeans-Jeffreys tidal theory, but it was fueled by the revival of a modified nebular hypothesis, developments in fields as diverse as double star astronomy and the measurement of stellar rotation periods, and—most surprising of all—by insistent claims that planetary systems, or their effects, had been actually observed (Table 3.1). Moreover, broader events in the field of cosmology conspired toward change also, events that Jeans himself could not ignore.

The implications of the revolution in cosmology of the 1920s and 1930s—a greatly enlarged galaxy, the existence of innumerable "island universes" full of stars, a universe expanding in space and expanded in time—are evident in Jeans' review of the subject of life on other worlds published in 1942. Having given a dim view of the chances of life on Mars and Venus, Jeans turned to the realm of the stars, and the origin of planetary systems. He pointed out that under present conditions in the universe the frequency of stellar encounters would be only one in 1018 years, so that for stars two billion years old, one star in 500 million might have planets. So far this was his old argument. But it was a sign of the times that he went on to say that though this seemed like a small fraction, in a universe with 10 billion galaxies, each with 100 billion stars, this minute fraction still represented two million million stars that might have planetary systems! Statistics—and the new cosmology—had caught up with Jeans, even if only 2000 of these systems might be located in our own galaxy. Straining the definition of "rare," Jeans was forced to conclude that "although planetary systems may be rare in space, their total number is far from insignificant" (Jeans 1942a).

Later that year, however, Jeans' view had undergone a much more radical change. In a letter to *Nature* of June 20, 1942, reacting to recent claims of serious dynamical problems arising for the tidal theory assuming the Sun was about its present size at time of encounter, Jeans asserted that the Sun was most likely comparable in size to the present orbit of Uranus or Neptune when an encounter took place. In a last-ditch effort to save the tidal theory from dynamical objections, Jeans was forced to increase greatly the size of the Sun at the time of supposed planetary formation, a concession that greatly increased its cross-section and by analogy the cross section of other suns. Not only did this address the dynamical objections in Jeans' opinion, it also led to another conclusion: that the total chance of planet formation was now one in six with such a size for the Sun. Thus, "there is no longer any need to strain the probabilities to account for the existence of the planets." And the final conclusion is one hardly expected from Jeans: "A far larger proportion of the stars than we have hitherto imagined must be accompanied by planets; life may be incomparably more abundant in the universe than we have thought" (Jeans 1942b). The whole exercise demonstrated the fragility of the argument, and the dangers of using equations whose parameters were not well-determined. For 25 years Jeans had epitomized the concept of the rarity of life in the universe. Now in the last years of his life, he recanted, and his death in 1946 left no substantial heirs to his theory.

Jeans' turnabout was just the beginning, and the cracks opening in the tidal theory in 1941–1942 were to become a breach through which the floodwaters of change would rush in the following year, when strong and independent observational claims were made for the existence of two planetary systems around nearby stars. Many astronomers were quick to draw general conclusions, especially in light of the new observations; as Henry Norris Russell wrote in 1943, "On the basis of this new [observational] evidence, it therefore appears probable that among the stars at large there may be a very large number which are attended by bodies as small as the planets of our own system. This is a radical change – indeed practically a reversal – of the view which was generally held a decade or two ago" (Russell 1943). Table 3.1 indicates how completely the change was, due in no small part to a change in theory.

Examples of the role of theory in the extraterrestrial life debate could be multiplied, but the lesson in this case seems to be that theories can play both positive and negative roles in getting at the truth, especially when empirical evidence is lacking. One thinks of Harold Urey's theory of a reducing atmosphere on the primitive Earth, based on his speculation that the solar nebula was largely hydrogen. This led to Stanley Miller's famous experiment in which amino acids were produced under such a simulated primitive Earth atmosphere, a result that in turn gave much hope to those who believed extraterrestrial life might be common. The nature of the primitive Earth atmosphere has since been called into question, leaving scientists oscillating between optimism and pessimism—one might almost say between hope and despair—undoubtedly also influenced by predispositions to one side or the other.

39.4 Testing History: Plenitude, Mediocrity, Anthropocentrism and Rare Earth

History demonstrates a mixed record of success for general arguments in the extraterrestrial life debate, including the uniformity of nature, the principle of plenitude, the principle of mediocrity, the Goldilocks argument for rare Earth, large number arguments, and the Fermi paradox. In closing we focus here briefly on three related arguments: plenitude, rare Earth, and the principle of mediocrity, with the idea that our current understanding of the number of exoplanets, the conditions for the origins of life, the conditions for life on planetary and satellite surfaces, and "The Great Silence" can be used to evaluate the validity of arguments in astrobiology over the last 50 years and more. The arguments of Ward and Brownlee (2000) in their book *Rare Earths*, makes such an evaluation in the light of history an important endeavor.

That general arguments can sometimes be illuminating is illustrated in the case of the principle of plentitude, which posits the fecundity of God or Nature and states that whatever God or Nature can do, they will do. In his classic volume *The Great Chain of Being*, the historian Arthur O. Lovejoy put it this way: that "no genuine potentiality of being can remain unfulfilled, that the extent and abundance of the.

creation must be as great as the possibility of existence and commensurate with the productive capacity of a 'perfect' and inexhaustible 'Source,' and that the world is better, the more things it contains" (Lovejoy 1936, 52). He posited this argument as the chief argument of the entire plurality of worlds debate through the nineteenth century, a claim now seen to be too simplistic.

That does not mean, however, that ideas of plenitude have not played a significant role throughout history, as Chap. 1 of this volume demonstrates. Nor does it mean that the same principle is not invoked even today, consciously or unconsciously, implicitly or explicitly. Despite skepticism prior to 1995, the last two decades have shown that a principle of plenitude does indeed apply in the case of exoplanets, which we now know exist in abundance around normal Sun-like stars. That the idea of plenitude has its limits is indicated by the fact that, although some planets are also found around exotic stars such as pulsars and in binary systems, they do not exist around all classes of stars. General metaphysical ideas such as plenitude must be mediated by sober physical reality and empirical findings. Whether or not this kind of mediated general argument carries over to life and intelligence remains to be seen—indeed that is the very question to be answered by more scientific means. But the principle of plenitude remains a kind of guiding argument for the optimists, and so far a useful one.

Other arguments, such as the "rare Earth" and related "Goldilocks" genre, do not fare so well when evaluated in terms of history. For example the natural philosopher William Whewell, who coined the term "scientist" in the mid-nineteenth century, used numerous Goldilocks arguments to "prove" that other worlds could not exist—that the Earth was indeed rare. His treatise *Of a Plurality of Worlds: an Essay*, which appeared anonymously in 1853, was the most learned, radical, and influential anti-pluralist treatise of the century (Crowe 1986). As with all participants in the debate, Whewell had his predispositions in the matter of extraterrestrial life. When it came to the compatibility of other worlds with Christianity, unlike Thomas Paine and others Whewell argued that it was other worlds, not Christianity, that should be rejected. To the argument that all the vast space must have some purpose, he countered that geology reveals human existence on Earth to be but a short "atom of time" compared to the age of the Earth; therefore why could not intelligence be confined to the "atom of space" that was the Earth?

Moreover, although the universe was indeed vast (about 3000 light years by his estimate), Whewell argued that the possible locales for inhabitants had been vastly overrated. With arguments that were plausible at the time he held that (1) all nebulae were gaseous rather than stellar systems, meaning that not nearly as many stars existed as some believed; (2) the analogy between the Sun and the stars may be less strong than many thought—all stars may not be similar to our Sun; (3) many stars were double stars, unsuitable for planets; and (4) there is no evidence of planets around other stars. In short, Whewell saw the analogies as greatly exaggerated in the case of other worlds. No longer was the Copernican implication that the planets were Earths a sufficiently precise argument; greater attention had to be given to the details of their physical conditions. But he was wrong that all nebulae are gaseous, and while there was no evidence at the time of other planets, today they are known to exist in abundance, even around double stars. Whewell was indeed correct that not all stars are similar to the Sun, but this argument was obliterated by the fact that we now know we live in a vastly larger universe that he conceived, harboring billions of stars even in a single galaxy.

All of Whewell's conclusions were based on uncertain evidence, but his arguments narrowing the number of habitats and claiming that everything had to be just right for life on those habitats were misguided. Today, with our knowledge of the extremities of life in deep sea hydrothermal vents, deep underground environments, and in conditions of extreme salinity, acidity, pressure, radiation, and almost every variable one can imagine, Goldilocks arguments seem even less relevant as a guiding principle. Such rare-Earth-type arguments are now known to be fallacious in light of new knowledge, and it seems likely that the modern rare Earth hypothesis (Ward and Brownlee 2000) will suffer a similar fate, as the Kepler spacecraft is already beginning to indicate. Of course, the rare Earth declaration depends on the definition of rare, and on the definition of Earths, especially with the discovery of the category of planets called "Super Earths." But with cases in our Solar System such as Europa, Callisto, and Ganymede, the very concept of a habitable zone has been revolutionized.

Related to the rare Earth argument is the insidious role of anthropocentrism, deployed consciously or unconsciously. 50 years after Whewell's treatise, at the beginning of the twentieth century, A. R. Wallace—the cofounder with Darwin of the theory of natural selection—used arguments similar to Whewell's to demonstrate the Earth had a favored place in the universe, that the sole purpose of the universe was to produce humans, and that humans were the only life in the universe. His influential work *Man's Place in the Universe: A Study of the Results of Scientific Research in Relation to the Unity or Plurality of* Worlds (1903) incorporated many of the biological problems that would be elaborated in ever more subtle form throughout the century. Although Wallace's book in some ways marks a signal advance in the debate about other worlds, its failure is marked by the dominance of the anthropocentric worldview over all other arguments. Convinced of the nearly central position of the Sun in the universe, Wallace first sought—and found—the significance of this fact in the uniqueness of life, and then adduced arguments in favor of the view that life was found beyond the Earth neither in our Solar System nor in others. 50 years after Whewell's treatise, Wallace confidently concluded that

"Our position in the material universe is special and probably unique, and … it is such as to lend support to the view, held by many great thinkers and writers today, that the supreme end and purpose of this vast universe was the production and development of the living soul in the perishable body of man" (Wallace 1903). Although professing a scientific approach, Wallace's book serves as a lesson on the limits of science when worldviews dominate empirical evidence (Dick 1996). It is a lesson the twenty-first century should take to heart.

The same lesson, however, needs to be applied to the other side of the argument. During the twentieth century the tug-of-war between anthropocentrism and other worlds was profoundly affected by radical changes in astronomical worldview. While it was still possible as the century began for scientists to argue for an anthropocentric universe based on the Earth's privileged physical position in the cosmos, by 1930 advances in astronomy had destroyed this argument. The resultant world view—an expanding universe of enormous dimensions in which the Solar System was at the periphery of one galaxy among millions—tipped the scales strongly toward the presumption of other worlds for the rest of the century. "The assumption of mediocrity" became an underlying current of thought favoring other inhabited worlds, superseding the assumption of uniqueness that had opposed it. The hopes for anthropocentrism at the beginning of the century, and its rapid demise thereafter, constitute one of the profound shifts in twentieth century thought. In this sense the proponents of extraterrestrial life therefore champion not only a scientific theory, but an entire philosophy. But a principle of mediocrity cannot be taken for granted any more than can an anthropocentric worldview. Only observation will provide the answer.

Whatever the scientific merits of the extraterrestrial life debate, the emotional issue of human status is inextricably linked to all discussions of inhabited worlds. Pluralism and anthropocentrism have long been locked in a deadly battle that has not been completely decided by the dawn of the twenty-first century. Committed anthropocentrists, whether for religious or other reasons, are likely to be the staunchest foes of pluralism, no matter what the evidence, and pluralists—whether they liked it or not—contributed significantly to the demise of anthropocentrism. For all the appeal to scientific argument, the continuing battle between anthropocentrism and other worlds pervades modern discussions of extraterrestrial life, and carries the Darwinian debate on the status of humanity into the universe at large.

39.5 Commentary 2020

This paper was written for an anthology edited by Douglas Vakoch (Vakoch 2013). For many years Vakoch held the memorable title of "Director of Interstellar Communications" at the SETI Institute. After leaving the SETI Institute, Vakoch was the founding President of METI International, organized to proactively message ETI (see Chap. 15), something the SETI Institute did not want to do. In addition to being a METI pioneer, Vakoch has edited many anthologies centered societal aspects of astrobiology (Vakoch 2011, 2013, 2014a, b).

References

Aitken, Robert G. 1938. "Is the Solar System Unique?"*Astronomical Society of the Pacific Leaflet No. 112*, 98–106: 105.

Bianciardi, G., J. Miller, Patricia A. Straat, Gilbert Levin. 2012. "Complexity Analysis of the Viking Labeled Release Experiments." *International Journal of Aeronautical and Space Sciences*, 13, 14–26.

Coleridge, Samuel Taylor. 1831. Remark made 18 December, 1831, in *Specimens of the Table Talk of Samuel Taylor Coleridge,* Henry N. Coleridge, ed.

Columbia Accident Investigation Board. 2003. *Report*, vol. 1, p. 195.

Crowe, Michael J. 1986. *The Extraterrestrial Life Debate, 1750-1900*. Cambridge: Cambridge University Press.

DeVaucouleurs, Gerard. 1954. *Physics of the planet mars: an introduction to areophysics*. London: Faber and Faber.

Dick, Steven J. 1996. *The Biological Universe: The Twentieth Century Extraterrestrial Life Debate and the Limits of Science*. Cambridge: Cambridge University Press.

Dick, Steven J. and James E. Strick. 2004. *The Living Universe: NASA and the Development of Astrobiology*, New Brunswick, NJ: Rutgers University Press.

DiGregorio, Barry E., Gilbert Levin, Patricia A. Stratt. 1997. *Mars: The Living Planet*. Berkeley, CA: Frog Books.

Eddington, Arthur S. 1929. *The Nature of the Physical World.* New York, Macmillan.

Ezell, Edward C. and Lin Ezell. 1984. *On Mars: Exploration of the Red Planet, 1958-1978* Washington, D.C.: NASA/Government Printing Office.

Gatewood, George and Heinrich Eichhorn. 1973. "An Unsuccessful Search for a Planetary Companion of Barnard's Star (BD +4 3561," *Astronomical Journal*, 78, 769–776.

Gould, Stephen J. 1989. *Wonderful Life: The Burgess Shale and the Nature of History*, New York: W. W. Norton.

Hegel, Georg W. F. 1832. *Lectures on the Philosophy of History*, Introduction, section 8, online at http://www.marxists.org/reference/archive/hegel/works/hi/history2.htm#II

Horowitz, Norman H. 1977. "The Search for Life on Mars." *Scientific American*, 237, 52–61: 61.

Horowitz, Norman H. 1986. *To Utopia and Back: The Search for Life in the Solar System.* New York: W. H. Freeman.

Huxley, Aldous. 1959. "Case of Voluntary Ignorance," in *Collected Essays*. New York: Harper.

Jeans, James. 1917. *The Motion of Tidally Distorted Masses*, London: Memoirs of the Royal Astronomical Society.

Jeans, James. 1919. *Problems of Cosmogony and Stellar Dynamics*. Cambridge: Cambridge University Press, p. 279, 290.

Jeans, James. 1923. *The Nebular Hypothesis and Modern Cosmogony*. Oxford: Oxford University Press, p. 30

Jeans, James. 1930. "Life and the Universe," in *The Universe Around Us*. Cambridge: Cambridge University Press, 335.

Jeans, James 1942a. "Is There Life on Other Worlds?" *Science*, 95, 589

Jeans, James. 1942b. "Origin of the Solar System," *Nature*, 149, 695.

Klein, Harold P. 1977. "The Viking Biological Investigation: General Aspects," *Journal of Geophysical Research*, 82 (Sept. 30, 1977), 4677–4680.

Kohler, Robert E. 1982. *From Medical Chemistry to Biochemistry*, Cambridge: Cambridge University Press.

Kuiper, Gerard P., ed. 1949. *The Atmospheres of the Earth and Planets: Papers Presented at the Fiftieth Anniversary Symposium of the Yerkes Observatory, September, 1947* Chicago, University of Chicago Press.

Kuiper, Gerard P. 1955. "On the Martian Surface Features," *Publications of the Astronomical Society of the Pacific*, 67 (October, 1955), 271–282.

Kumar, Shiv S. 1967. "On Planets and Black Dwarfs, *Icarus*, 6, 136–137.

Levin, Gilbert. 2011. Private communication.
Levin, Gilbert and Patricia Straat. 1976. "Viking Labeled Release Biology Experiment: Interim Results." *Science*, 194 (December, 1976), 1323.
Lovejoy, Arthur O. 1936. *The Great Chain of Being*, Cambridge, MA: Harvard University Press.
Marozzi, Justin. 2008. *The Way of Herodotus: Travels with the Man who Invented History* Cambridge, MA: Da Capo Press.
McKay, David, Everett K. Gibson, Jr., et al. 1996. "Search for Past Life on Mars: Possible Relic Biogenic Activity in Martian Meteorite ALH84001." *Science*, 273, 924–930.
Navarro-González, R., Vargas, E., de la Rosa, J., Raga, A.C., and McKay, C.P. 2010. "Reanalysis of the Viking Results Suggests Perchlorate and Organics at Mid-latitudes on Mars," *Journal of Geophysical Research*, 115: E12010.
Rea, D. G., T. Belsky, and Melvin Calvin. 1963. "Interpretation of the 3-to 4- Micron Infrared Spectrum of Mars," *Science*, 141, 923–237.
Rea, D. G., Brian O'Leary, and William M. Sinton. 1965. "The Origin of the 3.58 and 3.69-Micron Minima in the Infrared Spectrum," *Science*, 147, 1286–1288.
Russell, Henry N. 1935. *The Solar System and Its Origin*. New York.
Russell, Henry N. 1943. "Anthropocentrism's Demise." *Scientific American* (July, 1943), 18–19.
Sinton, William M. 1957. "Spectroscopic Evidence for Vegetation on Mars," *Astrophysical Journal*, 126 (September, 1957), 231–239.
Sinton, William M. 1959. "Further Evidence of Vegetation on Mars," *Science* (November 6, 1959), 1234.
Vakoch, Douglas, ed. 2011. *Communication with Extraterrestrial Intelligence*. Albany: SUNY Press.
Vakoch, Douglas, ed. 2013. *Astrobiology, History and Society: Life Beyond Earth and the Impact of Discovery*. Berlin-Heidelberg: Springer-Verlag.
Vakoch, Douglas, ed. 2014a. *Archaeology, Anthropology and Interstellar Communication*. Washington, D.C: NASA.
Vakoch, Douglas, ed. 2014b. *Extraterrestrial Altruism: Evolution and Ethics in the Cosmos*. Heidelberg and NewYork: Springer.
Van de Kamp, Peter. 1956. "Planetary Companions of Stars." *Vistas in Astronomy*, 2, 1040–1048.
Van de Kamp, Peter. 1963. "Astrometric Study of Barnard's Star from Plates Taken with the 24-Inch Sproul Refractor," *Astronomical Journal*, 68 (September, 1963), 515–521.
Van de Kamp, Peter. 1975. "Unseen Astrometric Companions of Stars," *Annual Reviews of Astronomy and Astrophysics*, 13, 295–333.
Van de Kamp, Peter. 1977. "Barnard's Star 1916-1976: A Sexagesimal Report," *Vistas in Astronomy*, 20, 501–521.
Wallace, Alfred R. 1903. *Man's Place in the Universe*. New York: McClure, Phillips and Co.
Ward, Peter and Donald Brownlee. 2000. *Rare Earth: Why Complex Life is Uncommon in the Universe,* New York: Springer-Verlag.

Chapter 40
Cosmology and Biology: An Entangled Web?

Abstract Even as astrobiologists search for life in the universe, scientists and philosophers alike are pondering the possibility of a deep and profound relationship between cosmology and biology. The universe appears in its very essence to be biocentric, in that the laws of nature and the physical constants are finely tuned for life, giving rise to what has been termed the *anthropic principle*. This in turn has given rise to the idea of an ensemble of universes, or multiverse, as an explanation for why we happen to be in a universe particularly suited for life. Exactly 20 years ago, in his book *Infinite in All Directions*, physicist Freeman Dyson speculated that the prospects are bright for a future-oriented science, joining together in a disciplined fashion the resources of biology and cosmology. Fine-tuning and the multiverse are two concepts that may prove central to this task. In such a *cosmic ecology*, life and intelligence would play a central role in the evolution of the universe, no less than its physical laws.

40.1 The Mysterious Universe: A Cosmological Enigma

12 years ago in my book *The Biological Universe* I proposed that, just as 400 years ago the geocentric and heliocentric world systems hung in the balance, in the late twentieth century the two reigning chief world views were the physical universe and the biological universe. This remains true today. I pointed out that these were two possible outcomes for cosmic evolution: that its ultimate endpoint consists primarily of matter such as planets, stars and galaxies (the physical universe), or its ultimate outcome is life, mind and intelligence (the biological universe). I framed the biological universe as an emerging cosmology, which I called the *biophysical cosmology*, similar to Barney Oliver's "biocosmology," a term he had used in his influential report Project Cyclops (Dick 1996, 34–335, 541–542, 1989, 13–17; Oliver 1973, 30).

The physical and biological universe are obviously connected in the sense that one cannot have biology before one has the physical substrate. But it now turns out that scientists and philosophers alike are increasingly aware of mysterious and deep connections between the physical universe of planets, stars, and galaxies, and the biological universe of life, mind and intelligence. When the British astronomer James Jeans

First published in *Proceedings of the 2008 Conference on the Society of Amateur Radio Astronomers*, June 29–July 2, 2008 (NRAO, Green Bank, West Virginia), pp. 1–15

S. J. Dick, *Space, Time, and Aliens*, https://doi.org/10.1007/978-3-030-41614-0_40

80 years ago wrote his book *The Mysterious Universe*, he was referring to the fact that the universe embodies mathematical laws, and that the human mind could comprehend this mathematical structure, a point later elaborated by Albert Einstein, Eugene Wigner and others.[1] The more we learn about the universe, the more we find that it is even more mysterious than we thought—that not only is the universe subject to mathematical laws, but also it is finely tuned for life. Or, to put it another way, the universe is *biofriendly*, as witnessed by our existence, and the possible existence of other life beyond Earth—the subject of astrobiology and SETI. As Winston Churchill said in another context, this situation "is a riddle, wrapped in a mystery, inside an enigma." "But perhaps," he added, "there is a key."[2] Churchill was talking about Russia's non-aggression pact with Germany, and the key was Russian national interest. Our key will be more difficult to discover, but the stakes are much greater, nothing less than the nature of the cosmos and its intimate relation to life. As cosmologists John Barrow and Frank Tipler wrote in their landmark volume *The Anthropic Cosmological Principle* (1986), "The realization that the possibility of biological evolution is strongly dependent upon the global structure of the Universe is truly surprising and perhaps provokes us to consider that the existence of life may be no more, but no less, remarkable than the existence of the Universe itself" (Barrow and Tipler 1986).

Already at the turn of the twentieth century there was some realization of the strange connections between the nature of the universe and the existence of life (Dick 2008). In the last chapter of his volume *Man's Place in the Universe, A. R.* Wallace—the cofounder with Darwin of the theory of natural selection—wrote in 1903:

> … I submit that the whole of the evidence I have here brought together leads to the conclusion that our earth is almost certainly the only inhabited planet in our solar system; and, further, that there is no inconceivability – no improbability even – in the conception that, in order to produce a world that should be precisely adapted in every detail for the orderly development of organic life culminating in man, such a vast and complex universe as that which we know exists around us, may have been absolutely required (Wallace 1903).[3]

Because he believed no life existed outside the solar system either, Wallace thus intimated that the universe was made for life on Earth, an anthropocentric and true anthropic principle (as we would now call it) in that it centered on humans: "the supreme end and purpose of this vast universe was the production and development of the living soul in the perishable body of man" (Wallace 1904, 304). Based on what was known at the turn of the twentieth century, there was some justification for this anthropocentric world view—the Solar System was still considered to be at or near the center of the observed universe (see Chap. 35, this volume).

The idea of a connection between the universe and life was also anticipated by Wallace's American contemporary, the Harvard biochemist Lawrence J. Henderson, who marveled at the fitness of the terrestrial environment for life, especially because of the properties of water. In the final paragraph to his book *The Fitness of the Environment*, Henderson extended his argument to the universe at large:

> The properties of matter and the course of cosmic evolution are now seen to be intimately related to the structure of the living being and to its activities; they become, therefore far more important in biology than previously suspected. For the whole evolutionary process, both cosmic and organic, is one, and the biologist may now rightly regard the universe in its very essence as biocentric (Henderson 1913; Fry 1996, 155–196).

In Henderson's formulation, the *biocentric principle* is a better term for the fitness of the universe for life, since it does not imply that terrestrial life forms, including humans, could be the only form of life. The universe was fine-tuned for life, including beings of higher consciousness, not only for humans, even if the only intelligent life in the universe we yet know is human. Others have gone even further and suggested that what is really indicated is a *complexity principle*, that the universe is fine-tuned for complexity, not just biology (Carr 2007a, 88).

Unfortunately, when these questions were once again pondered six decades later, the term "anthropic principle"was coined, almost by accident. This occurred in a path-breaking article on the subject by the British astronomer Brandon Carter (Carter, 1974), who first formulated what he called the weak anthropic principle: "what we can expect to observe must be restricted by the conditions necessary for our presence as observers" (Carter 2007).[4] Thus the gravitational constant is constrained by the fact that we exist, otherwise the conditions for life could never have arisen. Carter went on to expound a "strong anthropic principle," namely, "that the universe (and hence the fundamental parameters on which it depends) must be such as to admit the creation of observers within it at some stage," a much more problematic claim. Although such reasoning turned the deductive method on its head, Carter argued that it might be considered a kind of explanation for why our universe is the way it is. He later admitted that "anthropic" was an unfortunately limiting term.

Five years after Carter's paper, Bernard Carr and Martin Rees wrote a paper bringing together all the known constraints on the physical characteristics of the universe that seemed necessary for the emergence of life (Carr and Rees 1979). Over the last several decades the British Astronomer Sir Martin Rees, physicist and astrobiologist Paul Davies and many others have explored the ramifications of this possibly deep and profound relationship between biology and cosmology (Rees 1997a, 2001; Davies 1982, 1992, 2007). In his book *Just Six Numbers* Rees crystallized the mystery as encapsulated in the following six numbers (Rees 1997b, 2–4):

1. N, a ratio equal to ten to the 36th power (1 with 36 zeros), the strength of electrical forces that hold atoms together divided by the force of gravity between them. "If N had a few less zeroes, only a short-lived miniature universe could exist: no creatures could grow larger than insects, and there would be no time for biological evolution."
2. Epsilon, .007, defining the binding power of holding nuclei together, controlling the transmutation of hydrogen into the other elements of the periodic table. If Epsilon were .006 or .008, Rees points out, we could not exist because the proper elements would not exist.
3. Omega, the amount of material in our universe, including dark matter, a measure of gravity vs. expansion energy. If the universe expanded too slowly, no galaxies or stars would have formed; too fast and the universe would have collapsed long ago. "The initial expansion speed seems to have been finely tuned," Rees comments.
4. Lambda, a cosmic antigravity that controls the expansion rate of the universe. Its small value allowed cosmic evolution to take place.

5. Q, the ratio of two fundamental energies, which accounts for the seeds of all cosmic structures imprinted in the Big Bang, with a value of about 1/100,000.
6. D, the number of spatial dimensions in the world, equal to 3. Life could not exist if D were 2 or 4, Rees comments.

In addition to these numbers, there are fundamental constants at another level that are crucial to life, sometimes related to the numbers above, such as the Newtonian gravitational constant (G), Planck's constant (h), the charge of the electron and the mass of the proton (Barrow 2002; Barrow and Tipler 1986, 224).[5] So the bottom line is, we live in a *Goldilocks universe*, where conditions for life are not too bad and not too good, but just right.

There are a number of ways to explain why the universe is so biofriendly:

1. It's just a coincidence. For some this is enough, but for most people who have thought deeply about the problem, it's too good to be true and requires an explanation.
2. A supernatural intelligence fine-tuned the universe. This is better known as the *God Hypothesis*.
3. A highly evolved natural intelligence fine-tuned the universe. This possibility has received some attention, but not a great deal.
4. A multiverse—an ensemble of universes—exists, and we happen to live in a universe that is suitable for life. Such a multiverse could exist sequentially or simultaneously.

In the remainder of this paper, I will review these last three options, and in the process analyze the arguments and offer my own thoughts on the connections between the physical universe and the biological universe.

40.2 Supernatural Intelligence: The God Hypothesis

The God Hypothesis as related to our subject has a long history in the form of the design argument, the history of which has been written many times, including Chap. 2 of Barrow and Tipler's book *The Anthropic Cosmological Principle* (1986). In my first book on the extraterrestrial life debate, *Plurality of Worlds*, I showed how already in the late seventeenth and early eighteenth centuries, Newtonians intimately wedded the idea of an abundance of life beyond Earth to the munificence of an omnipotent God; this was one form of the widespread argument and ideology known as "natural theology" (Dick 1982, Chap. 6). The Englishman William Derham even wrote a well-known treatise *Astro-Theology: or, a Demonstration of the Being and Attributes of God, from a Survey of the Heavens* (1715), in which he utilized every variety of argument, from the magnitude of the universe to the size, shape and motion of its bodies, as evidence of the Deity. Though yet undiscovered, he favored the idea of inhabited planets and numerous solar systems as completely in line with this natural theology view of the world (Dick 1982, Chap. 6, 151–154). From William Paley's famous "watchmaker" argument in the late eighteenth century to John Polkinghorne in the twentieth century, this argument has been made again and again. Most recently Robert Collins has given it a more nuanced view (Barrow and Tipler 1986, 76–82; Polkinghorne 1994; Collins 2007, 459–480). The design argument is not one that most scientists embrace, and the "God of the gaps" argument is not even one that most theologians today embrace, history demonstrating that most "gaps" are filled in as science progress. It is, however, the core of the Intelligent Design movement that has made some headway in the United States by invoking the idea of "irreducible complexity," even if its proponents are often vague about the nature of the designing intelligence, not always pronouncing it as the canonical "God." (Numbers 2006).

While the God Hypothesis would be interesting if true, it remains by definition unprovable, since it is supernatural and thus outside the realm of science (which to its proponents is precisely the point). For those who seek a scientifically verifiable explanation based on the natural world, it is unsatisfactory.

40.3 Natural Intelligence

Returning from the supernatural to the realm of science, how about the idea of a "natural intelligence" as a designer? As Paul Davies has pointed out, this idea could be traced back as far as Plato, whose Demiurge had to work within the laws of nature to create the universe (Davies 2007, 201). Olaf Stapledon's Star Maker, in his famous science fiction novel by that name from the 1930s, would presumably also be in this tradition (Stapledon 1968). More recently the idea has been developed in the context of the extraterrestrial life debate. After all, most SETI proponents consider advanced extraterrestrial intelligence to be very old and wise; why not have the ultimate superintelligence, a product of natural selection, create the universe? This is exactly what the British astronomer Fred Hoyle proposed in his book *The Intelligent Universe* in 1983. Here Hoyle, a self-proclaimed atheist, concluded that the simplest explanation for the biofriendly universe was that a natural (not supernatural) super-intellect had engineered the universe to make it fit for carbon-based life and intelligence. Edward Harrison proposed a similar scenario in 1995, as did I in one of the Templeton conferences in 1998 (Hoyle 1983; Harrison 1995; Dick 2000). More recently James Gardner has elaborated this theme in a book with the same title as Hoyle's, following up on his book *Biocosm*. His *selfish Biocosm Hypothesis* is that life and intelligence are the primary cosmic phenomena, and that "the capacity of the universe to generate life and to evolve ever more capable intelligence is encoded as a hidden subtext to the basic laws and constants of nature," and the biofriendly laws of nature are equivalent to DNA in living creatures, "providing a recipe for development of a living universe and a blueprint for the construction of offspring (baby universes)" (Gardner 2003, 159; Gardner 2007a, 2007b).

Such thoughts raise the question of the nature of such a super-intelligence, a question necessarily speculative but that can be discussed based on what little we know about cultural evolution on Earth. In the context of SETI strategies, I have proposed by a series of arguments that extraterrestrial intelligence may be postbiological, so the superintelligence may be some form of artificial intelligence (Dick 2003). There are other possibilities, but in any case, in a universe 13.7 billion years old in which we treat astronomical and biological evolution as a known fact, we must take equally seriously cultural evolution. As we see on Earth, cultural evolution completely dominates other forms of evolution in terms of its rapidity. If extraterrestrial intelligence has been around for millions or billions of years, it too will have undergone cultural evolution, the ultimate form of which is the improvement of its own intelligence. A highly evolved natural intelligence is unlikely to be biological, least of all humanoid. As a natural intelligence, this explanation for the fine-tuning of the universe holds out the possibility of being proved at least in principle; as such it is very different from the God Hypothesis and the Intelligent Design movement.

40.4 The Multiverse

In contrast to the God Hypothesis and the Natural Intelligence explanations for the intimate relation between cosmology and biology, there is one explanation where the parameters are not "chosen" by a Designer, whether natural or supernatural. This is the so-called multiverse idea, "the possibility that there could be other universes (either connected or disconnected from ours) in which the constants of physics (and perhaps even the laws of nature) are different" (Carr 2007b, xv). The general idea of an ensemble of universes has been around for a long time, going back to the ancient Greek atomists, who believed in an infinite number of *kosmoi,* ordered worlds separated in space, some larger or smaller, some with life or devoid of life, and so on (Dick 1982, 6–11). The idea is also notable in the Leibnitzian discussion of the best of all possible worlds, and the more physically-grounded many worlds interpretation of quantum mechanics (Gale 1998, 195–212).[6] John Barrow cites the Cambridge biologist Charles Pantin as the first to articulate the many worlds approach in the modern sense. In 1965 Pantin wrote,

> …the properties of the material Universe are uniquely suitable for the evolution of living creatures … If we could know that our own Universe was only one of an indefinite number with varying properties we could perhaps invoke a solution analogous to the principle of Natural Selection, that only in certain Universes, which happen to include ours, are the conditions suitable for the existence of life, and unless that condition is fulfilled there will be no observers to note the fact. (Barrow 2002, 276)[7]

This is not the only example of how biology can lend fruitful concepts to astronomy.

The origin of the term "multiverse" is more recent, not becoming widespread until the 1990s. The word is not found in Paul Davies' early book on the subject, *The Accidental Universe* (1982), for example, where he spoke of "the many universes theory," nor in the 1986 Barrow and Tipler volume. The first use of the term I have seen is in the work of the Canadian philosopher John Leslie, whose 1989 book *Universes* begins its first chapter with the question "God or multiverse?"

Leslie points out that he borrowed the term from, of all places, *The Education of Henry Adams*, published in 1918, where Adams used the term in quite a different context (Leslie 1996, 1998).[8] But the term "multiverse" came into its own in the late 1990s after British Astronomer Royal Martin Rees championed the idea of multiple universes, and used the term in his books beginning as early as 1997. With the recent publication of a definitive set of papers edited by Bernard Carr and titled *Universe or Multiverse?* the term is assured a long existence, though the term "parallel universes" is also used (Carr 2007b).[9]

Logically speaking, a multiverse could be a large number of actually existing universes, or could refer to the philosophical problem of all possible worlds/universes, either actually existing or not all existing. The late Princeton philosopher David Lewis made a career of elucidating the philosophical problem of all possible worlds, and arguing that all possible worlds do exist in actuality. His book *On the Plurality of Worlds* (1986), although titled very similarly to my first book and published only 4 years after it, is very different in the sense that while it argues that all possible universes are real, it says very little about the only universe we know for sure is real—our own (Lewis 1986). Nevertheless, it stands as a monument to rigorous philosophical thinking, is widely known and cited in philosophical circles, and is in fact relevant to the multiverse idea from a foundational point of view.

But why raise such a seemingly non-empirical idea as the multiverse in the first place? From the point of view of science there are several reasons to think many universes might exist, though none of these rise to the level of firm evidence.[10]

1. Current cosmology indicates that the universe is infinite, and that universes may exist beyond our observable cosmic horizon, though it is open to question whether one should really classify these as different universes.[11]
2. Physicist Lee Smolin has argued in his book *The Life of the Cosmos*, the formation of black holes may spawn baby universes with different fundamental constants (Smolin 1997).[12] He even proposes a kind of cosmological natural selection, in which after many generations the production of black holes is maximized by selecting certain values of constants. In this view the fundamental constants provide the "variation" on which selection acts, just as variation of physical traits are the essence of Darwinian biological evolution. Even Smolin himself, however, has argued recently that this idea is neither falsifiable nor testable, though not everyone agrees (Smolin 2007a).
3. Some models of the inflationary theory of the Big Bang, a leading theory of the origin of our universe initiated by Alan Guth, indicate that inflationary processes happen more than once, leading to a scenario known as *chaotic* or *eternal inflation*. Such multiple inflations might lead to the creation of other universes. Eternal inflation is especially associated with the work of Andre Linde (2007, 127–149; Guth 1997).
4. Superstring theory hypothesizes "landscapes" with large numbers of vacua, which could constitute universes. These vacua arise because in string theory there are more than three dimensions, which are "compactified" into what we see as three dimensions. This compactification of dimensions can be done in many

ways, and our universe corresponds to just one such compactification. This idea is associated with Leonard Susskind (Susskind 2007). As Paul Davies says in his latest book *Cosmic Jackpot*, other universes "Might have ten rather than three species of neutrino, of five sorts of photon. Others might have only four quarks, or as many as forty. There could be worlds in which electromagnetism is stronger than the strong nuclear force, or where there are eight fundamental forces instead of four—and so on. Obviously our world is just one among a dizzying number of alternatives." (Davies 2007, 112). Of course, despite the large number of physicists working on it, the validity of string theory is widely questioned; Lee Smolin, for one, has enumerated its problems and associated methodology in his book *The Trouble with Physics* (Smolin 2007b).

5. Many universes could also be generated from cosmic branching according to what is known as the Everett-Wheeler interpretation of quantum theory, the original of several formulations of the many worlds interpretation of quantum mechanics (Everett 1957; Carr 2007, 450; Davies 2007, 228).[13]

Cosmologist Max Tegmark argues that the idea of the multiverse is grounded in well-tested theories such as relativity and quantum mechanics, and that it is empirical science because it makes predictions and can be falsified. The key question, he says, is not whether the multiverse exists, but how many levels it has. Accordingly, he has described four possible levels of multiverse, according to how different they would be from ours, ranging from those similar to ours but beyond our observable horizon, to those having different physical laws (see Table 40.1) (Tegmark 2003a, b). A Level I multiverse comprises those universes currently beyond our observable universe, which grows at the rate of 1 light year per year as light from farther away

Table 40.1 Tegmark's multiverse levels

Level	Theoretical basis	Proponents	Characteristics
I	Big Bang cosmology	Numerous	Universes beyond our light horizon. Same laws and dimensions; different initial conditions
II	Black Hole Generation	L. Smolin	Different physical constants; Different space-time dimensions, Different elementary particles
	Oscillating universe	J. Wheeler	ditto
	Chaotic Inflation theory	A. Linde	ditto
	String theory	Susskind	ditto
III	Quantum Mechanics	H. Everett, III	Multiverses in infinite
	Many worlds interpretation	J. Wheeler	Dimensional Hilbert space
IV	Plato vs. Aristotle	M. Tegmark	Different physical laws
	Pi in the sky	J. Barrow	Different fundamental equations
	Principle of fecundity	R. Nozick	
	Modal realism	D. Lewis	

Credit: From Tegmark (2003a)

reaches us. It is in the Level I multiverse that you are supposed to find your doppelganger, as all possible permutations are played out in the infinite space. The Level II multiverse, associated with black holes, inflation or string theory, might have different physical constants, different space-time dimensionality and different elementary particles. The Level III multiverse is the one associated with the many worlds interpretation of quantum mechanics, where all possibilities based on human action are played out. As Tegmark notes, whereas in the Level I multiverse your doppelganger exists in three-dimensional space, in Level III "they live on another quantum branch in infinite-dimensional Hilbert space" (Tegmark 2003a, b, 48). At Level IV, the multiverse is a form of radical Platonism in the sense that Plato's "realm of ideas" is considered to be reality. These multiverses and their characteristics are summarized in Table 40.1.

None of the theoretical mechanisms listed in Table 40.1, it should be noted, constitute direct evidence for the existence of a multiverse, though the fine-tunings for life might be considered indirect evidence. In this context it should also be noted that the ideas of a multiverse and the fine-tunings of the anthropic principle are separable; the fine-tunings could have nothing to do with a multiverse, and a multiverse could exist without the fine-tunings. Like the God Hypothesis, they are interesting if true, but also unprovable.

40.5 Philosophical Issues

The three primary explanations for the biofriendly universe beg an important question: are they verifiable? For those who believe in the supernatural, the God Hypothesis is acceptable, but not verifiable; even the design argument does not uniquely point to a supernatural explanation—that is precisely why some favor the multiverse hypothesis as an alternate explanation.[14] For those who believe many universes exist and we just happen to be in one that is biofriendly, the multiverse is acceptable, but almost by definition other universes beyond our own are not verifiable by normal scientific standards.[15] This leaves a highly evolved natural intelligence as the leading theory. As a natural entity in our own universe this would be verifiable in principle, though admittedly difficult. I therefore recommend that more attention be paid to cultural evolution in the cosmos, and its possible endpoints, of which postbiologicals are one manifestation. Creation by a natural intelligence raises intriguing possibilities, one of which has been played out in the movie The Matrix.

Lest we be accused of unbridled speculation, let us recall that we have been led to these conclusions by the empirically observed fine-tuning of the universe, inelegantly and erroneously named the anthropic principle. As others have noted, it is neither anthropic nor a principle, but by whatever name it is an enigma. But is it science? Bernard Carr in his book *Universe or Multiverse* admits the concept has earned the disdain of many physicists in the past. For example in his book *Perfect Symmetry* Heinz Pagels (who has thought deeply about these matters in his book *The Cosmic Code*) asserts "the influence of the anthropic principle on contemporary cosmological models has been sterile. It has explained nothing and it has even had a negative influence. I would opt for rejecting the anthropic principle as needless clutter in the conceptual repertoire of science" (Pagels 1985, 1982). However, in favor of the anthropic principle Carr cites a fundamental change in the epistemological status of multiverse proposal and the anthropic principle because of the many ways in which we now realize other universes could originate. Still, the discomfort remains and Carr says the reason is clear: "The idea is highly speculative and, from both a cosmological and a particle physics perspective, the reality of a multiverse is currently untestable. Indeed, it may always remain so, in the sense that astronomers may never be able to observe the other universes with telescopes and particle physicists may never be able to observe the extra dimensions with their accelerators" (Carr 2007b, 3, 14; Davies 2007, 204–205). He holds out hope for empirical verification only if the effects of other dimensions became visible at TeV scales.

There is another way in which the anthropic principle and its teleological underpinnings may be scientific, and it is encapsulated in the comment of Freeman Dyson: "I do not feel like an alien in this Universe. The more I examine the Universe and examine the details of its architecture, the more evidence I find that the Universe in some sense must have known we were coming" (Dyson 1979, 447). Could the universe have known we were coming? Physicist Paul Davies has discussed this possibility of "backwards causation" in the context of quantum theory. He points out some famous theories of physics involve backward causation: the Wheeler-Feynman theory of electrodynamics, the Hoyle-Narlikar theory of gravitation, and the Gellmann-Hartle-Hawking theory of quantum cosmology. Unfortunately, none of these, he hastens to add, have any experimental evidence (Davies 2007, 242–249).

In my view, in the end the concept of an advanced natural intelligence tuning our own universe tops the other two explanations in terms of verifiability according to normal scientific practice. But all three explanations for the biofriendly universe suffer from the same inherent weakness. If God fine-tuned the universe, who created God? If a natural intelligence did the fine tuning, it evolved from a more primitive intelligence and more primitive life, but what is the origin of that life? And if the multiverse explains fine tuning, who created the multiverse, or the first universe that spawned all others? Those questions lead to a broader, deeper and age-old philosophical problem: Why is there something rather than nothing?

40.6 Conclusions

20 years ago, the physicist Freeman Dyson called for the need to build bridges between biology and cosmology. "Only a few heretics … dare to express the view that the structure of the universe may not be unambiguously reducible to a problem of physics," he wrote in his iconoclastic book *Infinite in All Directions*. "Only a few romantics like me continue to hope that one day the links between biology and cosmology may be restored." He suggested further that

> The prospects are bright for a future-oriented science, joining together in a disciplined fashion the resources of biology and cosmology. When this new science has grown mature enough to differentiate itself clearly from the surrounding farrago of myth and fiction, it might call itself "cosmic ecology," the science of life in interaction with the cosmos as a whole. Cosmic ecology would look to the future rather than to the past for its subject matter, and would admit life and intelligence on an equal footing with general relativity as factors influencing the evolution of the universe. (Dyson 1988, 50–51).

Fine-tuning and the multiverse are two concepts that may prove central to this task. In such a cosmic ecology, life and intelligence do indeed play a central role in the evolution of the universe, no less than its physical laws. If the universe indeed "saw us coming," life may be its ultimate purpose for existing, returning to physics a kind of teleology that had been banned from science. This lends all the more importance to the current disciplines of astrobiology and SETI, which will determine just how abundant life is in the universe—or (perhaps in the very long run!) the multiverse. The biological universe may be not the result but the cause of the physical universe, part of its very fabric, and more surprisingly fundamental than we have ever realized.

40.7 Commentary 2020

This paper was first given as a keynote address at the Society of Amateur Radio Astronomers (SARA) at its meeting at the National Radio Astronomy Observatory (NRAO) in Green Bank, West Virginia in 2008. It appeared in the Proceedings of that meeting. As hinted in Chap. 35 on A. R. Wallace, the status of the anthropic principle remains a subject of hot debate (Siegfried 2019), even as its history is now much better known (Rubenstein 2014).

Notes

1. Einstein's statement was "The most incomprehensible thing about the world is that it is at all comprehensible." Wigner's thoughts on the subject are in (Wigner 1960).
2. Churchill's comment came in a radio address in 1939, 2 months after Russia's non-aggression compact with Germany.
3. This was the article on which Wallace's book was based.
4. Carter also went on to apply the anthropic principle to biology and quantum theory, in addition to cosmology(Carr 2007b, 285).
5. The constants of nature have been a fascinating problem since the "Large Number Coincidences" that Arthur S. Eddington and P. A. M Dirac studied in the 1930s. But these authors did not relate to the fine-tuning problem. See Miller (2005) and Douglas 1979).
6. Gale distinguishes three categories of multiple worlds: spatially multiple, temporally multiple, and other-dimensional multiple.
7. Pantin also added "But even if there were any conceivable way of testing such a hypothesis we should only have put off the problem of why, in all those Universes, our own should be possible?!" See also Barrow and Tipler (1986, 250).

8. Leslie's quote from Adams is "If he were obliged to insist on a Universe, he seemed driven to the Church," which he follows with his own paraphrase "so he opted for a 'multiverse' of largely or entirely separate worlds with very different characteristics." Adams does use the word "multiverse", but these worlds for Adams were not physical universes. Leslie refers to the 1931 edition of Adams, but it was first published in 1918. As the references in his volume show, Leslie is the philosopher who has written most on the subject of other universes, beginning with his ideas in the 1970s that the universe is "ethically required." See also the collection of articles in Leslie (1998)
9. Another important set of papers is found in a special issue of the *International Journal of Astrobiology*, on "Fine Tuning in Living Systems," volume 2, issue 2 (April, 2003).
10. There have been several taxonomies of the multiverse idea: Leslie, *Universes* (1996), pp. 6–7; Tegmark (2003a, b); and Kuhn (2007).
11. On the concept of cosmic horizon, and the differences between the observed universe, the observable universe, and the entire universe see, Davies (2007), pp. 30–40.
12. For his most recent analysis, which pans the anthropic principle, but not the multiverse, see Smolin (2007b)
13. In its "Copenhagen Interpretation" entry, the sometimes-accurate Wikipedia reports that according to a poll at a Quantum Mechanics workshop in 1997, "the Copenhagen interpretation is the most widely-accepted specific interpretation of quantum mechanics, followed by the many-worlds interpretation. Although current trends show substantial competition from alternative interpretations, throughout much of the twentieth century the Copenhagen interpretation had strong acceptance among physicists. Astrophysicist and science writer John Gribbin describes it as having fallen from primacy after the 1980s."
14. Carr (2007a, 16) points out that, while Cosmic Design should not be confused with Intelligent Design, "atheists might hope that the multiverse theory will have the same impact in the context of cosmic design as the theory of evolution did in the context of biological design."
15. The Jesuit astronomer William R. Stoeger offers the possibility that multiverses are testable through the concept of "retroduction," an idea proposed by the philosopher C. S. Peirce and supported by Notre Dame philosopher/historian Ernan McMullin. He defines retroduction as "inference based on the success or fruitfulness of an hypothesis in accounting for better understanding of a set of phenomena." Thus, if a hypothesis accomplishes the following, it may be true despite the inability to actually see its hidden properties (such as a multiverse): (1) account for empirical data (empirical adequacy); (2) provide long-term explanatory success and stimulate productive lines of further enquiry (theory fertility); and (3) bring together previously disparate domains of facts (unifying power) (Stoeger 2007, 445–457: 450–45). James Gardner uses "retrodiction" in his Selfish Biocosm Hypothesis Gardner (2003, 197–211).

References

Barrow, John D. 2002. *The Constants of Nature*. New York: Vintage Books

Barrow, John and Frank Tipler. 1986. *The Anthropic Cosmological Principle*. New York: Oxford University Press.

Carr, Bernard. 2007a. "The Anthropic Principle Revisited," in (Carr, 2007b, 88)

Carr, Bernard. 2007b. *Universe or Multiverse?* Cambridge: Cambridge University Press

Carr, Bernard J., and M. J. Rees. 1979. "The Anthropic Principle and the Structure of the Physical World," *Nature*, 278 (April 12, 1979), 605–612.

Carter, Brandon. 1974. "Large Number Coincidences and the Anthropic Principle in Cosmology," in M. S. Longair, ed. *Confrontation of Cosmological Theories with Observational Data*. Dordrecht, 291–298.

Collins, Robert. 2007. "The Multiverse Hypothesis: A Theistic Perspective," in Carr (2007b), pp. 459–480.

Davies, Paul. 1982. *The Accidental Universe*. Cambridge: Cambridge University Press
Davies, Paul. 1992. *The Mind of God*. New York: Simon and Schuster
Davies, Paul. 2007. *Cosmic Jackpot: Why Our Universe is Just Right for Life*. Houghton Mifflin: Boston and New York.
Dick, Steven J. 1982. *Plurality of Worlds: The Origins of the Extraterrestrial Life Debate from Democritus to Kant*. Cambridge University Press: Cambridge
Dick, Steven J. 1989. "The Concept of Extraterrestrial Intelligence - An Emerging Cosmology," *Planetary Report*, 9 (March-April, 1989), 13–17
Dick, Steven J. 1996. *The Biological Universe: The Twentieth Century Extraterrestrial Life Debate and the Limits of Science*. Cambridge: Cambridge University Press
Dick, Steven J. 2000. "Cosmotheology: Theological Implications of the New Universe," in Steven J. Dick, ed, *Many Worlds. The New Universe, Extraterrestrial Life and the Theological Implications*. Philadelphia and London: Templeton Foundation Press, pp. 191–210.
Dick, Steven J. 2003. Cultural Evolution, the Postbiological Universe and SETI, *International Journal of Astrobiology*, 2, 65–74
Dick, Steven J. 2008. "The Universe and Alfred Russel Wallace," in *Natural Selection and Beyond: The Intellectual Legacy of Alfred Russel Wallace*, Charles H. Smith and George W. Beccaloni, eds. Oxford: Oxford University Press
Douglas, A. V. 1979. The Life of Arthur Stanley Eddington. London: Thomas Nelson and Sons
Dyson, Freeman. 1979. *Reviews of Modern Physics*, 51, 447.
Dyson, Freeman. 1988. *Infinite in all Directions*. New York: Harper and Row.
Everett, III, Hugh. 1957. "Relative State Formulation of Quantum Mechanics," *Reviews of Modern Physics*, 29, no. 3 (July, 1957)
Fry, Iris. 1996. "On the Biological Significance of the Properties of Matter: L. J. Henderson's Theory of the Fitness of the Environment," *Journal of the History of Biology*, 29, 155–196.
Gale, George. 1998. "Cosmological Fecundity: Theories of Multiple Universes," in John Leslie, ed., *Modern Cosmology & Philosophy*. Prometheus Books: Amherst, pp. 195–212.
Gardner, James. 2003. *Biocosm: The New Scientific Theory of Evolution: Intelligent Life is the Architect of the Universe*. Makawao, Maui, Hawaii: Inner Ocean Publishing.
Gardner, James. 2007a. *The Intelligent Universe: AI, ET and the Emerging Mind of the Cosmos*. Franklin Lakes, N.J.: New Page Books
Gardner, James. 2007b. "The Physical Constants as Biosignature: An Anthropic Retrodiction of the Selfish Biocosm Hypothesis," in Gardner (2007a), 197–211
Guth, Alan. 1997. *The Inflationary Universe: The Quest for a New Theory of Cosmic Origins*. Reading, Mass.: Addison Wesley
Harrison, Edward W. 1995. "The Natural Selection of Universes Containing Intelligent Life," *Quarterly Journal of the Royal Astronomical Society*, 36, no. 3, 193
Henderson, Lawrence J. 1913. *The Fitness of the Environment: An Inquiry into the Biological Significance of the Properties of Matter*. Cambridge, Mass.: Macmillan, reprinted in 1970 with an introduction by George Wald, Gloucester, Mass: Peter Smith, p. 312.
Hoyle, Fred. 1983. *The Intelligent Universe: A New View of Creation and Evolution*. New York: Holt, Rinehart and Winston
Kuhn, Robert Lawrence. 2007. "Why This Universe? Toward a Taxonomy of Possible Explanations," *Skeptic*, vol. 13, no. 2, pp. 30–41.
Leslie, John. 1996. *Universes*. Routledge, London and New York, p. 1.
Leslie, John. 1998. *Modern Cosmology & Philosophy*. Prometheus Books: Amherst, NY.
Lewis, David. 1986. *On the Plurality of Worlds*. Oxford: Blackwell.
Linde, Andre. 2007. "The Inflationary Multiverse," in Carr (2007), 127–149.
Miller, Arthur I. 2005. *Empire of the Stars*. Boston: Houghton Mifflin, especially pp. 67–72 and 143–146
Numbers, Ron. 2006. *The Creationists: From Scientific Creationism to Intelligent Design*. Cambridge, Mass.: Harvard University Press
Oliver, Bernard. 1996. *Project Cyclops: A Design Study of a System for Detecting Extraterrestrial Intelligent Life*, 1973, reprinted, 1996.

Pagels, Heinz. 1982. *The Cosmic Code: Quantum Physics as the Language of Nature*. New York: Simon and Schuster

Pagels, Heinz. 1985. *Perfect Symmetry: The Search for the Beginning of Time*. New York: Simon and Schuster, p. 359.

Polkinghorne, J. 1994. *The Faith of a Physicist*. Princeton University Press

Rees, Martin. 1997a. *Before the Beginning: Our Universe and Others*. New York: Perseus Books

Rees, Martin. 1997b. *Just Six Numbers: The Deep Forces that Shape the Universe*. New York: Basic Books

Rees, Martin. 2001. *Our Cosmic Habitat*. Princeton, N.J.: Princeton University Press.

Rubenstein, Mary-Jane. 2014. *Worlds Without End: The Many Lives of the Multiverse*. New York, Columbia University Press.

Siegfried, Tom. 2019. *The Number of the Heavens: A History of the Multiverse and the Quest to Understand the Cosmos*. Cambridge, MA, Harvard University Press.

Smolin, Lee. 2007b. *The Trouble with Physics* (Boston and New York: Houghton and Mifflin, 2007)

Smolin, Lee. 1997. *The Life of the Cosmos*. New York: Oxford University Press

Smolin, Lee. 2007a. "Scientific Alternatives to the Anthropic Principle, in Carr (2007b), 323–366: 335–336.

Stapledon, Olaf. 1968. *Last and First Men* and *Star Maker*. New York.

Stoeger, William. 2007. "Are Anthropic Arguments, Involving Multiverses and Beyond, Legitimate?," in Carr (2007b)

Susskind, Leonard. 2007. "The Anthropic Landscape of String Theory," in Carr (2007b), 247–266.

Tegmark, Max. 2003a. "Parallel Universes," *Scientific American* (May, 2003) 41–51

Tegmark, Max. 2003b. "Parallel Universes," in *Science and Ultimate Reality: From Quantum to Cosmos*, J. D. Barrow, P.C.W. Davies, and C.L. Harper, eds. Cambridge: Cambridge University Press.

Wallace, A. R. 1903. "Man's Place in the Universe, as Indicated by the New Astronomy," *The Fortnightly Review*, March 1, 1903, 396. Online at http://www.wku.edu/~smithch/wallace/S602.htm.

Wallace, A. R. 1904. *Man's Place in the Universe: A Study of the Results of Scientific Research in Relation to the Unity or Plurality of Worlds*. New York: McClure, Phillips and Co., p. 306. Online at http://www.wku.edu/~smithch/wallace/S728-1.htm.

Wigner, Eugene. 1960. "The Unreasonable Effectiveness of Mathematics in the Natural Sciences," in *Communications in Pure and Applied Mathematics*, vol. 13, No. 1 (February 1960). New York: John Wiley & Sons, Inc. also at http://www.dartmouth.edu/~matc/MathDrama/reading/Wigner.html

Part VII
Envoi: Reflections on Humanity and the Cosmos

Part VII Frontispiece Statue of the brooding Giordano Bruno in the Piazza Campo de' Fiori, Rome. Here the Roman Inquisition burned the philosopher for heresy February 17, 1600, in part for his belief in an infinite number of inhabited worlds. Bruno held a cosmic perspective shared by many great thinkers

> *Look again at that dot. That's here. That's home. That's us. On it everyone you love, everyone you know, everyone you ever heard of, every human being who ever was, lived out their lives. The aggregate of our joy and suffering, thousands of confident religions, ideologies, and economic doctrines, every hunter and forager, every hero and coward, every creator and destroyer of civilization, every king and peasant, every young couple in love, every mother and father, hopeful child, inventor and explorer, every teacher of morals, every corrupt politician, every "superstar," every "supreme leader," every saint and sinner in the history of our species lived – on a mote of dust suspended in a sunbeam.*
>
> Carl Sagan

In one of the most famous passages from his prolific writings, Carl Sagan displays the cosmic perspective *par excellence*. It is a viewpoint that has been implicit in many of the articles in this volume, and in closing we examine this "view from a distant star" directly as a way of reflecting on humanity's future and on the role that astronomy and cosmology should play in our daily lives.

One person who held that perspective in the early twentieth century was the British philosopher Olaf Stapledon (1886–1950). In 1930 at the age of 44 Stapledon, a graduate of Oxford University in history and Liverpool University in philosophy, took up writing science fiction, in which aliens immediately played an essential role. In his novels *Last and First Men* (1930) and *Star Maker* (1937), political, religious and philosophical ideas dominate rather than the usual space opera adventure. Stapledon lived in an era when the immensity of the cosmos was well known, and his novels appropriately cover billions of years, the longest timescale of any science fiction novel. He knew of the work of Edwin Hubble (Chap. 36), and for his conception of the size of the cosmos he cited the astronomer William J. Luyten's *The Pageant of the Stars*. Still, "Immensity is not itself a good thing," Stapledon wrote.

> A living man is worth more than a lifeless galaxy. But immensity has indirect importance through its facilitation of mental richness and diversity [...] though spatial and temporal immensity of a cosmos have no intrinsic merit, they are the ground for psychical luxuriance, which we value. Physical immensity opens up the possibility of vast physical complexity, and this offers scope for complex minded organisms. (Stapledon 1968)

In this immense tapestry of infinite space and billions of years of time, Stapledon's characters were a variety of amazing and evolutionarily connected life forms. They became a fertile source of imagination for future science fiction authors, including Arthur C. Clarke an Stanislaw Lem, who were directly influenced by Stapledon and who continued his brand of what I would call philosophical science fiction. Through his novels Stapledon taught us to think long-term about space, time and aliens, and the richness of this thought over these time spans is still in many ways unsurpassed. *Last and First Men* and *Star Maker* are classic epics that contemplate the human future and our place in the universe over these immense scales of time.

Chapter 41, "Interstellar Humanity," is a takeoff from Stapledon's essay "Interplanetary Man," written in 1948 at the beginning of the atomic age. As its title suggests, Stapledon's article confines itself to our Solar System, already known at that time not to contain any complex life beyond our own Earth. Human destiny in Stapledon's worldview was therefore to spread into the Solar System, and his essay contemplated the implications of that scenario. By contrast, we do not yet know if intelligent life exists in interstellar space, and Chap. 41 examines our potential expansion into a cosmos that may well contain extraterrestrials—as Stapledon himself speculated in his novels. We conclude that the prospect of interstellar humanity during the next millennium is likely to have an effect on all branches of terrestrial endeavor, whether religion, philosophy, science, or the arts. The stage of human drama will be vastly expanded.

In Chap. 42—a number that is "the ultimate answer to the question of life, the universe, and everything," according to Douglas Adams in his *Hitchhiker's Guide to the Galaxy*—we show that this cosmic perspective has been present among certain writers even in ancient times, sometimes even those with a geocentric worldview. We choose four writers widely dispersed in space and time—the philosophers Boethius and Giordano Bruno and the astronomers Harlow Shapley and Carl Sagan to illustrate this cosmic perspective. Just as Boethius spoke of "the consolation of philosophy" in the early sixth century, so Urania, the Muse of astronomy, offers her own consolations. Those consolations include an expansive cosmological worldview that frames our daily lives, and a vision of unfolding cosmic evolution over 13.8 billion years, of which we are a product, though perhaps not the only biological (or postbiological) product.

The cosmic perspective, similar to but more robust than what Sagan long ago called "the cosmic connection," is widespread today and is growing even among theologians, philosophers, and a public otherwise too far removed from science. That perspective allows us to envision our place in the universe of space and time, offering an inspiring vision of both life and death.

References

Stapledon, Olaf. 1968. 'A Note on Magnitude,' in Stapledon, *Last and First Men* and *Star Maker*, New York: Dover Publications, 435.

Chapter 41
Interstellar Humanity

Abstract 50 years after Olaf Stapledon's landmark essay "Interplanetary Man?", we propose the coming era of "Interstellar Humanity." Over the next thousand years the domain of humanity will increasingly spread to the stars, a process that will alter our future in profound ways. At least three factors will in all probability drive this expansion: (1) increased understanding of cosmic evolution, changing our perception of ourselves and our place in the universe; (2) radio contact with extraterrestrial intelligence, bringing knowledge, wisdom, and problems of culture contact now unforeseen; and (3) interstellar space travel, transporting humanity's emissaries physically to at least the nearest stars. The consequences of these events are not predictable in detail, but may be studied by examining the lessons of cosmic evolution; by using history to analyze the reception of new world views and intellectual culture contacts on Earth; and by anticipating the likelihood of success in interstellar travel and its effects. The prospect of Interstellar Humanity during the next millennium is likely to have an effect on all branches of terrestrial endeavor, whether religion, philosophy, science or the arts. The stage of human drama will be vastly expanded.

41.1 Introduction

My title reflects an essay entitled "Interplanetary Man?" written 50 years ago by the British philosopher and science fiction writer Olaf Stapledon. That essay had its origin in a lecture delivered in 1948 at the British Interplanetary Society at the request of its President, Arthur C. Clarke. At the brink of the Space Age, Stapledon, best known for his *Last and First Men* (1930) and *Star Maker* (1937) epics, believed that humanity would spread throughout the Solar System over the next few centuries, barring the destruction of civilization made possible by the new atomic age. He realized that expansion of the human domain would be very different depending on whether or not the planets were inhabited; he concluded, in line with scientific thinking of his time, that this was very unlikely.

Humanity would therefore be left to colonize the planets of our system, most likely unencumbered by problems of extraterrestrial contact; the question that

---

First published in *Futures*, 32 (2000), 555–567

S. J. Dick, *Space, Time, and Aliens*, https://doi.org/10.1007/978-3-030-41614-0_41

exercised Stapledon was what humanity would do with the planets. Exploitation of physical resources, power over a larger physical domain, and the satisfaction of curiosity would all be part of it, but Stapledon believed the Solar System was destined for more than mere extension of life as it existed on Earth. Man should use other worlds not for the sole purpose of increasing luxury or power, but

> ...should avail himself of their resources in such ways as to advance the expression of the spirit in the life of mankind. He should use them so as to afford to every human being the greatest possible opportunity for developing and expressing his distinctively human capacity as an instrument of the spirit, as a centre of sensitive and intelligent awareness of the objective universe, as a centre of love of all lovely things, and of creative action for the spirit (Stapledon 1997, 218–241: 237–238).

But Stapledon had more in mind than a simply a wider scope for homo sapiens. He believed that through eugenics specially adapted human or quasi-human races should be developed.

> Thus the goal for the solar system would seem to be that it should become an interplanetary community of very diverse worlds each inhabited by its appropriate race of intelligent beings, its characteristic "humanity," and each contributing to the common experience its characteristic view of the universe. Through the pooling of this wealth of experience, through this "commonwealth of worlds," new levels of mental and spiritual development should become possible, levels at present quite inconceivable to man (Stapledon 1997, 238).

By "Interplanetary Humanity," Stapledon thus hoped for more than the mere spread of humanity through the Solar System; he hoped for a species physically, mentally, and spiritually improved over terrestrial humanity, privileged with new insights.

Such was the main message of Stapledon's best-known essay. In a postscript, however, he noted that intelligence in other planetary systems was also possible, no matter how unlikely it was in our own beyond the Earth. Unaware of the developments in radio astronomy, begun with Karl Jansky in 1933 and being developed in 1948 by Grote Reber and a very few others, Stapledon conjectured that "communication with intelligent races in even the remotest planetary systems may be effected by a highly developed technique of telepathy." Though he was (as far as we know!) wrong about the technique, Stapledon was very much in the spirit of today's tradition when he wrote "If, by one means or another, man does succeed in communicating with intelligent races in remote worlds, then the right aim will be to enter into mutual understanding and appreciation with them, for mutual enrichment and the further expression of the spirit. One can imagine some sort of cosmical community of worlds."

With the benefit of 50 more years in understanding of our universe, we can see that not only will at least parts of Stapledon's vision of interplanetary humanity likely be realized over the next centuries, but that over the next millennium homo sapiens will almost certainly become interstellar humanity. This will happen in three stages, each one filled with more potential—and more danger—than its predecessor. At level one we will have an increased awareness of our place in the universe by an increased understanding of cosmic evolution. This increased awareness is simply a matter of the advance of many branches of science, and is well underway at the end of the twentieth century. At level two we will have contact with extraterrestrial intelligence, most likely in the form of remote radio contact rather than physical contact. We are now on the brink of determining whether such intelligence is rare or abundant in the universe; we have only begun to think about the consequences. At level three, humanity will travel to the stars, physically spreading into the universe a genome constructed over 3.8 billion years of evolution on Earth, and perhaps by that time also altered along Stapledonian lines.

In all stages of interstellar humanity, the questions of cosmic purpose and human destiny will be paramount. Our place in the scheme of cosmic evolution, the outcome of contact with extraterrestrials, and our ability to travel to the stars, all bear strongly on the issue of our place in the universe and the ultimate fate of our species. Increased cosmic awareness, the exchange of information with extraterrestrials, and even the realization that intelligence is scarce as we physically expand into the galaxy, will alter our religions, philosophies, and the image of ourselves. Whether or not the "blind watchmaker" universe has an underlying purpose in cosmic evolution, whether our destiny is intertwined with alien sentient minds, and the nature of our long-term future are questions likely to be answered in the next millennium. All other questions pale in comparison.

41.2 Cosmic Evolution

At the end of the twentieth century more and more members of our species are increasingly aware of the new universe. At the beginning of the twentieth century, respected scientists could still believe that our Sun was located at the center of our galaxy, and that this galaxy—with a diameter of some 3600 light years—comprised the entire universe. Today we know that the universe extends some 12–15 billion light years, is populated with bizarre objects like pulsars, quasars and black holes, and is an Einsteinian space-time continuum that has no center. The Hubble Space Telescope, peering into the very depths of this universe, finds billions of galaxies similar and dissimilar to our own. The Earth seems an insignificant speck in the vastness of space.

At the same time this new universe of immense space is not static, but evolving, and presumably has been evolving for as long as 15 billion years. Following in the wake of the Darwinian revolution, the idea of cosmic evolution was already glimpsed at the beginning of the twentieth century, though the immensity of the time scale was not yet known. Percival Lowell's *The Evolution of Worlds* (1909) applied physical evolution to the solar system, while George Elery Hale's *The Study of Stellar Evolution* (1907) applied it to the stars. By 1929 Edwin Hubble gave an evolutionary dimension to the universe as a whole, by providing data on the recession rate of galaxies that showed the universe was expanding (Lowell 1909; Hale 1907; Hubble 1936). Today the Big Bang cosmology is believed to be the reason for that expansion, and the density of the universe appears to be such that the expansion will continue forever (Adams and Laughlin 1998).

But cosmic evolution applies to more than the physical universe. Harvard biochemist Lawrence J. Henderson not only grasped the essential idea of cosmic biological evolution in *The Fitness of the Environment* (1913), but also claimed a superior role for biology within it: "The properties of matter and the course of cosmic evolution are now seen to be intimately related to the structure of the living being and to its activities; they become, therefore, far more important in biology than has been previously suspected. For the whole evolutionary process, both cosmic and organic, is one, and the biologist may now rightly regard the universe in its very essence as biocentric." Volumes such as Harlow Shapley's *Of Stars and Men* (1958), Eric Chaisson's *Cosmic Dawn* (1981), and Carl Sagan's numerous writings (Sagan 1980) spread the idea that humanity must be placed in the context of cosmic evolution.

Today, cosmic biological evolution is the central assumption of NASA's Origins program, the touchstone for both astronomers and biologists, the starting point for numerous science fiction writers. The worldview of cosmic evolutionists from Henderson to the present is epitomized by Nobel biochemist Christian de Duve, who in *Vital Dust: Life as a Cosmic Imperative* (1995) concluded that "the universe is not the inert cosmos of the physicist, with a little life added for good measure. The universe is life, with the necessary infrastructure around; it consists foremost of trillions of biospheres generated and sustained by the rest of the universe."

This *biological universe*, as I have termed it elsewhere, is very different from a merely physical universe filled with lifeless planets, stars and galaxies (Dick 1996, 1998). As the new universe and cosmic evolution seep into the consciousness of homo sapiens through popular volumes such as those mentioned above, holds different meanings for different groups. Cosmic consciousness is expressed in many forms, some of them not embraced by scientists: UFO believers, space-oriented religious cults such as Heaven's Gate, and ever more elaborate (and often hostile) scenarios of science fiction. While this should warn us that human reactions to the new universe will not be monolithic, it should not prevent us from recognizing some undeniable underlying principles.

First, for all groups, the increased awareness of the new universe should dash any remaining hopes for an anthropocentric universe. Even though the idea that the universe was made for man survives in the form of the elegantly misnamed *anthropic principle*, in fact that principle is (in Henderson's terms) a *biocentric principle* that points to the abundance of life in the universe in many forms, rather than in the form of man. If life, mind and intelligence are the common outcomes of cosmic evolution rather than simply planets, stars and galaxies, then our religions, philosophies, and other human endeavors as currently formulated are too parochial. They will need to be significantly altered, expanded, or discarded in the coming millennium.

Secondly, quite apart from its nonanthropocentric consequences, cosmic evolution provides humanity a cosmic context in time. Our own planet is 4.5 billion years old; as we look into space we can now see solar systems only half a billion years old. Because of the finite speed of light, the more distant we look, the further back we are looking into our past, stretching back perhaps 15 billion years. But we are also looking into our future. Several generations of stars have already been born, lived and died, and by peering into space we see their fate, and our fate. We know that in 1.1 billion years our Sun will become too hot for life on Earth, that in 7 billion years it will become a full-fledged red giant engulfing the Earth, and that a few hundred million years later it will settle down to the dead end of stellar evolution known as a white dwarf. Though we are not accustomed to thinking in such time scales, cosmic evolution allows us to see the life of our species in perspective.

Finally, cosmic evolution teaches us that we are all "star stuff," in Sagan's colorful terminology. All elements except hydrogen and helium, including the biogenic elements and those found inside our bodies, were forged deep within the stars. Over the next millennium, as new generation space-based and ground-based telescopes carry out the "archaeology of the heavens," they will uncover our past and future. Even as they confirm the immensity of the universe and its blindingly hostile nature compared to our fragile biosphere, our human star stuff makes humanity a part of the cosmos in the same way that Copernicus made the Earth part of the cosmos 500 years ago. With this knowledge we should feel at home in the universe. We will increasingly understand it, to some extent control it, and perhaps have to share it, depending on whether or not life is abundant. In this context, the prevalence of extraterrestrial intelligence takes on peculiar importance.

41.3 Extraterrestrial Intelligence

Nothing could usher in a sustained interstellar humanity more forcefully or more quickly than contact with extraterrestrial intelligence. While the first half of the twentieth century still held hope for interplanetary intelligence, even Stapledon realized—and we are now virtually certain—that if extraterrestrial intelligence exists, it will only be found among the planets of other stars. The discovery of possible Martian fossils, of Europan oceans, of the ancient origin of terrestrial life under adverse conditions, and of current life on Earth under conditions much more extreme than previously believed, all point to life as a common emergent property of the physical universe. At the same time the discovery of numerous planetary systems populated by gas giants, presumably accompanied by terrestrial planets, provides life abundant sites on which to develop. Although the frequency of the evolution of intelligence is still problematic, electromagnetic SETI (Search for Extraterrestrial Intelligence) programs, if they are to succeed at all, will certainly succeed in the next thousand years. Alternative modes of contact are logically possible: Bracewell probes or Tough microprobes, a breakthrough in UFO studies, and the discovery of alien artifacts in the course of space exploration.

The consequences of the discovery of extraterrestrial intelligence in the next millennium depends strongly on the contact scenario. But we need not surrender because of this uncertainty. We may consider at least three approaches to studying the consequences of extraterrestrial contact. The first is science fiction. The effect of the physical presence of extraterrestrials on Earth in the mode of Arthur C. Clarke's *Childhood's End* (1953) is very different from the discovery of an alien artifact, as played out in Clarke's *Rendezvous with Rama* (1973) and its sequels. Both are again very different from scenarios of radio contact, as examined in James Gunn's *The Listeners* (1972) and Carl Sagan's *Contact* (1985). And all these scenarios are far removed from the vision of Fred Hoyle's *Black Cloud* (1957), where the intelligence is very different from the human form, or Stanislaw Lem's *Solaris* (1961), where the alien intelligence remains incomprehensible. Science fiction literature (some more elegantly than others) provides a rich source of thoughtful commentary on the consequences of alien contact. Though there is an overemphasis on alien invaders in the mode of H. G. Wells's *War of the Worlds* (1898), we do not in fact know which of the logical possibilities we will find in reality.

A second approach is grounded in the data of human experience and history. We may demonstrate this by taking the case of radio contact. Contrary to popular wisdom, radio contact with extraterrestrials—the object of SETI programs—will not be analogous to the physical contact of cultures on Earth. As has often been pointed out, physical culture contacts on Earth have most often been disastrous: witness Cortes and the Aztecs or Pizarro and the Incas or any number of other culture contacts. Rather, radio contact is analogous to intellectual culture contacts on Earth. Particularly apt is the analogy to the transmission of Greek knowledge to the Latin West via the Arabs in the twelfth and thirteenth centuries. The result was the Renaissance. Moreover, because contact among extraterrestrial civilizations is fundamentally delayed by the finite speed of light in direct proportion to the intervening distance, like the Greek and European civilizations, communications must take place across time.

If the decoding of an ET message results in a vast amount of knowledge, these analogies might break down. But a case can be made that deciphering a message could be a slow process perhaps continuing over generations, and thus more analogous to the translation of Greek knowledge. While both positive and negative analogies may be usefully applied, the nature of the message is the all-important unknown factor. Because of the age of the universe, however, extraterrestrial civilizations are likely to be much older than us, perhaps, it has been conjectured, holding the knowledge of the universe in an *Encyclopedia Galactica* (Dick 1995). By making contact we may become part of a Galactic Club—Stapledon's cosmical community of worlds.

In a more general sense we may use the trajectory of worldviews as an analogy for extraterrestrial contact. The belief in a biological universe—one in which cosmic evolution commonly ends in life and intelligence—is a kind of worldview similar in status to the Copernican world view or the Darwinian worldview. Like them, the biological universe as worldview strongly affects humanity's place in the universe. Like them, it is testable locally via spacecraft and globally via SETI programs. And like them it will go through certain stages of interpretation and explorations of implications. The rich literature in history of science examining these revolutions may therefore be used cautiously to examine the implications of the biological universe (Dick 1991, 1996).

The third mode of examining the effect of extraterrestrial intelligence is social science.

Albert Harrison's recent book, *After Contact: The Human Response to Extraterrestrial Life* (1997) has led the way in showing how fields such as psychology, sociology, and anthropology can be used as an aid to thinking about implications of contact. In particular he advocates a kind of systems approach, called *Living Systems Theory*, in which what we know about organisms, societies and supranational systems on Earth can be used to discuss the outer space analogues of aliens, alien civilizations and the galactic club. This is a promising approach to the study of implications, though it applies only to aliens and civilizations, not to low life forms and Martian nanofossils. The latter entails a shift in world view quite distinct from communicative extraterrestrials (Harrison 1997; Davies 1995).

Science fiction, terrestrial analogues in history, and Living Systems Theory are three fundamentally different ways to gauge the impact of extraterrestrial contact. All, it is true, make use of our terrestrial experience. But it is difficult to transcend that experience until contact is actually made. In any case, a systematic study using these three approaches should shed considerable light on our future over the next millennium. While the nature of the knowledge gained from extraterrestrial contact remains a wildcard, these approaches can nevertheless give us an outline of possible impacts. And we may predict that the exchange of information will shed light on the great problems of philosophy; if communication can be achieved, for example, the problem of objective knowledge will be raised to a new level of sophistication as the communicating intelligences share their knowledge. It is hard to imagine any event in the next millennium that could have greater consequences for human life and the future of our planet.

41.4 Interstellar Travel

Although some believe that interstellar travel is a physical impossibility because of distances and power requirements, history is not on the side of the pessimists. It is likely that their pessimism falls in the category of what Arthur C. Clarke has termed "failure of imagination," and that ways will be found to cross the ocean of space no less than the ingenuity of our predecessors carried them across the oceans of Earth. Indeed, the first full engineering study for an interstellar starship was published under the auspices of the British Interplanetary Society in 1978, 30 years after Stapledon's lecture. That vehicle was designed to travel to Barnard's star, at 6 light years one of the closest stars to our Sun. A feel for even nearby interstellar distances may be had by realizing that at a cruising speed 12% the speed of light, such a starship would take 50 years. The propulsion system was based on nuclear fusion, and required mining the helium-rich atmosphere of Jupiter to propel the 450-ton payload. Even then this was a flyby mission; no fuel would remain for deceleration from 12% the speed of light (Martin 1978).

Clearly interstellar travel will challenge humanity, and many practical problems await solution. It is possible, even probable, that over the next millennium propulsion systems such as a matter-antimatter drive or star sails will become commonplace. Whatever the case, interstellar travel will be the capstone to interstellar humanity. Daunting as the task seems, someday the hundreds of billions of stars in our own Galaxy will have been explored, and the nearby galaxies will beckon. We may safely predict, however, that the era of intergalactic humanity will not take place in the next millennium. At the speed of light, it would take a half-million light years to reach the nearest galaxy. Unless, that is, one is transported to distant regions of the universe via wormholes associated with black holes. But this, too, is unlikely in the next millennium. Humanity is likely to be preoccupied over the next millennium with roving our own galaxy, with its 100,000-light-year diameter (Crawford 1990; Halpern 1992).

Interstellar travel has an interesting connection with extraterrestrial life via the so-called Fermi paradox. Put most simply in terms of the question "Where are they?", the paradox asserts that given the long time scales of the universe, if extraterrestrials are abundant in the universe, they should have colonized the galaxy within a few million years and arrived on Earth long ago. Since we do not see them (barring arguments for the extraterrestrial hypothesis of UFOs), they do not exist. This argument has led some SETI leaders to assert that interstellar travel is not economical, and not the most efficient mode of interstellar contact.

They may be right, but quite aside from any merits of the Fermi Paradox, human interstellar travel over the next millennium will likely proceed. Exploration is an essential part of the human imperative, and while for economic and other reasons robot probes may precede human exploration, even the brief history of interplanetary exploration indicates that human exploration will follow. The Pioneer 10 and 11 and Voyager 1 and 2 spacecraft, vestiges of interplanetary humanity, are humanity's first emissaries to the stars, complete with messages for any extraterrestrials who chance to intercept them. At .005% the speed of light, the Voyagers will take 80,000 years to reach the distance of the nearest star, Alpha Centauri.

It is possible, though we believe not probable in light of what we now know about cosmic evolution, that homo sapiens is the only intelligent species in the Galaxy. There is precedent for this too in science fiction; perhaps its most famous story, Isaac Asimov's *Foundation* series, is carried out in a Galaxy devoid of extraterrestrials. One can hardly say that it is a less interesting universe; notably, the preservation of knowledge is its centerpiece, and the Encyclopedia Galactica its anchor. Strikingly, an Encyclopedia Galactica is a frequent theme of universes populated with extraterrestrials also, now containing all the knowledge of the Galactic Club. In either case, the era of interstellar humanity will not be "the end of history" or "the end of science" in the rather parochial view of Earthbound humanity posited at the end of the second millennium (Futuyama 1992; Horgan 1996).

With interstellar travel the human genome will physically be transported into space. Moreover, it is likely that the long distances required in interstellar travel will drive human intervention in that genome, so that humans, or their successor species, are engineered to withstand the rigors of the trip. Even if traditional ideas like suspended animation prove practical, genetic considerations are likely to play an important role in maximizing efficiency and success. Asimov reminds us that new sciences such as "psychohistory"—the statistical treatment of historical trends—will also play a role in the next millennium, though in ways intrinsically unforeseeable.

What are the consequences of interstellar travel? A glimpse of the struggles that interstellar travelers will face may be seen in Finney and Jones *Interstellar Migration and the Human Experience*. (Finney and Jones 1985). Again, it is based on analogy, with European exploration, Polynesian expansion across the Pacific, and the stillborn attempts of other societies at colonization. For those who succeed we may predict only in the most general terms that interstellar humanity will be highly dependent on technology, still struggling with the same problems of human nature that we face today, but will operate in a vastly expanded domain that incorporates a full knowledge of cosmic evolution and may involve interstellar politics in which the principals are extraterrestrial civilizations. Unlike radio contact, physical contact with extraterrestrial intelligence forces us to consider the negative analogies of physical culture contact on Earth. If we can survive such contact, the deepest questions of the species, pondered since our earliest human ancestors looked skyward and wondered about the stars, will be answered.

41.5 Cosmic Purpose and Human Destiny

In becoming interstellar humanity the question of cosmic purpose and human destiny will be paramount. Almost at the same time Stapledon wrote his essay on "Interplanetary Man," the French scientist LeComte du Noüy wrote a widely hailed volume entitled *Human Destiny* (1947). There he concluded that the simple protein molecule would have taken 243 orders of magnitude more time than the age of the Earth to form by chance, that life itself could therefore not have arisen by chance, and that "these consequences inevitably lead to the idea of God." The same theme pervaded the influential book *Chance and Necessity* (1971) of the French biologist Jacques Monod when he wrote "The universe was not pregnant with life, nor the biosphere with man. Our number came up in the Monte Carlo game" (du Noüy 1947; Monod 1971).

It was this kind of universe that led to pessimistic ideas of purpose and destiny. Nobel physicist Steven Weinberg, for example, wrote that "It is almost irresistible for humans to believe that we have some special relationship to the universe, that human life is not just a more-or-less farcical outcome of a chain of accidents reaching back to the first three minutes, that we were somehow built in from the beginning … It is hard to believe that all this is just a tiny part of an overwhelmingly hostile universe… The more the universe seems comprehensible, the more it also seems pointless" (Weinberg 1977).

But this conclusion did not take into account other possibilities or the biological universe. Since that time Stuart Kaufmann and others have pioneered the idea of self-organization; in his book *At Home in the Universe*, Kaufmann has suggested how life could have originated by the natural tendency of matter to self-organize. Under this scenario, Nobel biochemist Christian de Duve could write "I view this universe not as a cosmic joke, but as a meaningful entity - made in such a way to generate life and mind, bound to give birth to living beings able to discern truth, apprehend beauty, feel love, yearn after goodness, define evil, experience mystery"(De Duve 1995; Kauffman 1995). This is the view championed by many cosmic evolution pioneers, including Carl Sagan. The whole scenario of cosmic biological evolution, not taken into account by du Noüy, Monod or Weinberg, potentially places human life in a new context in terms of purpose and destiny. It is just such questions that coincide with the Stapledonian wish to use new worlds to advance the human spirit; to answer those questions of ultimate importance to interstellar humanity.

Human destiny depends on the course cosmic evolution has taken. With Stapledon, we may conclude that it will be very different if the Galaxy is inhabited, or if it is not. If it is, then we may join the Galactic Club, have access (and perhaps even add) to the Encyclopedia Galactica. But if it is not inhabited, then it is our destiny to fill the Galaxy with life; the Encyclopedia Galactica will be the history of the human race, or its genetic descendants. In the latter case interstellar migration and colonization are our future; interstellar travel, rather than extraterrestrial intelligence, predominates interstellar humanity. In that case, the Fermi Paradox will have proven true, though it is likely the search for life will go on as an important part of the exploration imperative.

41.6 Conclusions

Cosmic evolution, extraterrestrial intelligence and interstellar travel are three factors that will drive interstellar humanity in the next millennium. Those who take part directly in these ventures will certainly be affected, but so also will much of humanity. Consciousness of cosmic evolution is increasing and will continue to increase because of a fundamental human desire to answer the question that more than 130 years ago T. H. Huxley called "the question of questions for mankind – the problem which underlies all others and is more deeply interesting than any other – humanity's place in nature" (Huxley 1863, 71). Huxley was speaking in the context of Darwinian theory of evolution by natural selection; that question has now been extended to cosmic evolution, and therefore the problem of our place in the universe.

The search for extraterrestrial life is an essential part of that problem. Whereas Kepler, Galileo, Newton and their modern day successors demonstrated the role of physical law in the universe, the question at stake in the extraterrestrial life debate is whether an analogous "biological law" reigns throughout the universe, whether Darwinian natural selection is a universal phenomenon rather than simply a terrestrial one, whether there are other biologies, histories, religions, and philosophies beyond the Earth. The physical universe and the biological universe are fundamentally different worldviews. During the next millennium we will likely discover which one is true. The rest of the millennium will be spent in exploring the implications of this profound truth.

Looking back from the year 3000—some 40 generations hence—most of the historical and political issues that concern us now will have been forgotten. World War II will seem as distant as the Battle of Hastings does to us now. The geopolitical landscape will have transformed into an astropolitical landscape. Our science will seem quaint and embryonic. However, the desire to know better our place in the universe, to push the frontiers, to explore beyond one more barrier, will remain.

A final cautionary note is in order. The question mark in Stapledon's title referred not only to the accuracy of his prediction, but also to the survival of civilization in the new atomic age. Even in our age, we must still allow for the possibility that the hopes for interplanetary and interstellar man will be dashed by the follies of adolescent humanity, and that 1000 years from now some extraterrestrial archeological expedition will be picking through the remains of terrestrial man, whose follies exceeded its promise and cut short its brief history. As with the atomic age, the age of genetic engineering brings danger as well as opportunity. Even though the destruction of homo sapiens might constitute only a hiccup in the scheme of cosmic evolution (depending on the prevalance of intelligence elsewhere), as individuals each of us must do what we can to see that this future does not prevail. Even as we must do what we can to improve life on Earth, one way or another human destiny is among the stars. Human purpose remains elusive.

41.7 Commentary 2020

Interest remains high in the work of Olaf Stapledon, both philosophical and science fictional. The definitive biography remains Crossley (1994). In 1997 the same author published a collection of Stapledon's shorter works (Crossley 1997), including his essay on "Interplanetary Man." Dick (1996, 245–249) places Stapledon's work in the context of science fiction. Stapledon's archives are located at the University of Liverpool, https://web.archive.org/web/20051028102900/http://www.sfhub.ac.uk/Stapledon.htm and some of his digitized work, including his two main science fiction novels, can be accessed at https://www.torontopubliclibrary.ca/search.jsp?Erp=20&N=38537+38532+4294698632&view=grid

and at the Project Gutenberg site http://www.gutenberg.net.au/plusfifty-n-z.html#letterS

Not everyone will agree with the sentiments expressed in "Interplanetary Man?" or in this chapter on "Interstellar Humanity." Many would disagree that humanity should spread into either interplanetary or interstellar space. Those skeptics fall into at least two classes. One class would argue that we should not undertake space exploration of any kind until humanity has solved its considerable problems on Earth. In other words, never. Another group would argue that we will destroy the outer space environment in the same way we seem bent on destroying the environment on Earth. Call me an optimist, but I hope and believe we can learn the lessons of history, despite an admittedly sketchy record on this subject. Yet, perhaps the title of this chapter, like Stapledon's, should also have a question mark, both because of the dangers of the atomic age and the folly of our recent political history, but also because of qualms about genetic engineering. Genetic engineering or not, it also remains possible that interstellar humanity will be postbiological (Chap. 12).

Acknowledgements My thanks to Professors Albert Harrison and Allen Tough for their insightful comments.

References

Adams, Fred C. and Gregory Laughlin. "The Future of the Universe," *Sky and Telescope*, 96 (1998), 32–39

Chaisson, E. 1981. *Cosmic Dawn: The Origins of Matter and Life*. New York, Little Brown & Company.

Crawford, I. A. 1990. "Interstellar Travel: A Review for Astronomers, 31, *Quarterly Journal of the Royal Astronomical Society*, 377–400.

Crossley, Robert. 1994. *Olaf Stapledon: Speaking for the Future*, Syracuse University Press: New York

Crossley, Robert. 1997. *An Olaf Stapledon Reader*, ed. Robert Crossley, Syracuse University Press: Syracuse, New York.

Davies, Paul. (1995) *Are We Alone? Philosophical Implications of Discovery of Extraterrestrial Intelligence*. New York, Basic Books.

De Duve, Christian. 1995. *Vital Dust: Life as a Cosmic Imperative*, New York, Basic Books.

Dick, Steven J. 1991. "From the Physical World to the Biological Universe," in *Bioastronomy: The Search for Extraterrestrial Life*, J. Heidmann and M. J. Klein, eds. Berlin, Springer-Verlag, 356–363.

Dick, Steven J. 1995. "Consequences of Success in SETI: Lessons from the History of Science," in Shostak (1995), pp. 521–532.

Dick, Steven J. 1996. *The Biological Universe*. Cambridge, Cambridge University Press

Dick, Steven J. 1998. *Life on Other Worlds*. Cambridge, Cambridge University Press.

du Noüy, LeComte. 1947. *Human Destiny*. New York, Longmans, Green & Co.

Finney Ben R. and Eric M. Jones, eds.1985. *Interstellar Migration and the Human Experience*, Berkeley, University of California Press

Futuyama, Douglas. 1992. *The End of History and the Last Man*. New York, Free Press

Hale, George Elery. 1907. *The Study of Stellar Evolution*. Chicago, 2

Halpern, Paul. 1992. *Cosmic Wormholes: The Search for Interstellar Shortcuts*. New York, Dutton

Harrison, Albert. 1997. *After Contact: The Human Response to Extraterrestrial Life*, New York, Plenum

Henderson, L. J. 1913. *The Fitness of the Environment*. Cambridge, Mass. reprinted with an introduction by George Wald (Gloucester, Mass., 1979), p. 312;

Horgan, John. 1996. *The End of Science: Facing the Limits of Knowledge in the Twilight of the Scientific Age*. Reading, Mass., Addison-Wesley

Hubble, Edwin. 1936. *The Realm of the Nebulae*. New Haven, Yale University Press.

Huxley, T. H. 1863. *Man's Place in Nature*. 1863; Ann Arbor, Mich., Univ. Michigan Press, 1971 ed., 71

Kauffman, Stuart. 1995. *At Home in the Universe: The Search for the Laws of Self-Organization and Complexity*. New York and Oxford, Oxford University Press

Lowell, Percival. 1909. *The Evolution of Worlds*. New York, Macmillan Company

Martin, A. R. 1978. *Project Daedalus - The Final Report on the BIS Starship Study, Journal of the British Interplanetary Society*, Supplement

Monod, Jacques. 1971. *Chance and Necessity*. New York, Knopf, 144–146

Sagan, Carl. 1980. *Cosmos*, New York, Random House
Shapley, Harlow. 1958. *Of Stars and Men,* Boston, Beacon Press
Shostak, Seth. 1995. *Progress in the Search for Extraterrestrial Life*, ed. G. Seth Shostak. San Francisco: Astronomical Society of the Pacific
Stapledon, Olaf. 1997. "Interstellar Man?" in *An Olaf Stapledon Reader,* ed. Robert Crossley, Syracuse, N. Y., 218–241: 237–238.
Weinberg, Steven. 1977. *The First Three Minutes*. New York, 154–155.

Chapter 42
The Consolations of Astronomy and the Cosmic Perspective

Abstract Almost exactly 1500 years ago, the Roman statesman and philosopher Boethius wrote *The Consolation of Philosophy*, a conversation with Lady Philosophy in which he proclaimed the superiority of a life of the mind over wealth and fame. What does Urania, the Muse of Astronomy, have to offer? Here we argue that astronomy offers us a cosmological worldview within which we can frame our daily lives, providing a cosmic perspective sorely needed in the modern world. Boethius himself exhibited such a perspective as he sat in prison during the last days of his life, as did Giordano Bruno and many others, both in the realm of natural philosophy and natural science. Poets, philosophers, and scientists fostered it as the scientific revolution took root in the decades and centuries following Galileo, Kepler, and Newton, even as John Donne feared in the wake of the Copernican theory "'Tis all in pieces, all coherence gone." In the twentieth century, astronomers Harlow Shapley and Carl Sagan actively promoted a much more robust cosmic perspective, based on modern science and featuring many worlds, extraterrestrial intelligence, and an expansive image of cosmic evolution. The cosmic perspective, similar to but more robust than what Sagan called "the cosmic connection," is widespread today and is growing even among theologians, philosophers, and a public otherwise too far removed from science. One of the "consolations of astronomy" is that it allows us to envision our place in the universe of space and time, offering an inspiring vision of both life and death.

42.1 Introduction: On the Importance of Cosmological Worldviews

Whether we realize it or not, each of us harbors a complex variety of worldviews, accumulated through time and making us who we are (Vidal 2014). We can envision this complex as follows: cosmological and scientific worldviews preside at the apex of the hierarchy of worldviews each of us hold, followed by philosophical and religious worldviews, with cultural worldviews at the base of a broad pyramid (Fig. 42.1). Our cosmological and scientific worldviews are grounded in nature, and, I believe, should lay claim to grounding our overall worldviews in reality.

Published here for the first time.

S. J. Dick, *Space, Time, and Aliens*, https://doi.org/10.1007/978-3-030-41614-0_42

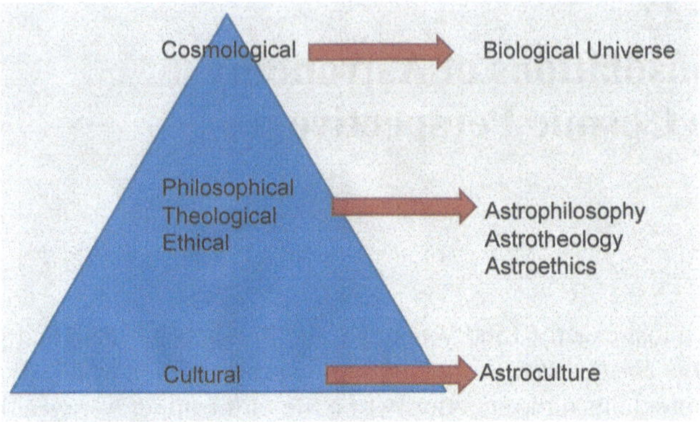

Fig. 42.1 A hierarchy of worldviews, in which cosmological worldviews trickle down and profoundly affect various levels of human culture. (From Dick 2018)

This is a controversial statement these days; while some might claim the primary direction of influence in Fig. 42.1 is upward because worldviews are socially constructed, I claim the primary influence is downward. Trickle-down economics demonstrably does not work, but I am arguing trickle-down worldviews do. Thus, cosmological and scientific worldviews should inform our worldviews at all levels, including the philosophical, theological and cultural.

Even if we confine ourselves to Western civilization, successive cosmological worldviews have been held following the first creation myths, ranging from the ancient atomists with their infinite number of worlds, to the Aristotelian/Ptolemaic cosmology with the Earth at its center, and the Copernican heliocentric worldview that made the Earth a planet and the planets potential Earths. The short-lived Cartesian system with its vortices first gave credence to solar systems beyond our own, and the Newtonian universe ruled by gravitational law often imagined other solar systems like ours. These cosmologies have risen and fallen with new discoveries, or in the case of the Newtonian universe, been subsumed in the worldview of Einstein's space-time and general relativity. Each culture, from the ancient Egyptian and Chinese to Native Americans and Europeans, has had its own history of worldviews. But today our worldviews have converged: in terms of spatial geography we have a galacto-eccentric worldview, the latter placing us in the far precincts of the Milky Way Galaxy in the midst of billions of galaxies. And in terms of time we see ourselves in the context of 13.8 billion years of cosmic evolution, as described in Chap. 8. As reflected in this volume, the discovery of life beyond Earth may result in a transition from our current physical cosmology to one based on a biological universe. As shown in Fig. 42.1 philosophical, religious, and ethical worldviews may be generalized and are in the process of transforming into astrophilosophy, astrotheology, and astroethics (Dick 2018). And the transformation of human culture to astroculture will be accelerated beyond the initial thrust given it by the Space Age.

Whether we realize it or not, these cosmological worldviews have historically played a major role in our everyday thinking. Dante's *Divine Comedy,* with its moral lessons played out in a cosmological worldview of nested spheres with the Earth at the center and the sphere of fixed stars at the periphery, provided the backdrop for medieval life. Within this ordered framework, both a mental and moral model as well as a physical system, all human actions took place. "In every age the human mind is deeply influenced by the accepted model of the universe," C. S. Lewis reminded us in his book on medieval and Renaissance literature, *The Discarded Image* (Lewis 1964, 222). The model, he insisted, "is also influenced by the prevailing temper of mind." As historian of science Nasser Zakariya has demonstrated in detail in his book *A Final Story: Science, Myth, and Beginnings*, the same is true in the modern world, where the sweeping scientific epic of cosmic evolution embraces the cosmological, the biological and the social, and provides the background for modern life—including the biological universe (Zakariya 2017).

The convergence of these cosmological worldviews has provided us with a cosmic perspective in both space and time. The epic of cosmic evolution is our story, our Genesis for the Third Millennium, and like most stories it holds lessons. Not everyone, of course, has been deeply influenced by these worldviews, nor explicitly expressed the extent of that influence. But throughout history certain individuals have implicitly and explicitly held and promoted the cosmic perspective, and it has become part of the cultural sea in which we all live. In this chapter we highlight only a few enlightened individuals by way of illustrating what we call "the consolations of astronomy," taking our cue from an influential work of the sixth-century philosopher Boethius, *The Consolation of Philosophy*.

42.2 Boethius and Bruno: A Tale of Two Martyrs

In the year 523 the Roman statesman and philosopher Anicius Manlius Severinus Boethius (c. 480–524) was imprisoned at a country estate near Pavia, Italy. Born shortly after the date typically taken as the end of the Roman empire (476 AD), at an early age he became a senator in what was left of the Western Roman empire and later entered the service of the Ostrogothic King Theoderic the Great (Marebon 2009; O'Donnell 2008, 117–118, 163–170). He rose through the ranks and by 522 held the office of *magister officiorum*, a post that made him a kind of foreign minister. He was, the historian James O'Donnell says, "proud of his family, ostentatious in his learning, and ambitious in every way … he was the second ranking civil dignitary of the imperial government … the most prominent citizen in Theoderic's realm." Boethius had his differences with Theoderic (sometimes spelled Theodoric); for starters he fought against the Arian heresy that Theoderic embraced from his capital at Ravenna. But more importantly he made it his life's goal to reconcile the classical Roman culture with the nascent Ostrogothic culture. He became perhaps too ambitious and influential, and may have harbored ambitions to become emperor and rule as Plato's philosopher king, since the aging Theoderic had no satisfactory successor.

Fig. 42.2 Boethius teaching his students, from a 1385 Italian manuscript of the *Consolation of Philosophy*. (Image in public domain)

In any case, in what O'Donnell calls "the unhappiest story of the age," at a time of great change and swirling alliances, Boethius became embroiled in political controversies that led to accusations of treason. Not one to remain idle, during the year before his execution he wrote *De Consolatione Philosophiae* (*The Consolation of Philosophy*). In this work, Boethius (Fig. 42.2) contemplated how evil could exist in a world governed by God, highlighted the transitory nature of money and power, argued for the superiority of a life of the mind, and defended himself from what he claimed were false accusations, even while still hinting at his political ambitions.

As Boethius sat in the prison, contemplating his life and likely early death, his despair was palpable (Boethius 1998, Book I, 39):

So sinks the mind in deep despair
And sight grows dim; when storms of life
Blow surging up the weight of care,
It banishes its inward light.
And turns in trust to the dark without

But his next lines reveal that Boethius was also contemplating matters celestial:

This was the man who once was free
To climb the sky with zeal devout
To contemplate the crimson sun,
The frozen fairness of the moon –

Astronomer once used in joy
To comprehend and to commune
With planets on their wandering ways.

Boethius was no astronomer in the modern observational sense, even if these and other passages in the *Consolation* reveal him to be a keen observer of the heavens. Rather, most importantly he had written books on music, astronomy, geometry and arithmetic, and perhaps translated at least parts of Euclid's *Elements* and Ptolemy's *Almagest*, translations that no longer survive but that were meant to further his goal of cultural transmission. Moreover, historian of science Stephen McCluskey states that "Boethius expressed his views on astronomy and the cosmos in many of his extant writings" (McCluskey 2012, 51). The *Consolation* itself is rife with astronomical passages, including this striking sentiment:

As you have heard from the demonstrations of the astronomers, in comparison to the vastness of the heavens it is agreed that the whole extent of the earth has the value of a mere point ... Further, as you have learned from the authority of Ptolemy, only about one-fourth of this so miniscule spot in the universe is the portion inhabited by animate creatures known to us (Boethius, Book II, Part VII)

Moreover, Boethius held there was a moral justification for studying the heavens, since it

Would mold both my actions and the principle of life in general in accordance with the model of heavenly order (McCluskey 2012, 51).

As McCluskey points out, Ptolemy had expressed a similar attitude in the Preface to his *Almagest*:

With regard to virtuous conduct in practical actions and character, astronomy, above all things, could make men see clearly; from the constancy, order, symmetry and calm which are associated with the divine, it makes it followers lovers of this divine beauty, accustoming them and reforming their natures, as it were, to a similar spiritual state (McCluskey 2012, 51)

And closer to his own time Boethius certainly knew of Macrobius's *Commentary on the Dream of Scipio* (Macrobius 1990) in which Macrobius (395–423) analyzed Cicero's *Somnium Sciponis*. In the latter, found in the sixth book of *De re publica*, Cicero has Scipio travelling through the planetary spheres, providing a cosmic perspective on Earthly affairs.

In short, it is clear that between his observations, his statements about a moral role for astronomy, and his scholarly goals, Boethius had acquired a cosmic perspective that pervaded his life and helped him through his year-long imprisonment, all within the geocentric worldview of sixth-century Italy.

The *Consolation of Philosophy* became one of the most famous works of the Middle Ages and early Renaissance, receiving translations by Geoffrey Chaucer and Queen Elizabeth I, among others. Boethius was canonized a saint in the Catholic Church in 1883, and was eulogized as recently as 2008 by Pope Benedict XVI (Benedict 2008). Visitors can still contemplate the mosaic wonders of Ravenna, wandering where Boethius once walked and worked, a man caught between the classical knowledge of the past and the harsh difficulties of the coming Middle Ages. And a man who knew his place in the cosmos.

A little more than a thousand years after the death of Boethius, Giordano Bruno (1548–1600) was born in Nola, near what is now Naples, Italy, 500 miles to the south of Pavia where Boethius had been executed. Once a Dominican monk, Bruno abandoned his order to spread radically new ideas on natural and moral philosophy. No small part of that natural philosophy was a strong belief in an infinite number of Earthlike worlds. Bruno led a wandering life, travelling to Paris, London, Oxford, Prague, Frankfurt, and various parts of Italy, often mixing in high political and scholarly circles due in part to his ideas and in part due to fame for his methods in the "art of memory." The year 1584 found him in Reformation England, and writing a detailed and systematic discussion of other worlds in the context of the Copernican theory, leavened by a much broader Hermetic philosophy that viewed Nature in magical and animistic terms (Yates 1964). In his *De l'infinito universo et mondi* (*On the Infinite Universe and Worlds*) Bruno wrote.

> For there is a single general space, a single vast immensity which we may freely call *Void*. In it are innumerable globes like this on which we live and grow. This space we declare to be infinite, since neither reason, convenience, possibility, sense-perception nor nature assign to it a limit. In it are an infinity of worlds of the same kind as our own. For there is no reason nor defect of nature's gift, either of active or of passive power, to hinder the existence other worlds throughout space, which is identical in natural character with our own space. (Bruno 1584; Singer 1968, 58–59)

This was not just a passing theory, but part of a broader vision. "The whole of Bruno's philosophy is based on his view of an infinite universe with an infinity of worlds," wrote one of his early biographers, Dorothea Singer (Singer 1968, 50). As she points out, his cosmological views were most influenced by two thinkers at opposite philosophical poles, the atheist Roman poet and atomist Lucretius, whose *De rerum natura (On the Nature of Things)* had been rediscovered in the previous century, and Nicholas of Cusa, a cardinal in the Catholic Church who sought in cosmology a reinforcement of his theological views. Hermetic philosophy notwithstanding, Bruno forcefully deployed the principle of plenitude, which in a universe full of atoms as Lucretius and his Greek predecessor Epicurus had proposed, led to an infinite universe with an infinite number of worlds (Dick 1982, 63–69; Gatti 1999, 117). The stars were just so many Suns, and our Earth was not unique.

This was the cosmic perspective that colored all of Bruno's other works and guided his daily life. As another of his biographers, Ingrid Rowland, wrote, "Bruno's enthusiasm for his real concern … was to show how the realization that we inhabit an infinite universe will transform every aspect of our lives." Living the Nolan philosophy, she wrote, "meant accepting the world as it was, which in turn implied accepting a much larger definition of life in which the earth, stars, and planets were also living things, infused with divinity." This was not the modern worldview of today, but it was a universe that demanded moral behavior, one that he described as "a mother, a nursemaid, a well-spring, light, love, a creation truly worthy of an omnipotent God" (Rowland 2008, 159, 174). Shorn of its supernatural elements, parts of Bruno's worldview are reminiscent of today's Gaia concept.

Other parts approach our worldview of a cosmic sea of suns and planets, some perhaps inhabited. But the larger point is Bruno's insistence by example that knowledge of the universe should affect our daily lives.

After a career of wandering Europe and writing books, Bruno was imprisoned by the Inquisition in 1592, first in Venice then in Rome, and burned at the stake in 1600 (Fig. 42.3). Many reasons undoubtedly precipitated this tragic event, none of them, it need hardly be said, justified. Aside from infinite worlds, the Church considered many of his ideas heretical, including the non-divinity of Christ. As a result many have portrayed him not as a martyr for science but as a heretic of Catholic doctrine, as if that justified his fate. Recent evidence, however, demonstrates conclusively that his belief in other worlds was a major reason for his execution, even if only one of several (Martinez 2016, 2018). At the same time it is by all accounts true that his personality did not help his causes. As one of his novelists has written with only slight artistic exaggeration "Bruno was no sweetheart. No supplicant ready to roll over and play dead. He was a man who rubbed everybody the wrong way, a man with a big mouth, a man who knocked everybody's ideas but his own, a man with all enemies and no friends" (Yanow 1998).

By the time of his imprisonment Bruno's books had already been written, and as he languished in prison at Castel Sant'Angelo in Rome and elsewhere for 8 years, one wonders what occupied his fertile mind. Perhaps he even contemplated the fate

Fig. 42.3 Bruno at the stake. (Image from the base of his statue at the Piazza Campo de' Fiori in Rome. Photo by Steven Dick)

of Boethius, whose work he certainly would have known. Like Boethius he believed in the superiority of the mind over petty politics and religion. Unlike Boethius he held the heliocentric view of Copernicus, in which the Earth was no longer the center of the universe. Boethius had been executed for treason against the state, Bruno would be executed for a kind of "treason" against Christianity as well as for his ideas of other worlds. Bruno is thus best seen not only as a martyr for science, but also as a martyr for freedom of thought. And this is why the brooding statue of Bruno at the Piazza Campo de' Fiori in Rome (Part VII Frontispiece), by design defiantly facing the Vatican, remains to this day a vibrant monument to the freedom of thought. In fact the entire piazza has become such a monument. Today visitors can dine under Bruno's watchful gaze at one of several restaurants on the piazza, stroll the Libreria Fahrenheit 451, and view scenes from his life beneath the statue, all surrounded by a colorful open market. While scholars and theologians of various persuasions argue about the relative weight of his offenses, perhaps we can all agree that burning at the stake for whatever reason is never a good idea. Boethius had died at age 44; Bruno was 51. The cosmos must have been among their last thoughts.

Cosmic perspectives became more common in the centuries following Bruno's death. Galileo's first use of the telescope 10 years later made other worlds real, and those with some imagination could visualize the view of the Earth from those worlds. The German intellectual historian Hans Blumenberg has spoken of "reflexive telescopics," imagining the Earth from an extraterrestrial viewpoint, and Dennis Danielson has addressed the subsequent relevance of this idea (Blumenberg 1987; Danielson 2013). The unleashing of the extraterrestrial gaze, which could be documented with many more examples, was one more step toward a cosmic perspective. The incipient cosmic worldviews of Boethius and Bruno would not be squashed.

42.3 Shapley and Sagan: A Tale of Two Astronomers

If we fast-forward to the twentieth century, we can highlight two prime examples of influential purveyors of the cosmic perspective that now permeates cultures around the world, not just in the Western context. One of them, Carl Sagan, became a household name due to his books, television appearances, and eloquence on a variety of societal issues. The other is less known among the general public, but deserves to be better known, not only for his astronomical accomplishments but also for his sustained attempts to relate cosmos to culture.

By the mid-twentieth century the cosmic perspective had become familiar to the general public, not only through reports of the astronomical discoveries of Hubble and other astronomers, discoveries that greatly enlarged our idea of the universe, but also through science fiction. Who can forget the eerie opening lines of H. G. Wells's 1898 *War of the Worlds*?

> No one would have believed in the last years of the 19th century that this world was being watched keenly and closely by intelligences greater than man's and yet as mortal as his own; that as men busied themselves about their various concerns they were scrutinized and

> studied, perhaps almost as narrowly as a man with a microscope might scrutinize the transient creatures that swarm and multiply in a drop of water. With infinite complacency men went to and fro over this globe about their little affairs, serene in their assurance of their empire over matter. It is possible that the infusoria under the microscope do the same. (Wells 1993, 51)

This is reflexive telescopics on steroids, not only an extraterrestrial gaze, but an extraterrestrial interaction with intelligent beings. One person greatly influenced by Wells was the British philosopher Olaf Stapledon (Crossley 1994, 196–198; 210–215), whose *Last and First Men* (1930) and *Star Maker* (1937) placed humanity in a cosmic context of billions of years of cosmic evolution (Chap. 41). As early as 1912 Stapledon had written.

> I think perhaps the greatest change that has come to the human mind since the middle ages is the conception of the infinite heavens, wherein the earth is a point, not the all-important centre. And as (in my view at least) the second beauty of the world is the beauty of the stars on a clear night, this change of conception is very important to the growth of humanity. (Crossley 1994, 229)

That attitude was exhibited repeatedly in his novels. Stapledon in turn influenced many science fiction writers, including Arthur C. Clarke, who wrote late in life that "Though more than sixty years have passed, I can still visualize the very shelf in the Minehead Public library where I discovered Olaf Stapledon's *Last and First Men.* No other book had a greater influence on my life" (Crossley 1994). For Clarke it was extraterrestrial life that gave perspective to humanity. From *Childhood's End* (1953) to *The City and the Stars* (1956) and *2001: A Space Odyssey* (1968), aliens were at the center of Clarke's cosmic perspective. During the twentieth century, that perspective became a primary theme in science fiction, one that inspired and influenced many scientists.

Both Shapley and Sagan had imbibed the cosmic perspective of these science fiction writers. Harvard Observatory Director Harlow Shapley (1885–1972) made pioneering contributions to astronomy, not least by showing that our Solar System was located at the periphery of the Milky Way Galaxy, which itself was embedded in a sea of galaxies. This was a confirmation and extension of the Copernican principle that the Earth was in no way central in the universe. A farm boy from Missouri, Shapley earned his way into astronomical stardom by sheer force of hard work, as he described in his memoir *Through Rugged Ways to the Stars* (Shapley 1969). Already in his 30s, Shapley achieved fame not only through his discovery of the location of our solar system in the galaxy, but also through his participation in "the Great Debate" of 1920 on the size of the universe and the nature of the "nebulae." In this debate Shapley took the wrong side in arguing that all nebulae, including those outside the galactic plane, were located inside our own galaxy. Edwin Hubble proved just the opposite only a few years later when he demonstrated that many of the nebulae were distant galaxies in their own right (see Chap. 36). But even large mistakes such as this are common in the frontier areas of astronomy, and Shapley's error was only a blip in his long career as Director of Harvard College Observatory from 1921–1952.

The technical aspects of astronomy, which is as far as most astronomers go in their professional careers, was only one aspect of Shapley's life, which deeply affected, and was affected by, culture and politics. As one historian of science put it, "Elucidating the broader and very human significance of the cosmic facts was this scientist's true calling. Bringing the cosmos to human culture was Harlow Shapley's lifelong mission" (Palmeri 2009). Among the books Shapley wrote to his end were *Of Stars and Men* (1958), *The View from a Distant Star* (1963), and *Beyond the Observatory* (1967). The phrase "the view from distant star" captures the cosmic perspective, found throughout that volume and his other works. Shapley pictured Earth as "Planet No. 3, in the family of a run-of-the-mill yellowish star, situated in the outer part of a typical galaxy that contains billions of typical stars" (Shapley 1958, 74–75). And he placed humanity in a cosmic context, not only in space but also in time:

> Mankind is made of star stuff, ruled by universal laws. The thread of cosmic evolution runs through his history, as through all phases of the universe – the microcosmos of atomic structures, molecular forms, and microscopic organisms, and the macrocosmos of higher organisms, planets, stars, and galaxies. Evolution is still proceeding in galaxies and man – to what end, we can only vaguely surmise. (Shapley 1963, v)

Shapley explained that all the elements of our bodies were made inside stars before being spewed out of supernova explosions and spread throughout the universe, including into the molecular clouds out of which our solar system and others formed over eons of cosmic evolution. As JoAnn Palmeri has written, Shapley popularized his cosmic evolutionary perspective "with missionary zeal" during the 1950s and 1960s. Not only did that perspective influence the general public, it also influenced scientists like Carl Sagan, who overlapped with Shapley in an informal way at Harvard until he was denied tenure in 1968. The evocative phrase "star stuff" was often repeated by Sagan and is sometimes attributed to him. But as the above quotation shows, it was a phrase popularized by Shapley at a time when Sagan was just beginning his career. In any case the phrase has become commonplace, an icon of the cosmic perspective.

Like Giordano Bruno and Harlow Shapley, during his storied career Carl Sagan (1934–1996) made other worlds and extraterrestrial life a hallmark of his thought (Davidson 1999; Poundstone 1999). Unlike Bruno, he would not be burned at the stake, although he would die a relatively early death of natural causes. Sagan, of course, inherited a much more robust scientific tradition that did Bruno, but was often criticized for his far-out thinking. Nonetheless his book *Intelligent Life in the Universe* (Shklovskii and Sagan 1966), written with Russian astrophysicist Joseph Shklovskii, became the Bible for a generation of scholars who pursued the quest for life beyond Earth. His *Cosmic Connection* became an inspiration for a much wider audience for whom it evoked the majesty of the cosmos and our place in it (Sagan 1973, 2000). Sagan was a popularizer *par excellence* at a time when popularization was not held in his esteem among most of his colleagues. At least in part for this reason, he was denied membership in the National Academy of Sciences. He died just as the first planets were being discovered beyond the solar system, a vindication of his ideas.

As is evident in his television series and book *Cosmos* (Sagan 1980), Sagan held the cosmic perspective in spades, and spread it to millions of people through a variety of books and media even in the era before social media. He knew that just as the voyages of discovery had changed the human worldview, so had space exploration, with its views of Earthrise from the Moon, of Earth as a "pale blue dot" as seen by the Voyager 1 spacecraft from the edge of the Solar System (Fig. 42.4), and with the increasing awareness that space science gives us our place in cosmic evolution. Again an extraterrestrial gaze on steroids, now enabled by the Space Age. In his book *Pale Blue Dot* (Sagan 1994) Sagan epitomized the cosmic perspective in a passage that has become famous. As he contemplated the Voyager image, he wrote

> Look again at that dot. That's here. That's home. That's us. On it everyone you love, everyone you know, everyone you ever heard of, every human being who ever was, lived out their lives. The aggregate of our joy and suffering, thousands of confident religions, ideologies, and economic doctrines, every hunter and forager, every hero and coward, every creator and destroyer of civilization, every king and peasant, every young couple in love, every mother and father, hopeful child, inventor and explorer, every teacher of morals, every corrupt politician, every "superstar," every "supreme leader," every saint and sinner in the history of our species lived – on a mote of dust suspended in a sunbeam.

Much more could be said about Sagan's role in fostering the cosmic perspective. The astronauts and cosmonauts have had similar, if not quite so poetic, sentiments, and many people now carry such thoughts in their minds vicariously, a kind of cosmological meme. That perspective was present already in thinkers such as Boethius, Bruno, and Shapley, but is now much enhanced by the Space Age. Though "the view from a distant star" seems not to have had widespread impact so far in fostering human understanding and peace, it is a perspective sorely needed today. We can only hope that it will have more impact in the future.

Fig. 42.4 Earth, described by Carl Sagan as a "Pale Blue Dot," as seen by Voyager 1 from a distance of more than 4 billion miles (6.4 billion kilometers), February 14, 1990. (This image inspired the title of Sagan's 1994 book, *Pale Blue Dot: A Vision of the Human Future in Space*. NASA/JPL-Caltech)

42.4 The Cosmic Perspective and the Consolation of Astronomy

It is easy to philosophize about such grand thoughts. But how does all this affect our daily lives? For starters, aside from the cosmological framework that humans have always had in one form or another, the cosmic perspective also provides a spiritual perspective that some see as akin to religion. Shapley stated that "stars … provide me with the awe, the reference, the poetry, the mystery, the beauty, the inspiration, the respect for and service to fellow man that form the basis of what seems to me to be the essence of religion" (Palmeri 2009, 493). As Palmeri has documented in detail, Shapley's religious views and activities in institutions such as the Institute for Religion in an Age of Science all stemmed from this core belief. And although Sagan was famously an atheist, he also drew a kind of spiritual sustenance from the cosmic perspective. His book *The Varieties of Scientific Experience: A Personal View of the Search for God*, has as its frontispiece a view of the Hubble Deep Field (similar to Fig. 18.5), the plethora of galaxies immediately putting humans in our proper place. Shorn of its supernaturalism, Sagan opined in that volume, "religion can provide in a very significant way, and without any mystical trappings, ethical standards for adults, stories for children, social organization for adolescents, ceremonials and rite of passage, history, literature, music, solace in times of bereavement, continuity with the past, and faith in the future" (Sagan 2006).

Others have also found spiritual sustenance in the epic of cosmic evolution. The new universe of the late twentieth century has spawned renewed analysis of the relation of humans to the cosmos, both inside and outside established religions. Biologist Ursula Goodenough, for one, argues in *The Sacred Depths of Nature* that cosmic evolution is a shared worldview capable of evoking an abiding religious response. "Any global tradition," she writes, "needs to begin with a shared worldview – a culture-independent, globally accepted consensus as to how things are" (Goodenough 1998). She finds this consensus in "our scientific account of Nature, an account that can be called The Epic of Evolution. The Big Bang, the formation of stars and planets, the origin and evolution of life on this planet, the advent of human consciousness and the resultant evolution of cultures – this is the story, the one story, that has the potential to unite us, because it happens to be true." Her elaboration of the religious implications fits well with religious naturalism, already a significant movement in the twenty-first century (Stone 2010).

This is the attitude that informs the principles of cosmotheology in Chap. 13, and also the attitude that Einstein espoused in his "cosmic religion," which he saw as "the strongest and noblest motive for scientific research" (Einstein 1930, 39; Jammer 1999, 75–84). "I cannot conceive of a God who rewards and punishes his creatures, or has a will of the type of which we are conscious in ourselves," Einstein wrote. "An individual who should survive his physical death is also beyond my comprehension, nor do I wish it otherwise… Enough for me the mystery of the eternity of life, and the inkling of the marvelous structure of reality, together with the single-hearted endeavor to comprehend a portion, be it never so tiny, of the reason that manifests itself in nature" (Einstein 1935, 5).

To be sure, the cosmic perspective will not solve our everyday problems. But it does put them into a broader perspective that makes them less burdensome to our remarkable but sometimes petty minds. Who can dwell unduly on their daily problems when galaxies are colliding, the Klingons are attacking, and extraterrestrial civilizations are rising and falling? Who among us cannot do with a little celestial contemplation, shorn of supernaturalism but full of wonder?

The colorful terminology of star stuff and starfolk evoked by Shapley, Sagan, and others with its full integration of humans into the cosmos, encourages us to be "at home in the universe" in the felicitous phrase used by several distinguished scientists in the late twentieth century (Kauffman 1995; Wheeler 1996). We now know that the atoms in our bodies were forged in nuclear reactions in stellar furnaces, spewed into the universe in supernovae explosions, and incorporated into our bodies through the long process of the evolution of life over the last 3.8 billion years on Earth. These, the life-giving elements, come from the stars.

The idea that we are made of star stuff should not only help us live more uplifting lives but also fosters an inspiring vision of death. We recognize that after death our bodily atoms will be dispersed once again through the universe, destined to become star stuff in a cycle of events that will end only with the death of the universe itself. We are part and parcel of the universe, and at the hour of our death when we return to the universe, the old phrase from the *Book of Common Prayer* based on Genesis—"earth to earth, ashes to ashes, dust to dust"—need only be slightly altered to "earth to earth, ashes to ashes, stardust to stardust," in order to be literally true. Or, to put a gloss on Genesis 3:19, "stardust thou art and unto stardust thou shalt return," in an eternal cycle. In the end we will all return to the stars, and they to us. That may be the ultimate message of the new universe and cosmic evolution. And the consolation of astronomy that Urania offers all of us.

References

Benedict XVI, Pope. 2008. "Boethius and Cassiodorus," http://w2.vatican.va/content/benedict-xvi/en/audiences/2008/documents/hf_ben-xvi_aud_20080312.html

Blumenberg, Hans. 1987. *The Genesis of the Copernican World*. Trans. Robert M. Wallace. Cambridge, MA: MIT Press

Boethius. 1998. *The Consolation of Philosophy*. London: The Folio Society

Bruno, Giordano. 1584. *De l'infinito universo et mondi*. Translated in Singer, 1968.

Crossley, Robert. 1994. *Olaf Stapledon: Speaking for the Future*, Syracuse University Press: New York

Danielson, Dennis. 2013. 'Early Modern ET, Reflexive Telescopics, and Their Relevance Today," in Vakoch (2013), 57–72.

Davidson, Keay. 1999. *Carl Sagan: A Life*. New York, John Wiley and Sons.

Dick, Steven J. 1982. *Plurality of Worlds: The Origins of the Extraterrestrial Life Debate from Democritus to Kant*. Cambridge: Cambridge University Press.

Dick, Steven J. 2018. *Astrobiology, Discovery, and Societal Impact*: Cambridge: Cambridge University Press.

Dick, Steven J. and Lupisella, M. L., editors. 2009. *Cosmos & Culture: Cultural Evolution in a Cosmic Context*. Washington, DC: NASA, online at http://history.nasa.gov/SP-4802.pdf

Einstein, Albert. 1935. *The World As I See It. London,* John Lane.
Einstein, Albert. 1930. "Religion and Science," in *Ideas and Opinions*. New York: Bonanza Books, 1954.
Gatti, Hilary. 1999. *Giordano Bruno and Renaissance Science*. Ithaca: Cornell University Press.
Goodenough, U. 1998. *The Sacred Depths of Nature*. Oxford, Oxford University Press.
Jammer, Max. 1999. *Einstein and Religion*. Princeton: Princeton University Press.
Kauffman, S. 1995. *At Home in the Universe: The Search for the Laws of Self- Organization and Complexity*. Oxford: Oxford University Press.
Kaylor, Noel Harold. 2012. "Introduction: The Life, Times, and Work of Boethius," in Noel Harold Kaylor, Jr. and Philip Edward Phillips (eds.), *A Companion to Boethius in the Middle Ages. Brill's companions to the Christian tradition,* volume 30, Leiden and Boston: Brill, pp. 1–46.
Lewis, C. S. 1964. *The Discarded Image*. Cambridge: Cambridge University Press
Macrobius. 1990. *Commentary on the Dream of Scipio by Macrobius*, William H. Stahl, trans. and ed. New York: Columbia University Press.
Marebon, John. 2009. *The Cambridge Companion to Boethius*. Cambridge: Cambridge University Press.
Martinez, Alberto A. 2016. "Giordano Bruno and the Heresy of Many Worlds," *Annals of Science*, 73, 4, pp. 345–374.
Martinez, Alberto A. 2018. *Burned Alive: Giordano Bruno, Galileo, and the Inquisition*. London, Reaktion Books.
McCluskey, Stephen C. 2012. "Boethius's Astronomy and Cosmology," in Noel Harold Kaylor, Jr. and Philip Edward Phillips (eds.), *A Companion to Boethius in the Middle Ages. Brill's companions to the Christian tradition,* volume 30, Leiden and Boston: Brill, pp. 47–73.
O'Donnell, James. 2008. *The Ruin of the Roman Empire: A New History*. New York: Harper Collins.
Palmeri, JoAnn. 2009. "Bringing Cosmos to Culture: Harlow Shapley and the Uses of Cosmic Evolution," in Dick and Lupisella, *Cosmos and Culture,* pp. 489–521.
Poundstone, William. 1999. *Carl Sagan: A Life in the Cosmos*. New York, Henry Holt and Company.
Rowland, Ingrid D. 2008. *Giordano Bruno: Philosopher Heretic*. New York: Ferrar, Straus and Giroux.
Sagan, Carl. 1973. *The Cosmic Connection*. New York: Doubleday
Sagan, Carl. 1980. *Cosmos*. New York: Random House.
Sagan, Carl. 1994. *Pale Blue Dot: A Vision of the Human Future in Space*. New York, Ballantine Books
Sagan, Carl. 2000. *Carl Sagan's Cosmic Connection: An Extraterrestrial Perspective*. Cambridge: Cambridge University Press.
Sagan, Carl. 2006. *The Varieties of Scientific Experience: A Personal View of the Search for God*. New York: Penguin Press.
Shapley, Harlow. 1958. *Of Stars and Men: The Human Response to an Expanding Universe*. Beacon Press, Boston, 104–114.
Shapley, Harlow. 1963. *The View from a Distant Star: Man's Future in the Universe*, New York: Basic Books.
Shapley, Harlow. 1969. *Through Rugged Ways to the Stars*. New York: Scribners.
Shklovskii, I. S. and Sagan, C. 1966. *Intelligent Life in the Universe*, Holden-Day, San Francisco.
Singer, Dorothea. 1968. *Giordano Bruno: His Life and Thought, with Annotated Translation of his Work On the Infinite Universe and Worlds*. New York: Greenwood.
Stone, Jerome A. 2010. *Religious Naturalism Today:* The *Rebirth of a Forgotten Alternative*. Albany: SUNY.
Vakoch, Douglas, ed. 2013. *Astrobiology, History and Society: Life Beyond Earth and the Impact of Discovery*. Berlin-Heidelberg: Springer-Verlag.
Vidal, Clément. 2014. *The Beginning and the End: The Meaning of Life in a Cosmological Perspective*. New York: Springer.

Wells, H. G. 1993. *A Critical Edition of the War of the Worlds*, Introduction and Notes by David Y. Hughes and Harry M. Geduld Bloomington: Indiana University Press.
Wheeler, J. 1996. *At Home in the Universe*. New York, Springer-Verlag.
Yanow, Morton Leonard. 1998. *The Nolan: Prisoner of the Inquisition*. New York: Crossroad Publishing Company.
Yates, Frances. 1964. *Giordano Bruno and the Hermetic Tradition*. New York: Vintage Books.
Zakariya, Nasser. 2017. *A Final Story: Science, Myth and Beginnings*. Chicago: University of Chicago Press.

Appendix A: Astrobiology Meets the United States Congress

Abstract Testimony before the full Committee on Science, Space, and Technology of the United States House of Representatives, December 13, 2013 at a session on "Astrobiology: Search for Biosignatures in our Solar System and Beyond."

Chairman Smith, Ranking Member Johnson, and members of the Committee, thank you for the opportunity to testify today on the subject of the past and future of astrobiology. I do so not as a practitioner in the field, but as a historian of science who for four decades has documented the debate over life beyond Earth, including in the context of NASA (Steven J. Dick and James E. Strick, *The Living Universe*: *NASA and the Development of Astrobiology* (Rutgers University Press, 2004). In that role I can say this is a subject rich in history and promise, and one that fascinates the American public. During my time as NASA Chief Historian, everywhere I went people of all ages wanted to know about life on other worlds. Astrobiology raises fundamental questions and evokes a sense of awe and wonder as we realize perhaps there *is* something new under our Sun, and the Suns of other worlds.

Key Discoveries in Astrobiology Over the Last Decade

The key discoveries in astrobiology over the last decade have evoked that sense of awe and wonder. High on the list must be must be the discovery of planets beyond our solar system, those so-called exoplanets that are the very first goal of the NASA Astrobiology Roadmap. Ground-based telescopes, as well as the Hubble and Spitzer space telescopes, have all contributed to these discoveries. But NASA's Kepler spacecraft has opened the floodgates. Twenty years ago no planets were known around Sun-like stars. As of the end of 2013 more than 1000 planets have been confirmed, and thousands more are awaiting confirmation. Smaller and smaller planets are being detected, including Super Earths and Earth-sized planets. Kepler-37b is

S. J. Dick, *Space, Time, and Aliens*, https://doi.org/10.1007/978-3-030-41614-0

only slightly larger than our Moon, one of at least three planets in its system. Kepler-62e and Kepler-62f are only about 50% larger than our Earth, and orbit in the habitable zone of their system. Scientists at the Kepler Science Conference last month reported that smaller planets are now being discovered at the most rapid rate, that most stars in our galaxy have at least one planet, and that one in five Sun-like stars are likely to have Earth-sized planets orbiting in their habitable zone.

A second highlight is the continued search for life in our solar system—Goal 2 of the Roadmap. Past spacecraft, including the Mars Global Surveyor, Mars Odyssey, the Mars Exploration Rovers (Spirit and Opportunity), the Mars Reconnaissance Orbiter, as well the Curiosity Rover, have demonstrated that Mars had enough liquid water in the past to be hospitable for life. In 2008 the Phoenix Lander detected the presence of shallow subsurface water, and only a few months ago researchers reported that Curiosity's first sample of Martian soil was composed of about 2% water. The Martian polar caps harbor large amounts of frozen water and carbon dioxide. Meanwhile, spacecraft have probed the icy moons of the outer solar system, including the Jovian moon Europa and the Saturnian satellite Enceladus. Europa almost certainly has an ocean under its icy crust, and Enceladus outdoes its Jovian counterpart by spouting jets of water vapor from beneath its icy surface. The still-ongoing Cassini/Huygens mission has found an atmosphere, believed to be rich in prebiotic organic compounds on the Saturnian moon Titan, and in 2006 discovered lakes of methane on the satellite. In a broader activity, just a few months ago Cassini captured an image of Earth, a pale blue dot against the darkness of space, as seen through the rings of Saturn.

Another of the highlights over the last decade has been to demonstrate further the tenacity of life in extreme environments—Goal 5 of the Astrobiology Roadmap. Life has been found in hydrothermal vents at high temperatures and pressures deep below the ocean; it has been found three kilometers below the ground employing radioactivity rather than photosynthesis for its metabolic processes; it has been found way above the boiling point of water in the brilliant hot spring of Yellowstone and way below its freezing point in the deserts of Antarctica, under conditions of extreme radiation, salinity, acidity and so on. The point is that life is much more tenacious than once thought, and so may arise on planets under conditions once thought unfavorable. Genomic analysis of these microorganisms continues to shed light on how they function.

Recently scientists have found evidence of microbial life in 3.48 billion year-old rocks in Australia, the oldest biosignatures yet found on Earth. These findings also feed into the origins of life debate, and if true, indicate that life arose relatively quickly after the late heavy bombardment of the Earth that ended about 3.8 billion years ago. These findings are likely to remain controversial over the next decade, similar to the Braiser-Schopf controversy that erupted in 2002 over the 3.45 billion year old Apex chert microfossils. Such controversy is an integral part of the scientific enterprise.

These are only some of the highlights of the numerous studies undertaken under the banner of astrobiology. More details are found in the Annual Reports of the NASA Astrobiology Institute (https://astrobiology.nasa.gov/nai/reports/annual-reports/). Uniting all these studies is the concept of cosmic evolution—the 13.8 billion year unfolding of the universe, resulting in galaxies, stars and planets—and in at least one case, life. Cosmic evolution provides the context for astrobiology, which among other things seeks to follow the evolution of organic compounds from the interstellar medium through protoplanetary disks to habitable planets, possibly including delivery by comets, and to understand how these organics gave rise to life on Earth.

Critical Issues and Challenges for Next Decade

The challenge now from an observational point of view is to classify and characterize the newly discovered planets, as well as to search for even smaller ones, especially those in the habitable zones of their parent stars. Over the next decade spacecraft such as the Transiting Exoplanet Survey Satellite (TESS) will search for rocky planets around stars, and the James Webb Space Telescope will further characterize these planets and their potential for life by searching for biosignatures in their atmospheres, Goal 7 of the Astrobiology Roadmap. Already in February of this year detection of carbon monoxide and water absorption lines were reported in a massive planet circling the star HR 8799, revealing the planet's chemical composition, atmospheric structure, and surface gravity (Konopacky et al., Science, 339, 1398). Ever more detailed spectra of ever more planetary atmospheres will surely be a major priority over the next decade. At the same time theoretical studies of the formation of circumstellar disks and planets will shed light on how other solar systems came to exist, illuminating their structure and how unique our solar system is. Numerous theoretical challenges exist, not least to explain how so many planets came to orbit so close to their parent stars.

Among the critical issues in the search for life will be a continued research program on past or present life on Mars, employing spacecraft such as MAVEN just launched 2 weeks ago. Mars Odyssey, Mars Express, the Mars Reconnaissance Orbiter, the Opportunity Rover, and the Mars Science Laboratory (Curiosity) will continue to return data as long as batteries and funding last. Meanwhile, Mars remains a planet of mystery in many ways. Claims of methane of possible biogenic origin have been made based on observations from Earth and from Mars orbit, but the Curiosity rover, using its laser spectrometer, has not detected any. Mars clearly had surface water in the past, as evidenced by channel and outflow features. Where and when did all the water go? Why did the climate of Mars change so remarkably? Results from the Phoenix lander in 2008 in the form of perchlorates on the Martian surface—a toxic compound of chlorine and oxygen—have even reopened the interpretation of the results of the biological experiments aboard the Viking landers from the late 1970s. These questions, and other, will likely be resolved over the next decade.

Research on the origins and limits of life on Earth will continue with the goal of elaborating a "universal biology" that applies not only to Earth but also to other planets. Today the debate is at the level of the molecular assembly of life, determining the geochemical steps that led to the origin of life on Earth. A major task will be to identify the origin of the first replicating molecules, which some researchers believe to be found in the "RNA world," in which RNA is able to both store information and catalyze reactions. In this scenario RNA was the predecessor to current life, based on DNA. Major gaps remain in this scenario, however, and the next decade could determine whether this, or another theory, comes to the fore. Origins of life work also incorporates the work of the late Carl Woese, who showed that single-celled organisms on Earth ("prokaryotes") actually consist of two distinct domains, archaea and bacteria, and that the archaea are actually closer to the third domain of life, eukarya (including plants and animals).

Work on biosignatures will continue in the atmospheres of other planets, but also in rocks and microbial ecosystems with environments analogous to early Earth. The results from studies of ancient biosignatures on Earth can be used to search for biomarkers on other planets.

One of the most appealing characteristics of astrobiology is that the discipline forces us to ask questions that put in perspective our place in the universe: What are life, consciousness, and intelligence in a universal context, and what are the metaphysical assumptions that underlie our understanding of these concepts? Is there a general theory of living systems, a universal biology as there is a universal physics? What are culture and civilization? What is our place in the 13.8 billion-year unfolding of cosmic evolution? Some of these questions bearing on consciousness and intelligence are beyond the scope of the current NASA astrobiology program, but they are nevertheless an important part of the search for life in the universe. Almost exactly 20 years ago, in the same session that saw the demise of the Superconducting Super Collider, the 103rd Congress terminated the NASA Search for Extraterrestrial Intelligence (SETI) program. In addition to a renewed search with the latest technology, the reinstatement of funding for SETI would allow a systematic examination of these intriguing questions. It would also repair the artificial programmatic divorce between the search for microbial and intelligent life, which, despite engaging different scientific communities, are part of the same research problem. And I believe SETI would be supported by the public, which as always is interested in life beyond Earth, whether microbial or intelligent.

The work described here is carried out by the 15 teams of the NASA Astrobiology Institute (NAI), plus individual grantees in NASA's Exobiology and Evolutionary Biology grant programs. In 2012 alone, NAI teams issued 172 project reports and 849 publications. These teams are guided by the NASA Astrobiology Roadmap, initially developed in 1998, and updated in 2003 and 2008. The 2008 Roadmap is now being updated as part of the 2014 Astrobiology Strategic Plan. The consolidation and expansion of astrobiology's goals over the last two decades demonstrate how the field is rapidly changing based on new ideas and new evidence. (see http://astrobiology.nasa.gov/roadmap/ for all three versions of the Roadmap).

Other institutions in the United States and around the world are also involved in such research. Astrobiology continues to become an ever-more robust discipline, defying past labels as a "science without a subject." That label is a misrepresentation of science. Every science is looking for its subject until it finds it, as in the case of planetary systems, the Higgs boson, and gravitational waves. From an epistemological point of view, the methods of astrobiology are as empirical as in any historical science such as astronomy or geology, though it is true that astrobiological observations and experiments are often especially difficult.

Astrobiology and Society

I would be remiss if I did not mention that among the issues and challenges for the next decade are those related to astrobiology and society. Indeed the Astrobiology Roadmap recognizes as one of its four implementation principles "a broad societal interest in its endeavors." Just as the American Association for the Advancement of Science and the National Science Foundation recognize the importance of studies of Science and Society, and just as some university programs across the country study the interactions of Biology and Society, including the ethical, legal and social implications of the Human Genome Project, so astrobiology raises profound questions with respect to the impact on society.

What will be the effect on our worldviews, philosophies and religions if we discover microbial or intelligent life beyond Earth? Are there useful analogies, such as the changes in worldview in the wake of the Copernican and Darwinian revolutions? A study of the nature of discovery indicates that the discovery of life beyond Earth is likely to be spread over an extended period of time, as in the debates over the Viking spacecraft results, and the ALH84001 Mars rock controversy. A second quite different critical issue in the domain of Astrobiology and Society may be formulated as follows: What are the ethical issues in pursuing the search for extraterrestrial life? How do we balance planetary protection and stewardship with the human exploration imperative, of which astrobiology is a significant part? NASA has for decades had a Planetary Protection program to address these issues, to ensure protection of "all of the planets all of the time." The gravity of the issues can hardly be over-emphasized, not only because of the real prospect of forward contamination of planets, but also because a single "Andromeda Strain" scenario would be enough to jeopardize the terrestrial biosphere.

These are the kinds of humanistic aspects of astrobiology I am studying as part of my time at the Library of Congress, where, in a related event a few weeks ago, we celebrated the opening of the papers of Carl Sagan, known worldwide for his work in astrobiology. Others are also studying these societal impact questions, especially since 2011 when the NASA Astrobiology Institute supported a Roadmap and a focus group on Astrobiology and Society (Margaret Race, Kathryn Denning et al., "Astrobiology and Society: Building an Interdisciplinary Research Community," *Astrobiology*, 12 (2012), 958–965).

Finally let me say that in my view astrobiology embodies the most important ideals of discovery and exploration. I like to quote Nobelist Baruch S. Blumberg—the first Director of the NASA Astrobiology Institute and the inspiration behind the Blumberg NASA/Library of Congress Chair in Astrobiology that I hold at the Library's Kluge Center, which brings together scholars and policymakers. Astrobiology, Dr. Blumberg said, counters the usual academic trends by drawing in numerous disciplines rather than compartmentalizing knowledge and increasing specialization. Moreover, it is in the best tradition of our species, and in the best American tradition dating back to Lewis and Clark to ask great questions, to explore our world and other worlds, to infuse our culture with new ideas, to be changed for the better because of that exploration, and to evoke that sense of awe and wonder as we discover the true place in the universe of our pale blue dot.

The full committee hearing, including the statements of NASA's senior scientist for astrobiology, Mary Voytek and MIT Professor Sara Seager, is available at

https://web.archive.org/web/20131130031810/http://science.house.gov/hearing/full-committee-hearing-astrobiology-search-biosignatures-our-solar-system-and-beyond

Appendix B: Sources and Permissions

Most of the chapters in this volume are reprinted from the sources listed below; other chapters are published here for the first time. My thanks to the copyright holders for permission to reprint the original articles. Illustration credits are given in the captions.

Chapter 1. Published as "Plurality of Worlds," in Norriss S. Hetherington, ed., Encyclopedia of Cosmology: Historical, Philosophical and Scientific Foundations of Modern Cosmology, Garland Publishing (New York and London, 1993), pp. 502–512.

Chapter 2. "The Twentieth Century Extraterrestrial Life Debate: Major Themes and Lessons Learned," in Astrobiology, History and Society: Life Beyond Earth and the Impact of Discovery, Douglas A. Vakoch, ed. (Springer: Heidelberg, 2013), pp. 133–175.

Chapter 3. "From the Physical World to the Biological Universe: Historical Developments Underlying SETI," *Bioastronomy: The Search for Extraterrestrial Life - The Exploration Broadens,* Proceedings of the Third International Symposium on Bioastronomy Held at Val Cenis, France, 18–23 June, 1990, Jean Heidmann and Michael Klein, eds. (Springer-Verlag, Berlin, 1991).

Chapter 4. "The Biophysical Cosmology: The Place of Bioastronomy in the History of Science," in *Astronomical and Biochemical Origins and the Search for Life in the Universe*, C.B. Cosmovici, S. Bowyer and D. Werthimer, eds. (Editrice Compositori, Bologna, 1997), 785–88.

Chapter 5. "The Biological Universe Revisited," in *The New Astronomy: Opening the Electromagnetic Window and Expanding our View of Planet Earth*, Wayne Orchiston, ed. (Springer: 2005), 15–26.

Chapter 6. "Back to the Future: SETI Before the Space Age," *The Planetary Report* (January–February, 1995), pp. 4–7.

Chapter 7. "The Drake Equation in Context," Introduction to *The Drake Equation: Estimating the Prevalence of Extraterrestrial Life Through the Ages,* Douglas Vakoch and Matthew Dowd, eds. (Cambridge University Press: Cambridge, 2015), pp. 1–20.

S. J. Dick, *Space, Time, and Aliens*, https://doi.org/10.1007/978-3-030-41614-0

Chapter 8. "Cosmic Evolution: History, Culture, and Human Destiny," Steven J. and Mark Lupisella, eds. *Cosmos & Culture: Cultural Evolution in a Cosmic Context*. Washington, DC: NASA, 2009, pp. 25–62

Chapter 9. "Consequences of Success in SETI: Lessons from the History of Science," *Progress in the Search for Extraterrestrial Life*, Seth Shostak, ed. (Proceedings of Santa Cruz meeting on SETI, August, 1993; ASP Conference series: San Francisco, 1995), 521–532.

Chapter 10. "Cultural Aspects of Astrobiology: A Preliminary Reconnaissance at the Turn of the Millennium", in *Bioastronomy 99: A New Era in Bioastronomy*, Guillermo Lemarchand and Karen Meech, eds. (San Francisco, 2000), pp. 649–659.

Chapter 11. "Anthropology and the Search for Extraterrestrial Intelligence," *Anthropology Today*, 22 (April, 2006), 3–7.

Chapter 12. Published as "Cultural Evolution, the Postbiological Universe, and SETI," *International Journal of Astrobiology*, 2 (2003), 65–74.

Chapter 13. "Toward a Constructive Naturalistic Cosmotheology," in Ted Peters, ed. *Astrotheology: Science and Theology Meet Extraterrestrial Life*. (Wipf and Stock, Cascade Books: Eugene, Oregon, 2018), 228–244.

Chapter 14. "Astroethics and Cosmocentrism," first published as a blog on the *Scientific American* website, August 8, 2018. https://blogs.scientificamerican.com/observations/astroethics-and-cosmocentrism/

Chapter 15. "Should We Message ET and is an Asiolomar Consultation Process Possible?" The first part of this chapter was a blog posted on the METI International website on December 9, 2015, available at http://meti.org/en/blog/should-we-message-et. The latter part of this chapter discussing Asilomar was first delivered at the International Space Development Conference (ISDC), held in St. Louis in 2017 and is published here for the first time.

Chapter 16. "Astrobiology and Society Comes of Age," Introduction to Kelly Smith, "*Social and Conceptual Issues in Astrobiology* (Oxford: Oxford University Press. 2019).

Chapter 17. "Exploring the Unknown: 50 Years of NASA History," in *50 Years of Space: A Global Perspective* (Universities Press India: Hyderabad, India, 2007).

Chapter 18. "Exploration, Discovery and Culture: NASA's Role in History," in Steven J. Dick, ed., *NASA's First 50 Years: Historical Perspectives* (NASA History Series, 2010), pp. 587–662.

Chapter 19. "Space, Time and Aliens: The Role of Imagination in Outer Space," in *Imagining Outer Space: European Astroculture in the Twentieth Century*, Alexander T. Geppert, ed. (Palgrave-MacMillan: New York, 2012), pp. 27–44.

Chapter 20. "The Impact of the Hubble Space Telescope," Introduction to Part 3 of *Hubble's Legacy: Reflections by Those Who Dreamed it, Built It, and Observed the Universe with It*," Roger Launius and David DeVorkin, eds. (Smithsonian Institution Scholarly Press: Washington, D.C., 2014), pp. 74–78.

Chapter 21. "The Decision to Cancel the Hubble Space Telescope Servicing Mission 4 (and its Reversal)," in *Hubble's Legacy: Reflections by Those Who Dreamed it, Built It, and Observed the Universe with It*," Roger Launius and David

DeVorkin, eds. (Smithsonian Institution Scholarly Press: Washington, D.C., 2014), pp.151–189.

Chapter 22. Reflections presented at a meeting on "French-American Relations in Space, 1957–1975" held in Paris December 8–9, 2005

Chapter 23. Published as "Astrometry," in *History of Astronomy: An Encyclopedia*, edited by J. A. Lankford (Garland Publishing, 1997), 47–60.

Chapter 24. "Pulkovo Observatory and the National Observatory Movement," *Inertial Coordinate Systems on the Sky*, Proceedings of International Astronomical Union Symposium 141, Leningrad, October 17–21, 1989, Jay H. Lieske and Victor K. Abalakin, eds. (Dordrecht, 1990), pp. 29–38.

Chapter 25. "John Quincy Adams, the Smithsonian Bequest, and the Origins of the U. S. Naval Observatory," *Journal for the History of Astronomy* , 22 (February, 1991), 31–44.

Chapter 26. (with Ian R. Bartky). "The First Time Balls," *Journal for the History of Astronomy*, 12 (October 1981), 155–64. (with Ian R. Bartky). And "The First North American Time Ball," *Journal for the History of Astronomy*, 13 (February 1982), 50–54.

Chapter 27. "The U. S. Naval Astronomical Expedition of James Melville Gilliss in the Southern Hemisphere, 1849–1852," Address given March 2016 at the Pontificia Universidad Católica Chile in Santiago, Chile, published here for the first time.

Chapter 28. "The American Transit of Venus Expeditions of 1874 and 1882," in D. W. Kurtz, ed., *Transits of Venus: New Views of the Solar System and the Galaxy*, IAU Colloquium 196 (Cambridge University Press, 2005), pp. 100–110.

Chapter 29. "Geodesy, Time, and the Markowitz Moon Camera Program: An Interwoven IGY Story," in *Globalizing Polar Science: Reconsidering the Social and Intellectual Implications of the International Polar and Geophysical Years,* edited by Roger D. Launius, James R. Fleming, and David H. DeVorkin (Palgrave MacMillan: New York, 2010)

Chapter 30. "Pluto, Discovery, and Classification in Astronomy," condensed from portions of *Discovery and Classification in Astronomy: Controversy and Consensus* (Cambridge: Cambridge University Press, 2013)

Chapter 31. "Astronomy's Three Kingdoms: A Comprehensive Classification for Astronomy," First published as "Astronomy's Three Kingdom System" (International Society for Knowledge Organization, *Encyclopedia of Knowledge Organization,* 2019).

Chapter 32. "The Discovery of Polar Motion and its Importance," First published as "Polar Motion: A Historical Overview on the Occasion of the Centennial of the International Latitude Service," in *Polar Motion: Historical and Scientific Problems*, S. J. Dick, D. McCarthy and B. Luzum, eds (San Francisco, 2000), 3–23

Chapter 33. "Observation and Interpretation of the Leonid Meteors over the Last Millennium," *Journal of Astronomical History and Heritage*, 1 (1998), 1–20.

Chapter 34. "Discovery and Exploration of the Moons of Mars," keynote address at the First International Conference on the Exploration of Phobos and Deimos, held November 5–7, 2007 at NASA Ames Research Center in Mountain View, California.

Chapter 35. "The Universe and Alfred Russel Wallace," in *Natural Selection & Beyond: The Intellectual Legacy of Alfred Russel Wallace*, Charles H. Smith and George Beccaloni, eds. (Oxford University Press, 2008), pp. 320–340.

Chapter 36. "Discovering a New Realm of the Universe: Hubble, Galaxies, and Classification," from Steven J. Dick, *Discovery and Classification in Astronomy: Controversy and Consensus* (Cambridge: Cambridge University Press, 2013).

Chapter 37. "The Philosophy of Astronomy, Cosmology, and Astrobiology: A Preliminary Reconnaissance," published here for the first time.

Chapter 38. "Critical Issues in the History, Philosophy and Sociology of Astrobiology," *Astrobiology, 12,* (2012), 906–927.

Chapter 39. From "The Twentieth Century Extraterrestrial Life Debate: Major Themes and Lessons Learned," in *Astrobiology, History and Society: Life Beyond Earth and the Impact of Discovery*, Douglas A. Vakoch, ed. (Springer: Heidelberg, 2013), pp. 133–175.

Chapter 40. "Cosmology and Biology," Proceedings of the 2008 Conference on the Society of Amateur Radio Astronomers, June 29-July 2, 2008 (NRAO, Green Bank, West Virginia), pp. 1–15

Chapter 41. "Interstellar Humanity," *Futures*, 32 (2000), 555–567.

Chapter 42. "The Consolations of Astronomy and the Cosmic Perspective," published here for the first time.

About the Author

Steven J. Dick is the former NASA Chief Historian and Director of the NASA History Office. He was the 2014 Baruch S. Blumberg NASA/Library of Congress Chair in Astrobiology at the Library of Congress's John W. Kluge Center. In 2013 he testified before the United States Congress on the subject of astrobiology. From 2011 to 2012 he held the Charles A. Lindbergh Chair in Aerospace History at the National Air and Space Museum. Prior to that, he was an astronomer and historian of science at the U.S. Naval Observatory for more than two decades. He is the author or editor of 23 books, including most recently *Discovery and Classification in Astronomy: Controversy and Consensus (Cambridge, 2013), The Impact of Discovering Life Beyond Earth (Cambridge, 2015), Astrobiology, Discovery, and Societal Impact (Cambridge, 2018), and Classifying the Cosmos: How We can Make Sense of the Celestial Landscape (Springer, 2019). Astrobiology, Discovery and Societal Impact* received the PROSE Award in Cosmology and Astronomy from the Association of American Publishers. In 2006, Dick received the LeRoy E. Doggett Prize from the American Astronomical Society for a career that has significantly influenced the field of the history of astronomy. He is the recipient of the Navy Meritorious Civilian Service Medal, the NASA Exceptional Service Medal, and two NASA Group Achievement Awards. In 2009, minor planet 6544 Stevendick was named in his honor. His reminiscences may be found in *Stardust Memories: A Dick Family History and Memoir,* privately published in 2019.

S. J. Dick, *Space, Time, and Aliens*, https://doi.org/10.1007/978-3-030-41614-0

Index

S. J. Dick, *Space, Time, and Aliens*, https://doi.org/10.1007/978-3-030-41614-0

Q

R

S

The manufacturer's authorised representative in the EU is Springer Nature Customer Service Centre GmbH, Europaplatz 3, 69115 Heidelberg, Germany. If you have any concerns regarding our products, please contact ProductSafety@springernature.com

Printed and bound by CPI Group (UK) Ltd, Croydon, CR0 4YY
30/09/2026
02225697-0019